More Than Just a Textbook

Log on to *geomconcepts.com* to...

- access your Online Student Edition from home so you don't need to bring your textbook home each night.
- link to Student Workbooks and Online Study Tools.

See mathematical concepts come to life
- Personal Tutor
- Concepts in Motion: Interactive Labs
- Concepts in Motion: Animations

Practice what you've learned
- Extra Examples
- Self-Check Quizzes
- Data Updates
- Career Data
- Vocabulary Review
- Chapter Tests
- Standardized Test Practice

Try these other fun activities
- Scavenger Hunts

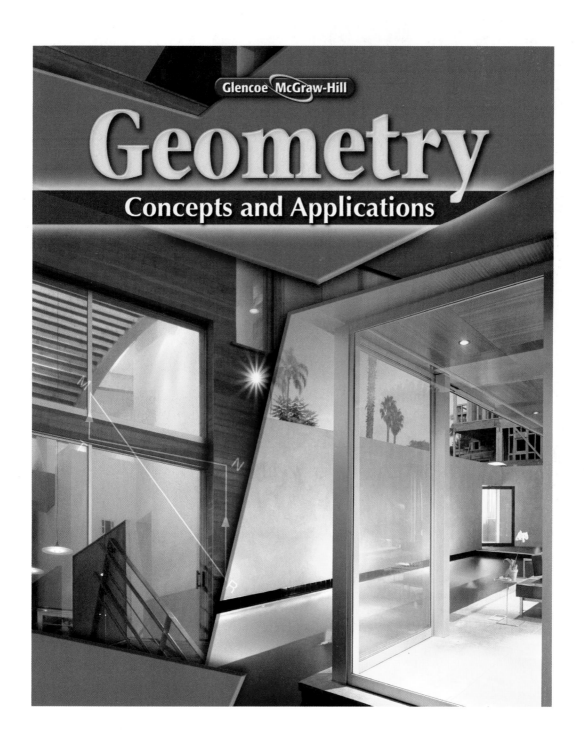

Glencoe McGraw-Hill

Geometry
Concepts and Applications

Glencoe

New York, New York   Columbus, Ohio   Chicago, Illinois

Visit the Glencoe Mathematics Internet Site for
Geometry: Concepts and Applications at

www.geomconcepts.com

You'll find:

- Concepts in Motion
- Personal Tutor
- Self-Check Quizzes
- Research Helps
- Data Updates
- Career Data
- Review Activities
- Test Practice
- Multilingual Glossary

 links to Web sites relevant to Problem-Solving Workshops, Investigations, Math In the Workplace features, exercises, and much more!

 Glencoe

The *McGraw-Hill* Companies

Send all inquiries to:
Glencoe/McGraw-Hill
8787 Orion Place
Columbus, OH 43240-4027

ISBN: 978-0-07-879914-3
MHID: 0-07-879914-7

Printed in the United States of America.

6 7 8 9 10 DOW 14 13 12 11 10

Dear Students,

Geometry: Concepts and Applications is designed to help you discover, learn, and apply geometry. You will be challenged to make connections from concrete examples to abstract concepts. The real-world photographs and realistic art will help you see geometry in your world. You will also have plenty of opportunities to review and use algebra concepts as you study geometry. And for those of you who love a good debate, you will find plenty of opportunities to flex your logical muscles.

We know that most of you haven't yet decided which careers you would like to pursue, so we've also included a little career guidance. This text offers real examples of how mathematics is used in many types of careers.

You may have to take an end-of-course exam for geometry and/or a proficiency test for graduation. When you enter the workforce, you may also have to take job placement tests. All of these tests include geometry problems. Because all of the major geometry concepts are covered in this text, this program will prepare you for all of those tests.

Each day, as you use **Geometry: Concepts and Applications,** you will see the practical value of geometry. You will grow to appreciate how often geometry is used in ways that relate directly to your life. You will have meaningful experiences that will prepare you for the future. If you don't already see the importance of geometry in your life, you soon will!

Sincerely,
The Authors

Jerry Cummins

Tinh Y. Ford

Margaret J. Kenney

Carol E. Malloy

Yvonne M. Mojica

P.S. To help you learn how to use your math book, use the Scavenger Hunt at geomconcepts.com. The Scavenger Hunt will help you learn where things are located in each chapter.

Contents in Brief

Jerry Cummins
Staff Development Specialist
Bureau of Education and
   Research
State of Illinois
President, National Council of
   Supervisors of Mathematics
Western Springs, IL

Tim Kanold
Superintendent
Former Director of
   Mathematics
Adlai Stevenson High School
Lincolnshire, IL

Margaret Kenney
Professor of Mathematics
Assistant to the Director,
   Mathematics Institute
Boston College
Chestnut Hill, MA

Carol Malloy
Assistant Professor of
   Mathematics Education
University of North Carolina
   at Chapel Hill
Chapel Hill, NC

Yvonne Mojica
Mathematics Teacher and
   Mathematics Department
   Chairperson
Verdugo Hills High School
Tujunga, CA

**Contributing Author
Dinah Zike**
Educational Consultant
Dinah-Might Activities, Inc.
San Antonio, TX

Academic Consultants and Teacher Reviewers

Each of the Academic Consultants read all 16 chapters, while each Teacher Reviewer read two chapters. The Consultants and Reviewers gave suggestions for improving the Student Edition and the Teacher's Wraparound Edition.

Academic Consultants

Patricia Beck
Mathematics Chair
Fort Bend ISD
Sugar Land, Texas

Judith Cubillo
Mathematics Department
  Chairperson
Northgate High School
Walnut Creek, California

Deborah A. Hutchens, Ed.D.
Graphing Calculator
  Consultant
Assistant Principal
Great Bridge Middle School
Chesapeake, Virginia

Nicki Hudson
Mathematics Teacher
West Linn High School
West Linn, Oregon

Don McGurrin
Senior Administrator for
  Secondary Mathematics
Wake County Public
  Schools
Raleigh, North Carolina

C. Vincent Pané, Ed.D.
Associate Professor
Mathematics Education
Molloy College
Rockville Centre, New York

Jane Wentzel
Secondary Mathematics
  Specialist
Fresno USD Mathematics
  Office
Fresno, California

Special thanks to Yuria Alban, District Mathematics Supervisor for the Dade County Public Schools in Miami, Florida, for helping to develop the philosophy of this program.

Teacher Reviewers

Patricia K. Bezona
Assistant Professor
Valdosta State University
Valdosta, Georgia

Denise J. Bodry
Mathematics Teacher
Aloha High School
Aloha, Oregon

Kimberly A. Brown
Mathematics Teacher
McGeehee High School
McGeehee, Arkansas

Karen A. Cannon
Mathematics Coordinator
Rockwood School District
Eureka, Missouri

Helen Carpini
Mathematics Department
  Chairperson
Middletown High School
Middletown, Connecticut

Donna L. Cooper
Mathematics Department
  Chairperson
Walter E. Stebbins High
  School
Riverside, Ohio

James D. Crawford
Instructional Coordinator,
  Mathematics
Manchester Memorial High
  School
Manchester, New Hampshire

David A. Crine
Mathematics Department
  Chairperson
Basic High School
Henderson, Nevada

Johnnie Ebbert
Department Chairperson
DeLand High School
DeLand, Florida

Astrid I. Ferony, Ph.D.
Mathematics Curriculum
  Representative
Papillion LaVista Public
  Schools
Papillion, Nebraska

Candace Frewin
Mathematics Teacher
East Lake High School
Tarpon Springs, Florida

Daniel A. Gandel
Mathematics Teacher
Hunter College
New York, New York

Glenn E. Gould
Mathematics Teacher
Liberty High School
Bealeton, Virginia

R. Emilie Greenwald
Mathematics Teacher
Worthington Kilbourne
    High School
Worthington, Ohio

Janice S. Hunter
Mathematics Teacher
Lexington High School
Lexington, South Carolina

Kathleen B. Jackson, Ed.D.
Mathematics Teacher
Phoenixville Area High School
Phoenixville, Pennsylvania

Linda T. Jones
Geometry Teacher
Robert S. Alexander
    High School
Douglasville, Georgia

Jim Keehn
Mathematics Teacher
Westview High School
Beaverton, Oregon

Wallace J. Mack
Mathematics Department
    Chairperson
Ben Davis High School
Indianapolis, Indiana

Kori N. Markle
Mathematics Teacher
Madison High School
Middletown, Ohio

Marilyn Martau
Mathematics Teacher
Lakewood High School
Lakewood, Ohio

Ross A. Martin
Mathematics Department
    Chairperson
Bethel High School
Bethel, Connecticut

Mary F. McPhaul
Mathematics Teacher
Maurice J. McDonough
    High School
Pomfret, Maryland

Jane E. Morey
Mathematics Department
    Chairperson
Washington High School
Sioux Falls, South Dakota

Rinda Olson
Mathematics Department
    Chairperson
Skyline High School
Idaho Falls, Idaho

Lawrence D. Patterson, Ed.D.
Curriculum Supervisor
Dorchester County Public
    Schools
Cambridge, Maryland

Dennis C. Preisser
Mathematics Department
    Chairperson
New Kensington-Arnold
    School District
New Kensington,
    Pennsylvania

Donna H. Preston
Mathematics Teacher
Thomas Worthington
    High School
Worthington, Ohio

Beverly Morris Sanderson
Mathematics Department
    Chairperson
Northwestern High School
Rock Hill, South Carolina

William Shutters
Mathematics Teacher
Urbandale High School
Urbandale, Iowa

Dora M. Swart
Mathematics Department
    Chairperson
W. F. West High School
Chehalis, Washington

Ronald R. Vervaecke
Mathematics Coordinator
Warren Consolidated Schools
Warren, Michigan

Dale I. Winters
Mathematics Teacher
Worthington Kilbourne
    High School
Worthington, Ohio

Michael J. Zelch
Mathematics Teacher
Worthington Kilbourne
    High School
Worthington, Ohio

Field Test Schools

Glencoe/McGraw-Hill wishes to thank the following schools that field tested
the pre-publication manuscript. They were instrumental in providing feedback
and verifying the effectiveness of this program.

Celina High School
Celina, Ohio

Worthington Kilbourne High School
Worthington, Ohio

Thomas Worthington High School
Worthington, Ohio

Table Of Contents

Lesson 1–2, page 12

Math In the Workplace

Standardized Test Practice

Hands-On Geometry 6

inter**NET** CONNECTION

Graphing Calculator Exploration 32

Chapter 2 Segment Measure and Coordinate Graphing............48

Chapter 3 Angles

Lesson 3–4, page 111

Lesson 4–2, page 154

Lesson 5-3, page 200

Chapter 6   More About Triangles...................226

Lesson 6–3, page 243

Chapter 7 Triangle Inequalities 274

Lesson 7–3, page 292

Chapter 8 Quadrilaterals 308

Math In the Workplace

Standardized Test Practice

Hands-On Geometry 312, 322, 328

*inter*NET CONNECTION

Graphing Calculator Exploration 316

Lesson 8–1, page 310

Chapter ⑨ Proportions and Similarity.............348

Lesson 9–5, page 379

Lesson 10–2, page 411

Lesson 11–1, page 460

Lesson 12-5, page 525

xix

Lesson 13–5, page 577

Chapter 14   Circle Relationships

Lesson 14–3, page 602

Chapter 15 Formalizing Proof 630

Investigation, page 667

Chapter 16   More Coordinate Graphing and Transformations 674

Lesson 16–4, page 692

Math In the Workplace

Standardized Test Practice

Hands-On Geometry

interNET CONNECTION

Graphing Calculator Exploration

Photo Graphic

Preparing for Standardized Test Success

The Preparing for Standardized Tests pages at the end of each chapter and at the end of the book have been created to help you get ready for the mathematics portions of your standardized tests. On these pages, you will find strategies for solving problems and test-taking advice to help you maximize your score.

It is important to remember that there are many different standardized tests given by schools and states across the country. Find out as much as you can about your test. Start by asking your teacher and counselor for any information, including practice materials, that may be available to help you prepare.

To help you get ready for these tests, do the Standardized Test Practice question in each lesson. Also review the concepts and techniques contained in the Preparing for Standardized Tests pages at the end of each chapter listed below. This will help you become familiar with the types of math questions that are asked on various standardized tests.

The Preparing for Standardized Tests pages in the Student Handbook review common types of problems on standardized tests and provide practice for each format.

The Preparing for Standardized Tests pages are part of a complete test preparation course offered in this text. The test items on these pages were written in the same style as those in state proficiency tests and standardized tests like ACT and SAT. The 16 topics are closely aligned with those tests, the geometry curriculum, and this text. These topics cover all of the types of problems you will see on these tests.

Chapter	Mathematics Topic	Pages
1	Number Concept Problems	46–47
2	More Number Concept Problems	86–87
3	Counting and Probability Problems	138–139
4	Data Analysis Problems	184–185
5	Statistics Problems	224–225
6	Algebra Problems	272–273
7	Algebra Word Problems	306–307
8	Coordinate Geometry Problems	346–347
9	Ratio and Proportion Problems	398–399
10	Triangle and Quadrilateral Problems	450–451
11	Function Problems	492–493
12	Angle, Line, and Arc Problems	544–545
13	Perimeter, Circumference, and Area Problems	582–583
14	Right Triangle and Trigonometry Problems	628–629
15	Solid Figure Problems	672–673
16	Systems of Equations and Polynomial Problems	714–715

With some practice, studying, and review, you will be ready for standardized test success. Good luck from Glencoe/McGraw-Hill!

Graphing Calculator Quick Reference Guide

Throughout this text, Graphing Calculator Explorations have been included so you can use technology to solve problems. These activities use the TI–83 Plus or TI–84 Plus graphing calculator.

The TI–83 Plus/TI–84 Plus has a wide variety of applications and features. If you are just beginning to use the calculator, you will not need to use all of its features. This page is designed to be a quick reference for the features you will need to use as you study from this text.

To open a geometry session: APPS ▼ (Name) ENTER
To quit a menu: ESC

Tool	Keystrokes
Alpha-Num	F5 2
Angle	F5 4 ▶ 3
Angle Bisector	F3 4
Area	F5 4 ▶ 2
Calculate	F5 6
Circle	F2 4
Clear	F5 7
Comment	F4 5
Compass	F3 6
Coordinates & Equation	F5 5
Dilation	F4 5
Distance & Length	F5 3 ▶ 1
Hide/Show	F5 1
Intersection Point	F2 ▶ 3
Line	F2 2

Tool	Keystrokes
Locus	F3 7
Midpoint	F3 5
Parallel Line	F3 2
Perpendicular Bisector	F3 3
Perpendicular Line	F3 1
Point	F2 1
Point on Object	F2 ▶ 2
Reflection	F4 2
Rotation	F4 4
Segment	F2 3
Slope	F5 4 ▶ 4
Symmetry	F4 1
Translation	F4 3
Triangle	F2 5

Reasoning in Geometry

What You'll Learn

Key Ideas

- Identify patterns and use inductive reasoning. *(Lesson 1–1)*
- Identify, draw models of, and use postulates about points, lines, and planes. *(Lessons 1–2 and 1–3)*
- Write statements in if-then form and write their converses. *(Lesson 1–4)*
- Use geometry tools. *(Lesson 1–5)*
- Use a four-step plan to solve problems. *(Lesson 1–6)*

Key Vocabulary

line *(p. 12)*
line segment *(p. 13)*
plane *(p. 14)*
point *(p. 12)*
ray *(p. 13)*

Why It's Important

Interior Design The goal of an interior designer is to make a room beautiful and functional. Designers listen carefully to their client's needs and preferences and put together a design plan and budget. The plan includes coordinating colors and selecting furniture, floor coverings, and window treatments.

Reasoning in geometry is used to solve real-life problems. You will use the four-step plan for problem solving to find the amount of wallpaper border an interior designer would need for a room in Lesson 1-6.

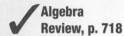

✓ Check Your Readiness

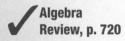

Algebra Review, p. 718

Evaluate each expression.

1. $2 \times 9 + 2 \times 3$ **2.** $2(4) + 2(7)$ **3.** $2(9) + 2(12)$

4. $2(14) + 2(18)$ **5.** 2×11 **6.** $9(10)$

7. 9×3 **8.** 9×8 **9.** $12(7)$

Algebra Review, p. 720

Solve each equation.

10. $10.1 - 0.2 = x$ **11.** $y = 2.6 - 1.4$ **12.** $n = 4.7 - 3.1$

13. $j = 100.4 - 94.9$ **14.** $1.43 + 0.84 = p$ **15.** $4.6 + 2.9 = n$

16. $0.8 + 1.3 = d$ **17.** $11.1 + 0.2 + 0.2 = t$ **18.** $h = 7.4(4.1)$

19. $m = 2.3(8.8)$ **20.** $(10.7)(15.5) = a$ **21.** $0.6(143.5) = g$

Algebra Review, p. 721

22. $q = \frac{5}{12} - \frac{1}{12}$ **23.** $\frac{7}{10} - \frac{1}{10} = t$ **24.** $\frac{2}{3} - \frac{1}{6} = w$

25. $y = \frac{11}{12} - \frac{1}{3}$ **26.** $b = \frac{3}{5} - \frac{1}{4}$ **27.** $\frac{4}{5} - \frac{1}{2} = c$

28. $v = \frac{4}{5} \cdot \frac{1}{3}$ **29.** $\frac{3}{5}\left(\frac{7}{8}\right) = d$ **30.** $z = \frac{5}{9} \cdot \frac{3}{4}$

FOLDABLES™
Study Organizer

Make this Foldable to help you organize your Chapter 1 notes. Begin with a sheet of $8\frac{1}{2}$" by 11" paper.

❶ **Fold** lengthwise in fourths.

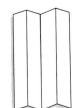

❷ **Draw** lines along the folds and label each column *sequences, patterns, conjectures,* and *conclusions*.

Reading and Writing As you read and study the chapter, record different sequences and describe their patterns. Also, record conjectures and state whether they are true or false; if false, provide at least one counterexample.

What You'll Learn

You'll learn to identify patterns and use inductive reasoning.

Why It's Important

Business Businesses look for patterns in data. *See Example 5.*

If you see dark, towering clouds approaching, you might want to take cover. Why? Even though you haven't heard a weather forecast, your past experience tells you that a thunderstorm is likely to happen. Every day you make decisions based on past experiences or patterns that you observe.

Rain clouds approaching

When you make a conclusion based on a pattern of examples or past events, you are using **inductive reasoning**. Originally, mathematicians used inductive reasoning to develop geometry and other mathematical systems to solve problems in their everyday lives.

You can use inductive reasoning to find the next terms in a sequence.

Example ❶ **Find the next three terms of the sequence 33, 39, 45, ...**

Study the pattern in the sequence.

33,   39,   45
     +6   +6

Each term is 6 more than the term before it. Assume that this pattern continues. Then, find the next three terms using the pattern of adding 6.

33,   39,   45,   51,   57,   63
     +6   +6   +6   +6   +6

The next three terms are 51, 57, and 63.

 **Your Turn**

Find the next three terms of each sequence.

a. 1.25, 1.45, 1.65, . . .   **b.** 13, 8, 3, . . .

c. 1, 3, 9, . . .   **d.** 32, 16, 8, . . .

Example 2

Find the next three terms of the sequence 1, 3, 7, 13, 21, ...

Notice the pattern 2, 4, 6, 8, ... To find the next terms in the sequence, add 10, 12, and 14.

1, 3, 7, 13, 21, 31, 43, 57
+2 +4 +6 +8 +10 +12 +14

The next three terms are 31, 43, and 57.

Your Turn

Find the next three terms of each sequence.

e. 10, 12, 15, 19, . . . **f.** 1, 2, 6, 24, . . .

Some patterns involve geometric figures.

Example 3

Draw the next figure in the pattern.

There are two patterns to study.

• First, the pattern with the squares (S) and triangles (T) is SSTTSS. The next figure should be a triangle (T).

• Next, the pattern with the colors white (W) and blue (B) is WBWBWB. The next figure should be white.

Therefore, the next figure should be a white triangle.

Your Turn

g.

Throughout this text, you will study many patterns and make conjectures. A **conjecture** is a conclusion that you reach based on inductive reasoning. In the following activity, you will make a conjecture about rectangles.

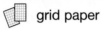 **Hands-On Geometry**

Materials: grid paper ✏ ruler

Step 1 Draw several rectangles on the grid paper. Then draw the diagonals by connecting each corner with its opposite corner.

Step 2 Measure the diagonals of each rectangle. Record your data in a table.

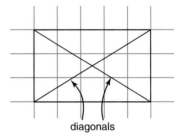

diagonals

Try These

1. **Make a conjecture** about the diagonals of a rectangle.

2. **Verify** your conjecture by drawing another rectangle and measuring its diagonals.

A conjecture is an educated guess. Sometimes it may be true, and other times it may be false. How do you know whether a conjecture is true or false? Try out different examples to test the conjecture. If you find one example that does not follow the conjecture, then the conjecture is false. Such a false example is called a **counterexample**.

Example 4

Number Theory Link

Akira studied the data in the table at the right and made the following conjecture.

The product of two positive numbers is always greater than either factor.

Factors		Product
2	8	16
5	15	75
20	38	760
54	62	3348

Find a counterexample for his conjecture.

The numbers $\frac{1}{2}$ and 10 are positive numbers.

However, the product of $\frac{1}{2}$ and 10 is 5, which is less than 10.

Therefore, the conjecture is false.

🌐**nline Personal Tutor at geomconcepts.com**

Businesses often look for patterns in data to find trends.

Example —**⑤**

Business Link

The following graph shows the revenue from the sale of waste management equipment in billions of dollars. Find a pattern in the graph and then make a conjecture about the revenue for 2005.

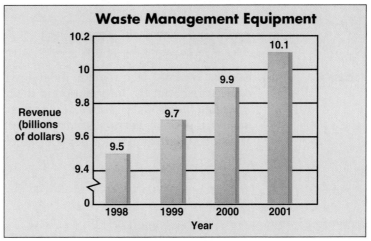

Source: Statistical Abstract of the United States

The graph shows an increase of 200 million dollars ($0.2 billion) each year from the sale of waste management equipment. In 2001, the revenue was 10.1 billion dollars.

2001	*2002*	*2003*	*2004*	*2005*
10.1	10.3	10.5	10.7	10.9

$+0.2$ $+0.2$ $+0.2$ $+0.2$

One possible conjecture is that the revenue in 2005 will grow to 10.9 billion dollars.

Check for Understanding

Communicating Mathematics

1. **Write** a definition of *conjecture*.
2. **Explain** how you can show that a conjecture is false.
3. **Writing Math** Write your own sequence of numbers. Then write a sentence that describes the pattern in the numbers.

Vocabulary

inductive reasoning
conjecture
counterexample

Guided Practice

🕒 **Getting Ready** **Tell how to find the next term in each pattern.**

Sample: 15, 18, 21, 24, . . .   **Solution:** Add 3.

4. 20, 26, 32, 38, . . .   5. 10, 7, 4, 1, . . .
6. 3, 6, 12, 24, . . .   7. 30, 31, 33, 36, . . .

Examples 1 & 2   **Find the next three terms of each sequence.**

8. 1, 3, 5, 7, . . .   9. 9, 6, 3, 0, . . .
10. 96, 48, 24, 12, . . .   11. 7, 8, 11, 16, . . .

Example 3

Draw the next figure in the pattern.

12.

13.

Example 4

14. **Number Theory** Jacqui made the following conjecture about the information in the table. *If the first number is negative and the second number is positive, the sum is always negative.* Find a counterexample for her conjecture.

Addends		Sum
−5	3	−2
−3	2	−1
−8	4	−4
−10	6	−4

Exercises

Practice

Homework Help	
For Exercises	See Examples
15–26, 33, 34	1, 2
27–32, 38	3
35, 37	4
36	5
Extra Practice	
See page 726.	

Find the next three terms of each sequence.

15. 5, 9, 13, 17, . . .

16. 12, 8, 4, 0, . . .

17. 12, 21, 30, 39, . . .

18. 1, 2, 4, 8, . . .

19. 3, 15, 75, 375, . . .

20. 2, −3, −8, −13, . . .

21. −1.4, 2.6, 6.6, 10.6, . . .

22. 6, 7, 9, 12, . . .

23. 13, 14, 16, 19, . . .

24. 20, 22, 26, 32, . . .

25. 10, 13, 19, 28, . . .

26. 10, 17, 31, 52, . . .

Draw the next figure in each pattern.

27.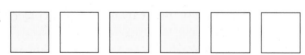

28.

29.

30.

31.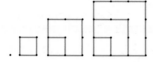

32.

33. Find the next term in the sequence $\frac{1}{2}, \frac{3}{2}, \frac{5}{2}$.

34. What operation would you use to find the next term in the sequence 3, 6, 12, 24, . . . ?

Applications and Problem Solving

35. Pets Find a counterexample for this statement: All dogs have spots.

36. Entertainment The graph shows the number of movie tickets sold yearly in the United States from 1993 to 2003. Predict the number of movie tickets that will be sold in 2006.

Math Online Data Update For the latest information about movie attendance, visit: geomconcepts.com

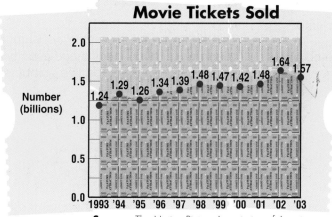

Source: The Motion Picture Association of America

37. Law Enforcement All fingerprint patterns can be divided into three main groups: arches, loops, and whorls.

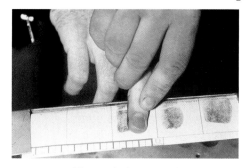

Fingerprinting procedure

Arch   Loop   Whorl

Name a situation that would provide a counterexample to this statement.

38. Critical Thinking Find the total number of small triangles in the eighth figure of the pattern.

Standardized Test Practice
(A) (B) (C) (D)

39. Multiple Choice Choose the expression that represents the value, in cents, of n nickels and d dimes. *(Algebra Review)*

(A) $n + d$
(B) $10n + 5d$
(C) $5n + 10d$
(D) $15nd$

To *Grandmother's House* WE Go!

Materials

 calculator

 colored pencils

Pascal's triangle is named for French mathematician Blaise Pascal (1623–1662).

Number Patterns in Pascal's Triangle

The triangular-shaped pattern of numbers below is called **Pascal's triangle**. Each row begins and ends with the number 1. Each other number is the sum of the two numbers above it.

row 0 → 1
row 1 → 1　1
row 2 → 1　2　1　　2 = 1 + 1
row 3 → 1　3　3　1　　3 = 1 + 2, 3 = 2 + 1
row 4 → 1　4　6　4　1　　4 = 1 + 3, 6 = 3 + 3, 4 = 3 + 1

Investigate

1. Copy row 0 through row 4 of Pascal's triangle on your paper.

   a. Complete row 5 through row 9 following the pattern.

   b. Find the sum of the numbers in each row. Examine the sums. What pattern do you see in the sums?

   c. Predict the sum of the numbers in row 10. Then check your answer by finding row 10 of Pascal's triangle and finding its sum.

2. The figure at the right shows how to find the sum of the diagonals of Pascal's triangle.

   a. Describe the pattern in the sums of the diagonals.

   b. Predict the sum of the next two diagonals.

The pattern in the sum of the diagonals is called the Fibonacci sequence.

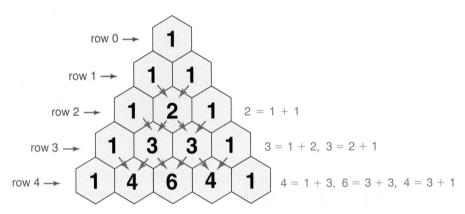

Suppose the grid at the right represents all of the streets between your house and your grandmother's house. You will start at your house and move down the grid to get to your grandmother's house.

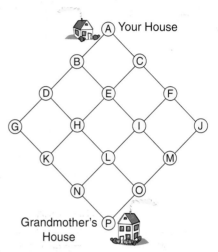
Your House

Grandmother's House

3. Copy the grid. How many different routes are there between each pair of points? Write your answers on the grid.

a. *A* and *B*   b. *A* and *C*
c. *A* and *D*   d. *A* and *E*
e. *A* and *F*   f. *A* and *G*
g. *A* and *H*   h. *A* and *I*

4. Explain how Pascal's triangle is related to the numbers on the grid.

5. Extend the pattern to find the number of routes between your house and your grandmother's house.

Extending the Investigation

In this extension, you will find more patterns in Pascal's triangle. Here are some suggestions.

- The figure at the right shows the pattern when multiples of 2 are shaded. Show the patterns when multiples of 3, 4, 5, and 6 are highlighted. (*Hint:* You will need to use at least 16 rows of Pascal's triangle.)

- Square numbers are 1, 4, 9, 16, ... Find two places where square numbers appear. (*Hint:* Look for the sum of adjacent numbers.)

- Find the powers of 11 in Pascal's triangle.

Presenting Your Conclusions

Here are some ideas to help you present your conclusions to the class.

- Write a paragraph about some of the patterns found in Pascal's triangle.
- Make a bulletin board that shows some of the visual patterns found in Pascal's triangle.

Investigation   For more information on Pascal's triangle, visit: www.geomconcepts.com

1-2 Points, Lines, and Planes

What You'll Learn
You'll learn to identify and draw models of points, lines, and planes, and determine their characteristics.

Why It's Important
Art Artists use points and lines to add shading to their drawings.
See Exercise 31.

Geometry is the study of points, lines, and planes and their relationships. Everything we see contains elements of geometry. Even the painting below is made entirely of small, carefully placed dots of color.

Georges Seurat, *Sunday Afternoon on the Island of La Grande Jatte*, 1884–1886

Each dot in the painting represents a point. A **point** is the basic unit of geometry. The shoreline in the painting represents part of a line. A **line** is a series of points that extends without end in two directions.

Reading Geometry

Points, lines, and planes are referred to as **undefined terms**.

Term	Description	Model
Point	• A point has no size. • Points are named using capital letters. • The points at the right are named *point A* and *point B*.	• *A* • *B*
Line	• A line is made up of an infinite number of points. • The arrows show that the line extends without end in both directions. • A line can be named with a single lowercase script letter or by two points on the line. • The line at the right is named *line AB*, *line BA*, or line ℓ. • The symbol for line *AB* is $\overleftrightarrow{AB}$.	*A* *B* ℓ

When we show the figure of a line, this is only a small part of a line.

1 **Name two points on line *m*.**

Two points are point *P* and point *Q*.

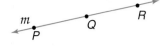

2 **Give three names for the line.**

Any two points on the line or the script letter can be used to name it. Three names are $\overleftrightarrow{PQ}$, $\overleftrightarrow{QR}$, and line *m*.

Your Turn

a. Name another point on line *m*.

b. Give two other names for line *m*.

Three points may lie on the same line, as in Example 1. These points are **collinear**. Points that do *not* lie on the same line are **noncollinear**.

Example

3 **Name three points that are collinear and three points that are noncollinear.**

D, *B*, and *C* are collinear.

A, *B*, and *C* are noncollinear.

Your Turn

c. Name three other points that are collinear.

d. Name three other points that are noncollinear.

Reading Geometry

The order of the letters can be switched with lines, but *not* with rays. Ray *DF* and ray *FD* are *not* the same ray.

Rays and line segments are parts of lines. A **ray** has a definite starting point and extends without end in one direction. The sun's rays represent a ray. A **line segment** has a definite beginning and end.

Term	Description	Model
Ray	• The starting point of a ray is called the **endpoint**. • A ray is named using the endpoint first, then another point on the ray. • The rays at the right are named *ray DF* and *ray CA*. • The symbol for ray *CA* is $\overrightarrow{CA}$.	*D* *F* *A* *C*

Term	Description	Model
Line Segment	• A line segment is part of a line containing two endpoints and all points between them. • A line segment is named using its endpoints. • The line segment at the right is named *segment BL* or *segment LB.* • The symbol for segment *BL* is $\overline{BL}$.	*B* *L*

In this text, we will refer to line segments as segments.

Name two segments and one ray.

Two segments are $\overline{PB}$ and $\overline{BC}$.
One ray is $\overrightarrow{PC}$.
$\overrightarrow{PB}$ is another name for $\overrightarrow{PC}$.

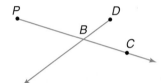

Your Turn

e. Name another segment and another ray.

The painting on page 12 was painted on a flat surface called a canvas. A canvas represents a plane. A **plane** is a flat surface that extends without end in all directions.

Term	Description	Model
Plane	• For any three noncollinear points, there is only one plane that contains all three points. • A plane can be named with a single uppercase script letter or by three noncollinear points. • The plane at the right is named plane *ABC* or plane *M*.	•*A* *M* •*B* •*C*

Whenever we draw a plane in this text, it is only a part of the whole plane. The whole plane continues in all directions.

Points that lie in the same plane are **coplanar**. Points that do *not* lie in the same plane are **noncoplanar**.

14   **Chapter 1** Reasoning in Geometry

Hands-On Geometry
Paper Folding

Materials: ☐ unlined paper

Step 1 Place points *A, B, C, D,* and *E* on a piece of paper as shown in the drawing.

Step 2 Fold the paper so that point *A* is on the crease.

Step 3 Open the paper slightly. The two sections of the paper represent different planes.

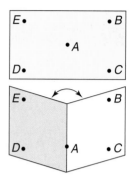

Try These

1. Name three points that are coplanar.
2. Name three points that are noncoplanar.
3. Name a point that is in both planes.

Check for Understanding

Communicating Mathematics

1. **Explain** the difference between a line and a segment.

2. **YOU Decide?** Joel says that $\overrightarrow{JL}$ and $\overrightarrow{LJ}$ name the same ray. Pat says they name different rays. Who is correct? Explain your reasoning.

Vocabulary

point
line
collinear
noncollinear
ray
line segment
plane
coplanar
noncoplanar

Guided Practice

Use the figure below to name an example of each term.

Example 1   3. point

Example 2   4. line

Example 4   5. ray

Example 4   6. segment

Example 3   7. **Maps** The map shows the state of Colorado. Name three cities that appear to be collinear. Name three cities that are *not* collinear.

Practice

Use the figure at the right to name examples of each term.

8. three points
9. two lines
10. three rays
11. three segments
12. point that is *not* on $\overline{AD}$
13. line that does *not* contain point E
14. ray with point A as the endpoint
15. segment with points E and F as its endpoints
16. three collinear points
17. three noncollinear points

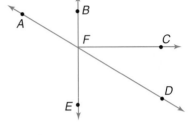

Homework Help	
For Exercises	See Examples
8, 18, 20	1
9, 24	2
10, 11, 14, 15, 21, 22, 25, 29, 31	4
12, 13, 16, 17, 27, 28	3

Extra Practice
See page 726.

Determine whether each model suggests a point, a line, a ray, a segment, or a plane.

18. the tip of a needle
19. a wall
20. a star in the sky
21. rules on notebook paper
22. a beam from a flashlight
23. a skating rink

Draw and label a figure for each situation described.

24. line ℓ
25. $\overline{CD}$
26. plane XYZ
27. collinear points A, B, and C
28. lines ℓ and m intersecting at point T
29. $\overrightarrow{BD}$ and $\overrightarrow{BE}$ so that point B is the only point common to both rays

Applications and Problem Solving

30. **Construction** The gable roof is the most common type of roof. It has two surfaces that meet at the top. The roof and the walls of the building are models of planes.

   a. Name a point that is coplanar with points C and D.
   b. Name a point that is noncoplanar with points R and S.
   c. Name one point that is in two different planes.

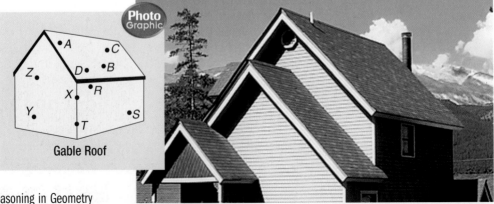

Gable Roof

31. Art Artists use segments to add shading to their drawings.

Hatching

a. How are the segments placed to create dark images?

b. How are the segments placed to create light images?

Fishing Boats on the Beach at Saintes-Maries-de-la-Mer, 1888
Vincent van Gogh

32. Critical Thinking Is the following statement *true* or *false*? Explain. *Two rays can have at most one point in common.*

Mixed Review

Find the next three terms of each sequence. *(Lesson 1–1)*

33. 5, 10, 20, 40, . . .

34. 112, 115, 118, 121, . . .

35. 1, −1, −3, −5, . . .

36. 1, 2, 4, 7, . . .

Standardized Test Practice

Ⓐ Ⓑ Ⓒ Ⓓ

37. Short Response Draw a figure that is a counterexample for the following conjecture: *All figures with four sides are squares.* *(Lesson 1–1)*

38. Multiple Choice Choose the figure that will continue the pattern. *(Lesson 1–1)*

Ⓐ   Ⓑ   Ⓒ   Ⓓ

Quiz 1   Lessons 1–1 and 1–2

Find the next three terms of each sequence. *(Lesson 1–1)*

1. 15, 11, 7, 3, . . .

2. 8, 10, 14, 20, . . .

Use the figure to name an example of each term. *(Lesson 1–2)*

3. line

4. point

5. ray

1-3 Postulates

What You'll Learn

You'll learn to identify and use basic postulates about points, lines, and planes.

Why It's Important

Architecture
Architects use points and lines in perspective drawings.
See page 23.

Geometry is built on statements called postulates. **Postulates** are statements in geometry that are accepted as true. The postulates in this lesson describe how points, lines, and planes are related.
*Another term for postulate is **axiom**.*

Postulate	Words	Models
1–1	Two points determine a unique line. *There is only one line that contains points P and Q.*	
1–2	If two distinct lines intersect, then their intersection is a point. *Lines ℓ and m intersect at point T.*	
1–3	Three noncollinear points determine a unique plane. *There is only one plane that contains points A, B, and C.*	

Examples

Points *D, E,* and *F* are noncollinear.

1 **Name all of the different lines that can be drawn through these points.**

There is only one line through each pair of points. Therefore, the lines that contain *D, E,* and *F,* taken two at a time, are $\overleftrightarrow{EF}, \overleftrightarrow{DF},$ and $\overleftrightarrow{DE}.$

2 **Name the intersection of $\overleftrightarrow{DE}$ and $\overleftrightarrow{EF}$.**

The intersection of $\overleftrightarrow{DE}$ and $\overleftrightarrow{EF}$ is point *E.*

Your Turn

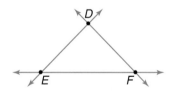

a. Points *Q, R, S,* and *T* are noncollinear. Name all of the different lines that can be drawn through these points.

b. Name the intersection of $\overleftrightarrow{QR}$ and $\overleftrightarrow{RS}$.

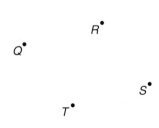

Example **3** **Name all of the planes that are represented in the figure.**

There are four points, *C*, *G*, *A*, and *H*.
There is only one plane that contains three
noncollinear points. Therefore, the planes
that can contain the points, taken three at
a time, are plane *ACG*, plane *GCH*, plane
GHA, and plane *AHC*.

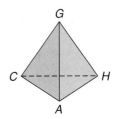

Reading
Geometry

When four points are
coplanar, you can name
the plane by choosing
any three points.

 Your Turn

c. Name all of the planes that are
represented in the figure.

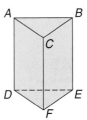

When two distinct lines intersect, they have only one point in common.
In the following activity, you will investigate what happens when two
planes intersect.

Hands-On Geometry
Paper Folding

Materials: 2 sheets of different-colored paper

scissors   tape

Step 1   Label one sheet of paper *M* and
the other *N*. Hold the two sheets of
paper together and cut a slit halfway
through both sheets.

Step 2   Turn the papers so that the two
slits meet and insert one sheet into the
slit of the other sheet. Use tape to hold
the two sheets together.

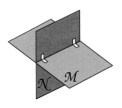

Try These
1. Draw two points, *D* and *E*, so they lie in both planes.
2. Draw the line determined by points *D* and *E*.
3. Describe the intersection of planes *M* and *N*.

| Postulate 1–4 | **Words:** If two distinct planes intersect, then their intersection is a line. |
| | **Model:** 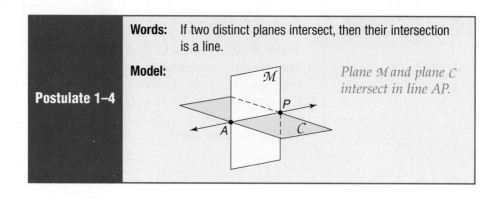 Plane M and plane C intersect in line AP. |

Example **4**

The figure shows the intersection of six planes. Name the intersection of plane *CDG* and plane *BCD*.

The intersection is $\overleftrightarrow{CD}$.

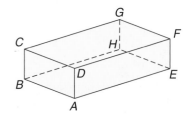

Your Turn

    d. Name two planes that intersect in $\overleftrightarrow{BA}$.

Check for Understanding

Communicating Mathematics

1. **Draw** plane *ABC* and plane *DEF* that intersect in $\overleftrightarrow{GH}$.

2. **State** the number of lines that are determined by two points.

3. **Writing Math** Write in your own words a sentence describing each postulate in this lesson. Include a diagram with each postulate.

> **Vocabulary**
>
> postulate

Guided Practice
Example 1

4. Points *X*, *Y*, and *Z* are noncollinear. Name all of the different lines that can be drawn through these points.

Refer to the figure at the right.

Example 2   5. Name the intersection of $\overleftrightarrow{DC}$ and $\overleftrightarrow{CB}$.

Example 3   6. Name all of the planes that are represented.

Example 4   7. Name the intersection of plane *ABC* and plane *ACD*.

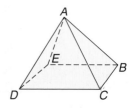

Example 3

8. **Photography** Cameras are often mounted on tripods to stabilize them. A tripod has three legs. Which postulate in the lesson guarantees that the tripod is stable?

Exercises

• • • • •   • • • • • • • • • • • • • • • •

Practice

Homework Help	
For Exercises	See Examples
9–11, 23, 25	1
12–14, 22	2
15–17, 26, 29, 30	3
18–21, 24, 27, 28, 32	4
Extra Practice	
See page 726.	

Name all of the different lines that can be drawn through each set of points.

9. A•   •C   •B

10. •D   •F   •E

11. •G   K•   •H   •J

Refer to Exercises 9–11. Name the intersection of each pair of lines.

12. $\overleftrightarrow{AC}$ and $\overleftrightarrow{AB}$

13. $\overleftrightarrow{FE}$ and $\overleftrightarrow{ED}$

14. $\overleftrightarrow{KJ}$ and $\overleftrightarrow{GK}$

Name all of the planes that are represented in each figure.

15.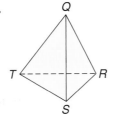

16.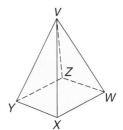

17.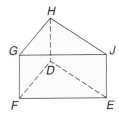

Refer to the figure at the right.

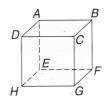

18. Name the intersection of plane ABC and plane BCG.

19. Name the intersection of plane DCG and plane HGF.

20. Name two planes that intersect in $\overleftrightarrow{AD}$.

21. Name two planes that intersect in $\overleftrightarrow{EF}$.

Determine whether each statement is *true* or *false*.

22. If two lines intersect in a point, then the point is in both lines.

23. More than one line can be drawn through two points.

24. Two planes can intersect in a line.

25. Two points determine two lines.

Determine whether each statement is *true* or *false*.

26. Three noncollinear points determine a plane.

27. If two planes intersect in a line, then the line is in both planes.

28. Two planes can intersect in a point.

29. It is possible for two lines to lie in the same plane.

30. Three planes can intersect in a point.

Applications and Problem Solving

31. Art In art, a line is *the path of a dot through space*. Using this definition, the figures at the right are *lines*. Explain why these curves are not called *lines* in geometry.

32. Buildings You can think of your classroom as a model of six planes: the ceiling, the floor, and the four walls.

a. Find two planes that intersect.

b. Find two planes that do *not* intersect.

c. Is it possible for three planes to intersect? If so, find the intersection.

33. Critical Thinking Three noncollinear points determine a plane. How many planes can contain three collinear points?

Mixed Review

Use the painting at the right to describe examples of each term. *(Lesson 1–2)*

34. point   **35.** line   **36.** plane

37. Explain how a ray is different from a line. *(Lesson 1–2)*

Standardized Test Practice
Ⓐ Ⓑ Ⓒ Ⓓ

38. Grid In Dyani shoots baskets every day to increase her free throw percentage. On Sunday, she shoots 20 free throws and plans to increase the number by 5 each day until Saturday. How many free throws will she shoot on Saturday? *(Lesson 1–1)*

39. Short Response Add two numbers to the data below so that the median does not change. *(Statistics Review)*

11, 13, 16, 12, 25, 8, 25, 33, 51

Irene Rice Pereira, *Untitled*, 1951

Math In the Workplace

Architect

Did you ever spend time building castles out of blocks or designing a dream home for your dolls? Then you might enjoy a career as an architect.

In addition to preparing blueprints and technical drawings, architects often prepare *perspective drawings* for their clients. The following steps show how to make a two-point perspective drawing of an office building.

Step 1: Draw a horizon line. Mark two *vanishing points*.

Step 2: Draw a vertical line, called the *key edge*. This will be a corner of the building.

Step 3: Connect the end of the key edge to each vanishing point. Draw the edges of the two visible sides of the building.

Step 4: Add details to the building.

1. Make a two-point perspective drawing of a building.
2. Do research about other kinds of perspective drawings.

FAST FACTS About Architects

Working Conditions
- generally work in a comfortable environment
- may be under great stress, working nights and weekends to meet deadlines

Education
- training period required after college degree
- knowledge of computer-aided design and drafting
- artistic ability is helpful, but not essential

Employment

Where Architects Are Employed

Architectural Firms 70%

Self-employed 30%

Source: Bureau of Labor Statistics

---

*inter*NET CONNECTION

Career Data For the latest information about a career as an architect, visit:

www.geomconcepts.com

1-4 Conditional Statements and Their Converses

What You'll Learn
You'll learn to write statements in if-then form and write the converses of the statements.

Why It's Important
Advertising Many advertisements are written in if-then form. *See Example 4.*

In mathematics, you will come across many **if-then statements**. For example, if a number is even, then the number is divisible by 2.

If-then statements join two statements based on a condition: a number is divisible by 2 only if the number is even. Therefore, if-then statements are also called **conditional statements.**

Conditional statements have two parts. The part following *if* is the **hypothesis.** The part following *then* is the **conclusion.**

> Hypothesis: a number is even.
> Conclusion: the number is divisible by 2.

How do you determine whether a conditional statement is true or false?

Conditional Statement	True or False?	Why?
If it is the fourth of July, then it is a holiday.	True	The statement is true because the conclusion follows from the hypothesis.
If an animal lives in the water, then it is a fish.	False	You can show that this statement is false by giving one counterexample. Whales live in water, but whales are mammals, not fish.

COncepts in MOtion
Animation
geomconcepts.com

In geometry, postulates are often written as if-then or conditional statements. You can easily identify the hypothesis and conclusion in a conditional statement.

Example

① **Identify the hypothesis and conclusion in this statement.**
If it is Saturday, then Elisa plays soccer.

Hypothesis: it is Saturday
Conclusion: Elisa plays soccer

Your Turn

a. If two lines intersect, then their intersection is a point.

There are different ways to express a conditional statement. The following statements all have the same meaning.

- If you are a member of Congress, then you are a U.S. citizen.
- All members of Congress are U.S. citizens.
- You are a U.S. citizen if you are a member of Congress.

U.S. Capitol, Washington, D.C.

Example ② **Write two other forms of this statement.**
If points are collinear, then they lie on the same line.

All collinear points lie on the same line.
Points lie on the same line if they are collinear.

Your Turn

b. If three points are noncollinear, then they determine a plane.

The **converse** of a conditional statement is formed by exchanging the hypothesis and the conclusion.

Example ③ **Write the converse of this statement.**
If a figure is a triangle, then it has three angles.

To write the converse, exchange the hypothesis and conclusion.

Conditional: If a figure is a triangle, then *it has three angles*.

Converse: If *a figure has three angles*, then it is a triangle.

Your Turn

c. If you are at least 16 years old, then you can get a driver's license.

If a conditional statement is true, is its converse always true?

Conditional: If a figure is a square, then it has four sides.
Converse: If a figure has four sides, then it is a square.

Look Back

Counterexample,
Lesson 1–1

The conditional statement is true. But there are many four-sided figures that are not squares. One counterexample is a rectangle. Therefore, the converse of this conditional is false.

Example 4

Advertising Link

Write the statement from the advertisement at the right in if-then form. Then write the converse of the statement.

It's OK to buy the wrong lipstick if you buy it in the right place.

CONNIE'S COSMETICS

MONEY BACK GUARANTEE
IT'S RISK FREE

Statement: It's OK to buy the wrong lipstick if you buy it in the right place.

If-then form: If you buy lipstick in the right place, then it's OK to buy the wrong lipstick.

Converse: If it's OK to buy the wrong lipstick, then you should buy it in the right place.

Check for Understanding

Communicating Mathematics

1. **Write** a conditional in which *there are clouds in the sky* is the hypothesis and *it may rain* is the conclusion.

2. **Explain** how to form the converse of a conditional.

> **Vocabulary**
> if-then statement
> conditional statement
> hypothesis
> conclusion
> converse

Guided Practice

Example 1

Identify the hypothesis and the conclusion of each statement.

3. If a figure is a quadrilateral, then it has four sides.

4. If a player misses three practices, he is off the team.

Example 2

Write two other forms of each statement.

5. You can vote if you are at least 18 years old.

6. All students who fight in school will be suspended.

Examples 3 & 4

Write the converse of each statement.

7. If it is raining, then the ground is wet.

8. If you cut class, you will be assigned a detention.

Example 2

9. **Biology** Write the if-then form of this statement. *All cats are mammals.*

Practice

Homework Help

For Exercises	See Examples
10–15	1
16–21	2
22–27, 29, 30	3
28	4

Extra Practice
See page 727.

Identify the hypothesis and the conclusion of each statement.

10. If the dog barks, it will wake the neighbors.

11. If a set of points has two endpoints, it is a line segment.

12. School will be cancelled if it snows more than six inches.

13. I will call my friend if I finish my homework.

14. All butterflies are arthropods.

15. All students should report to the gymnasium.

Write two other forms of each statement.

16. If the probability of an event is close to 1, it is very likely to happen.

17. If you eat fruits and vegetables, you will be healthy.

18. Your teeth will be whiter if you use a certain brand of toothpaste.

19. You'll win the race if you run the fastest.

20. All whole numbers are integers.

21. All people over age 18 can serve in the armed forces.

Write the converse of each statement.

22. If $2x = 20$, then $x = 10$.

23. If you play a musical instrument, you will do well in school.

24. The football team will play for the championship if it wins tonight.

25. You'll play softball if it stops raining.

26. All even numbers have a factor of 2.

27. All lines extend without end in two directions.

Applications and Problem Solving

28. Comics Write two conditional statements from the comic below in if-then form.

THE MIDDLETONS

29. Advertising Find an advertisement in a magazine or newspaper that contains an if-then statement. Write the converse of the statement.

30. **Number Theory** Consider this statement.
If two numbers are negative, then their product is positive.
   a. Write the converse of the statement.
   b. Determine whether the converse is *true* or *false*. If *false*, give a counterexample.

31. **Critical Thinking** The **inverse** of a conditional is formed by negating both the hypothesis and conclusion of the conditional.

   Conditional: If it is raining, then it is cloudy.
   Inverse: If it is *not* raining, then it is *not* cloudy.

   The **contrapositive** of a conditional is formed by negating both the hypothesis and conclusion of the *converse* of the conditional.

   Converse: If it is cloudy, then it is raining.
   Contrapositive: If it is *not* cloudy, then it is *not* raining.

   a. Write the inverse and contrapositive of this statement.
      If a figure has five sides, then it is a pentagon.
   b. Write a conditional. Then write its converse, inverse, and contrapositive.

Mixed Review

32. Determine whether the following statement is *true* or *false.*
If two planes intersect, then their intersection is a point. (*Lesson 1–3*)

Refer to the figure at the right. (*Lesson 1–2*)

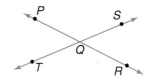

33. Name a ray.
34. Name a segment.
35. Name three collinear points.
36. Name three noncollinear points.

Exercises 33–36

Standardized Test Practice

Ⓐ Ⓑ Ⓒ Ⓓ

37. **Multiple Choice** Find the next term of the sequence
0, 3, 9, 18, ... (*Lesson 1–1*)

   Ⓐ 21   Ⓑ 30   Ⓒ 54   Ⓓ 162

Quiz 2   Lessons 1–3 and 1–4

1. Name the intersection of plane X and plane W. (*Lesson 1–3*)
2. Points R, S, and T are noncollinear. Name all the different lines that can be drawn through these points. (*Lesson 1–3*)

Exercises 3–5 refer to this statement.
If today is Monday, then I have band practice. (*Lesson 1–4*)

3. Identify the hypothesis.
4. Identify the conclusion.
5. Write the converse of the statement.

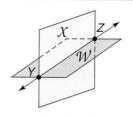

Exercise 1

www.geomconcepts.com/self_check_quiz

1-5 Tools of the Trade

What You'll Learn
You'll learn to use geometry tools.

Why It's Important
Landscaping
Landscapers use a compass-like device to draw large circles. *See Exercise 13.*

Industrial designers usually make rough sketches to begin new designs. Then they use drafting tools to make technical drawings of their plans. Some drafting tools are shown at the right.

As you study geometry, you will use some of these basic tools. The first tool is a straightedge. A **straightedge** is an object used to draw a straight line. A credit card, a piece of cardboard, or a ruler can serve as a straightedge. A straightedge is also used to check if a line is straight.

An *optical illusion* is a misleading image. Points, lines, and planes in geometry can be arranged to create such illusions.

Example	❶

Determine whether the sides of the triangle are straight.

Place a straightedge along each side of the triangle. You can see that the sides are straight.

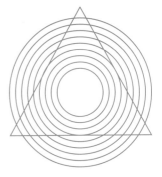

Your Turn

a. Use a straightedge to determine which of the three segments at the upper left forms a straight line with the segment at the lower right.

Lesson 1–5 Tools of the Trade **29**

A **compass** is another useful tool. A common use for a compass is drawing arcs and circles. *An arc is part of a circle.*

The figures below show two kinds of compasses.

A-shaped compass

Safety compass

The A-shaped compass usually has a point and a pencil or lead point. The two legs of the compass are on a hinge so they can be adjusted for different settings. This kind of compass is often used by engineers and cartographers, people who draw maps. One use of a compass is to compare lengths of segments.

Example

2 **Use a compass to determine which segment is longer, $\overline{AB}$ or $\overline{CD}$.**

Place the point of the compass on B and adjust the compass so that the pencil is on A.

Without changing the setting of the compass, place the point of the compass on C. The pencil point does not reach point D. Therefore, $\overline{CD}$ is longer.

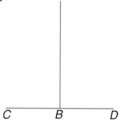

Your Turn

b. Which is longer, the segment from A to B or the segment from B to D?

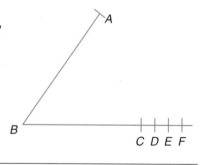

In geometry, you will draw figures using only a compass and a straightedge. These drawings are called **constructions**. No standard measurements are used in constructions. A construction is shown in Example 3.

 www.geomconcepts.com/extra_examples

Use a compass and straightedge to construct a six-sided figure.

First, use the compass to draw a circle. Then using the same compass setting, put the point on the circle and draw a small arc on the circle.

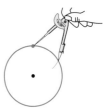

Move the compass point to the arc and then draw another arc along the circle. Continue doing this until there are six arcs.

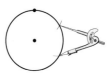

Use a straightedge to connect the points in order.

Another tool of the trade is patty paper. It can be used to do constructions such as finding a point in the middle of a segment, which is called a **midpoint**.

Hands-On Geometry
Paper Folding

Materials: patty paper straightedge

Step 1 Draw points *A* and *B* anywhere on a sheet of patty paper. Connect the points to form $\overline{AB}$.

Step 2 Fold the paper so that the endpoints lie on top of each other. Pinch the paper to make a crease on the segment.

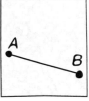

Step 1

Step 3 Open the paper and label the point where the crease intersects $\overline{AB}$ at *C*. *C* is the midpoint of $\overline{AB}$.

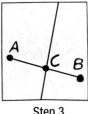

Step 3

Try This

Draw a circle on patty paper and cut it out. Fold it in half and then in half again. What do you call the point where the fold lines meet?

You can use graphing calculators and computers as tools in geometry. The following activity shows how to use a TI-83 Plus or TI-84 Plus to construct a triangle with three sides of equal length.

 Graphing Calculator Exploration

To open a geometry session, press $\boxed{\text{APPS}}$, select CabriJr and press $\boxed{\text{ENTER}}$.

Step 1 Open the $\boxed{\text{F2}}$ menu and select Segment. Draw a line segment for the first side of the triangle.

Step 2 Open the $\boxed{\text{F3}}$ menu and the Compass tool. The length of the segment you drew is the setting for the compass. Move your cursor to the segment and select it. The calculator will show a circle. Move the cursor so that the center of the circle is at one endpoint of the line segment and press $\boxed{\text{ENTER}}$.

Graphing Calculator Tutorial
See pp. 782–785.

Step 3 Repeat Step 2 and place the circle at the other endpoint of the line segment.

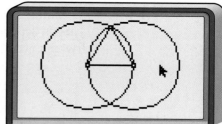

Step 4 Use the Segment tool on the $\boxed{\text{F2}}$ menu. Draw the line segments from each endpoint to the point of intersection of the two circles.

Try These

1. Open menu $\boxed{\text{F5}}$ and select Measure and then D. & Length. Then select each side of the triangle to find the lengths of the sides. Are the sides of the triangle all the same length?

2. Change the length of the original line segment. How is the triangle affected?

Check for Understanding

Communicating Mathematics

1. **Explain** the difference between a construction and other kinds of drawings.

2. **Name** four tools that are used in geometry.

3. Mario says that a straightedge and a ruler are the same. Curtis says they are different. Who is correct? Explain your reasoning.

Vocabulary

straightedge
compass
construction
midpoint

Use a straightedge or compass to answer each question.

4. Which segment on the upper left forms a straight line with the segment on the lower right?

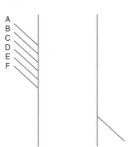

5. Which is greater, the height of the hat (from A to B) or the width of the hat (from C to D)?

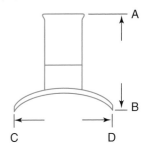

Example 3

6. **Design** Use a compass to make a design like the one shown at the right. (*Hint:* Draw large arcs from one "side" of the circle to the other.)

Exercises • • • • • • • • • • • • • • • •

Practice

Use a straightedge or compass to answer each question.

7. Which segment is longest?

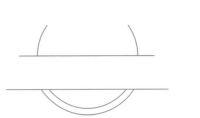

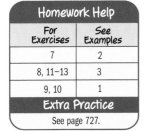

Homework Help

For Exercises	See Examples
7	2
8, 11–13	3
9, 10	1

Extra Practice

See page 727.

8. Which arc in the lower part of the figure goes with the upper arcs to form part of a circle?

9. Are the two horizontal segments straight or do they bend?

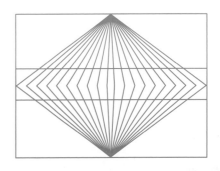

10. If extended, will $\overline{AB}$ intersect $\overline{CD}$ at C?

11. Use a compass to draw three different-sized circles that all have the same center.

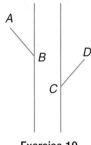

Exercise 10

Applications and Problem Solving

12. Sewing Explain how you could use a pencil and a long piece of string to outline a circular cloth for a round table.

13. Landscaping One way to mark off a circle for a bed of flowers is by using a measuring tape. From the center of the bed, extend the measuring tape a given distance and walk around the center, making marks on the ground. Explain how this method is similar to drawing a circle with a compass.

14. Critical Thinking In this text, you will be asked to make conjectures about geometric figures. Explain why you should not make conclusions about figures based only on their appearance. (*Hint: Think of optical illusions.*)

Mixed Review

Write the converse of each statement. *(Lesson 1–4)*

15. If a figure is a triangle, then it has three sides.

16. All whole numbers can be written as decimals.

17. You like the ocean if you are a surfer.

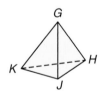
Standardized Test Practice
Ⓐ Ⓑ Ⓒ Ⓓ

18. Short Response Name all of the planes that are represented in the figure at the right. *(Lesson 1–3)*

19. Multiple Choice Which is *not* a name for this figure? *(Lesson 1–2)*

Ⓐ $\overleftrightarrow{JK}$ Ⓑ $\overleftrightarrow{\ell H}$ Ⓒ $\overleftrightarrow{KH}$ Ⓓ line ℓ

 www.geomconcepts.com/self_check_quiz

What You'll Learn

You'll learn to use a four-step plan to solve problems that involve the perimeters and areas of rectangles and parallelograms.

Why It's Important

Interior Design
Designers use formulas for perimeter and area to order materials.
See Exercise 11.

A useful measurement in geometry is perimeter. **Perimeter** is the distance around a figure. It is the sum of the lengths of the sides of the figure. The perimeter of the room shown at the right is found by adding.

$P = 12 + 9 + 6 + 6 + 18 + 15$ or 66

The perimeter of the room is 66 feet.

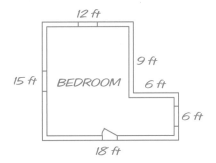

Some figures have special characteristics. For example, the opposite sides of a rectangle have the same length. This allows us to use a formula to find the perimeter of a rectangle. A **formula** is an equation that shows how certain quantities are related.

Perimeter of a Rectangle	**Words:**	The perimeter P of a rectangle is the sum of the measures of its sides. It can also be expressed as two times the length ℓ plus two times the width w.
	Symbols:	$P = \ell + w + \ell + w$ $P = 2\ell + 2w$ **Model:**

Example

Algebra Review
Evaluating Expressions, p. 718

1 **Find the perimeter of the rectangle.**

9 m

5 m

$P = 2\ell + 2w$ *Perimeter formula*
$P = 2(9) + 2(5)$ *Replace ℓ with 9 and w with 5.*
$P = 18 + 10$ or 28   *Simplify.*

The perimeter is 28 meters.

Your Turn

a. Find the perimeter of a rectangle with a length of 17 feet and a width of 8 feet.

Another important measure is area. The **area** of a figure is the number of square units needed to cover its surface. Two common units of area are the *square centimeter* and the *square inch*.

The area of the rectangle below can be found by dividing it into 20 unit squares.

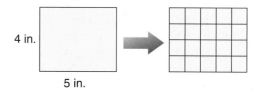

4 in.

5 in.

The area of a rectangle is also found by multiplying the length and the width.

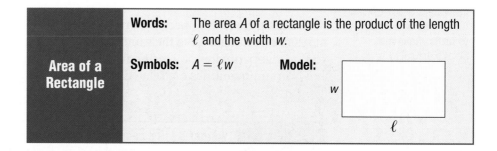

Area of a Rectangle

Words: The area A of a rectangle is the product of the length ℓ and the width w.

Symbols: $A = \ell w$ **Model:**

w

ℓ

Example ❷ **Find the area of the rectangle.**

14 in.

10 in.

$A = \ell w$ *Area formula*
$A = (14)(10)$ *Replace ℓ with 14*
$A = 140$ *and w with 10.*

The area of the rectangle is 140 square inches.

Your Turn

b. Find the area of a rectangle with a length of 8.2 meters and a width of 2.4 meters.

The opposite sides of a parallelogram also have the same length. The area of a parallelogram is closely related to the area of a rectangle.

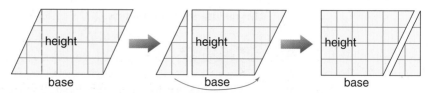

height   height   height

base   base   base

The area of a parallelogram is found by multiplying the base and height.

 www.geomconcepts.com/extra_examples

<table>
<tr><td rowspan="3">**Area of a Parallelogram**</td><td>**Words:** The area *A* of a parallelogram is the product of the base *b* and the height *h*.</td></tr>
<tr><td>**Symbols:** $A = bh$ **Model:**</td></tr>
</table>

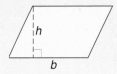

Example ③

Find the area of the parallelogram.

Algebra Review
Operations with Decimals, p. 720

$A = bh$ *Area formula*
$A = (5.2)(4)$ *Replace b with 5.2*
$A = 20.8$ *and h with 4.*

The area of the parallelogram is 20.8 square meters.

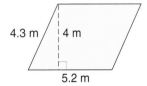

Your Turn

c. Find the area of a parallelogram with a height of 6 feet and a base of 8 feet.

Some mathematics problems can be solved by using a formula. Others can be solved by using a problem-solving strategy like finding a pattern or making a model. No matter what type of problem you need to solve, you can always use a **four-step plan**.

<table>
<tr><td rowspan="4">**Problem-Solving Plan**</td><td>1. **Explore** the problem.</td></tr>
<tr><td>2. **Plan** the solution.</td></tr>
<tr><td>3. **Solve** the problem.</td></tr>
<tr><td>4. **Examine** the solution.</td></tr>
</table>

Example ④

Interior Design Link

Julia wants to paint two rectangular walls of her bedroom. One bedroom wall is 15 feet long and 8 feet high. The other wall is 12 feet long and 8 feet high. She wants to put two coats of paint on the walls. She knows that 1 gallon of paint will cover about 350 square feet of surface. Will one gallon of paint be enough?

Explore   You know the dimensions of each wall. You also know that one gallon of paint covers about 350 square feet. You also know that she wants to use two coats of paint. You need to determine whether one gallon of paint is enough.

(continued on the next page)

Plan Since you need to find the total area that will be covered with paint, you can use the formula for the area of a rectangle. Find the total area of the two walls with two coats of paint. Then compare to 350 square feet.

Solve Area of first wall   Area of second wall

$A = \ell w$ $A = \ell w$

$A = (15)(8)$ $A = (12)(8)$

$A = 120$ $A = 96$

The total area of the two walls is 120 + 96 or 216 square feet.

Since Julia wants to use two coats of paint, she needs to cover 2 × 216 or 432 square feet of area. One gallon covers only 350 square feet, so one gallon will not be enough.

Examine Is the answer reasonable? The area of the first wall with two coats of paint is 2 × 120 or 240 square feet, which is more than one-half of 350 square feet. The answer seems reasonable.

Check for Understanding

Communicating Mathematics

1. **Draw and label** two rectangles and one parallelogram, each having an area of 12 square feet.
2. **Explain** the difference between *perimeter* and *area*.
3. **Name** the four steps of the four-step plan for problem solving.

Guided Practice

Getting Ready Find the area of each figure.

Sample:

3 in.
4 in.

Solution: The surface can be covered by 12 unit squares. The area is 4 × 3 or 12 square inches.

4.
4 cm
6 cm

5.
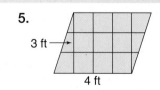
3 ft →
4 ft

Examples 1 & 2

Find the perimeter and area of each rectangle.

6.

6 ft

3 ft

7.
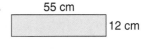
55 cm

12 cm

8. $\ell = 18$ cm, $w = 12$ cm

9. $\ell = 25$ ft, $w = 5$ ft

Example 3

10. Find the area of the parallelogram.

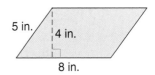
5 in.

4 in.

8 in.

Example 4

11. Interior Design An interior designer wants to order wallpaper border to place at the top of the walls in the room shown at the right. If one roll of border is 5 yards long, how many rolls of border should the designer order?

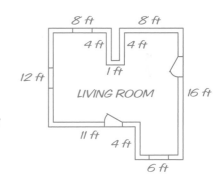
8 ft   8 ft

4 ft   4 ft

12 ft   1 ft

LIVING ROOM   16 ft

11 ft   4 ft

6 ft

Exercises

Practice

Homework Help	
For Exercises	**See Examples**
12–23, 30	1, 2
24–26, 29	3
27, 28, 31	2
Extra Practice	
See page 727.	

Find the perimeter and area of each rectangle.

12.
15 in.

4 in.

13.
10 m

10 m

14.
4 ft

10 ft

15.
1.6 mm

9.6 mm

16.
9 m

4.1 m

17.
6 mi

6 mi

Find the perimeter and area of each rectangle described.

18. $\ell = 12$ in., $w = 6$ in.

19. $\ell = 15$ ft, $w = 10$ ft

20. $\ell = 14$ m, $w = 4$ m

21. $\ell = 18$ cm, $w = 18$ cm

22. $\ell = 8.4$ mm, $w = 5$ mm

23. $\ell = 10$ mi, $w = 6.5$ mi

Find the area of each parallelogram.

24.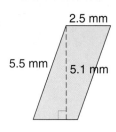

45 cm   40 cm

52 cm

25. 6 ft   5 ft

16 ft

26. 2.5 mm

5.5 mm   5.1 mm

27. Find the area of a rectangle with length 15 meters and width 2.3 meters.

28. The length of a rectangle is 24 inches, and the width of the rectangle is 18 inches. What is the area?

Applications and Problem Solving

29. Algebra   What is the base of a parallelogram with area 45 square yards and height 9 yards?

30. Algebra   What is the width of a rectangle with perimeter 18 centimeters and length 5 centimeters?

31. Remodeling   A remodeler charges $3.25 per square foot to refinish a wood floor. How much would it cost to refinish a wood floor that measures 30 feet by 18 feet?

32. Critical Thinking   A square is a rectangle in which all four sides have the same measure. Suppose *s* represents the measure of one side of a square.

a. Write a formula for the perimeter of a square.

b. Write a formula for the area of a square.

Mixed Review

33. Use a compass to determine which segment is longer, $\overline{AB}$ or $\overline{BC}$. *(Lesson 1–5)*

A   B   C

34. Advertising   A billboard reads *If you want an exciting vacation, come to Las Vegas.* Identify the hypothesis and conclusion of this statement. *(Lesson 1–4)*

Find the next three terms of each sequence.   *(Lesson 1–1)*

35. 13, 9, 5, 1, . . .

36. 50, 51, 53, 56, . . .

37. Multiple Choice   Which expression can be used to find the total cost of *b* bats and *g* gloves if a bat costs $50 and a glove costs $75? *(Algebra Review)*

Ⓐ $(50 + 75) \times (b + g)$

Ⓑ $(50 \times 75) + (b \times g)$

Ⓒ $50b + 75g$

Ⓓ $75b + 50g$

Math In the Workplace

Real Estate Agent

Do you like to work with people? Are you enthusiastic, well organized, and detail oriented? Then you may enjoy a career as a real estate agent. Real estate agents help people with an important event—buying and selling a home.

But before you can be a real estate agent, you must obtain a license. All states require prospective agents to pass a written test, which usually contains a section on real estate mathematics. Here are some typical questions.

1. Determine the total square footage of the kitchen and dinette in the blueprint.

2. How many square feet of concrete would be needed to construct a walk 7 feet wide around the outside corner of a corner lot measuring 50 feet by 120 feet?

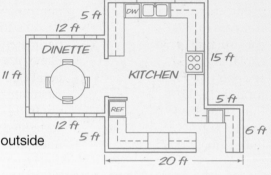

FAST FACTS About Real Estate Agents

Working Conditions
- growing number work from their homes because of advances in telecommunications
- work evenings and weekends to meet the needs of their clients

Education
- high school graduate
- 30 to 90 hours of classroom instruction about real estate mathematics and laws
- continuing education for license renewal

Earnings

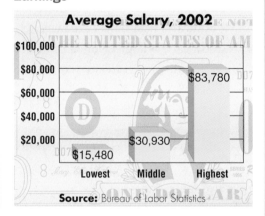

Average Salary, 2002

Lowest: $15,480
Middle: $30,930
Highest: $83,780

Source: Bureau of Labor Statistics

inter NET CONNECTION
www.geomconcepts.com

Career Data For up-to-date information about a career as a real estate agent, visit:

Understanding and Using the Vocabulary

After completing this chapter, you should be able to define each term, property, or phrase and give an example or two of each.

*inter***NET**
CONNECTION **Review Activities**
For more review activities, visit:
www.geomconcepts.com

Geometry

area *(p. 36)*
axiom *(p. 18)*
collinear *(p. 13)*
compass *(p. 30)*
construction *(p. 30)*
coplanar *(p. 14)*
endpoint *(p. 13)*
formula *(p. 35)*
four-step plan *(p. 37)*
line *(p. 12)*
line segment *(p. 13)*

midpoint *(p. 31)*
noncollinear *(p. 13)*
noncoplanar *(p. 14)*
Pascal's triangle *(p. 10)*
perimeter *(p. 35)*
plane *(p. 14)*
point *(p. 12)*
postulate *(p. 18)*
ray *(p. 13)*
straightedge *(p. 29)*
undefined terms *(p. 12)*

Logic

conclusion *(p. 24)*
conditional statement *(p. 24)*
conjecture *(p. 6)*
contrapositive *(p. 28)*
converse *(p. 25)*
counterexample *(p. 6)*
hypothesis *(p. 24)*
if-then statement *(p. 24)*
inductive reasoning *(p. 4)*
inverse *(p. 28)*

Choose the correct term to complete each sentence.

1. A (line, plane) is named using three noncollinear points.

2. The intersection of two planes is a (point, line).

3. The part following *if* in an if-then statement is called the (hypothesis, conclusion).

4. (Conjectures, Constructions) are special drawings created using only a compass and a straightedge.

5. The distance around a figure is called its (perimeter, area).

6. A conclusion reached using inductive reasoning is called a (hypothesis, conjecture).

7. A (line segment, ray) has a definite beginning and end.

8. (Hypotheses, Postulates) are facts about geometry that are accepted to be true.

9. A credit card or a piece of cardboard can serve as a (straightedge, compass).

10. It takes only one (converse, counterexample) to show that a conjecture is not true.

Skills and Concepts

Objectives and Examples	Review Exercises

• **Lesson 1–1** Identify patterns and use inductive reasoning.

The next figure in this pattern is ☐ .

Find the next three terms of each sequence.

11. 2, 3, 6, 11, . . .

12. 27, 21, 15, 9, . . .

Draw the next figure in the pattern.

13.

 www.geomconcepts.com/vocabulary_review

| **Objectives and Examples** | **Review Exercises** |

• **Lesson 1–2** Identify and draw models of points, lines, and planes and determine their characteristics.

$\overleftrightarrow{CB}$ is a line.
$\overrightarrow{EA}$ and $\overrightarrow{AD}$ are rays.
$\overline{CE}$ and $\overline{BA}$ are segments.
Points A, B, and D are collinear.

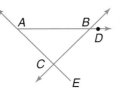

Use the figure to name examples of each term.

14. three segments
15. two rays
16. a line containing point P
17. three noncollinear points

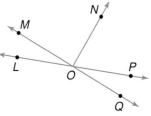

• **Lesson 1–3** Identify and use basic postulates about points, lines, and planes.

Two of the planes represented in this figure are planes ABF and ADE.

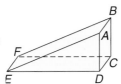

Planes ABC and BCF intersect at $\overleftrightarrow{BC}$.

18. Name the intersection of planes ADE and CDE.

19. Name three other planes represented in the figure.

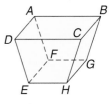

• **Lesson 1–4** Write statements in if-then form and write the converses of the statements.

Statement: All integers are rational numbers.

Write the statement in if-then form and then write its converse.

If-then: If a number is an integer, then it is a rational number.

Converse: If a number is a rational number, then it is an integer.

Identify the hypothesis and the conclusion of each statement.

20. If an animal has wings, then it is a bird.
21. All school buses are yellow.

Write two other forms of each statement.

22. Every cloud has a silver lining.
23. People who own pets live long lives.

Write the converse of each statement.

24. If the month is December, then it has 31 days.
25. All students like to play baseball.

Mixed Problem Solving
See pages 758–765.

Objectives and Examples

Review Exercises

• **Lesson 1–5** Use geometry tools.

In which figure
is the middle
segment longer?

a	b
_____	_____
_____	_____
_____	_____

The middle segment in figure a is longer.

26. Use a straightedge to
determine whether the
two heavy segments
are straight.

27. Use a compass to draw
two circles that have
two points of intersection.

• **Lesson 1–6** Use a four-step plan to solve
problems that involve the perimeters and
areas of rectangles and parallelograms.

Find the perimeter and area of a rectangle
with length 17 inches and width 5.5 inches.

Perimeter	**Area**
$P = 2\ell + 2w$	$A = \ell w$
$= 2(17) + 2(5.5)$	$= (17)(5.5)$
$= 34 + 11$ or 45	$= 93.5$

The perimeter is 45 inches, and the area is
93.5 square inches.

**Find the perimeter and area of each
rectangle.**

28.

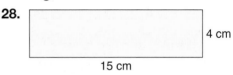

4 cm

15 cm

29. $\ell = 18$ ft, $w = 23$ ft
30. $\ell = 4.2$ m, $w = 1.5$ m

31. Find the base of a parallelogram with
height 9 centimeters and area 108 square
centimeters.

Application and Problem Solving

32. **Number Theory** Consider this statement.
*If a number is divisible by 6, it is also divisible
by 3.*
   a. Write the converse of the statement.
   b. Determine whether the converse is *true*
   or *false*. If false, give a counterexample.
   (Lesson 1–4)

33. **Construction** A rectangular patio
measures 15 feet by 18 feet. The patio will
be covered with square tiles measuring
1 foot on each side. If the tiles are $4.50
each, find the total cost of the tiles.
(Lesson 1–6)

34. **Retail Sales** A display of cereal boxes is stacked in the shape of a pyramid.
There are 4 boxes in the top row, 6 boxes in the next row, 8 boxes in the next
row, and so on. The display contains 7 rows of boxes. How many boxes are in
the seventh row? *(Lesson 1–1)*

1. **Explain** the difference between a drawing and a construction.
2. **Draw** a ray with endpoint A that also contains point B.
3. **Draw and label** a parallelogram that has an area of 24 square inches.
4. **Compare and contrast** lines and rays.

Find the next three terms of each sequence.

5. $1, 2, 4, 7, \ldots$

6. $-800, 400, -200, 100, \ldots$

7. $11, 15, 19, 23, \ldots$

For Exercises 8–11, refer to the figure at the right.

8. Name the intersection of $\overleftrightarrow{AB}$ and $\overleftrightarrow{CD}$.
9. Name the intersection of plane $\mathcal{J}$ and plane $\mathcal{L}$.
10. Name a point that is coplanar with points A and E.
11. Name three collinear points.

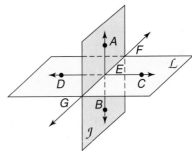

Determine whether each statement is *true* or *false*.
If false, replace the underlined word(s) to make a true statement.

12. The intersection of two planes is a point.
13. Two points determine a line.
14. A line segment has two endpoints.
15. Three collinear points determine a plane.

Write the converse of each statement. Then determine whether the converse is *true* or *false*. If false, give a counterexample.

16. If $x = 3$, then $x + 10 = 13$.
17. The sum of two odd numbers is an even number.
18. If you live in Vermont, then you live in the United States.

Use a straightedge or compass to answer each question.

19. Is the segment from A to B as long as the segment from C to D?

20. Which is longer, pencil L or R?

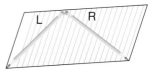

Find the perimeter and area of each rectangle described.

21. $\ell = 5$ mm, $w = 12$ mm
22. $\ell = 22$ ft, $w = 3$ ft
23. $\ell = 16$ m, $w = 14.25$ m

24. Find the area of a parallelogram with base 15 feet and height 10 feet.

25. **Agriculture** A sod farmer wants to fertilize and seed a rectangular plot of land 150 feet by 240 feet. A bag of fertilizer covers 5000 square feet, and a bag of grass seed covers 3000 square feet. How many bags of each does the farmer need to buy for this plot of land?

Number Concept Problems

All standardized tests contain numerical problems. You'll need to understand and apply these mathematical terms.

absolute value   decimals   divisibility
exponents   factors   fractions
integers   odd and even   positive and negative
prime numbers   roots   scientific notation

Problems on standardized tests often use these terms. Be sure you understand each term and read the problem carefully!

Example 1

Ricky earned about 3.6×10^4 dollars last year. If he worked 50 weeks during the year, how much did he earn per week?

A $72   B $180
C $720   D $7200

Hint Look for key terms, like *per week*. "Per" tells you to use division.

Solution You know the total amount earned in a year. You need to find the amount earned in one week. So divide the total amount by the number of weeks, 50. The total amount is written in scientific notation. Express this amount in standard notation. Then divide.

$$\frac{3.6 \times 10^4}{50} = \frac{3.6 \times 10,000}{50} \qquad 10^4 = 10,000$$

$$= \frac{3.6 \times \overset{200}{\cancel{10,000}}}{\underset{1}{\cancel{50}}} \qquad \begin{array}{l}\textit{Divide the numerator}\\\textit{and denominator by 50.}\end{array}$$

$$= 3.6 \times 200 \qquad \textit{Simplify.}$$

$$= 720 \qquad \textit{Multiply.}$$

The answer is C.

Example 2

What is the sum of the positive even factors of 12?

Hint Look for terms like *positive*, *even*, and *factor*.

Solution First, find all the factors of 12. To be sure you don't miss any factors, write all the *integers* from 1 to 12. Then cross out the numbers that are *not* factors of 12.

1 2 3 4 5̶ 6 7̶ 8̶ 9̶ 1̶0̶ 1̶1̶ 12

Reread the question. It asks for the sum of *even* factors. Circle the factors that are even numbers.

1 ②3 ④ 5̶ ⑥ 7̶ 8̶ 9̶ 1̶0̶ 1̶1̶ ⑫

Now add these factors to find the sum.

$$2 + 4 + 6 + 12 = 24$$

The answer is 24. Record it on the grid.

- Start with the *left* column.
- Write the answer in the boxes at the top. Write one digit in each column.
- Mark the correct oval in each column.
- *Never* grid a mixed number; change it to a fraction or a decimal.

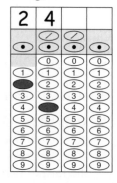

After you work each problem, record your answer on the answer sheet provided or on a sheet of paper.

Multiple Choice

1. The daily cost of renting a car is $25.00 plus $0.30 per mile driven. What is the cost of renting the car for one day and driving it 75 miles? (*Algebra Review*)

   Ⓐ $22.50   Ⓑ $27.50

   Ⓒ $47.50   Ⓓ $55.00

2. The product of a number and 1.85 is less than 1.85. Which of the following is the number? (*Algebra Review*)

   Ⓐ 1.5   Ⓑ 1

   Ⓒ 185   Ⓓ 0.75

3. Four students were asked to find the distance between their homes and school. Their responses were: 3.5 miles, $3\frac{3}{8}$ miles, $3\frac{3}{5}$ miles, and $3\frac{1}{3}$ miles. Which is the *greatest* distance? (*Algebra Review*)

   Ⓐ 3.5 miles   Ⓑ $3\frac{3}{8}$ miles

   Ⓒ $3\frac{3}{5}$ miles   Ⓓ $3\frac{1}{3}$ miles

4. If the pattern below continues, what will the 18th figure look like? (*Lesson 1–1*)

   Ⓐ △   Ⓑ ▢

   Ⓒ △   Ⓓ ▢

5. In 2003, about 746,000,000 CDs were shipped in the United States. What is another way of expressing the number 746,000,000? (*Algebra Review*)

   Ⓐ 7.46×10^8 Ⓑ 74.6×10^8

   Ⓒ 74.6 million   Ⓓ 74.6 billion

6. Which of the following expresses the prime factorization of 54? (*Algebra Review*)

   Ⓐ 9×6 Ⓑ $3 \times 3 \times 6$

   Ⓒ $3 \times 3 \times 2$ Ⓓ $3 \times 3 \times 3 \times 2$

   Ⓔ 5.4×10

7. If 8 and 12 are each factors of *K*, what is the value of *K*? (*Algebra Review*)

   Ⓐ 6   Ⓑ 24

   Ⓒ 8   Ⓓ 96

   Ⓔ It cannot be determined from the information given.

8. A rectangle has a perimeter of 38 feet and an area of 48 square feet. What are the dimensions of the rectangle? (*Lesson 1–6*)

   Ⓐ 4 ft × 12 ft   Ⓑ 6 ft × 8 ft

   Ⓒ 16 ft × 3 ft   Ⓓ 24 ft × 2 ft

Grid In

9. Dr. Cronheim has 379 milliliters of solution to use for a class experiment. She divides the solution evenly among the 24 students. If she has 19 milliliters of the solution left after the experiment, how many milliliters of the solution did she give to each student? (*Algebra Review*)

Extended Response

10. The altitude in Galveston, Texas, is about 10 feet. There are 5280 feet in a mile. (*Algebra Review*)

    Part A Explain how you can find this altitude in miles.

    Part B Give your answer both as a fraction and as a decimal to the nearest ten-thousandth.

Segment Measure and Coordinate Graphing

What You'll Learn

Key Ideas

- Find the distance between two points on a number line. *(Lesson 2–1)*

- Apply the properties of real numbers to the measures of segments. *(Lesson 2–2)*

- Identify congruent segments and find the midpoints of segments. *(Lesson 2–3)*

- Name and graph ordered pairs on a coordinate plane. *(Lesson 2–4)*

- Find the coordinates of the midpoint of a segment. *(Lesson 2–5)*

Key Vocabulary

coordinate plane *(p. 68)*

equation *(p. 57)*

graph *(p. 68)*

ordered pair *(p. 68)*

origin *(pp. 52, 68)*

Why It's Important

Science While most cats hunt by stalking and pouncing, a cheetah runs down its prey. Several characteristics make the cheetah built for speed. They have grooved pads on their feet for stopping, a flexible spine for quick turns, large nostrils and lungs for extra oxygen, and lightweight bones and small teeth for low body weight.

Segment measure and coordinate graphing are essential skills for the study of geometry. You will use a coordinate graph to analyze the speed of a cheetah and other animals in Lesson 2-4.

✓ Check Your Readiness

Solve each equation. Check your solution.

✓ Algebra
Review, p. 722

1. $5 + b = 9$

2. $12 + x = 21$

3. $34 = m + 15$

4. $a + 12 = 21$

5. $17 = q + 4$

6. $65 = d + 32$

7. $n - 6 = 3$

8. $16 = p - 5$

9. $42 - k = 27$

10. $10 - x = 4$

11. $16 - t = 5$

12. $2n - 4 = 8$

✓ Algebra
Review, p. 723

13. $3j + 2 = 23$

14. $14 - 2v = 6$

15. $11 + 7p = 39$

16. $5w + 2 = 27$

17. $26 = 3z + 2$

18. $36 - 5a = 1$

19. $\dfrac{x + 5}{2} = 9$

20. $\dfrac{y + 8}{2} = 10$

21. $\dfrac{y + (-1)}{2} = -3$

✓ Algebra
Review, p. 724

22. $c + 6 = 3c$

23. $6w = 3w + 12$

24. $2g = 5g - 6$

25. $8t + 6 = 10t - 4$

26. $37 - g = 4g + 2$

27. $18 - 3h = 26 - 5h$

28. $6f + 15 = 11f$

29. $q + 9 = 3q - 1$

30. $8r - 1 = 5r + 17$

Study Organizer

Make this Foldable to help you organize your Chapter 2 notes. Begin with a sheet of notebook paper.

❶ **Fold** lengthwise to the holes.

❷ **Cut** along the top line and then cut 10 tabs.

❸ **Label** each tab with a highlighted term from the chapter.

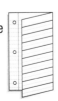

Reading and Writing Store your Foldable in a 3-ring binder. As you read and study the chapter, write definitions of important terms and examples under each tab.

Real Numbers and Number Lines

Numbers that share common properties can be *classified* or grouped into sets. Different sets of numbers can be shown on number lines.

Whole Numbers

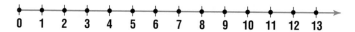

This figure shows the set of **whole numbers**. The whole numbers include 0 and the **natural**, or counting, **numbers**. The arrow to the right indicates that the whole numbers continue indefinitely. Zero is the least whole number.

Integers

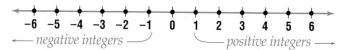

A number line can be used to represent the set of **integers**. Integers include 0, the positive integers, and the negative integers. The arrows indicate that the numbers go on forever in both directions.

Rational Numbers

A number line can also show **rational numbers**. A rational number is any number that can be written as a fraction, $\frac{a}{b}$, where a and b are integers and b cannot equal 0. The number line above shows some of the rational numbers between -2 and 2. In fact, there are infinitely many rational numbers between any two integers.

Rational numbers can also be represented by decimals.

$$\frac{3}{8} = 0.375 \qquad \frac{2}{3} = 0.666\ldots \qquad \frac{0}{5} = 0$$

Decimals may be **terminating** or **nonterminating**.

0.375 and 0.49 are terminating decimals.
0.666 . . . and -0.12345 . . . are nonterminating decimals.

The three periods following the digits in the nonterminating decimals indicate that there are infinitely many digits in the decimal.

Some nonterminating decimals have a repeating pattern.

0.171717 . . . repeats the digits 1 and 7 to the right of the decimal point.

A bar over the repeating digits is used to indicate a repeating decimal.

0.171717 . . . = $0.\overline{17}$

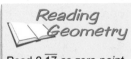
Each rational number can be expressed as a terminating decimal or a nonterminating decimal with a repeating pattern.

Irrational Numbers

Decimals that are nonterminating *and* do not repeat are called **irrational numbers**.

6.028716 . . . and 0.101001000 . . . appear to be irrational numbers.

Real Numbers

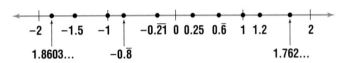

Real numbers include both rational and irrational numbers. The number line above shows some real numbers between −2 and 2.

Postulate 2–1 Number Line Postulate	Each real number corresponds to exactly one point on a number line. Each point on a number line corresponds to exactly one real number.

Examples

For each situation, write a real number with ten digits to the right of the decimal point.

1 a rational number less than 10 with a 3-digit repeating pattern

Sample answer:
5.1231231231 . . .

2 an irrational number between −4 and −2

Sample answer:
−2.6366366636 . . .

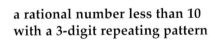

Your Turn

a. a rational number greater than −10 with a 2-digit repeating pattern

b. an irrational number between 1 and 2

The number that corresponds to a point on a number line is called the **coordinate** of the point. On the number line below, 10 is the coordinate of point A. The coordinate of point B is −4. Point C has coordinate 0 and is called the **origin**.

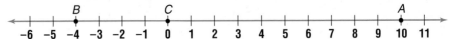

The distance between two points A and B on a number line is found by using the Distance and Ruler Postulates.

Postulate 2–2 **Distance** **Postulate**	**Words:** For any two points on a line and a given unit of measure, there is a unique positive real number called the **measure** of the distance between the points. **Model:** 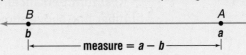

Postulate 2–3 **Ruler** **Postulate**	**Words:** Points on a line are paired with the real numbers, and the measure of the distance between two points is the positive difference of the corresponding numbers. **Model:**

Suppose you want to find the distance between points R and S on the number line below.

Reading Geometry

RS represents *the measure of the distance between points R and S.*

The measure of the distance between points R and S is the positive difference 11 − 3, or 8. The notation for the measure of the distance between two points is indicated by the capital letters representing the points. Since the measure from point S to point R is the same as from R to S, you can write RS = 8 or SR = 8.

Another way to calculate the measure of the distance is by using **absolute value**. The absolute value of a number is the number of units a number is from zero on the number line. In symbols, the absolute value is denoted by two vertical slashes.

$$SR = |11 - 3| \qquad\qquad RS = |3 - 11|$$
$$= |8| \qquad\qquad\qquad = |-8|$$
$$= 8 \qquad\qquad\qquad\quad = 8$$

| Example | 3 |

Example — **3** Use the number line to find *BE*.

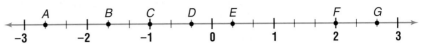

The coordinate of *B* is $-1\frac{2}{3}$, and the coordinate of *E* is $\frac{1}{3}$.

$BE = \left| -1\frac{2}{3} - \frac{1}{3} \right|$ *Subtract to find distance.*

$= |-2|$ or 2   *Absolute values are nonnegative.*

Algebra Review

Operations with Fractions, p. 721

Your Turn

Use the number line above to find each measure.

c. *CF* **d.** *AD* **e.** *BG*

Highways with their mile markers can represent number lines.

Real World

Example — **4**

Travel Link

Jamal traveled on I-71 from Grove City to Washington Courthouse. The Grove City entrance to I-71 is at the 100-mile marker, and the Washington Courthouse exit is at the 66-mile marker. How far did Jamal travel on I-71?

$|100 - 66| = |34|$ or 34   *Ruler Postulate*

Jamal traveled 34 miles on I-71.

Check for Understanding

Communicating Mathematics

1. **Explain** why a number line has arrows at each end.

2. **Write** a problem that can be solved by finding $|9 - 17|$. What is the value of $|9 - 17|$?

3. Consider 0.34, $0.3\overline{4}$, and $0.\overline{34}$.
   a. How are these numbers alike? How are they different?
   b. Which is greatest?
   c. How would you read each number?

Vocabulary

whole numbers
natural numbers
integers
rational numbers
terminating decimals
nonterminating decimals
irrational numbers
real numbers
coordinate
origin
measure
absolute value

4. Writing Math Copy and complete the diagram at the right. Give two examples of each type of number represented in the large rectangle. Write a paragraph describing how this diagram shows the relationship among different sets of numbers.

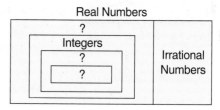

Real Numbers

?

Integers

?

?

Irrational Numbers

Guided Practice
Examples 1 & 2

For each situation, write a real number with ten digits to the right of the decimal point.

5. an irrational number between 1 and 2

6. a rational number greater than 10 with a 2-digit repeating pattern

Example 3

Use the number line to find each measure.

7. *CD*   8. *BF*   9. *EG*

Example 3

10. **Geography** In the Netherlands, the higher region of the Dunes protects the lower region of the Polders from the sea. The Dunes rise to 25 feet above sea level. The lowest point of the Polders is 22 feet below sea level.

a. Represent these two numbers on a number line.

b. Find the distance between these two points on the number line.

The Netherlands

Exercises

• • • •   • • • • • • • • • • • • • • •

Practice

For each situation, write a real number with ten digits to the right of the decimal point.

11. a rational number less than 0 with a 2-digit repeating pattern

12. an irrational number between 5 and 6

13. a rational number greater than 3 with a 4-digit repeating pattern

14. a rational number between −3.5 and −4 with a 3-digit repeating pattern

15. two irrational numbers between 0 and 1

16. an irrational number between −7 and −6.8

Homework Help

For Exercises	See Examples
11–16	1, 2
17–28, 31	3
29, 30	4

Extra Practice

See page 728.

Use the number line to find each measure.

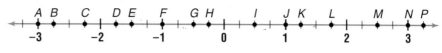

```
    A B   C   D E   F   G H     I   J K   L     M   N P
  ←─┼─┼─┼─┼─┼─┼─┼─┼─┼─┼─┼─┼─┼─┼─┼─┼─┼─┼─┼─┼─┼─┼─→
      -3      -2      -1      0      1      2      3
```

17. *AJ*	**18.** *AN*	**19.** *EG*	**20.** *IM*	**21.** *JK*
22. IN	**23.** *FK*	**24.** *AP*	**25.** *CK*	**26.** *HM*

27. Find the measure of the distance between *B* and *J*.

28. What is the measure of the distance between *D* and *L*?

Applications and Problem Solving

29. Sports   Hatsu is practicing on a rock-climbing range. Markers on the wall indicate the number of feet she has climbed. When Hatsu started, she reached for a handhold at the 6-foot marker. She is now reaching for the 22-foot marker.

   a. How much higher is the current handhold than the first one?

   b. If the highest handhold is at the 35-foot marker, how far does she need to climb?

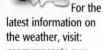
Math Online **Data Update**   For the latest information on the weather, visit: geomconcepts.com

30. Weather   The normal high and low temperatures for four cities for January are given in degrees Celsius. Find the measure of the difference between the two temperatures.

   a. Boston, 2°C, −6°C   **b.** San Francisco, 13°C, 6°C

   c. Chicago, −2°C, −11°C   **d.** Houston, 16°C, 4°C

31. Critical Thinking   Name two points that are 7 units from −5 on the number line. (*Hint:* Use a number line.)

Mixed Review

Find the perimeter and area of each rectangle.   *(Lesson 1–6)*

32.
```
┌──────────┐
│          │ 8 ft
│          │
└──────────┘
    8 ft
```

33.
```
┌──────────────┐
│              │ 6 cm
└──────────────┘
      10 cm
```

34.
```
┌────────┐
│        │ 13 m
│        │
│        │
└────────┘
   8 m
```

Name the tool needed to draw each figure.   *(Lesson 1–5)*

35. circle   **36.** straight line

Standardized Test Practice
Ⓐ Ⓑ Ⓒ Ⓓ

37. Multiple Choice   Monsa purchased shoes that were originally priced at $84.00. On that day, the store was having a 10% off sale. The sales tax was 7%. How much did Monsa pay for the shoes? *(Percent Review)*

    Ⓐ $70.31   Ⓑ $80.89   Ⓒ $85.93   Ⓓ $98.87

Given three collinear points on a line, one point is always between the other two points. In the figure below, point *B* is between points *A* and *C*.

Point *B* lies to the right of point *A* and to the left of point *C*. **Betweenness** is also defined in terms of distances.

Definition of Betweenness	**Words:**	Point *R* is between points *P* and *Q* if and only if *R*, *P*, and *Q* are collinear and $PR + RQ = PQ$.
	Model:	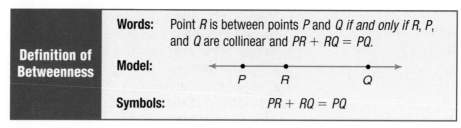
	Symbols:	$PR + RQ = PQ$

If and only if means that both the statement and its converse are true. Statements that include this phrase are called biconditionals.

Unless stated otherwise, betweenness and collinearity of points may be assumed if they are given in a figure.

Example —① **Points *A*, *B*, and *C* are collinear. If *AB* = 12, *BC* = 48, and *AC* = 36, determine which point is between the other two.**

Check to see which two measures add to equal the third.

$$12 + 36 = 48$$
$$BA + AC = BC$$

Therefore, *A* is between *B* and *C*.

Check: You can check by modeling the distances on a number line. Let 12 units = 1 inch.

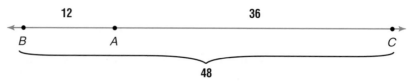

The solution checks.

Your Turn

a. Points *R*, *S*, and *T* are collinear. If *RS* = 42, *ST* = 17, and *RT* = 25, determine which point is between the other two.

Segment measures are real numbers. Let's review some of the properties of real numbers relating to equality.

Properties of Equality for Real Numbers	
Reflexive Property	For any number a, $a = a$.
Symmetric Property	For any numbers a and b, if $a = b$, then $b = a$.
Transitive Property	For any numbers a, b and c, if $a = b$ and $b = c$, then $a = c$.
Addition and Subtraction Properties	For any numbers a, b, and c, if $a = b$, then $a + c = b + c$ and $a - c = b - c$.
Multiplication and Division Properties	For any numbers a, b, and c, if $a = b$, then $a \cdot c = b \cdot c$, and if $c \neq 0$, then $\frac{a}{c} = \frac{b}{c}$.
Substitution Property	For any numbers a and b, if $a = b$, then a may be replaced by b in any equation.

A statement that includes the symbol = is an **equation** or equality. You can use equations to solve problems in geometry.

Example 2
Algebra Link

If $QS = 29$ and $QT = 52$, find ST.

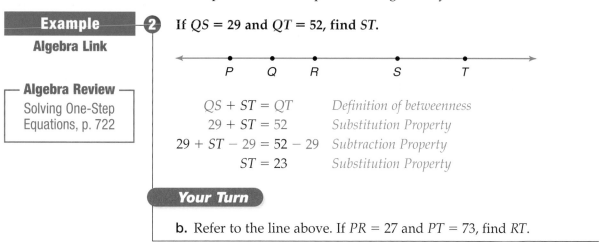

$QS + ST = QT$	*Definition of betweenness*
$29 + ST = 52$	*Substitution Property*
$29 + ST - 29 = 52 - 29$	*Subtraction Property*
$ST = 23$	*Substitution Property*

─ **Algebra Review** ─
Solving One-Step
Equations, p. 722

Your Turn

b. Refer to the line above. If $PR = 27$ and $PT = 73$, find RT.

Measurements, such as 10 centimeters and 4 inches, are composed of two parts: the measure and the **unit of measure**. The measure of a segment gives the number of units. When only measures are given in a figure in this text, you can assume that all of the measures in the figure have the same unit of measure.

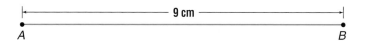

The measure of $\overline{AB}$ is 9, and $AB = 9$. The unit of measure is the centimeter. So, the measurement of $\overline{AB}$ is 9 centimeters.

The measurement of a segment is also called the *length* of the segment.

Example 3

Find the length of $\overline{XY}$ in centimeters and in inches.

Use a metric ruler to measure the segment. Put the 0 point at point X. *Caution: This point may not be at the end of the ruler.*
Then measure the distance to Y on the metric scale.

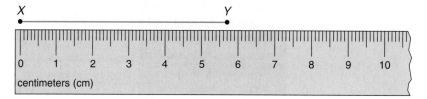

The length of $\overline{XY}$ is 5.7 centimeters.

Use a customary ruler to measure $\overline{XY}$ in inches. Put the 0 point at X and measure the distance to Y.

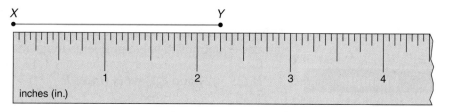

The length of $\overline{XY}$ is $2\frac{1}{4}$ inches.

The **precision** of a measurement depends on the smallest unit used to make the measurement. The **greatest possible error** is half the smallest unit used to make the measurement. The **percent of error** is found by comparing the greatest possible error with the measurement itself.

$$\text{percent of error} = \frac{\text{greatest possible error}}{\text{measurement}} \times 100\%$$

Compare the two measurements of $\overline{XY}$ in Example 3.

Centimeters	Inches
measurement: 5.7 cm or 57 mm	measurement: $2\frac{1}{4}$ (or 2.25) in.
precision: 1 mm	precision: $\frac{1}{16}$ in.
greatest possible error: 0.5 mm	greatest possible error: $\frac{1}{32}$ (or 0.03125) in.
percent of error: $\frac{0.5}{57} \times 100\%$ or about 0.88%	percent of error: $\frac{0.03125}{2.25} \times 100\%$ or about 1.39%

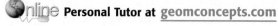 **Personal Tutor at geomconcepts.com**

Communicating Mathematics

1. **Write** a sentence that explains the difference between the *measure* and the *measurement* of a segment.

2. **Name** some units of measure for length.

3. **YOU Decide?** Jalisa says that the most precise measurement for a can of corn would be 2 pounds. Joseph says that 34 ounces is more precise. Who is correct, and why?

Vocabulary

betweenness
equation
measurement
unit of measure
precision
greatest possible error
percent of error

Guided Practice

Three segment measures are given. The three points named are collinear. Determine which point is between the other two.

Example 1

4. $TM = 21$, $MH = 37$, $TH = 16$ 5. $XZ = 36$, $YZ = 17$, $XY = 19$

Example 2

Refer to the line for Exercises 6–7.

6. If $AB = 23$ and $AD = 51$, find BD.

7. If $CD = 19$ and $AC = 38$, find AD.

Example 3

Find the length of each segment in centimeters and in inches.

8. ─────────────────   9. ─────────────────────

Example 2

10. **Travel**   Emilio drives on Route 40 from Little Rock to Nashville. He stops in Memphis for lunch. The distance from Little Rock to Memphis is 139 miles, and the distance from Little Rock to Nashville is 359 miles. How far does Emilio need to travel after lunch to reach Nashville?

Exercises

Practice

Three segment measures are given. The three points named are collinear. Determine which point is between the other two.

11. $AD = 25$, $ED = 33$, $AE = 58$ 12. $RS = 45$, $TS = 19$, $RT = 26$

13. $GH = 44$, $HK = 87$, $GK = 43$ 14. $PQ = 34$, $QR = 71$, $PR = 37$

15. $AB = 32$, $BC = 13.8$, $AC = 18.2$ 16. $WV = 27.6$, $VZ = 35.8$, $WZ = 8.2$

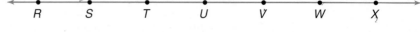

Homework Help	
For Exercises	See Examples
11–16	1
17–22, 31	2
23–29	3
Extra Practice	
See page 728.	

17. If $RS = 19$ and $RV = 71$, find SV.

18. If $UV = 17$ and $SU = 38$, find SV.

19. If $VX = 13$ and $SX = 30$, find SV.

20. If $TW = 81$ and $VW = 35$, find TV.

21. If $SW = 44.5$ and $SV = 37.1$, find VW.

22. If $TU = 15.9$ and $UW = 28.3$, find TW.

Find the length of each segment in centimeters and in inches.

23. ────────────

24. ──────

25. ─────────

26. ─────

27. ──────────────────────

28. ────────────────────

Applications and Problem Solving

29. Auto Mechanics   Lucille Treganowan is a grandmother with a weekly TV show on auto repair. She uses a socket wrench to tighten and loosen bolts on cars. Measure the distance across the head of each bolt in millimeters to find the size of socket needed for the bolt.

a.

b.

c.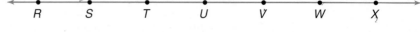

30. Clothing   The sizes of men's hats begin at $6\frac{1}{4}$ and go up by $\frac{1}{8}$ inch. How precise are the hat sizes?

31. Critical Thinking   If $AB = 5$, $BD = 14$, $CE = 19$, and $AE = 35$, find BC, CD, and DE.

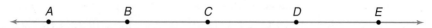

Mixed Review

Use the number line to find each measure. *(Lesson 2–1)*

32. *BG* **33.** *AF* **34.** *DE*

35. Photography The outer edges of a picture frame are 21 inches by 15 inches. The sides of the frame are 2 inches wide. *(Lesson 1–6)*

  a. Draw a picture to represent the frame. Label all the information presented in the problem.

  b. Find the area of a picture that will show in this frame.

36. Short Response Describe the intersection of two planes. *(Lesson 1–3)*

37. Multiple Choice Which point is collinear with *T* and *U*? *(Lesson 1–2)*

  Ⓐ *R* Ⓑ *S*
  Ⓒ *V* Ⓓ *W*

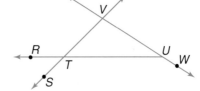

Quiz 1   Lessons 2–1 and 2–2

1. Write a rational number between 4 and 5 with a 3-digit repeating pattern. Name ten digits to the right of the decimal point. *(Lesson 2–1)*

2. Use the number line to find *PQ*. *(Lesson 2–1)*

3. Points *R*, *S*, and *T* are collinear. If *RS* = 71, *ST* = 55, and *RT* = 16, determine which point is between the other two. *(Lesson 2–2)*

4. Refer to the line below. If *AB* = 28 and *AC* = 44, find *BC*. *(Lesson 2–2)*

5. Find the length of the line segment in centimeters and in inches. *(Lesson 2–2)*

In geometry, two segments with the same length are called **congruent segments**.

Definition of Congruent Segments	Two segments are congruent if and only if they have the same length.

In the figures at the right, $\overline{AB}$ is congruent to $\overline{BC}$, and $\overline{PQ}$ is congruent to $\overline{RS}$. The symbol $\cong$ is used to represent congruence.

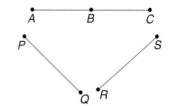

$$\overline{AB} \cong \overline{BC} \text{ and } \overline{PQ} \cong \overline{RS}$$

From the definition of congruent segments, we can also say $AB = BC$ and $PQ = RS$.

Example ──① **Use the number line to determine whether the statement is *true* or *false*. Explain your reasoning.**

$$\overline{DF} \text{ is congruent to } \overline{EG}.$$

Because $DF = 8$ and $EG = 7$, $DF \neq EG$. So, $\overline{DF}$ is not congruent to $\overline{EG}$, and the statement is false.

Your Turn

a. Is the statement $\overline{EG} \cong \overline{FH}$ true or false? Explain your reasoning.

Since congruence is related to the equality of segment measures, there are properties of congruence that are similar to the corresponding properties of equality. These statements are called **theorems**. Theorems are statements that can be justified by using logical reasoning.

We know that $AB = AB$. Therefore, $\overline{AB} \cong \overline{AB}$ and we can see that congruence is reflexive. You can make similar arguments to show congruence is symmetric and transitive.

Theorem	Words	Symbols
2–1	Congruence of segments is reflexive.	$\overline{AB} \cong \overline{AB}$
2–2	Congruence of segments is symmetric.	If $\overline{AB} \cong \overline{CD}$, then $\overline{CD} \cong \overline{AB}$.
2–3	Congruence of segments is transitive.	If $\overline{AB} \cong \overline{CD}$ and $\overline{CD} \cong \overline{EF}$, then $\overline{AB} \cong \overline{EF}$.

Example ──2 **Determine whether the statement is *true* or *false*. Explain your reasoning.**

$$\overline{JK} \text{ is congruent to } \overline{KJ}.$$

Congruence of segments is reflexive, so $\overline{JK} \cong \overline{JK}$. We know that $\overline{KJ}$ is another name for $\overline{JK}$. By substitution, $\overline{JK} \cong \overline{KJ}$. The statement is true.

Your Turn

b. If $\overline{AB} \cong \overline{CD}$ and $\overline{DC} \cong \overline{EF}$, then $\overline{AB} \cong \overline{EF}$.

There is a unique point on every segment called the **midpoint**. On the number line below, M is the midpoint of $\overline{ST}$. What do you notice about SM and MT?

Definition of Midpoint	Words:	A point M is the midpoint of a segment $\overline{ST}$ if and only if M is between S and T and $SM = MT$.
	Model:	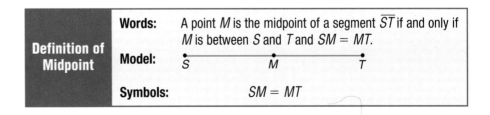
	Symbols:	$SM = MT$

The midpoint of a segment separates the segment into two segments of equal length. So, by the definition of congruent segments, the two segments are congruent.

Example **3** **Algebra Link**

In the figure, *C* is the midpoint of $\overline{AB}$. Find the value of *x*.

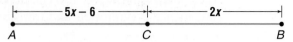

Explore   You are given a segment and its midpoint. You want to find the value of *x*.

Plan   Since *C* is the midpoint of $\overline{AB}$, *AC* = *CB*. Use this information to write an equation involving *x*, and solve for *x*.

Solve

$AC = CB$	*Definition of Midpoint*
$5x - 6 = 2x$	*Substitute.*
$5x - 6 - 5x = 2x - 5x$	*Subtract 5x from each side.*
$-6 = -3x$	*Simplify.*
$\dfrac{-6}{-3} = \dfrac{-3x}{-3}$	*Divide each side by −3.*
$2 = x$	*Simplify.*

Examine   Replace *x* with 2 to find *AC* and *CB*.

$AC = 5x - 6$	*Original equation*	$CB = 2x$	
$= 5(2) - 6$	*Substitution Property*	$= 2(2)$	
$= 4$	*Simplify.*	$= 4$	

Since *AC* = *CB*, *C* is the midpoint of $\overline{AB}$, and the answer is correct.

Your Turn

c. In the figure below, *W* is the midpoint of $\overline{XY}$. Find the value of *a*.

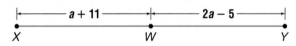

To **bisect** something means to separate it into two congruent parts. The midpoint of a segment bisects the segment because it separates the segment into two congruent segments. A point, line, ray, segment, or plane can also bisect a segment.

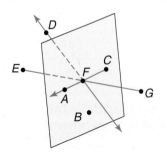

Point F bisects $\overline{EG}$.

$\overleftrightarrow{FD}$ bisects $\overline{EG}$.

$\overrightarrow{FA}$ bisects $\overline{EG}$.

$\overline{AC}$ bisects $\overline{EG}$.

Plane ABC bisects $\overline{EG}$.

The midpoint of the segment must be found to separate a segment into two congruent segments. If the segment is part of a number line, you can use arithmetic to find the midpoint. If there is no number line, you can use a construction to find the midpoint.

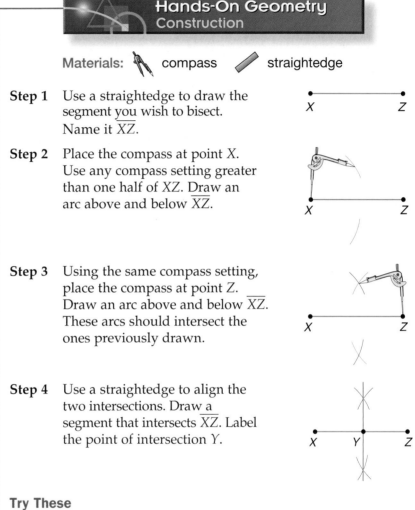

Hands-On Geometry
Construction

Materials: compass   straightedge

Step 1 Use a straightedge to draw the segment you wish to bisect. Name it $\overline{XZ}$.

Step 2 Place the compass at point X. Use any compass setting greater than one half of XZ. Draw an arc above and below $\overline{XZ}$.

Step 3 Using the same compass setting, place the compass at point Z. Draw an arc above and below $\overline{XZ}$. These arcs should intersect the ones previously drawn.

Step 4 Use a straightedge to align the two intersections. Draw a segment that intersects $\overline{XZ}$. Label the point of intersection Y.

Try These

1. Measure $\overline{XY}$ and $\overline{YZ}$. What can you conclude about point Y?
2. Fold $\overline{XZ}$ so that Z is over X. Does this confirm your conclusion in Exercise 1?
3. Can you make any other conjectures about the line segment that intersects $\overline{XZ}$?

Check for Understanding

Communicating Mathematics

1. **Draw** two diagrams or find two photographs that illustrate the use of congruent segments when building houses in the area where you live.

2. a. Explain why segment congruence is symmetric.

   b. Explain why segment congruence is transitive.

Guided Practice

Example 1

Use the number line to determine whether each statement is *true* or *false*. Explain your reasoning.

3. $\overline{AB}$ is congruent to $\overline{CD}$.   **4.** D is the midpoint of $\overline{CE}$.

Example 2

Determine whether each statement is *true* or *false*. Explain your reasoning.

   5. If $\overline{XY} \cong \overline{YZ}$, then Y is the midpoint of $\overline{ZY}$.

   6. If $\overline{RS} \cong \overline{CD}$, then $\overline{CD} \cong \overline{RS}$.

Example 3

   7. Algebra   In the figure below, M is the midpoint of $\overline{PQ}$. Find the value of x.

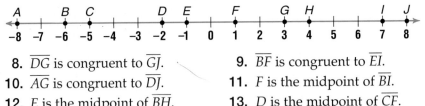

Exercises

• • • • • • • • • • • • • • • • • • •

Practice

Use the number line to determine whether each statement is *true* or *false*. Explain your reasoning.

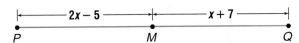

Homework Help	
For Exercises	**See Examples**
8–10	1
11–13, 21, 22	3
15	2
Extra Practice	
See page 728.	

8. $\overline{DG}$ is congruent to $\overline{GJ}$.   **9.** $\overline{BF}$ is congruent to $\overline{EI}$.

10. $\overline{AG}$ is congruent to $\overline{DJ}$.   **11.** F is the midpoint of $\overline{BI}$.

12. E is the midpoint of $\overline{BH}$.   **13.** D is the midpoint of $\overline{CF}$.

Determine whether each statement is *true* or *false*. Explain your reasoning.

14. If $XY = YZ$, then $\overline{XY} \cong \overline{YZ}$.

15. If $\overline{AB} \cong \overline{BC}$, $\overline{XY} \cong \overline{FG}$, and $\overline{BC} \cong \overline{FG}$, then $\overline{AB} \cong \overline{XY}$.

16. Every segment has only one bisector.

17. A plane can bisect a segment at an infinite number of points.

18. If $\overline{RS} \cong \overline{ST}$, then S is the midpoint of $\overline{RT}$.

19. If points D, E, and F are collinear and E is not between D and F, then F is between D and E.

20. Draw a segment like $\overline{MN}$ on your paper. Then use a compass and straightedge to bisect the segment.

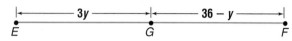

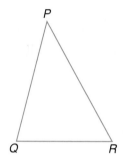

Applications and Problem Solving

21. Algebra In the figure below, G is the midpoint of $\overline{EF}$.

$$\overset{\longleftarrow\text{ }3y\text{ }\longrightarrow}{\underset{E}{\bullet}}\overset{\longleftarrow\text{ }36-y\text{ }\longrightarrow}{\underset{G}{\bullet}\quad\quad\underset{F}{\bullet}}$$

    a. Find the value of y.

    b. Find EG and GF.

    c. Find EF.

22. Science The center of mass of an object is the point where the object can be balanced in all directions. Draw the shape of a triangular object like the one at the right. Use the following steps to find its center of mass.

    a. Find the midpoint of each side of the triangle.

    b. Draw a segment between the midpoint of $\overline{QR}$ and P.

    c. Draw a segment between the midpoint of $\overline{PR}$ and Q.

    d. Draw a segment between the midpoint of $\overline{PQ}$ and R.

    e. The center of mass is the point where these three segments intersect. Label the center of mass C.

23. Critical Thinking In the figure below, C is any point between A and B, E is the midpoint of $\overline{AC}$, and F is the midpoint of $\overline{CB}$. Write a ratio comparing AB to EF.

$$\underset{A}{\bullet}\quad\underset{E}{\bullet}\quad\underset{C}{\bullet}\quad\quad\quad\quad\underset{F}{\bullet}\quad\quad\underset{B}{\bullet}$$

Mixed Review

Three segment measures are given. The three points named are collinear. Determine which point is between the other two. *(Lesson 2–2)*

24. $MN = 17$, $NP = 6.5$, $MP = 23.5$ **25.** $RS = 7.1$, $TR = 2.9$, $TS = 4.2$

26. Write an irrational number between 0 and -2 that has ten digits to the right of the decimal point. *(Lesson 2–1)*

Standardized Test Practice
Ⓐ Ⓑ Ⓒ Ⓓ

27. Grid In A soccer field is a rectangle that is 100 meters long and 73 meters wide. Find the area of the soccer field in square meters. *(Lesson 1–6)*

28. Multiple Choice
Solve $2y + 3 = 9$. *(Algebra Review)*
    Ⓐ 6   Ⓑ 5
    Ⓒ 4   Ⓓ 3

In coordinate geometry, grid paper is used to locate points. The plane of the grid is called the **coordinate plane**.

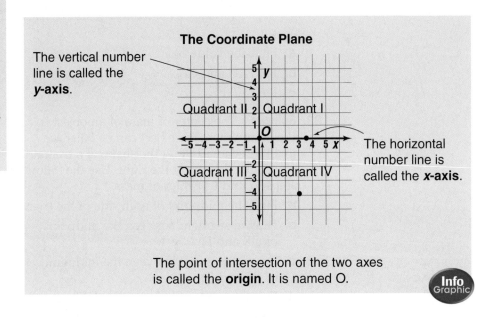

The Coordinate Plane

The vertical number line is called the **y-axis**.

Quadrant II   Quadrant I

Quadrant III   Quadrant IV

The horizontal number line is called the **x-axis**.

The point of intersection of the two axes is called the **origin**. It is named O.

Info Graphic

The two axes separate the plane into four regions called **quadrants**. Points can lie in one of the four quadrants or on an axis. The points on the *x*-axis to the right of the origin correspond to positive numbers. To the left of the origin, the points correspond to negative numbers. The points on the *y*-axis above the origin correspond to positive numbers. Below the origin, the points correspond to negative numbers.

An **ordered pair** of real numbers, called the **coordinates** of a point, locates a point in the coordinate plane. Each ordered pair corresponds to exactly one point in the coordinate plane.

The point in the coordinate plane is called the **graph** of the ordered pair. Locating a point on the coordinate plane is called *graphing* the ordered pair.

Postulate 2–4 Completeness Property for Points in the Plane	Each point in a coordinate plane corresponds to exactly one ordered pair of real numbers. Each ordered pair corresponds to exactly one point in a coordinate plane.

The figure at the right shows the graph of the ordered pair (5, 3). The first number, 5, is called the **x-coordinate**. It tells the number of units the point lies to the left or right of the origin. The second number, 3, is called the **y-coordinate**. It tells the number of units the point lies above or below the origin. What are the coordinates of the origin?

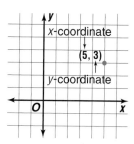

Examples

1 **Graph point A at (2, −3).**

Start at the origin. Move 2 units to the right. Then, move 3 units down. Label this point A.
The location of A at (2, −3) is also written as A(2, −3).

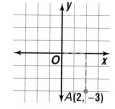

2 **Name the coordinates of points B and C.**

Point B is 2 units to the left of the origin and 4 units above the origin. Its coordinates are (−2, 4).

Point C is zero units to the left or right of the origin and 3 units below the origin. Its coordinates are (0, −3).

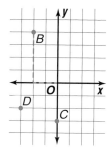

Your Turn

a. Graph point E at (−3, 4).
b. Name the coordinates of point D.

Hands-On Geometry

Materials: grid paper

Step 1 Draw lines representing the x-axis and y-axis on a piece of grid paper. Label the x-axis, y-axis, and the origin.

Step 2 Graph the points P(3, 4), Q(3, 0), R(3, −1), and S(3, −3).

(continued on the next page)

COncepts in MOtion
Interactive Lab
geomconcepts.com

Try These

1. What do you notice about the graphs of these points?

2. What do you notice about the *x*-coordinates of these points?

3. Name and graph three other points with an *x*-coordinate of 3. What do you notice about these points?

4. Write a general statement about ordered pairs that have the same *x*-coordinate.

5. Now graph $T(-4, -2)$, $U(0, -2)$, $V(1, -2)$, and $W(2, -2)$.

6. What do you notice about the graphs of these points?

7. What do you notice about the *y*-coordinates of these points?

8. Write a general statement about ordered pairs that have the same *y*-coordinate.

Horizontal and vertical lines in the coordinate plane have special characteristics. All lines can be described, or named, by equations. If a vertical line passes through (3, 4), then the *x*-coordinate of all points on the line is 3. If a horizontal line passes through $(-4, -2)$, then the *y*-coordinate of all points on the line is -2.

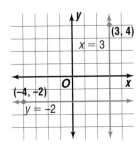

Theorem 2–4 summarizes this relationship for any vertical or horizontal line.

Theorem 2–4	If *a* and *b* are real numbers, a vertical line contains all points (x, y) such that $x = a$, and a horizontal line contains all points (x, y) such that $y = b$.

The equation of a vertical line is $x = a$, and the equation of a horizontal line is $y = b$.

Example — **3**

Algebra Link

Graph $y = 4$.

The graph of $y = 4$ is a horizontal line that intersects the *y*-axis at 4.

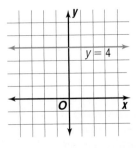

Your Turn

c. Graph $x = -3$.

Check for Understanding

Communicating Mathematics

1. **Describe** how an artist can use a grid to create a larger or smaller drawing.

2. **a.** **Graph** several points that form a horizontal line. Describe the common coordinate for each of these points.

   b. **Graph** several points that form a vertical line. Describe the common coordinate for each of these points.

3.  List at least five words that start with *quad*. Recall that the *x*-and *y*-axes divide the coordinate plane into four regions called *quadrants*. Consult a dictionary to see if all the words in your list relate to the number four.

Vocabulary
- coordinate plane
- *y*-axis
- *x*-axis
- quadrant
- origin
- ordered pair
- coordinates
- graph
- *x*-coordinate
- *y*-coordinate

Guided Practice

⏱ **Getting Ready** Name the *x*-coordinate and *y*-coordinate of each ordered pair.

Sample: $(-7, 2)$ **Solution:** $x = -7, y = 2$

4. $(0, -2)$ 5. $(-3, -6)$ 6. $(5, 8)$ 7. $(11, 0)$

Example 1 Draw and label a coordinate plane on a piece of grid paper. Then graph and label each point.

8. $M(-6, -2)$ 9. $J(2, 0)$ 10. $P(-5, 3)$

Example 2 Refer to the coordinate plane at the right. Name the ordered pair for each point.

11. A

12. B

13. C

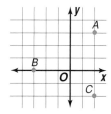

Exercises 11–13

Example 3 14. **Algebra** Graph $x = -5$.

Exercises

Practice

Draw and label a coordinate plane on a piece of grid paper. Then graph and label each point.

15. $T(0, -1)$ 16. $R(-2, -4)$ 17. $Q(5, 5)$

18. $C(0, 5)$ 19. $N(1, -5)$ 20. $S(3, 6)$

21. $G(-4, 0)$ 22. $L(-1, 4)$ 23. $F(6, -2)$

Homework Help

For Exercises	See Examples
24–31, 35	2
34	3
36, 37	1

Extra Practice
See page 729.

Name the ordered pair for each point on the coordinate plane at the right.

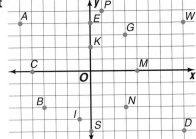

24. A

25. I

26. W

27. P

28. D

29. S

30. B

31. C

32. What point is located at (4, 0)?

33. Name the point at (3, −3).

Applications and Problem Solving

34. **Algebra** Graph $y = 6$.

35. **Geography** In geography, places are located using latitude (horizontal) and longitude (vertical) lines in much the same way as points are located in a coordinate plane.

 a. Name the city that is located at 30°N and 90°W.

 b. State the latitude and longitude of St. Petersburg. Round to the nearest ten.

 c. Suppose you are standing at 30°S and 20°E. Name the country you are visiting.

 d. State the latitude and longitude of the city or town where you live.

36. **Science** The average weight and top speeds of various animals are given below.

Animal	Avg. Weight (pounds)	Top Speed (miles per hour)
Cheetah	128	70
Chicken	7	9
Coyote	75	43
Fox	14	42
Horse	950	43
Polar Bear	715	35
Rabbit (domestic)	8	35

Sources: *Comparisons* and *The World Almanac*

 a. If the x-coordinate of an ordered pair represents the average weight and the y-coordinate represents the top speed, then (128, 70) would represent the cheetah. Write an ordered pair for each animal.

 b. Graph the ordered pairs.

 c. Look for patterns in the graph. Are larger animals usually faster or slower than smaller animals?

37. Critical Thinking   Graph $A(-3, -2)$ and $B(2, -2)$. Draw $\overline{AB}$. Find the coordinates of two other points that when connected with A and B would form a 5-by-3 rectangle.

Mixed Review

Use the number line to determine whether each statement is *true* or *false*. Explain your reasoning.   *(Lesson 2–3)*

38. D is the midpoint of $\overline{CE}$.   **39.** $\overline{AC} \cong \overline{CE}$

Refer to the line below for Exercises 40–42.   *(Lesson 2–2)*

40. If $XY = 14$ and $YZ = 27$, find XZ.
41. If $WX = 15$ and $WZ = 54$, find XZ.
42. If $WY = 21$ and $YZ = 21$, find WZ.

Standardized Test Practice
Ⓐ Ⓑ Ⓒ Ⓓ

43. Short Response   Write the following statement in if-then form.   *(Lesson 1–4)*

Students who do their homework will pass the course.

44. Multiple Choice   Charo walks 15 minutes the first day, 22 minutes the second day, and 29 minutes the third day. If she continues this pattern, how many minutes will Charo walk the fifth day?   *(Lesson 1–1)*

  Ⓐ 33 min   Ⓑ 36 min
  Ⓒ 39 min   Ⓓ 43 min

Quiz 2   Lessons 2–3 and 2–4

Use the number line to determine whether each statement is *true* or *false*. Explain your reasoning.   *(Lesson 2–3)*

1. $\overline{AC} \cong \overline{EF}$ **2.** $\overline{AB} \cong \overline{CE}$ **3.** D is the midpoint of $\overline{AF}$.

Draw and label a coordinate plane on a piece of grid paper. Then graph and label each point.   *(Lesson 2–4)*

4. $G(-2, 4)$ **5.** $H(0, -3)$

Investigation

What A View!

Materials

 centimeter cubes

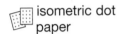 isometric dot paper

 ruler

Orthographic and Isometric Drawings

When an architect designs a building, he or she must communicate the design to the builder clearly. One way to do this is to supply detailed drawings of different views of the finished building.

Investigate

If you see a three-dimensional object from only one viewpoint, you may not know its true shape. Here are four views of a square pyramid. The two-dimensional views of the top, left, front, and right sides of an object are called an **orthographic drawing**.

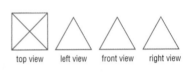

top view   left view   front view   right view

1. The orthographic drawing below represents the model at the right.

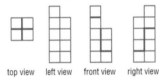

top view   left view   front view   right view

Step 1 The top view indicates two rows and two columns of different heights.

Step 2 The front view indicates that the left side is 5 blocks high and the right side is 3 blocks high. The dark segments indicate breaks in the surface.

Step 3 The right view indicates that the right front column is only one block high. The left front column is 4 blocks high. The right back column is 3 blocks high.

Make a model of a figure given the orthographic drawings below.

2.

top view   left view   front view   right view

3.

top view   left view   front view   right view

The view of a figure from a corner is called the **corner view**. You can use isometric dot paper to draw the corner view of a solid figure. This is called an **isometric drawing**. In this lesson, isometric dot paper will be used to draw and construct two-dimensional models of geometric solids.

left

front   right

CONcepts in MOtion

Interactive Lab
geomconcepts.com

4. To sketch a rectangular prism 2 units high, 5 units long, and 3 units wide using isometric dot paper, follow these steps.

Step 1   Mark the corner of the solid. Then draw 2 units down, 5 units to the left, and 3 units to the right.

Step 2   Draw a parallelogram for the top of the solid.

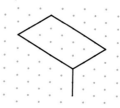

Step 3   Draw segments 2 units down from each vertex for the vertical edges.

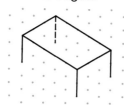

Step 4   Connect the corresponding vertices. Use dashed lines for the hidden edges. Shade the top of the solid.

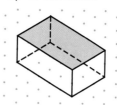

Sketch each solid using isometric dot paper.

5. rectangular prism 3 units high, 4 units long, and 5 units wide

6. cube 4 units on each edge

Extending the Investigation

In this extension, you will use isometric drawings to construct figures. Use centimeter cubes to construct a model of each figure.

1.

2.

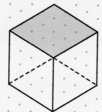

3.

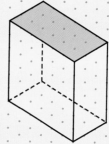

Presenting Your Conclusions

Here are some ideas to help you present your conclusions to the class.

- Make a poster that expalins how to make orthographic and isometric drawings
- Research the use of orthographic and isometric drawings in art, engineering, or architecture. Write a report about your findings. Include at least three specific ways in which they are used and a real-world example of each.

*inter*NET CONNECTION   **Investigation**   For more information on orthographic and isometric drawing, visit: www.geomconcepts.com

2-5 Midpoints

What You'll Learn
You'll learn to find the coordinates of the midpoint of a segment.

Why It's Important
Interior Design
Interior designers can determine where to place things by finding a midpoint. *See Exercise 34.*

The midpoint of a line segment, $\overline{AB}$, is the point C that bisects the segment.

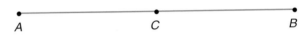

$$\overline{AC} \cong \overline{CB} \text{ (or } AC = CB)$$

You can use a number line to find the coordinates of the midpoint of a line segment.

Hands-On Geometry

Materials: grid paper scissors straightedge

Step 1 Draw a number line and mark the coordinates of the points from −10 to 10. Locate point A at −7 and point B at 5.

Step 2 Cut out $\overline{AB}$. Fold the segment so that points A and B are together. What is the coordinate of the midpoint of $\overline{AB}$?

Step 3 Find the sum of the coordinates of A and B. Divide the sum by 2.

Try These

1. How do the results of Steps 2 and 3 compare?

2. **a.** On a number line, locate point C with coordinate 2 and point D with coordinate 10. What is the coordinate of the midpoint of $\overline{CD}$?

   b. Find $(2 + 10) \div 2$.

   c. Compare your answers.

3. Repeat Exercise 2 with point E with coordinate −9 and point F with coordinate 1. What are the results?

4. **Make a conjecture** about the coordinate of the midpoint of a line segment on a number line.

In the activity above, you discovered that the coordinate of the midpoint of a segment on the number line equals the sum of the coordinates of the endpoints divided by 2.

	Words: On a number line, the coordinate of the midpoint of a segment whose endpoints have coordinates a and b is $\frac{a+b}{2}$.
Theorem 2–5 Midpoint Formula for a Number Line	**Model:** 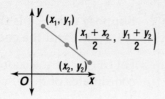

A similar relationship is true for the midpoint of a segment on a coordinate plane.

	Words: On a coordinate plane, the coordinates of the midpoint of a segment whose endpoints have coordinates (x_1, y_1) and (x_2, y_2) are $\left(\frac{x_1 + x_2}{2}, \frac{y_1 + y_2}{2}\right)$.
Theorem 2–6 Midpoint Formula for a Coordinate Plane	**Model:**

Example ➊ Find the coordinate of the midpoint of $\overline{RS}$.

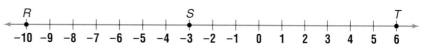

Use the Midpoint Formula to find the coordinate of the midpoint of $\overline{RS}$.

$\frac{a+b}{2} = \frac{-10 + (-3)}{2}$ *The coordinate of R is −10. So, a = −10.*
The coordinate of S is −3. So, b = −3.

$= \frac{-13}{2}$ or $-6\frac{1}{2}$ *Simplify.*

The coordinate of the midpoint is $-6\frac{1}{2}$.

Your Turn

a. Refer to the number line above. Find the coordinate of the midpoint of $\overline{RT}$.

2 Find the coordinates of M, the midpoint of $\overline{JK}$, given endpoints $J(2, -9)$ and $K(8, 3)$.

Use the Midpoint Formula to find the coordinates of M.

$$\left(\frac{x_1 + x_2}{2}, \frac{y_1 + y_2}{2}\right) = \left(\frac{2 + 8}{2}, \frac{-9 + 3}{2}\right) \qquad (x_1, y_1) = (2, -9)$$
$$(x_2, y_2) = (8, 3)$$
$$= \left(\frac{10}{2}, \frac{-6}{2}\right) \text{ or } (5, -3) \quad \textit{Simplify.}$$

The coordinates of M are $(5, -3)$.

Your Turn

b. Find the coordinates of N, the midpoint of $\overline{VW}$, given the endpoints $V(-4, -3)$ and $W(6, 11)$.

c. Find the coordinates of Q, the midpoint of $\overline{PR}$, given the endpoints $P(-5, 1)$ and $R(2, -8)$.

Algebra Link

3 Suppose $G(8, -9)$ is the midpoint of $\overline{FE}$ and the coordinates of E are $(18, -21)$. Find the coordinates of F.

Let (x_1, y_1) be the coordinates of F and let (x_2, y_2) or $(18, -21)$ be the coordinates of E. So, $x_2 = 18$ and $y_2 = -21$. Use the Midpoint Formula.

$$\left(\frac{x_1 + x_2}{2}, \frac{y_1 + y_2}{2}\right) = (8, -9)$$

x-coordinate of F		*y*-coordinate of F
$\dfrac{x_1 + x_2}{2} = 8$	*Midpoint formula*	$\dfrac{y_1 + y_2}{2} = -9$
$\dfrac{x_1 + 18}{2} = 8$	*Replace x_2 with 18 and y_2 with -21.*	$\dfrac{y_1 + (-21)}{2} = -9$
$\dfrac{x_1 + 18}{2}(2) = 8(2)$	*Multiply each side by 2.*	$\dfrac{y_1 - 21}{2}(2) = -9(2)$
$x_1 + 18 = 16$	*Simplify.*	$y_1 - 21 = -18$
$x_1 + 18 - 18 = 16 - 18$	*Add or subtract to isolate the variable.*	$y_1 - 21 + 21 = -18 + 21$
$x_1 = -2$	*Simplify.*	$y_1 = 3$

— Algebra Review —
Solving Multi-Step Equations, p. 723

The coordinates of F are $(-2, 3)$.

Your Turn

d. Suppose $K(-10, 17)$ is the midpoint of $\overline{IJ}$ and the coordinates of J are $(4, 12)$. Find the coordinates of I.

e. Suppose $S\left(3, -\dfrac{3}{4}\right)$ is the midpoint of $\overline{RT}$ and the coordinates of T are $(-2, 6)$. Find the coordinates of R.

COncepts in MOtion
Interactive Lab
geomconcepts.com

You can use a TI–83/84 Plus calculator to draw figures on a coordinate plane.

Graphing Calculator Tutorial
See pp. 782–785.

Graphing Calculator Exploration

Step 1   To display a coordinate plane, open the [F5] menu.

Step 2   Select Hide/Show and then select Axes.

The calculator will display a coordinate plane on which you can construct geometric figures.

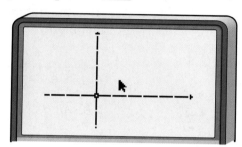

Try These

1. Use the Segment tool on the [F2] menu to construct a segment in Quadrant I. Select the Midpoint tool on the [F3] menu and construct the midpoint of the segment. Use the Coordinates & Equation tool on the [F5] menu to display the coordinates of the endpoints and midpoint of the segment. What do you notice about the coordinates of the midpoint?

2. Drag one endpoint of the segment into Quadrant III. How do the coordinates of the midpoint change as you do this?

3. Open the [F5] menu. Select Measure, then select Distance & Length to display the distance from the midpoint to each endpoint of the segment. How are the two distances related? What happens to these distances if you drag an endpoint of the segment?

Check for Understanding

Communicating Mathematics

1. **Graph** $A(1, 3)$ and $B(5, 1)$. Draw $\overline{AB}$. Use your graph to estimate the midpoint of $\overline{AB}$. Check your answer by using the Midpoint Formula.

2. **Explain** why it is correct to say that the coordinates of the midpoint of a segment are the means of the coordinates of the endpoints of the segment.

3. **You Decide** Fina wants to find the midpoint of a segment on a number line. She finds the length of the segment and divides by 2. She adds this number to the coordinate of the left endpoint to find the midpoint. Kenji says she should subtract the number from the coordinate of the right endpoint to find the midpoint. Who is correct? Explain your reasoning.

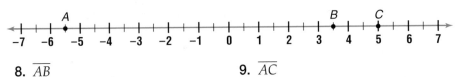

Getting Ready Find the mean of each pair of numbers.

Sample: −4 and 10	Solution: $\dfrac{-4+10}{2} = \dfrac{6}{2}$ or 3

4. 4 and 8 **5.** −2 and 6 **6.** 5 and −6 **7.** −4 and −10

Example 1

Use the number line to find the coordinate of the midpoint of each segment.

A   B   C
−7   −6   −5   −4   −3   −2   −1   0   1   2   3   4   5   6   7

8. $\overline{AB}$ **9.** $\overline{AC}$

Example 2

The coordinates of the endpoints of a segment are given. Find the coordinates of the midpoint of each segment.

10. $(-3, 6), (-5, -2)$ **11.** $(-8, 6), (1, -3)$ **12.** $(-3, 2), (6, 5)$

Example 3

13. Algebra Suppose $R(3, -5)$ is the midpoint of $\overline{PQ}$ and the coordinates of P are $(7, -2)$. Find the coordinates of Q.

Exercises

• • • • • • • • • • • • • • • • • • •

Practice

Use the number line to find the coordinate of the midpoint of each segment.

R   S   T   U   V   W   X
−7   −6   −5   −4   −3   −2   −1   0   1   2   3   4   5   6   7

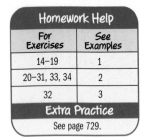

Homework Help	
For Exercises	See Examples
14–19	1
20–31, 33, 34	2
32	3

Extra Practice
See page 729.

14. $\overline{RV}$ **15.** $\overline{TW}$ **16.** $\overline{UX}$

17. $\overline{SU}$ **18.** $\overline{TX}$ **19.** $\overline{ST}$

The coordinates of the endpoints of a segment are given. Find the coordinates of the midpoint of each segment.

20. $(0, 4), (0, 0)$ **21.** $(-1, -2), (-3, -6)$

22. $(6, 0), (13, 0)$ **23.** $(4, 6), (-2, -3)$

24. $(-3, 2), (-5, 6)$ **25.** $(-1, -7), (6, 1)$

26. $(-8, 3), (6, -6)$ **27.** $(-18, 5), (-3, -16)$

28. $(a, b), (0, 0)$ **29.** $(a, b), (c, d)$

30. Find the midpoint of the segment that has endpoints at $(-1, 6)$ and $(-5, -18)$.

31. What is the midpoint of $\overline{ST}$ if the endpoints are $S(2a, 2b)$ and $T(0, 0)$?

32. Algebra Suppose $C(-4, 5)$ is the midpoint of $\overline{AB}$ and the coordinates of A are $(2, 17)$. Find the coordinates of B.

33. Travel Donte is traveling on I-70 in Kansas. He gets on the interstate at the 128-mile marker and gets off the interstate at the 184-mile marker to go to Russell. Which mile marker is the midpoint of his drive on I-70?

34. Interior Design Chapa is an interior designer. She has drawn a scale model of the first floor of a client's house. She plans to install a paddle fan in the ceiling at the midpoint of the diagonals of the great room. Name the coordinates of the location for the fan.

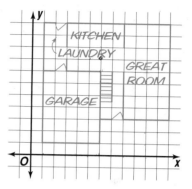

35. Critical Thinking Name the coordinates of the endpoints of five different segments with $M(6, 8)$ as the midpoint.

Mixed Review

Refer to the coordinate plane at the right. Name the ordered pair for each point. *(Lesson 2–4)*

36. G

37. H

38. J

39. K

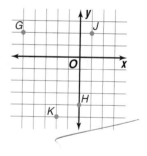

40. Algebra In the figure, C is the midpoint of $\overline{AB}$. Find the value of x. *(Lesson 2–3)*

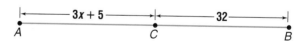

$$A \quad 3x + 5 \quad C \quad 32 \quad B$$

41. Short Response Name the intersection of plane DAC and plane EBF. *(Lesson 1–3)*

42. Short Response How would you describe any three points that lie in the same plane? *(Lesson 1–2)*

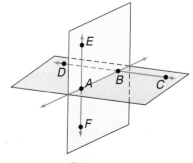

Exercise 41

Understanding and Using the Vocabulary

*inter***NET**
CONNECTION **Review Activities**
For more review activities, visit:
www.geomconcepts.com

After completing this chapter, you should be able to define each term, property, or phrase and give an example or two of each.

Geometry
betweenness *(p. 56)*
bisect *(p. 64)*
congruent segments *(p. 62)*
greatest possible error *(p. 58)*
measure *(p. 52)*
measurements *(p. 58)*
midpoint *(p. 63)*
percent of error *(p. 58)*
precision *(p. 58)*
theorems *(p. 62)*
unit of measure *(p. 58)*
vector *(p. 74)*

Algebra
absolute value *(p. 52)*
coordinate *(p. 52)*
coordinate plane *(p. 68)*
coordinates *(p. 68)*
equation *(p. 57)*
graph *(p. 68)*
integers *(p. 50)*
irrational numbers *(p. 51)*
natural numbers *(p. 50)*
nonterminating *(p. 50)*
ordered pair *(p. 68)*

origin *(pp. 52, 68)*
quadrants *(p. 68)*
rational numbers *(p. 50)*
real numbers *(p. 51)*
terminating *(p. 50)*
whole numbers *(p. 50)*
x-axis *(p. 68)*
x-coordinate *(p. 69)*
y-axis *(p. 68)*
y-coordinate *(p. 69)*

Choose the term or terms from the list above that best complete each statement.

1. The ___?___ numbers include 0 and the natural numbers.
2. A ___?___ is any number of the form $\frac{a}{b}$, where a and b are integers and b cannot equal zero.
3. Decimals that are nonterminating and do not repeat are called ___?___ numbers.
4. The number that corresponds to a point on a number line is called the ___?___ of the point.
5. The number of units from zero to a number on the number line is called its ___?___ .
6. The second component of an ordered pair is called the ___?___ .
7. Two segments are ___?___ if and only if they have the same length.
8. ___?___ are statements that can be justified using logical reasoning.
9. To ___?___ a segment means to separate it into two congruent segments.
10. The two axes separate a coordinate plane into four regions called ___?___ .

Skills and Concepts

Objectives and Examples

- **Lesson 2–1** Find the distance between two points on a number line.

Use the number line at the right to find *BE*.

$BE = |-3 - 1|$ *The coordinate of B is −3.*
$\quad = |-4|$ or 4   *The coordinate of E is 1.*

Review Exercises

Use the number line to find each measure.

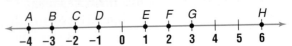

```
    A   B   C   D       E   F   G           H
 ───┼───┼───┼───┼───┼───┼───┼───┼───┼───┼───
   −4  −3  −2  −1   0   1   2   3   4   5   6
```

11. *AD*
12. *FH*
13. *CG*

 www.geomconcepts.com/vocabulary_review

Objectives and Examples

- **Lesson 2-2** Apply the properties of real numbers to the measure of segments.

If $XY = 39$ and $XZ = 62$, find YZ.

$XY + YZ = XZ$	*Def. of Betweenness*
$39 + YZ = 62$	*Substitution*
$39 + YZ - 39 = 62 - 39$	*Subtraction*
$YZ = 23$	*Simplify.*

- **Lesson 2-3** Identify congruent segments, and find the midpoints of segments.

Determine whether B is the midpoint of $\overline{AC}$.

Because $AB = 3$ and $BC = 2$, $AB \neq BC$. So, B is not the midpoint of AC.

- **Lesson 2-4** Name and graph ordered pairs on a coordinate plane.

Graph point B at $(-2, -3)$.

Start at the origin. Move 2 units to the left. Then, move 3 units down. Label this point B.

Review Exercises

Refer to the line for Exercises 14–15.

14. If $ST = 15$ and $SR = 6$, find RT.
15. If $SR = 6$ and $RT = 4.5$, find ST.

16. Find the length of the segment below in centimeters and in inches.

_____

Use the number line at the left to determine whether each statement is *true* or *false*. Explain your reasoning.
17. $\overline{BD} \cong \overline{EG}$
18. $\overline{AB} \cong \overline{DE}$
19. The midpoint of $\overline{AE}$ is C.

Determine whether each statement is *true* or *false*. Explain your reasoning.
20. If $\overline{RQ} \cong \overline{TP}$ and $\overline{RQ} \cong \overline{FG}$, then $\overline{TP} \cong \overline{FG}$.
21. $\overline{LM}$ is not congruent to $\overline{ML}$.
22. If points K, L, and M are collinear, then L is the midpoint of $\overline{KM}$.

Name the ordered pair for each point.
23. F
24. C
25. H
26. D

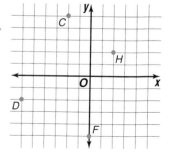

Draw and label a coordinate plane on a piece of grid paper. Then graph and label each point.
27. $A(5, 5)$
28. $B(0, 4)$
29. $E(-4, 0)$
30. $G(2, -2)$

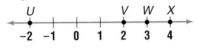

Mixed Problem Solving
See pages 758–765.

Objectives and Examples

- **Lesson 2-5** Find the coordinates of the midpoint of a segment.

Find the coordinates of M, the midpoint of $\overline{CD}$, given the endpoints $C(3, 1)$ and $D(9, 9)$.

Let $x_1 = 3$, $x_2 = 9$, $y_1 = 1$, and $y_2 = 9$.

$$\left(\frac{x_1 + x_2}{2}, \frac{y_1 + y_2}{2}\right) = \left(\frac{3 + 9}{2}, \frac{1 + 9}{2}\right)$$

$$= \left(\frac{12}{2}, \frac{10}{2}\right) \text{ or } (6, 5)$$

The coordinates of M are $(6, 5)$.

Review Exercises

Use the number line to find the coordinate of the midpoint of each segment.

```
       U                 V  W  X
  <----●--+--+--+--●--●--●---->
      -2 -1  0  1  2  3  4
```

31. $\overline{UW}$ **32.** $\overline{VX}$

The coordinates of the endpoints of a segment are given. Find the coordinates of the midpoint of each segment.

33. $(-1, -5), (3, -3)$ **34.** $(4, 7), (-1, 2)$

Applications and Problem Solving

35. Temperatures Temperatures on the planet Mars range from $-122°C$ to $31°C$. What is the difference between these two temperatures? *(Lesson 2–1)*

36. Geography The highest point in Asia is Mount Everest at 29,028 feet above sea level. The lowest point in Asia is the Dead Sea at 1312 feet below sea level. What is the vertical distance between these two points? *(Lesson 2–2)*

37. Environment The table at the right shows the mid-1990s Gross National Product (GNP) per person and municipal waste production for six countries. *(Lesson 2–4)*

   a. Graph the data. Let the x-coordinate of an ordered pair represent the GNP per person, and let the y-coordinate represent the number of kilograms of waste per person.

   b. Does the graph show that countries with a higher GNP per person generate more or less waste per person? Explain.

Country	GNP ($ per person)	Waste (kg per person)
United States	27,550	720
France	26,290	560
Japan	41,160	400
Mexico	2521	330
United Kingdom	19,020	490
Spain	14,160	370

Source: *Statistical Abstract of U.S.*

38. Algebra Suppose $K(3, -4)$ is the midpoint of $\overline{JL}$. The coordinates of J are $(-3, -2)$. Find the coordinates of L. *(Lesson 2–5)*

For each situation, write a real number with ten digits to the right of the decimal point.

1. a rational number less than -2 with a 3-digit repeating pattern

2. an irrational number between 3.5 and 4

Refer to the number line at the right.

3. *True* or *false*: $\overline{AD} \cong \overline{CE}$

4. What is the measure of $\overline{AF}$?

5. What is the midpoint of $\overline{CG}$?

Refer to the line at the right.

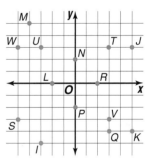

6. Find the length of $\overline{EI}$ in centimeters and in inches.

7. If $GH = 17$ and $FH = 23$, find FG.

8. If $FG = 28$ and $GH = 12$, find FH.

Name the ordered pair for each point in the coordinate plane at the right.

9. *M*   10. *P*   11. *V*

What point is located at each of the coordinates in the coordinate plane at the right?

12. $(-2, 0)$ 13. $(3, -4)$ 14. $(-5, -3)$

Exercises 9–14

The coordinates of the endpoints of a segment are given. Find the coordinates of the midpoint of each segment.

15. $(3, -8), (7, 2)$ 16. $(-4, 2), (-3, 1)$ 17. $(-11, 9), (3, -5)$

18. **Algebra**   In the figure at right, M is the midpoint of $\overline{LN}$. Find the value of x.

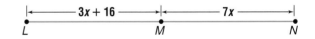

19. **Hardware**   Naomi purchased an extension ladder consisting of two 8-foot sections. When fully extended, the ladder measures 13 feet 7 inches. By how much do the two ladder sections overlap?

20. **Algebra**   Plot the points for the ordered pairs on grid paper. Connect the points in the given order with straight line segments. What shape is formed?   *(Lesson 2–4)*

$(0, 2), (1, 3), (2, 3), (3, 2), (3, 0), (2, -2), (1, -3), (0, -4), (-1, -3),$
$(-2, -2), (-3, 0), (-3, 2), (-2, 3), (-1, 3), (0, 2)$

More Number Concept Problems

Numerical problems on standardized tests can involve integers, fractions, decimals, percents, square roots, or exponents.

Many problems ask you to convert between fractions or decimals and percents. It's a good idea to memorize these common decimal-fraction-percent equivalents.

$0.01 = \frac{1}{100} = 1\%$ $\qquad$ $0.1 = \frac{1}{10} = 10\%$ $\qquad$ $0.2 = \frac{2}{10} = 20\%$

$0.25 = \frac{1}{4} = 25\%$ $\qquad$ $0.5 = \frac{1}{2} = 50\%$ $\qquad$ $0.75 = \frac{3}{4} = 75\%$

Test-Taking Tip
Remember the order of operations.
1. Parentheses
2. Exponents
3. Multiply, **D**ivide
4. Add, **S**ubtract
Please **E**xcuse **M**y **D**ear **A**unt **S**ally

Example 1

Evaluate the following expression.

$$[(9 - 5) \times 6] + 4^2 \div 4$$

Hint Begin inside the parentheses.

Solution Use the order of operations. Evaluate the expression inside the parentheses. Then evaluate the resulting expression inside the brackets.

$[(9 - 5) \times 6] + 4^2 \div 4$

$\quad = [4 \times 6] + 4^2 \div 4$ $\quad 9 - 5 = 4$

$\quad = 24 + 4^2 \div 4$ $\qquad 4 \times 6 = 24$

$\quad = 24 + 16 \div 4$ $\qquad 4^2 = 16$

$\quad = 24 + 4$ $\qquad\quad 16 \div 4 = 4$

$\quad = 28$ $\qquad\qquad 24 + 4 = 28$

The answer is 28.

Example 2

At a restaurant, diners get an "early bird" discount of 10% off their bill. If a diner orders a meal regularly priced at $18 and leaves a tip of 15% of the discounted meal, how much does she pay in total?

- **A** $13.50
- **B** $16.20
- **C** $18.63
- **D** $18.90
- **E** $20.70

Hint Be sure to read the question carefully.

Solution First, find the amount of the discount.

$$10\% \text{ of } \$18.00 = 0.10(18.00) \text{ or } \$1.8$$

Then subtract to find the cost of the discounted meal.

$$\$18.00 - \$1.80 = \$16.20$$

This is choice B, but it is *not* the answer to the question. You need to find the total cost of the meal plus the tip. Calculate the amount of the tip. 15% of $16.20 is $2.43.

The total amount paid is $16.20 + $2.43 or $18.63. The answer is C.

After you work each problem, record your answer on the answer sheet provided or on a sheet of paper.

Multiple Choice

1. Which is the correct order of the set of numbers from least to greatest? *(Algebra Review)*

$$-5, 4, 0, -\sqrt{22}, \sqrt{18}, 8$$

- **A** $-\sqrt{22}, \sqrt{18}, 0, 4, -5, 8$
- **B** $-\sqrt{22}, -5, 0, 4, 8, \sqrt{18}$
- **C** $-5, -\sqrt{22}, 0, 4, \sqrt{18}, 8$
- **D** $-5, -\sqrt{22}, 0, 4, 8, \sqrt{18}$

2. What are the coordinates of the intersection of $\overleftrightarrow{AB}$ and $\overleftrightarrow{CD}$? *(Lesson 2–4)*

- **A** $(-2, -2)$
- **B** $(-2, 2)$
- **C** $(-3, 0)$
- **D** $(0, -6)$

3. After   has been simplified to a single

fraction in lowest terms, what is the denominator? *(Algebra Review)*

- **A** 2
- **B** 3
- **C** 5
- **D** 9
- **E** 13

4. Talia is a travel agent. The agency gives a 7% bonus to any agent who sells at least $9000 in travel packages each month. If an average travel package is $855, how many packages must Talia sell to receive a bonus each month? *(Percent Review)*

- **A** 9 or more
- **B** 10 or more
- **C** 11 or more
- **D** less than 9

5. If n is an even integer, which must be an odd integer? *(Algebra Review)*

- **A** $3n - 2$
- **B** $3(n + 1)$
- **C** $n - 2$
- **D** $\frac{n}{3}$
- **E** n^2

6. Luke is making a model of our solar system. He has placed Venus and Mars in his model on the coordinate grid at the right. He wants to place the model of Earth at the midpoint of the segment connecting Venus and Mars. What will be the coordinates for the model of Earth? *(Lesson 2–5)*

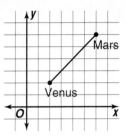

- **A** $(2, 2)$
- **B** $(3, 3)$
- **C** $(4, 4)$
- **D** $(5, 5)$

7. The length of the page in a textbook is $10\frac{7}{8}$ inches. The top and bottom margins total $1\frac{1}{16}$ inches. What is the length of the page inside the margins? *(Algebra Review)*

- **A** $8\frac{3}{16}$
- **B** $8\frac{13}{16}$
- **C** $9\frac{13}{16}$
- **D** $11\frac{15}{16}$

8. For a positive integer x, 10% of x% of 1000 equals— *(Percent Review)*

- **A** x.
- **B** $10x$.
- **C** $100x$.
- **D** $1000x$.
- **E** $10,000x$.

Grid In

9. Set S consists of all multiples of 3 between 11 and 31. Set T consists of all multiples of 5 between 11 and 31. What is one possible number in S but NOT in T? *(Algebra Review)*

Short Response

10. You must choose between two Internet providers. One charges a flat fee of $22 per month for unlimited usage, and the other charges a fee of $10.99 for 10 hours of use per month, plus $1.95 for each additional hour. Decide which provider would be more economical for you to use. *(Algebra Review)*

www.geomconcepts.com/standardized_test

CHAPTER

3

Angles

What You'll Learn

Key Ideas

- Name and identify parts of an angle. *(Lesson 3–1)*

- Measure, draw, classify angles, and find the bisector of an angle. *(Lessons 3–2 and 3–3)*

- Identify and use adjacent angles, linear pairs of angles, complementary and supplementary angles, and congruent and vertical angles. *(Lessons 3–4 to 3–6)*

- Identify, use properties of, and construct perpendicular lines and segments. *(Lesson 3–7)*

Key Vocabulary

acute angle *(p. 98)*

angle *(p. 90)*

obtuse angle *(p. 98)*

right angle *(p. 98)*

straight angle *(p. 90)*

Why It's Important

Hobbies Kites first appeared in China around 500 B.C. Silk and lightweight bamboo were used to make kites in rectangular shapes. After the invention of paper, exotic kites were made in a variety of shapes, such as dragons, birds, insects, and people.

Angles are used to describe relationships between real-life and mathematical objects. You will examine the angles in a flat kite in Lesson 3–1.

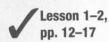
 ✓ Check Your Readiness

Use the figure to name examples of each term.

1. three points

2. three rays

3. a point that is not on $\overrightarrow{MS}$

4. a ray with point N as the endpoint

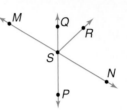

Solve each equation. Check your solution.

5. $4y = 180$

6. $65 + b = 90$

7. $8v = 168$

8. $180 = d + 25$

9. $90 - k = 12$

10. $180 - p = 142$

11. $90 = 4x + 26$

12. $16 + 2t = 84$

13. $3x + 18 = 180$

14. $180 - 2x = 66$

15. $5n - 45 = 120$

16. $6h + 18 = 90$

17. $52 + (8n - 4) = 160$

18. $2x + (4x - 16) = 110$

19. $75 + (3x + 12) = 180$

20. $8y + (10y - 16) = 128$

FOLDABLES™
Study Organizer

Make this Foldable to help you organize your Chapter 3 notes. Begin with a sheet of plain $8\frac{1}{2}$" by 11" paper.

❶ **Fold** in half lengthwise.

❷ **Fold** again in thirds.

❸ **Open** and cut along the second fold to make three tabs.

❹ **Label** as shown.

Reading and Writing As you read and study the chapter, explain and draw examples of points in the interior, exterior, and on an angle under the tabs. You may want to make another Foldable to record notes about the three types of angles, *right, acute,* and *obtuse.*

Opposite rays are two rays that are part of the same line and have only their endpoints in common.

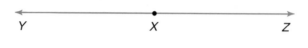

$\overrightarrow{XY}$ and $\overrightarrow{XZ}$ are opposite rays.

The figure formed by opposite rays is also referred to as a **straight angle**.

There is another case where two rays can have a common endpoint. This figure is called an **angle**. *Unless otherwise noted, the term "angle" in this book means a nonstraight angle.* Some parts of angles have special names. The common endpoint is called the **vertex**, and the two rays that make up the angle are called the **sides** of the angle.

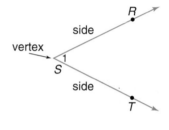

There are several ways to name the angle shown above.

Method	Symbol
1. Use the vertex and a point from each side. *The vertex letter is always in the middle.*	$\angle RST$ or $\angle TSR$
2. Use the vertex only. *If there is only one angle at a vertex, then the angle can be named with that vertex.*	$\angle S$
3. Use a number.	$\angle 1$

Definition of Angle		
Words:	An angle is a figure formed by two noncollinear rays that have a common endpoint.	
Model:	D E ⟨ 2 F	**Symbols:** $\angle DEF$ $\angle FED$ $\angle E$ $\angle 2$

**Name the angle in four ways.
Then identify its vertex and its sides.**

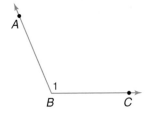

The angle can be named in four ways:
∠ABC, ∠CBA, ∠B, and ∠1.
Its vertex is point *B*. Its sides are $\overrightarrow{BA}$ and $\overrightarrow{BC}$.

Look Back

Naming Rays:
Lesson 1–2

Your Turn

a.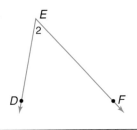

b.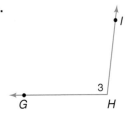

Look at the figure at the right. Three
angles have *P* as their vertex. Whenever
there is more than one angle at a given
vertex, use three points or a number to
name an angle as shown at the right.

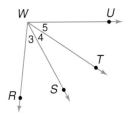

Example ➋

Name all angles having *W* as their vertex.

There are three distinct angles with vertex *W*:
∠3, ∠4, and ∠XWZ.

What other names are there for ∠3?
What other names are there for ∠4?
What other name is there for ∠XWZ?
Is there an angle that can be named ∠W?

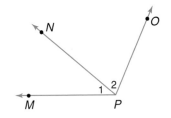

Your Turn

c.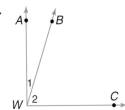

d.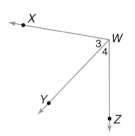

An angle separates a plane into three
parts: the **interior** of the angle, the **exterior**
of the angle, and the angle itself. In the
figure shown, point *W* and all other points
in the blue region are in the interior of the
angle. Point *V* and all other points in the
yellow region are in the exterior of the
angle. Points *X*, *Y*, and *Z* are on the angle.

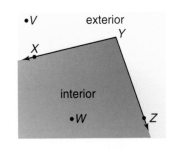

**Concepts
in MOtion**
Animation
geomconcepts.com

Tell whether each point is in the *interior*, *exterior*, or *on* the angle.

3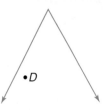

4

5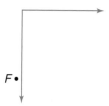

Point *D* is in the interior of the angle.

Point *E* is on the angle.

Point *F* is in the exterior of the angle.

Your Turn

e.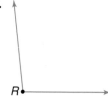

f.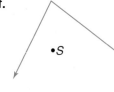

g.

Check for Understanding

Communicating Mathematics

1. **Sketch and label** an angle with sides $\overrightarrow{EF}$ and $\overrightarrow{EG}$.

2. **Draw** an angle *MNP* that has a point *Q* in the interior of the angle.

3. **Explain** why angle *PTR* cannot be labeled $\angle T$.

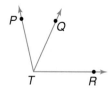

Vocabulary

opposite rays
straight angle
angle
vertex
sides
interior
exterior

Guided Practice

Examples 1 & 2

4. Name the angle in four ways. Then identify its vertex and its sides.

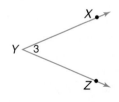

5. Name all angles having *K* as their vertex.

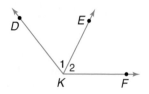

Tell whether each point is in the *interior*, *exterior*, or *on* the angle.

6.

7.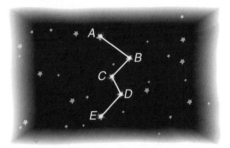

Examples 1 & 2

8. **Science** The constellation Cassiopeia is one of the 88 constellations in the sky.

 a. How many angles are formed by the arrangement of the stars in the constellation?

 b. Name each angle in two ways.

The constellation Cassiopeia

Exercises

Practice

Name each angle in four ways. Then identify its vertex and its sides.

Homework Help	
For Exercises	**See Examples**
9–11, 21, 23	1
12–14	2
15–20, 22	3–5
Extra Practice	
See page 729.	

9.

10.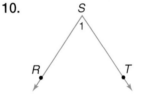

11.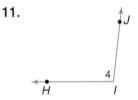

Name all angles having *J* as their vertex.

12.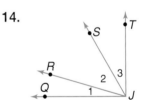

13.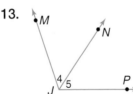

14.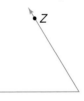

Tell whether each point is in the *interior*, *exterior*, or *on* the angle.

15.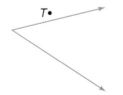

16.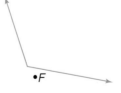

17.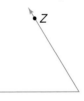

18.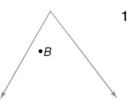

19.

20.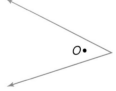

Determine whether each statement is *true* or *false*.

21. Angles may have four different names.

22. The vertex is in the interior of an angle.

23. The sides of $\angle ABC$ are $\overrightarrow{AB}$ and $\overrightarrow{BC}$.

24. Hobbies The oldest basic type of kite is called the *flat kite*.

  a. How many angles are formed by the corners of a flat kite?

  b. Name each angle in two ways.

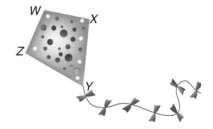

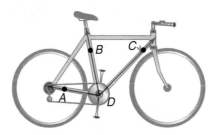

25. Design Bicycle manufacturers use angles when designing bicycles. Name each angle shown. Then identify the sides of each angle.

26. Critical Thinking Using three letters, how many different ways can the angle at the right be named? List them.

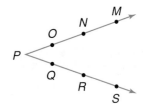

27. The coordinates of the endpoints of a segment are $(2, 3)$ and $(4, 5)$. Find the coordinates of the midpoint. *(Lesson 2–5)*

28. Draw and label a coordinate plane. Then graph and label point A at $(2, -3)$. *(Lesson 2–4)*

29. Interior Design Ke Min is planning to add a wallpaper border to his rectangular bathroom. How much border will he need if the length of the room is 8 feet and the width is 5 feet? *(Lesson 1–6)*

30. Use a compass and a straightedge to construct a five-sided figure. *(Lesson 1–5)*

31. Short Response Points P, Q, R, and S lie on a circle. List all of the lines that contain exactly two of these four points.

(Lesson 1–2)

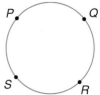

32. Multiple Choice Simplify $4y + 3(6 + 2y)$. *(Algebra Review)*

  Ⓐ $6y + 18$

  Ⓑ $9y + 9$

  Ⓒ $10y + 18$

  Ⓓ $18y + 18$

 www.geomconcepts.com/self_check_quiz

Math
In the Workplace

Astronomer

Do the stars in the night sky captivate you? If so, you may want to consider a career as an astronomer. In addition to learning about stars, galaxies, the sun, moon, and planets, astronomers study events such as *eclipses*.

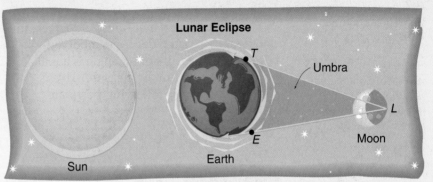

Lunar Eclipse

Umbra

T

L

E

Moon

Sun

Earth

A *total lunar eclipse* occurs when the moon passes totally into Earth's dark shadow, or *umbra*. Notice the angle that is formed by Earth's umbra.

1. Name the angle in three ways.
2. Identify the vertex and its sides.
3. Is the umbra in the *exterior*, in the *interior*, or *on* the angle?
4. Research lunar eclipses. Explain the difference between a lunar eclipse and a total lunar eclipse.

FAST FACTS About Astronomers

Working Conditions
- usually work in observatories
- may have to travel to remote locations
- may work long hours and nights

Education/Skills
- high school math and physical science courses
- college degree in astronomy or physics
- mathematical ability, computer skills, and the ability to work independently are essential

Employment

Where Astronomers Are Employed

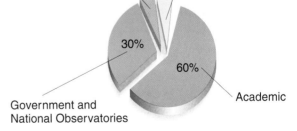

Other (Museums, Public Information) 5%

Business and Private Industry 5%

30%

60%

Government and National Observatories

Academic

---

*inter*NET
CONNECTION
www.geomconcepts.com

Career Data For the latest information on careers in astronomy, visit:

In geometry, angles are measured in units called **degrees**. The symbol for degree is °.

The angle shown measures 75 degrees. In the notation, there is no degree symbol with 75 because the measure of an angle is a real number with no unit of measure. This is summarized in the following postulate.

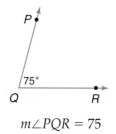

$m\angle PQR = 75$

	Words:	For every angle, there is a unique positive number between 0 and 180 called the *degree measure* of the angle.
Postulate 3–1 Angles Meaure Postulate	**Model:** 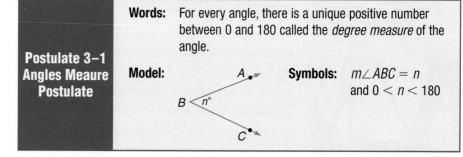	**Symbols:** $m\angle ABC = n$ and $0 < n < 180$

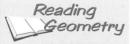
In this text, the term degree measure will be used in all appropriate theorems and postulates. Elsewhere we will refer to the degree measure of an angle as just measure.

You can use a **protractor** to measure angles and sketch angles of given measure.

Examples

1 Use a protractor to measure ∠DEF.

Step 1 Place the center point of the protractor on vertex E. Align the straightedge with side $\overrightarrow{EF}$.

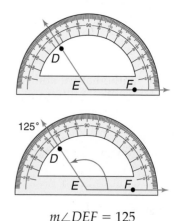

Step 2 Use the scale that begins with 0 at $\overrightarrow{EF}$. Read where the other side of the angle, $\overrightarrow{ED}$, crosses this scale.

$m\angle DEF = 125$

Angle *DEF* measures 125°.

2 Find the measures of ∠*BXE*, ∠*CXE*, and ∠*AXB*.

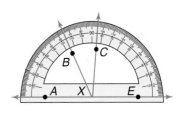

$m\angle BXE = 115$ $\overrightarrow{XE}$ *is at 0° on the right.*

$m\angle CXE = 85$ $\overrightarrow{XE}$ *is at 0° on the right.*

$m\angle AXB = 65$ $\overrightarrow{XA}$ *is at 0° on the left.*

Your Turn

a. Use a protractor to measure ∠*CDF*.

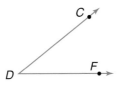

b. Find the measure of ∠*PQR*, ∠*PQS*, and ∠*PQT*.

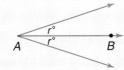

Just as Postulate 3–1 provides a way to measure angles, Postulate 3–2 describes the relationship between angle measures and numbers.

Postulate 3–2 **Protractor** **Postulate**	**Words:**	On a plane, given $\overrightarrow{AB}$ and a number r between 0 and 180, there is exactly one ray with endpoint A, extending on each side of $\overrightarrow{AB}$ such that the degree measure of the angle formed is r.
	Model:	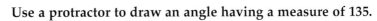

Example **3** Use a protractor to draw an angle having a measure of 135.

Step 1   Draw $\overrightarrow{YZ}$.

Step 2   Place the center point of the protractor on Y. Align the mark labeled 0 with the ray.

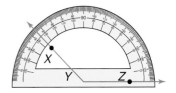

Step 3   Locate and draw point X at the mark labeled 135. Draw $\overrightarrow{YX}$.

COncepts in MOtion
Animation
geomconcepts.com

Your Turn

c. Use a protractor to draw an angle having a measure of 65.

Once the measure of an angle is known, the angle can be classified as one of three types of angles. These types are defined in relation to a right angle.

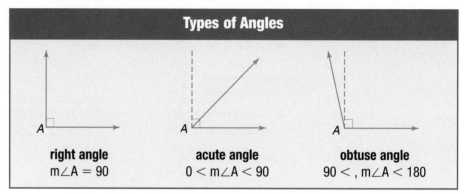

Types of Angles

right angle
$m\angle A = 90$

acute angle
$0 < m\angle A < 90$

obtuse angle
$90 < , m\angle A < 180$

Examples

Classify each angle as *acute*, *obtuse*, or *right*.

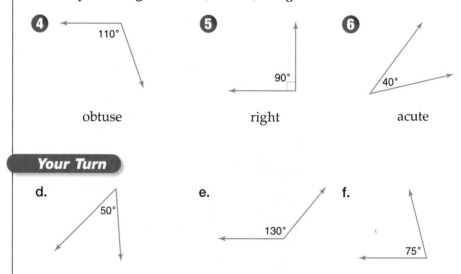

4 110°

obtuse

5 90°

right

6 40°

acute

Your Turn

d. 50°

e. 130°

f. 75°

Example

Algebra Link

7 The measure of $\angle B$ is 138. Solve for x.

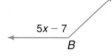

Explore   You know that $m\angle B = 138$
and $m\angle B = 5x - 7$.

Plan   Write and solve an equation.

Solve
$$138 = 5x - 7 \qquad \textit{Substitution}$$
$$138 + 7 = 5x - 7 + 7 \quad \textit{Add 7 to each side.}$$
$$145 = 5x \qquad \textit{Simplify.}$$
$$\frac{145}{5} = \frac{5x}{5} \qquad \textit{Divide each side by 5.}$$
$$29 = x \qquad \textit{Simplify.}$$

Examine   Since $m\angle B = 5x - 7$, replace x with 29.
$5(29) - 7 = 138$ and $m\angle B = 138$.

Algebra Review

Solving Multi-Step Equations, p. 723

To construct two angles of the same measure requires a compass and straightedge.

Step 1 Draw an angle like ∠P on your paper.

Step 2 Use a straightedge to draw a ray on your paper. Label its endpoint T.

Step 3 With P as the center, draw a large arc that intersects both sides of ∠P. Label the points of intersection Q and R.

Step 4 Using the same compass setting, put the compass at point T and draw a large arc that starts above the ray and intersects the ray. Label the point of intersection S.

Step 5 Place the point of the compass on R and adjust so that the pencil tip is on Q.

Step 6 Without changing the setting, place the compass at point S and draw an arc to intersect the larger arc you drew in Step 4. Label the point of intersection U.

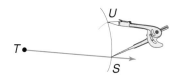

Step 7 Use a straightedge to draw $\overrightarrow{TU}$.

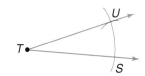

Try These

1. Cut out ∠QPR and ∠UTS and then compare them.
2. Do the two angles have the same measure? If so, write an equation.
3. Construct an angle whose measure is equal to the measure of ∠E.

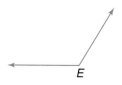

Communicating Mathematics

1. **Draw** an angle having a measure of 70 using a protractor.

2. **Draw** any angle. Then construct an angle whose measure is equal to the measure of the angle drawn.

3. ~~Writing Math~~ Write a few sentences describing how rulers and protractors are used in geometry.

Guided Practice

Examples 1, 2, 4–6

Use a protractor to find the measure of each angle. Then classify each angle as *acute*, *obtuse*, or *right*.

4. $m\angle PTR$ 5. $m\angle PTW$

6. $m\angle RTW$ 7. $m\angle PTQ$

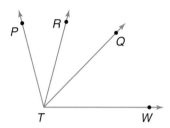

Examples 3–6

Use a protractor to draw an angle having each measurement. Then classify each angle as *acute*, *obtuse*, or *right*.

8. 45°   9. 115°

Example 7

10. **Algebra**   The measure of $\angle J$ is 84. Solve for y.

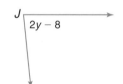

$2y - 8$

Exercises

Use a protractor to find the measure of each angle. Then classify each angle as *acute*, *obtuse*, or *right*.

11. $\angle AGD$ 12. $\angle CGD$

13. $\angle EGF$ 14. $\angle BGE$

15. $\angle CGF$ 16. $\angle EGC$

17. $\angle AGB$ 18. $\angle FGD$

19. $\angle BGF$ 20. $\angle BGC$

21. $\angle AGC$ 22. $\angle BGD$

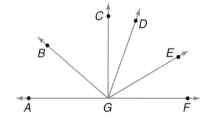

Homework Help

For Exercises	See Examples
11–22	1–2, 4–6
23–28	3, 4–6
29	1–6

Extra Practice
See page 730.

Use a protractor to draw an angle having each measurement. Then classify each angle as *acute*, *obtuse*, or *right*.

23. 42°   24. 155°   25. 26°

26. 95°   27. 75°   28. 138°

29. **Statistics**   The circle graph shows the enrollment in math courses at Hayes High School.

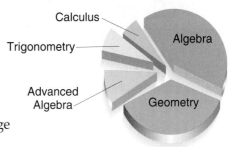

Hayes High School Enrollment in Math Courses

a. Use a protractor to find the measure of each angle of the circle graph.

b. Classify each angle as *acute*, *obtuse*, or *right*.

c. What is the greatest percentage that an acute angle could represent on a circle graph? Explain your reasoning.

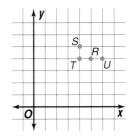

Math Online **Data Update**   For the latest information on school enrollment, visit: geomconcepts.com

30. **Sports**   In golf, the launch angle is the angle of a ball's initial flight path relative to horizontal. While most amateur golfers hit the ball at a 7° angle, professional golfers hit the ball at a 10° angle. A launch angle of 13° is optimal.

a. Draw a diagram that shows these launch angles.

b. Explain why an angle of 13° is optimal.

c. Explain why an angle of 30° is not optimal.

31. **Algebra**   The measure of ∠ABC is 6 more than twice the measure of ∠EFG. The sum of the measures of the two angles is 90. Find the measure of each angle.

32. **Critical Thinking**   Tell how a corner of a sheet of notebook paper could be used to classify an angle.

Mixed Review

33. Draw ∠XYZ that has a point W on the angle.   *(Lesson 3–1)*

34. Find the midpoint of a segment that has endpoints at (3, −5) and (−1, 1).   *(Lesson 2–5)*

35. What is the ordered pair for point R? *(Lesson 2–4)*

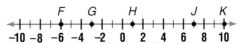

Standardized Test Practice
Ⓐ Ⓑ Ⓒ Ⓓ

36. **Extended Response**   Use the number line to determine whether H is the midpoint of $\overline{FJ}$. Explain your reasoning.   *(Lesson 2–3)*

37. **Short Response**   Write a sequence in which each term is 6 more than the previous term.   *(Lesson 1–1)*

Those Magical Midpoints

Materials

 straightedge

 compass

 scissors

Triangles, Quadrilaterals, and Midpoints

What happens when you find the midpoints of the sides of a three-sided figure and connect them to form a new figure? What if you connect the midpoints of the sides of a four-sided figure? Let's find out.

Investigate

1. A three-sided closed figure is called a **triangle**. Use paper and scissors to investigate the midpoints of the sides of a triangle.

   a. On a piece of paper, draw a triangle with all angles acute and all sides of different lengths.

   b. Use a compass to construct the midpoints of the three sides of your triangle. Connect the three midpoints as shown.

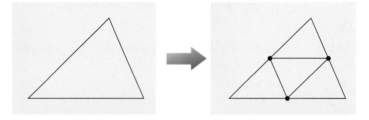

   c. Label the inner triangle 4. Label the outer triangles 1, 2, 3. Cut out each triangle. Compare the shape and size of the triangles.

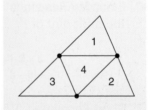

   d. What appears to be true about the four triangles?

2. A four-sided closed figure is called a **quadrilateral**. Use paper and scissors to investigate the midpoints of the sides of a quadrilateral.

   a. On a piece of paper, draw a large quadrilateral with all sides of different lengths.

   b. Use a compass to construct the midpoints of the four sides of your quadrilateral. Connect the four midpoints with line segments as shown.

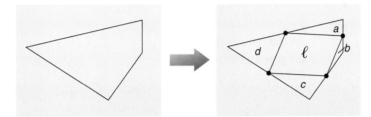

   c. Label the inner quadrilateral ℓ and the outer triangles a, b, c, and d. Cut out each triangle. Compare the shape and size of the triangles.

   d. Assemble all four triangles to cover quadrilateral ℓ completely. Sketch the arrangement on quadrilateral ℓ.

Extending the Investigation

In this extension, you will investigate other triangles and quadrilaterals and their midpoints.

Use paper and scissors or geometry software to complete these investigations.

1. Make a conjecture about the triangles formed when the midpoints of a triangle are connected. Test your conjecture on at least four triangles of different shapes and sizes. Include one triangle with a right angle and one with an obtuse angle.

2. Make a conjecture about the inner quadrilateral and the four triangles formed by connecting the midpoints of a quadrilateral. Test your conjecture on at least four quadrilaterals of different shapes and sizes. Include one quadrilateral with at least one right angle and one quadrilateral with at least one obtuse angle.

Presenting Your Conclusions

Here are some ideas to help you present your conclusions to the class.

- Make a poster that summarizes your results.
- Design an experiment using geometry software to test your conjectures about triangles, quadrilaterals, and the midpoints of their sides.

 Investigation For more information on midpoints and fractals, visit: www.geomconcepts.com

What You'll Learn
You'll learn to find the measure of an angle and the bisector of an angle.

Why It's Important
Sailing Angle measures can be used to determine sailing positions. *See Exercise 24.*

In the following activity, you will learn about the Angle Addition Postulate.

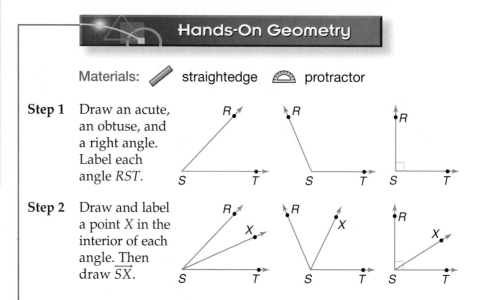

Hands-On Geometry

Materials: ✏ straightedge 📐 protractor

Step 1 Draw an acute, an obtuse, and a right angle. Label each angle *RST*.

Step 2 Draw and label a point *X* in the interior of each angle. Then draw $\overrightarrow{SX}$.

Step 3 For each angle, find $m\angle RSX$, $m\angle XST$, and $m\angle RST$.

Try These

1. For each angle, how does the sum of $m\angle RSX$ and $m\angle XST$ compare to $m\angle RST$?
2. **Make a conjecture** about the relationship between the two smaller angles and the larger angle.

The activity above leads to the following postulate.

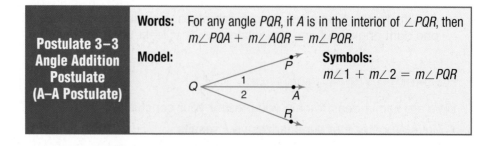

**Postulate 3–3
Angle Addition Postulate
(A–A Postulate)**

Words: For any angle *PQR*, if *A* is in the interior of $\angle PQR$, then $m\angle PQA + m\angle AQR = m\angle PQR$.

Model:

Symbols:
$m\angle 1 + m\angle 2 = m\angle PQR$

There are two equations that can be derived using Postulate 3–3.

$m\angle 1 = m\angle PQR - m\angle 2$ *These equations are true no matter where*
$m\angle 2 = m\angle PQR - m\angle 1$ *A is located in the interior of $\angle PQR$.*

1 **If $m\angle EFH = 35$ and $m\angle HFG = 40$, find $m\angle EFG$.**

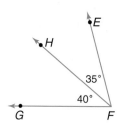

$$m\angle EFG = m\angle EFH + m\angle HFG$$
$$ = \quad 35 \quad + \quad 40 \qquad Substitution$$
$$ = 75 \qquad\qquad\qquad\quad Add.$$

So, $m\angle EFG = 75$.

2 **Find $m\angle 2$ if $m\angle XYZ = 86$ and $m\angle 1 = 22$.**

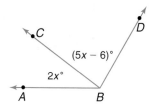

$$m\angle 2 = m\angle XYZ - m\angle 1$$
$$ = \quad 86 \quad - \quad 22 \qquad Substitution$$
$$ = 64 \qquad\qquad\qquad\quad Subtract.$$

So, $m\angle 2 = 64$.

Algebra Link

3 **Find $m\angle ABC$ and $m\angle CBD$ if $m\angle ABD = 120$.**

$$m\angle ABC + m\angle CBD = m\angle ABD \qquad Postulate\ 3\text{--}3$$
$$2x + (5x - 6) = 120 \qquad\qquad Substitution$$
$$7x \quad\ - 6 = 120 \qquad\qquad Combine\ like\ terms.$$
$$7x - 6 + 6 = 120 + 6 \qquad Add\ 6\ to\ each\ side.$$
$$7x = 126 \qquad\qquad\quad Simplify.$$
$$\frac{7x}{7} = \frac{126}{7} \qquad\qquad Divide\ each\ side\ by\ 7.$$
$$x = 18 \qquad\qquad\quad Simplify.$$

— Algebra Review —
Solving Multi-Step
Equations, p. 723

To find $m\angle ABC$ and $m\angle CBD$, replace x with 18 in each expression.

$$m\angle ABC = 2x \qquad\qquad\qquad m\angle CBD = 5x - 6$$
$$ = 2(18) \quad x = 18 \qquad\qquad = 5(18) - 6 \quad x = 18$$
$$ = 36 \qquad\qquad\qquad\qquad\ = 90 - 6\ \text{or}\ 84$$

So, $m\angle ABC = 36$ and $m\angle CBD = 84$.
Check: Is the sum of the measures 120?

Your Turn

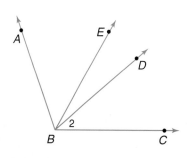

a. Find $m\angle ABC$ if $m\angle ABD = 70$ and $m\angle DBC = 43$.

b. If $m\angle EBC = 55$ and $m\angle EBD = 20$, find $m\angle 2$.

c. Find $m\angle ABD$ if $m\angle ABC = 110$ and $m\angle 2 = 36$.

Look Back

Bisector of a
Segment:
Lesson 2–3

Just as every segment has a midpoint that bisects the segment, every angle has a ray that bisects the angle. This ray is called an **angle bisector**.

Definition of an Angle Bisector	**Words:** The bisector of an angle is the ray with its endpoint at the vertex of the angle, extending into the interior of the angle. The bisector separates the angle into two angles of equal measure.
	Model: 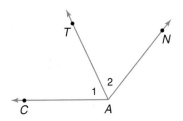 **Symbols:** $\overrightarrow{PW}$ is the bisector of $\angle P$. $m\angle QPW = m\angle WPR$

Example — ④ If $\overrightarrow{AT}$ bisects $\angle CAN$ and $m\angle CAN = 130$, find $m\angle 1$ and $m\angle 2$.

Since $\overrightarrow{AT}$ bisects $\angle CAN$, $m\angle 1 = m\angle 2$.

$m\angle 1 + m\angle 2 = m\angle CAN$	*Postulate 3–3*
$m\angle 1 + m\angle 2 = 130$	*Replace $m\angle CAN$ with 130.*
$m\angle 1 + m\angle 1 = 130$	*Replace $m\angle 2$ with $m\angle 1$.*
$2(m\angle 1) = 130$	*Combine like terms.*
$\dfrac{2(m\angle 1)}{2} = \dfrac{130}{2}$	*Divide each side by 2.*
$m\angle 1 = 65$	*Simplify.*

Since $m\angle 1 = m\angle 2$, $m\angle 2 = 65$.

Your Turn

d. If $\overrightarrow{JK}$ bisects $\angle RJT$ and $\angle RJT$ is a right angle, find $m\angle 1$ and $m\angle 2$.

The angle bisector of a given angle can be constructed using the following procedure.

Hands-On Geometry
Construction

Materials: compass 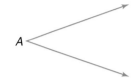 straightedge

Step 1 Draw an angle like ∠A on your paper.

Step 2 Place a compass at point A and draw a large arc that intersects both sides of ∠A. Label the points of intersection B and C.

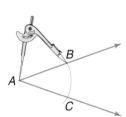

Step 3 With the compass at point B, draw an arc in the interior of ∠A.

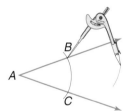

Step 4 Keeping the same compass setting, place the compass at point C. Draw an arc that intersects the arc drawn in Step 3. Label the point of intersection D.

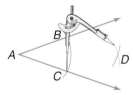

Step 5 Draw $\overrightarrow{AD}$.

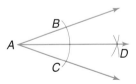

Try These

1. How does $m\angle BAD$ compare to $m\angle DAC$?
2. Name the bisector of ∠BAC.
3. Draw an angle like ∠Y on your paper. Then construct the angle bisector of ∠Y.

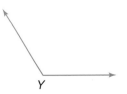

COncepts in MOtion
Animation
geomconcepts.com

Check for Understanding

Communicating Mathematics

1. **State** the Angle Addition Postulate in your own words.

2. **Draw** an acute angle and label it $\angle D$. Then construct the angle bisector and label it $\overrightarrow{DM}$.

3. **You Decide?** Josh says that you get two obtuse angles after bisecting an angle. Brandon disagrees. Who is correct, and why?

Guided Practice

⊙ **Getting Ready** Use the Angle Addition Postulate to solve each of the following.

Sample: If $m\angle 1 = 36$ and $m\angle 2 = 73$, find $m\angle 1 + m\angle 2$.

Solution: $m\angle 1 + m\angle 2 = 36 + 73$ or 109

4. If $m\angle 1 + m\angle 2 = 134$ and $m\angle 2 = 90$, find $m\angle 1$.

5. If $m\angle 1 + m\angle 2 = 158$ and $m\angle 1 = m\angle 2$, find $m\angle 1$.

6. If $m\angle 1 + m\angle 2 = 5x$ and $m\angle 1 = 2x + 1$, find $m\angle 2$.

Refer to the figure at the right.

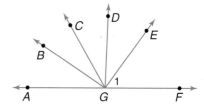

Example 1

7. If $m\angle AGB = 40$ and $m\angle BGC = 24$, find $m\angle AGC$.

Example 2

8. If $m\angle BGD = 52$ and $m\angle BGC = 24$, find $m\angle CGD$.

Example 4

9. If $\overrightarrow{GE}$ bisects $\angle CGF$ and $m\angle CGF = 116$, find $m\angle 1$.

Example 3

10. **Algebra** Find $m\angle PQT$ and $m\angle TQR$ if $m\angle PQT = x$, $m\angle TQR = 5x + 18$, and $m\angle PQR = 90$.

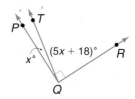

Exercises

• • • • • • • • • • • • • • • • • • • •

Practice

Refer to the figures at the right.

11. If $m\angle UZW = 77$ and $m\angle VZW = 35$, find $m\angle 1$.

12. Find $m\angle VZX$ if $m\angle VZW = 35$ and $m\angle WZX = 78$.

13. If $m\angle WZX = 78$ and $m\angle XZY = 25$, find $m\angle WZY$.

14. If $m\angle UZW = 76$ and $\overrightarrow{ZV}$ bisects $\angle UZW$, find $m\angle UZV$.

15. Find $m\angle KPM$ if $\overrightarrow{PM}$ bisects $\angle KPN$ and $m\angle KPN = 30$.

16. If $m\angle JPM = 48$ and $m\angle KPM = 15$, find $m\angle JPK$.

17. If $m\angle JPO = 126$ and $\overrightarrow{PN}$ bisects $\angle JPO$, find $m\angle NPO$.

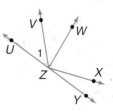

Exercises 11–14

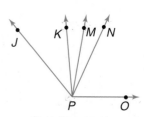

Exercises 15–17

Homework Help	
For Exercises	**See Examples**
11–20, 23, 24	1–3
22–24	4
Extra Practice	
See page 730.	

Refer to the figure at the right.

18. If $m\angle QSU = 38$ and $m\angle UST = 18$, find $m\angle QST$.

19. If RST is a right angle and $m\angle UST = 18$, find $m\angle RSU$.

20. Find $m\angle QSV$ if $m\angle TSU = 18$, $m\angle TSV = 24$, and $m\angle QSU = 38$.

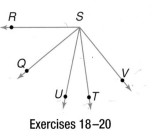

Exercises 18–20

21. If an acute angle is bisected, what type of angles are formed?

22. What type of angles are formed when an obtuse angle is bisected?

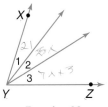

Exercise 23

Applications and Problem Solving

23. **Algebra** If $m\angle 1 = 21$, $m\angle 2 = 5x$, $m\angle 3 = 7x + 3$, and $m\angle XYZ = 18x$, find x.

24. **Sailing** The graph shows sailing positions. Suppose a sailboat is in the run position. How many degrees must the sailboat be turned so that it is in the close reach position?

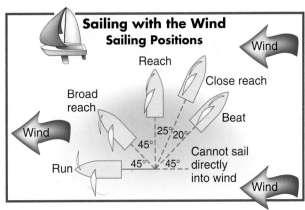

Source: Coast Guard

25. **Critical Thinking** What definition involving segments and points is similar to the Angle Addition Postulate?

Mixed Review

26. Use a protractor to measure $\angle ABC$. *(Lesson 3–2)*

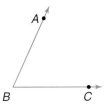

27. Name all angles having P as their vertex. *(Lesson 3–1)*

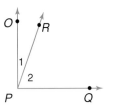

28. Points A, B, and C are collinear. If $AB = 12$, $BC = 37$, and $AC = 25$, determine which point is between the other two. *(Lesson 2–2)*

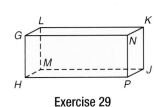

Exercise 29

Standardized Test Practice
Ⓐ Ⓑ Ⓒ Ⓓ

29. **Short Response** Name the intersection of plane GNK and plane PJK. *(Lesson 1–3)*

30. **Multiple Choice** A stock rose in price from $2.50 to $2.75 a share. Find the percent of increase in the price of the stock. *(Percent Review)*

     Ⓐ 10%   Ⓑ 9%   Ⓒ 0.1%   Ⓓ 0.09%

Adjacent Angles and Linear Pairs of Angles

What You'll Learn

You'll learn to identify and use adjacent angles and linear pairs of angles.

Why It's Important

Architecture
Adjacent angles and linear pairs are used in architecture.
See Example 6.

When you bisect an angle, you create two angles of equal measure. The two angles are called **adjacent angles**.

Angles 1 and 2 are examples of adjacent angles. They share a common ray.

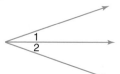

Definition of Adjacent Angles	**Words:**	Adjacent angles are angles that share a common side and have the same vertex, but have no interior points in common.
	Model:	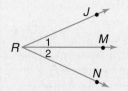 ∠1 and ∠2 are adjacent with the same vertex *R* and common side $\overrightarrow{RM}$.

Examples

Determine whether ∠1 and ∠2 are adjacent angles.

1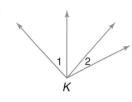

No. They have the same vertex *K*, but no common side.

2

Yes. They have the same vertex *P* and a common side with no interior points in common.

3

No. They do not have a common side or a common vertex.
The side of ∠1 is $\overrightarrow{TL}$.
The side of ∠2 is $\overrightarrow{ML}$.

Your Turn

a.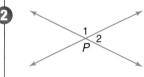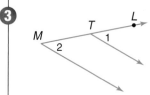

b.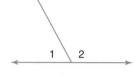

c.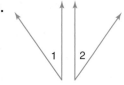

In Example 2, the noncommon sides of the adjacent angles form a straight line. These angles are called a **linear pair**.

Definition of Linear Pair	**Words:** Two angles form a linear pair if and only if they are adjacent and their noncommon sides are opposite rays.
	Model: 
	∠1 and ∠2 are a linear pair.

Examples

In the figure, $\overrightarrow{CM}$ and $\overrightarrow{CE}$ are opposite rays.

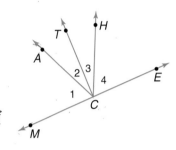

④ **Name the angle that forms a linear pair with ∠1.**

∠ACE and ∠1 have a common side $\overrightarrow{CA}$, the same vertex C, and opposite rays $\overrightarrow{CM}$ and $\overrightarrow{CE}$. So, ∠ACE forms a linear pair with ∠1.

⑤ **Do ∠3 and ∠TCM form a linear pair? Justify your answer.**

No, their noncommon sides are not opposite rays.

Your Turn

d. Name the angle that forms a linear pair with ∠MCH.
e. Tell whether ∠TCE and ∠TCM form a linear pair. Justify your answer.

Real World

Example

Architecture Link

⑥ **The John Hancock Center in Chicago, Illinois, contains many types of angles. Describe the highlighted angles.**

The angles are adjacent, and they form a linear pair.

You can use a TI–83/84 Plus graphing calculator to investigate how the angle bisectors for a linear pair are related.

Graphing Calculator Exploration

Graphing Calculator Tutorial

See pp. 782–785.

Step 1 Construct a line that passes through a point *P*. Use the Point on Object tool on the F2 menu to mark points *A* and *B* on opposite sides of point *P*. Use the Line tool on F2 to construct line *PC*.

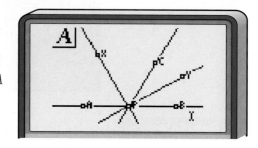

Step 2 Use the Angle Bisector tool on the F3 menu to construct the lines that bisect ∠*APC* and ∠*BPC*. Label points *X* and *Y* on these lines.

Step 3 Use the Measure Angle tool on the F5 menu to display the measure of ∠*XPY*.

Try These

1. What value does the calculator display for ∠*XPY*?
2. Use the Angle tool to display the measures of ∠*XPC* and ∠*CPY*. What is the sum of these measures?
3. Drag point *C*. Describe what happens to the angle measures.
4. **Make a conjecture** about the relationship between bisectors of a linear pair.

Check for Understanding

Communicating Mathematics

1. **Draw and label** two adjacent angles for which the sum of their measures is 90.

2. **Writing Math** Write a sentence explaining why you think the term *linear pair* is used to describe angles such as ∠1 and ∠*ACE* in Example 4.

Vocabulary

adjacent angles
linear pair

Guided Practice

Examples 1–5

Use the terms *adjacent angles*, *linear pair*, or *neither* to describe angles 1 and 2 in as many ways as possible.

3.
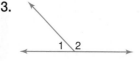

4.
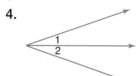

Examples 4 & 5

In the figure at the right, $\overrightarrow{UZ}$ and $\overrightarrow{UW}$ are opposite rays.

5. Name two angles that are adjacent to $\angle WUX$.

6. Which angle forms a linear pair with $\angle YUZ$?

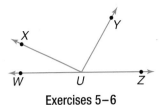

Exercises 5–6

Example 6

7. **Science** Describe the illustrated angles in the spider web.

Exercises

Practice

Homework Help	
For Exercises	See Examples
8–13	1–5
14–15	4–5
16, 18	1–2
17, 19, 21	4–5
20	1–2, 4–5
Extra Practice	
See page 730.	

Use the terms *adjacent angles, linear pair,* or *neither* to describe angles 1 and 2 in as many ways as possible.

8.

9.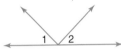

10.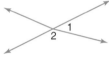

11.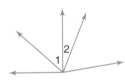

12.

13.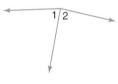

In the figure, $\overrightarrow{GA}$ and $\overrightarrow{GD}$, and $\overrightarrow{GB}$ and $\overrightarrow{GE}$ are opposite rays.

14. Which angle forms a linear pair with $\angle DGC$?

15. Do $\angle BGC$ and $\angle EGD$ form a linear pair? Justify your answer.

16. Name two angles that are adjacent to $\angle CGD$.

17. Name two angles that form a linear pair with $\angle BGD$.

18. Name three angles that are adjacent to $\angle AGB$.

19. Do $\angle CGE$ and $\angle CGB$ form a linear pair? Justify your answer.

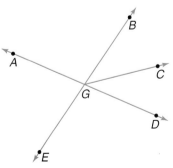

20. Plumbing   A plumber uses a T-fitting
to join three pieces of copper piping
as shown. Describe the type of angles
formed by the three pieces of pipe
and the fitting.

21. Flags   Sailors use international code flags to
communicate at sea. The flag shown represents
the letter z. How many linear pairs are in the design
of the flag?

22. Critical Thinking   How many pairs
of adjacent angles are in the design
of the window shown at the right?
Name them.

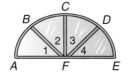

23.   ∠*ABC* is shown at the right. Find *m*∠2 if
m∠*ABC* = 87 and *m*∠1 = 19.   *(Lesson 3–3)*

24. Use a protractor to draw an 85° angle.
Then classify the angle.   *(Lesson 3–2)*

25. Draw ∠*ABC* that has point *T* in the
exterior of the angle.   *(Lesson 3–1)*

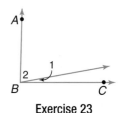

Exercise 23

26. Grid In   Find the measure of the
distance between *B* and *C*.   *(Lesson 2–1)*

27. Multiple Choice   Find the area of a rectangle with length 16 feet and
width 9 feet.   *(Lesson 1–6)*

    Ⓐ   50 ft²   Ⓑ   71 ft²   Ⓒ   86 ft²   Ⓓ   144 ft²

Quiz 1   Lessons 3–1 through 3–4

1. Name the angle in four ways. Then identify its vertex and
its sides.   *(Lesson 3–1)*

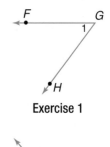

Exercise 1

**Use a protractor to draw an angle for each measurement. Then
classify each angle as *acute*, *obtuse*, or *right*.**   *(Lesson 3–2)*

2. 97°   **3.** 35°

4. Algebra   If *m*∠1 = 3*x*, *m*∠2 = 5*x*, and
m∠*ABC* = 96, find *x*.   *(Lesson 3–3)*

5. Use the terms *adjacent angles, linear pair,* or
neither to describe the pair of angles in as
many ways as possible.   *(Lesson 3–4)*

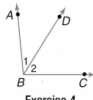

Exercise 4

Exercise 5

www.geomconcepts.com/self_check_quiz

Drafter

Do you like to draw? Does a career that involves drawing interest you? If so, you may enjoy a career as a drafter. Drafters prepare drawings and plans that are used to build everything from ordinary houses to space stations.

When preparing a drawing, drafters may use *drafting triangles* along with a *T-square* to draw various angles.

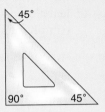

The diagram at the right shows how a drafter would use these tools to draw a 75° angle.

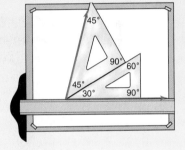

Draw a diagram that shows how a drafter would use drafting triangles and a T-square to draw each angle measure.

1. 105°   2. 150°   3. 135°

FAST FACTS About Drafters

Working Conditions
- usually work in a comfortable office
- sit at drafting tables or computer terminals
- may be susceptible to eyestrain, hand and wrist problems, and back discomfort

Education
- high school math, science, computer, design, and drafting courses
- postsecondary training in drafting at a technical school or community college

Employment
Where Drafters Are Employed

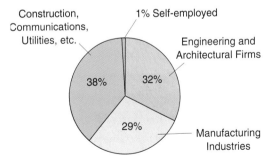

Construction, Communications, Utilities, etc. — 38%

1% Self-employed

Engineering and Architectural Firms — 32%

Manufacturing Industries — 29%

---

interNET
CONNECTION
www.geomconcepts.com

Career Data For the latest information on careers in drafting, visit:

Angles are all around us, even in nature. The veins of a maple leaf show a pair of **complementary angles**.

$$m\angle 1 + m\angle 2 = 90$$

Definition of Complementary Angles	**Words:** Two angles are complementary if and only if the sum of their degree measures is 90.
	Model: 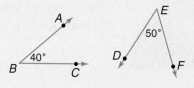
	Symbols: $m\angle ABC + m\angle DEF = 90$

If two angles are complementary, each angle is a *complement* of the other. For example, $\angle ABC$ is the complement of $\angle DEF$ and $\angle DEF$ is the complement of $\angle ABC$.

Complementary angles do not need to have a common side or even the same vertex. Some examples of complementary angles are shown.

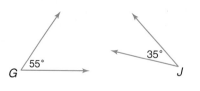

$$m\angle G + m\angle J = 90$$

$$m\angle PQR + m\angle RQS = 90$$

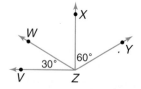

$$m\angle VZW + m\angle XZY = 90$$

If the sum of the measures of two angles is 180, they form a special pair of angles called **supplementary angles**.

<table>
<tr><td rowspan="3">Definition of Supplementary Angles</td><td>Words:</td><td>Two angles are supplementary if and only if the sum of their degree measures is 180.</td></tr>
<tr><td>Model:</td><td>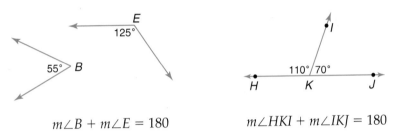</td></tr>
<tr><td>Symbols:</td><td>$m\angle MNP + m\angle RST = 180$</td></tr>
</table>

If two angles are supplementary, each angle is a *supplement* of the other. For example, $\angle MNP$ is the supplement of $\angle RST$ and $\angle RST$ is the supplement of $\angle MNP$.

Like complementary angles, supplementary angles do not need to have a common side or the same vertex. The figures below are examples of supplementary angles.

$m\angle B + m\angle E = 180$ $m\angle HKI + m\angle IKJ = 180$

$m\angle QVU + m\angle SVT = 180$

Example ①

Name a pair of adjacent complementary angles.

$m\angle STV + m\angle VTR = 90$, and they have the same vertex T and common side $\overrightarrow{TV}$ with no overlapping interiors.

So, $\angle STV$ and $\angle VTR$ are adjacent complementary angles.

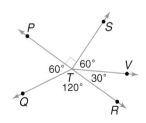

Your Turn

a. Name a pair of nonadjacent complementary angles.

2 Name a pair of nonadjacent supplementary angles.

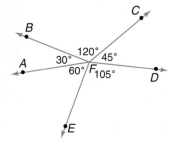

$m\angle BFC + m\angle AFE = 180$, and they have the same vertex F, but no common side.

So, $\angle BFC$ and $\angle AFE$ are nonadjacent supplementary angles.

3 **Find the measure of an angle that is supplementary to $\angle CFD$.**

Let x = the measure of the angle that is supplementary to $\angle CFD$.

$m\angle CFD + x = 180$	*Supplementary angles have a sum of 180.*
$45 + x = 180$	$m\angle CFD = 45$
$45 + x - 45 = 180 - 45$	*Subtract 45 from each side.*
$x = 135$	*Simplify.*

The measure of an angle that is supplementary to $\angle CFD$ is 135.

Your Turn

b. Name a pair of adjacent supplementary angles.

c. Find the measure of the angle that is complementary to $\angle QTR$.

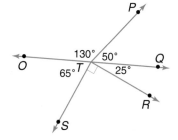

Algebra Link

4 **Angles A and B are complementary. If $m\angle A = x$ and $m\angle B = 5x$, find x. Then find $m\angle A$ and $m\angle B$.**

$m\angle A + m\angle B = 90$	*Definition of Complementary Angles*
$x + 5x = 90$	*Substitution*
$6x = 90$	*Combine like terms.*
$\dfrac{6x}{6} = \dfrac{90}{6}$	*Divide each side by 6.*
$x = 15$	*Simplify.*

─ **Algebra Review** ─
Solving One-Step
Equations, p. 722

Substitute the value of x into each expression.

$m\angle A = x$	$x = 15$	$m\angle B = 5x$	$x = 15$
$= 15$		$= 5(15)$ or 75	

So, $x = 15$, $m\angle A = 15$, and $m\angle B = 75$.

 Personal Tutor at geomconcepts.com

In the figure, $\angle WUX$ and $\angle XUY$ form a linear pair. Postulate 3–4 states that if two angles form a linear pair, the angles are supplementary.

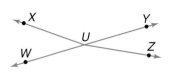

Postulate 3–4 Supplement Postulate	**Words:**	If two angles form a linear pair, then they are supplementary.
	Model:	B $110°\ /\ 70°$ A $\qquad$ D $\qquad$ C
	Symbols:	$m\angle ADB + m\angle BDC = 180$

Example

5 If $m\angle 1 = 57$ and $\angle 1$ and $\angle 2$ form a linear pair, find $m\angle 2$.

If $\angle 1$ and $\angle 2$ form a linear pair, then they are supplementary.

$$m\angle 1 + m\angle 2 = 180 \qquad \textit{Supplement Postulate and Definition of Supplementary Angles}$$
$$57 \ + m\angle 2 = 180 \qquad \textit{Replace } m\angle 1 \textit{ with } 57.$$
$$57 + m\angle 2 - 57 = 180 - 57 \qquad \textit{Subtract 57 from each side.}$$
$$m\angle 2 = 123 \qquad \textit{Simplify.}$$

So, $m\angle 2 = 123$.

Your Turn

d. If $m\angle 2 = 39$ and $\angle 1$ and $\angle 2$ form a linear pair, find $m\angle 1$.

Check for Understanding

Communicating Mathematics

1. **Draw** a pair of adjacent angles that are complementary and have the same measure. What is the measure of each angle?

2. **Explain** why an obtuse angle cannot have a complement.

3. **Tell** whether the angles shown are *complementary*, *supplementary*, or *neither*.

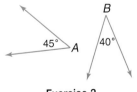

Exercise 3

Guided Practice

⊕ **Getting Ready**   Determine the measures of the complement and supplement of each angle.

| **Sample:** 62 | **Solution:** $90 - 62 = 28$; $180 - 62 = 118$ |

4. 38   **5.** 42   **6.** 79   **7.** 55

Examples 1–3

Refer to the figure at the right.

8. Name a pair of adjacent supplementary angles.

9. Name a pair of nonadjacent complementary angles.

10. Find the measure of an angle that is supplementary to ∠DGE.

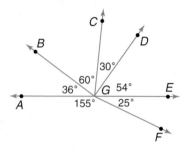

Example 4

11. **Algebra** Angles *G* and *H* are supplementary. If $m\angle G = x + 3$ and $m\angle H = 2x$, find the measure of each angle.

Example 5

12. Angles *XYZ* and *WYX* form a linear pair. If $m\angle WYX = 56$, what is $m\angle XYZ$?

Exercises • • • • • • • • • • • • • • • • • •

Practice

Refer to the figures at the right.

13. Name two pairs of complementary angles.

14. Find the measure of an angle that is supplementary to ∠HNM.

15. Name a pair of adjacent supplementary angles.

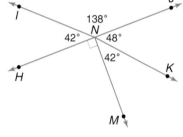

Exercises 13–15

Homework Help	
For Exercises	See Examples
13, 20	1
14, 19, 25–27, 30	3
15, 17	1, 2
16	1, 3
18, 21	2
22–24, 29	5
28	4
Extra Practice	
See page 731.	

16. Find the measure of an angle that is complementary to ∠VWU.

17. Name a pair of nonadjacent complementary angles.

18. Name two pairs of supplementary angles.

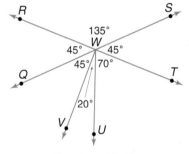

Exercises 16–18

19. Find the measure of an angle that is supplementary to ∠EGF.

20. Name a pair of adjacent complementary angles.

21. Name a pair of nonadjacent supplementary angles.

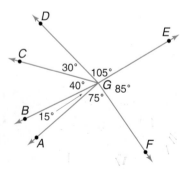

Exercises 19–21

22. If $\angle 1$ and $\angle 2$ form a linear pair and $m\angle 2 = 96$, find $m\angle 1$.

23. Find $m\angle 2$ if $\angle 1$ and $\angle 2$ form a linear pair and $m\angle 1 = 127$.

24. Angles ABC and DEF form a linear pair. If $m\angle DEF = 49$, what is $m\angle ABC$?

25. Can two acute angles be supplementary? Explain.

26. What kind of angle is the supplement of an acute angle?

27. What kind of angle is the supplement of a right angle?

Applications and Problem Solving

28. **Algebra** Angles 1 and 2 are complementary. If $m\angle 1 = 3x + 2$ and $m\angle 2 = 2x + 3$, find the measure of each angle.

29. **Algebra** Angles J and K are supplementary. Find the measures of the two angles if $m\angle J = x$ and $m\angle K = x - 60$.

30. **Carpentry** A carpenter uses a circular saw to cut a piece of lumber at a 145° angle. What is the measure of the other angle formed by the cut?

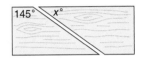

31. **Critical Thinking** Angles 1 and 2 are complementary, and $\angle 1$ and $\angle 3$ are also complementary. Describe the relationship that exists between $\angle 2$ and $\angle 3$.

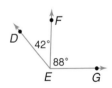

Mixed Review

32. Use the terms *adjacent angles*, *linear pair*, or *neither* to describe the pair of angles in as many ways as possible. *(Lesson 3–4)*

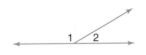

33. If $m\angle DEF = 42$ and $m\angle FEG = 88$, find $m\angle DEG$. *(Lesson 3–3)*

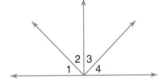

34. **Technology** A videotape cartridge has a length of 18.7 centimeters and a width of 10.3 centimeters. What is the perimeter of the cartridge? *(Lesson 1–6)*

Write the converse of each statement. *(Lesson 1–4)*

35. If it snows, then he will go skiing.

36. If she has 10 dollars, then she will go to the movies.

Standardized Test Practice
Ⓐ Ⓑ Ⓒ Ⓓ

37. **Multiple Choice** How many planes are represented in the figure? *(Lesson 1–3)*

Ⓐ 4
Ⓑ 5
Ⓒ 6
Ⓓ 7

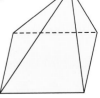

What You'll Learn
You'll learn to identify and use congruent and vertical angles.

Why It's Important
Quilting Congruent and vertical angles are often found in quilt patterns.
See Exercise 22.

Recall that congruent segments have the same measure. **Congruent angles** also have the same measure.

Definition of Congruent Angles	**Words:** Two angles are congruent if and only if they have the same degree measure.
	Model: 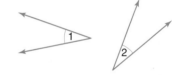 **Symbols:** $\angle A \cong \angle B$ if and only if $m\angle A = m\angle B$.

Reading Geometry

The notation $\angle A \cong \angle B$ is read as *angle A is congruent to angle B.*

If and only if means that if $m\angle 1 = m\angle 2$, then $\angle 1 \cong \angle 2$ and if $\angle 1 \cong \angle 2$, then $m\angle 1 = m\angle 2$.

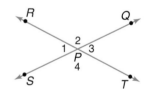

The arcs are used to show congruent angles.

In the figure at the right, $\overrightarrow{SQ}$ and $\overrightarrow{RT}$ intersect. When two lines intersect, four angles are formed. There are two pairs of nonadjacent angles. These pairs are called **vertical angles**.

Definition of Vertical Angles	**Words:** Two angles are vertical if and only if they are two nonadjacent angles formed by a pair of intersecting lines.
	Model: Vertical angles: $\angle 1$ and $\angle 3$ $\angle 2$ and $\angle 4$

Vertical angles are related in a special way. Suppose you cut out and fold a piece of patty paper twice as shown. Compare the angles formed. What can you say about the measures of the vertical angles?

These results are stated in the Vertical Angle Theorem.

Theorem 3–1 Vertical Angle Theorem	**Words:** Vertical angles are congruent.
	Model: 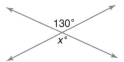 **Symbols:** $\angle 1 \cong \angle 3$ $\angle 2 \cong \angle 4$

Examples

Find the value of *x* in each figure.

1

The angles are vertical angles. So, the value of *x* is 130.

2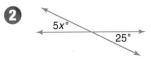

Since the angles are vertical angles, they are congruent.

$5x = 25 \quad \div 5 \quad x = 5$

So, the value of *x* is 5.

Your Turn

a.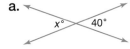

b.

🌐 **Personal Tutor at geomconcepts.com**

Suppose two angles are congruent. What do you think is true about their complements? What is true about their supplements? Draw several examples and make a conjecture.

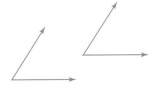

These results are stated in these theorems.

Theorem	Words	Models
3–2	If two angles are congruent, then their complements are congruent. *The measure of angles complementary to $\angle A$ and $\angle B$ is 30.*	60° A B 60° $\angle A \cong \angle B$
3–3	If two angles are congruent, then their supplements are congruent. *The measure of angles supplementary to $\angle 1$ and $\angle 4$ is 110.*	70° 4 \ 3 110° 110° 2 / 1 70° $\angle 1 \cong \angle 4$

Theorem	Words	Models
3–4	If two angles are complementary to the same angle, then they are congruent. ∠3 is complementary to ∠4. ∠5 is complementary to ∠4. ∠3 ≅ ∠5	
3–5	If two angles are supplementary to the same angle, then they are congruent. ∠1 is supplementary to ∠2. ∠3 is supplementary to ∠2. ∠1 ≅ ∠3	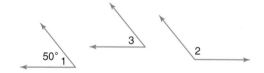

Examples

③ **Suppose ∠J ≅ ∠K and m∠K = 35. Find the measure of an angle that is complementary to ∠J.**

Since ∠J ≅ ∠K, their complements are congruent. The complement of ∠K is 90 − 35 or 55. So, the measure of an angle that is complementary to ∠J is 55.

④ **In the figure, ∠1 is supplementary to ∠2, ∠3 is supplementary to ∠2, and m∠1 = 50. Find m∠2 and m∠3.**

∠1 and ∠2 are supplementary. So, m∠2 = 180 − 50 or 130.
∠2 and ∠3 are supplementary. So, m∠3 = 180 − 130 or 50.

Your Turn

c. Suppose ∠A ≅ ∠B and m∠A = 52. Find the measure of an angle that is supplementary to ∠B.

d. If ∠1 is complementary to ∠3, ∠2 is complementary to ∠3, and m∠3 = 25, what are m∠1 and m∠2?

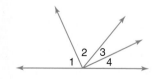

Suppose you draw two angles that are congruent and supplementary as shown at the right. What is true about the angles?

Theorem	Words	Models
3–6	If two angles are congruent and supplementary, then each is a right angle. *∠1 is supplementary to ∠2.* *m∠1 and m∠2 = 90.*	
3–7	All right angles are congruent.	$\angle A \cong \angle B$

Check for Understanding

Communicating Mathematics

1. **Construct** a pair of congruent angles.

2. **Explain** the difference between $m\angle F = m\angle G$ and $\angle F \cong \angle G$.

3. Keisha says that if $m\angle A = 45$ and $m\angle B = 45$, then it is correct to write $m\angle A \cong m\angle B$. Roberta disagrees. She says that it is correct to write $m\angle A = m\angle B$. Who is correct? Explain your reasoning.

> **Vocabulary**
>
> congruent angles
> vertical angles

Guided Practice

Examples 1 & 2

Find the value of x in each figure.

4.

28°   x°

5.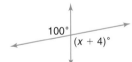

100° $(x + 4)°$

Refer to the figure at the right.

Example 3

6. If $m\angle BEC = 68$, what is the measure of an angle that is complementary to $\angle AED$?

Example 4

7. If $\angle 1$ is supplementary to $\angle 4$, $\angle 3$ is supplementary to $\angle 4$, and $m\angle 1 = 64$, what are $m\angle 3$ and $m\angle 4$?

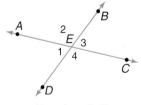

Exercises 6–7

Example 4

8. **Algebra** $\angle 1$ is complementary to $\angle 3$, and $\angle 2$ is complementary to $\angle 3$. If $m\angle 2 = 2x + 9$ and $m\angle 3 = 4x - 3$, find $m\angle 1$ and $m\angle 3$.

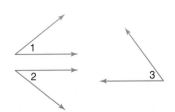

Lesson 3–6 Congruent Angles **125**

Exercises

Practice

Homework Help	
For Exercises	See Examples
9–14, 21	1–2
16, 17, 19, 20	3–4
18	4
Extra Practice	
See page 731.	

Find the value of x in each figure.

9.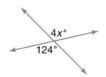

10.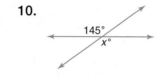

11.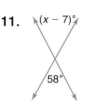

12.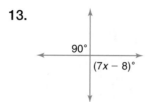

13.

14.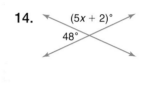

15. What is the measure of an angle that is supplementary to $\angle DEF$ if $\angle ABC \cong \angle DEF$?

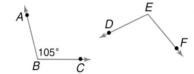

16. If $\angle 1$ is complementary to $\angle 2$, $\angle 3$ is complementary to $\angle 2$, and $m\angle 1 = 28$, what are $m\angle 2$ and $m\angle 3$?

17. If $\angle 2 \cong \angle 3$ and $m\angle 2 = 55$, find the measure of an angle that is supplementary to $\angle 3$.

18. If $\angle RST$ is supplementary to $\angle TSU$, $\angle VSU$ is supplementary to $\angle TSU$, and $m\angle TSU = 62$, find $m\angle RST$ and $m\angle VSU$.

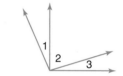

Exercises 17–18

19. Find the measure of an angle that is complementary to $\angle B$ if $\angle B \cong \angle E$ and $m\angle E = 43$.

20. If $\angle 1$ is complementary to $\angle 3$, $\angle 2$ is complementary to $\angle 3$, and $m\angle 1 = 42$, what are $m\angle 2$ and $m\angle 3$?

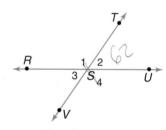

Exercises 19–20

Applications and Problem Solving

21. **Algebra** What is the value of x if $\angle AEC$ and $\angle DEB$ are vertical angles and $m\angle AEC = 27$ and $m\angle DEB = 3x - 6$?

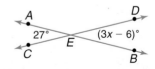

22. Quilting The quilt pattern shown is called the *Lone Star*. If ∠1 is supplementary to ∠2, ∠3 is supplementary to ∠2, and $m\angle 1 = 45$, what are $m\angle 2$ and $m\angle 3$?

23. Critical Thinking Show that Theorem 3–6 is true.

Mixed Review

24. Algebra Angles *G* and *H* are supplementary. If $m\angle G = x$ and $m\angle H = 4x$, what are $m\angle G$ and $m\angle H$? *(Lesson 3–5)*

25. Use the terms *adjacent angles, linear pair,* or *neither* to describe the relationship between ∠1 and ∠2. *(Lesson 3–4)*

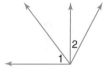

26. Draw an angle having a measure of 35°. *(Lesson 3–2)*

27. Short Response Write an irrational number between 2 and 3 that has ten digits to the right of the decimal point. *(Lesson 2–1)*

28. Multiple Choice Tamika is planning to install vinyl floor tiles in her basement. Her basement measures 20 feet by 16 feet. How many boxes of vinyl floor tile should she buy if one box covers an area of 20 square feet? *(Lesson 1–6)*

   Ⓐ 4   Ⓑ 12   Ⓒ 16   Ⓓ 20

Quiz 2   Lessons 3–5 and 3–6

1. Draw a pair of adjacent complementary angles. *(Lesson 3–5)*

2. If $m\angle 1 = 62$ and ∠1 and ∠2 form a linear pair, find $m\angle 2$. *(Lesson 3–5)*

3. Angles *J* and *K* are vertical angles. If $m\angle J = 37$, what is $m\angle K$? *(Lesson 3–6)*

Refer to the figure at the right. *(Lesson 3–6)*

4. If $m\angle AEB = 35$, what is the measure of an angle complementary to ∠*CED*?

5. If $m\angle 2 = 135$, find $m\angle 3$ and $m\angle 4$.

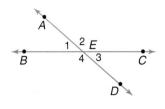

Lines that intersect at an angle of 90 degrees are **perpendicular lines**. In the figure below, lines $\overleftrightarrow{AB}$ and $\overleftrightarrow{CD}$ are perpendicular.

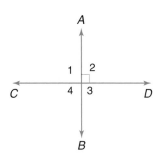

The square symbol where the two lines intersect indicates that the two lines are perpendicular. In the figure, four right angles are formed at the point of intersection.

Also, notice that the four pairs of adjacent angles $\angle 1$ and $\angle 2$, $\angle 2$ and $\angle 3$, $\angle 3$ and $\angle 4$, and $\angle 4$ and $\angle 1$ are supplementary. These adjacent angles also form linear pairs because the nonadjacent sides in each pair are opposite rays.

Definition of Perpendicular Lines	**Words:** Perpendicular lines are lines that intersect to form a right angle.
	Model: m **Symbols:** $m \perp n$ n

Because rays and segments are parts of lines, these too can be perpendicular. For rays or segments to be perpendicular, they must be part of perpendicular lines and they must intersect. In the figure at the right, $\overrightarrow{EC} \perp \overrightarrow{EA}$ and $\overline{CD} \perp \overline{AB}$.

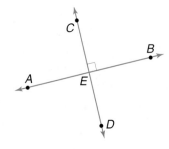

In the figure below, $\ell \perp m$. The following statements are true.

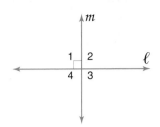

1. $\angle 1$ is a right angle. *Definition of Perpendicular Lines*
2. $\angle 1 \cong \angle 3$ *Vertical angles are congruent.*
3. $\angle 1$ and $\angle 4$ form a linear pair. *Definition of Linear Pair*
4. $\angle 1$ and $\angle 4$ are supplementary. *Linear pairs are supplementary.*
5. $\angle 4$ is a right angle. $m\angle 4 + 90 = 180, m\angle 4 = 90$
6. $\angle 4 \cong \angle 2$ *Vertical angles are congruent.*

These statements lead to Theorem 3–8.

Theorem 3–8	**Words:** If two lines are perpendicular, then they form four right angles. **Model:** 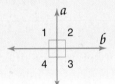 **Symbols:** $a \perp b$ $m\angle 1 = 90$ $m\angle 2 = 90$ $m\angle 3 = 90$ $m\angle 4 = 90$

Examples

In the figure, $\overleftrightarrow{OP} \perp \overleftrightarrow{MN}$ and $\overleftrightarrow{NP} \perp \overleftrightarrow{QS}$. Determine whether each of the following is *true* or *false*.

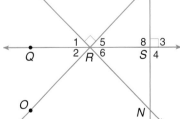

1 $\angle PRN$ is an acute angle.

False. Since $\overleftrightarrow{OP} \perp \overleftrightarrow{MN}$, $\angle PRN$ is a right angle.

2 $\angle 4 \cong \angle 8$

True. $\angle 4$ and $\angle 8$ are vertical angles, and vertical angles are congruent.

Your Turn

a. $m\angle 5 + m\angle 6 = 90$ **b.** $\overline{QR} \perp \overline{PR}$

Find $m\angle EGT$ **and** $m\angle TGC$ **if** $\overline{EG} \perp \overline{CG}$,
$m\angle EGT = 7x + 2$, **and** $m\angle TGC = 4x$.

Since $\overline{EG} \perp \overline{CG}$, $\angle EGC$ is a right angle.
So, $m\angle EGT + m\angle TGC = 90$.

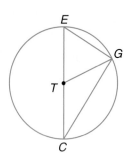

— **Algebra Review** —
Solving Multi-Step
Equations, p. 723

$m\angle EGT + m\angle TGC = 90$	*Definition of Perpendicular Lines*
$(7x + 2) + \quad 4x \quad = 90$	*Substitution*
$11x + 2 = 90$	*Combine like terms.*
$11x + 2 - 2 = 90 - 2$	*Subtract 2 from each side.*
$11x = 88$	*Simplify.*
$\dfrac{11x}{11} = \dfrac{88}{11}$	*Divide each side by 11.*
$x = 8$	*Simplify.*

To find $m\angle EGT$ and $m\angle TGC$, replace x with 8 in each expression.

$m\angle EGT = 7x + 2$
$\quad = 7(8) + 2$ or 58

$m\angle TGC = 4x$
$\quad = 4(8)$ or 32

The following activity demonstrates how to construct a line
perpendicular to a line through a point on the line.

Hands-On Geometry
Construction

Materials: compass straightedge

Step 1 Draw a line ℓ that contains a point T.

Step 2 Place the compass at point T.
Using the same compass setting,
draw arcs to the left and right of
T, intersecting line ℓ. Label these
points D and K.

Step 3 Open the compass to a setting
greater than $\overline{DT}$. Put the compass
at point D and draw an arc above
line ℓ.

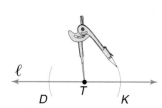

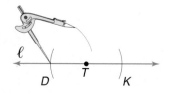

Step 4 Using the same compass setting, put the compass at point K and draw an arc to intersect the one previously drawn. Label the point of intersection S.

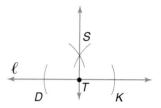

Step 5 Use a straightedge to draw $\overleftrightarrow{ST}$.

Try These

1. Find $m\angle DTS$ and $m\angle STK$.

2. Describe the relationship between $\overleftrightarrow{ST}$ and line ℓ.

3. Construct a line perpendicular to line n through point B.

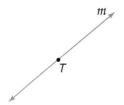

In the activity, you constructed a line through point T and perpendicular to line ℓ. Could you have constructed a different line through T that is perpendicular to line ℓ?

Think of a point T on line m. How many lines can be drawn through that given point? How many lines can be drawn that are perpendicular to line m? How many lines in a plane can be drawn that are perpendicular to line m and go through point T? The next theorem answers this question.

| **Theorem 3–9** | If a line m is in a plane and T is a point on m, then there exists exactly one line in that plane that is perpendicular to m at T. |

Check for Understanding

Communicating Mathematics

1. **Choose** the types of angles that are *not* formed by two perpendicular lines.

   a. vertical

   b. linear pair

   c. complementary

2. **Writing Math** Write a few sentences explaining why it is impossible for two perpendicular lines to form exactly one right angle.

> **Vocabulary**
>
> perpendicular

$\overrightarrow{AB} \perp \overrightarrow{CD}$ and $\overrightarrow{AB} \perp \overrightarrow{EF}$. Determine whether each of the following is *true* or *false*.

3. $m\angle 1 + m\angle 4 = 180$

4. $m\angle 1 = 90$

5. $\overline{EF} \perp \overline{BG}$

6. $m\angle AGE < m\angle 3$

Example 3

7. Algebra If $m\angle 3 = 2x + 6$ and $m\angle 4 = 2x$, find $m\angle 3$ and $m\angle 4$.

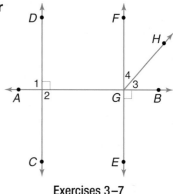

Exercises 3–7

Exercises

• • • • • • • • • • • • • • • •

$\overrightarrow{BN} \perp \overrightarrow{RT}$, $\overrightarrow{MN} \perp \overrightarrow{AB}$, and point T is the midpoint of $\overline{NB}$. Determine whether each of the following is *true* or *false*.

8. $\angle 5$ is a right angle.

9. $\overline{MO} \perp \overline{OR}$

10. $\angle 2 \cong \angle TON$

11. $\angle NOB \cong \angle MOA$

12. $\angle 1$ and $\angle 2$ are complementary.

13. $\angle AON$ and $\angle 3$ are supplementary.

14. $\overline{BT} \perp \overline{OT}$

15. $m\angle BOM + m\angle AOR = 180$

16. $\overline{NT} \cong \overline{BT}$

17. $m\angle BOM + m\angle 5 = 90$

18. $m\angle BTR = m\angle 5$

19. $m\angle 1 + m\angle TON \geq 90$

20. $\overleftrightarrow{AB}$ is the only line $\perp$ to $\overleftrightarrow{MN}$ at O.

21. If $m\angle 1 = 48$, what is $m\angle ROM$?

22. Name four right angles if $\overleftrightarrow{TK} \perp \overleftrightarrow{LY}$.

23. Name a pair of supplementary angles.

24. Name a pair of angles whose sum is 90.

Homework Help

For Exercises	See Examples
8–24, 26	1–2
25	3

Extra Practice
See page 731.

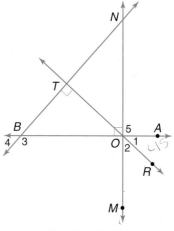

Exercises 8–21

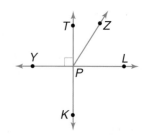

Exercises 22–24

25. Algebra If $\overrightarrow{OP} \perp \overrightarrow{OR}$, $\overrightarrow{OM}$ and $\overrightarrow{ON}$ are opposite rays, $m\angle NOP = 5x$, and $m\angle MOR = 2x - 1$, find $m\angle NOP$ and $m\angle MOR$.

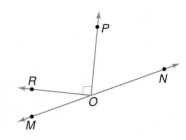

26. **Engineering** A site planner is preparing the layout for a new construction site.

   a. Which street appears to be perpendicular to Fair Avenue?

   b. Which streets appears to be perpendicular to Main Street?

27. **Modeling** Two planes are *perpendicular planes* if they form a right angle. Give a real-world example of two perpendicular planes.

28. **Critical Thinking** Refer to the figure below. Explain in writing, which lines, if any, are perpendicular.

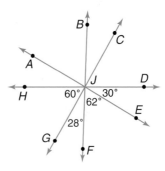

Mixed Review

29. Angles P and Q are vertical angles. If $m\angle P = 47$, what is $m\angle Q$? *(Lesson 3–6)*

30. **Algebra** Angles M and N are complementary. If $m\angle M = 3x$ and $m\angle N = 2x - 5$, find x. Then find $m\angle M$ and $m\angle N$. *(Lesson 3–5)*

31. Draw and label a coordinate plane. Then graph and label point C at $(-5, 3)$. *(Lesson 2–4)*

Standardized Test Practice
Ⓐ Ⓑ Ⓒ Ⓓ

32. **Short Response** Find the length of $\overline{RS}$ in centimeters and in inches. *(Lesson 2–2)*

R •————————————• S

33. **Multiple Choice** The graph shows the estimated number of satellite television subscribers in the United States over five years. Use the pattern in the graph to predict the number of satellite subscribers in 2005. *(Lesson 1–1)*

   Ⓐ 20 million
   Ⓑ 28 million
   Ⓒ 24 million
   Ⓓ 32 million

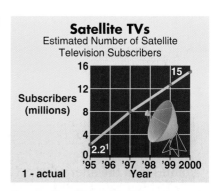

Source: Donaldson, Lufkin & Jenrette

Understanding and Using the Vocabulary

interNET
CONNECTION **Review Activities**
For more review activities, visit:
www.geomconcepts.com

After completing this chapter, you should be able to define each term, property, or phrase and give an example or two of each.

acute angle (p. 98)
adjacent angles (p. 110)
angle (p. 90)
angle bisector (p. 106)
complementary angles (p. 116)
congruent angles (p. 122)
degrees (p. 96)
exterior (p. 92)

interior (p. 92)
linear pair (p. 111)
obtuse angle (p. 98)
opposite rays (p. 90)
perpendicular (p. 128)
protractor (p. 96)
quadrilateral (p. 103)

right angle (p. 98)
sides (p. 90)
straight angle (p. 90)
supplementary angles (p. 116)
triangle (p. 102)
vertex (p. 90)
vertical angles (p. 122)

State whether each sentence is *true* or *false*. If false, replace the underlined word(s) to make a true statement.

1. Angles are measured in units called <u>degrees</u>.
2. In Figure 1, ∠2 and ∠3 are <u>complementary</u> angles.
3. A <u>compass</u> is used to find the measure of an angle.
4. In Figure 1, ∠3 is an <u>acute</u> angle.
5. In Figure 2, the two angles shown are <u>supplementary</u>.
6. In Figure 3, ∠5 and ∠6 are <u>vertical</u> angles.
7. Perpendicular lines intersect to form <u>obtuse</u> angles.
8. In Figure 3, A is called a <u>side</u> of ∠6.
9. In Figure 1, ∠1 and ∠4 form a <u>linear pair</u>.
10. In Figure 4, $\overrightarrow{KM}$ is the <u>vertex</u> of ∠JKL.

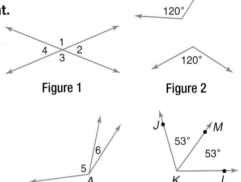

Figure 1   Figure 2

Figure 3   Figure 4

Skills and Concepts

Objectives and Examples

Review Exercises

• **Lesson 3–1** Name and identify parts of an angle.

This angle can be named in four ways:
∠XYZ, ∠ZYX, ∠Y, or ∠1.

The vertex is Y, and the sides are $\overrightarrow{YX}$ and $\overrightarrow{YZ}$.

Point A is in the interior of ∠XYZ.

Name each angle in four ways. Then identify its vertex and its sides.

11.   G
        5   H
     F

12.   S
             4   T
          U

13. Name all angles having P as their vertex.
14. Is Q in the *interior*, *exterior*, or on ∠3?

 www.geomconcepts.com/vocabulary_review

Objectives and Examples	**Review Exercises**

• **Lesson 3–2** Measure, draw, and classify angles.

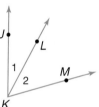

To find the measure of an angle, use a protractor.

The measure of ∠B is 125°.

Since 90 < m∠B < 180, ∠B is obtuse.

Use a protractor to find the measure of each angle. Then classify each angle as *acute*, *obtuse*, or *right*.

15. ∠MQP
16. ∠PQO
17. ∠LQN

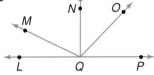

18. Use a protractor to draw a 65° angle.

Exercises 15–17

• **Lesson 3–3** Find the measure of an angle and the bisector of an angle.

Find m∠2 if m∠JKM = 74 and m∠1 = 28.

$m∠2 = m∠JKM - m∠1$
$\quad = \quad 74 \quad - 28$ or 46

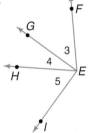

Refer to the figure at the right.

19. Find m∠FEH if m∠3 = 52 and m∠4 = 31.
20. If $\overrightarrow{EH}$ bisects ∠IEF and m∠HEF = 57, find m∠5.
21. If m∠GEI = 90 and m∠5 = 42, find m∠4.

• **Lesson 3–4** Identify and use adjacent angles and linear pairs of angles.

∠1 and ∠2 are adjacent angles. Since $\overrightarrow{XW}$ and $\overrightarrow{XZ}$ are opposite rays, ∠1 and ∠2 also form a linear pair. ∠1 and ∠3 are nonadjacent angles.

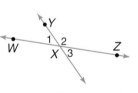

In the figure at the right, $\overrightarrow{TU}$ and $\overrightarrow{TS}$ are opposite rays.

22. Do ∠VTR and ∠UTV form a linear pair? Justify your answer.
23. Name two angles that are adjacent to ∠VTU.
24. Which angle forms a linear pair with ∠STR?

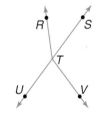

• **Lesson 3–5** Identify and use complementary and supplementary angles.

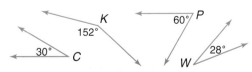

Since m∠C + m∠P = 90, ∠C and ∠P are complementary angles.
Since m∠K + m∠W = 180, ∠K and ∠W are supplementary angles.

Refer to the figure.

25. Name a pair of nonadjacent supplementary angles.
26. Name a pair of supplementary angles.
27. Find the measure of an angle that is supplementary to ∠KAJ.
28. Find the measure of an angle that is complementary to ∠DAS.

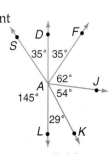

Mixed Problem Solving
See pages 758–765.

Objectives and Examples

Review Exercises

• **Lesson 3–6** Identify and use congruent and vertical angles.

If $m\angle 1 = 51$ and $\angle 2$ and $\angle 3$ are complementary, find $m\angle 3$.

$\angle 1$ and $\angle 2$ are vertical angles. So, $\angle 1 \cong \angle 2$.
$m\angle 1 = 51$. So, $m\angle 2 = 51$.
$\angle 2$ and $\angle 3$ are complementary. So,
$m\angle 3 = 90 - 51$ or 39.

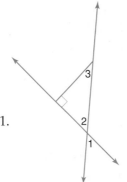

Find the value of x in each figure.

29.

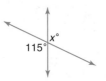

30.

Refer to the figures.

31. Find the measure of an angle that is complementary to $\angle R$ if $\angle R \cong \angle S$ and $m\angle S = 73$.

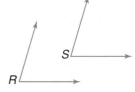

32. If $\angle 1$ is supplementary to $\angle 2$, $\angle 3$ is supplementary to $\angle 1$ and $m\angle 1 = 56$, what are $m\angle 2$ and $m\angle 3$?

• **Lesson 3–7** Identify, use properties of, and construct perpendicular lines and segments.

If $\overrightarrow{WY} \perp \overleftrightarrow{ZX}$, then the following are true.
1. $\angle WVZ$ is a right angle.
2. $\angle YVZ \cong \angle WVX$
3. $m\angle 1 + m\angle 2 = 90$

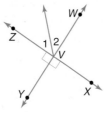

$\overrightarrow{BC} \perp \overrightarrow{AE}$ and $\overrightarrow{BF} \perp \overrightarrow{BD}$. **Determine whether each of the following is *true* or *false*.**

33. $\angle ABC$ is obtuse.
34. $m\angle FBD + m\angle ABC = 180$
35. $\angle DBF \cong \angle CBE$
36. $\overrightarrow{BD} \perp \overrightarrow{AE}$
37. $\angle 1 \cong \angle 3$

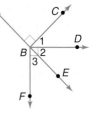

Applications and Problem Solving

38. Manufacturing A conveyor belt is set at a 25° angle to the floor of a factory. If this angle is increased, will the value of y increase or decrease? *(Lesson 3–2)*

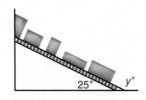

39. Nature In the picture of the snowflake, $\overrightarrow{FN}$ bisects $\angle AFL$ and $m\angle AFL = 120$. Find $m\angle 1$, $m\angle 2$, and $m\angle 3$. *(Lessons 3–3 & 3–6)*

Refer to the figures at the right.

1. Name a pair of opposite rays.
2. *True* or *false*: ∠CTE is adjacent to ∠ATC.
3. Name an angle congruent to ∠ATC.
4. Find the measure of an angle that is complementary to ∠FTE.
5. Name a pair of supplementary angles.
6. Name an angle that forms a linear pair with ∠ATB.
7. Find the measure of ∠ATE. Then classify the angle as *acute*, *right*, or *obtuse*.

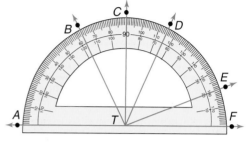

Exercises 1–7

8. Name ∠3 in two other ways.
9. If $\overrightarrow{FK}$ bisects ∠GFP and $m\angle 3 = 38$, find $m\angle KFP$.
10. If $m\angle GFB = 114$ and $m\angle BFT = 34$, find $m\angle GFT$.
11. If ∠PFK is supplementary to ∠KFB, ∠PFK is supplementary to ∠TFP, and $m\angle PFK = 33$, what is $m\angle KFB$ and $m\angle TFP$?

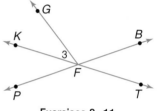

Exercises 8–11

12. Find the measure of an angle that is complementary to ∠C if ∠C ≅ ∠D and $m\angle D = 27$.
13. If ∠JKL and ∠CKD are vertical angles and $m\angle JKL = 35$, find $m\angle CKD$.

In the figure, $\overleftrightarrow{UV} \perp \overleftrightarrow{YW}$.

14. If $m\angle 2 = 44$, find $m\angle 1$.
15. Find $m\angle VYW + m\angle ZWY$.
16. *True* or *false*: $\overleftrightarrow{UV} \perp \overline{ZY}$
17. Find $m\angle UYX + m\angle XYW$.
18. Name two pairs of adjacent right angles.

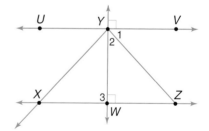

Exercises 14–18

19. **Sports** In pocket billiards, when a ball is hit so that no spin is produced, the angle at which the ball strikes the cushion is equal to the angle at which the ball rebounds off the cushion. That is, $m\angle 1 = m\angle 3$. If $m\angle 1 = 35$, find $m\angle 2$ and $m\angle 3$.

20. **Algebra** ∠G and ∠H are supplementary angles. If $m\angle G = 4x$ and $m\angle H = 7x + 15$, find the measure of each angle.

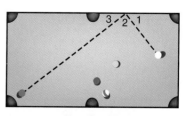

Exercise 19

www.geomconcepts.com/chapter_test

Counting and Probability Problems

Standardized tests usually include problems that ask you to count or calculate probabilities. You may need to know these concepts.

combinations   permutations   tree diagram
outcomes   probability

Test-Taking Tip

To solve counting problems, you can use arithmetic, make a list, draw a tree diagram, use permutations, use combinations, or draw a Venn diagram.

It's a good idea to memorize the definition of the probability of an event.

$$P(\text{event}) = \frac{\text{number of favorable outcomes}}{\text{total number of outcomes}}$$

Example 1

How many combinations of 5 flowers can you choose from one dozen different flowers?

  A 99
  B 396
  C 792
  D 1024

Solution   You need to find the combinations of 5 items out of 12. (These are *combinations*, not permutations, because the order of the flowers does not matter.) Calculate $C(12, 5)$, the number of combinations of 12 things taken 5 at a time.

$$C(12, 5) = \frac{P(12, 5)}{5!}$$

$$= \frac{12 \times 11 \times 10 \times 9 \times 8}{5 \times 4 \times 3 \times 2 \times 1}$$

Hint   Simplify numeric expressions when possible.

$$= \frac{\overset{1}{\cancel{12}} \times 11 \times \overset{1}{\cancel{10}} \times 9 \times 8}{\underset{1}{\cancel{5}} \times \underset{1}{\cancel{4}} \times \underset{1}{\cancel{3}} \times \underset{1}{\cancel{2}} \times 1}$$

$$= 792$$

The answer is choice C, 792.

Example 2

A box of donuts contains 3 plain, 5 cream-filled, and 4 chocolate donuts. If one of the donuts is chosen at random from the box, what is the probability that it will NOT be cream-filled?

Hint   If the probability of an event is p, then the probability of NOT an event is $1 - p$.

Solution   One method for solving this problem is to first find the total number of donuts in the box: $3 + 5 + 4 = 12$.

Then find the number of donuts that are NOT cream-filled. This is the sum of plain plus chocolate: $3 + 4 = 7$.

Calculate the probability of randomly selecting a donut that is NOT cream-filled.

$$\frac{\text{number of favorable outcomes} \rightarrow}{\text{total number of outcomes} \rightarrow} \frac{7}{12}$$

Another method is to find the probability of selecting a donut that *is* cream-filled, $\frac{5}{12}$. Then subtract this probability from 1.

$$1 - \frac{5}{12} = \frac{7}{12}$$

After you work each problem, record your answer on the answer sheet provided or on a sheet of paper.

Multiple Choice

1. How many ways can a family of 5 be seated in a theater if the mother sits in the middle? *(Statistics Review)*

   Ⓐ 120   Ⓑ 24   Ⓒ 15   Ⓓ 10

2. For a class play, student tickets cost $2 and adult tickets cost $5. A total of 30 tickets are sold. If the total sales must exceed $90, then what is the minimum number of adult tickets that must be sold? *(Algebra Review)*

   Ⓐ 7   Ⓑ 8   Ⓒ 9
   Ⓓ 10   Ⓔ 11

3. Andrew's family wants to fence in a 40-meter by 75-meter rectangular area on their ranch. How much fencing should they buy? *(Lesson 1–6)*

   Ⓐ 115 m   Ⓑ 230 m
   Ⓒ 1500 m   Ⓓ 3000 m

4. A coin was flipped 20 times and came up heads 10 times and tails 10 times. If the first and the last flips were both heads, what is the greatest number of consecutive heads that could have occurred? *(Statistics Review)*

   Ⓐ 1   Ⓑ 2   Ⓒ 8
   Ⓓ 9   Ⓔ 10

5. A suitcase designer determines the longest item that could fit in a particular suitcase to be $\sqrt{360}$ centimeters. Which is equivalent to this value? *(Algebra Review)*

   Ⓐ $6\sqrt{10}$ Ⓑ $10\sqrt{6}$
   Ⓒ 36   Ⓓ 180

6. $-|-7| - |-5| - 3|-4| =$ *(Algebra Review)*

   Ⓐ -24 Ⓑ -11 Ⓒ 0
   Ⓓ 13   Ⓔ 24

7. A rope is used to stake a tent pole as shown. Which could be the measure of an angle that is supplementary to the angle that the rope makes with the ground? *(Lesson 3–5)*

   Ⓐ 45°
   Ⓑ 75°
   Ⓒ 90°
   Ⓓ 125°

8. Of the 16 people waiting for the subway, 12 have briefcases, 8 have overcoats, and 5 have both briefcases and overcoats. The other people have neither. How many people have just a briefcase? *(Statistics Review)*

   Ⓐ 10   Ⓑ 7   Ⓒ 6   Ⓓ 3

Grid In

9. Celine made a basket 9 out of 15 times. Based on this, what would be the odds *against* her making a basket the next time she shoots? Write as a fraction. *(Statistics Review)*

Extended Response

10. Spin the two spinners and add the numbers. If the sum is even, you get one point; if odd, your partner gets one point. *(Statistics Review)*

Part A Use a tree diagram to find the probability of getting an even number. Explain why this makes sense.

Part B How could you change the spinners so that the probability of getting an even number equals the probability of getting an odd number?

Parallels

What You'll Learn

Key Ideas

- Describe relationships among lines, parts of lines, and planes. *(Lesson 4–1)*

- Identify relationships among angles formed by two parallel lines and a transversal. *(Lessons 4–2 and 4–3)*

- Identify conditions that produce parallel lines and construct parallel lines. *(Lesson 4–4)*

- Find the slopes of lines and use slope to identify parallel and perpendicular lines. *(Lesson 4–5)*

- Write and graph equations of lines. *(Lesson 4–6)*

Key Vocabulary

linear equation *(p. 174)*

parallel lines *(p. 142)*

skew lines *(p. 143)*

slope *(p. 168)*

transversal *(p. 148)*

Why It's Important

Sports College football has been an American tradition since the 1800s. American football, soccer, and rugby are all derived from games played in ancient Greece and Rome. Now, college football attracts over 40 million people to games and is watched by millions more on television each year.

Parallel lines are often used as part of building and road construction. You will determine how the parallel lines of a football field can be marked in Lesson 4–4.

✔ Check Your Readiness

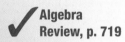

Lesson 1–2, pp. 12–17

Use the figure to name examples of each term.

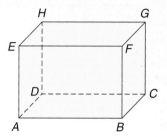

1. segment

2. segment with point H as an endpoint

3. a point that is *not* in plane *ABF*

4. two segments that *do not* intersect

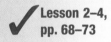

Algebra Review, p. 719

Find the value of each expression.

5. $\frac{6-2}{5-3}$

6. $\frac{12-3}{6-3}$

7. $\frac{10-9}{9-5}$

8. $\frac{-3-(-2)}{4-2}$

9. $\frac{-4-1}{5-3}$

10. $\frac{-10-(-7)}{3-0}$

11. $\frac{-6-(-4)}{-2-(-1)}$

12. $\frac{4-(-4)}{4-9}$

13. $\frac{-6-1}{0-2}$

Lesson 2–4, pp. 68–73

Draw and label a coordinate plane on a piece of grid paper. Then graph and label each point.

14. $A(2, 4)$

15. $B(-1, -3)$

16. $C(2, 0)$

17. $D(-3, 1)$

18. $E(0, -1)$

19. $F(4, -4)$

FOLDABLES™ Study Organizer

Make this Foldable to help you organize your Chapter 4 notes. Begin with four sheets of plain $8\frac{1}{2}$" by 11" paper.

❶ **Fold** in half along the width.

❷ **Open** and fold up the bottom to form a pocket.

❸ **Repeat** steps 1 and 2 three times and glue all four pieces together.

❹ **Label** each pocket with a lesson name. Use the last two pockets for vocabulary. Place an index card in each pocket.

Parallels

Reading and Writing As you read and study the chapter, write the main ideas, examples of theorems, postulates, and definitions on the index cards.

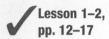

What You'll Learn

You'll learn to describe relationships among lines, parts of lines, and planes.

Why It's Important

Construction

Carpenters use parallel lines and planes in the construction of furniture.
See Exercise 11.

Suppose you could measure the distance between the columns of a building at various points. You would find that the distance remains the same at all points. The columns are parallel.

In geometry, two lines in a plane that are always the same distance apart are **parallel lines**. No two parallel lines intersect, no matter how far you extend them.

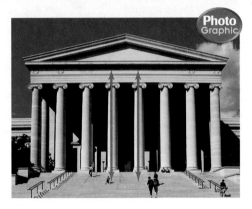

National Gallery of Art, Washington, D.C.

Definition of Parallel Lines	Two lines are parallel if and only if they are in the same plane and do not intersect.

Since segments and rays are parts of lines, they are considered parallel if the lines that contain them are parallel.

Reading Geometry

Read the symbol ∥ as *is parallel to.*

Arrowheads are often used in figures to indicate parallel lines.

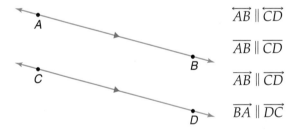

$$\overleftrightarrow{AB} \parallel \overleftrightarrow{CD}$$

$$\overline{AB} \parallel \overline{CD}$$

$$\overrightarrow{AB} \parallel \overrightarrow{CD}$$

$$\overrightarrow{BA} \parallel \overrightarrow{DC}$$

Planes can also be parallel. The shelves in a bookcase are examples of parts of planes. The shelves are the same distance apart at all points, and do not appear to intersect. They are parallel. In geometry, planes that do not intersect are called **parallel planes**.

plane *PSR* ∥ plane *JML*
plane *JMS* ∥ plane *KLR*
plane *PJK* ∥ plane *SML*

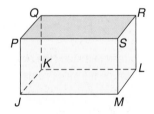

Recall that plane PSR refers to the plane containing points P, S, and R.

Sometimes lines that do not intersect are not in the same plane. These lines are called **skew lines**.

Definition of Skew Lines	Two lines that are not in the same plane are skew if and only if they do not intersect.

Segments and rays can also be skew if they are contained in skew lines. In the figure, $\overline{AX}$ and $\overline{BC}$ are skew segments. They are parts of noncoplanar lines that do not intersect. $\overline{AX}$ and $\overline{XZ}$ intersect at X. They are not skew segments.

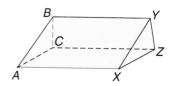

Examples

Name the parts of the rectangular prism shown below. Assume segments that look parallel are parallel.

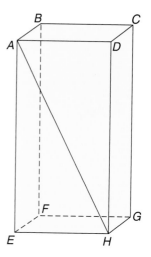

1 **all planes parallel to plane *ABC***

Plane *ABC* is parallel to plane *EFG*.

2 **all segments that intersect $\overline{AB}$**

$\overline{BC}$, $\overline{AD}$, $\overline{AH}$, $\overline{AE}$, and $\overline{BF}$ intersect $\overline{AB}$.

3 **all segments parallel to $\overline{FG}$**

$\overline{BC}$, $\overline{AD}$, and $\overline{EH}$ are parallel to $\overline{FG}$.

4 **all segments skew to $\overline{EF}$**

$\overline{CG}$, $\overline{DH}$, $\overline{AD}$, $\overline{BC}$, and $\overline{AH}$ are skew to $\overline{EF}$.

Your Turn

Name the parts of the figure above.

a. all planes parallel to plane *ABF*

b. all segments that intersect $\overline{DH}$

c. all segments parallel to $\overline{CD}$

d. all segments skew to $\overline{AB}$

Communicating Mathematics

1. **Draw and label** two parallel lines, ℓ and m. Indicate that the lines are parallel by using the arrowhead symbol.

2. **Describe** a real-world example or model of parallel lines.

3. **Writing Math** Sketch the diagram shown below. Then describe and explain the relationship between lines ℓ and m.

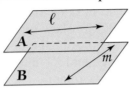

Planes A and B are parallel.

Guided Practice

Examples 2–4

Describe each pair of segments in the prism as *parallel*, *skew*, or *intersecting*.

4. $\overline{KN}, \overline{HL}$
5. $\overline{JM}, \overline{ML}$
6. $\overline{JM}, \overline{KH}$

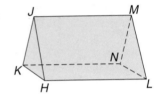

Examples 1, 3, & 4

Name the parts of the cube shown at the right.

7. all planes parallel to plane WXQ
8. all segments parallel to $\overline{PQ}$
9. all segments skew to $\overline{PS}$
10. all pairs of parallel planes

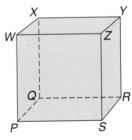

Examples 2–4

11. **Carpentry** A carpenter is constructing a chair like the one shown at the left. Describe a pair of parts that are parallel, a pair that intersect, and a pair that are skew.

Practice

Homework Help

For Exercises	See Examples
12–21, 24–27, 29–31	2–4
22, 23, 28	1
32–39	1–4

Extra Practice
See page 732.

Describe each pair of segments in the prism as *parallel*, *skew*, **or** *intersecting*.

12. $\overline{BE}$, $\overline{CF}$

13. $\overline{AD}$, $\overline{BE}$

14. $\overline{AD}$, $\overline{EF}$

15. $\overline{BC}$, $\overline{EF}$

16. $\overline{AB}$, $\overline{DE}$

17. $\overline{AB}$, $\overline{CF}$

18. $\overline{AB}$, $\overline{BC}$

19. $\overline{AD}$, $\overline{BC}$

20. $\overline{BE}$, $\overline{BC}$

21. $\overline{BC}$, $\overline{DE}$

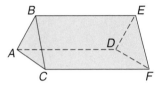

Name the parts of the cube shown at the right.

22. six planes

23. all pairs of parallel planes

24. all segments parallel to $\overline{EH}$

25. all segments skew to $\overline{GH}$

26. all segments parallel to $\overline{AE}$

27. all segments skew to $\overline{BF}$

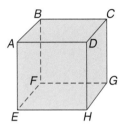

Name the parts of the pyramid shown at the right.

28. all pairs of intersecting planes

29. all pairs of parallel segments

30. all pairs of skew segments

31. all sets of three segments that intersect in a common point

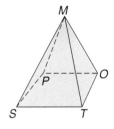

Draw and label a figure to illustrate each pair.

32. congruent parallel segments

33. parallel segments not congruent

34. segments not parallel or congruent

35. skew segments

36. segments not intersecting or skew

37. parallel planes

38. intersecting planes

39. parallel rays

Complete each sentence with *sometimes*, *always*, or *never*.

40. Skew lines ___?___ intersect.

41. Skew lines are ___?___ parallel.

42. Two parallel lines ___?___ lie in the same plane.

43. Two lines in parallel planes are ___?___ skew.

44. Two lines that have no points in common are ___?___ parallel.

45. If two lines are parallel, then they ___?___ lie in the same plane.

Applications and Problem Solving

46. **Interior Design** The shower stall shown in the diagram is formed by a series of intersecting planes. Name two skew segments in the diagram.

47. **Construction** The Empire State Building, built in 1930–1931, is 102 stories and reaches a height of 1250 feet. Suppose the stories represent parallel planes equal distances apart. What is the approximate distance between floors?

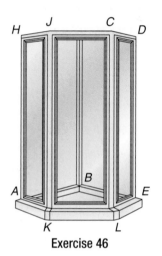

Exercise 46

48. **Sports** Describe how skis on a skier's feet could *intersect*, be *parallel*, or be *skew*.

49. **Graphic Arts** The comic artist has used parallel lines two ways in the *B.C.* comic below. Explain the two uses of parallel lines in the comic.

B.C.   **by johnny hart**

50. Critical Thinking Plane *A* is parallel to plane *B*, and plane *B* is parallel to plane *C*. Is Plane *A* parallel to Plane *C*? Write *yes* or *no*, and explain your answer. Then describe something in your school building that illustrates your response.

Mixed Review

51. In the figure at the right, $\overline{BE} \perp \overline{AD}$. *(Lesson 3–7)*

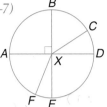

   a. Name four right angles.

   b. Name a pair of supplementary angles.

   c. Name two pairs of angles whose sum is 90.

52. If $\angle JKL \cong \angle PQR$, $m\angle PQR = 4x + 5$, and $m\angle JKL = 5x - 12$, what is $m\angle PQR$? *(Lesson 3–6)*

Refer to the figure at the right.
(Lesson 3–3)

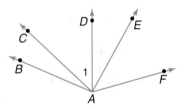

53. If $m\angle CAE = 78$ and $m\angle DAE = 30$, find $m\angle 1$.

54. Find $m\angle DAF$ if $m\angle DAE = 30$ and $m\angle EAF = 75$.

55. Time Do the hands of a clock at 12:20 P.M. form an acute, obtuse, or right angle? *(Lesson 3–2)*

56. Cartography On a map of Ohio, Cincinnati is located at (3, 5), and Massillon is located at (17, 19). If Columbus is halfway between the two cities, what ordered pair describes its position? *(Lesson 2–5)*

Cincinnati, Ohio

57. It is 4.5 blocks from Carlos' house to Matt's house. From Matt's house to Keisha's house is 10.5 blocks, and from Keisha's house to Carlos' house is 6 blocks. If they live on the same street, whose house is between the other two? *(Lesson 2–2)*

Standardized Test Practice
Ⓐ Ⓑ Ⓒ Ⓓ

58. Short Response Name all the planes that can contain three of the points *J*, *K*, *L*, and *M*, if no three points are collinear and the points do not all lie in the same plane. *(Lesson 1–3)*

59. Multiple Choice 12 is what percent of 40? *(Percent Review)*

  Ⓐ 0.3%   Ⓑ $3\frac{1}{3}\%$ Ⓒ 30%   Ⓓ 33%

In geometry, a line, line segment, or ray that intersects two or more lines at different points is called a **transversal**. $\overleftrightarrow{AB}$ is an example of a transversal. It intersects lines ℓ and m. Note all of the different angles that are formed at the points of intersection.

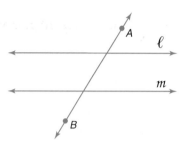

| **Definition of Transversal** | In a plane, a line is a transversal if and only if it intersects two or more lines, each at a different point. |

The lines cut by a transversal may or may not be parallel.

Parallel Lines

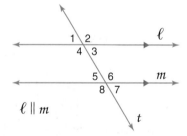

$\ell \parallel m$

t is a transversal for ℓ and m.

Nonparallel Lines

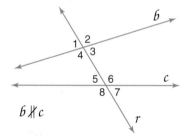

$b \nparallel c$

r is a transversal for b and c.

When a transversal intersects two lines, eight angles are formed, as shown in the figures above. These angles are given special names.

Interior angles lie between the two lines.

$\angle 3, \angle 4, \angle 5, \angle 6$

Alternate interior angles are on opposite sides of the transversal.

$\angle 3$ and $\angle 5$, $\angle 4$ and $\angle 6$

Consecutive interior angles are on the same side of the transversal.

$\angle 3$ and $\angle 6$, $\angle 4$ and $\angle 5$

Exterior angles lie outside the two lines.

$\angle 1, \angle 2, \angle 7, \angle 8$

Alternate exterior angles are on opposite sides of the transversal.

$\angle 1$ and $\angle 7$, $\angle 2$ and $\angle 8$

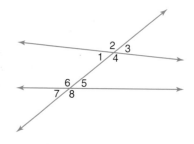

Look Back

Vertical Angles:
Lesson 3–6

Identify each pair of angles
as *alternate interior, alternate
exterior, consecutive interior,*
or *vertical.*

1 ∠2 and ∠8

∠2 and ∠8 are exterior angles on
opposite sides of the transversal, so
they are *alternate exterior* angles.

2 ∠1 and ∠6

∠1 and ∠6 are interior angles on the same side of the transversal, so
they are *consecutive interior* angles.

Your Turn

a. ∠2 and ∠4   **b.** ∠4 and ∠6

In the following activity, you will investigate the relationships among
the angles formed when a transversal intersects two parallel lines.

Hands-On Geometry

Materials: lined paper   ✏ straightedge

◗ protractor

*Note: Save this
drawing to use again
in Lesson 4–3.*

Step 1 Use a straightedge to darken any
two horizontal lines on a piece of
lined paper.

Step 2 Draw a transversal for the lines and
label the angles 1 through 8. Use a
protractor to find the measure of
each angle.

Try These

1. Compare the measures of the alternate interior angles.
2. What is the sum of the measures of the consecutive interior angles?
3. Repeat Steps 1 and 2 above two more times by darkening different
   pairs of horizontal lines on your paper. Make the transversals
   intersect the lines at a different angle each time.
4. Do the interior angles relate to each other the same way for each pair
   of lines?
5. Compare the measures of the alternate exterior angles in each
   drawing.

The results of the activity suggest the theorems stated below.

Theorem	Words	Models and Symbols
4–1 **Alternate Interior Angles**	If two parallel lines are cut by a transversal, then each pair of alternate interior angles is congruent.	$\angle 4 \cong \angle 6$ $\angle 3 \cong \angle 5$
4–2 **Consecutive Interior Angles**	If two parallel lines are cut by a transversal, then each pair of consecutive interior angles is supplementary.	$m\angle 3 + m\angle 6 = 180$ $m\angle 4 + m\angle 5 = 180$
4–3 **Alternate Exterior Angles**	If two parallel lines are cut by a transversal, then each pair of alternate exterior angles is congruent.	$\angle 1 \cong \angle 7$ $\angle 2 \cong \angle 8$

Look Back

Supplementary
Angles:
Lesson 3–5

You can use these theorems to find the measures of angles.

Examples — ❸ **In the figure, $p \parallel q$, and r is a transversal. If $m\angle 5 = 28$, find $m\angle 8$.**

$\angle 5$ and $\angle 8$ are alternate exterior angles, so by Theorem 4–3 they are congruent. Therefore, $m\angle 8 = 28$.

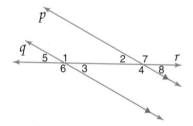

❹ **In the figure, $\overleftrightarrow{AB} \parallel \overleftrightarrow{CD}$, and t is a transversal. If $m\angle 8 = 68$, find $m\angle 6$, $m\angle 7$, and $m\angle 9$.**

$\angle 6$ and $\angle 8$ are consecutive interior angles, so by Theorem 4–2 they are supplementary.

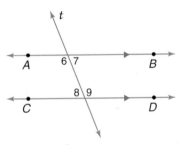

$$
\begin{aligned}
m\angle 6 + m\angle 8 &= 180 && \textit{Supplementary angles} \\
m\angle 6 + 68 &= 180 && \textit{Replace } m\angle 8 \textit{ with 68.} \\
m\angle 6 + 68 - 68 &= 180 - 68 && \textit{Subtract 68 from each side.} \\
m\angle 6 &= 112 && \textit{Simplify.}
\end{aligned}
$$

$\angle 7$ and $\angle 8$ are alternate interior angles, so by Theorem 4–1 they are congruent. Therefore, $m\angle 7 = 68$.

Look Back

Linear Pairs:
Lesson 3–4;
Supplement Postulate:
Lesson 3–5

∠6 and ∠9 are alternate interior angles, so by Theorem 4–1 they are congruent. Thus, $m\angle 9 = 112$.

Your Turn

Refer to the figure in Example 3. Find the measure of each angle.
c. ∠1   d. ∠2   e. ∠3   f. ∠4

Example ⑤

Algebra Link

— Algebra Review —

Solving Equations with
the Variable on Both
Sides, p. 724

In the figure, $s \parallel t$, and m is a transversal. Find $m\angle EBF$.

By Theorem 4–1, $\angle ABC$ is congruent to $\angle BCD$.

$$
\begin{array}{ll}
m\angle ABC = m\angle BCD & \textit{Congruent angles have equal measures.} \\
3x - 5 = 4x - 29 & \textit{Substitution} \\
3x - 5 - 3x = 4x - 29 - 3x & \textit{Subtract 3x from each side.} \\
-5 = x - 29 & \textit{Simplify.} \\
-5 + 29 = x - 29 + 29 & \textit{Add 29 to each side.} \\
24 = x & \textit{Simplify.}
\end{array}
$$

The measure of $\angle ABC$ is $3x - 5$.

$$
\begin{array}{ll}
m\angle ABC = 3x - 5 & \\
m\angle ABC = 3(24) - 5 & \textit{Replace x with 24.} \\
= 72 - 5 \text{ or } 67 &
\end{array}
$$

$\angle EBF$ and $\angle ABC$ are vertical angles and are therefore congruent.

$$
\begin{array}{ll}
m\angle EBF = m\angle ABC & \textit{Congruent angles have equal measures.} \\
= 67 & \textit{Substitution}
\end{array}
$$

Your Turn

g. Find $m\angle GCH$.

Check for Understanding

Communicating Mathematics

1. **Explain** why ∠2 and ∠3 must be congruent.

2. **Describe** two different methods you could use to find $m\angle 3$ if $m\angle 1 = 130$.

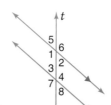

┌─ **Vocabulary** ─┐

transversal
interior angles
alternate interior angles
consecutive interior angles
exterior angles
alternate exterior angles

3. **Name** each transversal and the lines it intersects in the figure at the right.

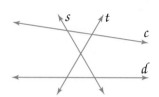

Guided Practice

Identify each pair of angles as *alternate interior*, *alternate exterior*, *consecutive interior*, or *vertical*.

Examples 1 & 2

4. ∠3 and ∠7   5. ∠1 and ∠5
6. ∠2 and ∠8   7. ∠2 and ∠3

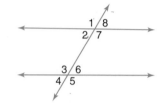

Examples 3 & 4

a ∥ b, and *h* is a transversal. If *m*∠1 = 48, find the measure of each angle. Give a reason for each answer.

8. ∠2   9. ∠3
10. ∠4   11. ∠7

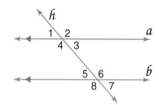

Example 5

12. **Algebra** In the figure at the right, *r ∥ t*, and *d* is a transversal. Find *m*∠1 and *m*∠2.

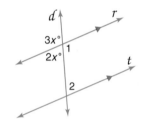

Exercises • • • • • • • • • • • • • • • • • • •

Practice

Identify each pair of angles as *alternate interior*, *alternate exterior*, *consecutive interior*, or *vertical*.

Homework Help	
For Exercises	See Examples
13–24	1, 2
25–36	3, 4
37–39	5
Extra Practice	
See page 732.	

13. ∠1 and ∠5   14. ∠2 and ∠10
15. ∠5 and ∠15   16. ∠11 and ∠3
17. ∠1 and ∠4   18. ∠16 and ∠8
19. ∠5 and ∠6   20. ∠10 and ∠14
21. ∠15 and ∠14   22. ∠8 and ∠2
23. ∠12 and ∠14   24. ∠9 and ∠13

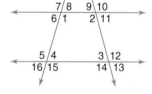

Find the measure of each angle.
Give a reason for each answer.

25. ∠1   26. ∠4
27. ∠6   28. ∠5

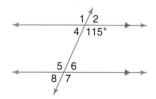

Find the measure of each angle. Give a reason for each answer.

29. ∠9   30. ∠12

31. ∠13   32. ∠15

33. ∠19   34. ∠21

35. ∠22   36. ∠24

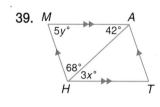

Find the values of x and y.

37.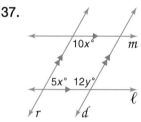

38.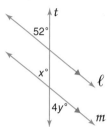

39.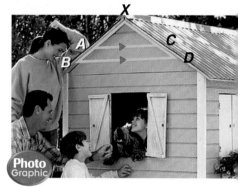

Reading Geometry

Double arrowheads indicate a second pair of parallel lines.

Applications and Problem Solving

40. **Road Maps** Trace and label a section of a road map that illustrates two parallel roads intersected by a transversal road or railroad. Use a protractor to measure the angles formed by the intersections on the map. How does this drawing support Theorems 4–1, 4–2, and 4–3?

41. **Construction** The roof at the right intersects the parallel lines of the siding. Which angles are congruent?

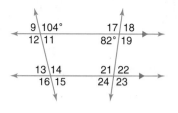

Exercise 41

42. **Critical Thinking** In the figure at the right, explain why you can conclude that ∠1 ≅ ∠4, but you cannot tell whether ∠3 is congruent to ∠2.

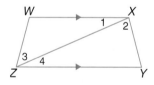

Mixed Review

Draw and label a cube. Name the following pairs. *(Lesson 4–1)*

43. parallel segments

44. intersecting segments

45. skew segments

46. In the figure at the right, $\overline{AD} \perp \overline{CD}$. If $m\angle ADB = 23$ and $m\angle BDC = 3y - 2$, find y. *(Lesson 3–7)*

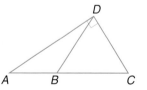

Draw and label a coordinate plane on a piece of grid paper. Then graph and label each point. *(Lesson 2–4)*

47. $G(-5, -1)$ 48. $H(3, -2)$ 49. $J(4, 0)$

50. **Short Response** Write a sequence of five numbers that follows the pattern +1, +3, +5, ... *(Lesson 1–1)*

When Does a Circle Become a Line?

Materials

 globe

 two large rubber bands

 removable tape

 ruler

 protractor

 scissors

Spherical Geometry

The geometry you have been studying in this text is called *Euclidean geometry*. It was named for a famous Greek mathematician named Euclid (325 B.C.–265 B.C.). There are, however, other types of geometry. Let's take a look at *spherical geometry*. Spherical geometry is one form of *non-Euclidean geometry*.

Investigate

1. Use a globe, two large rubber bands, and the steps below to investigate lines on a sphere.

   Latitude Lines

   Longitude Lines

   a. In Euclidean and spherical geometry, points are the same. A point is just a location that can be represented by a dot.

   b. In Euclidean geometry, you can represent a plane by a sheet of paper. Remember that the paper is only part of the plane. The plane goes on forever in all directions. In spherical geometry, a plane is a sphere. The sphere is **finite**, that is, it does not go on forever. The globe will represent a plane in spherical geometry for this investigation.

   c. If possible, place a large rubber band on the globe covering the equator. The equator is known as a **line** in spherical geometry. In Euclidean geometry, a line extends without end in both directions. In spherical geometry, a line is finite. In spherical geometry, a line is defined as a **great circle**, which divides a sphere into two congruent halves. On the globe, the equator is also called a **line of latitude**.

   d. Place a second large rubber band on the globe so that it extends over both the North and South Poles. Position the band so that it is also a great circle. On the globe, a line like this is also called a **line of longitude**.

   e. In Euclidean geometry you learned that when two lines intersect, they have only one point in common. Look at the rubber bands on your globe. How many points do these two lines have in common?

2. Use the globe, removable tape, and the steps below to investigate angle measures in spherical and Euclidean planes.

This works best if you cut the tape into strips about one-eighth inch wide.

a. Select two points on the equator. Select another point close to the North Pole. Use three pieces of removable tape to form a triangle as shown. Use a protractor to estimate the measure of each angle of the triangle. Record your results.

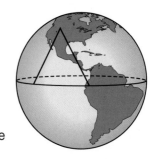

b. Carefully remove the tape from the sphere. Use the three strips to form a triangle on a sheet of paper. Use a protractor to estimate the measure of each angle of the triangle. Record the results.

c. You have formed two triangles with sides of the same length. The first was on the spherical plane. The second was on the Euclidean plane. How do the angle measures of the two triangles compare?

Extending the Investigation

In this extension, you will investigate lines in both Euclidean and spherical geometry by using a globe or geometry drawing software.

1. Determine whether all lines of longitude on the globe are lines in spherical geometry.

2. Determine whether all lines of latitude on the globe are lines in spherical geometry.

3. In Chapter 3, you learned the theorem that states: if two lines are perpendicular, then they form four right angles. Is this theorem true for spherical geometry? Explain your reasoning and include a sketch.

4. **Make a conjecture** about angle measures of triangles in Euclidean and spherical geometry. Use at least three different-sized triangles to support your idea.

Presenting Your Conclusions

Here are some ideas to help you present your conclusions to the class.

• Make a poster comparing a point, a line, and a plane in Euclidean geometry and spherical geometry. Include diagrams or sketches.

• Research parallel lines in spherical geometry. Write a paragraph to report your findings.

• Make a video demonstrating your findings in the project.

• Pair up with another group. Have a debate in which one group is in favor of Euclidean geometry, and the other is in favor of spherical geometry.

 Investigation For more information on non-Euclidean geometry, visit: www.geomconcepts.com

What You'll Learn

You'll learn to identify the relationships among pairs of corresponding angles formed by two parallel lines and a transversal.

Why It's Important

Design City planners use corresponding angles.
See Exercise 30.

When a transversal crosses two lines, the intersection creates a number of angles that are related to each other. Note $\angle 7$ and $\angle 8$ below. Although one is an exterior angle and the other is an interior angle, both lie on the same side of the transversal.

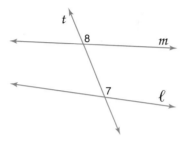

Angles 7 and 8 are called **corresponding angles**. They have different vertices. In the figure, three other pairs of corresponding angles are formed.

Example

1 Lines p and r are cut by transversal t.
Name two pairs of corresponding angles.

$\angle 1$ and $\angle 5$
$\angle 4$ and $\angle 7$

Your Turn

Refer to the figure above.

a. Name two other pairs of corresponding angles.

As with interior and exterior angles, there is a special relationship between corresponding angles when the transversal intersects parallel lines.

Recall that in Lesson 4–2, you discovered that if parallel lines are cut by a transversal, then each pair of alternate interior angles is congruent, and that each pair of consecutive interior angles is supplementary. Using the drawings you made for the Hands-On Geometry activity on page 149, measure the corresponding angles. What do you notice?

Postulate 4–1 Corresponding Angles	**Words:**	If two parallel lines are cut by a transversal, then each pair of corresponding angles is congruent.
	Model:	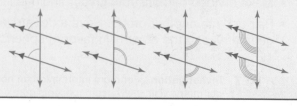

2 In the figure, $\ell \parallel m$, and a is a transversal. Which angles are congruent to $\angle 1$? Explain your answers.

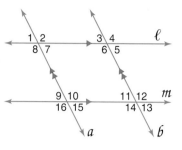

$\angle 1 \cong \angle 7$ *Vertical angles are congruent.*
$\angle 1 \cong \angle 9$ *Postulate 4–1*
$\angle 7 \cong \angle 15$ *Postulate 4–1*

Therefore, $\angle 1 \cong \angle 7 \cong \angle 9 \cong \angle 15$.

3 Find the measure of $\angle 10$ if $m\angle 1 = 62$.

$m\angle 1 = m\angle 9$, so $m\angle 9 = 62$.
$\angle 9$ and $\angle 10$ are a linear pair, so they are supplementary.

$$m\angle 9 + m\angle 10 = 180 \qquad \textit{Definition of supplementary angles}$$
$$62 \;\; + m\angle 10 = 180 \qquad \textit{Replace } m\angle 9 \textit{ with 62.}$$
$$62 + m\angle 10 - 62 = 180 - 62 \qquad \textit{Subtract 62 from each side.}$$
$$m\angle 10 = 118 \qquad \textit{Simplify.}$$

Your Turn

In the figure above, assume that b is also a transversal.

b. Which angles are congruent to $\angle 1$?
c. Find the measure of $\angle 5$ if $m\angle 14 = 98$.

 Personal Tutor at geomconcepts.com

Thus far, you have learned about the relationships among several kinds of special angles.

Concept Summary	Types of angle pairs formed when a transversal cuts two parallel lines.	
	Congruent alternate interior alternate exterior corresponding	**Supplementary** consecutive interior

You can use corresponding angles to prove the relationship of a perpendicular transversal to two parallel lines. In the figure, $r \parallel s$ and transversal c is perpendicular to r.

$\angle 1$ is a right angle.   *Definition of perpendicular lines*
$m\angle 1 = 90$ *Definition of right angle*
$\angle 1 \cong \angle 2$ *Postulate 4–1*
$m\angle 1 = m\angle 2$ *Definition of congruent angles*
$90 = m\angle 2$ *Substitution*
$\angle 2$ is a right angle.   *Definition of right angle*
$c \perp s$ *Definition of perpendicular lines*

This relationship leads to the following theorem.

Theorem 4–4 **Perpendicular** **Transversal**	If a transversal is perpendicular to one of two parallel lines, it is perpendicular to the other.

Example **④**

Algebra Link

In the figure, $p \parallel q$, and transversal r is perpendicular to q. If $m\angle 2 = 3x - 6$, find x.

$p \perp r$ *Theorem 4–4*

$\angle 2$ is a right angle.   *Definition of perpendicular lines*

$m\angle 2 = 90$ *Definition of right angles*

$$m\angle 2 = 3x - 6 \qquad \textit{Given}$$
$$90 = 3x - 6 \qquad \textit{Replace } m\angle 2 \textit{ with 90.}$$
$$90 + 6 = 3x - 6 + 6 \qquad \textit{Add 6 to each side.}$$
$$96 = 3x \qquad \textit{Simplify.}$$
$$\frac{96}{3} = \frac{3x}{3} \qquad \textit{Divide each side by 3.}$$
$$32 = x \qquad \textit{Simplify.}$$

— **Algebra Review** —

Solving Multi-Step
Equations, p. 723

Your Turn

d. Refer to the figure above. Find x if $m\angle 2 = 2(x + 4)$.

Check for Understanding

Communicating
Mathematics

1a. Identify two pairs of corresponding angles.

b. Explain why $\angle 6 \cong \angle 4$.

2. **You Decide!** Kristin says that $\angle 2$ and $\angle 3$ must be supplementary. Pedro disagrees. Who is correct, and why?

Vocabulary

corresponding angles

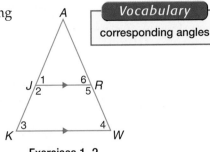

Exercises 1–2

3. Draw a pair of parallel lines cut by a transversal so that one pair of corresponding angles has the given measure. (Use a straightedge and protractor.)

a. 35   **b.** 90   **c.** 105   **d.** 140

158   Chapter 4 Parallels

⏱ Getting Ready Find the value of *x*.

Sample: $5x - 9 = 2x$	Solution: $5x - 2x = 9$
	$3x = 9$
	$x = 3$

4. $12x = 8x + 1$ **5.** $3x + 6 = 4x - 7$ **6.** $x - 10 + 7x = 180$

Example 2

In the figure, *s* ∥ *t* and *c* ∥ *d*. Name all angles congruent to the given angle. Give a reason for each answer.

7. ∠1

8. ∠5

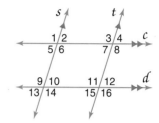

Examples 3 & 4

Find the measure of each numbered angle.

9.

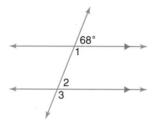

10.

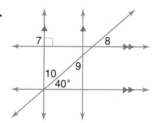

Examples 2 & 3

11. Farming The road shown in the diagram divides a rectangular parcel of land into two parts. If $m\angle 6 = 52$, find $m\angle 1$.

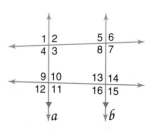

Example 4

12. Algebra If $m\angle 10 = 4x - 5$ and $m\angle 12 = 3x + 8$, find *x*, $m\angle 10$, and $m\angle 11$.

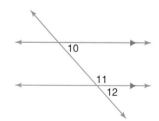

Exercises

In the figure, *a* ∥ *b*. Name all angles congruent to the given angle. Give a reason for each answer.

13. ∠2   **14.** ∠3

15. ∠8   **16.** ∠9

17. ∠12   **18.** ∠14

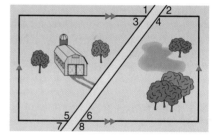

Homework Help

For Exercises	See Examples
13–18, 29, 30	2
19–24	3
25–28	4

Extra Practice
See page 732.

Find the measure of each numbered angle.

19.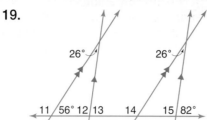

20.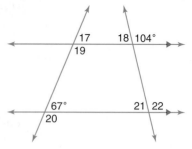

21.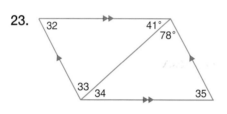

22.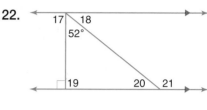

23.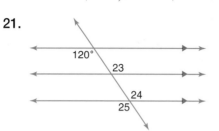

24.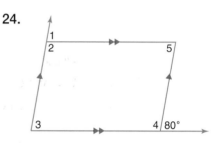

25. If $m\angle 4 = 2x + 7$ and $m\angle 8 = 3x - 13$, find x, $m\angle 4$, and $m\angle 8$.

26. If $m\angle 8 = 14x - 56$ and $m\angle 6 = 6x$, find x, $m\angle 8$, and $m\angle 6$.

27. If $m\angle 1 = 5x + 8$ and $m\angle 4 = 12x + 2$, find x, $m\angle 1$, and $m\angle 4$.

28. If $m\angle 6 = 5x + 25$ and $m\angle 7 = 3x - 5$, find x, $m\angle 6$, and $m\angle 7$.

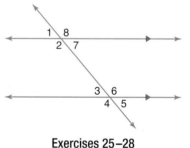

Exercises 25–28

Applications and Problem Solving

29. Flag Design Trace the drawing of the flag of the Bahamas. Assume segments that appear to be parallel are parallel. Make a conjecture about which angles you can conclude are congruent and explain your reasoning. Check your conjecture by measuring the angles.

Flag of the Bahamas

30. City Planning In New York City, roads running parallel to the Hudson River are named avenues, and those running perpendicular to the river are named streets. What is the measure of the angle formed at the intersection of a street and an avenue?

31. Critical Thinking In the figure at the right, why can you conclude that ∠6 and ∠4 are congruent, but you cannot state that ∠6 and ∠2 are congruent?

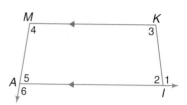

Mixed Review

32. Find the measure of ∠1.
(Lesson 4–2)

33. Name all pairs of parallel lines.
(Lesson 4–1)

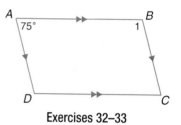

Exercises 32–33

Draw and label a figure for each situation described. *(Lesson 1–2)*

34. line ℓ

35. $\overline{AC}$

36. plane *FGH*

37. lines *p* and *q* intersecting at point *R*

Standardized Test Practice
Ⓐ Ⓑ Ⓒ Ⓓ

38. Grid In Austin practices the flute 9 minutes the first day, 10 minutes the second day, 12 minutes the third day, and 15 minutes the fourth day. If he continues this pattern, how many minutes will Austin practice the sixth day? *(Lesson 1–1)*

39. Multiple Choice Simplify $\frac{4r^9}{2r^3}$. *(Algebra Review)*

Ⓐ $2r^3$ Ⓑ $2r^{12}$ Ⓒ $\frac{1}{2}r^{12}$ Ⓓ $2r^6$

Quiz 1 — Lesson 4–1 through 4–3

1. Give examples of parallel lines and transversals as they are used in home design. *(Lesson 4–1)*

In the figure, $m\angle 2 = 56$. Find the measure of each angle.
(Lesson 4–2)

2. ∠3

3. ∠7

4. ∠4

5. Algebra In the figure, $a \parallel b$ and transversal *t* is perpendicular to *a*. If $m\angle 9 = 2x + 8$, find *x*.
(Lesson 4–3)

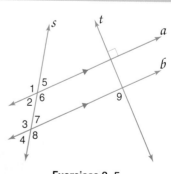

Exercises 2–5

What You'll Learn

You'll learn to identify conditions that produce parallel lines and to construct parallel lines.

Why It's Important

Maintenance
Groundskeepers use parallel lines when marking the yardage lines on football fields. *See Exercise 8.*

We can use geometry to prove that lines are parallel. We can also use geometry to construct parallel lines.

Hands-On Geometry
Construction

Materials: straightedge   compass

Step 1 Use a straightedge to draw a line ℓ and a point P, not on ℓ.

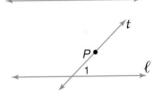

Step 2 Draw a line t through P that intersects line ℓ. Label $\angle 1$ as shown.

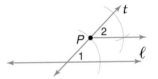

Step 3 Use a compass and a straightedge to construct an angle congruent to $\angle 1$ at P. Label this angle 2.

Step 4 Extend the side of $\angle 2$ to form line m.

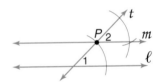

Try These

1. Identify the special angle pair name for $\angle 1$ and $\angle 2$.
2. Use a ruler to measure the distance between lines ℓ and m at several places. **Make a conjecture** about the relationship between the lines.

Look Back

Constructing Congruent Angles: Lesson 3–2

CONcepts in MOtion

Animation
geomconcepts.com

This activity illustrates a postulate that helps to prove two lines are parallel. This postulate is the converse of Postulate 4–1.

Postulate 4–2		
	Words:	In a plane, if two lines are cut by a transversal so that a pair of corresponding angles is congruent, then the lines are parallel.
	Model:	**Symbols:** If $\angle 1 \cong \angle 2$, then $a \parallel b$.

You can use Postulate 4–2 to find the angle measures of corresponding angles so that two lines are parallel.

Example

Engineering Link

1 **The intersection at the right is called a *trumpet interchange.* If $m\angle 1 = 13x - 8$ and $m\angle 2 = 12x + 4$, find x so that $\ell \parallel m$.**

From the figure, you know that $\angle 1$ and $\angle 2$ are corresponding angles. So, according to Postulate 4–2, if $m\angle 1 = m\angle 2$, then $\ell \parallel m$.

$m\angle 1 = m\angle 2$	*Corresponding angles*
$13x - 8 = 12x + 4$	*Substitution*
$13x - 8 - 12x = 12x + 4 - 12x$	*Subtract 12x from each side.*
$x - 8 = 4$	*Simplify.*
$x - 8 + 8 = 4 + 8$	*Add 8 to each side.*
$x = 12$	*Simplify.*

$m\angle 1 = 13x - 8$ $m\angle 2 = 12x + 4$
$= 13(12) - 8$ or 148 $= 12(12) + 4$ or 148

Your Turn

Refer to the figure in Example 1.

a. Find $m\angle 3$ and name the type of angle pair formed by $\angle 2$ and $\angle 3$.

b. Make a conjecture about the relationship between $\angle 2$ and $\angle 3$ that must be true for ℓ to be parallel to m.

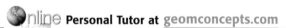

Personal Tutor at geomconcepts.com

Example 1 illustrates four additional methods for proving that two lines are parallel. These are stated as Theorems 4–5, 4–6, 4–7, and 4–8.

Theorem	Words	Models and Symbols
4–5	In a plane, if two lines are cut by a transversal so that a pair of alternate interior angles is congruent, then the two lines are parallel.	If $\angle 1 \cong \angle 2$, then $a \parallel b$.
4–6	In a plane, if two lines are cut by a transversal so that a pair of alternate exterior angles is congruent, then the two lines are parallel.	If $\angle 3 \cong \angle 4$, then $a \parallel b$.

Theorem	Words	Models and Symbols
4–7	In a plane, if two lines are cut by a transversal so that a pair of consecutive interior angles is supplementary, then the two lines are parallel.	If $m\angle 5 + m\angle 6 = 180$, then $a \parallel b$.
4–8	In a plane, if two lines are perpendicular to the same line, then the two lines are parallel.	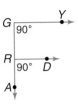 If $a \perp t$ and $b \perp t$, then $a \parallel b$.

So, we now have five ways to prove that two lines are parallel.

Concept Summary	• Show that a pair of corresponding angles is congruent. • Show that a pair of alternate interior angles is congruent. • Show that a pair of alternate exterior angles is congruent. • Show that a pair of consecutive interior angles is supplementary. • Show that two lines in a plane are perpendicular to a third line.

Examples

2 Identify the parallel segments in the letter Z.

$\angle ABC$ and $\angle BCD$ are alternate interior angles.
$m\angle ABC = m\angle BCD$ *Both angles measure 40°.*
$\overline{AB} \parallel \overline{CD}$ *Theorem 4–5*

Your Turn

c. Identify any parallel segments in the letter F. Explain your reasoning.

Algebra Link

3 Find the value of x so $\overline{BE} \parallel \overline{TS}$.

$\overline{ES}$ is a transversal for $\overline{BE}$ and $\overline{TS}$.
$\angle BES$ and $\angle EST$ are consecutive interior angles. If $m\angle BES + m\angle EST = 180$, then $\overline{BE} \parallel \overline{TS}$ by Theorem 4–7.

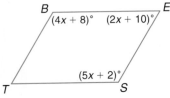

Algebra Review

Solving Multi-Step Equations, p. 723

$$m\angle BES + m\angle EST = 180$$
$$(2x + 10) + (5x + 2) = 180 \quad \text{Replace } m\angle BES \text{ with } 2x + 10$$
$$\text{and } m\angle EST \text{ with } 5x + 2.$$
$$(2x + 5x) + (10 + 2) = 180 \quad \text{Combine like terms.}$$
$$7x \quad + \quad 12 \quad = 180 \quad \text{Add.}$$
$$7x + 12 - 12 = 180 - 12 \quad \text{Subtract 12 from each side.}$$
$$\frac{7x}{7} = \frac{168}{7} \quad \text{Divide each side by 7.}$$
$$x = 24 \quad \text{Simplify.}$$

Thus, if $x = 24$, then $\overline{BT} \parallel \overline{TS}$.

Your Turn

d. Refer to the figure in Example 3. Find the value of x so $\overline{BT} \parallel \overline{ES}$.

Check for Understanding

Communicating Mathematics

1. **Explain** why $\overline{CA} \nparallel \overline{RT}$ in the figure at the right.

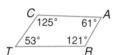

parallel postulate

2. **Describe** two situations in your own life in which you encounter parallel lines. How could you guarantee that the lines are parallel?

3. **Writing Math** Write a step-by-step argument to show that Theorem 4–6 is true.

Guided Practice

Examples 1 & 3

Find x so that $a \parallel b$.

4.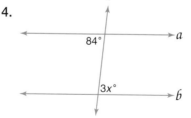

5.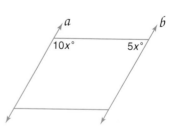

Example 2

Name the pairs of parallel lines or segments.

6.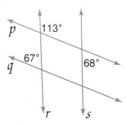

7.

Lesson 4–4 Proving Lines Parallel **165**

8. **Sports** The yardage lines on a football field are parallel. Explain how the grounds crew could use Theorem 4–8 to know where to paint the yardage lines.
Example 3

Exercises

Practice

Homework Help	
For Exercises	See Examples
9–14	3
15–20	2
21	2, 3
22, 23	1

Extra Practice
See page 733.

Find x so that a ∥ b.

9.
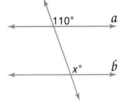

10.
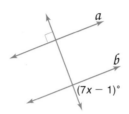

11.
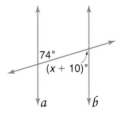

12.
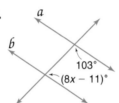

13.
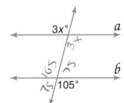

14.
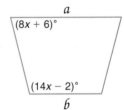

Name the pairs of parallel lines or segments.

15.
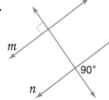

16.
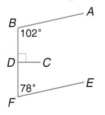

17.
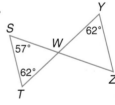

18.
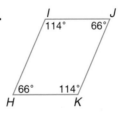

19.
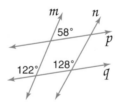

20.
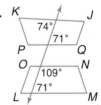

21. Refer to the figure at the right.
   a. Find x so that $\overline{AC} \parallel \overline{DE}$.
   b. Using the value that you found in part a, determine whether lines *AB* and *CD* are parallel.

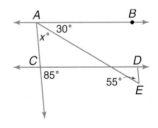

22. Solar Energy The figure at the right shows how the sun's rays reflect off special mirrors to provide electricity.

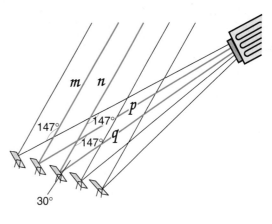

    a. Are the rays from the sun, lines *m* and *n*, parallel lines? Explain.

    b. Are the reflected rays, lines *p* and *q*, parallel lines? Explain.

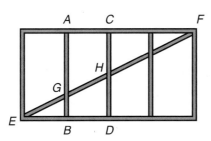

23. Construction Carpenters use parallel wall studs in building supports for walls. Describe three ways a carpenter could guarantee that the wall studs $\overline{AB}$ and $\overline{CD}$ are parallel.

24. Critical Thinking In the Hands-On Geometry activity on page 162, you constructed a line through a point *P* parallel to a line ℓ. In 1795, Scottish mathematician John Playfair (1748–1819) provided the modern version of Euclid's famous **Parallel Postulate**.

> If there is a line and a point not on the line, then there exists exactly one line through the point that is parallel to the given line.

Explain the meaning of *exactly one* in the postulate. If you try to draw two lines parallel to a given line through a point not on the line, what happens?

Mixed Review

Find the measure of each numbered angle.

25. ∠1   *(Lesson 4–3)*

26. ∠2   *(Lesson 4–2)*

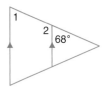

Determine whether each statement is *true* or *false*. Explain your reasoning. *(Lesson 2–3)*

27. If $LM = MJ$, then $\overline{LM} \cong \overline{MJ}$.

28. If $\overline{XY} \cong \overline{YZ}$, $\overline{ST} \cong \overline{PQ}$, and $\overline{YZ} \cong \overline{PQ}$, then $\overline{XY} \cong \overline{ST}$.

29. Multiple Choice The high temperature in Newport on January 12 was 6°C. The low temperature was −7°C. Find the range of the temperatures in Newport on this date. *(Lesson 2–1)*

    Ⓐ 1°C   Ⓑ 6.5°C   Ⓒ 7°C   Ⓓ 13°C

What You'll Learn

You'll learn to find the slopes of lines and use slope to identify parallel and perpendicular lines.

Why It's Important

Architecture
Architects use slope to determine the steepness of stairways.
See Exercise 9.

The steepness of a line is called the **slope**. Slope is defined as the ratio of the *rise*, or vertical change, to the *run*, or horizontal change, as you move from one point on the line to another.

You can use two points on a line to find its slope. Consider the slope of $\overrightarrow{AE}$.

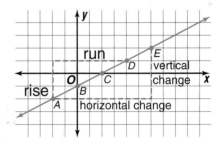

Points Used	A and C	A and D	A and E
Slope	$\dfrac{\text{rise}}{\text{run}} = \dfrac{2}{4}$ or $\dfrac{1}{2}$	$\dfrac{\text{rise}}{\text{run}} = \dfrac{3}{6}$ or $\dfrac{1}{2}$	$\dfrac{\text{rise}}{\text{run}} = \dfrac{4}{8}$ or $\dfrac{1}{2}$

Notice that the slope of $\overrightarrow{AE}$ is always $\frac{1}{2}$, regardless of the points chosen. The slope of a line is the same, or constant, everywhere along the line. This means that the choice of points used to find the slope of a line does not affect the slope. You can find the slope of a line as follows.

Definition of Slope	**Words:** The slope *m* of a line containing two points with coordinates (x_1, y_1) and (x_2, y_2) is given by the formula $\text{slope} = \dfrac{\text{difference of the } y\text{-coordinates}}{\text{difference of the corresponding } x\text{-coordinates}}.$ **Symbols:** $m = \dfrac{y_2 - y_1}{x_2 - x_1}$, where $x_2 \neq x_1$

The slope of a vertical line, where $x_1 = x_2$, is undefined.

Examples

Find the slope of each line.

①

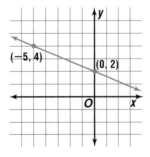

$m = \dfrac{2 - 4}{0 - (-5)}$ or $-\dfrac{2}{5}$

The slope is $-\dfrac{2}{5}$.

②

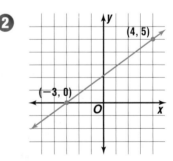

$m = \dfrac{5 - 0}{4 - (-3)}$ or $\dfrac{5}{7}$

The slope is $\dfrac{5}{7}$.

Algebra Review
Operations with Integers, p. 719

③ the line through points at
(−4, 3) and (2, 3)

$$m = \frac{3 - 3}{2 - (-4)} \text{ or } 0$$

The slope is 0.

④ the line through points at
(3, 4) and (3, −2)

$$m = \frac{-2 - 4}{3 - 3} \text{ or } -\frac{6}{0}$$

The slope is undefined.

Your Turn   **Find the slope of each line.**

a.

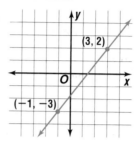

b. the line through points at
(−1, 3) and (1, −3)

Examples 1–3 suggest that a line with a negative slope seems to be going downhill, a line with a positive slope seems to be going uphill, and a line with a zero slope is a horizontal line. As shown in Example 4, the slope of a vertical line is undefined.

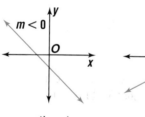

negative slope   positive slope   0 slope   undefined slope

**COncepts
in MOtion**
Interactive Lab
geomconcepts.com

Hands-On Geometry

Materials: 📄 grid paper   ✏️ straightedge   📐 protractor

Step 1 On a piece of grid paper, graph points A(−2, 0) and B(−3, −4). Using a straightedge, draw $\overleftrightarrow{AB}$.

Step 2 Graph points C(4, 4) and D(3, 0). Draw $\overleftrightarrow{CD}$. Using the definition of slope, find the slopes of $\overleftrightarrow{AB}$ and $\overleftrightarrow{CD}$.

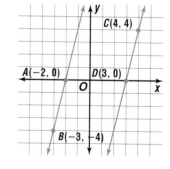

Try These

1. Measure ∠BAD and ∠ADC. What is true of these measures?
2. What special pair of angles do ∠BAD and ∠ADC form?
3. What is true of $\overleftrightarrow{AB}$ and $\overleftrightarrow{CD}$?

🖥 www.geomconcepts.com/extra_examples

Lesson 4–5 Slope   **169**

This activity illustrates a special characteristic of parallel lines.

| Postulate 4–3 | Two distinct nonvertical lines are parallel if and only if they have the same slope. |

All vertical lines are parallel.

You can use a TI–83/84 Plus calculator to study slopes of perpendicular lines.

Graphing Calculator Tutorial

See pp. 782–785.

Graphing Calculator Exploration

Step 1   First open a new session by choosing New from the F1 menu.

Step 2   Press F5 , select Hide/Show, and choose Axes.

Step 3   Draw a pair of nonvertical perpendicular lines on the coordinate plane.

Step 4   Open the F5 menu and select Measure and then Slope to find the slope of each of the two perpendicular lines that you drew.

Step 5   Use the Calculate feature on the F5 menu to calculate the product of the slopes of the perpendicular lines.

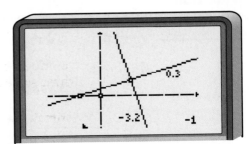

Try These

1. What number did you obtain as the product of the slopes of the perpendicular lines?
2. Go to the point that you used to draw the first of your perpendicular lines. Drag this point to a different location. Describe what happens to the slopes of the perpendicular lines and to the product of the slopes.
3. Drag on one end of the first of your perpendicular lines. Describe what happens to the slopes and their product.

This activity illustrates a special characteristic of perpendicular lines.

| Postulate 4–4 | Two nonvertical lines are perpendicular if and only if the product of their slopes is −1. |

Example **5**

Science Link

A dragonfly has two sets of wings. Given $A(-2, -2)$, $B(1, 2)$, $C(-3, 6)$, and $D(5, 0)$, prove that the second set of wings is perpendicular to the body. In other words, show that $\overleftrightarrow{AB} \perp \overleftrightarrow{CD}$.

First, find the slopes of $\overleftrightarrow{AB}$ and $\overleftrightarrow{CD}$.

$$\text{slope of } \overleftrightarrow{AB} = \frac{2 - (-2)}{1 - (-2)}$$

$$= \frac{4}{3}$$

$$\text{slope of } \overleftrightarrow{CD} = \frac{0 - 6}{5 - (-3)}$$

$$= \frac{-6}{8} \text{ or } -\frac{3}{4}$$

The product of the slopes for $\overleftrightarrow{AB}$ and $\overleftrightarrow{CD}$ is $\frac{4}{3} \cdot -\frac{3}{4}$ or -1. So, $\overleftrightarrow{AB} \perp \overleftrightarrow{CD}$, and the second set of wings is perpendicular to the body.

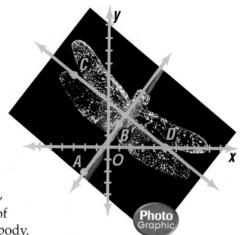

Photo Graphic

Your Turn

c. Given $P(-2, 2)$, $Q(2, 1)$, $R(1, -1)$, and $S(5, -2)$, prove that $\overleftrightarrow{PQ} \parallel \overleftrightarrow{RS}$.

Online **Personal Tutor at** geomconcepts.com

Check for Understanding

Communicating Mathematics

1. Describe a line whose slope is 0 and a line whose slope is undefined.

Vocabulary

slope

2. Estimate the slope of line ℓ shown at the right. Explain how you determined your estimate.

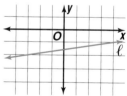

3. You Decide Sang Hee claims that a line with a slope of 2 is steeper than a line with a slope of $\frac{1}{4}$. Emily claims that a slope of $\frac{1}{4}$ is steeper than a slope of 2. Who is correct? Use a coordinate drawing to support your answer.

Guided Practice

Examples 1–4

Find the slope of each line.

4.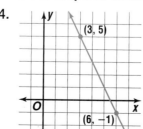

5.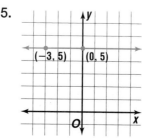

6. the line through points at $(-5, -2)$ and $(1, 2)$

Example 5

Given each set of points, determine if $\overleftrightarrow{PQ}$ and $\overleftrightarrow{RS}$ are *parallel*, *perpendicular*, or *neither*.

7. $P(-9, 2)$, $Q(2, -9)$, $R(9, 5)$, $S(0, -4)$

8. $P(6, 1)$, $Q(4, 0)$, $R(3, -5)$, $S(7, -3)$

Example 5

9. Construction Some building codes require the slope of a stairway to be no steeper than 0.88, or $\frac{22}{25}$. The stairs in Amad's house measure 11 inches deep and 6 inches high. Do the stairs meet the code requirements? Explain.

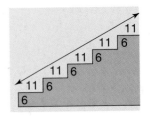

Exercises

Practice

Find the slope of each line.

10.

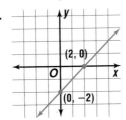

11.

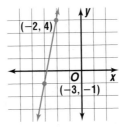

12.

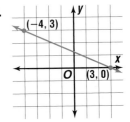

Homework Help	
For Exercises	See Examples
10–15	1, 2
16–18	3–4
19–24	5
25–28	3
Extra Practice	
See page 733.	

13.

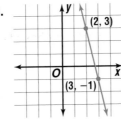

14.

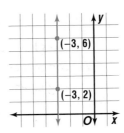

15.

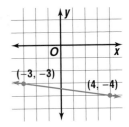

16. the line through $\left(\frac{1}{2}, 5\right)$ and $\left(2\frac{1}{2}, 1\right)$

17. the line through $(1, 3.5)$ and $(6, 3.5)$

18. the line through $(-1, -7)$ and $(3, -1)$

Given each set of points, determine if $\overleftrightarrow{JK}$ and $\overleftrightarrow{LM}$ are *parallel*, *perpendicular*, or *neither*.

19. $J(-4, 11)$, $K(-6, 3)$, $L(7, 7)$, $M(6, 3)$

20. $J(6, 9)$, $K(4, 6)$, $L(0, 8)$, $M(3, 6)$

21. $J(-8, 1)$, $K(-5, -8)$, $L(0, 10)$, $M(3, 11)$

22. $J(6, 3)$, $K(-7, 3)$, $L(-4, -5)$, $M(1, -5)$

23. $J(-1, 5)$, $K(2, -3)$, $L(7, 9)$, $M(2, 6)$

24. $J(3, -2)$, $K(5, -9)$, $L(6, 4)$, $M(4, -3)$

25. Find the slope of the line passing through points at $(-7, 4)$ and $(2, 9)$.

26. $C(r, -5)$ and $D(5, 3)$ are two points on a line. If the slope of the line is $\frac{2}{3}$, find the value of r.

27. **Sports**   Refer to the application at the beginning of the lesson. Find the *run* of a hill with a 36-foot *rise* if the hill has a slope of 0.02.

28. **Construction**   To be efficient, gutters should drop $\frac{1}{4}$ inch for every 4 feet that they run toward a downspout. What is the desired slope of a gutter?

29. **Critical Thinking**   Use slope to determine if $A(2, 4)$, $B(5, 8)$, $C(13, 2)$, and $D(10, -2)$ are the vertices of a rectangle. Explain.

Mixed Review

30. Name the pairs of parallel segments.   *(Lesson 4–4)*

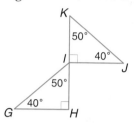

31. Find the measure of each numbered angle.   *(Lesson 4–3)*

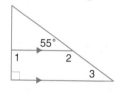

In the figure, $\overrightarrow{XA}$ and $\overrightarrow{XD}$ are opposite rays.
(Lesson 3–4)

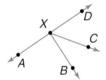

32. Which angle forms a linear pair with $\angle AXB$?

33. Name all pairs of adjacent angles.

**Standardized
Test Practice**

34. **Multiple Choice**   What is the hypothesis in the following statement?
(Lesson 1–4)

Angles are congruent if they have the same measure.

(A) congruent

(B) they have the same measure

(C) angles are congruent

(D) not congruent

Quiz 2   Lessons 4–4 and 4–5

Find *x* so that *a* ∥ *b*.   *(Lesson 4–4)*

1.

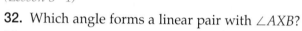

$85°$

$(x + 15)°$

2.

$140°$

$(6x - 2)°$

3.

$137°$

$(4x - 7)°$

4. **Music**   On the panpipe at the right, find the slope of $\overleftrightarrow{GH}$.
(Lesson 4–5)

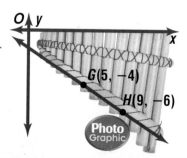

$G(5, -4)$

$H(9, -6)$

5. Given $P(-6, 1)$, $Q(6, 4)$, $R(3, 4)$, and $S(2, 8)$, determine if $\overleftrightarrow{PQ}$ and $\overleftrightarrow{RS}$ are *parallel*, *perpendicular*, or *neither*.
(Lesson 4–5)

The equation $y = 2x - 1$ is called a **linear equation** because its graph is a straight line. We can substitute different values for x in the equation to find corresponding values for y, as shown in the table at the right.

x	$2x - 1$	y	(x, y)
0	$2(0) - 1$	-1	$(0, -1)$
1	$2(1) - 1$	1	$(1, 1)$
2	$2(2) - 1$	3	$(2, 3)$
3	$2(3) - 1$	5	$(3, 5)$

We can then graph the ordered pairs (x, y). Notice that there is one line that contains all four points. There are many more points whose ordered pairs are solutions of $y = 2x - 1$. These points also lie on the line.

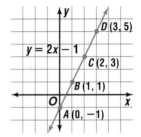

To find the slope of this line, choose any two points on the line, such as $B(1, 1)$ and $C(2, 3)$.

$$m = \frac{y_2 - y_1}{x_2 - x_1} \quad \textit{Definition of slope}$$

$$= \frac{3 - 1}{2 - 1} \quad \textit{Replace } y_2 \textit{ with 3, and } y_1 \textit{ with 1.}$$
$$\textit{Replace } x_2 \textit{ with 2, and x1 with 1.}$$

$$= \frac{2}{1} \text{ or } 2$$

Now look at the graph of $y = 2x - 1$. The y-value of the point where the line crosses the y-axis is -1. This value is called the **y-intercept** of the line.

$$y = 2x - 1$$
$$\textit{slope} \quad \quad \textit{y-intercept}$$

Most linear equations can be written in the form $y = mx + b$. This form is called the **slope-intercept form**.

Slope-Intercept Form	An equation of the line having slope m and y-intercept b is $y = mx + b$.

Name the slope and *y*-intercept of the graph of each equation.

1 $y = \frac{1}{2}x + 5$ **2** $y = 3$ **3** $x = -2$

The slope is $\frac{1}{2}$.

The *y*-intercept
is 5.

$y = 0x + 3$

The slope is 0.
The *y*-intercept
is 3.

The graph is a vertical
line.

The slope is undefined.
There is no *y*-intercept.

4 $2x - 3y = 18$

Rewrite the equation in slope-intercept form by solving for *y*.

$$2x - 3y = 18 \qquad \textit{Original equation}$$
$$2x - 3y - 2x = 18 - 2x \qquad \textit{Subtract 2x from each side.}$$
$$-3y = 18 - 2x \qquad \textit{Simplify.}$$
$$\frac{-3y}{-3} = \frac{18 - 2x}{-3} \qquad \textit{Divide each side by } -3.$$
$$y = -6 + \frac{2}{3}x \qquad \textit{Simplify.}$$
$$y = \frac{2}{3}x - 6 \qquad \textit{Write in slope-intercept form.}$$

The slope is $\frac{2}{3}$. The *y*-intercept is -6.

Your Turn

a. $y = -x + 8$ **b.** $x + 4y = 24$
c. $y = -1$ **d.** $x = 10$

5 **Graph $2x + y = 3$ using the slope and *y*-intercept.**

First, rewrite the equation in slope-intercept form.

$$2x + y = 3 \qquad \textit{Original equation}$$
$$2x + y - 2x = 3 - 2x \qquad \textit{Subtract 2x from each side.}$$
$$y = 3 - 2x \qquad \textit{Simplify.}$$
$$y = -2x + 3 \qquad \textit{Write in slope-intercept form.}$$

The *y*-intercept is 3. So, the point at (0, 3)
must be on the line. Since the slope is -2, or
$\frac{-2}{1}$, plot a point by using a *rise* of -2 units
(down) and a *run* of 1 unit (right). Draw a
line through the points.

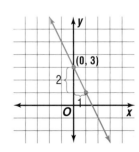

Your Turn

e. Graph $-x + 3y = 9$ using the slope and *y*-intercept.

The graphs of lines ℓ, m, and t are shown at the right.

$\ell:\ y = 2x - 3$

$m:\ y = 2x + 6$

$t:\ y = -\dfrac{1}{2}x + 3$

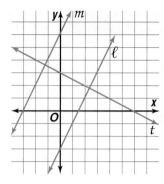

Notice that ℓ appears to be parallel to m. You can verify this by Postulate 4–3, since the slopes of ℓ and m are the same. Also, t appears to be perpendicular to ℓ and m. You can verify this by Postulate 4–4, since the product of the slope of t and the slope of ℓ or m is -1. Thus, the slope-intercept form can be used to write the equation of another line.

Example ⑥ **Write an equation of the line parallel to the graph of $y = 2x - 5$ that passes through the point at (3, 7).**

Explore   The equation of the line will be in the form $y = mx + b$. You need to find the slope m and the y-intercept b.

Plan   Because the lines are parallel, they must have the same slope. The slope of the given line is 2, so the slope m of the line parallel to the graph of $y = 2x - 5$ must be 2.

To find b, use the ordered pair (3, 7) and substitute for m, x, and y in the slope-intercept form.

Solve

$$
\begin{aligned}
y &= mx + b && \textit{Slope-intercept form} \\
7 &= 2(3) + b && \textit{Replace } m \textit{ with 2 and } (x, y) \textit{ with } (3, 7). \\
7 &= 6 + b && \textit{Multiply.} \\
7 - 6 &= 6 + b - 6 && \textit{Subtract 6 from each side.} \\
1 &= b && \textit{Simplify.}
\end{aligned}
$$

The value of b is 1. So, the equation of the line is $y = 2x + 1$.

Examine   The graphs of $y = 2x - 5$ and $y = 2x + 1$ both have the same slope, 2, so the lines are parallel. Since $7 = 2(3) + 1$, the graph of $y = 2x + 1$ passes through the point at (3, 7). The solution satisfies the conditions given.

Your Turn

Write each equation in slope-intercept form.

f. Write an equation of the line parallel to the graph of $3x + y = 6$ that passes through the point at (1, 4).

g. Write an equation of the line perpendicular to the graph of $y = \dfrac{1}{4}x + 5$ that passes through the point at $(-3, 8)$.

Communicating Mathematics

1. **Explain** why $y = mx + b$ is called the *slope-intercept form* of the equation of a line.

2. **Explain** how you know that lines ℓ, m, and n are parallel. Then name the y-intercept for each line.

> **Vocabulary**
>
> linear equation
> y-intercept
> slope-intercept form

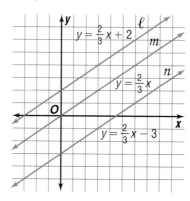

Guided Practice

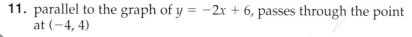

Getting Ready Solve each equation for y.

Sample: $8x + 4y = 1$

Solution: $4y = 1 - 8x$

$\frac{4y}{4} = \frac{1 - 8x}{4}$ or $y = \frac{1}{4} - 2x$

3. $y - 6x = -3$
4. $x + 7y = 14$
5. $5x - 3y = 9$

Examples 1–4

Name the slope and y-intercept of the graph of each equation.

6. $y = 2x + 6$
7. $3x + 2y = 8$

Example 5

Graph each equation using the slope and y-intercept.

8. $y = -x + 4$
9. $2x - 5y = 10$

Example 6

Write an equation of the line satisfying the given conditions.

10. slope = 4, goes through the point at $(-1, 3)$

11. parallel to the graph of $y = -2x + 6$, passes through the point at $(-4, 4)$

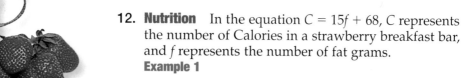

12. **Nutrition** In the equation $C = 15f + 68$, C represents the number of Calories in a strawberry breakfast bar, and f represents the number of fat grams.
    Example 1
    a. What is the slope of the line? What does it represent?
    b. What is the y-intercept? What does it represent?

Exercises

Practice

Name the slope and y-intercept of the graph of each equation.

13. $y = 9x + 1$

14. $7x + y = 12$

15. $3x - 2y = 18$

16. $x = 6$

17. $y = 5$

18. $3x + 4y = 2$

Homework Help	
For Exercises	See Examples
13–18	1–3
19–24, 31–34	5
25–30	6
35	1, 5
36	1, 4
Extra Practice	
See page 733.	

Graph each equation using the slope and y-intercept.

19. $y = 3x - 5$

20. $y = -x + 6$

21. $y + 7x = 4$

22. $-\frac{1}{2}x + 2y = 9$

23. $\frac{1}{3}x - y = 2$

24. $4x - 3y = -6$

Write an equation of the line satisfying the given conditions.

25. slope = 3, goes through the point at $(-1, 4)$

26. parallel to the graph of $y = -2x - 6$, passes through the point at $(4, 4)$

27. parallel to the graph of $4x + y = 9$, passes through the point at $(0, -5)$

28. parallel to the x-axis, passes through the point at $(-3, -6)$

29. slope is undefined, passes through the point at $(-3, 7)$

30. perpendicular to the graph of $y = \frac{1}{2}x - 3$, passes through the point at $(5, -4)$

Choose the correct graph of lines a, b, c, or d for each equation.

31. $y = 2x + 1$

32. $y = 2x + 3$

33. $y = -2x - 1$

34. $y = -2x + 3$

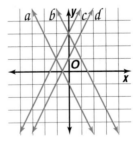

Applications and Problem Solving

Math Data Online Update For the latest information on telephone rates, visit: geomconcepts.com

35. Communication One telephone company's charges are given by the equation $y = 0.40x + 0.99$, where y represents the total cost in dollars for a telephone call and x represents the length of the call in minutes.

a. Make a table of values showing what a telephone call will cost after 0, 1, 2, 3, 4, and 5 minutes.

b. Graph the values in your table.

c. What is the slope of the line? What does it represent?

d. What is the y-intercept of the line? What does it represent?

36. **Sports** The graph at the right shows the distance a baseball can be hit when it is pitched at different speeds.

   a. What is the *y*-intercept? What does this value represent?

   b. Estimate the slope.

   c. Write an equation of the line.

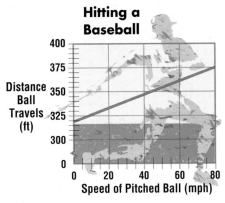

Hitting a Baseball

Distance Ball Travels (ft)

Speed of Pitched Ball (mph)

37. **Critical Thinking** Explain how you could find an equation of a line if you are only given the coordinates of two points on the line.

Mixed Review

The graph at the right shows the estimated cost of wireless phone use from 1998 to 2003. *(Lesson 4–5)*

38. Which section of the graph shows when the greatest change occurred? How does its slope compare to the rest of the graph?

39. Describe the slope of a graph showing an increase in cost.

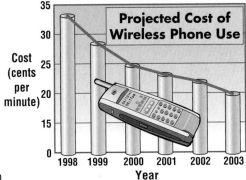

Projected Cost of Wireless Phone Use

Cost (cents per minute)

Year

Source: The Strategies Group

40. Find *x* so that $a \parallel b$. *(Lesson 4–4)*

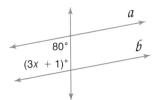

a

80°

b

$(3x + 1)°$

41. Haloke used the design shown below in a quilt she made for her grandmother. Name all the angles with *Q* as a vertex. *(Lesson 3–1)*

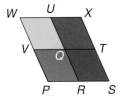

42. **Algebra** What is the length of a rectangle with area 108 square inches and width 8 inches? *(Lesson 1–6)*

43. **Extended Response** Just by looking, which segment appears to be longer, $\overline{AJ}$ or $\overline{KR}$? Use a ruler to measure the two segments. What do you discover? *(Lesson 1–5)*

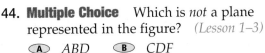

A —— K —— J

R

44. **Multiple Choice** Which is *not* a plane represented in the figure? *(Lesson 1–3)*

   Ⓐ *ABD*   Ⓑ *CDF*

   Ⓒ *BFE*   Ⓓ *ADF*

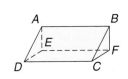

Understanding and Using the Vocabulary

After completing this chapter, you should be able to define each term, property, or phrase and give an example or two of each.

*inter***NET**
CONNECTION **Review Activities**
For more review activities, visit:
www.geomconcepts.com

Geometry

alternate exterior angles *(p. 148)*
alternate interior angles *(p. 148)*
consecutive interior angles *(p. 148)*
corresponding angles *(p. 156)*
exterior angles *(p. 148)*
finite *(p. 154)*
great circle *(p. 154)*

interior angles *(p. 148)*
line *(p. 154)*
line of latitude *(p. 154)*
line of longitude *(p. 154)*
parallel lines *(p. 142)*
parallel planes *(p. 142)*
skew lines *(p. 143)*
transversal *(p. 148)*

Algebra

linear equation *(p. 174)*
slope *(p. 168)*
slope-intercept form *(p. 174)*
y-intercept *(p. 174)*

Choose the letter of the term that best describes each set of angles or lines.

1. ∠2 and ∠7
2. ∠1, ∠3, ∠6, ∠8
3. ∠5 and ∠1
4. lines *m* and *n*
5. ∠7 and ∠4
6. line *q*
7. ∠2, ∠4, ∠5, ∠7
8. ∠1 and ∠6
9. lines *q* and *p*
10. lines *p* and *m*

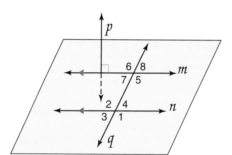

a. alternate exterior angles
b. alternate interior angles
c. consecutive interior angles
d. corresponding angles
e. exterior angles
f. interior angles
g. parallel lines
h. skew lines
i. perpendicular lines
j. transversal

Skills and Concepts

Objectives and Examples

• **Lesson 4–1** Describe relationships among lines, parts of lines, and planes.

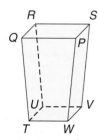

In the prism, plane *QRS* is parallel to plane *TUV*.

$\overline{TW}$ and $\overline{UV}$ are parallel to $\overline{RS}$.
$\overline{QR}, \overline{TU}, \overline{RS}, \overline{UV}$ intersect $\overline{RU}$.

$\overline{RS}, \overline{SP}, \overline{UV}$, and $\overline{WV}$ are skew to $\overline{QT}$.

Review Exercises

Describe each pair of segments in the prism as *parallel*, *skew*, or *intersecting*.

11. $\overline{HB}, \overline{FD}$
12. $\overline{BC}, \overline{AG}$
13. $\overline{EC}, \overline{HE}$

Name the parts of the prism.

14. six planes
15. all segments skew to $\overline{GE}$
16. all segments parallel to $\overline{GF}$

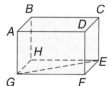

Exercises 11–16

 www.geomconcepts.com/vocabulary_review

Objectives and Examples

Review Exercises

• **Lesson 4–2** Identify the relationships among pairs of interior and exterior angles formed by two parallel lines and a transversal.

∠3 and ∠8 are alternate interior angles, so they are congruent.

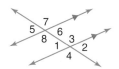

∠1 and ∠8 are consecutive interior angles, so they are supplementary.

∠4 and ∠7 are alternate exterior angles, so they are congruent.

Identify each pair of angles as *alternate interior, alternate exterior, consecutive interior,* or *vertical.*

17. ∠1 and ∠6
18. ∠4 and ∠2
19. ∠3 and ∠8
20. ∠7 and ∠3

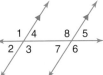

Exercises 17–24

If *m*∠1 = 124, find the measure of each angle. Give a reason for each answer.

21. ∠3
22. ∠4
23. ∠6
24. ∠8

• **Lesson 4–3** Identify the relationships among pairs of corresponding angles formed by two parallel lines and a transversal.

If *s* ∥ *t*, which angles are congruent to ∠2?

∠2 ≅ ∠5 *Vertical angles are congruent.*
∠2 ≅ ∠3 *Postulate 4–1*
∠3 ≅ ∠7 *Vertical angles are congruent.*

Therefore, ∠5, ∠3, and ∠7 are congruent to ∠2.

Name all angles congruent to the given angle. Give a reason for each answer.

25. ∠2
26. ∠5
27. ∠10

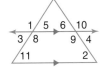

Find the measure of each numbered angle.

28.

29.

• **Lesson 4–4** Identify conditions that produce parallel lines and construct parallel lines.

Find *x* so that $\overline{JK} \parallel \overline{MN}$.

∠*JLM* and ∠*LMN* are alternate interior angles. If ∠*JLM* ≅ ∠*LMN*, then $\overline{JK} \parallel \overline{MN}$.
$m\angle JLM = m\angle LMN$
$63 = 5x - 2$ *Substitution*
$65 = 5x$ *Add 2 to each side.*
$13 = x$ *Divide each side by 5.*

Find *x* so that *c* ∥ *d*.

30.

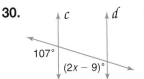

31.

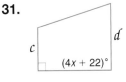

Name the pairs of parallel lines or segments.

32.

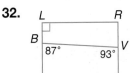

33.

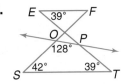

Objectives and Examples

- **Lesson 4–5** Find the slopes of lines and use slope to identify parallel and perpendicular lines.

Given $A(9, 2)$, $D(6, 3)$, $F(-5, 7)$, and $S(-4, 10)$, is $\overleftrightarrow{AD} \perp \overleftrightarrow{FS}$, $\overleftrightarrow{AD} \parallel \overleftrightarrow{FS}$, or neither?

slope of $\overleftrightarrow{AD}$

$m = \dfrac{3-2}{6-9}$

$= \dfrac{1}{-3}$ or $-\dfrac{1}{3}$

slope of $\overleftrightarrow{FS}$

$m = \dfrac{10-7}{-4-(-5)}$

$= \dfrac{3}{1}$ or 3

Since $\left(-\dfrac{1}{3}\right)\left(\dfrac{3}{1}\right) = -1$, $\overleftrightarrow{AD} \perp \overleftrightarrow{FS}$.

Review Exercises

Find the slope of each line.

34. a

35. b

36. c

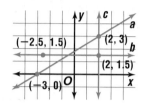

Given each set of points, determine if $\overleftrightarrow{GH}$ and $\overleftrightarrow{PQ}$ are *parallel*, *perpendicular*, or *neither*.

37. $G(-4, 3)$, $H(10, -9)$, $P(8, 6)$, $Q(1, 0)$

38. $G(11, 0)$, $H(3, -7)$, $P(4, 16)$, $Q(-4, 9)$

- **Lesson 4–6** Write and graph equations of lines.

Slope-intercept form:

$$y = mx + b$$

slope ⟶   ⟵ y-intercept

To graph a linear equation using slope and y-intercept, rewrite the equation by solving for y.

Name the slope and y-intercept of the graph of each equation. Then graph the equation.

39. $y = -\dfrac{3}{5}x + 1$ **40.** $2x + 5y = 20$

41. Write an equation of the line perpendicular to the graph of $y = -2x + 4$ that passes through the point at $(-8, 3)$.

Applications and Problem Solving

42. Gymnastics Describe bars ℓ and m on the equipment below as *parallel*, *skew*, or *intersecting*. *(Lesson 4–1)*

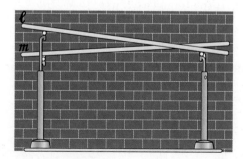

43. Construction A flight of stairs is parallel to the overhead ceiling. If $x = 122$, find y. *(Lesson 4–4)*

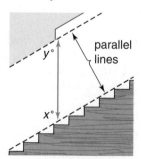

Choose the letter on the right that identifies each angle pair.

1. $\angle 5$ and $\angle 16$
2. $\angle 9$ and $\angle 3$
3. $\angle 10$ and $\angle 13$
4. $\angle 11$ and $\angle 9$
5. $\angle 2$ and $\angle 7$

a. corresponding angles
b. consecutive interior angles
c. alternate interior angles
d. alternate exterior angles
e. vertical angles

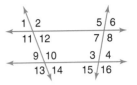

Describe each pair of segments or planes in the prism as *parallel*, *skew*, or *intersecting*.

6. $\overline{IJ}$ and $\overline{EF}$
7. plane *JBH* and plane *IGH*
8. $\overline{CG}$ and $\overline{AB}$
9. plane *ADF* and plane *HBC*
10. Name all segments parallel to $\overline{AD}$.
11. Name all segments that intersect $\overline{BH}$.

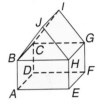

In the figure, *q* ‖ *r*.

12. If $m\angle 7 = 88$, find the measure of each numbered angle.
13. If $\angle 1$ is a right angle, name two pairs of perpendicular lines.
14. If $m\angle 5 = 5x$ and $m\angle 7 = 2x + 63$, find $m\angle 5$.

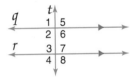

Name the parallel segments formed if the following angles are congruent.

15. $\angle 2$ and $\angle 6$
16. $\angle 3$ and $\angle 4$
17. If $m\angle DCB = 12x - 3$ and $m\angle 7 = 8x + 3$, find x so that $\overline{AD} \parallel \overline{BC}$.

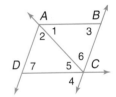

Given each set of points, find the slopes of $\overleftrightarrow{MN}$ and $\overleftrightarrow{UV}$. Then determine whether the two lines are *parallel*, *perpendicular*, or *neither*.

18. $M(-3, -5)$, $N(-3, 4)$, $U(7, -2)$, $V(4, -5)$
19. $M(2, 12)$, $N(-8, 9)$, $U(8, 4)$, $V(-2, 1)$
20. $M(4, 8)$, $N(9, 2)$, $U(0, 3)$, $V(-6, -2)$

21. Graph the equation $3x + y = 1$ using the slope and *y*-intercept.

22. Find and graph the equation of the line with slope -1 passing through the point at $(2, 3)$.

23. Find the equation of the line parallel to the *y*-axis and passing through the point at $(-4, 7)$.

24. **Meteorology**   The weather symbol at the right represents a heavy thunderstorm. If $\overline{AB}$ and $\overline{CD}$ are parallel segments and $m\angle B = 85$, find $m\angle C$.

25. **Civil Engineering**   A public parking lot is constructed so that all parking stalls in an aisle are parallel. If $m\angle KLA = 43$, find $m\angle PKR$.

 www.geomconcepts.com/chapter_test

Data Analysis Problems

Proficiency tests usually include several problems on interpreting graphs and creating graphs from data. The SAT and ACT may include a few questions on interpreting graphs. Often a graph is used to answer two or more questions.

You'll need to understand these ways of representing data: bar graphs, circle graphs, line graphs, stem-and-leaf plots, histograms, and frequency tables.

> **Test-Taking Tip**
>
> If a problem includes a graph, first look carefully at its axis labels and units. Then read the question(s) about the graph.

Example 1

The table below shows the number of employees who earn certain hourly wages at two different companies.

Hourly Wage ($)	Number of Employees	
	Company 1	Company 2
6.00	3	4
6.50	5	6
7.00	6	7
7.50	4	5
8.00	1	2

Construct a double-bar graph to show the number of employees at each company who earn the given hourly wages.

> **Hint** Decide what quantities will be shown on each axis of your graph. Label the axes.

Solution Draw the double-bar graph. Put the wages on the horizontal axis and the number of employees on the vertical axis.

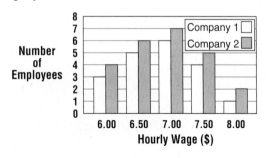

Example 2

Sara's Driving Speed, Saturday Afternoon

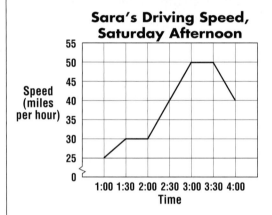

For what percent of the time was Sara driving 40 miles per hour or faster?

- **A** 20%
- **B** 25%
- **C** $33\frac{1}{3}\%$
- **D** 40%
- **E** 50%

> **Hint** Don't mix units, like *hours* and *minutes*. In this case, use hours.

Solution The total time Sara drove is 3 hours (from 1:00 until 4:00). She drove 40 miles per hour or faster from 2:30 to 4:00, or $1\frac{1}{2}$ hours. The fraction of the time she drove 40 mph or faster is $\dfrac{1\frac{1}{2}}{3}$ or $\frac{1}{2}$. The equivalent percent is 50%. So, the answer is E.

After you work each problem, record your answer on the answer sheet provided or on a sheet of paper.

Multiple Choice

1. What is a reasonable conclusion from the information given? *(Statistics Review)*

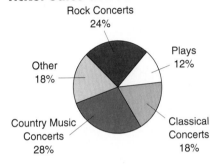

Ticket Sales at Bob's Ticket Outlet

Rock Concerts 24%
Plays 12%
Other 18%
Classical Concerts 18%
Country Music Concerts 28%

   (A) Fewer than 20 customers purchased classical concert tickets last month.
   (B) Plays are more popular than country music concerts at Bob's.
   (C) Rock concerts are less popular than country music concerts at Bob's.
   (D) Bob doesn't sell rap concert tickets.

2. $(-2)^3 + (3)^{-2} + \frac{8}{9} =$ *(Algebra Review)*
   (A) -7
   (B) $-1\frac{7}{9}$
   (C) $\frac{8}{9}$
   (D) $1\frac{7}{9}$
   (E) 12

3. Kendra has 80 CDs. If 40% are jazz CDs and the rest are blues CDs, how many blues CDs does she have? *(Percent Review)*
   (A) 32
   (B) 40
   (C) 42
   (D) 48
   (E) 50

4. Cody has 3 pairs of jeans and 4 sweatshirts. How many combinations of 1 pair of jeans and 1 sweatshirt can he wear? *(Statistics Review)*
   (A) 3
   (B) 4
   (C) 7
   (D) 12

5. If $3x + 7 = 28$, what is x? *(Algebra Review)*
   (A) 4
   (B) 5
   (C) 6
   (D) 7

6. $\overleftrightarrow{FB}$ and $\overleftrightarrow{EC}$ are parallel, and the measure of $\angle CXD$ is 62. Find the measure of $\angle FYX$. *(Lesson 4–2)*

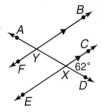

   (A) 28
   (B) 62
   (C) 90
   (D) 118

7. In a stem-and-leaf plot of the data in the chart, which numbers would be the best choice for the stems? *(Statistics Review)*

Average Temperature (°F)					
52	55	49	51	57	66
62	61	61	73	78	76
69	58	59	74	78	79

   (A) 5–7
   (B) 4–7
   (C) 0–7
   (D) 0–9

8. What is the y-intercept of line MN? *(Lesson 4–6)*

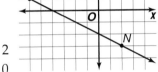

   (A) -4
   (B) 2
   (C) -2
   (D) 0

Grid In

9. Six cards are numbered 0 through 5. Two are selected without replacement. What is the probability that their sum is 4? *(Statistics Review)*

Extended Response

10. The numbers of visitors each day to a history museum during the month of June are shown below. *(Statistics Review)*

    11 19 8   7   18 43 22 18 14 21 19
    41 61 36 16 16 14 24 31 64 29 24
    27 33 31 71 89 61 41 34

    Part A Construct a frequency table for the data.

    Part B Construct a histogram that represents the data.

Triangles and Congruence

What You'll Learn

Key Ideas

- Identify parts of triangles and classify triangles by their parts. *(Lesson 5–1)*

- Use the Angle Sum Theorem. *(Lesson 5–2)*

- Identify translations, reflections, and rotations and their corresponding parts. *(Lesson 5–3)*

- Name and label corresponding parts of congruent triangles. *(Lesson 5–4)*

- Use the SSS, SAS, ASA, and AAS tests for congruence. *(Lessons 5–5 and 5–6)*

Key Vocabulary

congruent triangles *(p. 203)*

right triangle *(p. 188)*

triangle *(p. 188)*

Why It's Important

Social Studies The Hopi people have lived on three isolated mesas in what is now northern Arizona for more than a thousand years. Oraibi, a Hopi village on Third Mesa, is the oldest continuously inhabited village in the United States. The Hopi live in multilevel adobe or stone villages called pueblos.

Triangles are the simplest of the polygons. You will determine how triangles are used to create the design on Hopi pottery in Lesson 5–3.

Study these lessons to improve your skills.

✓Check Your Readiness

Use a protractor to draw an angle having each measurement. Then classify each angle as *acute*, *obtuse*, or *right*.

✓ **Lesson 3–2,** pp. 96–101

1. $52°$
2. $145°$
3. $18°$
4. $90°$
5. $75°$
6. $98°$

✓ **Lesson 3–5,** pp. 116–121

Determine the measures of the complement and supplement of each angle.

7. $34°$
8. $12°$
9. $44°$
10. $78°$
11. $66°$
12. $5°$

Solve each equation. Check your solution.

✓ **Algebra Review,** p. 722

13. $114 + n = 180$
14. $58 + x = 90$
15. $5m = 90$
16. $180 = 12g$
17. $90 - k = 23$
18. $180 - q = 121$

✓ **Algebra Review,** p. 723

19. $90 = 4b - 18$
20. $48 + 3g = 90$
21. $8y - 16 = 180$
22. $12c + 6 = 90$
23. $(n - 4) + n = 180$
24. $2x + 4x + 6x = 180$

FOLDABLES ™
Study Organizer

Make this Foldable to help you organize your Chapter 5 notes. Begin with 3 sheets of plain $8\frac{1}{2}$" by 11" paper.

❶ **Fold** in half lengthwise.

❷ **Fold** the top to the bottom.

❸ **Open** and cut along the second fold to make two tabs.

❹ **Label** each tab as shown.

Reading and Writing As you read and study the chapter, write what you learn about the two methods of classifying triangles under the tabs.

Optical art is a form of abstract art that creates special effects by using geometric patterns. The design at the right looks like a spiral staircase, but it is made mostly of triangles.

In geometry, a **triangle** is a figure formed when three noncollinear points are connected by segments. Each pair of segments forms an angle of the triangle. The **vertex** of each angle is a vertex of the triangle.

Triangles are named by the letters at their vertices. Triangle *DEF*, written △*DEF*, is shown below.

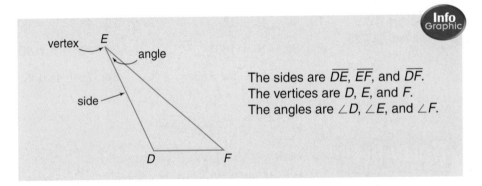

The sides are $\overline{DE}$, $\overline{EF}$, and $\overline{DF}$.
The vertices are *D*, *E*, and *F*.
The angles are ∠*D*, ∠*E*, and ∠*F*.

In Chapter 3, you classified angles as acute, obtuse, or right. Triangles can also be classified by their angles. All triangles have at least two acute angles. The third angle is either acute, obtuse, or right.

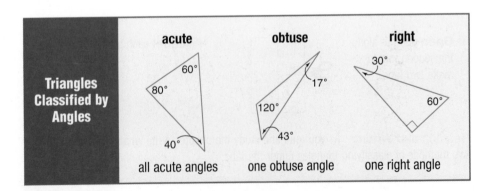

Triangles Classified by Angles	acute	obtuse	right
	60° 80° 40° all acute angles	17° 120° 43° one obtuse angle	30° 60° one right angle

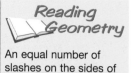

Reading
Geometry

An equal number of
slashes on the sides of
a triangle indicate that
those sides are
congruent.

Triangles can also be classified by their sides.

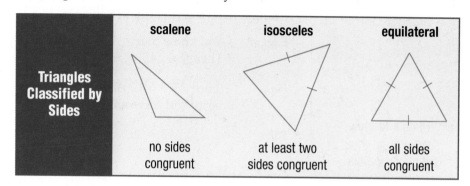

	scalene	isosceles	equilateral
Triangles Classified by Sides	no sides congruent	at least two sides congruent	all sides congruent

Since all sides of an equilateral triangle are congruent, then at least two
of its sides are congruent. So, *all equilateral triangles are also isosceles triangles.*

Some parts of isosceles triangles have special names.

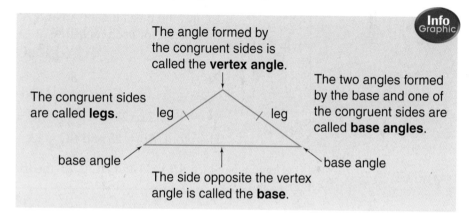

The angle formed by
the congruent sides is
called the **vertex angle**.

The congruent sides
are called **legs**.

leg   leg

The two angles formed
by the base and one of
the congruent sides are
called **base angles**.

base angle

base angle

The side opposite the vertex
angle is called the **base**.

Examples

Classify each triangle by its angles and by its sides.

1

△EFG is a right
isosceles triangle.

2

△ABC is an acute
equilateral triangle.

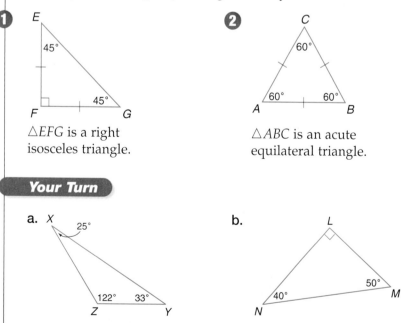

Your Turn

a.

b.

Example	**3**	**Find the measures of** $\overline{AB}$ **and** $\overline{BC}$ **of isosceles**
Algebra Link		**triangle** ABC **if** $\angle A$ **is the vertex angle.**

A diagram shows triangle ABC with apex A, left side labeled $5x - 7$, right side labeled 23, base BC labeled $3x - 5$ with vertices B and C.

Explore	You know that $\angle A$ is the vertex angle. Therefore, $\overline{AB} \cong \overline{AC}$.

Plan	Since $\overline{AB} \cong \overline{AC}$, $AB = AC$. You can write and solve an equation.

Algebra Review
Solving Multi-Step
Equations, p. 723

Solve		
	$AB = AC$	*Definition of congruent segments*
	$5x - 7 = 23$	*Substitution*
	$5x - 7 + 7 = 23 + 7$	*Add 7 to each side.*
	$5x = 30$	*Simplify.*
	$\dfrac{5x}{5} = \dfrac{30}{5}$	*Divide each side by 5.*
	$x = 6$	*Simplify.*

To find the measures of $\overline{AB}$ and $\overline{AC}$, replace x with 6 in the expression for each measure.

AB	**BC**
$AB = 5x - 7$	$BC = 3x - 5$
$= 5(6) - 7$	$= 3(6) - 5$
$= 30 - 7 \text{ or } 23$	$= 18 - 5 \text{ or } 13$

Therefore, $AB = 23$ and $BC = 13$.

Examine	Since $AB = 23$ and $AC = 23$, the triangle is isosceles.

Check for Understanding

Communicating Mathematics

1. **Draw** a scalene triangle.
2. **Sketch and label** an isosceles triangle in which the vertex angle is $\angle X$ and the base is $\overline{YZ}$.
3. Is an equilateral triangle also an isosceles triangle? **Explain** why or why not.

Vocabulary

triangle
vertex
equilateral
isosceles
scalene

Guided Practice

Examples 1 & 2

Classify each triangle by its angles and by its sides.

4.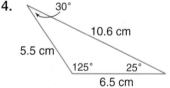

5.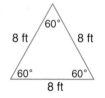

6.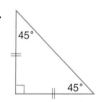

Example 3

7. **Algebra** $\triangle ABC$ is an isosceles triangle with base $\overline{BC}$. Find AB and BC.

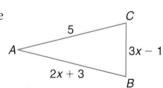

Practice

Classify each triangle by its angles and by its sides.

8.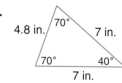

4.8 in. 70° 7 in. 70° 40° 7 in.

9.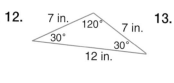

11 m 60° 60° 11 m 11 m 60°

10.

5.7 cm 65° 6 cm 60° 55° 6.3 cm

11.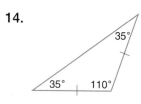

7.7 m 40° 6.4 m 10 m 50°

12.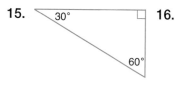

7 in. 120° 7 in. 30° 30° 12 in.

13.

80° 50° 50°

14.

35° 35° 110°

15.

30° 60°

16.

60° 60° 60°

17. Triangle *XYZ* has angles that measure 30°, 60°, and 90°. Classify the triangle by its angles.

Make a sketch of each triangle. If it is not possible to sketch the figure, write *not possible*.

18. acute isosceles

19. right equilateral

20. obtuse and *not* isosceles

21. right and *not* scalene

22. obtuse equilateral

Applications and Problem Solving

23. **Architecture** Refer to the photo at the right. Classify each triangle by its angles and by its sides.

    a. △*ABC*

    b. △*ACD*

    c. △*BCD*

24. **Art** Refer to the optical art design on page 188. Classify the triangles by their angles and by their sides.

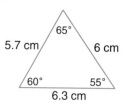

Alcoa Office Building, San Francisco, CA

25. Quilting Classify the triangles that are used in the quilt blocks.

a.

Ohio Star

b.

Duck's Foot in the Mud

26. Algebra △DEF is an equilateral triangle in which $ED = x + 5$, $DF = 3x - 3$, and $EF = 2x + 1$.

    a. Draw and label △DEF.

    b. Find the measure of each side.

27. Algebra Find the measure of each side of isosceles triangle ABC if ∠A is the vertex angle and the perimeter of the triangle is 20 meters.

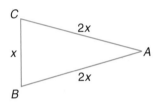

28. Critical Thinking Numbers that can be represented by a triangular arrangement of dots are called *triangular numbers*. The first four triangular numbers are 1, 3, 6, and 10.

Find the next two triangular numbers.

Mixed Review

Write an equation in slope-intercept form of the line with the given slope that passes through the given point. (*Lesson 4–6*)

29. $m = -3, (0, 4)$ **30.** $m = 0, (0, -2)$ **31.** $m = -2, (-2, 1)$

Find the slope of the lines passing through each pair of points. (*Lesson 4–5*)

32. $(5, 7), (4, 5)$ **33.** $(8, 4), (-2, 4)$ **34.** $(5, -2), (5, 1)$

35. Sports In the Olympic ski-jumping competition, the skier tries to make the angle between his body and the front of his skis as small as possible. If a skier is aligned so that the front of his skis makes a 20° angle with his body, what angle is formed by the tail of the skis and his body? (*Lesson 3–5*)

36. Multiple Choice Use the number line to find DA. (*Lesson 2–1*)

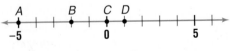

    Ⓐ -10 Ⓑ -6 Ⓒ 6 Ⓓ 10

www.geomconcepts.com/self_check_quiz

5-2 Angles of a Triangle

What You'll Learn
You'll learn to use the Angle Sum Theorem.

Why It's Important
Construction
Builders use the measure of the vertex angle of an isosceles triangle to frame buildings.
See Exercise 21.

If you measure and add the angles in any triangle, you will find that the sum of the angles have a special relationship. Cut and fold a triangle as shown below. Make a conjecture about the sum of the angle measures of a triangle.

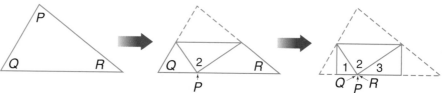

You can use a graphing calculator to verify your conjecture.

Graphing Calculator Exploration

Graphing Calculator Tutorial
See pp. 782–785.

Step 1 Use the Triangle tool on the [F2] menu. Move the pencil cursor to each location where you want a vertex and press [ENTER]. The calculator automatically draws the sides. Label the vertices *A*, *B*, and *C*.

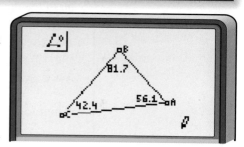

Step 2 Use the Angle tool under Measure on the [F5] menu to measure each angle.

Try These
1. Determine the sum of the measures of the angles of your triangle.
2. Drag any vertex to a different location, measure each angle, and find the sum of the measures.
3. Repeat Exercise 2 several times.
4. **Make a conjecture** about the sum of the angle measures of any triangle.

The results of the activities above can be stated in the Angle Sum Theorem.

Concepts in MOtion
Animation
geomconcepts.com

Theorem 5–1 Angle Sum Theorem	**Words:** The sum of the measures of the angles of a triangle is 180.
	Model:  **Symbols:** $x + y + z = 180$

You can use the Angle Sum Theorem to find missing measures in triangles.

Examples **❶** **Find $m\angle T$ in $\triangle RST$.**

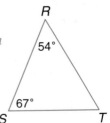

$$
\begin{array}{lll}
m\angle R + m\angle S + m\angle T = 180 & \textit{Angle Sum Theorem} \\
54 \ + \ 67 \ + m\angle T = 180 & \textit{Substitution} \\
121 + m\angle T = 180 & \textit{Add.} \\
121 - 121 + m\angle T = 180 - 121 & \textit{Subtract 121} \\
m\angle T = 59 & \textit{from each side.}
\end{array}
$$

> **Algebra Review**
> Solving One-Step
> Equations, p. 722

❷ **Find the value of each variable in $\triangle DCE$.**

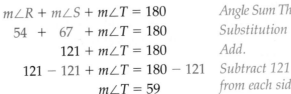

$\angle ACB$ and $\angle DCE$ are vertical angles. Vertical angles are congruent, so $m\angle ACB = m\angle DCE$. Therefore, $x = 85$.

Now find the value of y.

$$
\begin{array}{lll}
m\angle D + m\angle DCE + m\angle E = 180 & \textit{Angle Sum Theorem} \\
55 \ + \ \ 85 \ \ + \ y \ = 180 & \textit{Substitution} \\
140 + y = 180 & \textit{Add.} \\
140 - 140 + y = 180 - 140 & \textit{Subtract 140 from each side.} \\
y = 40 & \textit{Simplify.}
\end{array}
$$

Therefore, $x = 85$ and $y = 40$.

Your Turn

a. Find $m\angle L$ in $\triangle MNL$ if $m\angle M = 25$ and $m\angle N = 25$.
b. Find the value of each variable in the figure at the right.

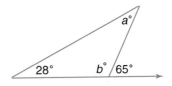

You can use the Angle Sum Theorem to discover a relationship between the acute angles of a right triangle. In $\triangle RST$, $\angle R$ is a right angle.

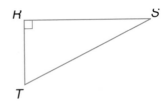

$$
\begin{array}{lll}
m\angle R + m\angle T + m\angle S = 180 & \textit{Angle Sum Theorem} \\
90 + m\angle T + m\angle S = 180 & \textit{Substitution} \\
90 - 90 + m\angle T + m\angle S = 180 - 90 & \textit{Subtract 90 from each side.} \\
m\angle T + m\angle S = 90 & \textit{Simplify.}
\end{array}
$$

Look Back

Complementary
Angles: Lesson 3–5

By the definition of complementary angles, $\angle T$ and $\angle S$ are complementary. This relationship is stated in the following theorem.

  www.geomconcepts.com/extra_examples

<table>
<tr>
<td rowspan="3">Theorem 5–2</td>
<td colspan="3">Words: The acute angles of a right triangle are complementary.</td>
</tr>
<tr>
<td>Model:</td>
<td></td>
<td>Symbols: $x + y = 90$</td>
</tr>
</table>

<table>
<tr>
<td>Example
3
Algebra Link</td>
<td>

Find $m\angle A$ and $m\angle B$ in right triangle ABC.

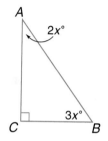

$m\angle A + m\angle B = 90$	*Theorem 5–2*
$2x \;\; + \;\; 3x \;\;\;\; = 90$	*Substitution*
$5x = 90$	*Combine like terms.*
$\dfrac{5x}{5} = \dfrac{90}{5}$	*Divide each side by 5.*
$x = 18$	*Simplify.*

Now replace x with 18 in the expression for each angle.

$\angle A$	$\angle B$
$m\angle A = 2x$	$m\angle B = 3x$
$= 2(18)$ or 36	$= 3(18)$ or 54

</td>
</tr>
</table>

An **equiangular triangle** is a triangle in which all three angles are congruent. You can use the Angle Sum Theorem to find the measure of each angle in an equiangular triangle.

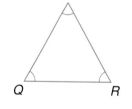

Triangle PQR is an equiangular triangle. Since $m\angle P = m\angle Q = m\angle R$, the measure of each angle of $\triangle PQR$ is $180 \div 3$ or 60.

This relationship is stated in Theorem 5–3.

<table>
<tr>
<td rowspan="3">Theorem 5–3</td>
<td colspan="3">Words: The measure of each angle of an equiangular triangle is 60.</td>
</tr>
<tr>
<td>Model:</td>
<td></td>
<td>Symbols: $x = 60$</td>
</tr>
</table>

Check for Understanding

1. **Choose** the numbers that are not measures of the three angles of a triangle.

   a. 10, 20, 150

   b. 30, 60, 90

   c. 40, 70, 80

   d. 45, 55, 80

2. **Explain** how to find the measure of the third angle of a triangle if you know the measures of the other two angles.

3. Writing Math Is it possible to have two obtuse angles in a triangle? Write a few sentences explaining why or why not.

Guided Practice

Examples 1 & 2

Find the value of each variable.

4.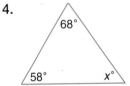

5.

6.

Example 3

7. **Algebra** The measures of the angles of a triangle are $2x$, $3x$, and $4x$. Find the measure of each angle.

Exercises

Practice

Find the value of each variable.

8.

9.

10.

11.

12.

13.

14.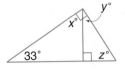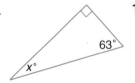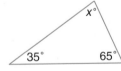

15.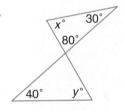

16.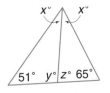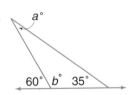

Find the measure of each angle in each triangle.

17.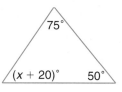
75°
(x + 20)° 50°

18.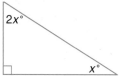
2x°
x°

19.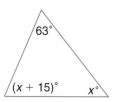
63°
(x + 15)° x°

20. The measure of one acute angle of a right triangle is 25. Find the measure of the other acute angle.

Applications and Problem Solving

21. **Construction** The roof lines of many buildings are shaped like the legs of an isosceles triangle. Find the measure of the vertex angle of the isosceles triangle shown at the right.

22. **Algebra** The measures of the angles of a triangle are $x + 5$, $3x + 14$, and $x + 11$. Find the measure of each angle.

23. **Critical Thinking** If two angles of one triangle are congruent to two angles of another triangle, what is the relationship between the third angles of the triangles? Explain your reasoning.

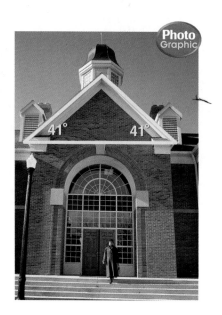

Mixed Review

24. The perimeter of $\triangle GHI$ is 21 units. Find GH and GI. *(Lesson 5–1)*

25. State the slope of the lines perpendicular to the graph of $y = 3x - 2$. *(Lesson 4–6)*

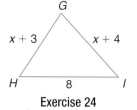
G
x + 3   x + 4
H   8   I
Exercise 24

Identify each pair of angles as *alternate interior*, *alternate exterior*, *consecutive interior*, or *vertical*. *(Lesson 4–2)*

26. $\angle 1, \angle 5$

27. $\angle 9, \angle 11$

28. $\angle 2, \angle 3$

29. $\angle 7, \angle 15$

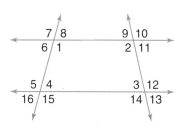
7 8   9 10
6 1   2 11

5 4   3 12
16 15   14 13

Standardized Test Practice
Ⓐ Ⓑ Ⓒ Ⓓ

30. **Short Response** Points X, Y, and Z are collinear, and $XY = 45$, $YZ = 23$, and $XZ = 22$. Locate the points on a number line. *(Lesson 2–2)*

What You'll Learn

You'll learn to identify translations, reflections, and rotations and their corresponding parts.

Why It's Important

Art Artists use motion geometry to make designs. *See Example 6.*

We live in a world of motion. Geometry helps us define and describe that motion. In geometry, there are three fundamental types of motion: **translation, reflection,** and **rotation.**

Translation	Reflection	Rotation
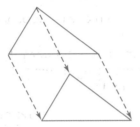	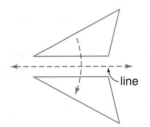	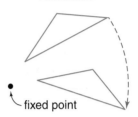
In a **translation**, you slide a figure from one position to another without turning it. Translations are sometimes called *slides*.	In a **reflection**, you flip a figure over a line. The new figure is a mirror image. Reflections are sometimes called *flips*.	In a **rotation**, you turn the figure around a fixed point. Rotations are sometimes called *turns*.

When a figure is translated, reflected, or rotated, the lengths of the sides of the figure do not change.

Examples

Identify each motion as a *translation, reflection,* or *rotation.*

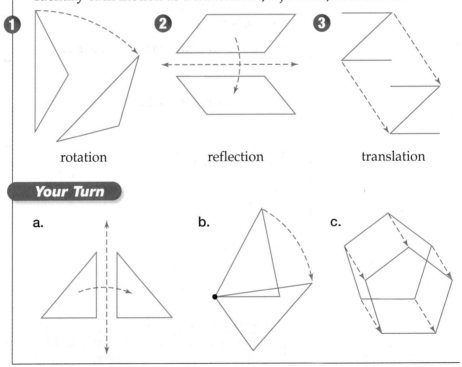

❶ rotation   ❷ reflection   ❸ translation

Your Turn

a.   b.   c.

The figure below shows a translation.

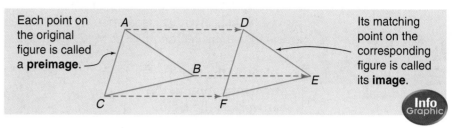

Each point on the original figure is called a **preimage**.

Its matching point on the corresponding figure is called its **image**.

Each point on the preimage can be paired with exactly one point on its image, and each point on the image can be paired with exactly one point on the preimage. This one-to-one correspondence is an example of a **mapping**.

The symbol → is used to indicate a mapping. In the figure, $\triangle ABC \rightarrow \triangle DEF$. In naming the triangles, the order of the vertices indicates the corresponding points.

Reading Geometry

Read $\triangle ABC \rightarrow \triangle DEF$ as *triangle ABC maps to triangle DEF*.

Preimage		Image		Preimage		Image
A	→	D		$\overline{AB}$	→	$\overline{DE}$
B	→	E		$\overline{BC}$	→	$\overline{EF}$
C	→	F		$\overline{AC}$	→	$\overline{DF}$

This mapping is called a **transformation**.

Examples

In the figure, $\triangle XYZ \rightarrow \triangle ABC$ by a reflection.

④ Name the image of ∠X.

$$\triangle XYZ \rightarrow \triangle ABC$$

∠X corresponds to ∠A.

So, ∠A is the image of ∠X.

⑤ Name the side that corresponds to $\overline{AB}$.

Point A corresponds to point X.

$$\triangle XYZ \rightarrow \triangle ABC$$

Point B corresponds to point Y.

So, $\overline{AB}$ corresponds to $\overline{XY}$.

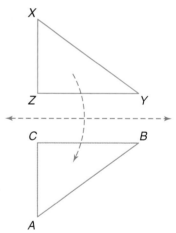

Your Turn

In the figure, $\triangle LMN \rightarrow \triangle QRS$ by a rotation.

d. Name the image of ∠M.

e. Name the angle that corresponds to ∠S.

f. Name the image of $\overline{LM}$.

g. Name the side that corresponds to $\overline{LN}$.

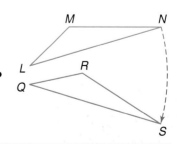

Translations, reflections, and rotations are all **isometries**. An isometry is a movement that does not change the size or shape of the figure being moved. Artists often use isometries in designs. One of the most famous artists to use this technique was M. C. Escher.

Example ⑥

Art Link

Identify the type of transformation in the artwork at the right.

Each figure can be moved to match another without turning or flipping. Therefore, the motion is a translation.

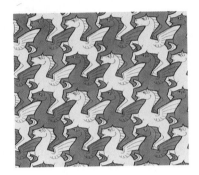

M. C. Escher, *Pegasus*

Check for Understanding

Communicating Mathematics

1. **Explain** the difference between a translation and a rotation.

2.  Suppose $\triangle ABC \rightarrow \triangle RST$. Antonio says that $\angle C$ corresponds to $\angle T$. Keisha says she needs to see the drawing to know which angles correspond. Who is correct? Explain your reasoning.

> **Vocabulary**
>
> translation
> reflection
> rotation
> transformation
> preimage
> isometry
> image
> mapping

Guided Practice

Identify each motion as a *translation*, *reflection*, or *rotation*.

Examples 1–3

3.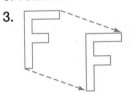

4.

5.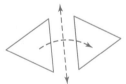

Examples 4 & 5

In the figure at the right, $\triangle XYZ \rightarrow \triangle RST$.

6. Name the image of $\overline{XY}$.

7. Name the angle that corresponds to $\angle R$.

Example 6

8. **Native American Designs** The design below was found on food bowls that were discovered in the ruins of an ancient Hopi pueblo. Identify the transformations in the design.

Exercises ·

Practice

Homework Help

For Exercises	See Examples
9–17	1–3
18–24	4, 5
25–27	6

Extra Practice

See page 734.

Identify each motion as a *translation*, *reflection*, or *rotation*.

9.

10.
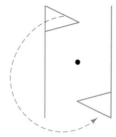

11.

12.
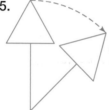

13.

14.

15.
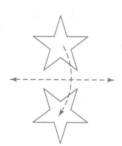

16.

17.
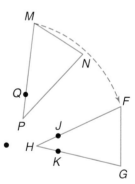

In the figure at the right, $\triangle MNP \rightarrow \triangle FGH$.

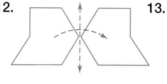

18. Which angle corresponds to $\angle N$?

19. Which side corresponds to $\overline{MN}$?

20. Name the angle that corresponds to $\angle H$.

21. Name the image of point Q.

22. Name the side that corresponds to $\overline{GH}$.

23. Name the image of $\overline{PQ}$.

24. If $\triangle ABC \rightarrow \triangle PQR$, which angle corresponds to $\angle R$?

Applications and Problem Solving

25. **Engines** *Cams* are important parts of engines because they change motion from one direction to another. As the cam turns around, the pistons move up and down. Identify the transformation that occurs in the cams.

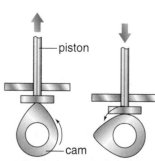

M. C. Escher, *Flying Fish*

26. **Art** The figure at the left shows an untitled work by M. C. Escher. Identify the type of transformation used to complete the work.

27. **Critical Thinking** The transformation below is called a *glide reflection*. How is this transformation different from a translation, reflection, and rotation?

Mixed Review

28. The measure of one acute angle of a right triangle is 30. Find the measure of the other acute angle. *(Lesson 5–2)*

29. **Algebra** $\triangle XYZ$ is an equilateral triangle in which $XY = 2x + 2$, $YZ = x + 7$, and $XZ = 4x - 8$. Find the measure of each side. *(Lesson 5–1)*

Draw a figure for each pair of planes or segments. *(Lesson 4–1)*

30. parallel planes   31. skew segments   32. intersecting planes

Standardized Test Practice

Ⓐ Ⓑ Ⓒ Ⓓ

33. **Multiple Choice** Which ordered pair represents the intersection of line t and line m? *(Lesson 2–4)*

Ⓐ (2, 3)

Ⓑ (−2, −3)

Ⓒ (2, −3)

Ⓓ (−2, 3)

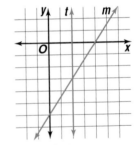

Quiz 1   Lessons 5–1 through 5–3

Classify each triangle by its angles and by its sides. *(Lesson 5–1)*

1.
45°
45°

2.
20°
6 in.
9 in.
25°
135°
4 in.

3.
60°
60°   60°

4. **Algebra** The measures of the angles of a triangle are $2x$, $5x$, and $5x$. Find the measure of each angle. *(Lesson 5–2)*

5. Identify the motion as a *translation*, *reflection*, or *rotation*. *(Lesson 5–3)*

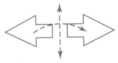

5-4 Congruent Triangles

What You'll Learn
You'll learn to identify corresponding parts of congruent triangles.

Why It's Important
Crafts The pieces of fabric used to make a quilt are congruent to a template.
See Exercise 27.

You've learned that congruent segments have the same length and congruent angles have the same degree measure. In the following activity, you will learn about congruent triangles.

Hands-On Geometry

Materials: grid paper   ✂ scissors   ✏ straightedge

Step 1 On a piece of grid paper, draw two triangles like the ones below. Label the vertices as shown.

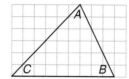

Step 2 Cut out the triangles. Put one triangle over the other so that the parts with the same measures match up.

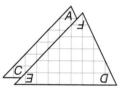

Try These

1. Identify all of the pairs of angles and sides that match or correspond.
2. Triangle *ABC* is congruent to △*FDE*. What is true about their corresponding sides and angles?

Reading Geometry

Arcs are used to show which angles are congruent. Slash marks are used to show which sides are congruent.

If a triangle can be translated, rotated, or reflected onto another triangle so that all of the vertices correspond, the triangles are **congruent triangles**. The parts of congruent triangles that "match" are called **corresponding parts**.

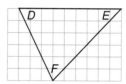

In the figure, △*ABC* ≅ △*FDE*. As in a mapping, the order of the vertices indicates the corresponding parts.

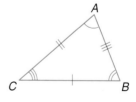

Congruent Angles	Congruent Sides
∠*A* ≅ ∠*F*	$\overline{AB} \cong \overline{FD}$
∠*B* ≅ ∠*D*	$\overline{BC} \cong \overline{DE}$
∠*C* ≅ ∠*E*	$\overline{AC} \cong \overline{FE}$

These relationships help to define congruent triangles.

Definition of Congruent Triangles (CPCTC)	If the corresponding parts of two triangles are congruent, then the two triangles are congruent.
	If two triangles are congruent, then the corresponding parts of the two triangles are congruent.

CPCTC is an abbreviation for Corresponding Parts of Congruent Triangles are Congruent.

Examples ❶ **If △PQR ≅ △MLN, name the congruent angles and sides. Then draw the triangles, using arcs and slash marks to show the congruent angles and sides.**

First, name the three pairs of congruent angles by looking at the order of the vertices in the statement △PQR ≅ △MLN.

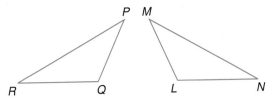

So, ∠P ≅ ∠M, ∠Q ≅ ∠L, and ∠R ≅ ∠N.

Since *P* corresponds to *M*, and *Q* corresponds to *L*, $\overline{PQ} \cong \overline{ML}$.
Since *Q* corresponds to *L*, and *R* corresponds to *N*, $\overline{QR} \cong \overline{LN}$.
Since *P* corresponds to *M*, and *R* corresponds to *N*, $\overline{PR} \cong \overline{MN}$.

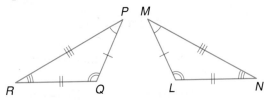

❷ **The corresponding parts of two congruent triangles are marked on the figure. Write a congruence statement for the two triangles.**

List the congruent angles and sides.

∠I ≅ ∠K $\overline{IH} \cong \overline{KH}$
∠G ≅ ∠J $\overline{GH} \cong \overline{JH}$
∠GHI ≅ ∠JHK $\overline{GI} \cong \overline{JK}$

The congruence statement can be written by matching the vertices of the congruent angles. Therefore, △IGH ≅ △KJH.

Your Turn

The corresponding parts of two congruent triangles are marked on the figure. Write a congruence statement for the two triangles.

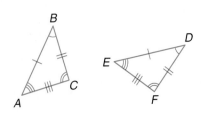

www.geomconcepts.com/extra_examples

Example ③ **Algebra Link**

△*RST* is congruent to △*XYZ*. Find the value of *n*.

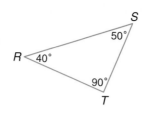
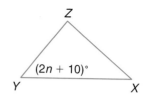

Algebra Review
Solving Multi-Step
Equations, p. 723

Since △*RST* ≅ △*XYZ*, the corresponding parts are congruent.

$$m\angle S = m\angle Y$$
$$50 = 2n + 10 \qquad Substitution$$
$$50 - 10 = 2n + 10 - 10 \qquad Subtract\ 10\ from\ each\ side.$$
$$40 = 2n \qquad Simplify.$$
$$\frac{40}{2} = \frac{2n}{2} \qquad Divide\ each\ side\ by\ 2.$$
$$20 = n \qquad Simplify.$$

🌐nline **Personal Tutor at** geomconcepts.com

Check for Understanding

Communicating Mathematics

1. **Explain** what it means when one triangle is congruent to another.
2. **Describe** how transformations are used to determine whether triangles are congruent.

Vocabulary
congruent triangles
corresponding parts

Guided Practice

⏱ **Getting Ready** **If △*ABC* ≅ △*DEF*, name the corresponding side or angle.**

Sample: ∠*B*   **Solution:** ∠*B* corresponds to ∠*E*.

3. ∠*F*   4. ∠*A*   5. $\overline{AC}$ 6. $\overline{EF}$

Example 1

7. If △*XYZ* ≅ △*EDF*, name the congruent angles and sides. Then draw the triangles, using arcs and slash marks to show the congruent angles and sides.

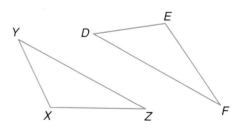

Example 2

Complete each congruence statement.

8.

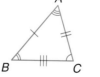

△*ABC* ≅ △ ___?___

9.

△*CBA* ≅ △ ___?___

Example 3
10. **Algebra** △RQP is congruent to △ONM. Find the value of x.

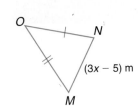

(3x − 5) m

P 70 m Q

Exercises

Practice

For each pair of congruent triangles, name the congruent angles and sides. Then draw the triangles, using arcs and slash marks to show the congruent angles and sides.

Homework Help	
For Exercises	See Examples
11, 12, 19–23, 26, 27	1
13–18	2
24, 25	3

Extra Practice
See page 735.

11.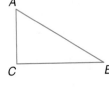

△ACB ≅ △EFD

12.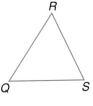

△QRS ≅ △TUV

Complete each congruence statement.

13.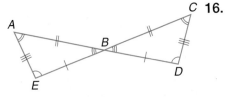

△BAD ≅ △ ___?___

14.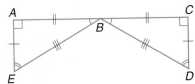

△BCD ≅ △ ___?___

15.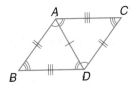

△AEB ≅ △ ___?___

16.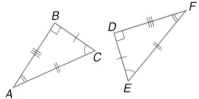

△ ___?___ ≅ △DFE

17.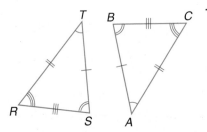

△RTS ≅ △ ___?___

18.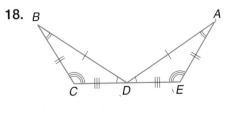

△AED ≅ △ ___?___

If △BCA ≅ △GFH, name the part that is congruent to each angle or segment.

19. ∠F   **20.** $\overline{BA}$ **21.** ∠A   **22.** $\overline{FG}$ **23.** ∠G

24. If △PRQ ≅ △YXZ, m∠P = 63, and m∠Q = 57, find m∠X.

Applications and Problem Solving

25. Algebra   If △DEF ≅ △HEG, what is the value of *x*?

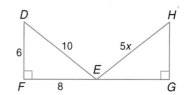

26. Landscaping   Two triangular gardens have the same size and shape. The landscaper needed 24 feet of fencing for one garden. How much fencing is needed for the second garden? Explain your reasoning.

27. Crafts   Many quilts are designed using triangles. Quilters start with a template and trace around the template, outlining the triangles to be cut out. Explain why the triangles are congruent.

template   fabric triangles

28. Critical Thinking   Determine whether each statement is *true* or *false*. If *true*, explain your reasoning. If *false*, show a counterexample.

a. If two triangles are congruent, their perimeters are equal.

b. If two triangles have the same perimeter, they are congruent.

Mixed Review

Identify each motion as a *translation*, *reflection*, or *rotation*.
(Lesson 5–3)

29. 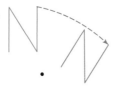   **30.**    **31.**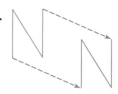

32. Communication   A support cable called a guy wire is attached to a utility pole to give it stability. Safety regulations require a minimum angle of 30° between the pole and the guy wire. Determine the measure of the angle between the guy wire and the ground.   *(Lesson 5–2)*

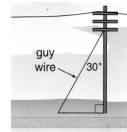

Standardized Test Practice
Ⓐ Ⓑ Ⓒ Ⓓ

33. Short Response If m∠R = 45, classify ∠R as *acute, right,* or *obtuse*.   *(Lesson 3–2)*

34. Multiple Choice   Choose the *false* statement. *(Lesson 1–3)*

Ⓐ   Two points determine two lines.

Ⓑ   A line contains at least two points.

Ⓒ   Three points that are not on the same line determine a plane.

Ⓓ   If two planes intersect, then their intersection is a line.

Investigation

Take a Shortcut

Materials

 patty paper

 scissors

 straightedge

Introducing the Congruence Postulates

Is it possible to show that two triangles are congruent without showing that all six pairs of corresponding parts are congruent? Let's look for a shortcut.

Investigate

1. Use patty paper to investigate three pairs of congruent sides.

   a. Draw a triangle on a piece of patty paper.

   b. Copy the sides of the triangle onto another piece of patty paper and cut them out.

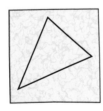 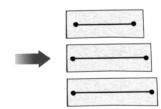

   c. Arrange the pieces so that they form a triangle.

   d. Is this triangle congruent to the original triangle? Explain your reasoning.

   e. Try to form another triangle. Is it congruent to the original triangle?

   f. Can three pairs of congruent sides be used to show that two triangles are congruent?

2. Use patty paper to investigate three pairs of congruent angles.

   a. Draw a triangle on a piece of patty paper.

   b. Copy each angle of the triangle onto a separate piece of patty paper and cut them out. Extend each ray of each angle to the edge of the patty paper.

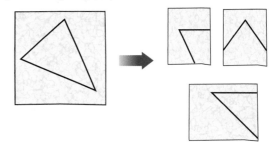

   c. Arrange the pieces so that they form a triangle.

   d. Is this triangle congruent to the original triangle? Explain your reasoning.

   e. Try to form another triangle. Is this triangle congruent to the original triangle?

   f. Can three pairs of congruent angles be used to show that two triangles are congruent?

Extending the Investigation

In this investigation, you will determine which three pairs of corresponding parts can be used to show that two triangles are congruent.

Use patty paper or graphing software to investigate these six cases. (You have already investigated the first two.)

1. three pairs of congruent sides
2. three pairs of congruent angles
3. two pairs of congruent sides and the pair of congruent angles between them
4. two pairs of congruent sides and one pair of congruent angles *not* between them
5. two pairs of congruent angles and the pair of congruent sides between them
6. two pairs of congruent angles and one pair of congruent sides *not* between them

Presenting Your Conclusions

Here are some ideas to help you present your conclusions to the class.

- Make a poster that summarizes your results.
- Make a model with straws that illustrates why certain pairs of corresponding parts cannot be used to show that two triangles are congruent. Be sure to show counterexamples.

 Investigation For more information on the congruence postulates, visit: www.geomconcepts.com

What You'll Learn

You'll learn to use the SSS and SAS tests for congruence.

Why It's Important

Construction
Architects add strength to their buildings by using triangles for support. *See Exercise 7.*

Triangles are common in construction, because triangles, unlike squares, maintain thair shape under stress. You can see this yourself if you use straws and a string to make a triangle and a four-sided figure.

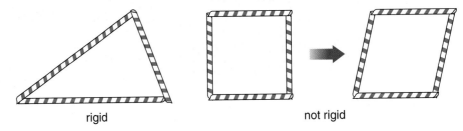

rigid   not rigid

This rigidity hints at an underlying geometric concept: a triangle with three sides of a set length has exactly one shape.

Hands-On Geometry
Construction

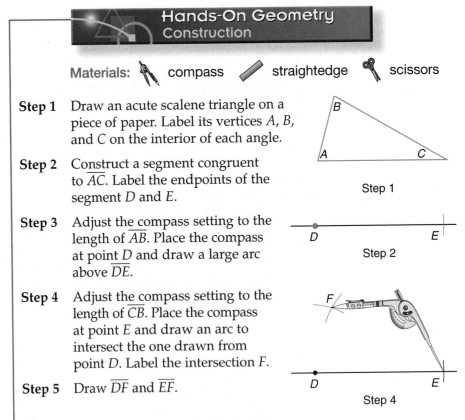

Materials: compass   straightedge   scissors

Step 1 Draw an acute scalene triangle on a piece of paper. Label its vertices *A*, *B*, and *C* on the interior of each angle.

Step 2 Construct a segment congruent to $\overline{AC}$. Label the endpoints of the segment *D* and *E*.

Step 3 Adjust the compass setting to the length of $\overline{AB}$. Place the compass at point *D* and draw a large arc above $\overline{DE}$.

Step 4 Adjust the compass setting to the length of $\overline{CB}$. Place the compass at point *E* and draw an arc to intersect the one drawn from point *D*. Label the intersection *F*.

Step 5 Draw $\overline{DF}$ and $\overline{EF}$.

Try These

1. Label the vertices of $\triangle DEF$ on the interior of each angle. Then cut out the two triangles. **Make a conjecture**. Are the triangles congruent?
2. If the triangles are congruent, write a congruence statement.
3. **Verify** your conjecture with another triangle.

In the previous activity, you constructed a congruent triangle by using only the measures of its sides. This activity suggests the following postulate.

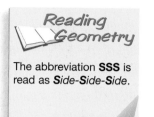
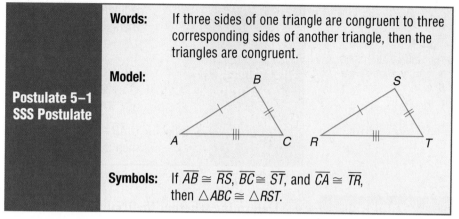

Postulate 5–1 SSS Postulate

Words: If three sides of one triangle are congruent to three corresponding sides of another triangle, then the triangles are congruent.

Model:

Symbols: If $\overline{AB} \cong \overline{RS}$, $\overline{BC} \cong \overline{ST}$, and $\overline{CA} \cong \overline{TR}$, then $\triangle ABC \cong \triangle RST$.

Example

1 In two triangles, $\overline{PQ} \cong \overline{ML}$, $\overline{PR} \cong \overline{MN}$, and $\overline{RQ} \cong \overline{NL}$. Write a congruence statement for the two triangles.

Draw a pair of congruent triangles. Identify the congruent parts with slashes. Label the vertices of one triangle.

Use the given information to label the vertices of the second triangle.

By SSS, $\triangle PQR \cong \triangle MLN$.

Your Turn

a. In two triangles, $\overline{ZY} \cong \overline{FE}$, $\overline{XY} \cong \overline{DE}$, and $\overline{XZ} \cong \overline{DF}$. Write a congruence statement for the two triangles.

In a triangle, the angle formed by two given sides is called the **included angle** of the sides.

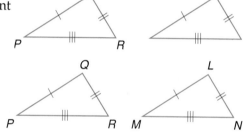

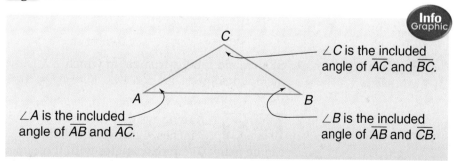

$\angle C$ is the included angle of $\overline{AC}$ and $\overline{BC}$.

$\angle A$ is the included angle of $\overline{AB}$ and $\overline{AC}$.

$\angle B$ is the included angle of $\overline{AB}$ and $\overline{CB}$.

Using the SSS Postulate, you can show that two triangles are congruent if their corresponding sides are congruent. You can also show their congruence by using two sides and the included angle.

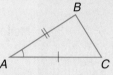
Postulate 5–2 SAS Postulate	**Words:**	If two sides and the included angle of one triangle are congruent to the corresponding sides and included angle of another triangle, then the triangles are congruent.
	Model:	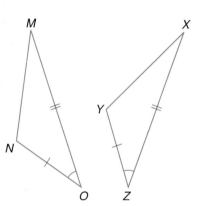
	Symbols:	If $\overline{AB} \cong \overline{RS}$, $\angle A \cong \angle R$, and $\overline{AC} \cong \overline{RT}$, then $\triangle ABC \cong \triangle RST$.

Example **2** **Determine whether the triangles shown at the right are congruent. If so, write a congruence statement and explain why the triangles are congruent. If not, explain why not.**

There are two pairs of congruent sides, $\overline{NO} \cong \overline{YZ}$ and $\overline{MO} \cong \overline{XZ}$. There is one pair of congruent angles, $\angle O \cong \angle Z$, which is included between the sides.

Therefore, $\triangle MNO \cong \triangle XYZ$ by SAS.

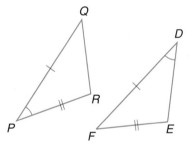

Your Turn

b. Determine whether the triangles shown at the right are congruent by SAS. If so, write a congruence statement and tell why the triangles are congruent. If not, explain why not.

COncepts in MOtion

Animation

geomconcepts.com

Check for Understanding

Communicating Mathematics

1. **Sketch and label** a triangle in which $\angle X$ is the included angle of $\overline{YX}$ and $\overline{ZX}$.

2. **You Decide?** Karen says that there is only one triangle with sides of 3 inches, 4 inches, and 5 inches. Mika says that there can be many different triangles with those measures. Who is correct? Explain your reasoning.

Guided Practice

Write a congruence statement for each pair of triangles represented.

Example 1 3. $\overline{RT} \cong \overline{UW}$, $\overline{RS} \cong \overline{UV}$, $\overline{TS} \cong \overline{WV}$ 4. $\overline{AB} \cong \overline{GH}$, $\overline{BC} \cong \overline{HI}$, $\angle B \cong \angle H$

Example 2 Determine whether each pair of triangles is congruent. If so, write a congruence statement and explain why the triangles are congruent.

5.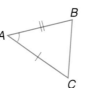

6.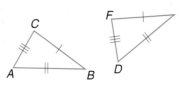

7. **Construction** Most roofs on residential buildings are made of triangular roof trusses. Explain how the SSS postulate guarantees that the triangles in the roof truss will remain rigid. **Example 1**

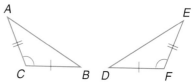

Web member — Gusset — Upper chord
12
5
Lower chord

Exercises

Practice

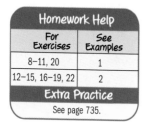

Homework Help	
For Exercises	See Examples
8–11, 20	1
12–15, 16–19, 22	2
Extra Practice	
See page 735.	

Write a congruence statement for each pair of triangles represented.

8. $\overline{JK} \cong \overline{MN}$, $\overline{LK} \cong \overline{ON}$, $\angle K \cong \angle N$

9. $\overline{CB} \cong \overline{EF}$, $\overline{CA} \cong \overline{ED}$, $\overline{BA} \cong \overline{FD}$

10. $\overline{XY} \cong \overline{CA}$, $\overline{XZ} \cong \overline{CB}$, $\angle X \cong \angle C$

11. $\overline{GH} \cong \overline{RT}$, $\overline{GI} \cong \overline{RS}$, $\overline{HI} \cong \overline{TS}$

Determine whether each pair of triangles is congruent. If so, write a congruence statement and explain why the triangles are congruent.

12.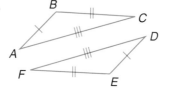

13.

14.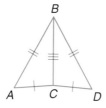

15.

Use the given information to determine whether the two triangles are congruent by SAS. Write *yes* or *no*.

16. $\angle A \cong \angle D$, $\overline{AB} \cong \overline{DE}$, $\overline{BC} \cong \overline{EF}$

17. $\overline{EF} \cong \overline{CA}$, $\overline{BC} \cong \overline{ED}$, $\angle C \cong \angle E$

18. $\overline{BC} \cong \overline{DF}$, $\overline{BA} \cong \overline{EF}$, $\angle B \cong \angle F$

19. $\overline{AB} \cong \overline{DF}$, $\overline{CA} \cong \overline{DE}$, $\angle C \cong \angle F$

20. **Carpentry** Suppose you are building a rectangular bookcase. How could you provide additional support so that the back of the bookcase won't shift?

21. **Landscaping** When small trees are planted, they are usually supported with a wooden stake as shown at the right. Explain how the stake provides support against the wind.

22. **Critical Thinking** Name the additional corresponding part needed to prove that the triangles below are congruent by SAS.

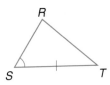

Exercise 21

Mixed Review

23. If $\triangle PQR \cong \triangle CAB$, $m\angle P = 45$, and $m\angle R = 38$, find $m\angle A$. *(Lesson 5–4)*

24. **Word Processing** The button in some computer programs makes the indicated change in the position of the word "Hello." Identify the change as a *rotation*, *reflection*, or *translation*. *(Lesson 5–3)*

The coordinates of the endpoints of a segment are given. Find the coordinates of the midpoint of each segment. *(Lesson 2–5)*

25. $(-1, -2)$, $(-3, -8)$ 26. $(4, 8)$, $(-3, -4)$ 27. $(0, 0)$, (x, y)

28. **Multiple Choice** Express 0.0025 in scientific notation. *(Algebra Review)*
Ⓐ 2.5×10^3 Ⓑ 2.5×10^4 Ⓒ 2.5×10^{-3} Ⓓ 2.5×10^{-4}

Quiz 2   Lessons 5–4 and 5–5

1. **Design** Which triangles in the figure appear to be congruent? *(Lesson 5–4)*

2. If $\triangle XYZ \cong \triangle RST$, which angle is congruent to $\angle S$? *(Lesson 5–4)*

3. In two triangles, $\overline{XZ} \cong \overline{BC}$, $\overline{YZ} \cong \overline{AC}$, and $\overline{YX} \cong \overline{AB}$. Write a congruence statement for the two triangles. *(Lesson 5–5)*

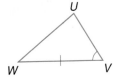

Exercise 1

Determine whether each pair of triangles is congruent. If so, write a congruence statement and explain why the triangles are congruent. *(Lesson 5–5)*

4.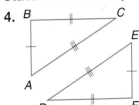

5.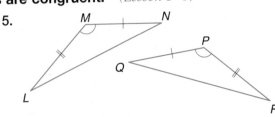

www.geomconcepts.com/self_check_quiz

5-6 ASA and AAS

What You'll Learn

You'll learn to use the ASA and AAS tests for congruence.

Why It's Important

Surveying Surveyors use the ASA Postulate when setting up sight markers.
See Exercise 10.

The side of a triangle that falls between two given angles is called the **included side** of the angles. It is the one side common to both angles.

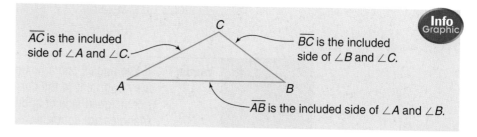

$\overline{AC}$ is the included side of $\angle A$ and $\angle C$.

$\overline{BC}$ is the included side of $\angle B$ and $\angle C$.

$\overline{AB}$ is the included side of $\angle A$ and $\angle B$.

Info Graphic

You can show that two triangles are congruent by using two angles and the included side of the triangles.

Reading Geometry

The abbreviation **ASA** is read as *Angle-Side-Angle*.

Postulate 5-3
ASA Postulate

Words: If two angles and the included side of one triangle are congruent to the corresponding angles and included side of another triangle, then the triangles are congruent.

Model:

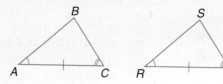

Symbols: If $\angle A \cong \angle R$, $\overline{AC} \cong \overline{RT}$, and $\angle C \cong \angle T$, then $\triangle ABC \cong \triangle RST$.

Example ❶ In $\triangle PQR \setminus \triangle KJL$, $\angle R \cong \angle K$, $\overline{RQ} \cong \overline{KL}$, and $\angle Q \cong \angle L$. **Write a congruence statement for the two triangles.**

Begin by drawing a pair of congruent triangles. Mark the congruent parts with arcs and slashes. Label the vertices of one triangle P, Q, and R.

Locate K and L on the unlabeled triangle in the same positions as R and Q. The unassigned vertex must be J.

Therefore, $\triangle PQR \cong \triangle JLK$ by ASA.

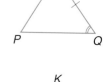

Your Turn

a. In $\triangle DEF$ and $\triangle LMN$, $\angle D \cong \angle N$, $\overline{DE} \cong \overline{NL}$, and $\angle E \cong \angle L$. Write a congruence statement for the two triangles.

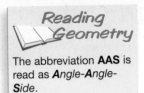
The Angle-Angle-Side Theorem is called a theorem because it can be derived from the ASA Postulate. In AAS, the S is *not* between the two given angles. Therefore, the S indicates a side that is not included between the two angles.

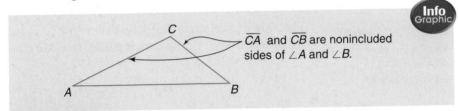

$\overline{CA}$ and $\overline{CB}$ are nonincluded sides of $\angle A$ and $\angle B$.

Theorem 5–4 AAS Theorem	**Words:**	If two angles and a nonincluded side of one triangle are congruent to the corresponding two angles and nonincluded side of another triangle, then the triangles are congruent.
	Model:	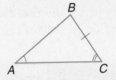
	Symbols:	If $\angle A \cong \angle R$, $C \cong T$, and $\overline{BC} \cong \overline{ST}$, then $\triangle ABC \cong \triangle RST$.

Example ➋ $\triangle ABC$ and $\triangle EDF$ each have one pair of sides and one pair of angles marked to show congruence. What other pair of angles must be marked so that the two triangles are congruent by AAS?

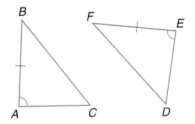

If $\angle B$ and $\angle F$ are marked congruent, then $\overline{AB}$ and $\overline{EF}$ would be included sides. However, AAS requires the nonincluded sides. Therefore, $\angle C$ and $\angle D$ must be marked congruent.

Your Turn

b. $\triangle DEF$ and $\triangle LMN$ each have one pair of sides and one pair of angles marked to show congruence. What other pair of angles must be marked so that the two triangles are congruent by AAS?

c. What other pair of angles must be marked so that the two triangles are congruent by ASA?

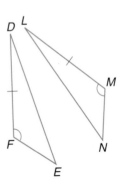

www.geomconcepts.com/extra_examples

Determine whether each pair of triangles is congruent by SSS, SAS, ASA, or AAS. If it is not possible to prove that they are congruent, write *not possible*.

3

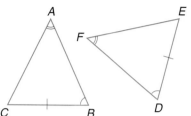

There are two pairs of congruent angles, $\angle A \cong \angle F$ and $\angle B \cong \angle D$. There is one pair of corresponding congruent sides, $\overline{CB} \cong \overline{ED}$, which is *not included* between the angles.

Therefore, $\triangle ABC \cong \triangle FDE$ by AAS.

4

There are two pairs of congruent sides, $\overline{MN} \cong \overline{RP}$ and $\overline{NO} \cong \overline{RQ}$. There is one pair of congruent angles, $\angle M \cong \angle P$, which is *not included* between the sides.

Since SSA is *not* a test for congruence, it is *not possible* to show the triangles are congruent from this information.

Your Turn

d.

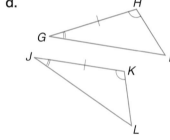

e.

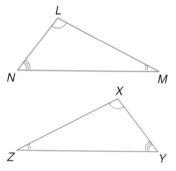

Onl**i**ne **Personal Tutor at** geomconcepts.com

Check for Understanding

Communicating Mathematics

1. **Sketch and label** triangle XYZ in which $\overline{XZ}$ is an included side. Then name the two angles $\overline{XZ}$ is between.

> **Vocabulary**
> included side

2. **Explain** how you could construct a triangle congruent to a given triangle using ASA.

3. *Writing Math* Write a few sentences explaining the SSS, SAS, ASA, and AAS tests for congruence. Give an example of each.

Guided Practice

Example 1

Write a congruence statement for each pair of triangles represented.

4. In $\triangle DEF$ and $\triangle RST$, $\angle D \cong \angle R$, $\angle E \cong \angle T$, and $\overline{DE} \cong \overline{RT}$.

5. In $\triangle ABC$ and $\triangle XYZ$, $\angle A \cong \angle X$, $\angle B \cong \angle Y$, and $\overline{BC} \cong \overline{YZ}$.

Example 2

Name the additional congruent parts needed so that the triangles are congruent by the postulate or theorem indicated.

6. ASA

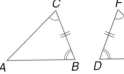

7. AAS

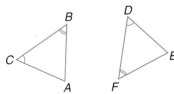

Examples 3 & 4

Determine whether each pair of triangles is congruent by SSS, SAS, ASA, or AAS. If it is not possible to prove that they are congruent, write *not possible*.

8.

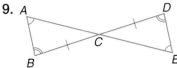

9.

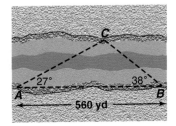

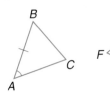

•**10. Surveying** Two surveyors 560 yards apart sight a marker C on the other side of a canyon at angles of 27° and 38°. What will happen if they repeat their measurements from the same positions on another day? Explain your reasoning.
Example 1

Surveying land •·········

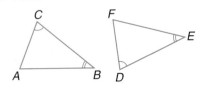

Exercises

· · · · · ● · · · · · · · · · · · · · · ·

Practice

Write a congruence statement for each pair of triangles represented.

11. In $\triangle QRS$ and $\triangle TUV$, $\angle Q \cong \angle T$, $\angle S \cong \angle U$, and $\overline{QS} \cong \overline{TU}$.

12. In $\triangle ABC$ and $\triangle DEF$, $\overline{AC} \cong \overline{ED}$, $\angle C \cong \angle D$, and $\angle B \cong \angle F$.

13. In $\triangle RST$ and $\triangle XYZ$, $\angle S \cong \angle X$, $\overline{ST} \cong \overline{XZ}$, and $\angle T \cong \angle Z$.

14. In $\triangle MNO$ and $\triangle PQR$, $\angle M \cong \angle P$, $\angle N \cong \angle R$, and $\overline{NO} \cong \overline{RQ}$.

Name the additional congruent parts needed so that the triangles are congruent by the postulate or theorem indicated.

Homework Help	
For Exercises	See Examples
11–14	1
15–18	2
19–22, 23	3, 4
Extra Practice	
See page 735.	

15. ASA

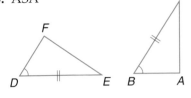

16. AAS

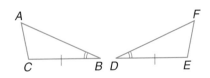

17. AAS

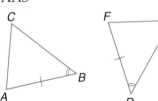

18. ASA

Determine whether each pair of triangles is congruent by SSS, SAS, ASA, or AAS. If it is not possible to prove that they are congruent, write *not possible*.

19.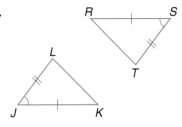

20.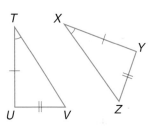

21.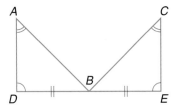

22.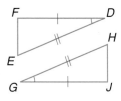

Applications and Problem Solving

23. **Math History** The figure shows how the Greek mathematician Thales (624 B.C.–547 B.C.) determined the distance from the shore to enemy ships during a war. He sighted the ship from point P and then duplicated the angle at $\angle QPT$. The angles at point Q are right angles. Explain why QT represents the distance from the shore to the ship.

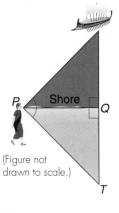

(Figure not drawn to scale.)

24. **Critical Thinking** In $\triangle RST$ and $\triangle UVW$, $\angle R \cong \angle U$, $\angle S \cong \angle V$, and $\overline{RT} \cong \overline{UW}$. So, $\triangle RST \cong \triangle UVW$ by AAS. Prove $\triangle RST \cong \triangle UVW$ by ASA.

Mixed Review

25. In two triangles, $\overline{MN} \cong \overline{PQ}$, $\overline{MO} \cong \overline{PR}$, and $\overline{NO} \cong \overline{QR}$. Write a congruence statement for the two triangles and explain why the triangles are congruent. *(Lesson 5–5)*

If $\triangle HRT \cong \triangle MNP$, complete each statement. *(Lesson 5–4)*

26. $\angle R \cong$ ___?___

27. $\overline{HT} \cong$ ___?___

28. $\angle P \cong$ ___?___

Standardized Test Practice
Ⓐ Ⓑ Ⓒ Ⓓ

29. **Multiple Choice** The graph shows the sales of athletic and sports equipment from 1995 to 2002. Between which two years was the percent of increase the greatest? *(Statistics Review)*

Ⓐ 1996 to 1997
Ⓑ 1998 to 1999
Ⓒ 1999 to 2000
Ⓓ 2000 to 2001

Athletic and Sports Equipment Sales (billions)

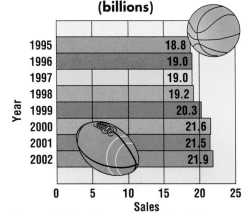

Year	Sales
1995	18.8
1996	19.0
1997	19.0
1998	19.2
1999	20.3
2000	21.6
2001	21.5
2002	21.9

Source: National Sporting Goods Association

Understanding and Using the Vocabulary

After completing this chapter, you should be able to define each term, property, or phrase and give an example or two of each.

acute triangle *(p. 188)*	included side *(p. 215)*	rotation *(p. 198)*
base *(p. 189)*	isometry *(p. 200)*	scalene triangle *(p. 189)*
base angles *(p. 189)*	isosceles triangle *(p. 189)*	transformation *(p. 199)*
congruent triangle *(p. 203)*	legs *(p. 189)*	translation *(p. 198)*
corresponding parts *(p. 203)*	mapping *(p. 199)*	triangle *(p. 188)*
equiangular triangle *(p. 195)*	obtuse triangle *(p. 188)*	vertex *(p. 188)*
equilateral triangle *(p. 189)*	preimage *(p. 199)*	vertex angle *(p. 189)*
image *(p. 199)*	reflection *(p. 198)*	
included angle *(p. 211)*	right triangle *(p. 188)*	

State whether each sentence is *true* or *false*. If false, replace the underlined word(s) to make a true statement.

1. Triangles can be classified by their angles and sides.

2. An isosceles triangle has two vertex angles.

3. The sum of the measures of the angles of a triangle is 360°.

4. An equiangular triangle is defined as a triangle with three congruent sides.

5. The acute angles of a right triangle are supplementary.

6. SSS, SAS, ASA, and AAS are ways to show that two triangles are congruent.

7. A translation is an example of a transformation.

8. An equilateral triangle is also an isosceles triangle.

9. AAS refers to two angles and their included side.

10. Reflections are sometimes called *turns*.

Skills and Concepts

Objectives and Examples	**Review Exercises**

• **Lesson 5–1** Identify the parts of triangles and classify triangles by their parts.

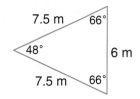

The triangle is acute and isosceles.

Classify each triangle by its angles and by its sides.

11.

12.

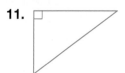

 www.geomconcepts.com/vocabulary_review

Objectives and Examples	**Review Exercises**

• **Lesson 5–2** Use the Angle Sum Theorem.

Find $m\angle A$ in $\triangle ABC$.

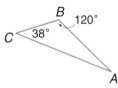

$m\angle A + m\angle B + m\angle C = 180$
$m\angle A + 120 + 38 = 180$
$m\angle A + 158 = 180$
$m\angle A + 158 - 158 = 180 - 158$
$m\angle A = 22$

Find the value of each variable.

13.

14.

15.

• **Lesson 5–3** Identify translations, reflections, and rotations and their corresponding parts.

$\triangle ABC \rightarrow \triangle RST$ by a translation.

$\angle R$ is the image of $\angle A$.

$\overline{BC}$ corresponds to $\overline{ST}$.

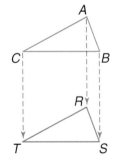

Suppose $\triangle ABE \rightarrow \triangle CBD$.

16. Name the angle that corresponds to $\angle D$.
17. Name the image of $\angle ABE$.
18. Name the image of $\overline{AE}$.
19. Identify the transformation that occurred in the mapping.

• **Lesson 5–4** Name and label corresponding parts of congruent triangles.

Write a congruence statement for the two triangles.

$\triangle ABC \cong \triangle DEF$

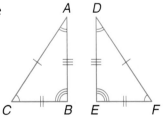

Complete each congruence statement.

20.

21.

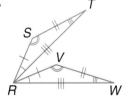

$\triangle MNO \cong \triangle\underline{\quad?\quad}$ $\triangle RST \cong \triangle\underline{\quad?\quad}$

Mixed Problem Solving
See pages 758–765.

Objectives and Examples

Review Exercises

• **Lesson 5–5** Use the SSS and SAS tests for congruence.

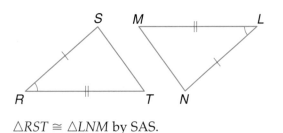

△RST ≅ △LNM by SAS.

Determine whether each pair of triangles is congruent. If so, write a congruence statement and explain why the triangles are congruent.

22. 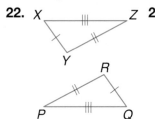 **23.**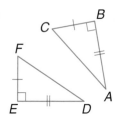

• **Lesson 5–6** Use the ASA and AAS tests for congruence.

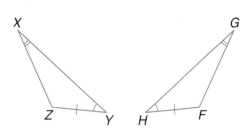

△XYZ ≅ △GHF by AAS.

Determine whether each pair of triangles is congruent by SSS, SAS, ASA, or AAS. If it is not possible to prove that they are congruent, write *not possible*.

24. 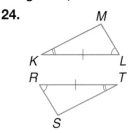 **25.**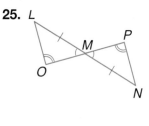

Applications and Problem Solving

26. Maps Classify the triangle by its sides. (*Lesson 5–1*)

27. Algebra Find the measure of ∠A in ∠ABC. (*Lesson 5–2*)

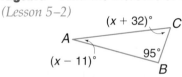

28. Construction The W-truss is the most widely used of light wood trusses. Identify two pairs of triangles in the truss below that appear to be congruent. (*Lesson 5–4*)

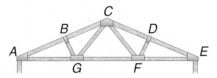

Choose the letter of the description that best matches each term.

1. scalene triangle
2. right triangle
3. isosceles triangle
4. acute triangle
5. equilateral triangle
6. equiangular triangle

a. has a right angle
b. all sides are congruent
c. no sides are congruent
d. has a vertex angle
e. all angles are acute
f. all angles are congruent

Find the value of each variable.

7.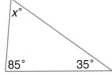

8.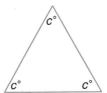

9.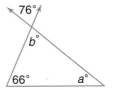

Identify each motion as a translation, reflection, or rotation.

10.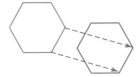

11.

12.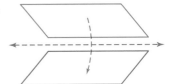

Complete each congruence statement.

13.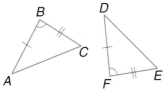

$\triangle ABC \cong \triangle$ ___?___

14.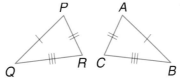

$\triangle$ ___?___ $\cong \triangle ABC$

15.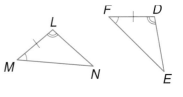

$\triangle$ ___?___ $\cong \triangle FDE$

16. In $\triangle CDE$, identify the included angle for sides $\overline{CD}$ and $\overline{EC}$.

Determine whether each pair of triangles is congruent by SSS, SAS, ASA, or AAS. If it is not possible to prove that they are congruent, write *not possible*.

17.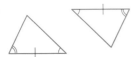

18.

19.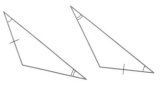

20. **Sports** The sail for a sailboat looks like a right triangle. If the angle at the top of the sail measures 54°, what is the measure of the acute angle at the bottom?

Statistics Problems

On some standardized tests, you will calculate the mean, median, and mode of a data set. You will also choose the most appropriate measure for a data set. On the SAT and ACT, you will apply the concept of the mean to solve problems.

$$\text{mean} = \frac{\text{sum of the numbers}}{\text{number of numbers}}$$

median = middle number of a set arranged in numerical order

mode = the number(s) that occurs most often

Example 1

The heights of ten National Champion Trees are listed in the table below. What is the median, in feet, of the heights?

Tree	Height (ft)	Tree	Height (ft)
American Beech	115	Loblolly Pine	148
Black Willow	76	Pinyon Pine	69
Coast Douglas Fir	329	Sugar Maple	87
Coast Redwood	313	Sugar Pine	232
Giant Sequoia	275	White Oak	79

Hint If there is no single middle number, find the median by calculating the mean of the two middle values.

Solution To find the median, first list the heights in numerical order.

69 76 79 87 115 148 232 275 313 329

Since there are ten numbers, there is no middle number. The two numbers in the middle are 115 and 148. Calculate the mean of these two numbers.

$$\frac{115 + 148}{2} = \frac{263}{2} \text{ or } 131\frac{1}{2}$$

The median is $131\frac{1}{2}$ feet.

Example 2

If the average of five numbers is 32 and the average of two of the numbers is 20, then what is the *sum* of the remaining three numbers?

- (A) 12
- (B) 40
- (C) $46\frac{2}{3}$
- (D) 120
- (E) 140

Hint Use the formula for mean to calculate the sum of the numbers.

Solution On the SAT, *average* is the same as *mean*. First find the sum of the five numbers. Then use the formula for the mean. You know the average (32) and the number of numbers (5).

$$32 = \frac{\text{sum of the five numbers}}{5}$$

$$5 \cdot 32 = 5 \cdot \frac{\text{sum of the five numbers}}{5}$$

$$160 = \text{sum of the five numbers}$$

Use the same method to find the sum of the two numbers.

$$20 = \frac{\text{sum of the two numbers}}{2}$$

$$40 = \text{sum of the two numbers}$$

You can find the sum of the other three numbers by subtracting: (sum of the five numbers) − (sum of the two numbers) = 160 − 40 or 120. The answer is D.

After you work each problem, record your answer on the answer sheet provided or on a sheet of paper.

Multiple Choice

1. Mr. Mendosa obtained estimates for painting from five companies. The estimates were $950, $850, $995, $1000, and $950. What is the mode of these estimates? *(Statistics Review)*

   (A) $150 (B) $949 (C) $950 (D) $995

2. $\sqrt{64 + 36} = ?$ *(Algebra Review)*

   (A) 10 (B) 14 (C) 28
   (D) 48 (E) 100

3. Jared's study group recorded the time they spent on math homework one day. Here are the results (in minutes): 30, 29, 32, 25, 36, 20, 30, 26, 56, 45, 33, and 34. What was the median time spent? *(Statistics Review)*

   (A) 20 min (B) 25 min
   (C) 30 min (D) 31 min

4. The figure below shows an example of a— *(Lesson 5–3)*

   (A) dilation. (B) reflection.
   (C) rotation. (D) translation.

5. Yoshi wants to buy a sweater priced at $59.95. If the sales tax rate is 6%, which is the best estimate of the tax paid on the sweater? *(Percent Review)*

   (A) $3.00 (B) $3.60
   (C) $4.00 (D) $4.20

6. How many even integers are there between 2 and 100, not including 2 and 100? *(Algebra Review)*

   (A) 98 (B) 97 (C) 50
   (D) 49 (E) 48

7. Jenny recorded high temperatures every day for a week. The temperatures, in degrees Fahrenheit, were 48, 55, 60, 55, 52, 47, and 40. What was the mean temperature? *(Statistics Review)*

   (A) 51 (B) 52 (C) 55 (D) 60

8. What is the value of x in the figure? *(Lesson 5–2)*

   (A) 10 (B) 18
   (C) 27 (D) 63

   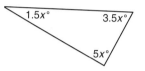

Grid In

9. There are 24 fish in an aquarium. If $\frac{1}{8}$ of them are tetras and $\frac{2}{3}$ of the remaining fish are guppies, how many guppies are there? *(Algebra Review)*

Extended Response

10. The table shows the percent of new passenger cars imported into the United States by country of origin in 2003. *(Statistics Review)*

Percent of New Passenger Cars Imported into U.S. by Country of Origin	
Country	New Cars (percent)
Canada	27
Germany	17
Japan	28
Mexico	10
Korea	7
Other	11

Source: Bureau of Census, Foreign Trade Division

Part A Make a circle graph to show the data. Label each section of the graph with the percent of imported cars.

Part B The total value of cars imported was about $114 billion. Use this information to determine the value of cars imported from outside North America.

More About Triangles

What You'll Learn

Key Ideas

- Identify and construct special segments in triangles.
  (Lessons 6–1 to 6–3)

- Identify and use properties of isosceles triangles.
  (Lesson 6–4)

- Use tests for congruence of right triangles. *(Lesson 6–5)*

- Use the Pythagorean Theorem and its converse.
  (Lesson 6–6)

- Find the distance between two points on the coordinate plane.
  (Lesson 6–7)

Key Vocabulary

hypotenuse *(p. 251)*

leg *(p. 251)*

Pythagorean Theorem *(p. 256)*

Why It's Important

Construction There are reports that ancient Egyptian surveyors used a rope tool to lay out right triangles. They would do this to restore property lines that were washed away by the annual flooding of the Nile River. The need to manage their property led them to develop a mathematical tool.

Right triangles are often used in modern construction. You will investigate a right triangular tool called a *builder's square* in Lesson 6–5.

✓ Check Your Readiness

Classify each triangle by its angles and sides.

1.

2.

3.
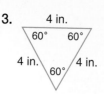

Find the value of each variable.

4.

5.

6.

Solve each equation. Check your solution.

7. $7b + 1 = 50$

8. $10k - 4 = 56$

9. $6r - 8 = 2r$

10. $2n + 1 = n + 6$

11. $8x - 1 = 7x + 9$

12. $15c + 1 = 16c - 6$

13. $12y - 3 = 8y + 13$

14. $9f + 11 = 14f + 1$

FOLDABLES™
Study Organizer

Make this Foldable to help you organize your Chapter 6 notes. Begin with four sheets of notebook paper.

❶ **Fold** each sheet in half along the width. Then cut along the crease.

❷ **Staple** the eight half-sheets together to form a booklet.

❸ **Cut** seven lines from the bottom of the top sheet, six from the second sheet, and so on.

❹ **Label** each tab with a lesson number. The last tab is for vocabulary.

Reading and Writing   As you read and study the chapter, use each page to write the main ideas, theorems, and examples for each lesson.

What You'll Learn
You'll learn to identify and construct medians in triangles.

Why It's Important
Travel Medians can be used to find the distance between two places.
See Exercise 22.

In a triangle, a **median** is a segment that joins a vertex of the triangle and the midpoint of the side opposite that vertex. In the figures below, a median of each triangle is shown in red.

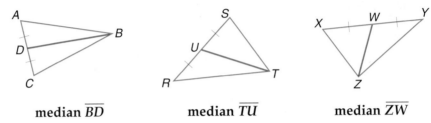

median $\overline{BD}$ median $\overline{TU}$ median $\overline{ZW}$

A triangle has three medians. You can use a compass and a straightedge to construct a median of a triangle.

Hands-On Geometry
Construction

Look Back

Construct the Bisector of a Segment:
Lesson 2–3

Materials:  compass   straightedge

Step 1 Draw a triangle like △ABC.

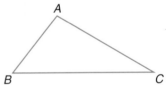

Step 2 The side opposite vertex A is $\overline{BC}$. Find the midpoint of $\overline{BC}$ by constructing the bisector of $\overline{BC}$. Label the midpoint M.

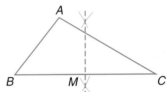

Step 3 Use a straightedge to draw $\overline{AM}$. $\overline{AM}$ is the median of △ABC drawn from vertex A.

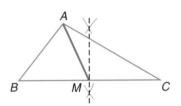

Try These

1. Draw another triangle ABC. Construct the median of △ABC from vertex B.
2. Construct the median of △ABC from vertex C.
3. Are there any other medians that can be drawn?
4. How many medians does a triangle have?

1 In △EFG, $\overline{FN}$ is a median.
Find *EN* if *EG* = 11.

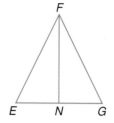

$\overline{FN}$ is a median. So, *N* is the midpoint of $\overline{EG}$.
Since *EG* = 11, *EN* = $\frac{1}{2}$ · 11 or 5.5.

Your Turn

In △MNP, $\overline{MC}$ and $\overline{ND}$ are medians.

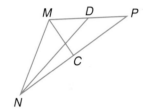

a. What is *NC* if *NP* = 18?
b. If *DP* = 7.5, find *MP*.

Algebra Link

2 In △RST, $\overline{RP}$ and $\overline{SQ}$ are medians. If *RQ* = 7*x* − 1, *SP* = 5*x* − 4,
and *QT* = 6*x* + 9, find *PT*.

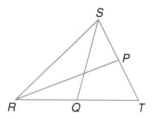

Since $\overline{RP}$ and $\overline{SQ}$ are medians, *Q* and *P* are midpoints.
Use the given values for *RQ* and *QT* to first solve for *x*.

RQ = *QT*	*Definition of Median*
7*x* − 1 = 6*x* + 9	*Substitution*
7*x* − 1 + 1 = 6*x* + 9 + 1	*Add 1 to each side.*
7*x* = 6*x* + 10	*Simplify.*
7*x* − 6*x* = 6*x* + 10 − 6*x*	*Subtract 6x from each side.*
x = 10	*Simplify.*

— **Algebra Review** —
Solving Multi-Step
Equations, p. 723

Next, use the values of *x* and *SP* to find *PT*.

SP = *PT*	*Definition of Median*
5*x* − 4 = *PT*	*Substitution*
5(10) − 4 = *PT*	*Replace x with 10.*
46 = *PT*	*Simplify.*

The medians of triangle JKM, $\overline{JR}$, $\overline{KP}$, and $\overline{MQ}$, intersect at a common point called the **centroid**. When three or more lines or segments meet at the same point, the lines are **concurrent**.

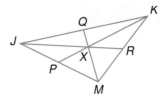

There is a special relationship between the length of the segment from the vertex to the centroid and the length of the segment from the centroid to the midpoint. Use the following diagrams to make a conjecture about the relationship between the two lengths.

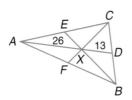

$AX = 26$
$XD = 13$

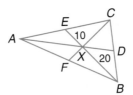

$BX = 20$
$XE = 10$

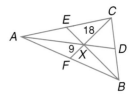

$CX = 18$
$XF = 9$

| Theorem 6–1 | **Words:** | The length of the segment from the vertex to the centroid is twice the length of the segment from the centroid to the midpoint. |
| | **Model:** | 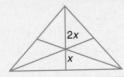 |

Examples

3 In △XYZ, $\overline{XP}$, $\overline{ZN}$, and $\overline{YM}$ are medians.
Find ZQ if QN = 5.

Since $QN = 5$, $ZQ = 2 \cdot 5$ or 10.

4 If XP = 10.5, what is QP?

Since $XP = 10.5$, $QP = \frac{1}{3} \cdot 10.5$ or 3.5.

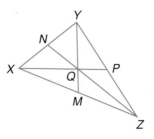

Your Turn

In △ABC, $\overline{CD}$, $\overline{BF}$, and $\overline{AE}$ are medians.

c. If CG = 14, what is DG?
d. Find the measure of $\overline{BF}$ if GF = 6.8.

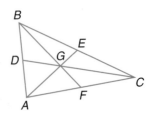

Check for Understanding

Communicating Mathematics

1. **Explain** how to draw a median of a triangle.

2. **Draw** a figure that shows three concurrent segments.

3. Kim says that the medians of a triangle are always the same length. Hector says that they are never the same length. Who is correct? Explain your reasoning.

Guided Practice

Example 1

4. In $\triangle XYZ$, $\overline{YW}$ is a median. What is XW if $XZ = 17$?

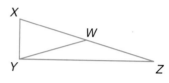

Example 2

5. **Algebra** In $\triangle ABC$, $\overline{BX}$, $\overline{CZ}$, and $\overline{AY}$ are medians. If $AX = 3x - 9$, $XC = 2x - 4$, and $ZB = 2x + 1$, what is AZ?

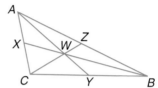

Examples 3 & 4

In $\triangle DEF$, $\overline{DS}$, $\overline{FR}$, and $\overline{ET}$ are medians.

6. Find EV if $VT = 5$.

7. If $FR = 20.1$, what is the measure of $\overline{VR}$?

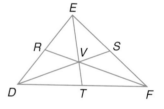

Exercises

Practice

In $\triangle TUV$, $\overline{TE}$, $\overline{UD}$, and $\overline{VC}$ are medians.

8. Find EV if $UV = 24$.

9. If $TC = 8$, find TU.

10. What is TD if $TV = 29$?

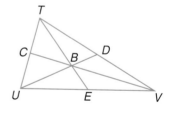

Homework Help	
For Exercises	See Examples
8–18, 22	3, 4
20, 21	2
Extra Practice	
See page 736.	

In $\triangle MNP$, $\overline{MY}$, $\overline{PX}$, and $\overline{NZ}$ are medians.

11. Find the measure of $\overline{WY}$ if $MW = 22$.

12. What is NW if $ZW = 10$?

13. If $PW = 13$, what is WX?

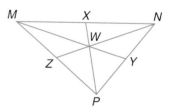

In $\triangle FGH$, $\overline{FJ}$, $\overline{HI}$, and $\overline{GK}$ are medians.

14. What is XK if $GK = 13.5$?

15. If $FX = 10.6$, what is the measure of $\overline{XJ}$?

16. Find HX if $HI = 9$.

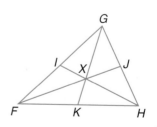

17. If $\overline{BD}$ and $\overline{CF}$ are medians of $\triangle ABC$ and $CE = 17$, what is EF?

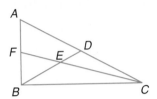

18. In $\triangle RST$, $\overline{RP}$ and $\overline{SQ}$ are medians. Find RU if $UP = 7.3$.

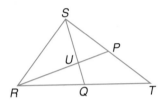

19. Draw a triangle with vertices R, S, and T. Then construct the medians of the triangle to show that they are concurrent.

Applications and Problem Solving

20. Algebra In $\triangle EFG$, $\overline{GP}$, $\overline{FM}$, and $\overline{EN}$ are medians. If $EM = 2x + 3$ and $MG = x + 5$, what is x?

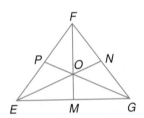

21. Algebra $\overline{RU}$, $\overline{SV}$, and $\overline{TW}$ are medians of $\triangle RST$. What is the measure of $\overline{RW}$ if $RV = 4x + 3$, $WS = 5x - 1$, and $VT = 2x + 9$?

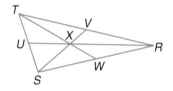

22. Travel On a map, the points representing the towns of Sandersville, Waynesboro, and Anderson form a triangle. The point representing Thomson is the centroid of the triangle. Suppose Washington is halfway between Anderson and Sandersville, Louisville is halfway between Sandersville and Waynesboro, and the distance from Anderson to Thomson is 75 miles. What is the distance from Thomson to Louisville?

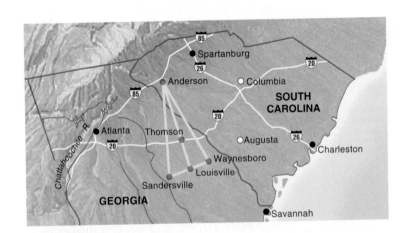

23. Critical Thinking Draw a triangle on a piece of cardboard and cut it out. Draw only one median on the cardboard. How can you find the centroid without using the other two medians? Place the point of a pencil on the centroid you found. Does the triangle balance on your pencil? Why?

24. In △*MNP* and △*RST*, ∠*M* ≅ ∠*R* and $\overline{MP}$ ≅ $\overline{RT}$. Name the additional congruent angles needed to show that the triangles are congruent by ASA. *(Lesson 5–6)*

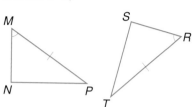

25. In △*DEF* and △*GHJ*, $\overline{DE}$ ≅ $\overline{GH}$, $\overline{EF}$ ≅ $\overline{HJ}$, and ∠*F* ≅ ∠*J*. Tell whether the triangles are congruent by SAS. Explain. *(Lesson 5–5)*

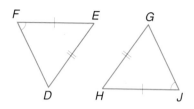

26. Fitness   A rower is designed to simulate rowing. In the diagram shown, notice that the base of the rower, along with the lower portion of the oar, and the hydraulic resistance bar form a triangle. Suppose the measures of two of the angles are 15 and 100. What is the measure of the third angle?   *(Lesson 5–2)*

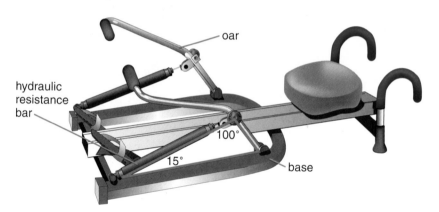

27. Entertainment   In the equation $y = 5x + 3$, y represents the total cost of a trip to the aquarium for x people in a car. Name the slope and y-intercept of the graph of the equation and explain what each value represents.   *(Lesson 4–6)*

Standardized Test Practice
Ⓐ Ⓑ Ⓒ Ⓓ

28. Grid In   Find x so that $a \parallel b$.   *(Lesson 4–4)*

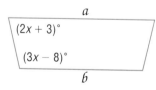

29. Multiple Choice   The toaster shown at the right is formed by a series of intersecting planes. Which names a pair of skew segments?   *(Lesson 4–1)*

Ⓐ $\overline{BC}$ and $\overline{EF}$ Ⓑ $\overline{BE}$ and $\overline{CH}$
Ⓒ $\overline{CD}$ and $\overline{DG}$ Ⓓ $\overline{EH}$ and $\overline{AD}$

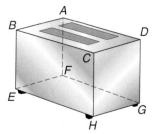

Altitudes and Perpendicular Bisectors

What You'll Learn

You'll learn to identify and construct altitudes and perpendicular bisectors in triangles.

Why It's Important

Construction
Carpenters use perpendicular bisectors and altitudes when framing roofs.
See Exercise 22.

In geometry, an **altitude** of a triangle is a perpendicular segment with one endpoint at a vertex and the other endpoint on the side opposite that vertex. The altitude $\overline{AD}$ is perpendicular to side $\overline{BC}$.

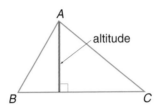

In the following activity, you will construct an altitude of a triangle.

Hands-On Geometry
Construction

Materials: compass   straightedge

Step 1   Draw a triangle like $\triangle ABC$.

Step 2   Place the compass point at B and draw an arc that intersects side $\overline{AC}$ in two points. Label the points of intersection D and E.

Step 3   Place the compass point at D and draw an arc below $\overline{AC}$. Using the same compass setting, place the compass point at E and draw an arc to intersect the one drawn.

Step 4   Use a straightedge to align the vertex B and the point where the two arcs intersect. Draw a segment from vertex B to side $\overline{AC}$. Label the point of intersection F.

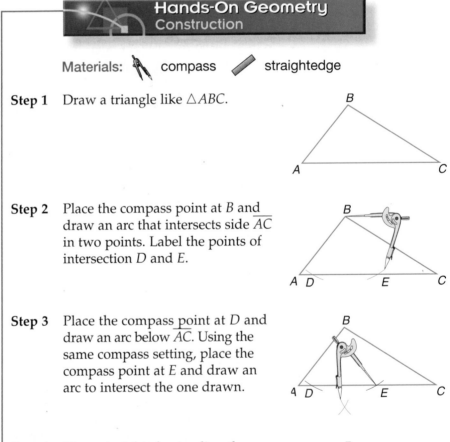

Try These

1. What can you say about $\overline{BF}$?
2. Does $\triangle ABC$ have any other altitudes? If so, construct them.
3. **Make a conjecture** about the number of altitudes in a triangle.

An altitude of a triangle may not always lie inside the triangle.

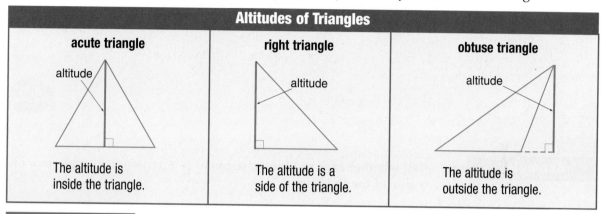

Altitudes of Triangles

acute triangle	right triangle	obtuse triangle
The altitude is inside the triangle.	The altitude is a side of the triangle.	The altitude is outside the triangle.

Examples

Tell whether each red segment is an altitude of the triangle.

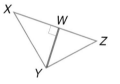 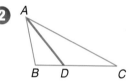 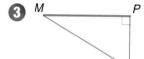

❶ $\overline{YW} \perp \overline{XZ}$, Y is a vertex, and W is on the side opposite Y. So, $\overline{YW}$ is an altitude of the triangle.

❷ $\overline{AD}$ is not a perpendicular segment. So, $\overline{AD}$ is *not* an altitude of the triangle.

❸ $\overline{MP} \perp \overline{NP}$, M is a vertex, and P is on the side opposite M. So, $\overline{MP}$ is an altitude of the triangle.

Your Turn

a. 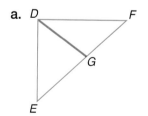 b. 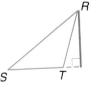 c.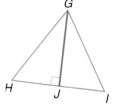

In this text, we will assume that a perpendicular bisector can be a line or a segment contained in that line.

Another special line in a triangle is a perpendicular bisector. A perpendicular line or segment that bisects a side of a triangle is called the **perpendicular bisector** of that side. Line m is a perpendicular bisector of side $\overline{BC}$.

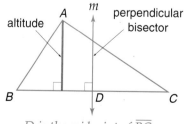

D is the midpoint of $\overline{BC}$.

In some triangles, the perpendicular bisector and the altitude are the same. If the perpendicular bisector of a side contains the opposite vertex, then the perpendicular bisector is also an altitude.

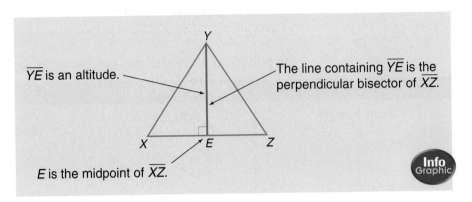

$\overline{YE}$ is an altitude.

The line containing $\overline{YE}$ is the perpendicular bisector of $\overline{XZ}$.

E is the midpoint of $\overline{XZ}$.

Info Graphic

Examples

Tell whether each red line or segment is a perpendicular bisector of a side of the triangle.

4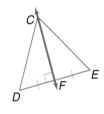

$\overleftrightarrow{CF} \perp \overline{DE}$ and F is the midpoint of $\overline{DE}$. So, $\overleftrightarrow{CF}$ is a perpendicular bisector of side $\overline{DE}$ in $\triangle CDE$.

5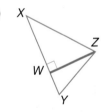

$\overline{ZW} \perp \overline{XY}$ but W is *not* the midpoint of $\overline{XY}$. So, $\overline{ZW}$ is *not* a perpendicular bisector of side $\overline{XY}$ in $\triangle XYZ$.

6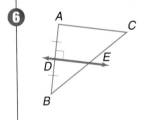

$\overleftrightarrow{DE} \perp \overline{AB}$ and D is the midpoint of $\overline{AB}$. So, $\overleftrightarrow{DE}$ is a perpendicular bisector of side $\overline{AB}$ in $\triangle ABC$.

Your Turn

d.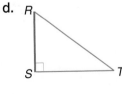

e.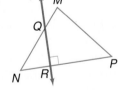

f.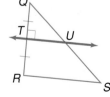

Example 7
Music Link

A balalaika is a stringed musical instrument that has a triangular body. Balalaikas are commonly played when performing Russian songs and dance music. A three-stringed balalaika is shown at the right. Tell whether string B is an *altitude*, a *perpendicular bisector*, *both*, or *neither*.

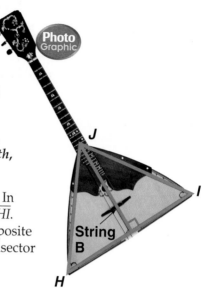

String B contains the midpoint of $\overline{HI}$. In addition, string B is perpendicular to $\overline{HI}$. Since it also contains the vertex, J, opposite $\overline{HI}$, string B is both a perpendicular bisector and an altitude.

Check for Understanding

Communicating Mathematics

1. **Draw** a right triangle. Then construct all of the altitudes of the triangle.

2. **Draw** a triangle like $\triangle ABC$. Then use a segment bisector construction to construct the perpendicular bisector of $\overline{AC}$.

3. **Writing Math** Compare and contrast altitudes and perpendicular bisectors.

> **Vocabulary**
> altitude
> perpendicular bisector

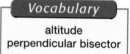

Exercise 2

Guided Practice

Examples 1–6

For each triangle, tell whether the red segment or line is an *altitude*, a *perpendicular bisector*, *both*, or *neither*.

4.
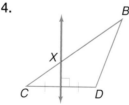

5.
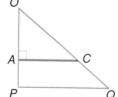

6.
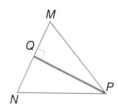

Example 7

7. **Camping** The front of a pup tent is shaped like a triangle. Tell whether the roof pole is an *altitude*, a *perpendicular bisector*, *both*, or *neither*.

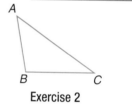

Practice

For each triangle, tell whether the red segment or line is an *altitude*, a *perpendicular bisector*, *both*, or *neither*.

8.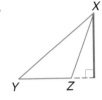

9.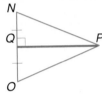

10.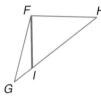

11.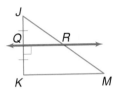

12.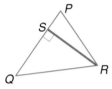

13.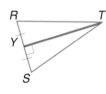

14.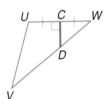

15.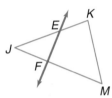

16.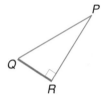

17. Name a perpendicular bisector in △*ABC*.

18. Tell whether $\overline{AC}$ is a *perpendicular bisector*, an *altitude*, *both*, or *neither*.

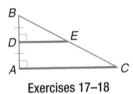

Exercises 17–18

19. In △*DEF*, $\overrightarrow{GH}$ is the perpendicular bisector of $\overline{EF}$. Is it possible to construct other perpendicular bisectors in △*DEF*? Make a conjecture about the number of perpendicular bisectors of a triangle.

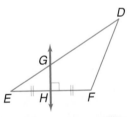

Applications and Problem Solving

20. Architecture The Transamerica building in San Francisco is triangular in shape. Copy the triangle onto a sheet of paper. Then construct the perpendicular bisector of each side.

Transamerica Building

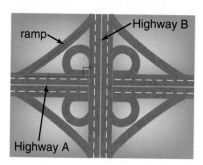

ramp

Highway B

Highway A

21. Transportation There are four major types of highway interchanges. One type, a cloverleaf interchange, is shown. Notice that each ramp along with sections of the highway form a triangle. Tell whether highway A is an *altitude*, a *perpendicular bisector*, *both*, or *neither*.

22. Construction The most common type of design for a house roof is a gable roof. The illustration shows the structural elements of a gable roof.

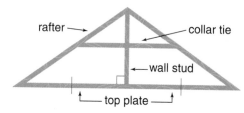

rafter → ← collar tie
← wall stud
└ top plate ┘

    a. Which structural element is a perpendicular bisector?

    b. Tell whether the top plate is an altitude. Explain your reasoning.

    c. Tell whether the collar tie is a perpendicular bisector. Explain your reasoning.

23. Critical Thinking Draw two types of triangles in which the altitude is on the line that forms the perpendicular bisector. Identify the types of triangles drawn, and draw the altitude and perpendicular bisector for each triangle.

Mixed Review

24. Algebra In $\triangle ABC$, $\overline{BZ}$, $\overline{CX}$, and $\overline{AY}$ are medians. If $BY = x - 2$ and $YC = 2x - 10$, find the value of x. *(Lesson 6–1)*

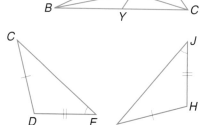

25. Determine whether $\triangle CDE$ and $\triangle GHJ$ are congruent by SSS, SAS, ASA, or AAS. If it is not possible to prove that they are congruent, write *not possible*. *(Lessons 5–5 & 5–6)*

26. Find the slope of the line passing through points at $(2, 3)$ and $(-2, 4)$. *(Lesson 4–5)*

Standardized Test Practice
Ⓐ Ⓑ Ⓒ Ⓓ

27. Short Response Draw the next figure in the pattern. *(Lesson 1–1)*

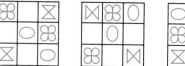

28. Multiple Choice Andrew is buying a pair of sunglasses priced at $18.99. What is the total cost of the sunglasses if he needs to pay a sales tax of 6%? Round to the nearest cent. *(Percent Review)*

    Ⓐ $19.10   Ⓑ $19.94   Ⓒ $20.13   Ⓓ $20.54

Recall that the bisector of an angle is a ray that separates the angle into two congruent angles.

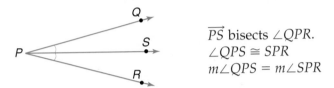

$\overrightarrow{PS}$ bisects $\angle QPR$.
$\angle QPS \cong SPR$
$m\angle QPS = m\angle SPR$

An **angle bisector** of a triangle is a segment that separates an angle of the triangle into two congruent angles. One of the endpoints of an angle bisector is a vertex of the triangle, and the other endpoint is on the side opposite that vertex.

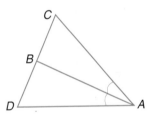

$\overline{AB}$ is an angle bisector of $\triangle DAC$.
$\angle DAB \cong \angle CAB$
$m\angle DAB = m\angle CAB$

Just as every triangle has three medians, three altitudes, and three perpendicular bisectors, every triangle has three angle bisectors.

Special Segments in Triangles			
Segment	• altitude	• perpendicular bisector	• angle bisector
Type	• line segment	• line • line segment	• ray • line segment
Property	from the vertex, a line perpendicular to the opposite side	bisects the side of a triangle	bisects the angle of a triangle

An angle bisector of a triangle has all of the characteristics of any angle bisector. In $\triangle FGH$, $\overline{FJ}$ bisects $\angle GFH$.

1. $\angle 1 \cong \angle 2$, so $m\angle 1 = m\angle 2$.

2. $m\angle 1 = \frac{1}{2}(m\angle GFH)$ or $2(m\angle 1) = m\angle GFH$

3. $m\angle 2 = \frac{1}{2}(m\angle GFH)$ or $2(m\angle 2) = m\angle GFH$

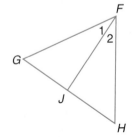

1 In △MNP, $\overline{MO}$ bisects ∠NMP. If $m\angle 1 = 33$, find $m\angle 2$.

Since $\overline{MO}$ bisects ∠NMP, $m\angle 1 = m\angle 2$.

Since $m\angle 1 = 33$, $m\angle 2 = 33$.

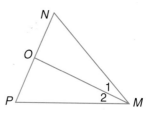

2 In △PQR, $\overline{QS}$ bisects ∠PQR. If $m\angle PQR = 70$, what is $m\angle 2$?

$m\angle 2 = \frac{1}{2}(m\angle PQR)$ *Definition of bisector*

$m\angle 2 = \frac{1}{2}(70)$ *Substitution*

$m\angle 2 = 35$ *Multiply.*

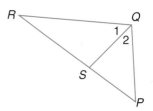

3 In △DEF, $\overline{EG}$ bisects ∠DEF. If $m\angle 1 = 43$, find $m\angle DEF$.

$m\angle DEF = 2(m\angle 1)$ *Definition of bisector*
$m\angle DEF = 2(43)$ *Substitution*
$m\angle DEF = 86$ *Multiply.*

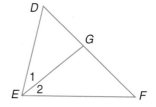

Your Turn

In △ABC, $\overline{AD}$ bisects ∠BAC.

a. If $m\angle 1 = 32$, find $m\angle 2$.
b. Find $m\angle 1$ if $m\angle BAC = 52$.
c. What is $m\angle CAB$ if $m\angle 1 = 28$?

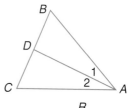

Algebra Link

4 In △RST, $\overline{SU}$ is an angle bisector. Find $m\angle UST$.

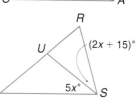

Algebra Review
Solving Multi-Step Equations, p. 723

$m\angle UST = m\angle RSU$ *Definition of bisector*
$5x = 2x + 15$ *Substitution*
$5x - 2x = 2x + 15 - 2x$ *Subtract 2x from each side.*
$3x = 15$ *Simplify.*
$\frac{3x}{3} = \frac{15}{3}$ *Divide each side by 3.*
$x = 5$ *Simplify.*

So, $m\angle UST = 5(5)$ or 25.

Check for Understanding

Communicating
Mathematics

1. **Describe** an angle bisector of a triangle.

2. **Draw** an acute scalene triangle. Then use a compass and straightedge to construct the angle bisector of one of the angles.

Vocabulary

angle bisector

Guided Practice

Examples 1–3

In △DEF, $\overline{EG}$ bisects ∠DEF, and $\overline{FH}$ bisects ∠EFD.

3. If $m\angle 4 = 24$, what is $m\angle DEF$?

4. Find $m\angle 2$ if $m\angle 1 = 36$.

5. What is $m\angle EFD$ if $m\angle 1 = 42$?

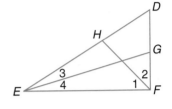

Example 4

6. **Algebra** In △XYZ, $\overline{ZW}$ bisects ∠YZX. If $m\angle 1 = 5x + 9$ and $m\angle 2 = 39$, find x.

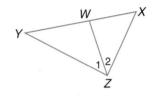

Exercises

Practice

In △ABC, $\overline{BD}$ bisects ∠ABC, and $\overline{AE}$ bisects ∠BAC.

7. If $m\angle 1 = 55$, what is $m\angle ABC$?

8. Find $m\angle 3$ if $m\angle BAC = 38$.

9. What is $m\angle 4$ if $m\angle 3 = 22$?

10. Find $m\angle 2$ if $m\angle ABC = 118$.

11. What is $m\angle BAC$ if $m\angle 3 = 20$?

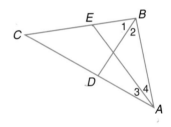

Homework Help

For Exercises	See Examples
7, 11, 15, 17	3
8, 10, 13, 14, 16, 20	2
9, 12	1
18	4
19	2, 3

Extra Practice
See page 736.

In △MNP, $\overline{NS}$ bisects ∠MNP, $\overline{MR}$ bisects ∠NMP, and $\overline{PQ}$ bisects ∠MPN.

12. Find $m\angle 4$ if $m\angle 3 = 31$.

13. If $m\angle MPN = 34$, what is $m\angle 6$?

14. What is $m\angle 3$ if $m\angle NMP = 64$?

15. Find $m\angle MNP$ if $m\angle 1 = 44$.

16. What is $m\angle 2$ if ∠MNP is a right angle?

17. In △XYZ, $\overline{YW}$ bisects ∠XYZ. What is $m\angle XYZ$ if $m\angle 2 = 62$?

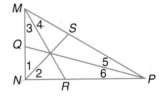

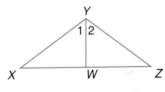

Applications and
Problem Solving

18. **Algebra** In △DEF, $\overline{EC}$ is an angle bisector. If $m\angle CEF = 2x + 10$ and $m\angle DEC = x + 25$, find $m\angle DEC$.

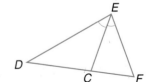

19. **Engineering** One type of bridge, a *cable-stayed bridge* is shown. Notice that the *cable stay anchorage* is an angle bisector of each triangle formed by the cables called *stays* and the roadway.

   a. Suppose $m\angle ABC = 120$, what is $m\angle 2$?

   b. Suppose $m\angle 4 = 48$, what is $m\angle DEF$?

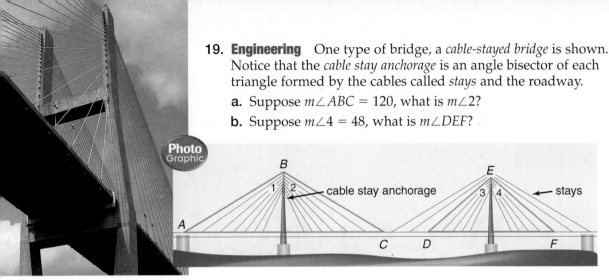

cable stay anchorage

stays

Tallmadge Bridge, Savannah, Georgia

Mixed Review

20. **Critical Thinking** What kind of angles are formed when you bisect an obtuse angle of a triangle? Explain.

21. Tell whether the red segment in $\triangle ABC$ is an *altitude*, a *perpendicular bisector*, *both*, or *neither*. *(Lesson 6–2)*

22. In $\triangle MNP$, $\overline{MC}$, $\overline{NB}$, and $\overline{PA}$ are medians. Find PD if $DA = 6$. *(Lesson 6–1)*

23. **Algebra** The measures of the angles of a triangle are $x + 2$, $4x + 3$, and $x + 7$. Find the measure of each angle. *(Lesson 5–2)*

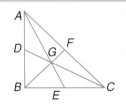

Exercise 22

Standardized Test Practice
Ⓐ Ⓑ Ⓒ Ⓓ

24. **Short Response** Triangle *DEF* has sides that measure 6 feet, 6 feet, and 9 feet. Classify the triangle by its sides. *(Lesson 5–1)*

25. **Multiple Choice** Multiply $2r + s$ by $r - 3s$. *(Algebra Review)*

   Ⓐ $2r^2 - 3s^2$ Ⓑ $2r^2 - 5rs - 3s^2$ Ⓒ $2r^2 - 3rs$ Ⓓ $-6r^2s^2$

Quiz 1   Lessons 6–1 through 6–3

In $\triangle ABC$, $\overline{AE}$, $\overline{BF}$, and $\overline{CD}$ are medians. *(Lesson 6–1)*

1. Find GE if $AG = 9$.   **2.** What is BF if $BG = 5$?

3. If $DG = 12$, what is the measure of $\overline{DC}$?

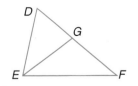

4. Tell whether $\overline{PO}$ is an *altitude*, a *perpendicular bisector*, *both*, or *neither*. *(Lesson 6–2)*

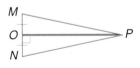

5. Algebra In $\triangle DEF$, $\overline{EG}$ is an angle bisector. If $m\angle DEG = 2x + 7$ and $m\angle GEF = 4x - 1$, find $m\angle GEF$. *(Lesson 6–3)*

Investigation

What a CIRCLE !

Materials

 ruler

 protractor

 compass

Circumcenter, Centroid, Orthocenter, and Incenter

Is there a relationship between the perpendicular bisectors of the sides of a triangle, the medians, the altitudes, and the angle bisectors of a triangle? Let's find out!

Investigate

1. Use construction tools to locate some interesting points on a triangle.

   a. Draw a large acute scalene triangle.

   b. On a separate sheet of paper, copy the following table.

Description of Points	Label of Points
midpoints of the three sides (3)	
circumcenter (1)	
centroid (1)	
intersection points of altitudes with the sides (3)	
orthocenter (1)	
midpoints of segments from orthocenter to each vertex (3)	
incenter (1)	
midpoint of segment joining circumcenter and orthocenter (1)	

   c. Construct the perpendicular bisector of each side of your triangle. Label the midpoints *E*, *F*, and *G*. Record these letters in your table. The **circumcenter** is the point where the perpendicular bisectors meet. Label this point *J* and record it. To avoid confusion, erase the perpendicular bisectors, but not the circumcenter.

   d. Draw the medians of your triangle. The point where the medians meet is the *centroid*. Label this point *M* and record it in your table. Erase the medians, but not the centroid.

e. Construct the altitudes of the triangle. Label the points where the altitudes intersect the sides *N*, *P*, and *Q*. Record these points. The point where the altitudes meet is the **orthocenter**. Label this point *S* and record it. Erase the altitudes, but not the orthocenter.

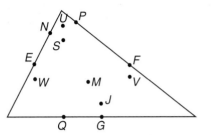

f. Draw three segments, each having the orthocenter as one endpoint and a vertex of your triangle as the other endpoint. Find the midpoint of each segment. Label the midpoints *U*, *V*, and *W* and record these points in your table. Erase the segments.

g. Construct the bisector of each angle of the triangle. The point where the angle bisectors meet is the **incenter**. Label this point *X* and record it. Erase the angle bisectors, but not the incenter.

2. You should now have 13 points labeled. Follow these steps to construct a special circle, called a **nine-point circle**.

a. Locate the circumcenter and orthocenter. Draw a line segment connecting these two points. Bisect this line segment. Label the midpoint *Z* and record it in the table. Do not erase this segment.

b. Draw a circle whose center is point *Z* and whose radius extends to a midpoint of the side of your triangle. How many of your labeled points lie on or very close to this circle?

c. Extend the segment drawn in Step 2a. This line is called the **Euler** (OY-ler) **line**. How many points are on the Euler line?

Extending the Investigation

In this extension, you will determine whether a special circle exists for other types of triangles.

Use paper and construction tools to investigate these cases.

1. an obtuse scalene triangle   **2.** a right triangle   **3.** an equilateral triangle

Presenting Your Conclusions

Here are some ideas to help you present your conclusions to the class.

• Make a booklet of your constructions. For each triangle, include a table in which all of the points are recorded.

• Research Leonhard Euler. Write a brief report on his contributions to mathematics, including the nine-point circle.

 Investigation For more information on the nine-point circle, visit: www.geomconcepts.com

Recall from Lesson 5–1 that an isosceles triangle has at least two congruent sides. The congruent sides are called **legs**. The side opposite the vertex angle is called the **base**. In an isosceles triangle, there are two base angles, the vertices where the base intersects the congruent sides.

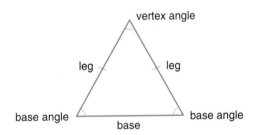

You can use a TI–83/84 Plus graphing calculator to draw an isosceles triangle and study its properties.

Graphing Calculator Exploration

Graphing Calculator Tutorial
See pp. 782–785.

Step 1 Draw a circle using the Circle tool on the F2 menu. Label the center of the circle *A*.

Step 2 Use the Triangle tool on the F2 menu to draw a triangle that has point *A* as one vertex and its other two vertices on the circle. Label these vertices *B* and *C*.

Step 3 Use the Hide/Show tool on menu F5 to hide the circle. Press the CLEAR key to quit the F7 menu. The figure that remains on the screen is isosceles triangle *ABC*.

Try These

1. Tell how you can use the measurement tools on F5 to check that △*ABC* is isosceles. Use your method to be sure it works.

2. Use the Angle tool on F5 to measure ∠*B* and ∠*C*. What is the relationship between ∠*B* and ∠*C*?

3. Use the Angle Bisector tool on F3 to bisect ∠*A*. Use the Intersection Point tool on F2 to mark the point where the angle bisector intersects $\overline{BC}$. Label the point of intersection *D*. What is point *D* in relation to side $\overline{BC}$?

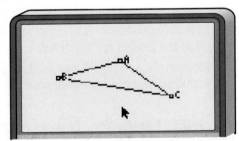

4. Use the Angle tool on ⌈F5⌉ to find the measures of ∠ADB and ∠ADC.

5. Use the Distance & Length tool on ⌈F5⌉ to measure $\overline{BD}$ and $\overline{CD}$. What is the relationship between the lengths of $\overline{BD}$ and $\overline{CD}$?

6. Is $\overline{AD}$ part of the perpendicular bisector of $\overline{BC}$? Explain.

The results you found in the activity are expressed in the following theorems.

Theorem	Words	Models	Symbols
6-2 **Isosceles Triangle Theorem**	If two sides of a triangle are congruent, then the angles opposite those sides are congruent.		If $\overline{AB} \cong \overline{AC}$, then ∠C ≅ ∠B.
6-3	The median from the vertex angle of an isosceles triangle lies on the perpendicular bisector of the base and the angle bisector of the vertex angle.		If $\overline{AB} \cong \overline{AC}$ and $\overline{BD} \cong \overline{CD}$, then $\overline{AD} \perp \overline{BC}$ and ∠BAD ≅ ∠CAD.

Example ──① **Find the value of each variable in isosceles triangle *DEF* if $\overline{EG}$ is an angle bisector.**

First, find the value of *x*.
Since △*DEF* is an isosceles triangle, ∠*D* ≅ ∠*F*. So, *x* = 49.

Now find the value of *y*.
By Theorem 6-3, $\overline{EG} \perp \overline{DF}$. So, *y* = 90.

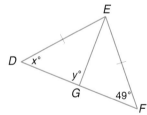

Your Turn

For each triangle, find the values of the variables.

a.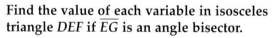

b.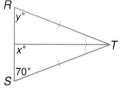

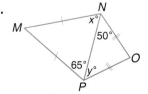

Suppose you draw two congruent acute angles on two pieces of patty paper and then rotate one of the angles so that one pair of rays overlaps and the other pair intersects.

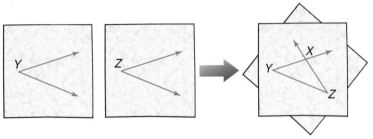

What kind of triangle is formed?   What is true about angles Y and Z? What is true about the sides opposite angles Y and Z? Is the converse of Theorem 6–2 true?

Theorem 6–4 Converse of Isosceles Triangle Theorem	**Words:** If two angles of a triangle are congruent, then the sides opposite those angles are congruent.
	Model: **Symbols:** If $\angle B \cong \angle C$, then $\overline{AC} \cong \overline{AB}$.

Example

Algebra Link

2 In $\triangle ABC$, $\angle A \cong \angle B$ and $m\angle A = 48$. Find $m\angle C$, AC, and BC.

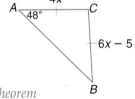

First, find $m\angle C$. You know that $m\angle A = 48$. Since $\angle A \cong \angle B$, $m\angle B = 48$.

Algebra Review

Solving Multi-Step Equations, p. 723

$m\angle A + m\angle B + m\angle C = 180$	*Angle Sum Theorem*
$48 + 48 + m\angle C = 180$	*Replace $m\angle A$ and $m\angle B$ with 48.*
$96 + m\angle C = 180$	*Add.*
$96 - 96 + m\angle C = 180 - 96$	*Subtract 96 from each side.*
$m\angle C = 84$	*Simplify.*

Next, find AC. Since $\angle A \cong \angle B$, Theorem 6–4 states that $\overline{BC} \cong \overline{AC}$.

$BC = AC$	*Definition of Congruent Segments*
$6x - 5 = 4x$	*Replace AC with $4x$ and BC with $6x - 5$.*
$6x - 5 - 6x = 4x - 6x$	*Subtract $6x$ from each side.*
$-5 = -2x$	*Simplify.*
$\dfrac{-5}{-2} = \dfrac{-2x}{-2}$	*Divide each side by -2.*
$2.5 = x$	*Simplify.*

By replacing x with 2.5, you find that $AC = 4(2.5)$ or 10 and $BC = 6(2.5) - 5$ or 10.

 Personal Tutor at geomconcepts.com

In Chapter 5, the terms *equiangular* and *equilateral* were defined. Using Theorem 6-4, we can now establish that equiangular triangles are equilateral.

$\triangle ABC$ is equiangular.
Since $m\angle A = m\angle B = m\angle C$,
Theorem 6-4 implies that
$BC = AC = AB$.

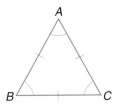

Theorem 6-5	A triangle is equilateral if and only if it is equiangular.

Check for Understanding

Communicating Mathematics

1. **Draw** an isosceles triangle. Label it $\triangle DEF$ with base $\overline{DF}$. Then state four facts about the triangle.

2. **Explain** why equilateral triangles are also equiangular and why equiangular triangles are also equilateral.

Guided Practice

Example 1

3.

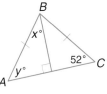

4.

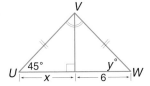

Example 2

5. **Algebra** In $\triangle MNP$, $\angle M \cong \angle P$ and $m\angle M = 37$. Find $m\angle P$, MQ, and PQ.

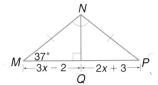

Exercises

Practice

For each triangle, find the values of the variables.

6.

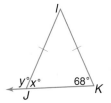

7.

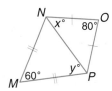

8.

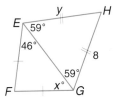

Homework Help	
For Exercises	See Examples
6-14, 17, 18	1
15, 16	2
Extra Practice	
See page 737.	

9.

10.

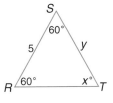

11.

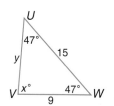

12. In △DEF, $\overline{DE} \cong \overline{FE}$. If $m\angle D = 35$, what is the value of x?

13. Find the value of y if $\overline{EN} \perp \overline{DF}$.

14. In △DMN, $\overline{DM} \cong \overline{MN}$. Find $m\angle DMN$.

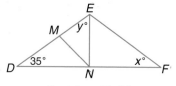

Exercises 12–14

Applications and Problem Solving

15. Algebra In △ABC, $\overline{AB} \cong \overline{AC}$. If $m\angle B = 5x - 7$ and $m\angle C = 4x + 2$, find $m\angle B$ and $m\angle C$.

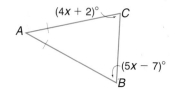

16. Algebra In △RST, $\angle S \cong \angle T$, $m\angle S = 70$, $RT = 3x - 1$, and $RS = 7x - 17$. Find $m\angle T$, RT, and RS.

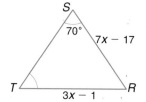

17. Advertising A business logo is shown.
  a. What kind of triangle does the logo contain?
  b. If the measure of angle 1 is 110, what are the measures of the two base angles of that triangle?

18. Critical Thinking Find the measures of the angles of an isosceles triangle such that, when an angle bisector is drawn, two more isosceles triangles are formed.

Mixed Review

19. In △JKM, $\overline{JQ}$ bisects $\angle KJM$. If $m\angle KJM = 132$, what is $m\angle 1$?
(Lesson 6–3)

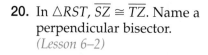

20. In △RST, $\overline{SZ} \cong \overline{TZ}$. Name a perpendicular bisector.
(Lesson 6–2)

21. Graph and label point H at $(-4, 3)$ on a coordinate plane.
(Lesson 2–4)

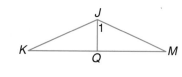

Exercise 20

22. Short Response Marcus used 37 feet of fencing to enclose his triangular garden. What is the length of each side of the garden? *(Lesson 1–6)*

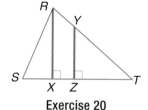

23. Short Response Write a sequence in which each term is 7 less than the previous term. *(Lesson 1–1)*

 www.geomconcepts.com/self_check_quiz

What You'll Learn
You'll learn to use tests for congruence of right triangles.

Why It's Important
Construction
Masons use right triangles when building brick, block, and stone structures.
See Exercise 20.

In a right triangle, the side opposite the right angle is called the **hypotenuse**. The two sides that form the right angle are called the **legs**.

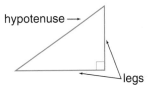

In triangles *ABC* and *DEF*, the two right angles are congruent. Also, the corresponding legs are congruent. So, the triangles are congruent by SAS.

Reading Geometry

The abbreviation LL is read as *Leg-Leg*.

Since right triangles are special cases of triangles, the SAS test for congruence can be used to establish the following theorem.

Theorem 6–6 LL Theorem	**Words:**	If two legs of one right triangle are congruent to the corresponding legs of another right triangle, then the triangles are congruent.
	Model:	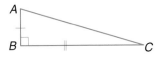 △*ABC* ≅ △*DEF*

Suppose the hypotenuse and an acute angle of the triangle on the left are congruent to the hypotenuse and acute angle of the triangle on the right.

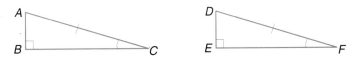

Since the right angles in each triangle are congruent, the triangles are congruent by AAS. The AAS Theorem leads to Theorem 6–7.

Theorem 6–7 HA Theorem

Words: If the hypotenuse and an acute angle of one right triangle are congruent to the hypotenuse and corresponding angle of another right triangle, then the triangles are congruent.

Model:

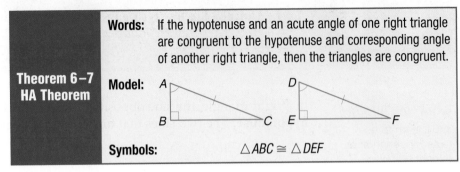

Symbols: △ABC ≅ △DEF

Suppose a leg and an acute angle of one triangle are congruent to the corresponding leg and acute angle of another triangle.

Case 1

The leg is included between the acute angle and the right angle.

Case 2

The leg is not included between the acute angle and the right angle.

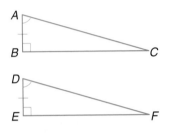
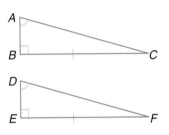

The right angles in each triangle are congruent. In Case 1, the triangles are congruent by ASA. In Case 2, the triangles are congruent by AAS. This leads to Theorem 6–8.

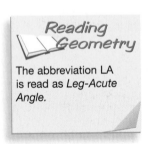
Theorem 6–8 LA Theorem

Words: If one leg and an acute angle of a right triangle are congruent to the corresponding leg and angle of another right triangle, then the triangles are congruent.

Model:

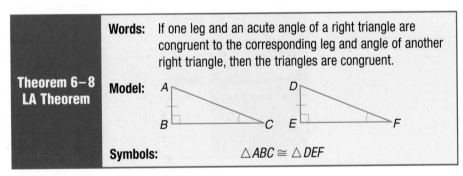

Symbols: △ABC ≅ △DEF

The following postulate describes the congruence of two right triangles when the hypotenuse and a leg of the two triangles are congruent.

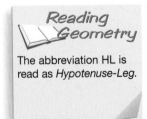
Postulate 6–1 HL Postulate

Words: If the hypotenuse and a leg of one right triangle are congruent to the hypotenuse and corresponding leg of another right triangle, then the triangles are congruent.

Model:

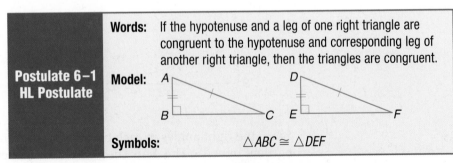

Symbols: △ABC ≅ △DEF

www.geomconcepts.com/extra_examples

Determine whether each pair of right triangles is congruent by LL, HA, LA, or HL. If it is not possible to prove that they are congruent, write *not possible*.

1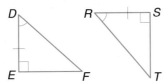

There is one pair of congruent acute angles, $\angle D \cong \angle R$. There is one pair of congruent legs, $\overline{DE} \cong \overline{RS}$.

So, $\triangle DEF \cong \triangle RST$ by LA.

2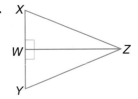

There is one pair of congruent legs, $\overline{YZ} \cong \overline{NP}$. The hypotenuses are congruent, $\overline{XZ} \cong \overline{MP}$.

So, $\triangle XYZ \cong \triangle MNP$ by HL.

Your Turn

a.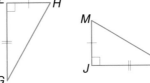

b.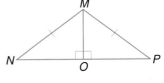

Check for Understanding

Communicating Mathematics

1. **Tell** which test for congruence is used to establish the LL Theorem.

2. **Write** a few sentences explaining the LL, HA, LA, and HL tests for congruence. Give an example of each.

Vocabulary
hypotenuse
legs

Guided Practice

Determine whether each pair of right triangles is congruent by LL, HA, LA, or HL. If it is not possible to prove that they are congruent, write *not possible*.

Examples 1 & 2

3.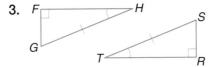

4.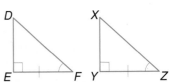

Examples 1 & 2

5. Which test for congruence proves that $\triangle DEF \cong \triangle XYZ$?

6. **Sports** A creel is a wicker basket used for holding fish. On the creel shown, the straps form two right triangles. Explain how $\triangle XYW \cong \triangle XZW$ by the HL Postulate if $\overline{XY} \cong \overline{XZ}$.

Exercises

• • • • • • • • • • • • • • • • • •

Practice

Homework Help

For Exercises	See Examples
7, 16, 21	1
8, 9, 15, 17, 19	2

Extra Practice
See page 737.

Determine whether each pair of right triangles is congruent by LL, HA, LA, or HL. If it is not possible to prove that they are congruent, write *not possible*.

7.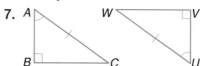

8.

9.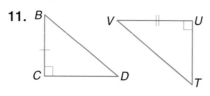

10.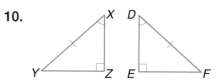

11.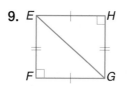

12.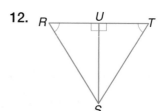

Given $\triangle ABC$ and $\triangle DEF$, name the corresponding parts needed to prove that $\triangle ABC \cong \triangle DEF$ by each theorem.

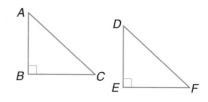

13. LL

14. HA

15. HL

16. LA

Name the corresponding parts needed to prove the triangles congruent. Then complete the congruence statement and name the theorem used.

17.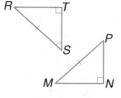

18.

19.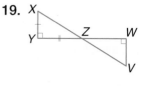

$\triangle BAC \cong \triangle\underline{\ \ ?\ \ }$ $\triangle RTS \cong \triangle\underline{\ \ ?\ \ }$ $\triangle XYZ \cong \triangle\underline{\ \ ?\ \ }$

20. **Construction**   Masons use bricks, concrete blocks, and stones to build various structures with right angles. To check that the corners are right angles, they use a tool called a *builder's square*. Which pair of builder's squares are congruent by the LL Theorem?

a.    b.   c.

21. **Sports**   There are many types of hurdles in track and field. One type, a steeplechase hurdle, is shown. Notice that the frame contains many triangles. In the figure, $\overline{AC}$ bisects $\angle BAD$, and $\overline{AC} \perp \overline{BD}$. What theorem can you use to prove $\triangle ABC \cong \triangle ADC$? Explain.

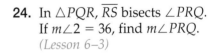

Math **Data**
online **Update**
For the latest information on track and field, visit:
geomconcepts.com

22. **Critical Thinking**   Postulates SAS, ASA, and SSS require three parts of a triangle be congruent to three parts of another triangle for the triangles to be congruent. Explain why postulates LA, LL, HA, and HL require that only two parts of a triangle be congruent to two parts of another triangle in order for the triangles to be congruent.

Mixed Review

23. Find the value of each variable in triangle ABC if $\overline{AD}$ is an angle bisector.   *(Lesson 6–4)*

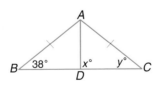

24. In $\triangle PQR$, $\overline{RS}$ bisects $\angle PRQ$. If $m\angle 2 = 36$, find $m\angle PRQ$.   *(Lesson 6–3)*

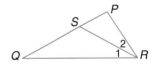

25. If $\triangle ABC \cong \triangle XYZ$, $m\angle A = 40$, and $m\angle C = 65$, find $m\angle Y$.   *(Lesson 5–4)*

**Standardized
Test Practice**
Ⓐ Ⓑ Ⓒ Ⓓ

26. **Short Response**   Find an equation of the line parallel to the graph of $y = -3x + 4$ that passes through $(1, -5)$.   *(Lesson 4–6)*

27. **Multiple Choice**   In the figure, $m \parallel n$, and q is a transversal. If $m\angle 2 = 70$, what is $m\angle 7$?   *(Lesson 4–2)*

Ⓐ 55   Ⓑ 70

Ⓒ 110   Ⓓ 140

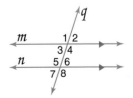

The stamp shown was issued in 1955 by Greece to honor the 2500th anniversary of the Pythagorean School. Notice the triangle bordered on each side by a checkerboard pattern. Count the number of small squares in each of the three larger squares.

The relationship among 9, 16, and 25 forms the basis for the **Pythagorean Theorem**. It can be illustrated geometrically.

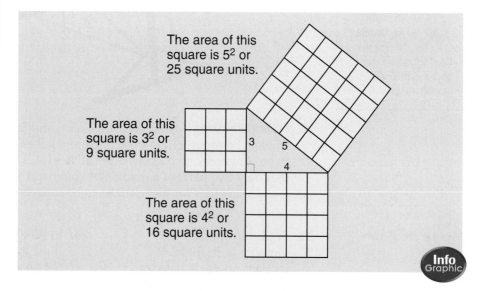

The area of this square is 5^2 or 25 square units.

The area of this square is 3^2 or 9 square units.

The area of this square is 4^2 or 16 square units.

The sides of the right triangle have lengths of 3, 4, and 5 units. The area of the larger square is equal to the total area of the two smaller squares.

$$5^2 = 3^2 + 4^2$$

$$25 = 9 + 16$$

This relationship is true for *any* right triangle.

Theorem 6–9 Pythagorean Theorem	**Words:** In a right triangle, the square of the length of the hypotenuse c is equal to the sum of the squares of the lengths of the legs a and b.
	Model: **Symbols:** $c^2 = a^2 + b^2$

If two measures of the sides of a right triangle are known, the Pythagorean Theorem can be used to find the measure of the third side.

Examples ❶ **Find the length of the hypotenuse of the right triangle.**

$c^2 = a^2 + b^2$ *Pythagorean Theorem*
$c^2 = 15^2 + 8^2$ *Replace a with 15 and b with 8.*
$c^2 = 225 + 64$ *$15^2 = 225, 8^2 = 64$*
$c^2 = 289$ *Simplify.*
$c = \sqrt{289}$ *Take the square root of each side.*
2nd [√] 289 ENTER *17*
$c = 17$

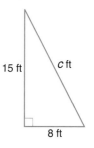

The length of the hypotenuse is 17 feet.

Reading Geometry

Always check to be sure that *c* represents the length of the longest side.

❷ **Find the length of one leg of a right triangle if the length of the hypotenuse is 14 meters and the length of the other leg is 6 meters.**

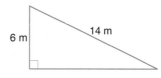

$c^2 = a^2 + b^2$ *Pythagorean Theorem*
$14^2 = 6^2 + b^2$ *Replace c with 14 and a with 6.*
$196 = 36 + b^2$ *$14^2 = 196, 6^2 = 36$*
$196 - 36 = 36 + b^2 - 36$ *Subtract 36 from each side.*
$160 = b^2$ *Simplify.*
$\sqrt{160} = b$ *Take the square root of each side.*
2nd [√] 160 ENTER *12.64911064*

To the nearest tenth, the length of the leg is 12.6 meters.

Your Turn

Find the missing measure in each right triangle.

a.

b.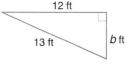

If *c* is the measure of the hypotenuse, find each missing measure. Round to the nearest tenth, if necessary.

c. $a = 7, b = ?, c = 25$

d. $a = ?, b = 10, c = 20$

Example **3** In pitched roof construction, carpenters build the roof with rafters, one piece at a time. The rise, the run, and the rafter form a right triangle. The rise and run are the legs, and the rafter is the hypotenuse. Find the rafter length for the roof shown at the right. Round to the nearest tenth.

Carpentry Link

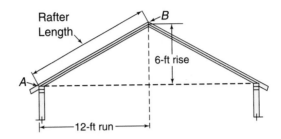

Explore   You know the rise is 6 feet and the run is 12 feet. You need to find the length of the rafter.

Plan   Let $a = 6$ and $b = 12$. Use the Pythagorean Theorem to find c, the hypotenuse.

Solve

$c^2 = a^2 + b^2$ *Pythagorean Theorem*

$c^2 = 6^2 + 12^2$ *Replace a with 6 and b with 12.*

$c^2 = 36 + 144$ *$6^2 = 36, 12^2 = 144$*

$c^2 = 180$ *Simplify.*

$c = \sqrt{180}$ *Take the square root of each side.*

$c \approx 13.4$ *Use a calculator.*

The length of the rafter is about 13.4 feet.

Examine   Since $10^2 = 100$ and $15^2 = 225$, $\sqrt{180}$ is between 10 and 15. Also, the length of the hypotenuse, 13.4 feet, is longer than the length of either leg.

 Personal Tutor at geomconcepts.com

You can use the converse of the Pythagorean Theorem to test whether a triangle is a right triangle.

Theorem 6–10 Converse of the Pythagorean Theorem	If c is the measure of the longest side of a triangle, a and b are the lengths of the other two sides, and $c^2 = a^2 + b^2$, then the triangle is a right triangle.

Example ④ The lengths of the three sides of a triangle are 5, 7, and 9 inches. Determine whether this triangle is a right triangle.

Since the longest side is 9 inches, use 9 as c, the measure of the hypotenuse.

$c^2 = a^2 + b^2$ *Pythagorean Theorem*
$9^2 \stackrel{?}{=} 5^2 + 7^2$ *Replace c with 9, a with 5, and b with 7.*
$81 \stackrel{?}{=} 25 + 49$ $9^2 = 81, 5^2 = 25, 7^2 = 49$
$81 \neq 74$ *Add.*

Since $c^2 \neq a^2 + b^2$, the triangle is *not* a right triangle.

Your Turn

The measures of three sides of a triangle are given. Determine whether each triangle is a right triangle.

e. 20, 21, 28   **f.** 10, 24, 26

Check for Understanding

Communicating Mathematics

1. **State** the Pythagorean Theorem.

2. **Explain** how to find the length of a leg of a right triangle if you know the length of the hypotenuse and the length of the other leg.

3. **Writing Math** Write a few sentences explaining how you know whether a triangle is a right triangle if you know the lengths of the three sides.

Guided Practice

⊕ **Getting Ready**

Find each square root. Round to the nearest tenth, if necessary.

Sample 1: $\sqrt{25}$ **Solution:** [2nd] [√] 25 [ENTER] 5
Sample 2: $\sqrt{32}$ **Solution:** [2nd] [√] 32 [ENTER] 5.656854249 ≈ 5.7

4. $\sqrt{64}$ **5.** $\sqrt{54}$ **6.** $\sqrt{126}$ **7.** $\sqrt{121}$ **8.** $\sqrt{196}$ **9.** $\sqrt{87}$

Example 1

Find the missing measure in each right triangle. Round to the nearest tenth, if necessary.

10.
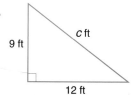
9 ft, c ft, 12 ft

11.
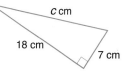
c cm, 18 cm, 7 cm

Example 2

If c is the measure of the hypotenuse, find each missing measure. Round to the nearest tenth, if necessary.

12. $a = 30, c = 34, b = ?$ **13.** $a = 7, b = 4, c = ?$

Example 4

The lengths of three sides of a triangle are given. Determine whether each triangle is a right triangle.

14. 9 mm, 40 mm, 41 mm

15. 9 ft, 16 ft, 20 ft

Example 3

16. Find the length of the diagonal of a rectangle whose length is 8 meters and whose width is 5 meters.

Exercises •

Practice

Homework Help

For Exercises	See Examples
17-28, 36-38	1–3
29-35, 40	4

Extra Practice
See page 737.

Find the missing measure in each right triangle. Round to the nearest tenth, if necessary.

17.
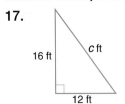

18.
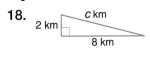

19.
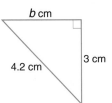

20.
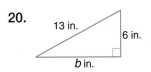

21.
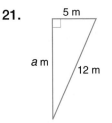

22.
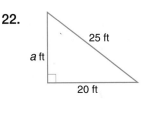

If *c* is the measure of the hypotenuse, find each missing measure. Round to the nearest tenth, if necessary.

23. $a = 6, b = 3, c = ?$

24. $b = 10, c = 11, a = ?$

25. $c = 29, a = 20, b = ?$

26. $a = \sqrt{5}, c = \sqrt{30}, b = ?$

27. $a = \sqrt{7}, b = \sqrt{9}, c = ?$

28. $a = \sqrt{11}, c = \sqrt{47}, b = ?$

The lengths of three sides of a triangle are given. Determine whether each triangle is a right triangle.

29. 11 in., 12 in., 16 in.

30. 11 cm, 60 cm, 61 cm

31. 6 ft, 8 ft, 9 ft

32. 6 mi, 7 mi, 12 mi

33. 45 m, 60 m, 75 m

34. 1 mm, 1 mm, $\sqrt{2}$ mm

35. Is a triangle with measures 30, 40, and 50 a right triangle? Explain.

36. Find the length of the hypotenuse of a right triangle if the lengths of the legs are 6 miles and 11 miles. Round to the nearest tenth if necessary.

37. Find the measure of the perimeter of rectangle *ABCD* if $OB = OC$, $AO = 40$, and $OB = 32$.

Exercise 37

Applications and Problem Solving

38. **Entertainment** Television sets are measured by the diagonal length of the screen. A 25-inch TV set has a diagonal that measures 25 inches. If the height of the screen is 15 inches, how wide is the screen?

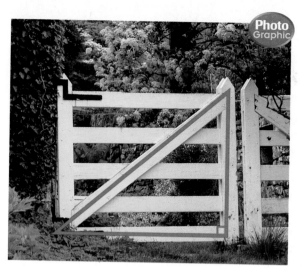

39. **Carpentry** Find the length of a diagonal brace for a rectangular gate that is 5 feet by 4 feet. Round to the nearest tenth.

40. **Critical Thinking** A **Pythagorean triple** is a group of three whole numbers that satisfies the equation $a^2 + b^2 = c^2$, where c is the measure of the hypotenuse. Some common Pythagorean triples are listed below.

   3, 4, 5   9, 12, 15   8, 15, 17   7, 24, 25

   a. List three other Pythagorean triples.

   b. Choose any whole number. Then multiply the whole number by each number of one of the Pythagorean triples you listed. Show that the result is also a Pythagorean triple.

Mixed Review

41. Which right angle test for congruence can be used to prove that $\triangle RST \cong \triangle XYZ$? *(Lesson 6–5)*

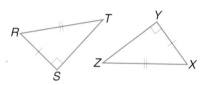

42. **Algebra** In $\triangle DEF$, $\angle D \cong \angle E$ and $m\angle E = 17$. Find $m\angle F$, DF, and FE. *(Lesson 6–4)*

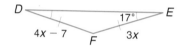

43. Draw an acute scalene triangle. *(Lesson 5–1)*

44. In the figure shown, lines m and n are cut by transversal q. Name two pairs of corresponding angles. *(Lesson 4–3)*

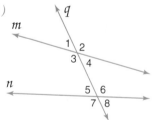

Draw an angle having the given measure. *(Lesson 3–2)*

45. 126

46. 75

47. **Multiple Choice** Which shows the graph of $N(2, -3)$? *(Lesson 2–4)*

Ⓐ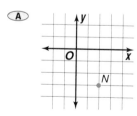

Ⓑ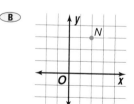

Ⓒ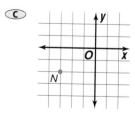

Ⓓ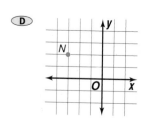

What You'll Learn

You'll learn to find the distance between two points on the coordinate plane.

Why It's Important

Transportation
Knowing how to find the distance between two points can help you determine distance traveled.
See Example 3.

In Lesson 2–1, you learned how to find the distance between two points on a number line. In this lesson, you will learn how to find the distance between two points on the coordinate plane.

When two points lie on a horizontal line or a vertical line, the distance between the two points can be found by subtracting one of the coordinates. In the coordinate plane below, points *A* and *B* lie on a horizontal line, and points *C* and *D* lie on a vertical line.

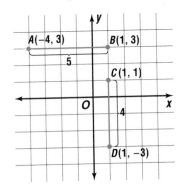

The distance between *A* and *B* is $|-4 - 1|$ or 5.

The distance between *C* and *D* is $|-3 - 1|$ or 4.

In the following activity, you will learn how to find the distance between two points that do not lie on a horizontal or vertical line.

Hands-On Geometry

Look Back

Graphing Ordered Pairs:
Lesson 2–4

Materials:  grid paper ✏ straightedge

Step 1 Graph $A(-3, 1)$ and $C(2, 3)$.

Step 2 Draw a horizontal segment from *A* and a vertical segment from *C*. Label the intersection *B* and find the coordinates of *B*.

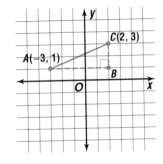

Try These

1. What is the measure of the distance between *A* and *B*?
2. What is the measure of the distance between *B* and *C*?
3. What kind of triangle is $\triangle ABC$?
4. If *AB* and *BC* are known, what theorem can be used to find *AC*?
5. What is the measure of $\overline{AC}$?

In the activity, you found that $(AC)^2 = (AB)^2 + (BC)^2$. By taking the square root of each side of the equation, you find that $AC = \sqrt{(AB)^2 + (BC)^2}$.

AC = measure of the distance between points A and C
AB = difference of the x-coordinates of A and C
BC = difference of the y-coordinates of A and C

This formula can be generalized to find the distance between any two points.

Theorem 6–11 **Distance** **Formula**	**Words:**	If d is the measure of the distance between two points with coordinates (x_1, y_1) and (x_2, y_2), then $d = \sqrt{(x_2 - x_1)^2 + (y_2 - y_1)^2}$.
	Model:	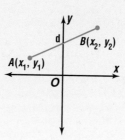

Example **1**   Use the Distance Formula to find the distance between $J(-8, 6)$ and $K(1, -3)$. Round to the nearest tenth, if necessary.

Use the Distance Formula. Replace (x_1, y_1) with $(-8, 6)$ and (x_2, y_2) with $(1, -3)$.

$d = \sqrt{(x_2 - x_1)^2 + (y_2 - y_1)^2}$	*Distance Formula*
$JK = \sqrt{[1 - (-8)]^2 + (-3 - 6)^2}$	*Substitution*
$JK = \sqrt{(9)^2 + (-9)^2}$	*Subtract.*
$JK = \sqrt{81 + 81}$	$9^2 = 81, (-9)^2 = 81$
$JK = \sqrt{162}$	*Add.*
$JK \approx 12.7$	*Simplify.*

Your Turn

Find the distance between each pair of points. Round to the nearest tenth, if necessary.

a. $M(0, 3), N(0, 6)$ **b.** $G(-3, 4), H(5, 1)$

You can use the Distance Formula to determine whether a triangle is isosceles given the coordinates of its vertices.

2 **Determine whether $\triangle ABC$ with vertices $A(-3, 2)$, $B(6, 5)$, and $C(3, -1)$ is isosceles.**

An isosceles triangle has at least two congruent sides. Use the Distance Formula to find the measures of the sides of $\triangle ABC$. Then determine if any two are equal.

Hint: Draw a picture on a coordinate plane.

$$AB = \sqrt{[6 - (-3)]^2 + (5 - 2)^2}$$
$$= \sqrt{9^2 + 3^2}$$
$$= \sqrt{81 + 9}$$
$$= \sqrt{90}$$

$$BC = \sqrt{(3 - 6)^2 + (-1 - 5)^2}$$
$$= \sqrt{(-3)^2 + (-6)^2}$$
$$= \sqrt{9 + 36}$$
$$= \sqrt{45}$$

$$AC = \sqrt{[3 - (-3)]^2 + (-1 - 2)^2}$$
$$= \sqrt{6^2 + (-3)^2}$$
$$= \sqrt{36 + 9}$$
$$= \sqrt{45}$$

$\overline{BC}$ and $\overline{AC}$ have equal measures. Therefore, $\triangle ABC$ is isosceles.

Transportation Link

3 **Lena takes the bus from Mill's Market to the Candle Shop. Mill's Market is 3 miles west and 2 miles north of Blendon Park. The Candle Shop is 2 miles east and 4 miles south of Blendon Park. How far is the Candle Shop from Mill's Market?**

Let Mill's Market be represented by (x_1, y_1) and the Candle Shop by (x_2, y_2). Then $x_1 = -3$, $y_1 = 2$, $x_2 = 2$, and $y_2 = -4$.

$$d = \sqrt{(x_2 - x_1)^2 + (y_2 - y_1)^2}$$
$$d = \sqrt{[2 - (-3)]^2 + (-4 - 2)^2}$$
$$d = \sqrt{(5)^2 + (-6)^2}$$
$$d = \sqrt{25 + 36}$$
$$d = \sqrt{61}$$
$$d \approx 7.8$$

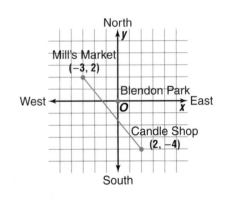

The Candle Shop is about 7.8 miles from Mill's Market.

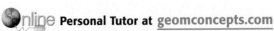

Online Personal Tutor at geomconcepts.com

Communicating Mathematics

1. **State** the Distance Formula for points represented by (x_1, y_1) and (x_2, y_2).

2. **Name** the theorem that is used to determine the Distance Formula in the coordinate plane.

3. **You Decide** Ana says that to find the distance from $A(-3, 2)$ to $B(-7, 5)$, you must evaluate the expression $\sqrt{[-7 - (-3)]^2 + (5 - 2)^2}$. Emily disagrees. She says that you must evaluate the expression $\sqrt{[-3 - (-7)]^2 + (2 - 5)^2}$. Who is correct? Explain your answer.

Guided Practice

⊕ Getting Ready Find the value of each expression.

Sample: $(-7 + 4)^2 + [3 - (-6)]^2$

Solution: $(-7 + 4)^2 + [3 - (-6)]^2 = (-3)^2 + [3 + 6]^2$
$$= (-3)^2 + 9^2$$
$$= 9 + 81 \text{ or } 90$$

4. $(6 + 2)^2 + (-5 + 3)^2$

5. $[-2 + (-3)]^2 + (2 + 3)^2$

6. $[-5 - (-6)]^2 + (4 - 2)^2$

Example 1

Find the distance between each pair of points. Round to the nearest tenth, if necessary.

7. $E(1, 2), F(3, 4)$

8. $R(-6, 0), S(-2, 0)$

9. $P(5, 6), Q(-3, 1)$

Example 2

10. Determine whether $\triangle FGH$ with vertices $F(-2, 1)$, $G(1, 6)$, and $H(4, 1)$ is isosceles.

11. **Hiking** Tamika and Matthew are going to hike from Cedar Creek Cave to the Ford Nature Center. Cedar Creek Cave is located 3 kilometers west of the ranger's station. The Ford Nature Center is located 2 kilometers east and 4 kilometers north of the ranger's station. **Example 3**

a. Draw a diagram on a coordinate grid to represent this situation.

b. What is the distance between Cedar Creek Cave and Ford Nature Center?

Exercises

· · · · · · · · · · · · · · · · · · · ·

Practice

Homework Help

For Exercises	See Examples
12–23	1
24–25	2
27–29	3

Extra Practice
See page 738.

Find the distance between each pair of points. Round to the nearest tenth, if necessary.

12. $A(5, 0)$, $B(12, 0)$

13. $M(2, 3)$, $N(5, 7)$

14. $D(-1, -2)$, $E(-3, -4)$

15. $X(-4, 0)$, $Y(3, -3)$

16. $P(-6, -4)$, $Q(6, -8)$

17. $T(6, 4)$, $U(2, 2)$

18. $B(0, 0)$, $C(-5, 6)$

19. $G(-6, 8)$, $H(-6, -4)$

20. $J(-3, -2)$, $K(3, 1)$

21. $S(-6, -4)$, $T(-3, -7)$

22. Find the distance between $A(-1, 5)$ and $C(3, 5)$.

23. What is the distance between $E(-3, -1)$ and $F(4, -2)$?

24. Is $\triangle MNP$ with vertices $M(1, 4)$, $N(-3, -2)$, and $P(4, -3)$ an isosceles triangle? Explain.

25. Determine whether $\triangle RST$ with vertices $R(1, 5)$, $S(-1, 1)$, and $T(5, 4)$ is scalene. Explain.

26. Triangle FGH has vertices $F(2, 4)$, $G(0, 2)$, and $H(3, -1)$. Determine whether $\triangle FGH$ is a right triangle. Explain.

Applications and Problem Solving

27. Gardening At Memorial Flower Garden, the rose garden is located 25 yards west and 30 yards north of the gazebo. The herb garden is located 35 yards west and 15 yards south of the gazebo.

   a. Draw a diagram on a coordinate grid to represent this situation.

   b. How far is the herb garden from the rose garden?

   c. What is the distance from the rose garden to the gazebo?

28. Communication To set long-distance rates, telephone companies superimpose an imaginary coordinate plane over the United States. Each ordered pair on this coordinate plane represents the location of a telephone exchange. The phone company calculates the distances between the exchanges in miles to establish long-distance rates. Suppose two exchanges are located at (53, 187) and (129, 71). What is the distance between these exchanges to the nearest mile? The location units are in miles.

29. Critical Thinking In $\triangle ABC$, the coordinates of the vertices are $A(2, 4)$, $B(-3, 6)$, and $C(-5, -2)$. To the nearest tenth, what is the measure of the median drawn from A to $\overline{BC}$? Include a drawing on a coordinate plane of the triangle and the median.

Mixed Review

30. Music The frame of the music stand shown contains several triangles. Find the length of the hypotenuse of right triangle MSC if the length of one leg is 10 inches and the length of the other leg is 8.5 inches. Round to the nearest tenth, if necessary. *(Lesson 6–6)*

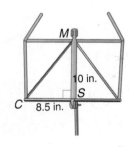

31. Which right triangle test for congruence can be used to prove that $\triangle BCD \cong \triangle FGH$? (*Lesson 6–5*)

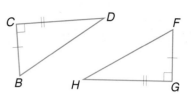

32. Identify the motion shown as a translation, reflection, or rotation. (*Lesson 5–3*)

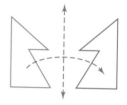

33. Short Response Classify the angle shown as *acute*, *obtuse*, or *right*. (*Lesson 3–2*)

34. Multiple Choice The line graph shows the retail coffee sales in the United States from 1999 to 2002. Estimate how much more coffee was sold in 2002 than in 2001. (*Statistics Review*)

Ⓐ $1,000,000

Ⓑ $10,000,000

Ⓒ $100,000,000

Ⓓ $1,000,000,000

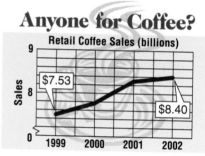

Anyone for Coffee?

Retail Coffee Sales (billions)

$7.53

$8.40

Source: Specialty Coffee Association of America

Quiz 2   Lessons 6–4 through 6–7

1. Find the value of x in $\triangle ABC$ if $AD \perp BC$. (*Lesson 6–4*)

2. Determine whether the pair of right triangles is congruent by LL, HA, LA, or HL. If it is not possible to prove that they are congruent, write *not possible*. (*Lesson 6–5*)

3. Find c in triangle MNP. Round to the nearest tenth, if necessary. (*Lesson 6–6*)

4. Landscape Tyler is planning to build a triangular garden. The lengths of the sides of the garden are 12 feet, 9 feet, and 15 feet. Will the edges of the garden form a right triangle? (*Lesson 6–6*)

5. Is $\triangle JKL$ with vertices at $J(2, 4)$, $K(-1, -1)$, and $L(5, -1)$ an isosceles triangle? Explain. (*Lesson 6–7*)

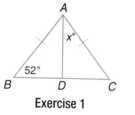

Exercise 1

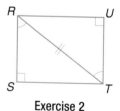

Exercise 2

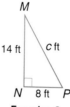

Exercise 3

Understanding and Using the Vocabulary

After completing this chapter, you should be able to define each term, property, or phrase and give an example or two of each.

*inter*NET
CONNECTION **Review Activities**
For more review activities, visit:
www.geomconcepts.com

altitude *(p. 234)*
angle bisector *(p. 240)*
base *(p. 246)*
centroid *(pp. 230, 244)*
circumcenter *(p. 244)*
concurrent *(p. 230)*

Euler line *(p. 245)*
hypotenuse *(p. 251)*
incenter *(p. 245)*
leg *(pp. 246, 251)*
median *(p. 228)*

nine-point circle *(p. 245)*
orthocenter *(p. 245)*
perpendicular bisector *(p. 235)*
Pythagorean Theorem *(p. 256)*
Pythagorean triple *(p. 261)*

State whether each sentence is *true* or *false*. If false, replace the underlined word(s) to make a true statement.

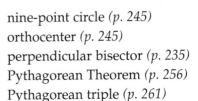

Figure 1   Figure 2

1. In Figure 1, $\overline{AB}$, $\overline{AC}$, and $\overline{AD}$ are <u>concurrent</u>.
2. The point where all of the <u>altitudes</u> of a triangle intersect is called the centroid.
3. In Figure 1, $\overline{AD}$ is a(n) <u>altitude</u> of $\triangle ABC$.
4. In Figure 2, $\overline{JM}$ is a(n) <u>median</u> of $\triangle JKL$.
5. In Figure 2, $(JK)^2 + (JL)^2 = (KL)^2$ by the Pythagorean Theorem.
6. In Figure 2, $\overline{JK}$ is a(n) <u>hypotenuse</u> of $\triangle JKL$.
7. In a(n) <u>acute</u> triangle, one of the altitudes lies outside the triangle.
8. In Figure 3, $\overleftrightarrow{EG}$ is a(n) angle bisector of $\overline{HF}$ in $\triangle FHI$.
9. In Figure 4, $\overline{XV}$ is a(n) <u>perpendicular bisector</u> of $\overline{YZ}$.
10. The side opposite the right angle of a right triangle is called the <u>leg</u>.

Figure 3   Figure 4

Skills and Concepts

Objectives and Examples	Review Exercises

• **Lesson 6–1**   Identify and construct medians in triangles.

In $\triangle PRV$, $\overline{PS}$, $\overline{VQ}$, and $\overline{RT}$ are medians. Find PW if $WS = 7.5$.

Since $WS = 7.5$, $PW = 2(7.5)$ or 15.

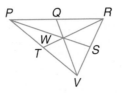

Refer to $\triangle PRV$ at the left for Exercises 11–14.

11. Find TW if $WR = 12$.
12. If $PQ = 14.5$, find QR.
13. What is the measure of $\overline{QW}$ if $WV = 11$?
14. If $PV = 20$, find TV.
15. In $\triangle CRT$, $\overline{TQ}$ and $\overline{MR}$ are medians. If $MC = 5x$, $TM = x + 16$, and $CQ = 8x + 6$, find QR.

 www.geomconcepts.com/vocabulary_review

Objectives and Examples

• **Lesson 6–2** Identify and construct altitudes and perpendicular bisectors in triangles.

In $\triangle DCE$, $\overline{CG}$ is an altitude, and $\overleftrightarrow{HF}$ is the perpendicular bisector of side DE.

$\overline{SQ}$ is both an altitude and a perpendicular bisector of $\triangle RST$.

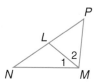

• **Lesson 6–3** Identify and use angle bisectors in triangles.

In $\triangle PMN$, $\overline{ML}$ is an angle bisector of $\angle PMN$. If $m\angle 1 = 55$, find $m\angle PMN$.

$m\angle PMN = 2(m\angle 1)$
$\qquad\quad = 2(55)$ or 110

• **Lesson 6–4** Identify and use properties of isosceles triangles.

If $\triangle BAJ$ is isosceles and $\overline{AK}$ bisects $\angle BAJ$, then the following statements are true.

$\angle B \cong \angle J \qquad \overline{AK} \perp \overline{BJ} \qquad \overline{AK}$ bisects $\overline{BJ}$.

• **Lesson 6–5** Use tests for congruence of right triangles.

Determine if $\triangle ABC \cong \triangle EFG$.
$\angle A \cong \angle E$ and $\overline{BC} \cong \overline{FG}$

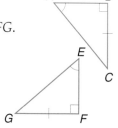

By the LA Theorem,
$\triangle ABC \cong \triangle EFG$.

Review Exercises

For each triangle, tell whether the red segment or line is an *altitude*, a *perpendicular bisector*, *both*, or *neither*.

16. 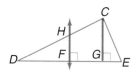 **17.**

18. 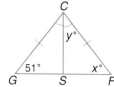 **19.**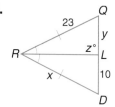

In $\triangle WXV$, $\overline{XY}$ bisects $\angle WXV$, $\overline{UV}$ bisects $\angle XVW$, and $\overline{WZ}$ bisects $\angle XWV$.

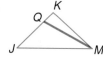

20. If $m\angle 2 = 38$, what is $m\angle 1$?

21. Find $m\angle 3$ if $m\angle WVX = 62$.

22. If $m\angle WXV = 70$ and $m\angle 2 = 3x - 4$, find the value of x.

Find the values of the variables.

23. 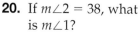 **24.**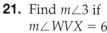

Determine whether each pair of right triangles is congruent by LL, HA, LA, or HL. If it is not possible to prove that they are congruent, write *not possible*.

25. 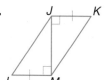 **26.**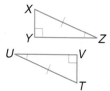

Mixed Problem Solving
See pages 758–765.

Objectives and Examples

Review Exercises

• **Lesson 6–6** Use the Pythagorean Theorem and its converse.

Find the value of b in $\triangle XYZ$.

$$a^2 + b^2 = c^2$$
$$16^2 + b^2 = 34^2$$
$$256 + b^2 = 1156$$
$$256 + b^2 - 256 = 1156 - 256$$
$$b^2 = 900$$
$$b = \sqrt{900} \text{ or } 30$$

If c is the measure of the hypotenuse, find each missing measure. Round to the nearest tenth, if necessary.

27. $a = 16, b = 12, c = ?$
28. $b = 5, c = 14, a = ?$

The lengths of three sides of a triangle are given. Determine whether each triangle is a right triangle.

29. 40 cm, 42 cm, 58 cm
30. 13 ft, 36 ft, 38 ft

• **Lesson 6–7** Find the distance between two points on the coordinate plane.

Use the Distance Formula to find the distance between $A(-3, 7)$ and $B(2, -5)$.

$$AB = \sqrt{(x_2 - x_1)^2 + (y_2 - y_1)^2}$$
$$AB = \sqrt{[2 - (-3)]^2 + (-5 - 7)^2}$$
$$AB = \sqrt{(5)^2 + (-12)^2}$$
$$AB = \sqrt{25 + 144}$$
$$AB = \sqrt{169} \text{ or } 13$$

Find the distance between each pair of points. Round to the nearest tenth, if necessary.

31. $J(-8, 2), K(0, -4)$
32. $A(3, -7), B(1, -9)$
33. $Y(-5, -2), X(6, -2)$

34. Determine whether $\triangle LMN$ with vertices $L(-1, 4), M(5, 1),$ and $N(2, -2)$ is isosceles. Explain.

Applications and Problem Solving

35. Music A *metronome* is a device used to mark exact time using a regularly repeated tick. The body of a metronome resembles an isosceles triangle. In the picture shown at right, is the shaded segment an *altitude*, *perpendicular bisector*, *both*, or *neither* of $\triangle MNP$?
(Lesson 6–4)

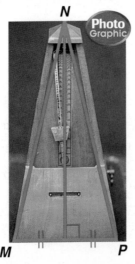

36. Sports Kimiko is parasailing 350 feet away from the boat that pulls her. Suppose she is lifted 400 feet into the air. Find the length of the rope used to keep her attached to the boat. Round to the nearest foot. *(Lesson 6–6)*

400 ft

350 ft

In △ABC, $\overline{AN}$, $\overline{BP}$, and $\overline{CM}$ are medians.

1. If $PE = 4$, find EB.
2. Find NB if $CB = 12$.
3. If $AE = 5$, what is EN?
4. If $AM = 2x + 3$, $MB = x + 5$, and $CP = 7x - 6$, find AC.

Exercises 1–4

For each triangle, tell whether the red segment or line is an *altitude*, a *perpendicular bisector*, *both*, or *neither*.

5.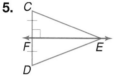

6.

7.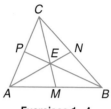

In the figure, $\overline{BA}$ bisects $\angle CBD$, $\overline{CG}$ bisects $\angle BCD$, and $\overline{DF}$ bisects $\angle CDB$.

8. Find $m\angle CBD$ if $m\angle ABC = 18$.
9. What is $m\angle BCG$ if $m\angle BCD = 54$?
10. If $m\angle BDF = 3x$ and $m\angle FDC = x + 20$, find x.

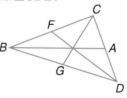

For each triangle, find the value of the variables.

11.

12.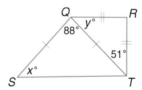

13.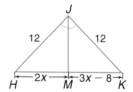

Determine whether each pair of right triangles is congruent by LL, HA, LA, or HL.

14.

15.

16.

17. Find the length of one leg of a right triangle to the nearest tenth if the length of the hypotenuse is 25 meters and the length of the other leg is 18 meters.

18. The measures of three sides of a triangle are 12, 35, and 37. Determine whether this triangle is a right triangle.

19. What is the distance between $R(-7, 13)$ and $S(1, -2)$?

20. **Games** The scoring area for the game of shuffleboard is an isosceles triangle. Suppose the measure of the vertex angle is 40. What are the measures of the two base angles?

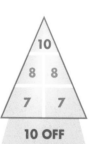

Exercise 20

Algebra Problems

Standardized test problems often ask you to simplify expressions, evaluate expressions, and solve equations.

You may want to review the rules for exponents. For any numbers a and b, and all integers m and n, the following are true.

$$(a^m)^n = a^{mn}$$

$$a^m a^n = a^{m+n}$$

$$(ab)^m = a^m b^m$$

$$\frac{a^m}{a^n} = a^{m-n}$$

Example 1

Evaluate $x^2 - 3x + 4$ if $x = -2$.

(A) 2   (B) 6   (C) 12   (D) 14

Hint Work carefully when combining negative integers.

Solution Replace x with -2. Then perform the operations. Remember the rules for operations using negative numbers.

$$\begin{aligned}
x^2 - 3x + 4 &= (-2)^2 - 3(-2) + 4 \\
&= 4 - (-6) + 4 \\
&= 4 + 6 + 4 \\
&= 14
\end{aligned}$$

The answer is D.

Example 2

For which of the following values of x is $\dfrac{x^2}{x^3}$ the LEAST?

(A) 1   (B) −1   (C) −2

(D) −3   (E) −4

Hint Make problems as simple as possible by simplifying expressions.

Solution Notice that the expression with exponents can be simplified. Start by simplifying it.

$$\frac{x^2}{x^3} = \frac{\overset{1}{\cancel{x^2}}}{\underset{x^1}{\cancel{x^3}}} = \frac{1}{x}$$

Now it is easy to evaluate the expression. The problem asks which value makes the expression the smallest. If you substitute each value for x in the expression, which gives you the least or smallest number?

It's easy to check each value. A is 1, B is −1, C is $-\frac{1}{2}$, D is $-\frac{1}{3}$, and E is $-\frac{1}{4}$. Which of these numbers is the least? Think of a number line.

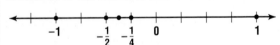

Since −1 is the least, the answer is B.

After you work each problem, record your answer on the answer sheet provided or on a sheet of paper.

Multiple Choice

1. Evaluate $x - 3(2) - 4$ if $x = 24$.
(Algebra Review)
- Ⓐ 2
- Ⓑ 6
- Ⓒ 12
- Ⓓ 14

2. Property tax is 2% of the assessed value of a house. How much would the property tax be on a house with an assessed value of $80,000? *(Percent Review)*
- Ⓐ $100
- Ⓑ $160
- Ⓒ $1000
- Ⓓ $1600
- Ⓔ $10,000

3. Carl has been practicing basketball free throws. Which statement is best supported by the data shown in the graph?
(Statistics Review)

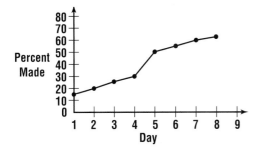

- Ⓐ By Day 10, Carl should be shooting 80%.
- Ⓑ Carl made a total of 60 shots on Day 8.
- Ⓒ Carl's performance improved most between Days 7 and 8.
- Ⓓ Carl's performance improved most between Days 4 and 5.

4. What is the measure of each base angle of an isosceles triangle in which the vertex angle measures 80°? *(Lesson 5–2)*

- Ⓐ 30°
- Ⓑ 50°
- Ⓒ 80°
- Ⓓ 100°

5. The average of x numbers is 16. If the sum of the x numbers is 64, what is the value of x? *(Statistics Review)*
- Ⓐ 3
- Ⓑ 4
- Ⓒ 8
- Ⓓ 16
- Ⓔ 48

6. The lengths of the sides of a triangle are consecutive even integers. The perimeter of the triangle is 48 inches. What are the lengths of the sides? *(Lesson 1–6)*
- Ⓐ 12, 14, 16
- Ⓑ 14, 16, 18
- Ⓒ 15, 16, 17
- Ⓓ 16, 18, 20

7. A box of 36 pens contains 12 blue pens, 14 red pens, and 10 black pens. Three students each successively draw a pen at random from the box and then replace it. If the first two students each draw and then replace a red pen, what is the probability that the third students does *not* draw a red pen?
(Statistics Review)
- Ⓐ $\frac{1}{3}$
- Ⓑ $\frac{7}{18}$
- Ⓒ $\frac{11}{18}$
- Ⓓ $\frac{11}{17}$

8. What is the solution of the inequality $-2 \leq 4 + x$? *(Algebra Review)*
- Ⓐ $x \geq -6$
- Ⓑ $x \geq 6$
- Ⓒ $x \leq -2$
- Ⓓ $x \geq 2$

Grid In

9. What is the mean of the ten numbers below?
(Statistics Review)

$-820, -65, -32, 0, 1, 2, 3, 32, 65, 820$

Short Response

10. Evaluate $x^2 - 1$ for the first eight prime numbers. If you delete the value of $x^2 - 1$ for $x = 2$, what pattern do you see in the other results? (*Hint:* Look at the greatest common factors.) Show your work. Describe the pattern you observed.
(Algebra Review)

CHAPTER 7

Triangle Inequalities

What You'll Learn

Key Ideas

- Apply inequalities to segment and angle measures. *(Lesson 7–1)*

- Identify exterior angles and remote interior angles of a triangle and use the Exterior Angle Theorem. *(Lesson 7–2)*

- Identify the relationships between the sides and angles of a triangle. *(Lesson 7–3)*

- Identify and use the Triangle Inequality Theorem. *(Lesson 7–4)*

Key Vocabulary

exterior angle *(p. 282)*

inequality *(p. 276)*

remote interior angles *(p. 282)*

Why It's Important

Entertainment Early moviemakers created illusions by stopping and restarting the camera or splicing film. Later, blue-screen and split-screen compositing allowed two or more separately shot scenes to be merged. Now computers make it possible to add or alter elements for perfect special effects.

Triangle inequalities can be used to analyze how objects are arranged. You will investigate how triangles are used to create a movie special effect in Lesson 7–2.

Study **these lessons to improve your skills.**

✔ Check Your Readiness

✔ **Lesson 2–1,**
pp. 50–53

Use the number line to find each measure.

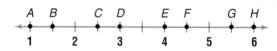

1. AD 2. DE 3. DH

4. AC 5. BF 6. GH

✔ **Algebra**
Review, p. 725

Solve each inequality. Graph the solution on a number line.

7. $n + 7 < 12$ 8. $14 + p < 32$ 9. $55 + r < 90$

10. $18 > d - 4$ 11. $w - 15 > 15$ 12. $22 - t > 15$

✔ **Lesson 3–5,**
pp. 116–121

Refer to the figure.

13. Name two pairs of adjacent angles.

14. Name a pair of supplementary angles.

15. Find the measure of $\angle CFB$.

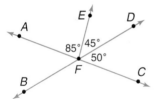

Study Organizer

Make this Foldable to help you organize your Chapter 7 notes. Begin with a sheet of notebook paper.

❶ **Fold** lengthwise to the holes.

❷ **Cut** along the top line and then cut four tabs.

❸ **Label** each tab with an inequality symbol.

Reading and Writing Store your Foldable in a 3-ring binder. As you read and study the chapter, describe each inequality symbol and give examples of its use under each tab.

Segments, Angles, and Inequalities

Why It's Important
Construction
Relationships between segment measures and angle measures are important in construction.
See Examples 3 & 4.

The Comparison Property of Numbers is used to compare two line segments of unequal measures. The property states that given two unequal numbers a and b, either $a < b$ or $a > b$. The same property is also used to compare angles of unequal measures. (Recall that measures of angles are real numbers.)

T•———— 2 cm ————•U

V•———————— 4 cm ————————•W

The length of $\overline{TU}$ is less than the length of $\overline{VW}$, or $TU < VW$.

133° J 60° K

The measure of $\angle J$ is greater than the measure of $\angle K$, or $m\angle J > m\angle K$.

The statements $TU < VW$ and $m\angle J > m\angle K$ are called **inequalities** because they contain the symbol $<$ or $>$. We can write inequalities to compare measures since measures are real numbers.

Postulate 7–1 Comparison Property	**Words:**	For any two real numbers, a and b, exactly one of the following statements is true.
	Symbols:	$a < b$ $\qquad a = b \qquad a > b$

Example **1**

Look Back

Finding Distance on a Number Line: Lesson 2–1

Replace ● with $<$, $>$, or $=$ to make a true sentence.
SL ● RL

SL ● RL
$2 - (-5)$ ● $2 - (-3)$ *Subtract to find distance.*
$7 > 5$ *Simplify.*

Your Turn

a. ND ● RD

b. SR ● DN

The results from Example 1 illustrate the following theorem.

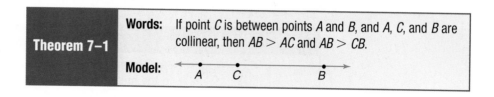

A similar theorem for comparing angle measures is stated below. This theorem is based on the Angle Addition Postulate.

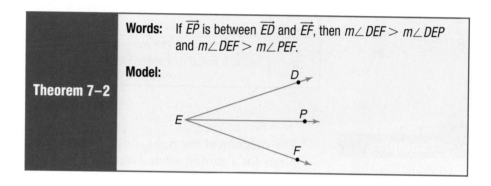

We can use Theorem 7–2 to solve the following problem.

Example ②
Music Link

The graph shows the portion of music sales for each continent. Replace ● with <, >, or = to make a true sentence.

$$m\angle SCI \bullet m\angle UCI$$

Since $\overline{CS}$ is between $\overline{CU}$ and $\overline{CI}$, then by Theorem 7–2, $m\angle SCI < m\angle UCI$.

Making Music

[circle graph] Europe 119° Asia 79° U M Other 40° S North America 122° C I

Source: International Federation of the Phonographic Industry

Check:

$m\angle SCI \overset{?}{\lessgtr} m\angle UCI$

$40 \overset{?}{\lessgtr} 79 + 40$ *Replace $m\angle SCI$ with 40 and $m\angle UCI$ with 79 + 40.*

$40 < 119$ ✓

Math Data
Online Update
For the latest information on world music sales, visit:
geomconcepts.com

Your Turn

c. $m\angle MCS \bullet m\angle ICM$ **d.** $m\angle UCM \bullet m\angle ICM$

Inequalities comparing segment measures or angle measures may also include the symbols listed in the table below.

Symbol	Statement	Words	Meaning
≠	$MN \neq QR$	The measure of $\overline{MN}$ is not equal to the measure of $\overline{QR}$.	$MN < QR$ or $MN > QR$
≤	$m\angle E \leq m\angle J$	The measure of angle E is less than or equal to the measure of angle J.	$m\angle E < m\angle J$ or $m\angle E = m\angle J$
≥	$PF \geq KD$	The measure of $\overline{PF}$ is greater than or equal to the measure of $\overline{KD}$.	$PF > KD$ or $PF = KD$
≰	$ZY \nleq LN$	The measure of $\overline{ZY}$ is not less than or equal to the measure of $\overline{LN}$.	$ZY > LN$
≱	$m\angle A \ngeq m\angle B$	The measure of angle A is not greater than or equal to the measure of angle B.	$m\angle A < m\angle B$

Examples

Construction Link

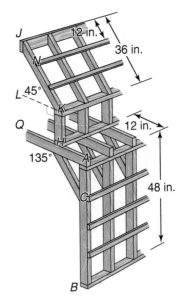

The diagram at the right shows the plans for a garden arbor. Use the diagram to determine whether each statement is *true* or *false*.

3 $AB \leq JK$

$48 \leq 36$ *Replace AB with 48 and JK with 36.*

This is false because 48 is not less than or equal to 36.

4 $m\angle LKN \ngeq m\angle LKH$

$45 \ngeq 90$ *Replace $m\angle LKN$ with 45 and $m\angle LKH$ with 90.*

This is true because 45 is not greater than or equal to 90.

Your Turn

e. $NK \neq HA$

f. $m\angle QHC \nleq m\angle JKH$

There are many useful properties of inequalities of real numbers that can be applied to segment and angle measures. Two of these properties are illustrated in the following example.

Example ⑤

Gemology Link

Diamonds are cut at angles that will create maximum sparkle. In the diamond at the right, $m\angle Q < m\angle N$. If each of these measures were multiplied by 1.2 to give a different type of cut, would the inequality still be true?

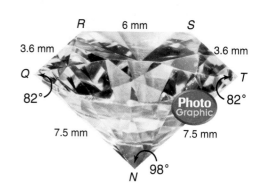

$$m\angle Q < m\angle N$$
$$82 < 98 \qquad \textit{Replace } m\angle Q \textit{ with 82 and } m\angle N \textit{ with 98.}$$
$$82 \cdot 1.2 \overset{?}{<} 98 \cdot 1.2 \qquad \textit{Multiply each side by 1.2.}$$
$$98.4 < 117.6 \quad \checkmark$$

Therefore, the original inequality still holds true.

— **Algebra Review** —
Solving Inequalities, p. 725

Your Turn

g. Suppose each side of the diamond was decreased by 0.9 millimeter. Write an inequality comparing the lengths of $\overline{TN}$ and $\overline{RS}$.

Example 5 demonstrates how the multiplication and subtraction properties of inequalities for real numbers can be applied to geometric measures. These properties, as well as others, are listed in the following table.

Property	Words	Example
Transitive Property	For any numbers a, b, and c, **1.** if $a < b$ and $b < c$, then $a < c$. **2.** if $a > b$ and $b > c$, then $a > c$.	If $6 < 7$ and $7 < 10$, then $6 < 10$. If $9 > 5$ and $5 > 4$, then $9 > 4$.
Addition and Subtraction Properties	For any numbers a, b, and c, **1.** if $a < b$, then $a + c < b + c$ and $a - c < b - c$. **2.** if $a > b$, then $a + c > b + c$ and $a - c > b - c$.	$1 < 3$ $1 < 3$ $1 + 8 < 3 + 8$ $1 - 8 < 3 - 8$ $9 < 11$ $-7 < -5$ *Write an example for part 2.*
Multiplication and Division Properties	For any numbers a, b, and c, **1.** if $c > 0$ and $a < b$, then $ac < bc$ and $\frac{a}{c} < \frac{b}{c}$. **2.** if $c > 0$ and $a > b$, then $ac > bc$ and $\frac{a}{c} > \frac{b}{c}$.	$12 < 18$ $12 < 18$ $12 \cdot 2 < 18 \cdot 2$ $\frac{12}{2} < \frac{18}{2}$ $24 < 36$ $6 < 9$ *Write an example for part 2.*

Communicating Mathematics

1. **Translate** the statement $m\angle J \nleq m\angle T$ into words two different ways. Then draw and label a pair of angles that shows the statement is true.

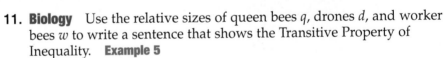

> **Vocabulary**
>
> inequality

2. M is the midpoint of $\overline{AB}$, and P is the midpoint of $\overline{MB}$. The length of $\overline{MP}$ is greater than 7.
   a. **Make a drawing** of $\overline{AB}$ showing the location of points M and P.
   b. **Write** an inequality that represents the length of $\overline{AB}$.

3. **You Decide** Mayuko says that if $a > 7$ and $b < 7$, then $a > b$. Lisa says that $a < b$. Who is correct? Explain your reasoning.

Guided Practice

> ⊖ **Getting Ready** State whether the given number is a possible value of *n*.
>
> **Sample:** $n \nleq 15$; 11
>
> **Solution:** n cannot be less than or equal to 15. So, 11 is not a possible value.

4. $n \neq 0$; -4 5. $n > 86$; 80   6. $n \ngeq 23$; 23

Examples 1 & 2

Replace each ● with <, >, or = to make a true sentence.
7. KP ● PL
8. $m\angle JPL$ ● $m\angle KPM$

Examples 3 & 4

Determine if each statement is *true* or *false*.
9. $JP \neq PM$
10. $m\angle KPM \geq m\angle LPK$

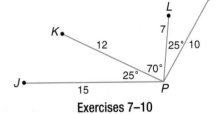

Exercises 7–10

Queen   Drone   Worker

11. **Biology** Use the relative sizes of queen bees q, drones d, and worker bees w to write a sentence that shows the Transitive Property of Inequality. **Example 5**

Exercises

Practice

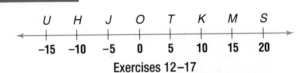

Exercises 12–17

Replace each ● with <, >, or = to make a true sentence.
12. MT ● JT 13. HK ● OK 14. JU ● OS

Determine if each statement is *true* or *false*.
15. $MH \geq JS$ 16. $HT \leq TM$ 17. $KH \nleq UK$

Homework Help	
For Exercises	See Examples
12–20	1, 2
21–28	3, 4
31–32	5
Extra Practice	
See page 738.	

**Lines *BE*, *FC*, and *AD* intersect at *G*.
Replace each ● with <, >, or = to
make a true sentence.**

18. $m\angle BGC$ ● $m\angle AGC$

19. $m\angle BGC$ ● $m\angle FGE$

20. $m\angle AGC$ ● $m\angle CGE$

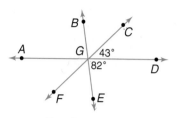

Exercises 18–28

Determine if each statement is *true* or *false*.

21. $m\angle AGF \geq m\angle DGC$ 22. $m\angle DGB \nleq m\angle BGC$

23. $m\angle AGE \neq m\angle BGD$ 24. $m\angle BGC \ngeq m\angle FGE$

25. $m\angle FGE \cdot 2 = m\angle BGC \cdot 2$ 26. $\dfrac{m\angle AGE}{4} < \dfrac{m\angle BGE}{4}$

27. $m\angle DGE - 15 > m\angle CGD - 15$

28. $m\angle CGE + m\angle BGC < m\angle FGE + m\angle BGC$

29. If $JK = 58$ and $GH = 67 - 3b$, what values of b make $JK \geq GH$?

30. If $m\angle Q = 62$ and $m\angle R = 44 - 3y$, what values of y make $m\angle Q < m\angle R$?

Applications and Problem Solving

31. **Algebra** If $m\angle 1 = 94$, $m\angle 2 = 16 - 5x$, and $m\angle 1 = m\angle 2 + 10$, find the value of x.

32. **Art** Important factors in still-life drawings are reference points and distances. The objects at the right are set up for a still-life drawing. If the artist moves the objects apart so that all the measures are increased by 3 centimeters, is the statement $MS < SD$ *true* or *false*? Explain.

33. **Critical Thinking** If $r < s$ and $p < q$, is it true that $rp < sq$? Explain. (*Hint:* Look for a counterexample.)

Exercise 32

Mixed Review

Find the distance between each pair of points. (*Lesson 6–7*)

34. $C(1, 5)$ and $D(-3, 2)$ 35. $L(0, -9)$ and $M(8, -9)$

36. The lengths of three sides of a triangle are 4 feet, 6 feet, and 9 feet. Is the triangle a right triangle? (*Lesson 6–6*)

37. **Construction** Draw an isosceles right triangle. Then construct the three angle bisectors of the triangle. (*Lesson 6–3*)

38. Name all angles congruent to the given angle. (*Lesson 4–3*)

    a. $\angle 2$ **b.** $\angle 7$ **c.** $\angle 8$

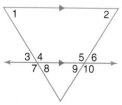

Exercise 38

Standardized Test Practice
Ⓐ Ⓑ Ⓒ Ⓓ

39. **Multiple Choice**
Solve $-3y + 2 < 17$. (*Algebra Review*)

    Ⓐ $y < -5$ Ⓑ $y > 18$

    Ⓒ $y > -5$ Ⓓ $y < 16$

In the figure at the right, recall that ∠1, ∠2, and ∠3 are *interior angles* of △PQR. Angle 4 is called an **exterior angle** of △PQR. An exterior angle of a triangle is an angle that forms a linear pair with one of the angles of the triangle.

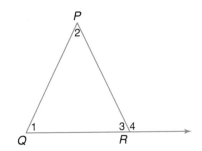

In △PQR, ∠4 is an exterior angle at R because it forms a linear pair with ∠3. **Remote interior angles** of a triangle are the two angles that do *not* form a linear pair with the exterior angle. In △PQR, ∠1 and ∠2 are the remote interior angles with respect to ∠4.

Each exterior angle has corresponding remote interior angles. How many exterior angles does △XYZ below have?

Vertex	Exterior Angle	Remote Interior Angles
X	∠4	∠2 and ∠3
X	∠9	∠2 and ∠3
Y	∠5	∠1 and ∠3
Y	∠6	∠1 and ∠3
Z	∠7	∠1 and ∠2
Z	∠8	∠1 and ∠2

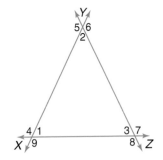

Notice that there are two exterior angles at each vertex and that those two exterior angles have the same remote interior angles. Also observe that an exterior angle is never a vertical angle to an angle of the triangle.

Real World

Example ➊
Design Link

In the music stand, name the remote interior angles with respect to ∠1.

Angle 1 forms a linear pair with ∠2. Therefore, ∠3 and ∠4 are remote interior angles with respect to ∠1.

Your Turn

a. In the figure above, ∠2 and ∠3 are remote interior angles with respect to what angle?

You can investigate the relationships among the interior and exterior angles of a triangle.

Hands-On Geometry

Materials: ▬ straightedge   ⌒ protractor

Step 1 Use a straightedge to draw and label △RPN. Extend side $\overline{RN}$ through K to form the exterior angle 4.

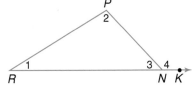

Step 2 Measure the angles of the triangle and the exterior angle.

Step 3 Find $m\angle1 + m\angle2$.

Step 4 Make a table like the one below to record the angle measures.

$m\angle1$	$m\angle2$	$m\angle1 + m\angle2$	$m\angle4$
31	103	134	134

Try These

1. Draw other triangles and collect the same data. Record the data in your table.
2. Do you see any patterns in your data? **Make a conjecture** that describes what you see.

The relationship you investigated in the activity above suggests the following theorem.

Theorem 7–3 Exterior Angle Theorem	**Words:**	The measure of an exterior angle of a triangle is equal to the sum of the measures of its two remote interior angles.
	Model:	 Y 2 1 3 4 X Z
	Symbols:	$m\angle4 = m\angle1 + m\angle2$

2 If $m\angle 2 = 38$ and $m\angle 4 = 134$, what is $m\angle 5$?

Examples 2–3

$m\angle 4 = m\angle 2 + m\angle 5$	*Exterior Angle Theorem*
$134 = 38 + m\angle 5$	*Replace $m\angle 4$ with 134 and $m\angle 2$ with 38.*
$134 - 38 = 38 + m\angle 5 - 38$	*Subtract 38 from each side.*
$96 = m\angle 5$	*Simplify.*

3 If $m\angle 2 = x + 17$, $m\angle 3 = 2x$, and $m\angle 6 = 101$, find the value of x.

Algebra Review

Solving Multi-Step
Equations, p. 723

$m\angle 6 = m\angle 2 + m\angle 3$	*Exterior Angle Theorem*
$101 = (x + 17) + 2x$	*Replace $m\angle 6$ with 101, $m\angle 2$ with $x + 17$, and*
$101 = 3x + 17$	$m\angle 3$ with $2x$.
$101 - 17 = 3x + 17 - 17$	*Subtract 17 from each side.*
$84 = 3x$	*Simplify.*
$\dfrac{84}{3} = \dfrac{3x}{3}$	*Divide each side by 3.*
$28 = x$	*Simplify.*

Your Turn

Refer to the figure above.

b. What is $m\angle 1$ if $m\angle 3 = 46$ and $m\angle 5 = 96$?

c. If $m\angle 2 = 3x$, $m\angle 3 = x + 34$, and $m\angle 6 = 98$, find the value of x.
Then find $m\angle 3$.

Personal Tutor at geomconcepts.com

There are two other theorems that relate to the Exterior Angle Theorem. In the triangle at the right, $\angle QRS$ is an exterior angle, and $\angle S$ and $\angle T$ are its remote interior angles. The Exterior Angle Theorem states that

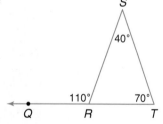

$$m\angle QRS = m\angle S + m\angle T.$$

In $\triangle RST$, you can see that the measure of $\angle QRS$ is greater than the measures of both $\angle S$ and $\angle T$, because $110 > 40$ and $110 > 70$. This suggests Theorem 7–4.

<table>
<tr>
<td rowspan="2">**Theorem 7–4**
Exterior Angle
Inequality
Theorem</td>
<td>**Words:**</td>
<td colspan="2">The measure of an exterior angle of a triangle is greater than the measure of either of its two remote interior angles.</td>
</tr>
<tr>
<td>**Model:**</td>
<td>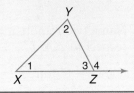</td>
<td>**Symbols:**
$m\angle 4 > m\angle 1$
$m\angle 4 > m\angle 2$</td>
</tr>
</table>

Example ▶ **4** **Name two angles in $\triangle MAL$ that have measures less than 90.**

$\angle MLC$ is a 90° exterior angle. $\angle M$ and $\angle A$ are its remote interior angles. By Theorem 7–4, $m\angle MLC > m\angle 1$ and $m\angle MLC > m\angle 2$. Therefore, $\angle 1$ and $\angle 2$ have measures less than 90.

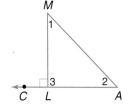

Your Turn

d. Name two angles in $\triangle VWX$ that have measures less than 74.

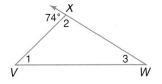

The results of Example 4 suggest the following theorem about the angles of a right triangle.

<table>
<tr>
<td>**Theorem 7–5**</td>
<td>If a triangle has one right angle, then the other two angles must be acute.</td>
</tr>
</table>

Check for Understanding

Communicating Mathematics

1. Draw a triangle and extend all of the sides. Identify an exterior angle at each of the vertices.

2. Trace $\triangle ABC$ on a blank piece of paper and cut out the triangle. Tear off corners with $\angle C$ and $\angle A$, and use the pieces to show that the Exterior Angle Theorem is true. Explain.

Vocabulary

exterior angle
remote interior angle

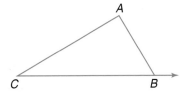

3. **You Decide?** Maurice says that the two exterior angles at the same vertex of a triangle are always congruent. Juan says it is impossible for the angles to be congruent. Who is correct? Explain your reasoning.

Guided Practice

Example 1 4. Name two remote interior angles with respect to $\angle AKL$.

Example 2 5. If $m\angle 3 = 65$ and $m\angle 5 = 142$, what is $m\angle 2$?

Example 3 6. If $m\angle 1 = 2x - 26$, $m\angle 3 = x$, and $m\angle 4 = 37$, find the value of x.

Example 4 7. Replace ● with $<$, $>$, or $=$ to make a true sentence.

$$m\angle 3 \bullet m\angle 1$$

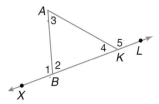

Exercises 4–7

Example 4 8. **Interior Design** Refer to the floor tile at the right.

a. Is $\angle 1$ an exterior angle of $\triangle ABC$? Explain.

b. Which angle must have a measure greater than $\angle 5$?

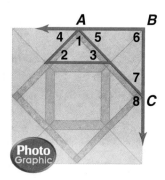

Exercises

Practice

Name the following angles.

9. an exterior angle of $\triangle SET$

10. an interior angle of $\triangle SCT$

11. a remote interior angle of $\triangle TCE$ with respect to $\angle JET$

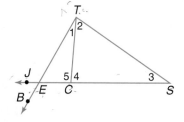

Homework Help	
For Exercises	**See Examples**
9–11	1
12–17, 24	2
22	3
Extra Practice	
See page 738.	

Find the measure of each angle.

12. $\angle 4$
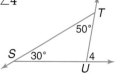

13. $\angle J$

14. $\angle A$
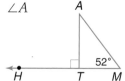

15. Find the value of x.

16. Find $m\angle C$.

17. Find $m\angle Y$.

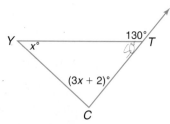

Exercises 15–17

Replace each ● with <, >, or = to make a true sentence.

18.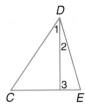

$m\angle 3 ● m\angle 1$

19.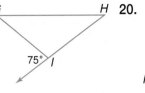

$m\angle G ● 75$

20.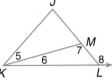

$m\angle 8 ● m\angle 6 + m\angle 7$

21. Write a relationship for $m\angle BAC$ and $m\angle ACD$ using <, >, or =.

22. Find the value of x.

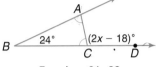

Exercises 21–22

Applications and Problem Solving

23. **Botany** The feather-shaped leaf at the right is called a *pinnatifid*. In the figure, does $x = y$? Explain.

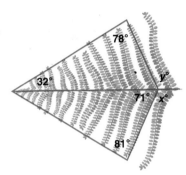

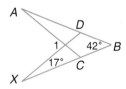

24. **Entertainment** For the 1978 movie *Superman*, the flying scenes were filmed using angled mirrors as shown in the diagram at the left. Find x, the measure of the angle made by the two-way mirror and the camera projection.

25. **Critical Thinking** If $\triangle ABC \cong \triangle XBD$, find the measure of $\angle 1$.

Mixed Review

26. **Transportation** Corning, Red Bluff, and Redding are California cities that lie on the same line, with Red Bluff in the center. Write a sentence using <, >, or = to compare the distance from Corning to Redding *CR* and the distance from Corning to Red Bluff *CB*. *(Lesson 7–1)*

27. Determine whether $\triangle XYZ$ with vertices $X(-2, 6)$, $Y(6, 4)$, and $Z(0, -2)$ is an isosceles triangle. Explain. *(Lesson 6–7)*

Find the perimeter and area of each rectangle. *(Lesson 1–6)*

28. $\ell = 12$ feet, $w = 16$ feet

29. $\ell = 3.5$ meters, $w = 1.2$ meters

Standardized Test Practice
Ⓐ Ⓑ Ⓒ Ⓓ

30. **Multiple Choice** What is the solution to $60 \leq 9r - 21 \leq 87$? *(Algebra Review)*

　Ⓐ $-9 \leq r \leq -12$　Ⓑ $9 \leq r \leq 12$　Ⓒ $9 \geq r \geq 12$　Ⓓ $12 \leq r \leq 9$

Linguine Triangles Hold the Sauce!

Materials

☐ unlined paper

 ruler

 protractor

 uncooked linguine

Measures of Angles and Sides in Triangles

What happens to the length of a side of a triangle as you increase the measure of the angle opposite that side? How does this change in angle measure affect the triangle? In this investigation, you will use linguine noodles to explore this relationship.

Investigate

1. Use uncooked linguine to investigate three different triangles. First, break a piece of linguine into two 3-inch lengths.

   a. Using a protractor as a guide, place the two 3-inch pieces of linguine together to form a 30° angle. Break a third piece of linguine so its length forms a triangle with the first two pieces. Trace around the triangle and label it Triangle 1. Measure and record the length of the third side of the triangle.

   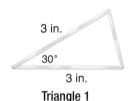
   3 in.
   30°
   3 in.
   Triangle 1

Hint: Use small pieces of modeling clay or tape to hold the linguine pieces together.

   b. Using a protractor, place the two 3-inch pieces of linguine together to form a 60° angle. Break another piece of linguine and use it to form a triangle with the first two pieces. Trace around the triangle and label it Triangle 2. Measure and record the length of the third side of your triangle.

   c. Using a protractor, place the two 3-inch pieces of linguine together to form a 90° angle. Break another piece of linguine and use it to form a triangle with the first two pieces. Trace around the triangle and label it Triangle 3. Measure and record the length of the third side of the triangle.

   d. As the angle opposite the third side of the triangle increases, what happens to the measure of the third side?

2. Break four pieces of linguine so that you have the following lengths: 2 inches, 3 inches, 4 inches, and 5 inches.

   a. Use a protractor to form a 40° angle between the 2-inch piece and the 3-inch piece as shown at the right. Break a third piece of linguine to form a triangle. Trace around the triangle and label it Triangle 4. Record the measure of angle 1 shown in the figure.

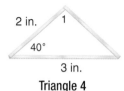

2 in. 1

40°

3 in.

Triangle 4

   b. In the linguine triangle from Step 2a, replace the 3-inch piece with the 4-inch piece. Keep the angle measure between the pieces 40°. Break a third piece of linguine to form a triangle. Trace around the triangle and label it Triangle 5. Record the measure of angle 1.

   c. In the linguine triangle from Step 2b, replace the 4-inch piece with the 5-inch piece. Keep the angle measure between the pieces 40°. Break a third piece of linguine to form a triangle. Trace around the triangle and label it Triangle 6. Record the measure of angle 1.

   d. In the three triangles that you formed, each contained a 40° angle. One side remained 2 inches long, but the other side adjacent to the 40° angle increased from 3 to 4 to 5 inches. As that side increased in length, what happened to the measure of angle 1?

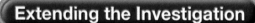

Extending the Investigation

In this extension, you will further investigate the relationship between the measures of the sides and angles in triangles.

Use linguine, geometry drawing software, or a graphing calculator to investigate these questions.

1. What happens to the length of the third side of a triangle as the angle between the other two sides ranges from 90° to 150°?

2. What happens to the measure of an angle of a triangle as you increase the length of the side opposite that angle?

Presenting Your Conclusions

Here are some ideas to help you present your conclusions to the class.

- Make a display or poster of your findings in this investigation.

- Write a description of the steps to follow to complete this investigation using geometry drawing software or a graphing calculator.

interNET **Investigation** For more information on triangle inequalities,
CONNECTION visit: www.geomconcepts.com

What You'll Learn
You'll learn to identify the relationships between the sides and angles of a triangle.

Why It's Important
Surveying Triangle relationships are important in undersea surveying.
See Example 2.

Florists often use triangles as guides in their flower arrangements. There are special relationships between the side measures and angle measures of each triangle. You will discover these relationships in the following activity.

Suppose in triangle ABC, the inequality $AC > BC$ holds true. Is there a similar relationship between the angles $\angle B$ and $\angle A$, which are across from those sides?

Graphing Calculator Tutorial
See pp. 782–785.

Graphing Calculator Exploration

Step 1 Use the Triangle tool on [F2] to draw and label $\triangle ABC$.

Step 2 Select Measure from the [F5] menu. Then use the Distance & Length tool and the Angle tool on [F6] to display the measures of the sides and angles of $\triangle ABC$.

Try These

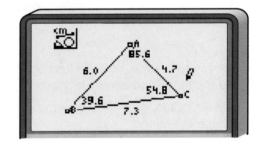

1. Refer to the triangle drawn using the steps above.
   a. What is the measure of the largest angle in your triangle?
   b. What is the measure of the side opposite the largest angle?
   c. What is the measure of the smallest angle in your triangle?
   d. What is the measure of the side opposite the smallest angle?

2. Drag vertex A to a different location.
   a. What are the lengths of the longest and shortest sides of the new triangle?
   b. What can you conclude about the measures of the angles of a triangle and the measures of the sides opposite these angles?

3. Use the Perpendicular Bisector tool on [F3] to draw the perpendicular bisector of side AB. Drag vertex C very close to the perpendicular bisector. What do you observe about the measures of the sides and angles?

The observations you made in the previous activity suggest the following theorem.

<table>
<tr><td rowspan="2">Theorem 7–6</td><td>Words:</td><td colspan="2">If the measures of three sides of a triangle are unequal, then the measures of the angles opposite those sides are unequal in the same order.</td></tr>
<tr><td>Model:</td><td>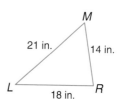</td><td>Symbols:
$PL < MP < LM$
$m\angle M < m\angle L < m\angle P$</td></tr>
</table>

The converse of Theorem 7–6 is also true.

<table>
<tr><td rowspan="2">Theorem 7–7</td><td>Words:</td><td colspan="2">If the measures of three angles of a triangle are unequal, then the measures of the sides opposite those angles are unequal in the same order.</td></tr>
<tr><td>Model:</td><td>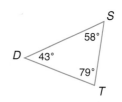</td><td>Symbols:
$m\angle W < m\angle J < m\angle K$
$JK < KW < WJ$</td></tr>
</table>

Example —❶ In △LMR, list the angles in order from least to greatest measure.

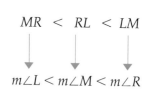

First, write the segment measures in order from least to greatest.

$$MR < RL < LM$$

Then, use Theorem 7–6 to write the measures of the angles opposite those sides in the same order.

$$m\angle L < m\angle M < m\angle R$$

The angles in order from least to greatest measure are $\angle L$, $\angle M$, and $\angle R$.

Your Turn

a. In △DST, list the sides in order from least to greatest measure.

Example **2**

Surveying Link

Scientists are developing automated robots for underwater surveying. These undersea vehicles will be guided along by sonar and cameras. If △NPQ represents the intended course for an undersea vehicle, which segment of the trip will be the longest?

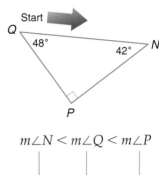

First, write the angle measures in order from least to greatest.

$$m\angle N < m\angle Q < m\angle P$$

Then, use Theorem 7–7 to write the measures of the sides opposite those angles in the same order.

$$PQ \ < \ NP \ < \ QN$$

So, $\overline{QN}$, the first segment of the course, will be the longest.

Your Turn

Undersea Robot Vehicle, *Oberon*

b. If △*ABC* represents a course for an undersea vehicle, which turn will be the sharpest—that is, which angle has the least measure?

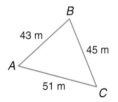

Example 2 illustrates an argument for the following theorem.

Theorem 7–8	**Words:** In a right triangle, the hypotenuse is the side with the greatest measure.
	Model: *W* 3 ⟋ 5 *X* — 4 — *Y* **Symbols:** *WY* > *YX* *WY* > *XW*

Check for Understanding

Communicating Mathematics

1. **Name** the angle opposite $\overline{ZH}$ in △*GHZ*.

2. **Choose** the correct value for *x* in △*GHZ* without using the Pythagorean Theorem: 14, 16, or 20. Explain how you made your choice.

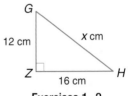

Exercises 1–2

3. **Writing Math** Identify the shortest segment from point P to line ℓ. Write a conjecture in your journal about the shortest segment from a point to a line.

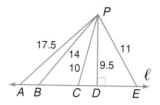

Guided Practice
Example 1

4. List the angles in order from least to greatest measure.

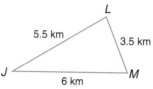

5. List the sides in order from least to greatest measure.

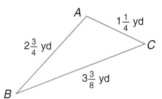

Example 2

6. Identify the angle with the greatest measure.

7. Identify the side with the greatest measure.

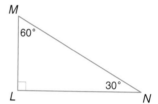

8. **Driving** The road sign indicates that a steep hill is ahead.

   a. Use a ruler to measure the sides of $\triangle STE$ to the nearest millimeter. Then list the sides in order from least to greatest measure.

   b. List the angles in order from least to greatest measure. **Example 2**

Lombard Street, San Francisco

Exercises

Practice

List the angles in order from least to greatest measure.

9.

10.

11.

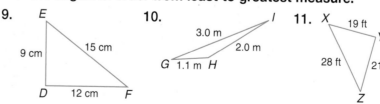

Homework Help

For Exercises	See Examples
9–11, 15–17, 22, 23	1
12–14, 18–20, 21, 24	2

Extra Practice
See page 739.

List the sides in order from least to greatest measure.

12.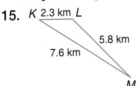

13.

14.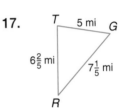

Identify the angle with the greatest measure.

15.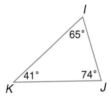

16.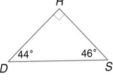

17.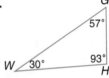

Identify the side with the greatest measure.

18.

19.

20.

21. In $\triangle PRS$, $m\angle P = 30$, $m\angle R = 45$, and $m\angle S = 105$. Which side of $\triangle PRS$ has the greatest measure?

22. In $\triangle WQF$, $WQ > QF > FW$. Which angle of $\triangle WQF$ has the greatest measure?

Applications and Problem Solving

23. Archaeology Egyptian carpenters used a tool called an *adze* to smooth and shape wooden objects. Does $\angle E$, the angle the copper blade makes with the handle, have a measure less than or greater than the measure of $\angle G$, the angle the copper blade makes with the work surface? Explain.

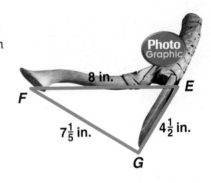

24. Maps Two roads meet at an angle of 50° at point *A*. A third road from *B* to *C* makes an angle of 45° with the road from *A* to *C*. Which intersection, *A* or *B*, is closer to *C*? Explain.

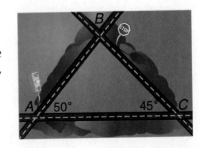

25. Critical Thinking In an obtuse triangle, why is the longest side opposite the obtuse angle?

Mixed Review

26. The measures of two interior angles of a triangle are 17 and 68. What is the measure of the exterior angle opposite these angles? *(Lesson 7–2)*

27. Algebra If $m\angle R = 48$ and $m\angle S = 2x - 10$, what values of x make $m\angle R \geq m\angle S$? *(Lesson 7–1)*

Complete each congruence statement. *(Lesson 5–4)*

28.

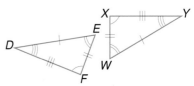

$\triangle MLK \cong \triangle$ ___?___

29.

$\triangle YXW \cong \triangle$ ___?___

Standardized Test Practice
Ⓐ Ⓑ Ⓒ Ⓓ

30. Short Response Sketch at least three different quilt patterns that could be made using transformations of the basic square shown at the right. Identify each transformation. *(Lesson 5–3)*

Quiz   Lessons 7–1 through 7–3

Replace each ● with <, >, or = to make a true sentence. *(Lesson 7–1)*

1. $JA \bullet ST$ **2.** $m\angle JST \bullet m\angle STN$

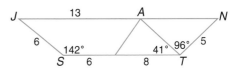

Find the measure of each angle. *(Lesson 7–2)*

3. $\angle 2$

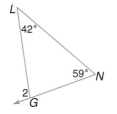

4. $\angle D$

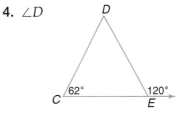

5. Geography Perth, Darwin, and Sydney are three cities in Australia. Which two of the cities are the farthest apart? *(Lesson 7–3)*

7-4 Triangle Inequality Theorem

What You'll Learn
You'll learn to identify and use the Triangle Inequality Theorem.

Why It's Important
Aviation Pilots use triangle inequalities when conducting search-and-rescue operations.
See page 301.

Can you always make a triangle with any three line segments? For example, three segments of lengths 1 centimeter, 1.5 centimeters, and 3 centimeters are given. According to the Triangle Inequality Theorem, it is not possible to make a triangle with the three segments. Why? The sum of any two sides of a triangle has to be greater than the third side.

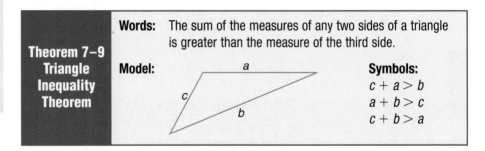

Theorem 7–9 Triangle Inequality Theorem

Words: The sum of the measures of any two sides of a triangle is greater than the measure of the third side.

Model:

Symbols:
$c + a > b$
$a + b > c$
$c + b > a$

You can use the Triangle Inequality Theorem to verify the possible measures for sides of a triangle.

Examples

Determine if the three numbers can be measures of the sides of a triangle.

❶ **5, 7, 4**

$5 + 7 > 4$ *yes*
$5 + 4 > 7$ *yes*
$7 + 4 > 5$ *yes*

All possible cases are true. Sides with these measures can form a triangle.

❷ **11, 3, 7**

$11 + 3 > 7$ *yes*
$11 + 7 > 3$ *yes*
$7 + 3 > 11$ *no*

All possible cases are not true. Sides with these measures cannot form a triangle.

Your Turn

a. Determine if 16, 10, and 5 can be measures of the sides of a triangle.

The next example shows another way you can use the Triangle Inequality Theorem.

Example ③ **Suppose △XYZ has side $\overline{YX}$ that measures 10 centimeters and side $\overline{XZ}$ that measures 7 centimeters. What are the greatest and least possible whole-number measures for $\overline{YZ}$?**

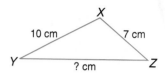

Explore   Cut one straw 10 centimeters long and another straw 7 centimeters long. Connect the two straws with a pin to form a moveable joint.

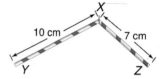

Plan   Lay the straws on a flat surface along a ruler. Hold the end representing point Y at the 0 point on the ruler.

Solve   With your other hand, push point X toward the ruler. When X is touching the ruler, the measure is about 17 centimeters. So the greatest measure possible for $\overline{YZ}$ is just less than 17. Now slide the end representing point Z toward the 0 point on the ruler. Just left of 3 centimeters, the point Z can no longer lie along the ruler. So the least possible measure is just greater than 3.

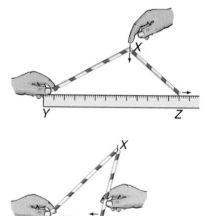

Therefore, $\overline{YZ}$ can be as long as 16 centimeters and as short as 4 centimeters.

Examine   Notice that $16 < 10 + 7$ and $4 > 10 - 7$.

Your Turn

b. What are the greatest and least possible whole-number measures for the third side of a triangle if the other two sides measure 8 inches and 3 inches?

🌐nline **Personal Tutor at** geomconcepts.com

Example 3 shows that the measure of an unknown side of a triangle must be less than the sum of the measures of the two known sides and greater than the difference of the measures of the two known sides.

The Grecian catapult at the right was used for siege warfare during the time of ancient Greece. If the two ropes are each 4 feet long, find x, the range of the possible distances between the ropes.

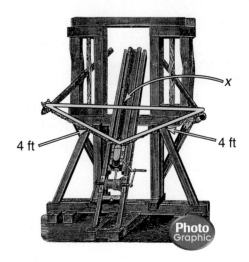

4 ft   4 ft

Let x be the measure of the third side of the triangle.

x is greater than the difference of the measures of the two other sides.	x is less than the sum of the measures of the two other sides.
$x > 4 - 4$ $\quad$ $x > 0$	$x < 4 + 4$ $\quad$ $x < 8$

The measure of the third side is greater than 0 but less than 8. This can be written as $0 < x < 8$.

Your Turn

c. If the measures of two sides of a triangle are 9 and 13, find the range of possible measures of the third side.

Check for Understanding

Communicating Mathematics

1. **Select** a possible measure for the third side of a triangle if its other two sides have measures 17 and 9.

2. **State** three inequalities that relate the measures of the sides of the triangle.

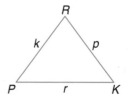

3. **Writing Math** Draw a triangle in your journal and explain why the shortest distance between two points is a straight line.

Determine if the three numbers can be measures of the sides of a triangle. Write *yes* or *no*. Explain.

Examples 1 & 2

4. 15, 8, 29

5. 100, 100, 8

Example 4

If two sides of a triangle have the following measures, find the range of possible measures for the third side.

6. 17, 8

7. 40, 62

Example 3

8. Birds If $\angle FGH$ in the flock of migrating geese changes, what are the greatest and least possible whole number values of x?

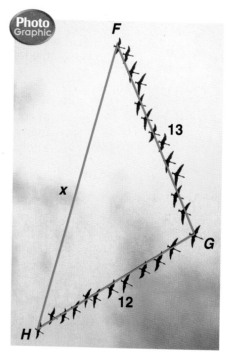

Exercise 8

Exercises

Practice

Determine if the three numbers can be measures of the sides of a triangle. Write *yes* or *no*. Explain.

9. 7, 12, 8

10. 6, 7, 13

11. 1, 2, 3

12. 9, 10, 14

13. 5, 10, 20

14. 60, 70, 140

For Exercises	See Examples
Homework Help	
9–14	1, 2
15–26	3, 4
Extra Practice	
See page 739.	

If two sides of a triangle have the following measures, find the range of possible measures for the third side.

15. 12, 8

16. 2, 7

17. 21, 22

18. 5, 16

19. 44, 38

20. 81, 100

21. The sum of KL and KM is greater than ___?___ .

22. If $KM = 5$ and $KL = 3$, then LM must be greater than ___?___ and less than ___?___ .

23. Determine the range of possible values for x if $KM = x$, $KL = 61$, and $LM = 83$.

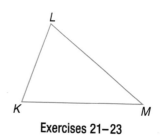

Exercises 21–23

Applications and Problem Solving

24. **Design** Some kitchen planners design kitchens by drawing a triangle and placing an appliance at each vertex. If the distance from the refrigerator to the sink is 6 feet and the distance from the sink to the range is 5 feet, what are possible distances between the refrigerator and the range?

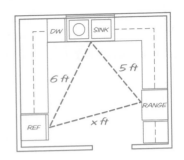

25. **History** Early Egyptians made triangles using a rope with knots tied at equal intervals. Each vertex of the triangle had to be at a knot. How many different triangles could you make with a rope with exactly 13 knots as shown below? Sketch each possible triangle.

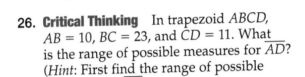

26. **Critical Thinking** In trapezoid *ABCD*, *AB* = 10, *BC* = 23, and *CD* = 11. What is the range of possible measures for $\overline{AD}$? (*Hint*: First find the range of possible measures for $\overline{AC}$.)

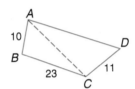

Mixed Review

27. **Art** The drawing at the right shows the geometric arrangement of the objects in the painting *Apples and Oranges*. In each triangle, list the sides in order from least to greatest length. (*Lesson 7–3*)

a. △*LMN*   b. △*UVW*   c. △*BCD*

Paul Cezanne, *Apples and Oranges*

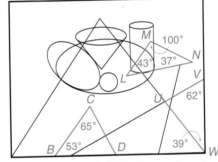

Exercises 27–28

28. What is the measure of the exterior angle at *D*? (*Lesson 7–2*)

29. **Camping** When Kendra's tent is set up, the front of the tent is in the shape of an isosceles triangle. If each tent side makes a 75° angle with the ground, what is the measure of the angle at which the sides of the tent meet? (*Lesson 6–5*)

Standardized Test Practice
Ⓐ Ⓑ Ⓒ Ⓓ

30. **Grid In** Find the value of *x* in the figure at the right. (*Lesson 3–6*)

$(x - 7)°$

$140°$

31. **Multiple Choice** Points *J*, *K*, and *L* are collinear, with *K* between *J* and *L*. If *KL* = $6\frac{1}{3}$ and *JL* = $16\frac{2}{5}$, what is the measure of $\overline{JK}$? (*Lesson 2–2*)

Ⓐ 10   Ⓑ $10\frac{1}{15}$ Ⓒ $22\frac{1}{2}$ Ⓓ $22\frac{11}{15}$

www.geomconcepts.com/self_check_quiz

Pilot

In search-and-rescue operations, direction findings are used to locate emergency radio beacons from a downed airplane. When two search teams from different locations detect the radio beacon, the directions of the radio signals can pinpoint the position of the plane.

Suppose search teams S and T have detected the emergency radio beacon from an airplane at point A. Team T measures the direction of the radio beacon signal 52° east of north. Team S measures the direction of the radio beacon signal 98° east of north and the direction of Team T 150° east of north.

1. Find the measure of each angle.
   a. 1
   b. 2
   c. 3
2. Which search team is closer to the downed airplane?

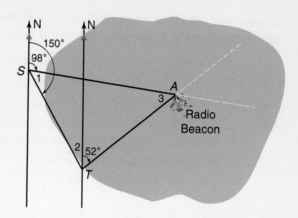

FAST FACTS About Pilots

Working Conditions
- often have irregular schedules and odd hours
- does not involve much physical effort, but can be mentally stressful
- must be alert and quick to react

Education
- commercial pilot's license
- 250 hours flight experience
- written and flying exams
- Most airlines require at least two years of college, including mathematics courses essential for navigation techniques.

Employment

Pilot Certificates, 2000

- Student 15%
- Private 40%
- Commercial 20%
- Other 25%

Source: *Federal Aviation Administration*

---

*inter*NET
CONNECTION   **Career Data**   For the latest information on a career as a pilot, visit:
www.geomconcepts.com

Understanding and Using the Vocabulary

After completing this chapter, you should be able to define each term, property, or phrase and give an example or two of each.

interNET
CONNECTION **Review Activities**
For more review activities, visit:
www.geomconcepts.com

exterior angle *(p. 282)*   inequality *(p. 276)*   remote interior angles *(p. 282)*

Determine whether each statement is *true* or *false*. If the statement is false, replace the underlined word or phrase to make it true.

1. The expression $4y - 9 \le 5$ is an example of an <u>equation</u>.
2. In Figure 1, $\angle 3$, $\angle 5$, and $\angle 8$ are <u>exterior</u> angles.
3. $CM \ge BQ$ means the length of $\overline{CM}$ is <u>less than</u> the length of $\overline{BQ}$.
4. A <u>remote interior</u> angle of a triangle is an angle that forms a linear pair with one of the angles of the triangle.
5. The Triangle Inequality Theorem states that the sum of the measures of any two sides of a triangle is <u>greater than</u> the measure of the third side.
6. In Figure 1, $m\angle 7 = m\angle 5 + m\angle 8$ by the <u>Interior</u> Angle Theorem.
7. $m\angle Z < m\angle Y$ means the measure of angle Z is <u>less than or equal to</u> the measure of angle Y.
8. In Figure 1, the exterior angles at K are $\angle 6$, $\angle 9$, and $\angle BKD$.
9. In Figure 2, $EF + FG$ is <u>equal to</u> EG.
10. In Figure 2, if $FG = 5$ and $EF = 9$, a possible measure for $\overline{EG}$ is <u>13.9</u>.

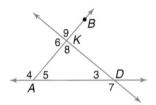

Figure 1

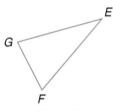

Figure 2

Skills and Concepts

Objectives and Examples	**Review Exercises**

• Lesson 7–1 Apply inequalities to segment and angle measures.

$LP > LN$
$LP > NP$

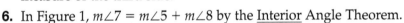

$m\angle GBK > m\angle GBH$
$m\angle GBK > m\angle HBK$

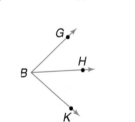

Replace each ● with <, >, or = to make a true sentence.

11. $m\angle FRV$ ● $m\angle FRM$
12. JR ● RF
13. FV ● FM
14. $m\angle JRV$ ● $m\angle MRF$

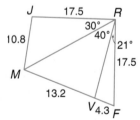

Exercises 11–16

Determine if each statement is *true* or *false*.

15. $FM \ne JR$ 16. $m\angle JRF \ge m\angle VRJ$

 www.geomconcepts.com/vocabulary_review

Objectives and Examples

- **Lesson 7–2** Identify exterior angles and remote interior angles of a triangle.

Interior angles of △UVW are ∠2, ∠4, and ∠5.

Exterior angles of △UVW are ∠1, ∠3, and ∠6.

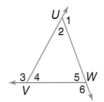

The remote interior angles of △UVW with respect to ∠1 are ∠4 and ∠5.

- **Lesson 7–2** Use the Exterior Angle Theorem.

If $m\angle 1 = 75$ and $m\angle 4 = 35$, find $m\angle 3$.

∠1 and ∠4 are remote interior angles of △CDN with respect to ∠3.

$m\angle 3 = m\angle 1 + m\angle 4$ *Exterior Angle Theorem*
$m\angle 3 = 75\ +\ 35$ *Substitution*
$m\angle 3 = 110$

- **Lesson 7–3** Identify the relationships between the sides and angles of a triangle.

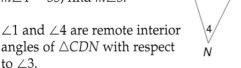

$m\angle S < m\angle R < m\angle O$
$OR\ <\ SO\ <\ RS$

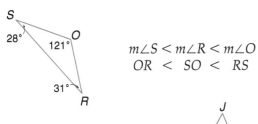

$JK\ <\ KN < NJ$
$m\angle N < m\angle J < m\angle K$

Review Exercises

Name the angles.

17. an exterior angle of △QAJ
18. all interior angles of △ZAQ
19. a remote interior angle of △QZJ with respect to ∠1
20. a remote interior angle of △ZAQ with respect to ∠2

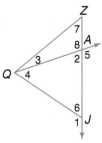

Exercises 17–20

Find the measure of each angle.

21. $m\angle PHF$ **22.** $m\angle RYK$

23. Replace ● with <, >, or = to make a true sentence.
$m\angle E$ ● 108
24. Find the value of x.
25. Find $m\angle B$.
26. Find $m\angle E$.

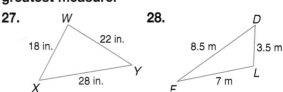

Exercises 23–26

List the angles in order from least to greatest measure.

27.   **28.**

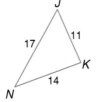

Identify the side with the greatest measure.

29.   **30.**

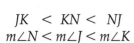

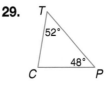

Mixed Problem Solving
See pages 758–765.

Objectives and Examples

- **Lesson 7–4** Identify and use the Triangle Inequality Theorem.

Determine if 15, 6, and 7 can be the measures of the sides of a triangle.

By the Triangle Inequality Theorem, the following inequalities must be true.

$$15 + 6 > 7 \quad \textit{yes}$$
$$15 + 7 > 6 \quad \textit{yes}$$
$$6 + 7 > 15 \quad \textit{no}$$

Since all possible cases are not true, sides with these measures cannot form a triangle.

Review Exercises

Determine if the three numbers can be measures of the sides of a triangle. Write *yes* or *no*. Explain.

31. 12, 5, 13
32. 27, 11, 39
33. 15, 45, 60

If two sides of a triangle have the following measures, find the range of possible measures for the third side.

34. 2, 9
35. 10, 30
36. 34, 18

Applications and Problem Solving

37. History The Underground Railroad used quilts as coded directions. In the quilt block shown below, the right triangles symbolize flying geese, a message to follow these birds north to Canada. If $m\angle FLG = 135$ and $m\angle LSG = 6x - 18$, find the value of x. *(Lesson 7–2)*

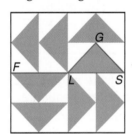

38. Theater A theater has spotlights that move along a track in the ceiling 16 feet above the stage. The lights maintain their desired intensity for up to 30 feet. One light is originally positioned directly over center stage C. At what distance d from C will the light begin to lose its desired intensity? *(Lesson 7–4)*

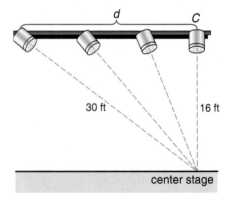

39. Problem Solving *True* or *false*: $TA = KT$. Explain. *(Lesson 7–3)*

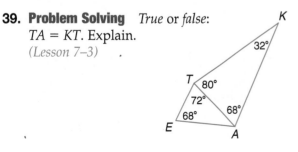

Replace each ● with <, >, or = to make a true sentence.

1. BK ● JK

2. $m\angle DJK$ ● $m\angle BDK$

3. $m\angle BJD$ ● $m\angle DKF$

4. JF ● DF

5. BD ● KF

6. $m\angle JDF$ ● $m\angle FDK$

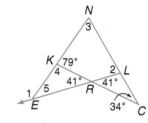

Determine if each statement is *true* or *false*.

7. $m\angle KFD > m\angle JKD$

8. $BK \geq DF$

9. $m\angle BDF \not\cong m\angle DKF$

10. $JF \neq BD$

Exercises 1–10

11. Name all interior angles of $\triangle NLE$.

12. Name an exterior angle of $\triangle KNC$.

13. Name a remote interior angle of $\triangle KRE$ with respect to $\angle KRL$.

14. Find $m\angle 2$.

15. Find $m\angle 5$.

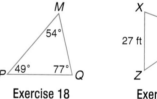

Exercises 11–17

Replace each ● with <, >, or = to make a true sentence.

16. $m\angle 3$ ● $m\angle RLC$

17. $m\angle 2 + m\angle 3$ ● $m\angle 1$

18. In $\triangle MPQ$, list the sides in order from least to greatest measure.

19. In $\triangle XYZ$, identify the angle with the greatest measure.

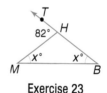

Exercise 18   Exercise 19

20. In $\triangle BTW$, $m\angle B = 36$, $m\angle T = 84$, and $m\angle W = 60$. Which side of $\triangle BTW$ has the greatest measure?

21. Is it possible for 3, 7, and 11 to be the measures of the sides of a triangle? Explain.

22. In $\triangle FGW$, $FG = 12$ and $FW = 19$. If $GW = x$, determine the range of possible values for x.

23. Algebra If $m\angle THM = 82$, find the value of x.

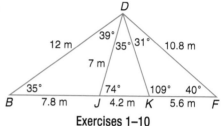

Exercise 23

24. Language The character below means *mountain* in Chinese. The character is enlarged on a copy machine so that it is 3 times as large as shown. Write a relationship comparing CD and EG in the enlarged figure using <, >, or =.

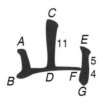

25. Storage Jana is assembling a metal shelving unit to use in her garage. The unit uses triangular braces for support, as shown in the diagram below. Piece r is 60 inches long and piece v is 25 inches long. Find the range of possible lengths for piece t before all the pieces are permanently fastened together.

Algebra Word Problems

You will need to write equations and solve word problems on most standardized tests.

The most common types of word problems involve consecutive integers, total cost, ages, motion, investments, or coins.

Example 1

Lin's Sundae Shoppe has a make-it-yourself sundae bar. A bowl of ice cream costs \$2. Each topping costs \$0.25. Which of the following equations shows the relationship between t, the number of toppings added, and C, the cost of the sundae?

A $C = 2 + 0.25t$ B $C = 2(t + 0.25)$

C $C = 0.25(2 + t)$ D $C = 2 + \frac{t}{0.25}$

Hint Write the equation and then compare it to the answer choices.

Solution Translate the words into algebra. The total cost is the cost of the ice cream and the toppings. Each topping costs \$0.25. The word *each* tells you to multiply.

Cost	equals	cost of ice cream	plus	\$0.25 per topping.
C	$=$	2	$+$	$0.25t$

$$C = 2 + 0.25t$$

The answer is A.

Example 2

Steve ran a 12-mile race at an average speed of 8 miles per hour. If Adam ran the same race at an average speed of 6 miles per hour, how many minutes longer than Steve did Adam take to complete the race?

A   9   B   12   C   16

D   24   E   30

Hint Be careful about units like hours and minutes.

Solution Read the question carefully. You need to find a number of minutes, not hours. The phrase "longer than" means you will probably subtract.

Use the formula for motion.

$$\text{distance} = \text{rate} \times \text{time or } d = rt$$

Solve this equation for t: $t = \frac{d}{r}$.

For Steve's race, $t = \frac{12}{8}$ or $1\frac{1}{2}$ hours.

For Adam's race, $t = \frac{12}{6}$ or 2 hours.

The question asks how many minutes longer did Adam take. Adam took $2 - 1\frac{1}{2}$ or $\frac{1}{2}$ hour longer. Since $\frac{1}{2}$ hour is 30 minutes, the answer is E.

After you work each problem, record your answer on the answer sheet provided or on a piece of paper.

Multiple Choice

1. In order for a student to be eligible for financial aid at a certain school, a student's parents must have a combined annual income of less than $32,000. If f is the father's income and m is the mother's income, which sentence represents the condition for financial aid? *(Algebra Review)*
   - (A) $f + m < \$32{,}000$
   - (B) $f + m > \$32{,}000$
   - (C) $f - m < \$32{,}000$
   - (D) $2f < \$32{,}000$

2. If the sum of two consecutive odd integers is 56, then the greater integer equals— *(Algebra Review)*
   - (A) 25.
   - (B) 27.
   - (C) 29.
   - (D) 31.
   - (E) 33.

3. The distance an object covers when it moves at a constant speed, or rate, is given by the formula $d = rt$, where d is distance, r is rate, and t is time. How far does a car travel in $2\frac{1}{2}$ hours moving at a constant speed of 60 miles per hour? *(Algebra Review)*
   - (A) 30 mi
   - (B) 60 mi
   - (C) 150 mi
   - (D) 300 mi

4. If 3 more than x is 2 more than y, what is x in terms of y? *(Algebra Review)*
   - (A) $y - 5$
   - (B) $y - 1$
   - (C) $y + 1$
   - (D) $y + 5$
   - (E) $y + 6$

5. The annual salaries for the eight employees in a small company are $12,000, $14,500, $14,500, $18,000, $21,000, $27,000, $38,000, and $82,000. Which of these measures of central tendency would make the company salaries seem as large as possible? *(Statistics Review)*
   - (A) mean
   - (B) median
   - (C) mode
   - (D) range

6. Shari's test scores in Spanish class are 73, 86, 91, and 82. She needs at least 400 points to earn a B. Which inequality describes the number of points p Shari must yet earn in order to receive a B? *(Algebra Review)*
   - (A) $p - 332 > 400$
   - (B) $p - 332 > 400$
   - (C) $p + 332 \geq 400$
   - (D) $400 - p \geq 332$

7. In $\triangle ABC$, $\angle A \cong \angle B$, and $m\angle C$ is twice the measure of $\angle B$. What is the measure, in degrees, of $\angle A$? *(Lesson 5-2)*

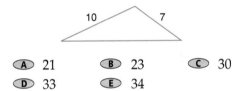

   - (A) 30
   - (B) 40
   - (C) 45
   - (D) 90

8. Which *cannot* be the perimeter of the triangle shown below? *(Lesson 7–4)*

   10   7

   - (A) 21
   - (B) 23
   - (C) 30
   - (D) 33
   - (E) 34

Grid In

9. A car repair service charges $36 per hour plus the cost of the parts. If Ken is charged $70.50 for repairs that took 1.5 hours, what was the cost in dollars and cents of the parts used? *(Algebra Review)*

Extended Response

10. Mei Hua is buying a $445 television set that is on sale for 30% off. The sales tax in her state is 6%. She reasons that she will save 30%, then pay 6%, so the total savings from the price listed will be 24%. She then calculates her price as $445 − 0.24($445). *(Percent Review)*

   Part A Calculate what answer she gets.

   Part B Is she right? If so, why? If not, why not, and what is the correct answer?

Quadrilaterals

What You'll Learn

Key Ideas

- Identify parts of quadrilaterals and find the sum of the measures of the interior angles of a quadrilateral. *(Lesson 8-1)*

- Identify and use the properties of parallelograms. *(Lessons 8-2 and 8-3)*

- Identify and use the properties of rectangles, rhombi, squares, trapezoids, and isosceles trapezoids. *(Lessons 8-4 and 8-5)*

Key Vocabulary

parallelogram *(p. 316)*

quadrilateral *(p. 310)*

rectangle *(p. 327)*

rhombus *(p. 327)*

square *(p. 327)*

trapezoid *(p. 333)*

Why It's Important

Art The work of architect Filippo Brunelleschi, designer of the famed cathedral in Florence, Italy, led to a mathematical theory of perspective. He probably developed his theories to help him render architectural drawings. In learning the mathematics of perspective, Renaissance painters were able to depict figures more fully and realistically than artists from the Middle Ages.

Quadrilaterals are used in construction and architecture. You will investigate trapezoids in perspective drawings in Lesson 8-5.

✔Check Your Readiness

Find the value of each variable.

1.

2.

3.
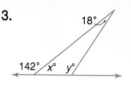

Find the measure of each angle.
Give a reason for each answer.

4. ∠1   5. ∠2

6. ∠3   7. ∠4

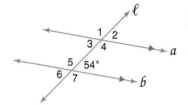

For each triangle, find the values of the variables.

8.

9.

10.

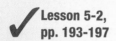

FOLDABLES™
Study Organizer

Make this Foldable to help you organize your Chapter 8 notes. Begin with three sheets of grid paper.

❶ **Fold** each sheet in half from top to bottom.

❷ **Cut** along each fold. Staple the six half-sheets together to form a booklet.

❸ **Cut** five tabs. The top tab is 3 lines wide, the next tab is 6 lines wide, and so on.

❹ **Label** each of the tabs with a lesson number.

Reading and Writing As you read and study the chapter, fill the journal with terms, diagrams, and theorems for each lesson.

What You'll Learn
You'll learn to identify parts of quadrilaterals and find the sum of the measures of the interior angles of a quadrilateral.

Why It's Important
City Planning
City planners use quadrilaterals in their designs.
See Exercise 36.

The building below was designed by Laurinda Spear. Different quadrilaterals are used as faces of the building.

Centre for Innovative Technology, Fairfax and Louden Counties, Virginia

A **quadrilateral** is a closed geometric figure with four sides and four vertices. The segments of a quadrilateral intersect only at their endpoints. Special types of quadrilaterals include squares and rectangles.

Quadrilaterals	*Not* Quadrilaterals

Quadrilaterals are named by listing their vertices in order. There are many names for the quadrilateral at the right. Some examples are quadrilateral *ABCD*, quadrilateral *BCDA*, or quadrilateral *DCBA*.

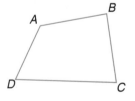

Any two sides, vertices, or angles of a quadrilateral are either **consecutive** or **nonconsecutive**.

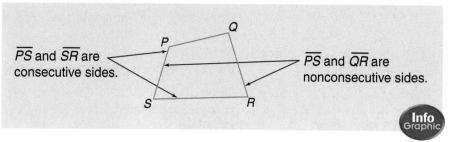

$\overline{PS}$ and $\overline{SR}$ are consecutive sides.

$\overline{PS}$ and $\overline{QR}$ are nonconsecutive sides.

Info Graphic

Segments that join nonconsecutive vertices of a quadrilateral are called **diagonals**.

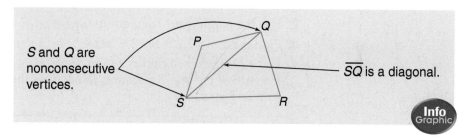

S and Q are nonconsecutive vertices.

$\overline{SQ}$ is a diagonal.

Info Graphic

Examples

Refer to quadrilateral *ABLE*.

1 Name all pairs of consecutive angles.

$\angle A$ and $\angle B$, $\angle B$ and $\angle L$, $\angle L$ and $\angle E$, and $\angle E$ and $\angle A$ are consecutive angles.

2 Name all pairs of nonconsecutive vertices.

A and *L* are nonconsecutive vertices.
B and *E* are nonconsecutive vertices.

3 Name the diagonals.

$\overline{AL}$ and $\overline{BE}$ are the diagonals.

Your Turn

Refer to quadrilateral *WXYZ*.

a. Name all pairs of consecutive sides.
b. Name all pairs of nonconsecutive angles.
c. Name the diagonals.

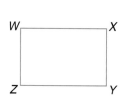

Look Back

Angle Sum Theorem: Lesson 5–2

In Chapter 5, you learned that the sum of the measures of the angles of a triangle is 180. You can use this result to find the sum of the measures of the angles of a quadrilateral.

Materials: / straightedge △ protractor

Step 1 Draw a quadrilateral like the one at the right. Label its vertices *A*, *B*, *C*, and *D*.

Step 2 Draw diagonal $\overline{AC}$. Note that two triangles are formed. Label the angles as shown.

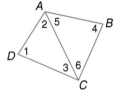

Try These

1. Use the Angle Sum Theorem to find $m\angle 1 + m\angle 2 + m\angle 3$.
2. Use the Angle Sum Theorem to find $m\angle 4 + m\angle 5 + m\angle 6$.
3. Find $m\angle 1 + m\angle 2 + m\angle 3 + m\angle 4 + m\angle 5 + m\angle 6$.
4. Use a protractor to find $m\angle 1$, $m\angle DAB$, $m\angle 4$, and $m\angle BCD$. Then find the sum of the angle measures. How does the sum compare to the sum in Exercise 3?

You can summarize the results of the activity in the following theorem.

Theorem 8–1	**Words:** The sum of the measures of the angles of a quadrilateral is 360.
	Model: **Symbols:** $a + b + c + d = 360$

Example

④ **Find the missing measure in quadrilateral *WXYZ*.**

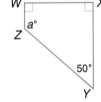

$$m\angle W + m\angle X + m\angle Y + m\angle Z = 360 \quad \textit{Theorem 8–1}$$
$$90 + 90 + 50 + a = 360 \quad \textit{Substitution}$$
$$230 + a = 360 \quad \textit{Add.}$$
$$230 - 230 + a = 360 - 230 \quad \textit{Subtract 230 from each side.}$$
$$a = 130 \quad \textit{Simplify.}$$

Therefore, $m\angle Z = 130$.

Your Turn

d. Find the missing measure if three of the four angle measures in quadrilateral *ABCD* are 50, 60, and 150.

 Personal Tutor at geomconcepts.com

Example

Algebra Link

⑤ **Find the measure of ∠U in quadrilateral *KDUC* if *m∠K* = 2x, *m∠D* = 40, *m∠U* = 2x and *m∠C* = 40.**

$m\angle K + m\angle D + m\angle U + m\angle C = 360$		*Theorem 8–1*
$2x + 40 + 2x + 40 = 360$		*Substitution*
$4x + 80 = 360$		*Add.*
$4x + 80 - 80 = 360 - 80$		*Subtract 80 from each side.*
$4x = 280$		*Simplify.*
$\dfrac{4x}{4} = \dfrac{280}{4}$		*Divide each side by 4.*
$x = 70$		*Simplify.*

Since $m\angle U = 2x$, $m\angle U = 2 \cdot 70$ or 140.

Algebra Review
Solving Multi-Step
Equations, p. 723

Your Turn

e. Find the measure of ∠B in quadrilateral *ABCD* if $m\angle A = x$, $m\angle B = 2x$, $m\angle C = x - 10$, and $m\angle D = 50$.

Check for Understanding

Communicating Mathematics

1. Sketch and label a quadrilateral in which $\overline{AC}$ is a diagonal.

2. Writing Math Draw three figures that are *not* quadrilaterals. Explain why each figure is *not* a quadrilateral.

Vocabulary

quadrilateral
consecutive
nonconsecutive
diagonal

Guided Practice

⏱ **Getting Ready** **Solve each equation.**

Sample: $120 + 55 + 45 + x = 360$ **Solution:** $220 + x = 360$
$x = 140$

3. $130 + x + 50 + 80 = 360$ **4.** $90 + 90 + x + 55 = 360$
5. $28 + 72 + 134 + x = 360$ **6.** $x + x + 85 + 105 = 360$

Refer to quadrilateral *MQPN* for Exercises 7–9.

Example 1   **7.** Name a pair of consecutive angles.
Example 2   **8.** Name a pair of nonconsecutive vertices.
Example 3   **9.** Name a diagonal.

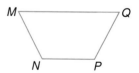

Example 4   **Find the missing measure in each figure.**

10.

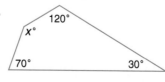

11.

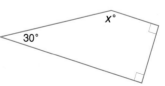

Lesson 8–1 Quadrilaterals   **313**

Example 5

12. **Algebra** Find the measure of ∠A in quadrilateral BCDA if m∠B = 60, m∠C = 2x + 5, m∠D = x, and m∠A = 2x + 5.

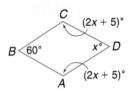

Exercises

· · · · · ● · · · · · · · · · · · · · · ●

Practice

Homework Help	
For Exercises	See Examples
17, 19	3
18	1
21	2
22–27, 35	4, 5
36	1–3
Extra Practice	
See page 739.	

Refer to quadrilaterals QRST and FGHJ.

13. Name a side that is consecutive with $\overline{RS}$.

14. Name the side opposite $\overline{ST}$.

15. Name a pair of consecutive vertices in quadrilateral QRST.

16. Name the vertex that is opposite S.

17. Name the two diagonals in quadrilateral QRST.

18. Name a pair of consecutive angles in quadrilateral QRST.

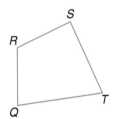

Exercises 13–18

19. Name a diagonal in quadrilateral FGHJ.

20. Name a pair of nonconsecutive sides in quadrilateral FGHJ.

21. Name the angle opposite ∠F.

Exercises 19–21

Find the missing measure(s) in each figure.

22.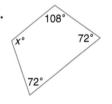

23.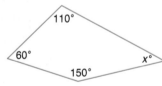

24.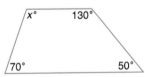

25.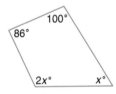

26.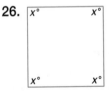

27.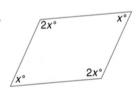

28. Three of the four angle measures in a quadrilateral are 90, 90, and 125. Find the measure of the fourth angle.

Use a straightedge and protractor to draw quadrilaterals that meet the given conditions. If none can be drawn, write *not possible*.

29. exactly two acute angles
30. exactly four right angles
31. exactly four acute angles
32. exactly one obtuse angle
33. exactly three congruent sides
34. exactly four congruent sides

Applications and Problem Solving

35. **Algebra**   Find the measure of each angle in quadrilateral *RSTU* if $m\angle R = x$, $m\angle S = x + 10$, $m\angle T = x + 30$, and $m\angle U = 50$.

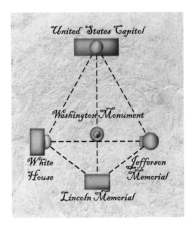

36. **City Planning**   Four of the most popular tourist attractions in Washington, D.C., are located at the vertices of a quadrilateral. Another attraction is located on one of the diagonals.

   a. Name the attractions that are located at the vertices.

   b. Name the attraction that is located on a diagonal.

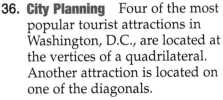

37. **Critical Thinking**   Determine whether a quadrilateral can be formed with strips of paper measuring 8 inches, 4 inches, 2 inches, and 1 inch. Explain your reasoning.

Mixed Review

Determine whether the given numbers can be the measures of the sides of a triangle. Write *yes* or *no*.   *(Lesson 7–4)*

38. 6, 4, 10
39. 2.2, 3.6, 5.7
40. 3, 10, 13.6

41. In $\triangle LNK$, $m\angle L < m\angle K$ and $m\angle L > m\angle N$. Which side of $\triangle LNK$ has the greatest measure?   *(Lesson 7–3)*

Name the additional congruent parts needed so that the triangles are congruent by the indicated postulate or theorem.   *(Lesson 5–6)*

42. ASA

43. AAS

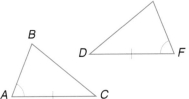

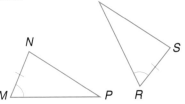

Standardized Test Practice
Ⓐ Ⓑ Ⓒ Ⓓ

44. **Multiple Choice**   The total number of students enrolled in public colleges in the U.S. is expected to be about 12,646,000 in 2005. This is a 97% increase over the number of students enrolled in 1970. About how many students were enrolled in 1970?   *(Algebra Review)*

   Ⓐ 94,000
   Ⓑ 6,419,000
   Ⓒ 12,267,000
   Ⓓ 24,913,000

What You'll Learn

You'll learn to identify and use the properties of parallelograms.

What You'll Learn

You'll learn to identify and use the properties of parallelograms.

Why It's Important

Carpentry
Carpenters use the properties of parallelograms when they build stair rails.
See Exercise 28.

A **parallelogram** is a quadrilateral with two pairs of parallel sides. A symbol for parallelogram $ABCD$ is $\square ABCD$. In $\square ABCD$ below, $\overline{AB}$ and $\overline{DC}$ are parallel sides. Also, $\overline{AD}$ and $\overline{BC}$ are parallel sides. The parallel sides are congruent.

Graphing Calculator Exploration

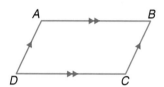

Graphing Calculator Tutorial

See pp. 782–785.

Step 1 Use the Segment tool on the **F2** menu to draw segments AB and AD that have a common endpoint A. Be sure the segments are not collinear. Label the endpoints.

Step 2 Use the Parallel Line tool on the **F3** menu to draw a line through point B parallel to $\overline{AD}$. Next, draw a line through point D parallel to $\overline{AB}$.

Step 3 Use the Intersection Point tool on the **F2** menu to mark the point where the lines intersect. Label this point C. Use the Hide/Show tool on the **F5** menu to hide the lines.

Step 4 Finally, use the Segment tool to draw $\overline{BC}$ and $\overline{DC}$. You now have a parallelogram whose properties can be studied with the calculator.

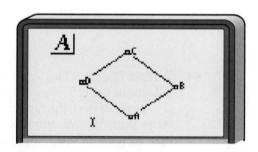

Try These

1. Use the Angle tool under Measure on the **F5** menu to **verify** that the opposite angles of a parallelogram are congruent. Describe your procedure.

2. Use the Distance & Length tool under Measure on the **F5** menu to **verify** that the opposite sides of a parallelogram are congruent. Describe your procedure.

3. Measure two pairs of consecutive angles. **Make a conjecture** as to the relationship between consecutive angles in a parallelogram.

4. Draw the diagonals of □*ABCD*. Label their intersection *E*. Measure $\overline{AE}$, $\overline{BE}$, $\overline{CE}$, and $\overline{DE}$. **Make a conjecture** about the diagonals of a parallellogram.

The results of the activity can be summarized in the following theorems.

Theorem	Words	Models and Symbols
8–2	Opposite angles of a parallelogram are congruent.	$\angle A \cong \angle C, \angle B \cong \angle D$
8–3	Opposite sides of a parallelogram are congruent.	$\overline{AB} \cong \overline{DC}, \overline{AD} \cong \overline{BC}$
8–4	The consecutive angles of a parallelogram are supplementary.	$m\angle A + m\angle B = 180$ $m\angle A + m\angle D = 180$

Using Theorem 8-4, you can show that the sum of the measures of the angles of a parallelogram is 360.

Examples

In □*PQRS*, *PQ* = 20, *QR* = 15, and *m*∠*S* = 70.

1 Find *SR* and *SP*.

$\overline{SR} \cong \overline{PQ}$ and $\overline{SP} \cong \overline{QR}$ *Theorem 8–3*

SR = *PQ* and *SP* = *QR* *Definition of congruent segments*

SR = 20 and *SP* = 15 *Replace PQ with 20 and QR with 15.*

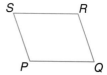

2 Find $m\angle Q$.

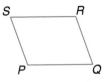

$\angle Q \cong \angle S$ *Theorem 8–2*

$m\angle Q = m\angle S$ *Definition of congruent angles*

$m\angle Q = \; 70$ *Replace $m\angle S$ with 70.*

3 Find $m\angle P$.

$m\angle S + m\angle P = 180$ *Theorem 8–4*

$70 \; + m\angle P = 180$ *Replace $m\angle S$ with 70.*

$70 - 70 + m\angle P = 180 - 70$ *Subtract 70 from each side.*

$m\angle P = 110$ *Simplify.*

Your Turn

In □$DEFG$, $DE = 70$, $EF = 45$, and $m\angle G = 68$.

a. Find GF.

b. Find DG.

c. Find $m\angle E$.

d. Find $m\angle F$.

The result in Theorem 8–5 was also found in the Graphing Calculator Exploration.

Theorem 8–5	**Words:** The diagonals of a parallelogram bisect each other.
	Model: 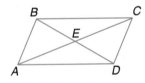
	Symbols: $\overline{AE} \cong \overline{EC}$, $\overline{BE} \cong \overline{ED}$

Example **4** In □$ABCD$, if $AC = 56$, find AE.

Theorem 8–5 states that the diagonals of a parallelogram bisect each other. Therefore, $\overline{AE} \cong \overline{EC}$ or $AE = \frac{1}{2}(AC)$.

$AE = \frac{1}{2}(AC)$ *Definition of bisect*

$AE = \frac{1}{2}(56)$ or **28**   *Replace AC with 56.*

Your Turn

e. If $DE = 11$, find DB.

A diagonal separates a parallelogram into two triangles. You can use the properties of parallel lines to find the relationship between the two triangles. Consider ▱ABCD with diagonal $\overline{AC}$.

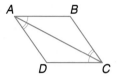

Look Back

Alternate Interior Angles: Lesson 4–2; ASA: Lesson 5–6

1. $\overline{DC} \parallel \overline{AB}$ and $\overline{AD} \parallel \overline{BC}$ *Definition of parallelogram*
2. $\angle ACD \cong \angle CAB$ and   *If two parallel lines are cut by a transversal,*
    $\angle CAD \cong \angle ACB$ *alternate interior angles are congruent.*
3. $\overline{AC} \cong \overline{AC}$ *Reflexive Property*
4. $\triangle ACD \cong \triangle CAB$ *ASA*

This property of the diagonal is illustrated in the following theorem.

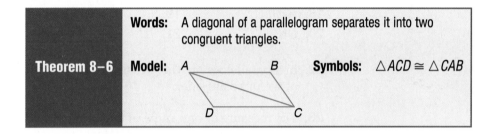

	Words: A diagonal of a parallelogram separates it into two congruent triangles.
Theorem 8–6	**Model:** **Symbols:** $\triangle ACD \cong \triangle CAB$

Check for Understanding

Communicating Mathematics

1. **Name** five properties that all parallelograms have.

2. **Draw** parallelogram *MEND* with diagonals *MN* and *DE* intersecting at *X*. Name four pairs of congruent segments.

3. Karen and Tai know that the measure of one angle of a parallelogram is 50°. Karen thinks that she can find the measures of the remaining three angles without a protractor. Tai thinks that is not possible. Who is correct? Explain your reasoning.

Guided Practice

Examples 1–3

Find each measure.

4. $m\angle S$ 5. $m\angle P$
6. *MP*   7. *PS*

Example 4

8. Suppose the diagonals of ▱*MPSA* intersect at point *T*. If *MT* = 15, find *MS*.

Example 1

9. **Drafting** Three parallelograms are used to produce a three-dimensional view of a cube. Name all of the segments that are parallel to the given segment.
    a. $\overline{AB}$ b. $\overline{BE}$ c. $\overline{DG}$

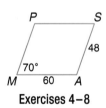

Exercises 4–8

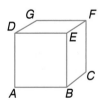

Practice

Find each measure.

10. $m\angle A$
11. $m\angle B$
12. AB
13. BC

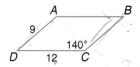

In the figure, $OE = 19$ and $EU = 12$. Find each measure.

14. LE
15. JO
16. $m\angle OUL$
17. $m\angle OJL$
18. $m\angle JLU$
19. EJ
20. OL
21. JU

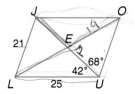

22. In a parallelogram, the measure of one side is 7. Find the measure of the opposite side.

23. The measure of one angle of a parallelogram is 35. Determine the measures of the other three angles.

Determine whether each statement is *true* or *false*.

24. The diagonals of a parallelogram are congruent.

25. In a parallelogram, when one diagonal is drawn, two congruent triangles are formed.

26. If the length of one side of a parallelogram is known, the lengths of the other three sides can be found without measuring.

Applications and Problem Solving

27. **Art** The Escher design below is based on a parallelogram. You can use a parallelogram to make a simple Escher-like drawing. Change one side of the parallelogram and then slide the change to the opposite side. The resulting figure is used to make a design with different colors and textures.

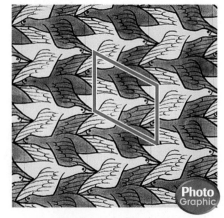

M. C. Escher, *Study of Regular Division of the Plane with Birds*

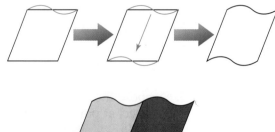

Make your own Escher-like drawing.

28. **Carpentry** The part of the stair rail that is outlined forms a parallelogram because the spindles are parallel and the top railing is parallel to the bottom railing. Name two pairs of congruent sides and two pairs of congruent angles in the parallelogram.

29. **Critical Thinking** If the measure of one angle of a parallelogram increases, what happens to the measure of its adjacent angles so that the figure remains a parallelogram?

Mixed Review

The measures of three of the four angles of a quadrilateral are given. Find the missing measure. *(Lesson 8–1)*

30. 55, 80, 125

31. 74, 106, 106

32. If the measures of two sides of a triangle are 3 and 7, find the range of possible measures of the third side. *(Lesson 7–4)*

33. **Short Response** Drafters use the MIRROR command to produce a mirror image of an object. Identify this command as a *translation*, *reflection*, or *rotation*. *(Lesson 5–3)*

34. **Multiple Choice** If $m\angle XRS = 68$ and $m\angle QRY = 136$, find $m\angle XRY$.
(Lesson 3–5)

Ⓐ 24   Ⓑ 44
Ⓒ 64   Ⓓ 204

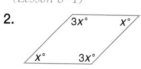

Quiz 1   Lessons 8–1 and 8–2

Find the missing measure(s) in each figure. *(Lesson 8–1)*

1.
$60°$ $x°$ $58°$
$79°$

2.
$3x°$ $x°$
$x°$ $3x°$

3. **Algebra** Find the measure of $\angle R$ in quadrilateral $RSTW$ if $m\angle R = 2x$, $m\angle S = x - 7$, $m\angle T = x + 5$, and $m\angle W = 30$. *(Lesson 8–1)*

In $\square DEFG$, $m\angle E = 63$ and $EF = 16$. Find each measure. *(Lesson 8–2)*

4. $m\angle D$

5. DG

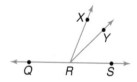

Theorem 8-3 states that the opposite sides of a parallelogram are congruent. Is the converse of this theorem true? In the figure below, $\overline{AB}$ is congruent to $\overline{DC}$ and $\overline{AD}$ is congruent to $\overline{BC}$.

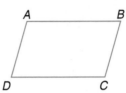

You know that a parallelogram is a quadrilateral in which both pairs of opposite sides are parallel. If the opposite sides of a quadrilateral are congruent, then is it a parallelogram?

In the following activity, you will discover other ways to show that a quadrilateral is a parallelogram.

Hands-On Geometry

Materials: 🖉 straws ✂ scissors 〰 pipe cleaners 📏 ruler

Step 1 Cut two straws to one length and two straws to a different length.

Step 2 Insert a pipe cleaner in one end of each straw. Connect the pipe cleaners at the ends to form a quadrilateral.

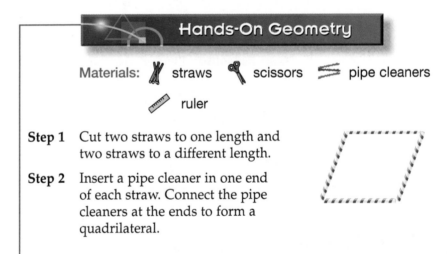

Try These

1. How do the measures of opposite sides compare?
2. Measure the distance between the top and bottom straws in at least three places. Then measure the distance between the left and right straws in at least three places. What seems to be true about the opposite sides?
3. Shift the position of the sides to form another quadrilateral. Repeat Exercises 1 and 2.
4. What type of quadrilateral have you formed? Explain your reasoning.

This activity leads to Theorem 8–7, which is related to Theorem 8–3.

| **Theorem 8–7** | **Words:** | If both pairs of opposite sides of a quadrilateral are congruent, then the quadrilateral is a parallelogram. |
| | **Model:** | 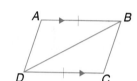 **Symbols:** $\overline{RS} \cong \overline{UT}$, $\overline{RU} \cong \overline{ST}$ |

You can use the properties of congruent triangles and Theorem 8–7 to find other ways to show that a quadrilateral is a parallelogram.

Example ❶ In quadrilateral *ABCD*, with diagonal *BD*, $\overline{AB} \parallel \overline{CD}$, $\overline{AB} \cong \overline{CD}$. Show that *ABCD* is a parallelogram.

Explore   You know $\overline{AB} \parallel \overline{CD}$ and $\overline{AB} \cong \overline{CD}$. You want to show that *ABCD* is a parallelogram.

Plan   One way to show *ABCD* is a parallelogram is to show $\overline{AD} \cong \overline{CB}$. You can do this by showing $\triangle ABD \cong \triangle CDB$. Make a list of statements and their reasons.

Look Back

Alternate Interior Angles: Lesson 4–2

Solve **1.** $\angle ABD \cong \angle CDB$ *If two $\parallel$ lines are cut by a transversal, then each pair of alternate interior angles is $\cong$.*

2. $\overline{BD} \cong \overline{BD}$ *Reflexive Property*
3. $\overline{AB} \cong \overline{CD}$ *Given*
4. $\triangle ABD \cong \triangle CDB$ *SAS*
5. $\overline{AD} \cong \overline{CB}$ *CPCTC*
6. *ABCD* is a *Theorem 8–7*
parallelogram.

Your Turn

In quadrilateral *PQRS*, $\overline{PR}$ and $\overline{QS}$ bisect each other at *T*. Show that *PQRS* is a parallelogram by providing a reason for each step.

a. $\overline{PT} \cong \overline{TR}$ and $\overline{QT} \cong \overline{TS}$
b. $\angle PTQ \cong \angle RTS$ and $\angle STP \cong \angle QTR$
c. $\triangle PQT \cong \triangle RST$ and $\triangle PTS \cong \triangle RTQ$
d. $\overline{PQ} \cong \overline{RS}$ and $\overline{PS} \cong \overline{RQ}$
e. *PQRS* is a parallelogram.

These examples lead to Theorems 8–8 and 8–9.

Theorem	Words	Models and Symbols
8–8	If one pair of opposite sides of a quadrilateral is parallel and congruent, then the quadrilateral is a parallelogram.	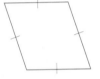 $\overline{AB} \cong \overline{DC}, \overline{AB} \parallel \overline{DC}$
8–9	If the diagonals of a quadrilateral bisect each other, then the quadrilateral is a parallelogram.	 $\overline{AE} \cong \overline{EC}, \overline{BE} \cong \overline{ED}$

Examples

Determine whether each quadrilateral is a parallelogram. If the figure is a parallelogram, give a reason for your answer.

2

The figure has two pairs of opposite sides that are congruent. The figure is a parallelogram by Theorem 8–7.

3

The figure has two pairs of congruent sides, but they are *not* opposite sides. The figure is *not* a parallelogram.

Your Turn

f.

g.

Check for Understanding

1. **Draw** a quadrilateral that meets each set of conditions and is *not* a parallelogram.

   a. one pair of parallel sides

   b. one pair of congruent sides

   c. one pair of congruent sides and one pair of parallel sides

2. Writing Math List four methods you can use to determine whether a quadrilateral is a parallelogram.

Determine whether each quadrilateral is a parallelogram. Write *yes* or *no*. If *yes*, give a reason for your answer.

Examples 2 & 3

3.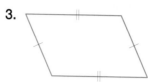

4.

Example 1

5. In quadrilateral *ABCD*, $\overline{BA} \parallel \overline{CD}$ and $\angle DBC \cong \angle BDA$. Show that quadrilateral *ABCD* is a parallelogram by providing a reason for each step.

    a. $\overline{BC} \parallel \overline{AD}$

    b. *ABCD* is a parallelogram.

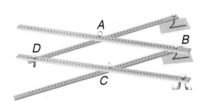

Examples 2 & 3

6. In the figure, $\overline{AD} \cong \overline{BC}$ and $\overline{AB} \cong \overline{DC}$. Which theorem shows that quadrilateral *ABCD* is a parallelogram?

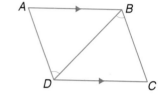

Exercises

$360°$

Practice

Determine whether each quadrilateral is a parallelogram. Write *yes* or *no*. If *yes*, give a reason for your answer.

Homework Help	
For Exercises	See Examples
7–12, 14–16	2, 3
13	1
17	2

Extra Practice
See page 740.

7.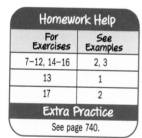

8.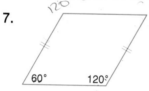

9.

10.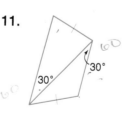

11.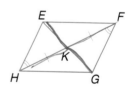

12.

13. In quadrilateral *EFGH*, $\overline{HK} \cong \overline{KF}$ and $\angle KHE \cong \angle KFG$. Show that quadrilateral *EFGH* is a parallelogram by providing a reason for each step.

    a. $\angle EKH \cong \angle FKG$

    b. $\triangle EKH \cong \triangle GKF$

    c. $\overline{EH} \cong \overline{GF}$

    d. $\overline{EH} \parallel \overline{GF}$

    e. *EFGH* is a parallelogram.

14. Explain why quadrilateral *LMNT* is a parallelogram. Support your explanation with reasons as shown in Exercise 13.

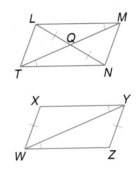

15. Determine whether quadrilateral *XYZW* is a parallelogram. Give reasons for your answer.

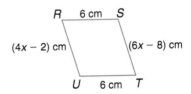

Applications and Problem Solving

16. **Algebra** Find the value for *x* that will make quadrilateral *RSTU* a parallelogram.

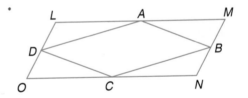

17. **Quilting** Faith Ringgold is an African-American fabric artist. She used parallelograms in the design of the quilt at the left. What characteristics of parallelograms make it easy to use them in quilts?

18. **Critical Thinking** Quadrilateral *LMNO* is a parallelogram. Points *A*, *B*, *C*, and *D* are midpoints of the sides. Is *ABCD* a parallelogram? Explain your reasoning.

Faith Ringgold, *#4 The Sunflowers Quilting Bee at Arles*

Mixed Review

In □*ABCD*, *m*∠*D* = 62 and *CD* = 45. Find each measure. *(Lesson 8–2)*

19. *m*∠*B*

20. *m*∠*C*

21. *AB*

22. **Drawing** Use a straightedge and protractor to draw a quadrilateral with exactly two obtuse angles. *(Lesson 8–1)*

23. Find the length of the hypotenuse of a right triangle whose legs are 7 inches and 24 inches. *(Lesson 6–6)*

Standardized Test Practice
Ⓐ Ⓑ Ⓒ Ⓓ

24. **Grid In** In order to "curve" a set of test scores, a teacher uses the equation $g = 2.5p + 10$, where *g* is the curved test score and *p* is the number of problems answered correctly. How many points is each problem worth? *(Lesson 4–6)*

25. **Short Response** Name two different pairs of angles that, if congruent, can be used to prove *a* ∥ *b*. Explain your reasoning. *(Lesson 4–4)*

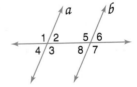

www.geomconcepts.com/self_check_quiz

In previous lessons, you studied the properties of quadrilaterals and parallelograms. Now you will learn the properties of three other special types of quadrilaterals: **rectangles**, **rhombi**, and **squares**. The following diagram shows how these quadrilaterals are related.

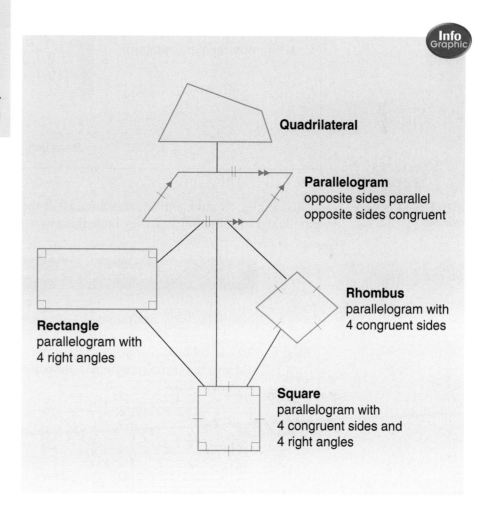

Quadrilateral

Parallelogram
opposite sides parallel
opposite sides congruent

Rhombus
parallelogram with
4 congruent sides

Rectangle
parallelogram with
4 right angles

Square
parallelogram with
4 congruent sides and
4 right angles

Notice how the diagram goes from the most general quadrilateral to the most specific one. Any four-sided figure is a quadrilateral. But a parallelogram is a special quadrilateral whose opposite sides are parallel. The opposite sides of a square are parallel, so a square is a parallelogram. In addition, the four angles of a square are right angles, and all four sides are equal. A rectangle is also a parallelogram with four right angles, but its four sides are not equal.

Both squares and rectangles are special types of parallelograms. The best description of a quadrilateral is the one that is the most specific.

Example

Art Link

1

Identify the parallelogram that is outlined in the painting at the right.

Parallelogram *ABCD* has four right angles, but the four sides are not congruent. It is a rectangle.

Your Turn

a. Identify the parallelogram.

Diana Ong, *Blue, Red, and Yellow Faces*

Reading Geometry

Rhombi is the plural of *rhombus*.

Rectangles, rhombi, and squares have all of the properties of parallelograms. In addition, they have their own properties.

Hands-On Geometry

Materials: dot paper ruler protractor

Step 1 Draw a rhombus on isometric dot paper. Draw a square and a rectangle on rectangular dot paper. Label each figure as shown below.

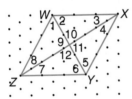
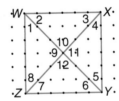
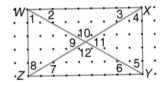

Step 2 Measure $\overline{WY}$ and $\overline{XZ}$ for each figure.

Step 3 Measure ∠9, ∠10, ∠11, and ∠12 for each figure.

Step 4 Measure ∠1 through ∠8 for each figure.

Try These

1. For which figures are the diagonals congruent?

2. For which figures are the diagonals perpendicular?

3. For which figures do the diagonals bisect a pair of opposite angles?

www.geomconcepts.com/extra_examples

The results of the previous activity can be summarized in the following theorems.

Theorem	Words	Models and Symbols
8–10	The diagonals of a rectangle are congruent.	$\overline{AC} \cong \overline{BD}$
8–11	The diagonals of a rhombus are perpendicular.	$\overline{AC} \perp \overline{BD}$
8–12	Each diagonal of a rhombus bisects a pair of opposite angles.	$m\angle 1 = m\angle 2,\ m\angle 3 = m\angle 4,$ $m\angle 5 = m\angle 6,\ m\angle 7 = m\angle 8$

CΟncepts in MΟtion
Animation
geomconcepts.com

A square is defined as a parallelogram with four congruent angles and four congruent sides. This means that a square is not only a parallelogram, but also a rectangle and a rhombus. Therefore, all of the properties of parallelograms, rectangles, and rhombi hold true for squares.

Examples

2 **Find XZ in square $XYZW$ if $YW = 14$.**

A square has all of the properties of a rectangle, and the diagonals of a rectangle are congruent. So, $\overline{XZ}$ is congruent to $\overline{YW}$, and $XZ = 14$.

3 **Find $m\angle YOX$ in square $XYZW$.**

A square has all the properties of a rhombus, and the diagonals of a rhombus are perpendicular. Therefore, $m\angle YOX = 90$.

Your Turn

b. Name all segments that are congruent to $\overline{WO}$ in square $XYZW$. Explain your reasoning.

c. Name all the angles that are congruent to $\angle XYO$ in square $XYZW$. Explain your reasoning.

Online **Personal Tutor at** geomconcepts.com

Check for Understanding

Communicating Mathematics

1. **Draw** a quadrilateral that is a rhombus but not a rectangle.

2. **Compare and contrast** the definitions of rectangles and squares.

3. Eduardo says that every rhombus is a square. Teisha says that every square is a rhombus. Who is correct? Explain your reasoning.

Guided Practice

🕐 **Getting Ready**   **Which quadrilaterals have each property?**

Sample: All angles are right angles.   **Solution:** square, rectangle

4. The opposite angles are congruent.

5. The opposite sides are congruent.

6. All sides are congruent.

Example 1

Identify each parallelogram as a *rectangle, rhombus, square,* or *none of these.*

7.

8.

Examples 2 & 3

Use square *FNRM* or rhombus *STPK* to find each measure.

9. *AR*   10. *MA*

11. *m∠FAN*   12. *TP*

13. *PB*   14. *m∠KTP*

Example 1

15. **Sports** Basketball is played on a court that is shaped like a rectangle. Name two other sports that are played on a rectangular surface and two sports that are played on a surface that is not rectangular.

Exercises

Practice

Identify each parallelogram as a *rectangle, rhombus, square,* or *none of these.*

16.

17.

18.

Homework Help

For Exercises	See Examples
16–21	1
22–38	2, 3

Extra Practice
See page 740.

19.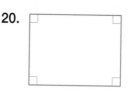

20.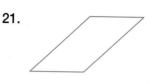

21.

Use square *SQUR* or rhombus *LMPY* to find each measure.

22. *EQ*　　　　　　**23.** *EU*

24. *SU*　　　　　　**25.** *RQ*

26. *m∠SEQ*　　　　**27.** *m∠SQU*

28. *m∠SQE*　　　　**29.** *m∠RUE*

30. *ZP*　　　　　　**31.** *YM*

32. *m∠LMP*　　　　**33.** *m∠MLY*

34. *m∠YZP*　　　　**35.** *YL*

36. *YP*　　　　　　**37.** *m∠LPM*

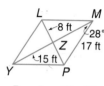

Exercises 22–29

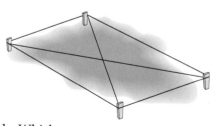

Exercises 30–37

38. Which quadrilaterals have diagonals that are perpendicular?

The Venn diagram shows relationships among some quadrilaterals. Use the Venn diagram to determine whether each statement is *true* or *false*.

39. Every square is a rhombus.

40. Every rhombus is a square.

41. Every rectangle is a square.

42. Every square is a rectangle.

43. All rhombi are parallelograms.

44. Every parallelogram is a rectangle.

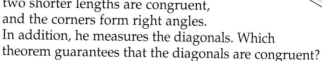

Quadrilaterals

Applications and Problem Solving

45. Algebra The diagonals of a square are $(x + 8)$ feet and $3x$ feet. Find the measure of the diagonals.

46. Carpentry A carpenter is starting to build a rectangular deck. He has laid out the deck and marked the corners, making sure that the two longer lengths are congruent, the two shorter lengths are congruent, and the corners form right angles. In addition, he measures the diagonals. Which theorem guarantees that the diagonals are congruent?

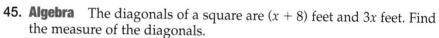

47. Critical Thinking Refer to rhombus *PLAN*.

　a. Classify △*PLA* by its sides.

　b. Classify △*PEN* by its angles.

　c. Is △*PEN* ≅ △*AEL*? Explain your reasoning.

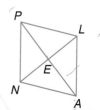

Determine whether each quadrilateral is a parallelogram. State *yes* or *no*. If *yes*, give a reason for your answer. *(Lesson 8–3)*

48.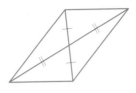

49.

50.

Determine whether each statement is *true* or *false*. *(Lesson 8–2)*

51. If the measure of one angle of a parallelogram is known, the measures of the other three angles can be found without using a protractor.

52. The diagonals of every parallelogram are congruent.

53. The consecutive angles of a parallelogram are complementary.

54. **Extended Response**   Write the converse of this statement. *(Lesson 1–4)*   *If a figure is a rectangle, then it has four sides.*

55. **Multiple Choice**   If x represents the number of homes with televisions in Dallas, which expression represents the number of homes with televisions in Atlanta? *(Algebra Review)*

   Ⓐ $x - 221$

   Ⓑ $x + 221$

   Ⓒ $x - 2035$

   Ⓓ $x + 2035$

Homes with Televisions
(thousands)

Area	Homes
New York	7376
Los Angeles	5402
Chicago	3399
Philadelphia	2874
San Francisco	2441
Boston	2392
Dallas-Ft. Worth	2256
Washington, DC	2224
Atlanta	2035
Detroit	1923

Source: Nielsen Media Research

Quiz 2   Lessons 8–3 and 8–4

Determine whether each quadrilateral is a parallelogram. State *yes* or *no*. If *yes*, give a reason for your answer. *(Lesson 8–3)*

1.

2.

Refer to rhombus *BTLE*. *(Lesson 8–4)*

3. Name all angles that are congruent to $\angle BIE$.
4. Name all segments congruent to $\overline{IE}$.
5. Name all measures equal to BE.

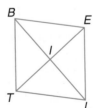

www.geomconcepts.com/self_check_quiz

Many state flags use geometric shapes in their designs. Can you find a quadrilateral in the Maryland state flag that has exactly one pair of parallel sides?

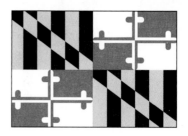

Maryland state flag

A **trapezoid** is a quadrilateral with exactly one pair of parallel sides. The parallel sides are called **bases**. The nonparallel sides are called **legs**.

Study trapezoid *TRAP*.

$\overline{TR} \parallel \overline{PA}$ $\overline{TR}$ and $\overline{PA}$ are the bases.

$\overline{TP} \not\parallel \overline{RA}$ $\overline{TP}$ and $\overline{RA}$ are the legs.

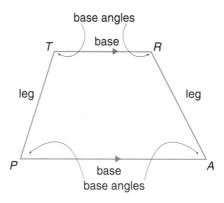

Each trapezoid has two pairs of **base angles**. In trapezoid *TRAP*, $\angle T$ and $\angle R$ are one pair of base angles; $\angle P$ and $\angle A$ are the other pair.

Example ①
Art Link

Artists use *perspective* to give the illusion of depth to their drawings. In perspective drawings, vertical lines remain parallel, but horizontal lines gradually come together at a point. In trapezoid *ZOID*, name the bases, the legs, and the base angles.

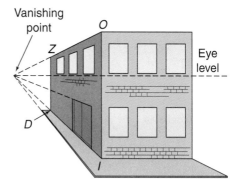

Bases $\overline{ZD}$ and $\overline{OI}$ are parallel segments.
Legs $\overline{ZO}$ and $\overline{DI}$ are nonparallel segments.
Base Angles $\angle Z$ and $\angle D$ are one pair of base angles; $\angle O$ and $\angle I$ are the other pair.

The **median** of a trapezoid is the segment that joins the midpoints of its legs. In the figure, $\overline{MN}$ is the median.

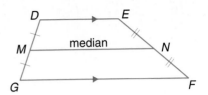

Theorem 8–13	**Words:** The median of a trapezoid is parallel to the bases, and the length of the median equals one-half the sum of the lengths of the bases.
	Model: 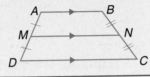 **Symbols:** $\overline{AB} \parallel \overline{MN}$, $\overline{DC} \parallel \overline{MN}$ $MN = \frac{1}{2}(AB + DC)$

Example ➋ Find the length of median MN in trapezoid $ABCD$ if $AB = 12$ and $DC = 18$.

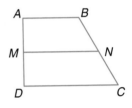

$MN = \frac{1}{2}(AB + DC)$ *Theorem 8–13*

$\quad = \frac{1}{2}(12 + 18)$ *Replace AB with 12 and DC with 18.*

$\quad = \frac{1}{2}(30)$ or 15 *Simplify.*

The length of the median of trapezoid $ABCD$ is 15 units.

Your Turn

a. Find the length of median MN in trapezoid $ABCD$ if $AB = 20$ and $DC = 16$.

Look Back

Isosceles Triangle: Lesson 6–4

If the legs of a trapezoid are congruent, the trapezoid is an **isosceles trapezoid**. In Lesson 6–4, you learned that the base angles of an isosceles triangle are congruent. There is a similar property for isosceles trapezoids.

Theorem 8–14	**Words:** Each pair of base angles in an isosceles trapezoid is congruent.
	Model:  **Symbols:** $\angle W \cong \angle X$, $\angle Z \cong \angle Y$

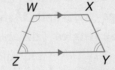 www.geomconcepts.com/extra_examples

Find the missing angle measures in isosceles trapezoid *TRAP*.

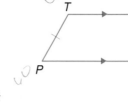

Find $m\angle P$.

$\angle P \cong \angle A$ *Theorem 8–14*

$m\angle P = m\angle A$ *Definition of congruent*

$m\angle P = 60$ *Replace $m\angle A$ with 60.*

Find $m\angle T$. Since *TRAP* is a trapezoid, $\overline{TR} \parallel \overline{PA}$.

$m\angle T + m\angle P = 180$ *Consecutive interior angles are supplementary.*

$m\angle T + 60 = 180$ *Replace $m\angle P$ with 60.*

$m\angle T + 60 - 60 = 180 - 60$ *Subtract 60 from each side.*

$m\angle T = 120$ *Simplify.*

Find $m\angle R$.

$\angle R \cong \angle T$ *Theorem 8–14*

$m\angle R = m\angle T$ *Definition of congruent*

$m\angle R = 120$ *Replace $m\angle T$ with 120.*

Look Back

Consecutive Interior
Angles:
Lesson 4–2

Your Turn

b. The measure of one angle in an isosceles trapezoid is 48. Find the measures of the other three angles.

In this chapter, you have studied quadrilaterals, parallelograms, rectangles, rhombi, squares, trapezoids, and isosceles trapezoids. The Venn diagram illustrates how these figures are related.

- The Venn diagram represents all quadrilaterals.

- Parallelograms and trapezoids do not share any characteristics except that they are both quadrilaterals. This is shown by the nonoverlapping regions in the Venn diagram.

- Every isosceles trapezoid is a trapezoid. In the Venn diagram, this is shown by the set of isosceles trapezoids contained in the set of trapezoids.

- All rectangles and rhombi are parallelograms. Since a square is both a rectangle and a rhombus, it is shown by overlapping regions.

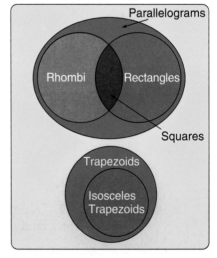

**C◌ncepts
in M◌tion**

Interactive Lab

geomconcepts.com

Check for Understanding

Communicating Mathematics

1. **Draw** an isosceles trapezoid and label the legs and the bases.

2. **Explain** how the length of the median of a trapezoid is related to the lengths of the bases.

3. _Writing Math_ Copy and complete the following table. Write _yes_ or _no_ to indicate whether each quadrilateral always has the given characteristics.

Vocabulary

trapezoid
bases
legs
base angles
median
isosceles trapezoid

Characteristics	Parallelogram	Rectangle	Rhombus	Square	Trapezoid
Opposite sides are parallel.					
Opposite sides are congruent.					
Opposite angles are congruent.					
Consecutive angles are supplementary.					
Diagonals bisect each other.					
Diagonals are congruent.					
Diagonals are perpendicular.					
Each diagonal bisects two angles.					

Guided Practice

Example 1

4. In trapezoid _QRST_, name the bases, the legs, and the base angles.

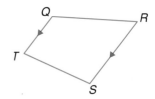

Example 2

Find the length of the median in each trapezoid.

5.

23 ft

51 ft

6.

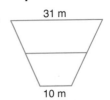

31 m

10 m

Example 3

7. Trapezoid _ABCD_ is isosceles. Find the missing angle measures.

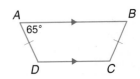

Example 3

8. Construction A hip roof slopes at the ends of the building as well as the front and back. The front of this hip roof is in the shape of an isosceles trapezoid. If one angle measures 30°, find the measures of the other three angles.

Exercises

Practice

Homework Help	
For Exercises	See Examples
8–20, 22	3
9–11, 29	1
12–17, 21, 30	2
Extra Practice	
See page 741.	

For each trapezoid, name the bases, the legs, and the base angles.

9.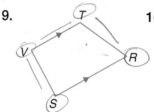

10.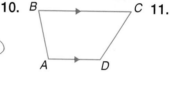

11.

Find the length of the median in each trapezoid.

12.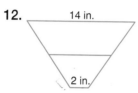

13.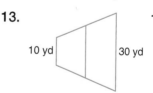

14.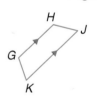

15.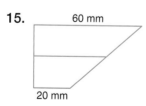

16.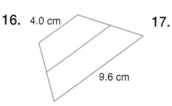

17.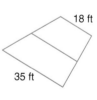

Find the missing angle measures in each isosceles trapezoid.

18.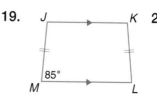

19.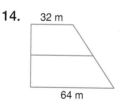

20.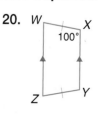

21. Find the length of the shorter base of a trapezoid if the length of the median is 34 meters and the length of the longer base is 49 meters.

22. One base angle of an isosceles trapezoid is 45°. Find the measures of the other three angles.

Determine whether it is possible for a trapezoid to have the following conditions. Write *yes* or *no*. If *yes*, draw the trapezoid.

23. three congruent sides   **24.** congruent bases

25. four acute angles   **26.** two right angles

27. one leg longer than either base

28. two congruent sides, but not isosceles

Applications and Problem Solving

29. Bridges   Explain why the figure outlined on the Golden Gate Bridge is a trapezoid.

30. Algebra   If the sum of the measures of the bases of a trapezoid is 4*x*, find the measure of the median.

31. Critical Thinking   A sequence of trapezoids is shown. The first three trapezoids in the sequence are formed by 3, 5, and 7 triangles.

   a. How many triangles are needed for the 10th trapezoid?

   b. How many triangles are needed for the *n*th trapezoid?

Mixed Review

Name all quadrilaterals that have each property.   *(Lesson 8–4)*

32. four right angles   **33.** congruent diagonals

34. Algebra   Find the value for *x* that will make quadrilateral *ABCD* a parallelogram.   *(Lesson 8–3)*

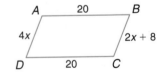

Standardized Test Practice
Ⓐ Ⓑ Ⓒ Ⓓ

35. Extended Response   Draw and label a figure to illustrate that $\overline{JN}$ and $\overline{LM}$ are medians of $\triangle JKL$ and intersect at *I*.   *(Lesson 6–1)*

36. Multiple Choice   In the figure, $AC = 60$, $CD = 12$, and *B* is the midpoint of $\overline{AD}$. Choose the correct statement.   *(Lesson 2–5)*

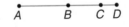

   Ⓐ $BC > CD$ Ⓑ $BC < CD$
   Ⓒ $BC = CD$ Ⓓ   There is not enough information.

   www.geomconcepts.com/self_check_quiz

Designer

Are you creative? Do you find yourself sketching designs for new cars or the latest fashion trends? Then you may like a career as a designer. Designers organize and design products that are visually appealing and serve a specific purpose.

Many designers specialize in a particular area, such as fashion, furniture, automobiles, interior design, and textiles. Textile designers design fabric for garments, upholstery, rugs, and other products, using their knowledge of textile materials and geometry. Computers—especially intelligent pattern engineering (IPE) systems—are widely used in pattern design.

1. Identify the geometric shapes used in the textiles shown above.
2. Design a pattern of your own for a textile.

FAST FACTS About Fashion Designers

Working Conditions
- vary by places of employment
- overtime work sometimes required to meet deadlines
- keen competition for most jobs

Education
- a 2- or 4-year degree is usually needed
- computer-aided design (CAD) courses are very useful
- creativity is crucial

Earnings

Median Hourly Wage in 2001

| $10 | $20 | $30 | $40 | $50 | $60 | $70 |

Source: Bureau of Labor Statistics

---

interNET CONNECTION **Career Data** For the latest information on a career as a designer, visit:
www.geomconcepts.com

Go Fly a Kite!

Materials

 unlined paper

 compass

 straightedge

 protractor

ruler

Kites

A kite is more than just a toy to fly on a windy day. In geometry, a **kite** is a special quadrilateral that has its own properties.

Investigate

1. Use paper, compass, and straightedge to construct a kite.

   a. Draw a segment about six inches in length. Label the endpoints *I* and *E*. Mark a point on the segment. The point should *not* be the midpoint of $\overline{IE}$. Label the point *X*.

   b. Construct a line that is perpendicular to $\overline{IE}$ through *X*. Mark point *K* about two inches to the left of *X* on the perpendicular line. Then mark another point, *T*, on the right side of *X* so that $\overline{KX} \cong \overline{XT}$.

   c. Connect points *K*, *I*, *T*, and *E* to form a quadrilateral. *KITE* is a kite. Use a ruler to measure the lengths of the sides of *KITE*. What do you notice?

   d. Write a definition for a kite. Compare your definition with others in the class.

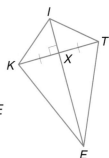

2. Use compass, straightedge, protractor, and ruler to investigate kites.

    a. Use a protractor to measure the angles of *KITE*. What do you notice about the measures of opposite and consecutive angles?

    b. Construct at least two more kites. Investigate the measures of the sides and angles.

    c. Can a kite be parallelogram? Explain your reasoning.

Extending the Investigation

In this extension, you will investigate kites and their relationship to other quadrilaterals. Here are some suggestions.

1. Rewrite Theorems 8–2 through 8–6 and 8–10 through 8–12 so they are true for kites.

2. Make a list of as many properties as possible for kites.

3. Build a kite using the properties you have studied.

Presenting Your Conclusions

Here are some ideas to help you present your conclusions to the class.

• Make a booklet showing the differences and similarities among the quadrilaterals you have studied. Be sure to include kites.

• Make a video about quadrilaterals. Cast your actors as the different quadrilaterals. The script should help viewers understand the properties of quadrilaterals.

 Investigation For more information on kites, visit: www.geomconcepts.com

Understanding and Using the Vocabulary

After completing this chapter, you should be able to define each term, property, or phrase and give an example or two of each.

*inter***NET**
CONNECTION **Review Activities**
For more review activities, visit:
www.geomconcepts.com

base angles *(p. 333)*
bases *(p. 333)*
consecutive *(p. 311)*
diagonals *(p. 311)*
isosceles trapezoid *(p. 334)*

kite *(p. 340)*
legs *(p. 333)*
median *(p. 334)*
midsegment *(p. 334)*
nonconsecutive *(p. 311)*

parallelogram *(p. 316)*
quadrilateral *(p. 310)*
rectangle *(p. 327)*
rhombus *(p. 327)*
square *(p. 327)*
trapezoid *(p. 333)*

Choose the term from the list above that best completes each statement.

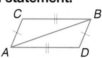

Figure 1   **Figure 2**

1. In Figure 1, *ACBD* is best described as a(n) ___?___ .
2. In Figure 1, $\overline{AB}$ is a(n) ___?___ of quadrilateral *ACBD*.
3. Figure 2 is best described as a(n) ___?___ .
4. The parallel sides of a trapezoid are called ___?___ .
5. Figure 3 is best described as a(n) ___?___ .
6. Figure 4 is best described as a(n) ___?___ .
7. In Figure 4, $\angle M$ and $\angle N$ are ___?___ .
8. A(n) ___?___ is a quadrilateral with exactly one pair of parallel sides.
9. A parallelogram with four congruent sides and four right angles is a(n) ___?___ .
10. The ___?___ of a trapezoid is the segment that joins the midpoints of each leg.

Figure 3   **Figure 4**

Skills and Concepts

Objectives and Examples	**Review Exercises**

- **Lesson 8–1** Identify parts of quadrilaterals and find the sum of the measures of the interior angles of a quadrilateral.

  The following statements are true about quadrilateral *RSVT*.

  - $\overline{RT}$ and $\overline{TV}$ are consecutive sides.
  - *S* and *T* are opposite vertices.
  - The side opposite $\overline{RS}$ is $\overline{TV}$.
  - $\angle R$ and $\angle T$ are consecutive angles.
  - $m\angle R + m\angle S + m\angle V + m\angle T = 360$

11. Name one pair of nonconsecutive sides.
12. Name one pair of consecutive angles.
13. Name the angle opposite $\angle M$.
14. Name a side that is consecutive with $\overline{AY}$.

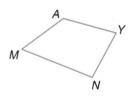

Find the missing measure(s) in each figure.

15.   16.

 www.geomconcepts.com/vocabulary_review

Objectives and Examples

Review Exercises

• **Lesson 8–2** Identify and use the properties of parallelograms.

If *JKML* is a parallelogram, then the following statements can be made.

$\overline{JK} \parallel \overline{LM}$ $\overline{JL} \parallel \overline{KM}$
$\angle JLM \cong \angle JKM$ $\angle LJK \cong \angle KML$
$\overline{JK} \cong \overline{LM}$ $\overline{JL} \cong \overline{KM}$
$\overline{JN} \cong \overline{NM}$ $\overline{LN} \cong \overline{NK}$
$\triangle JLM \cong \triangle MKJ$ $\triangle LJK \cong \triangle KML$
$m\angle LJK + m\angle JKM = 180$

In the parallelogram, *CG* = 4.5 and *BD* = 12. Find each measure.

17. *FD*
18. *BF*
19. $m\angle CBF$
20. $m\angle BCD$
21. *BG*
22. *GF*

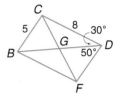

Exercises 17–22

23. In a parallelogram, the measure of one angle is 28. Determine the measures of the other angles.

• **Lesson 8–3** Identify and use tests to show that a quadrilateral is a parallelogram.

You can use the following tests to show that a quadrilateral is a parallelogram.

Theorem 8–7   Both pairs of opposite sides are congruent.
Theorem 8–8   One pair of opposite sides is parallel and congruent.
Theorem 8–9   The diagonals bisect each other.

Determine whether each quadrilateral is a parallelogram. Write *yes* or *no*. If *yes*, give a reason for your answer.

24.

25.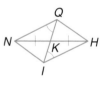

26. In quadrilateral *QNIH*, $\angle NQI \cong \angle QIH$ and $\overline{NK} \cong \overline{KH}$. Explain why quadrilateral *QNIH* is a parallelogram. Support your explanation with reasons.

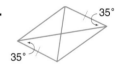

• **Lesson 8–4** Identify and use the properties of rectangles, rhombi, and squares.

rectangle   rhombus   square

Theorem 8–10   The diagonals of a rectangle are congruent.
Theorem 8–11   The diagonals of a rhombus are perpendicular.
Theorem 8–12   Each diagonal of a rhombus bisects a pair of opposite angles.

Identify each parallelogram as a *rectangle*, *rhombus*, *square*, or *none of these*.

27.

28.

29.

30.

Mixed Problem Solving
See pages 758–765.

Objectives and Examples

- **Lesson 8–5** Identify and use the properties of trapezoids and isosceles trapezoids.

If quadrilateral *BVFG* is an isosceles trapezoid, and $\overline{RT}$ is the median, then each is true.

$\overline{BV} \parallel \overline{GF}$ $\overline{BG} \cong \overline{VF}$

$\angle G \cong \angle F$ $\angle B \cong \angle V$

$RT = \frac{1}{2}(BV + GF)$

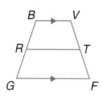

Review Exercises

31. Name the bases, legs, and base angles of trapezoid *CDJH* where $\overline{SP}$ is the median.

Exercises 33–34

32. If *CD* = 27 yards and *HJ* = 15 yards, find *SP*.

Find the missing angle measures in each isosceles trapezoid.

33.

34.

Applications and Problem Solving

35. Recreation Diamond kites are one of the most popular kites to fly and to make because of their simple design. In the diamond kite, $m\angle K = 135$ and $m\angle T = 65$. The measure of the remaining two angles must be equal in order to ensure a diamond shape. Find $m\angle I$ and $m\angle E$. *(Lesson 8–1)*

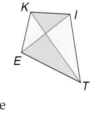

37. Car Repair To change a flat tire, a driver needs to use a device called a jack to raise the corner of the car. In the jack, $AB = BC = CD = DA$. Each of these metal pieces is attached by a hinge that allows it to pivot. Explain why nonconsecutive sides of the jack remain parallel as the tool is raised to point *F*. *(Lesson 8–3)*

36. Architecture The Washington Monument is an *obelisk*, a large stone pillar that gradually tapers as it rises, ending with a pyramid on top. Each face of the monument under the pyramid is a trapezoid. The monument's base is about 55 feet wide, and the width at the top, just below the pyramid, is about 34 feet. How wide is the monument at its median? *(Lesson 8–5)*

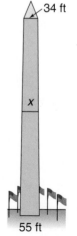

1. Name a diagonal in quadrilateral *FHSW*.
2. Name a side consecutive with $\overline{SW}$.
3. Find the measure of the missing angle in quadrilateral *FHSW*.
4. In □*XTRY*, find *XY* and *RY*.
5. Name the angle that is opposite ∠*XYR*.
6. Find *m*∠*XTR*.
7. Find *m*∠*TRY*.
8. If *TV* = 32, find *TY*.
9. In square *GACD*, if *DA* = 14, find *BC*.
10. Find *m*∠*DBC*.

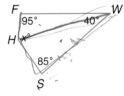

Exercises 1–3

Exercises 4–8   Exercises 9–10

Determine whether each quadrilateral is a parallelogram. Write *yes* or *no*. If yes, give a reason for your answer.

11.   12.   13.   14.

Identify each figure as a *quadrilateral*, *parallelogram*, *rhombus*, *rectangle*, *square*, *trapezoid*, or *none of these*.

15.   16.   17.   18.

19. Determine whether quadrilateral *ADHT* is a parallelogram. Support your answer with reasons.
20. In rhombus *WQTZ*, the measure of one side is 18 yards, and the measure of one angle is 57. Determine the measures of the other three sides and angles.
21. *NP* is the median of isosceles trapezoid *JKML*. If $\overline{JK}$ and $\overline{LM}$ are the bases, *JK* = 24, and *LM* = 44, find *NP*.

Exercise 19

Identify each statement as *true* or *false*.

22. All squares are rectangles.   23. All rhombi are squares.

24. **Music**   A series of wooden bars of varying lengths are arranged in the shape of a quadrilateral to form an instrument called a xylophone. In the figure, $\overline{XY} \parallel \overline{WZ}$, but $\overline{XW} \nparallel \overline{YZ}$. What is the best description of quadrilateral *WXYZ*?

25. **Algebra**   Two sides of a rhombus measure 5*x* and 2*x* + 18. Find *x*.

Coordinate Geometry Problems

Standardized tests often include problems that involve points on a coordinate grid. You'll need to identify the coordinates of points, calculate midpoints of segments, find the distance between points, and identify intercepts of lines and axes.

Be sure you understand these concepts.

axis	coordinates	distance	intercept
line	midpoint	ordered pair	

Example 1

In the figure at the right, which of the following points lies within the shaded region?

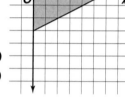

A $(-1, 1)$ B $(1, -2)$

C $(4, 3)$ D $(5, -4)$

E $(7, 0)$

Hint Try to eliminate impossible choices in multiple-choice questions.

Solution Notice that the shaded region lies in the quadrant where x is positive and y is negative. Look at the answer choices. Since x must be positive and y must be negative for a point within the region, you can eliminate choices A, C, and E.

Plot the remaining choices, B and D, on the grid. You will see that $(1, -2)$ is inside the region and $(5, -4)$ is not. So, the answer is B.

Example 2

A segment has endpoints at $P(-2, 6)$ and $Q(6, 2)$.

Part A Draw segment PQ.

Part B Explain how you know whether the midpoint of segment PQ is the same as the y-intercept of segment PQ.

Hint You may be asked to draw points or segments on a grid. Be sure to use labels.

Solution

Part A

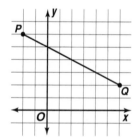

Part B Use the Midpoint Formula.

$$\left(\frac{x_1 + x_2}{2}, \frac{y_1 + y_2}{2}\right)$$

The midpoint of $\overline{PQ}$ is $\left(\frac{-2 + 6}{2}, \frac{6 + 2}{2}\right)$ or $(2, 4)$. The y-intercept is $(0, 5)$. So they are not the same point.

After you work each problem, record your
answer on the answer sheet provided or on
a sheet of paper.

Multiple Choice

1. The graph of
   $y = -\frac{1}{2}x + 1$ is
   shown. What is the
   x-intercept?
   (*Algebra Review*)

   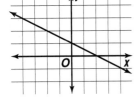

   Ⓐ 2   Ⓑ 1
   Ⓒ 0   Ⓓ −1

2. A team has 8 seniors, 7 juniors, 3 sophomores,
   and 2 freshmen. What is the probability that a
   player selected at random is *not* a junior or a
   freshman? (*Statistics Review*)

   Ⓐ $\frac{9}{20}$ Ⓑ $\frac{11}{20}$ Ⓒ $\frac{13}{20}$ Ⓓ $\frac{9}{11}$

3. A cubic inch is about 0.000579 cubic feet. How
   is this expressed in scientific notation?
   (*Algebra Review*)

   Ⓐ 5.79×10^{-4} Ⓑ 57.9×10^{-6}
   Ⓒ 57.9×10^{-4} Ⓓ 579×10^{-6}

4. Joey has at least one quarter, one dime, one
   nickel, and one penny. If he has twice as
   many pennies as nickels, twice as many
   nickels as dimes, and twice as many dimes
   as quarters, what is the least amount of
   money he could have? (*Algebra Review*)

   Ⓐ $0.41   Ⓑ $0.64   Ⓒ $0.71
   Ⓓ $0.73   Ⓔ $2.51

5. On a floor plan, two consecutive corners of
   a room are at (3, 15) and (18, 2). The
   architect places a window in the center of
   the wall containing these two points. What
   are the coordinates of the center of the
   window? (*Lesson 2–6*)

   Ⓐ (8.5, 10.5)   Ⓑ (10.5, 8.5)
   Ⓒ (17, 21)   Ⓓ (21, 17)

6. Find the distance between (−2, 1) and
   (1, −3). (*Lesson 6–7*)

   Ⓐ 3   Ⓑ 4   Ⓒ 5
   Ⓓ 6   Ⓔ 7

7. The graph shows a store's sales of greeting
   cards. The average price of a greeting card
   was $2. Which is the best estimate of the
   total sales during the 4-month period?
   (*Algebra Review*)

   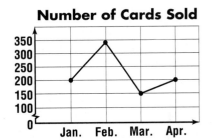

   Number of Cards Sold

   Ⓐ   less than $1000
   Ⓑ   between $1000 and $2000
   Ⓒ   between $2000 and $3000
   Ⓓ   between $3000 and $4000

8. At a music store, the price of a CD is
   three times the price of a cassette tape. If
   40 CDs were sold for a total of $480, and
   the combined sales of CDs and cassette
   tapes totaled $600, how many cassette
   tapes were sold? (*Algebra Review*)

   Ⓐ 4   Ⓑ 12   Ⓒ 30   Ⓓ 120

Short Response

9. Two segments with lengths 3 feet and
   5 feet form two sides of a triangle. Draw a
   number line that shows possible lengths
   for the third side. (*Lesson 7–4*)

Extended Response

10. Make a bar graph for the data below.
    (*Statistics Review*)

Destination	Frequency
Circle Center shopping district	ＨＴ |||
Indianapolis Children's Museum	ＨＴ ＨＴ ||
RCA Dome	ＨＴ ＨＴ ＨＴ	
Indianapolis 500	ＨＴ	
Indianapolis Art Museum	|||

Proportions and Similarity

What You'll Learn

Key Ideas

- Use ratios and proportions to solve problems. *(Lesson 9–1)*

- Identify similar polygons and use similarity tests for triangles. *(Lessons 9–2 and 9–3)*

- Identify and use the relationships between proportional parts of triangles. *(Lessons 9–4 to 9–6)*

- Identify and use proportional relationships of similar triangles. *(Lesson 9–7)*

Key Vocabulary

polygon *(p. 356)*

proportion *(p. 351)*

ratio *(p. 350)*

similar polygons *(p. 356)*

Why It's Important

Mechanics The pit crew of a racing team is responsible for making sure the car is prepared for the driver. To maintain the car, crew members must understand the workings of each part of the complex gear system in the engine.

Proportions are used to compare sizes using ratios. You will investigate gear ratios in Lesson 9–1.

Study these lessons to improve your skills.

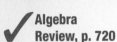
Algebra Review, p. 720

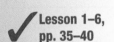
Algebra Review, p. 721

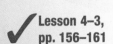
Lesson 4–3, pp. 156–161

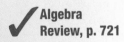
Lesson 1–6, pp. 35–40

Solve each equation.

1. $y = 12(4.5)$

2. $6.1(3.3) = p$

3. $n = 13.3 \div 3.5$

4. $11.16 \div 0.9 = q$

5. $d = \dfrac{2}{3} \cdot \dfrac{3}{4}$

6. $3\left(1\dfrac{1}{2}\right) = f$

7. $x = \dfrac{3}{8} \div \dfrac{1}{4}$

8. $7\dfrac{1}{2} \div 30 = w$

In the figure, $n \parallel o$. Name all angles congruent to the given angle. Give a reason for each answer.

9. $\angle 1$

10. $\angle 4$

11. $\angle 5$

12. $\angle 16$

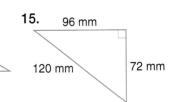

Find the perimeter of each triangle.

13.

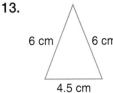

6 cm   6 cm

4.5 cm

14.

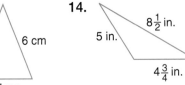

5 in.

$8\dfrac{1}{2}$ in.

$4\dfrac{3}{4}$ in.

15.

96 mm

120 mm

72 mm

FOLDABLES™
Study Organizer

Make this Foldable to help you organize your Chapter 9 notes. Begin with a sheet of notebook paper.

❶ **Fold** lengthwise to the holes.

❷ **Cut** along the top line and then cut 10 tabs.

❸ **Label** each tab with important terms.

Reading and Writing   Store the Foldable in a 3–ring binder. As you read and study the chapter, write definitions and examples of important terms under each tab.

Using Ratios and Proportions

What You'll Learn
You'll learn to use ratios and proportions to solve problems.

Why It's Important
Medicine Nurses solve proportions to determine the amount of medication to give a patient.
See Exercise 45.

In 2000, about 180 million tons of solid waste was created in the United States. Paper made up about 72 million tons of this waste. The **ratio** of paper waste to total solid waste is 72 to 180. This ratio can be written in the following ways.

72 to 180   72:180 $\dfrac{72}{180}$ 72 ÷ 180

Definition of Ratio	**Words:**	A ratio is a comparison of two numbers by division.
	Symbols:	a to b, $a:b$, $\dfrac{a}{b}$, or $a \div b$, where $b \neq 0$

All ratios should be written in simplest form. Because all fractions can be written as decimals, it is sometimes useful to express ratios in decimal form.

Examples

Reading Geometry

Read 45:340 as *45 to 340*.

Write each ratio in simplest form.

1 $\dfrac{45}{340}$

$\dfrac{45}{340} = \dfrac{45 \div 5}{340 \div 5}$ *Divide the numerator and denominator by 5.*

$\qquad = \dfrac{9}{68}$ *Simplify.*

2 **six days to two weeks**

To write this as a ratio, the units of measure must be the same. Write both using days. There are seven days in one week, so two weeks equal 14 days. The ratio is $\dfrac{6}{14}$ or $\dfrac{3}{7}$.

Your Turn

a. $\dfrac{18}{24}$

b. 10 kilometers to 20,000 meters

A **proportion** is an equation that shows two equivalent ratios.

$$\frac{20}{30} = \frac{2}{3}$$

Every proportion has two **cross products**. In the proportion above, the terms 20 and 3 are called the **extremes**, and 30 and 2 are called the **means**. The cross products are 20(3) and 30(2). The cross products are always equal in a proportion.

$$\frac{20}{30} = \frac{2}{3}$$

$$20(3) = 30(2)$$
extremes   means
$$60 = 60$$

Theorem 9–1 **Property of** **Proportions**	**Words:**	For any numbers a and c and any nonzero numbers b and d, if $\frac{a}{b} = \frac{c}{d}$, then $ad = bc$. Likewise, if $ad = bc$, then $\frac{a}{b} = \frac{c}{d}$.
	Numbers:	If $\frac{5}{10} = \frac{1}{2}$, then $5(2) = 10(1)$. If $5(2) = 10(1)$, then $\frac{5}{10} = \frac{1}{2}$.

You can use cross products to solve equations in proportion form. Remember that you should always check your solution in the original proportion.

Example

Algebra Link

┌─ **Algebra Review** ─┐
Solving Multi-Step
Equations, p. 723
└──────────────────┘

3 Solve $\frac{15}{35} = \frac{3}{2x + 1}$.

$\frac{15}{35} = \frac{3}{2x + 1}$	*Original equation*
$15(2x + 1) = 35(3)$	*Cross Products*
$30x + 15 = 105$	*Distributive Property*
$30x + 15 - 15 = 105 - 15$	*Subtract 15 from each side.*
$30x = 90$	*Simplify.*
$\frac{30x}{30} = \frac{90}{30}$	*Divide each side by 30.*
$x = 3$	*Simplify.*

Check:

$\frac{15}{35} = \frac{3}{2x + 1}$	*Original equation*
$\frac{15}{35} \stackrel{?}{=} \frac{3}{2(3) + 1}$	*Replace x with 3.*
$\frac{15}{35} \stackrel{?}{=} \frac{3}{7}$	*2(3) + 1 = 7*
$15(7) \stackrel{?}{=} 35(3)$	*Cross Products*
$105 = 105 \checkmark$	

The solution is 3.

(continued on the next page)

Your Turn

c. Solve $\frac{3}{b} = \frac{15}{60}$.

d. Solve $\frac{x-2}{4} = \frac{4}{8}$.

Proportions can help you solve real-life problems. The ratios in the proportions must be written in the same order.

Example ④

Automotive Link

In an engine, the volume of the cylinder changes as the piston moves up and down. The compression ratio is the expanded volume of the cylinder to the compressed volume of the cylinder. One car with a V6 engine has a compression ratio of 9 to 1. If the expanded volume of the cylinder is 22.5 cubic inches, find its compressed volume.

$$\begin{array}{c} expanded\ volume \rightarrow \\ compressed\ volume \rightarrow \end{array} \frac{9}{1} = \frac{22.5}{x} \begin{array}{c} \leftarrow compressed\ volume \\ \leftarrow expanded\ volume \end{array}$$

$$9(x) = 1(22.5) \quad Cross\ Products$$

$$x = 2.5 \quad Divide\ each\ side\ by\ 9.$$

The cylinder's compressed volume is 2.5 cubic inches.

Online Personal Tutor at geomconcepts.com

Check for Understanding

Communicating Mathematics

1. **Write** two ratios that form a proportion and two ratios that do not form a proportion.

2. **Explain** how you would solve the proportion $\frac{14}{21} = \frac{x}{24}$.

3. Lawanda says that if $\frac{7}{8} = \frac{x}{y}$, then $\frac{8}{7} = \frac{y}{x}$. Paul disagrees. Who is correct? Explain your reasoning.

Vocabulary
ratio
proportion
cross products
extremes
means

Guided Practice

 **Getting Ready** Write each ratio as a fraction in simplest form.

Sample: 6 ounces to 12 ounces

Solution: $\dfrac{6\ \cancel{oz}}{12\ \cancel{oz}} = \dfrac{6}{12}$

$= \dfrac{6 \div 6}{12 \div 6}$ or $\dfrac{1}{2}$

4. 7 feet to 3 feet

5. 3 grams to 11 grams

6. 16 cm to 5 cm

7. 21 miles to 16 miles

8. 15 km to 5 km

9. 6 meters to 10 meters

Examples 1 & 2

Write each ratio in simplest form.

10. $\dfrac{4}{2}$

11. $\dfrac{72}{100}$

12. 3 millimeters to 1 centimeter

Example 3

Solve each proportion.

13. $\dfrac{x}{3} = \dfrac{12}{18}$

14. $\dfrac{6}{2x} = \dfrac{15}{30}$

15. $\dfrac{7}{3} = \dfrac{3x - 1}{6}$

Example 4

16. Mechanics The gear ratio is the number of teeth on the driving gear to the number of teeth on the driven gear. If the gear ratio is 5:2 and the driving gear has 35 teeth, how many teeth does the driven gear have?

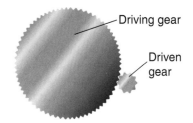

Driving gear

Driven gear

Exercises

Practice

Write each ratio in simplest form.

17. $\dfrac{2}{10}$

18. $\dfrac{8}{12}$

19. $\dfrac{10}{22}$

20. $\dfrac{18}{36}$

21. $\dfrac{45}{21}$

22. $\dfrac{40}{12}$

23. 44 centimeters to 2 meters

24. 6 inches to 2 feet

25. 6 quarts to 1 pint

26. 3 liters to 300 milliliters

Homework Help	
For Exercises	**See Examples**
17–26, 42	1, 2
27–41, 43–46	3, 4
Extra Practice	
See page 741.	

Solve each proportion.

27. $\dfrac{2}{5} = \dfrac{4}{x}$

28. $\dfrac{12}{5} = \dfrac{x}{10}$

29. $\dfrac{36}{3} = \dfrac{12}{x}$

30. $\dfrac{3x}{27} = \dfrac{2}{9}$

31. $\dfrac{14}{x - 1} = \dfrac{7}{4}$

32. $\dfrac{1}{x + 3} = \dfrac{3}{29}$

33. $\dfrac{5}{9} = \dfrac{5}{x - 3}$

34. $\dfrac{x + 2}{16} = \dfrac{7}{4}$

35. $\dfrac{30 - x}{x} = \dfrac{3}{2}$

36. If $3:x = 18:24$, find the value of x.

37. If 3 to 4 and $5x - 1$ to 12 form a proportion, what is the value of x?

38. Refer to the triangles below.

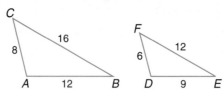

a. Write the ratio of *AB* to *DE*.

b. Write the ratio of *AC* to *DF*.

c. Do the two ratios form a proportion? Explain.

If $a = 3$, $b = 2$, $c = 6$, and $d = 4$, determine whether each pair of ratios forms a proportion.

39. $\dfrac{b}{a} = \dfrac{d}{c}$

40. $\dfrac{c}{b} = \dfrac{d}{a}$

41. $\dfrac{a + b}{b} = \dfrac{c + d}{d}$

Applications and Problem Solving

42. Money The average number of days Americans worked in a recent year to pay taxes is shown below.

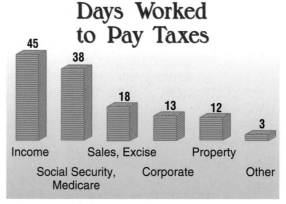

Days Worked to Pay Taxes

Source: Tax Foundation

a. Write a ratio of the number of days worked to pay income taxes to the total number of days in a year.

b. Write a ratio of the number of days worked to pay property taxes to the number of days to pay sales and excise taxes.

43. Environment If 2500 square feet of grass supplies enough oxygen for a family of four, how much grass is needed to supply oxygen for a family of five?

44. Science Light travels approximately 1,860,000 miles in 10 seconds. How long will it take light to travel the 93,000,000 miles from the sun to Earth?

45. Medicine Antonio is a nurse. A doctor tells him to give a patient 60 milligrams of acetaminophen. Antonio has a liquid medication that contains 240 milligrams of acetaminophen per 10 milliliters of medication. How many milliliters of the medication should he give the patient?

46. Critical Thinking The **geometric mean** between two positive numbers a and b is the positive number x in $\frac{a}{x} = \frac{x}{b}$.

a. Solve $\frac{4}{x} = \frac{x}{9}$ to find the geometric mean between 4 and 9.

b. In any right triangle XYZ, if $\overline{YW}$ is an altitude to the hypotenuse, then XY is the geometric mean between XZ and XW, and YZ is the geometric mean between XZ and WZ. If $XZ = 16$ and $XW = 4$, find XY.

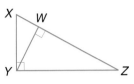

Mixed Review

47. Find the length of median $\overline{AB}$ in trapezoid $JKLM$ if $JM = 14$ inches and $KL = 21$ inches. *(Lesson 8–5)*

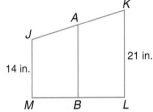

48. Puzzles A crossword puzzle is made up of various parallelograms. Identify the parallelogram that is outlined as a *rectangle, rhombus, square,* or *none of these.* If it is more than one of these, list all that apply. *(Lesson 8–4)*

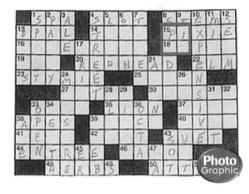

49. Determine if the numbers 8, 15, and 17 can be measures of the sides of a triangle. *(Lesson 7–4)*

50. In the triangle shown, $m\angle 5 = 9x$, $m\angle 4 = 6x + 2$, and $m\angle 2 = 92$. Find the values of x, $m\angle 5$, and $m\angle 4$. *(Lesson 7–2)*

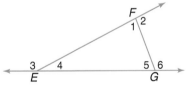

For each triangle, tell whether the red segment or line is an *altitude*, a *perpendicular bisector*, *both*, or *neither*. *(Lesson 6–2)*

51.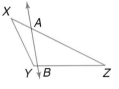

52.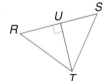

53.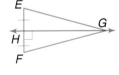

Standardized Test Practice
Ⓐ Ⓑ Ⓒ Ⓓ

54. Short Response According to the building code in Plainfield, Connecticut, the slope of a stairway cannot be steeper than 0.82. The stairs in Troy's home measure 10.5 inches deep and 7.5 inches high. Do the stairs in his home meet the code requirements? Explain. *(Lesson 4–5)*

What You'll Learn

You'll learn to identify similar polygons.

Why It's Important

Construction
Contractors use drawings that are similar to the actual building.
See Example 3.

A **polygon** is a closed figure in a plane formed by segments called **sides**. It is a general term used to describe a geometric figure with at least three sides. Polygons that are the same shape but not necessarily the same size are called **similar polygons**.

The two triangles shown below are similar. For naming similar polygons, the vertices are written in order to show the corresponding parts. The symbol for similar is ~.

$$\triangle RST \sim \triangle VWX$$

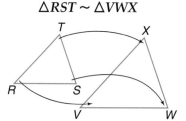

Corresponding Angles	Corresponding Sides
$\angle R \leftrightarrow \angle V$	$\overline{RS} \leftrightarrow \overline{VW}$
$\angle S \leftrightarrow \angle W$	$\overline{ST} \leftrightarrow \overline{WX}$
$\angle T \leftrightarrow \angle X$	$\overline{TR} \leftrightarrow \overline{XV}$

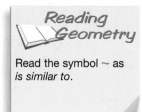

Reading Geometry

Read the symbol ~ as *is similar to*.

Recall that in congruent figures, corresponding angles and sides are congruent. In similar figures, corresponding angles are congruent, and the measures of corresponding sides have equivalent ratios, or are proportional.

$$\angle R \cong \angle V, \angle S \cong \angle W, \angle T \cong \angle X, \text{ and } \frac{RS}{VW} = \frac{ST}{WX} = \frac{TR}{XV}$$

Definition of Similar Polygons	**Words:** Two polygons are similar if and only if their corresponding angles are congruent and the measures of their corresponding sides are proportional.
	Model:
	$$\frac{AB}{EF} = \frac{BC}{FG} = \frac{CD}{GH} = \frac{DA}{HE} \text{ and}$$ $$\angle A \cong \angle E, \angle B \cong \angle F,$$ $$\angle C \cong \angle G, \angle D \cong \angle H$$
	Symbols: polygon *ABCD* ~ polygon *EFGH*

Example ① Determine if the polygons are similar. Justify your answer.

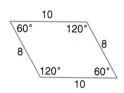

Since $\frac{4}{8} = \frac{5}{10} = \frac{4}{8} = \frac{5}{10}$, the measures of the sides of the polygons are proportional. However, the corresponding angles are not congruent. The polygons are not similar.

Your Turn

a.

b.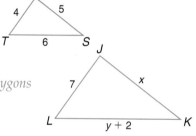

Concepts in MOtion
Animation
geomconcepts.com

Knowing several measures of two similar figures may allow you to find the measures of missing parts.

Example ② **Algebra Link**

Find the values of x and y if $\triangle RST \sim \triangle JKL$.

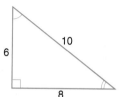

Use the corresponding order of the vertices to write proportions.

$\dfrac{RS}{JK} = \dfrac{ST}{KL} = \dfrac{TR}{LJ}$ *Definition of similar polygons*

$\dfrac{5}{x} = \dfrac{6}{y + 2} = \dfrac{4}{7}$ *Substitution*

Write the proportion that can be solved for x.

$$\frac{5}{x} = \frac{4}{7}$$

$5(7) = x(4)$ *Cross Products*

$35 = 4x$

$\dfrac{35}{4} = \dfrac{4x}{4}$ *Divide each side by 4.*

$8\frac{3}{4} = x$ *Simplify.*

Now write the proportion that can be solved for y.

$$\frac{6}{y + 2} = \frac{4}{7}$$

$6(7) = (y + 2)4$ *Cross Products*

$42 = 4y + 8$ *Distributive Property*

$42 - 8 = 4y + 8 - 8$ *Subtract 8 from each side.*

$34 = 4y$

$\dfrac{34}{4} = \dfrac{4y}{4}$ *Divide each side by 4.*

$8\frac{1}{2} = y$ *Simplify.*

— **Algebra Review** —
Solving Multi-Step Equations, p. 723

Therefore, $x = 8\frac{3}{4}$ and $y = 8\frac{1}{2}$.

(continued on the next page)

 www.geomconcepts.com/extra_examples

c. Find the values of x and y if $\triangle ABC \sim \triangle DEF$.

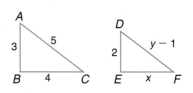

Scale drawings are often used to represent something that is too large or too small to be drawn at actual size. Contractors use scale drawings called *blueprints* to represent the floor plan of a house to be constructed. The blueprint and the floor plan are similar.

Example **3**

Construction Link

In the blueprint, 1 inch represents an actual length of 16 feet. Find the actual dimensions of the living room.

On the blueprint, the living room is $1\frac{1}{2}$ inches long and $1\frac{1}{4}$ inches wide. Use proportions to find the actual dimensions.

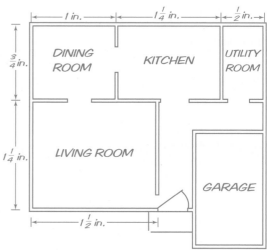

$$\begin{array}{ll} \textit{blueprint} \rightarrow \\ \textit{actual} \rightarrow \end{array} \frac{1 \text{ in.}}{16 \text{ ft}} = \frac{1\frac{1}{2} \text{ in.}}{x \text{ ft}} \begin{array}{l} \leftarrow \textit{blueprint} \\ \leftarrow \textit{actual} \end{array}$$

$$1(x) = 16\left(1\frac{1}{2}\right) \quad \textit{Cross Products}$$

$$x = 24$$

$$\begin{array}{ll} \textit{blueprint} \rightarrow \\ \textit{actual} \rightarrow \end{array} \frac{1 \text{ in.}}{16 \text{ ft}} = \frac{1\frac{1}{4} \text{ in.}}{y \text{ ft}} \begin{array}{l} \leftarrow \textit{blueprint} \\ \leftarrow \textit{actual} \end{array}$$

$$1(y) = 16\left(1\frac{1}{4}\right) \quad \textit{Cross Products}$$

$$y = 20$$

The actual dimensions of the living room are 24 feet by 20 feet.

Your Turn

d. Use the blueprint to find the actual dimensions of the kitchen.

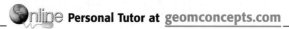
Online Personal Tutor at geomconcepts.com

Check for Understanding

1. **Compare and contrast** congruent polygons and similar polygons.

2. **Explain** how to find an actual distance using a scale drawing.

3. **Draw** two similar pentagons on grid paper. Label the vertices of the pentagons. Name the corresponding angles and sides. Write a proportion for the measures of the sides.

> **Vocabulary**
>
> polygon
> sides
> similar polygons
> scale drawing

Guided Practice
Example 1

Determine whether each pair of polygons is similar. Justify your answer.

4.

15
12   12
15

10
9   9
10

5.

Example 2

Each pair of polygons is similar. Find the values of x and y.

6.

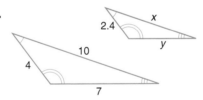

7.

    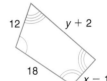

Example 3

8. **Construction** Refer to Example 3. Find the actual dimensions of the utility room.

Exercises

Practice

Determine whether each pair of polygons is similar. Justify your answer.

9.

9
9   9
9

7
7   7
7

yes

10.

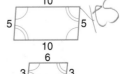

yes

11.

14
4   4
14

8
2   2
8

yes

Homework Help	
For Exercises	**See Examples**
9–14, 21, 22	1
15–20	2
23–26	3
Extra Practice	
See page 741.	

Determine whether each pair of polygons is similar. Justify your answer.

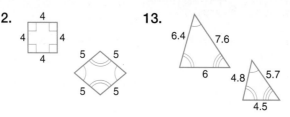

12.

13.

14.

Each pair of polygons is similar. Find the values of *x* and *y*.

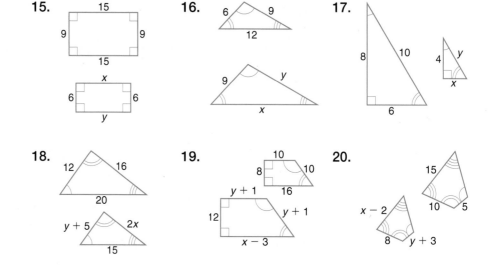

15.

16.

17.

18.

19.

20.

Determine whether each statement is *always*, *sometimes*, or *never* true.

21. Similar polygons are also congruent.

22. Congruent polygons are also similar.

Applications and Problem Solving

23. **Sports**   A soccer field is 91 meters by 46 meters. Make a scale drawing of the field if 1 millimeter represents 1 meter.

24. **Publishing**   Mi-Ling is working on the school yearbook. She must reduce a photo that is 4 inches wide by 5 inches long to fit in a space 3 inches wide. How long will the reduced photo be?

25. **Automotive Design**   Tracy is drawing a scale model of a car she is designing. If $\frac{1}{4}$ inch on the drawing represents 28 inches, find each measurement on the actual car.

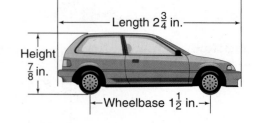

Length $2\frac{3}{4}$ in.

Height $\frac{7}{8}$ in.

Wheelbase $1\frac{1}{2}$ in.

a. length

b. height

c. wheelbase

26. Travel Each year, many tourists visit Madurodam in the Netherlands. Madurodam is a miniature town where 1 meter represents 25 meters. How high is a structure in Madurodam that represents a building that is actually 30 meters high?

Madurodam, Netherlands

27. Critical Thinking Marquis is doing a report on Wyoming. The state is approximately a rectangle measuring 362 miles by 275 miles. If Marquis wishes to draw the largest possible map of Wyoming on an $8\frac{1}{2}$-inch by 11-inch piece of paper, how many miles should one inch represent?

Mixed Review

28. Find the value of y if $4:y = 16:36$. *(Lesson 9–1)*

29. Trapezoid *TAHS* is isosceles. Find $m\angle T$, $m\angle H$, and $m\angle A$. *(Lesson 8–5)*

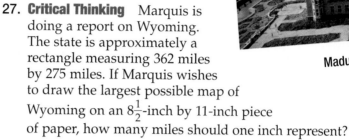

30. In quadrilateral *QRST*, diagonal *QS* bisects diagonal *RT*. Is *QRST* a parallelogram? Explain. *(Lesson 8–3)*

Standardized Test Practice
Ⓐ Ⓑ Ⓒ Ⓓ

31. Short Response In right triangle *ASP*, $m\angle S = 90$. Which side has the greatest measure? *(Lesson 7–3)*

32. Multiple Choice Which triangle is not obtuse? *(Lesson 5–1)*

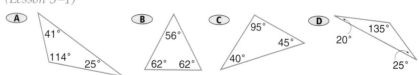

Quiz 1   Lessons 9–1 and 9–2

Solve each proportion. *(Lesson 9–1)*

1. $\dfrac{2}{x} = \dfrac{8}{12}$

2. $\dfrac{18}{4x} = \dfrac{3}{2}$

3. $\dfrac{x+4}{3} = \dfrac{25}{5}$

Each pair of polygons is similar. Find the values of x and y. *(Lesson 9–2)*

4.

5.

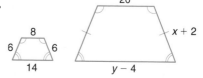

What You'll Learn

You'll learn to use AA, SSS, and SAS similarity tests for triangles.

Why It's Important

Surveying Surveyors use similar triangles to measure distances that cannot be measured directly. *See Exercise 5.*

The Bank of China building in Hong Kong is one of the ten tallest buildings in the world. Designed by American architect I. M. Pei, the outside of the 70-story building is sectioned into triangles, which are meant to resemble the trunk of a bamboo plant. Some of the triangles are similar, as shown below.

In previous chapters, you learned several basic tests for determining whether two triangles are congruent. Recall that each congruence test involves only three corresponding parts of each triangle. Likewise, there are tests for similarity that will not involve all the parts of each triangle.

Hands-On Geometry

Materials: ruler protractor

Step 1 On a sheet of paper, use a ruler to draw a segment 2 centimeters in length. Label the endpoints of the segment A and B.

A 2 cm B

Step 2 Use a protractor to draw an angle at A so that $m\angle A = 87$. Draw an angle at B so that $m\angle B = 38$. Extend the sides of $\angle A$ and $\angle B$ so that they intersect to form a triangle. Label the third vertex C.

Step 3 Now draw a segment 4 centimeters in length. Label the endpoints D and E.

D 4 cm E

Step 4 Use a protractor to draw an angle at D so that $m\angle D = 87$. Draw an angle at E so that $m\angle E = 38$. Extend the sides of $\angle D$ and $\angle E$ to form a triangle. Label the third vertex F.

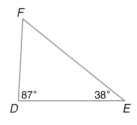

Try These

1. What is $m\angle C$? What is $m\angle F$?
2. Use a ruler to find BC, CA, EF, and FD.
3. Find $\frac{AB}{DE}$, $\frac{BC}{EF}$, and $\frac{CA}{FD}$.
4. Are the triangles similar? Why or why not?

The activity suggests Postulate 9–1.

Postulate 9–1 **AA Similarity**	**Words:**	If two angles of one triangle are congruent to two corresponding angles of another triangle, then the triangles are similar.
	Model:	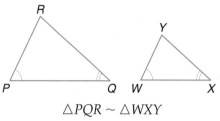
	Symbols:	If $\angle A \cong \angle D$ and $\angle B \cong \angle E$, then $\triangle ABC \sim \triangle DEF$.

In $\triangle PQR$ and $\triangle WXY$, if $\angle P \cong \angle W$ and $\angle Q \cong \angle X$, then the triangles are similar. By definition of similar polygons, there are four other parts of the triangles that are related.

$$\angle R \cong \angle Y \qquad \frac{PQ}{WX} = \frac{QR}{XY} = \frac{RP}{YW} \qquad \triangle PQR \sim \triangle WXY$$

Two other tests are used to determine whether two triangles are similar.

Theorem 9–2 **SSS Similarity**	**Words:**	If the measures of the sides of a triangle are proportional to the measures of the corresponding sides of another triangle, then the triangles are similar.
	Model:	
	Symbols:	If $\frac{AB}{DE} = \frac{BC}{EF} = \frac{CA}{FD}$, then $\triangle ABC \sim \triangle DEF$.

*A **fractal** is a geometric figure that is created by repeating the same process over and over. One characteristic of a fractal is that it has a self-similar shape. The Sierpinski Triangle shown below is an example of a fractal.*

Theorem 9–3 **SAS Similarity**	**Words:**	If the measures of two sides of a triangle are proportional to the measures of two corresponding sides of another triangle and their included angles are congruent, then the triangles are similar.
	Model:	
	Symbols:	If $\frac{AB}{DE} = \frac{BC}{EF}$, and $\angle B \cong \angle E$, then $\triangle ABC \sim \triangle DEF$.

1 Determine whether the triangles are similar. If so, tell which similarity test is used and complete the statement.

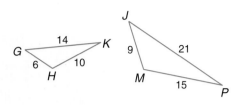

$$\triangle GHK \sim \triangle \underline{\quad ? \quad}$$

Since $\frac{6}{9} = \frac{10}{15} = \frac{14}{21}$, the triangles are similar by SSS Similarity. Therefore, $\triangle GHK \sim \triangle JMP$.

2 Find the value of x.

To see the corresponding parts more easily, flip $\triangle QRN$ so that it is in the same position as $\triangle XWT$.

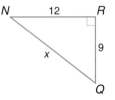

Since $\angle R$ and $\angle W$ are right angles, they are congruent. We also know that $\frac{12}{4} = \frac{9}{3}$. Therefore, $\triangle QRN \sim \triangle XWT$ by SAS Similarity. Use the definition of similar polygons to find x.

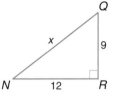

$$\frac{QR}{XW} = \frac{NQ}{TX} \qquad \textit{Definition of Similar Polygons}$$

$$\frac{9}{3} = \frac{x}{5} \qquad \textit{QR = 9, XW = 3, NQ = x, TX = 5}$$

$$9(5) = 3(x) \qquad \textit{Cross Products}$$

$$45 = 3x \qquad \textit{Simplify.}$$

$$\frac{45}{3} = \frac{3x}{3} \qquad \textit{Divide each side by 3.}$$

$$15 = x \qquad \textit{Simplify.}$$

Your Turn

a. Determine whether the triangles are similar. If so, tell which similarity test is used and complete the statement.

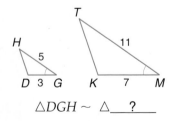

$$\triangle DGH \sim \triangle \underline{\quad ? \quad}$$

b. Find the values of x and y if the triangles are similar.

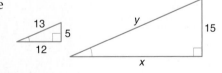

COncepts in MOtion

Interactive Lab

geomconcepts.com

Similar triangles can be used to find the length of an object that is difficult to measure directly.

Example ③
Landscaping Link

Editon is landscaping a yard. To see how well a tree shades an area of the yard, he needs to know the tree's height. The tree's shadow is 18 feet long at the same time that Editon's shadow is 4 feet long. If Editon is 6 feet tall, how tall is the tree?

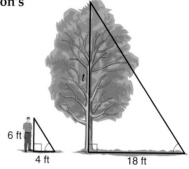

Draw a diagram. The rays of the sun form congruent angles with the ground. Both Editon and the tree form right angles with the ground. Therefore, the triangles in the diagram are similar by AA Similarity. Use the similar triangles to find the height of the tree t.

$$\begin{array}{ll} \text{Editon's height} \rightarrow & \dfrac{6}{t} = \dfrac{4}{18} \quad \leftarrow \text{Editon's shadow} \\ \text{tree's height} \rightarrow & \quad\quad\quad \leftarrow \text{tree's shadow} \end{array}$$

$$6(18) = t(4) \qquad \text{Cross Products}$$
$$108 = 4t \qquad \text{Simplify.}$$
$$\frac{108}{4} = \frac{4t}{4} \qquad \text{Divide each side by 4.}$$
$$27 = t \qquad \text{Simplify.}$$

The tree is 27 feet tall.

Online **Personal Tutor at** geomconcepts.com

Check for Understanding

Communicating Mathematics

1. **Sketch and label** two similar right triangles ABC and DEF with right angles at C and F. Let the measures of angles A and D be 30. Name the corresponding sides that are proportional.

2. Refer to Example 2.
   a. If the sides of $\triangle XWT$ form a Pythagorean triple, what is true about the sides of $\triangle QRN$?
   b. Why is this true?

Guided Practice
Example 1

3. Determine whether the triangles are similar. If so, tell which similarity test is used and complete the statement.

$$\triangle XYZ \sim \triangle \underline{\quad?\quad}$$

Example 2

4. Find the values of x and y.

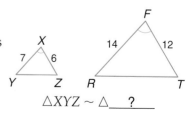

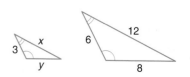

Example 3

5. **Surveying** Syreeta Coleman is a surveyor. To find the distance across Muddy Pond, she forms similar triangles and measures distances as shown at the right. What is the distance across Muddy Pond?

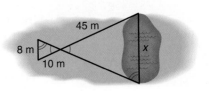

Exercises • • • • • • • • • • • • • • • • • •

Practice

Determine whether each pair of triangles is similar. If so, tell which similarity test is used and complete the statement.

Homework Help	
For Exercises	See Examples
6–8	1
9–11	2
14–16	2, 3
Extra Practice	
See page 742.	

6.

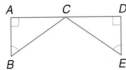

$\triangle RST \sim \triangle$ _____?_____

7.

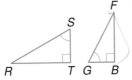

$\triangle XYZ \sim \triangle$ _____?_____

8.

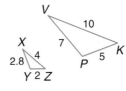

$\triangle MNQ \sim \triangle$ _____?_____

Find the value of each variable.

9.

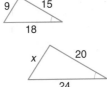

10.

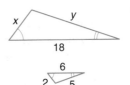

11.

Give a reason for each statement in Exercises 12–13.

12. If $\angle B \cong \angle E$ and $\angle A$ and $\angle D$ are right angles, show that $\dfrac{BC}{EC} = \dfrac{AB}{DE}$.

a. $\angle B \cong \angle E$

b. $\angle A$ and $\angle D$ are right angles.

c. $\angle A \cong \angle D$

d. $\triangle ABC \sim \triangle DEC$

e. $\dfrac{BC}{EC} = \dfrac{AB}{DE}$

13. If $\overline{JK} \parallel \overline{GH}$, show that $\dfrac{FJ}{FG} = \dfrac{FK}{FH}$.

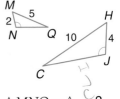

a. $\overline{JK} \parallel \overline{GH}$

b. $\angle 1 \cong \angle 2$

c. $\angle F \cong \angle F$

d. $\triangle FJK \sim \triangle FGH$

e. $\dfrac{FJ}{FG} = \dfrac{FK}{FH}$

14. Road Construction The state highway
department is considering the possibility
of building a tunnel through the mountain
from point *A* to point *B*. Surveyors provided
the map at the right. How long would the
tunnel be?

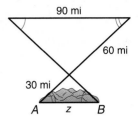

15. Construction The pitch of a roof
is the ratio of the rise to the run.
The Ace Construction Company
is building a garage that is 24 feet
wide. If the pitch of the roof is to be
1:3, find the rise of the roof.

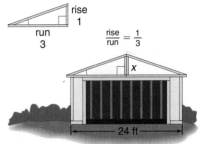

16. Architecture Maria is visiting Washington, D.C. She wants to know
the height of the Washington Monument. The monument's shadow
is 111 feet at the same time that Maria's shadow is 1 foot. Maria is
5 feet tall.

   a. Draw a figure to represent the problem. Label all known distances.

   b. Outline two similar triangles in red.

   c. Determine the height of the Washington Monument.

17. Critical Thinking A *primitive Pythagorean triple* is a set of whole
numbers that satisfies the equation $a^2 + b^2 = c^2$ and has no common
factors except 1. A *family of Pythagorean triples* is a primitive triple and
its whole number multiples. How are the triangles represented by a
family of Pythagorean triples related? Explain.

Mixed Review

18. Scale Drawings A window measures 8 feet by 3 feet. Make a scale
drawing of the window if $\frac{1}{4}$ inch represents 1 foot. *(Lesson 9–2)*

19. Write the ratio *10 months to 5 years* in simplest form. *(Lesson 9–1)*

20. Find *x* in the figure shown. *(Lesson 8–1)*

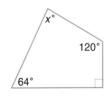

**Standardized
Test Practice**
Ⓐ Ⓑ Ⓒ Ⓓ

21. Short Response In $\triangle ACE$, $\overline{AC} \cong \overline{AE}$.
If $m\angle C = 7x + 2$, and $m\angle E = 8x - 8$,
what is $m\angle C$ and $m\angle E$? *(Lesson 6–4)*

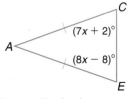

22. Multiple Choice 4 is what percent of 20? *(Percent Review)*

   Ⓐ 16%   Ⓑ 18%   Ⓒ 20%   Ⓓ 22%

Proportional Parts and Triangles

In $\triangle PQR$, $\overline{ST} \parallel \overline{QR}$, and $\overline{ST}$ intersects the other two sides of $\triangle PQR$. Note the shape of $\triangle PST$. Are $\triangle PQR$ and $\triangle PST$ similar?

Since $\angle 1$ and $\angle 2$ are congruent corresponding angles and $\angle P \cong \angle P$, $\triangle PST \sim \triangle PQR$. Why?

By Postulate 9–1, if two angles are congruent, then the two triangles are similar.

This characteristic of a line parallel to a side of a triangle is expressed in Theorem 9–4.

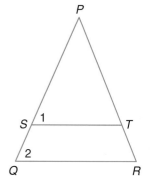

Theorem 9–4	**Words:**	If a line is parallel to one side of a triangle and intersects the other two sides, then the triangle formed is similar to the original triangle.
	Model:	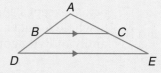
	Symbols:	If $\overline{BC} \parallel \overline{DE}$, then $\triangle ABC \sim \triangle ADE$.

You can use Theorem 9–4 to write proportions.

Example

1 Complete the proportion $\dfrac{SV}{SR} = \dfrac{?}{RT}$.

Since $\overline{VW} \parallel \overline{RT}$, $\triangle SVW \sim \triangle SRT$.

Therefore, $\dfrac{SV}{SR} = \dfrac{VW}{RT}$.

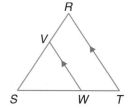

Your Turn

a. Use the triangle above to complete the proportion $\dfrac{ST}{SW} = \dfrac{SR}{?}$.

You can use proportions to solve for missing measures in triangles.

In the figure, $\overline{JK} \parallel \overline{GH}$. Find the value of x.

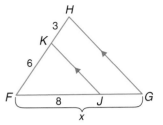

Explore You are given a triangle with a line parallel to one side of the triangle. You need to find the measure of $\overline{FG}$.

Plan You know $\triangle FJK \sim \triangle FGH$ by Theorem 9–4. Use this information to write a proportion and solve for x.

Solve
$\dfrac{FK}{FH} = \dfrac{FJ}{FG}$ *Definition of Similar Polygons*

$\dfrac{6}{9} = \dfrac{8}{x}$ *FK = 6, FH = 6 + 3 or 9, FJ = 8, FG = x*

$6(x) = 9(8)$ *Cross Products*

$6x = 72$ *Simplify.*

$\dfrac{6x}{6} = \dfrac{72}{6}$ *Divide each side by 6.*

$x = 12$ *Simplify.*

- **Algebra Review** -
Solving One-Step
Equations, p. 722

Examine Check the proportion by substituting 12 for x.

$\dfrac{6}{9} = \dfrac{8}{x}$ *Original Proportion*

$\dfrac{6}{9} \stackrel{?}{=} \dfrac{8}{12}$ *Substitution*

$6(12) \stackrel{?}{=} 9(8)$ *Cross Products*

$72 = 72$ ✓

Your Turn

b. In the figure, $\overline{AB} \parallel \overline{PR}$. Find the value of x.

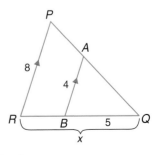

What other relationship occurs when a line is parallel to one side of a triangle and intersects the other two sides?

Hands-On Geometry

Materials:　□ lined paper　⌒ protractor　╱ ruler

Step 1 On a piece of lined paper, pick a point on one of the lines and label it *A*. Use a straightedge and protractor to draw ∠*A* so that *m*∠*A* < 90 and only the vertex lies on the line.

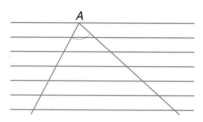

Step 2 Extend one side of ∠*A* down six lines. Label this point *G*. Do the same for the other side of ∠*A*. Label this point *M*. Now connect points *G* and *M* to form △*AGM*.

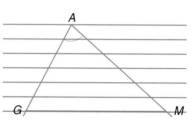

Step 3 Label the points where the horizontal rules intersect $\overline{AG}$, *B* through *F*, as shown. Label those points where the horizontal rules intersect $\overline{AM}$, *H* through *L*.

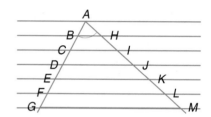

Try These

1. Measure $\overline{AC}$, $\overline{CG}$, $\overline{AD}$, $\overline{DG}$, $\overline{AE}$, $\overline{EG}$, $\overline{AI}$, $\overline{IM}$, $\overline{AJ}$, $\overline{JM}$, $\overline{AK}$, and $\overline{KM}$.
2. Calculate and compare the following ratios.

    a. $\frac{AC}{CG}$ and $\frac{AI}{IM}$　　**b.** $\frac{AD}{DG}$ and $\frac{AJ}{JM}$　　**c.** $\frac{AE}{EG}$ and $\frac{AK}{KM}$

3. What can you conclude about the lines through the sides of △*AGM* and parallel to $\overline{GM}$?

The activity above suggests Theorem 9–5.

Theorem 9–5	**Words:** If a line is parallel to one side of a triangle and intersects the other two sides, then it separates the sides into segments of proportional lengths.
	Model: 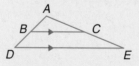
	Symbols: If $\overline{BC} \parallel \overline{DE}$, then $\frac{AB}{BD} = \frac{AC}{CE}$.

You can use a TI–83 Plus/TI–84 Plus graphing calculator to verify Theorem 9–5.

Graphing Calculator Tutorial
See pp. 782–785.

Graphing Calculator Exploration

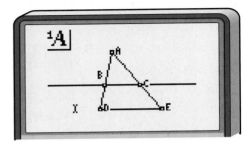

Step 1 Select the Triangle tool on [F2]. Draw and label triangle *DAE*.

Step 2 Next, use the Point on Object tool on [F2] to draw a point between *D* and *A* on side $\overline{DA}$. Label the point *B*.

Step 3 Use the Parallel Line tool on [F3] to draw a line through *B* parallel to side $\overline{DE}$.

Step 4 Use the Intersection Point tool on [F2] to mark the point where the line intersects side $\overline{EA}$. Label the point *C*. You now have a figure that you can use to verify Theorem 9–5.

Try These

1. Use the Distance & Length tool on [F5] to find the length of $\overline{AB}$, $\overline{BD}$, $\overline{AC}$, and $\overline{CE}$.
2. Use the Calculate tool on [F5] to calculate the value of $AB \div BD$ and $AC \div CE$. What can you say about these values?
3. Describe what happens to the segment lengths from Exercise 1 and the ratios from Exercise 2 when you drag point *B* closer to *A*.

Example ❸
Algebra Link

In the figure, $\overline{RS} \parallel \overline{UW}$. Find the value of *x*.

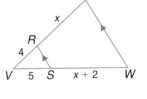

$$\frac{VR}{RU} = \frac{VS}{SW} \qquad \textit{Theorem 9–5}$$

$$\frac{4}{x} = \frac{5}{x + 2} \qquad \begin{array}{l}VR = 4,\ RU = x,\\ VS = 5,\ SW = x + 2\end{array}$$

$$4(x + 2) = x(5) \qquad \textit{Cross Products}$$

$$4x + 8 = 5x \qquad \textit{Distributive Property}$$

$$4x + 8 - 4x = 5x - 4x \quad \textit{Subtract 4x from each side.}$$

$$8 = x \qquad \textit{Simplify.}$$

Your Turn

c. In the figure, $\overline{GH} \parallel \overline{BC}$. Find the value of *x*.

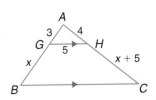

Check for Understanding

Communicating
Mathematics

1. **Explain** why $\triangle NRT \sim \triangle NPM$.

2. **Write** four proportions for the figure.

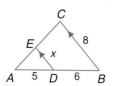

Exercises 1–2

3. 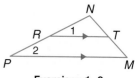 Casey uses the proportion $\frac{5}{6} = \frac{x}{8}$ to solve for x in the figure.
Jacob says she should use the proportion $\frac{5}{11} = \frac{x}{8}$.
Who is correct? Explain your reasoning.

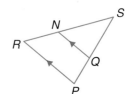

Guided Practice

Example 1

Complete each proportion.

4. $\dfrac{NQ}{RP} = \dfrac{?}{SR}$

5. $\dfrac{SN}{?} = \dfrac{SQ}{QP}$

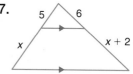

Examples 2 & 3

Find the value of each variable.

6.

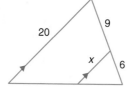

7.

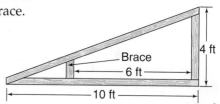

Example 2

8. **Construction** A roof rafter is shown at the right. Find the length of the brace.

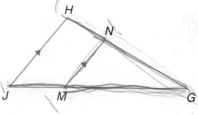

Brace
6 ft
4 ft
10 ft

Exercises

Practice

Complete each proportion.

9. $\dfrac{GN}{GH} = \dfrac{GM}{?}$

10. $\dfrac{GN}{NH} = \dfrac{GM}{?}$

11. $\dfrac{NM}{HJ} = \dfrac{?}{GJ}$

12. $\dfrac{?}{MN} = \dfrac{GH}{GN}$

13. $\dfrac{GJ}{?} = \dfrac{GH}{GN}$

14. $\dfrac{?}{NH} = \dfrac{GM}{MJ}$

Homework Help	
For Exercises	See Examples
9–14	1
15–24	2, 3
Extra Practice	
See page 742.	

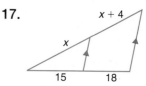

Find the value of each variable.

15.

16.

17.

Find the value of each variable.

18.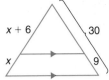

$x + 6$ 30

x 9

19.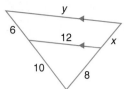

y

6 12 x

10 8

20.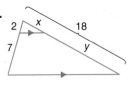

2 x 18

7 y

21. Find AD, AB, AE, and AC.

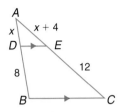

A

x $x + 4$

D E

8 12

B C

Applications and Problem Solving

22. **Surveying** Antoine wants to find the distance across Heritage Lake. According to his measurements, what is the distance across the lake?

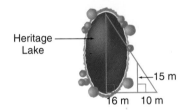

Heritage Lake

15 m

16 m 10 m

23. **Construction** Hannah is building a sawhorse. According to the diagram, how long should she make the brace?

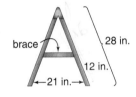

brace 28 in.

12 in.

21 in.

24. **Forestry** Ranger Lopez wants to know how tall the tree is that she planted five years ago. She walks away from the tree until the end of her shadow and the tree's shadow coincide. Use her measurements to determine the height of the tree.

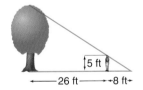

5 ft

26 ft 8 ft

25. **Critical Thinking** $\triangle JLN$ is equilateral. If $\overline{KM} \parallel \overline{JN}$, is $\triangle KLM$ equilateral? Explain.

L

K M

J N

Mixed Review

26. Draw and label two triangles that are similar by SAS Similarity. *(Lesson 9–3)*

27. **Sports** A volleyball court measures 30 feet by 60 feet. Make a scale drawing of the court if 1 centimeter represents 12 feet. *(Lesson 9–2)*

28. If $m\angle 1 = 67$ and $\angle 1$ and $\angle 2$ form a linear pair, find $m\angle 2$. *(Lesson 3–5)*

29. **Short Response** Graph point M with coordinates $(-4, -3)$. *(Lesson 2–4)*

30. **Multiple Choice** Solve $3h - 5 = 13$. *(Algebra Review)*

Ⓐ -4 Ⓑ -2 Ⓒ -7 Ⓓ 6

What You'll Learn
You'll learn to use proportions to determine whether lines are parallel to sides of triangles.

Why It's Important
Building Carpenters can use proportions to make sure boards are parallel to other boards.
See Exercise 1.

Jodie Rudberg is a carpenter. She is building the framework for a roof. How can she be sure the collar tie is parallel to the joist?

You know that if a line is parallel to one side of a triangle and intersects the other two sides, then it separates the sides into segments of proportional lengths (Theorem 9–5). The converse of this theorem is also true.

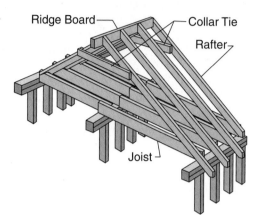

Ridge Board — Collar Tie

Rafter

Joist

Theorem 9–6	**Words:**	If a line intersects two sides of a triangle and separates the sides into corresponding segments of proportional lengths, then the line is parallel to the third side.

Model:

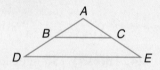

Symbols: If $\frac{AB}{BD} = \frac{AC}{CE}$, then $\overline{BC} \parallel \overline{DE}$.

Example

① Determine whether $\overline{TU} \parallel \overline{RS}$.

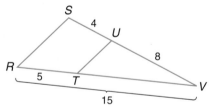

First, find TV.

$$RT + TV = RV$$
$$5 + TV = 15 \qquad \text{Replace } RT \text{ with 5 and } RV \text{ with 15.}$$
$$5 + TV - 5 = 15 - 5 \quad \text{Subtract 5 from each side.}$$
$$TV = 10$$

Then, determine whether $\frac{RT}{TV}$ and $\frac{SU}{UV}$ form a proportion.

$$\frac{RT}{TV} \stackrel{?}{=} \frac{SU}{UV}$$
$$\frac{5}{10} \stackrel{?}{=} \frac{4}{8} \qquad RT = 5, TV = 10, SU = 4, \text{ and } UV = 8$$
$$5(8) \stackrel{?}{=} 10(4) \quad \text{Cross Products}$$
$$40 = 40 \quad \checkmark$$

Using Theorem 9–6, $\overline{TU} \parallel \overline{RS}$.

In each figure, determine whether $\overline{MN} \parallel \overline{KL}$.

a.

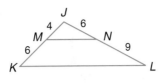

b.

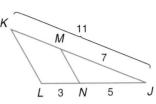

Consider the special case at the right.
S is the midpoint of $\overline{PQ}$ and T is the
midpoint of $\overline{PR}$. Because $\frac{5}{5} = \frac{7}{7}$,
$\overline{ST} \parallel \overline{QR}$ by Theorem 9–6. Therefore,
$\triangle PST \sim \triangle PQR$ by Theorem 9–4. Using
the definition of similar polygons,
$\frac{ST}{QR} = \frac{PS}{PQ}$. But $\frac{PS}{PQ} = \frac{5}{10}$, so $\frac{ST}{QR} = \frac{5}{10}$ or $\frac{1}{2}$.

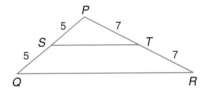

The general conclusion is stated in Theorem 9–7.

Theorem 9–7	**Words:**	If a segment joins the midpoints of two sides of a triangle, then it is parallel to the third side, and its measure equals one-half the measure of the third side.
	Model:	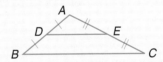
	Symbols:	If *D* is the midpoint of $\overline{AB}$ and *E* is the midpoint of $\overline{AC}$, then $\overline{DE} \parallel \overline{BC}$ and $DE = \frac{1}{2}BC$.

Examples

Algebra Link

X, *Y*, and *Z* are midpoints of the sides of $\triangle UVW$. Complete each
statement.

2 If $VW = 6a$, then $XY = $ ___?___ .

$XY = \frac{1}{2}VW$ *Theorem 9–7*

$XY = \frac{1}{2}(6a)$ *Replace VW with 6a.*

$XY = 3a$ *Multiply.*

If $VW = 6a$, then $XY = 3a$.

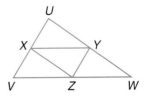

(continued on the next page)

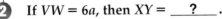

③ If $m\angle WZY = 2b + 1$, then $m\angle WVU = $ ___?___ .

By Theorem 9–6, $\overline{YZ} \parallel \overline{UV}$. Since $\overline{YZ}$ and $\overline{UV}$ are parallel segments cut by transversal $\overline{VW}$, $\angle WZY$ and $\angle WVU$ are congruent corresponding angles.

If $m\angle WZY = 2b + 1$, then $m\angle WVU = 2b + 1$.

Your Turn

c. $\overline{XZ} \parallel$ ___?___

d. If $YZ = c$, then $UV = $ ___?___ .

Check for Understanding

Communicating Mathematics

1. **Describe** how Ms. Rudberg can determine if the collar tie is parallel to the joist in the application at the beginning of the lesson.

2. **Draw** a triangle. Find the midpoints of two sides of the triangle and draw a segment between the two midpoints. Measure this segment and the third side of the triangle. Which theorem is confirmed?

Guided Practice

Example 1

In each figure, determine whether $\overline{GH} \parallel \overline{EF}$.

3.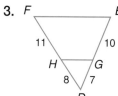

4.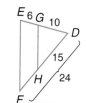

Examples 2 & 3

A, B, and C are the midpoints of the sides of $\triangle MNP$. Complete each statement.

5. $\overline{MP} \parallel$ ___?___

6. If $BC = 14$, then $MN = $ ___?___ .

7. If $m\angle MNP = s$, then $m\angle BCP = $ ___?___ .

8. If $MP = 18x$, then $AC = $ ___?___ .

9. **Communication** Ships signal each other using an international flag code. There are over 40 signaling flags, including the one at the right. The line that divides the white and blue portions of the flag intersects two sides of the flag at their midpoints. If the longer side of the blue portion is 48 inches long, find the length of the line dividing the white and blue portions of the flag. **Example 2**

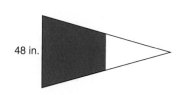

48 in.

Practice

In each figure, determine whether $\overline{EF} \parallel \overline{YZ}$.

Homework Help	
For Exercises	See Examples
10–17	1
18, 19, 22, 23, 26–31	2
20, 21, 24, 25	3
Extra Practice	
See page 742.	

10.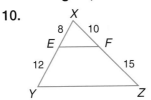

11.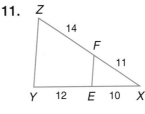

12.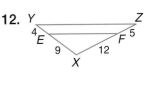

13.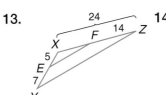

14.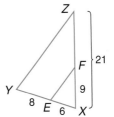

15.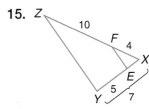

**R, S, and T are the midpoints of the sides of $\triangle GHJ$.
Complete each statement.**

16. $\overline{RS} \parallel$ ___?___

17. $\overline{GH} \parallel$ ___?___

18. If $RS = 36$, then $HJ =$ ___?___ .

19. If $GH = 44$, then $ST =$ ___?___ .

20. If $m\angle JST = 57$, then $m\angle JGH =$ ___?___ .

21. If $m\angle GHJ = 31$, then $m\angle STJ =$ ___?___ .

22. If $GJ = 10x$, then $RT =$ ___?___ .

23. If $ST = 20y$, then $GH =$ ___?___ .

24. If $m\angle HGJ = 4a$, then $m\angle TSJ =$ ___?___ .

25. If $m\angle JTS = 8b$, then $m\angle JHG =$ ___?___ .

26. If $RS = 12x$, then $HT =$ ___?___ .

27. If $GR = x + 5$, then $ST =$ ___?___ .

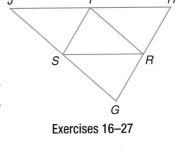

Exercises 16–27

28. A, B, and C are the midpoints of the sides of $\triangle DEF$.
    a. Find DE, EF, and FD.
    b. Find the perimeter of $\triangle ABC$.
    c. Find the perimeter of $\triangle DEF$.
    d. Find the ratio of the perimeter of
        $\triangle ABC$ to the perimeter of $\triangle DEF$.

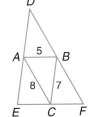

29. M is the midpoint of $\overline{GH}$, J is the midpoint
of $\overline{MG}$, N is the midpoint of $\overline{GI}$, and K is the
midpoint of $\overline{NG}$. If HI is 24, find MN and JK.

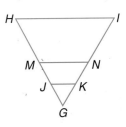

Applications and Problem Solving

30. Recreation Ryan is painting the lines on a shuffleboard court. Find AB.

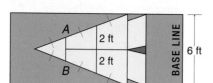

31. Algebra Find the value of x.

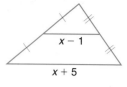

32. Critical Thinking $ABCD$ is a quadrilateral. E is the midpoint of $\overline{AD}$, F is the midpoint of $\overline{DC}$, G is the midpoint of $\overline{AB}$, and H is the midpoint of $\overline{BC}$.

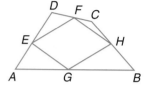

    a. What can you say about $\overline{EF}$ and $\overline{GH}$? Explain. (*Hint:* Draw diagonal $\overline{AC}$.)

    b. What kind of figure is *EFHG*?

Mixed Review

33. In the figure shown, $\overline{RS} \parallel \overline{NP}$. Find the value of x. *(Lesson 9–4)*

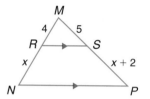

Determine whether the triangles shown are similar. If so, tell which similarity test is used and complete the statement. *(Lesson 9–3)*

34.

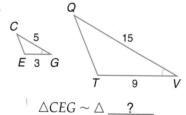

$\triangle CEG \sim \triangle \underline{\quad ? \quad}$

35.

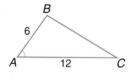

$\triangle \underline{\quad ? \quad} \sim \triangle DFE$

36. In $\square MTAH$, $MA = 46$. Find MC. *(Lesson 8–2)*

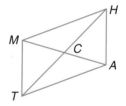

37. Grid In Find the measure of $\angle P$ in quadrilateral *GEPN* if $m\angle G = 130$, $m\angle E = 2x$, $m\angle P = 3x + 10$, and $m\angle N = 95$. *(Lesson 8–1)*

38. Multiple Choice What is the y-intercept of the graph of the equation $y = \frac{1}{3}x + 2$? *(Lesson 4–6)*

  Ⓐ 3   Ⓑ $\frac{1}{3}$ Ⓒ -2 Ⓓ 2

Math In the Workplace

Carpenter

Did you know that carpenters make up the largest group of skilled workers employed in the building trades? Carpenters cut, fit, and assemble wood and other materials in the construction of structures such as houses, buildings, and highways.

When constructing houses, carpenters use trusses to frame the roof. There are many types of trusses. One type, the *scissors truss*, is shown below.

1. Determine whether $\overline{XY} \parallel \overline{BC}$. Explain your reasoning.

2. Complete the following: $\triangle AXY \sim \triangle$ ___?___ .

FAST FACTS About Carpenters

Working Conditions
- generally work outdoors
- work may be strenuous
- can change employers each time a job is completed
- may risk injury from slips or falls, working with rough materials, and using tools and power equipment

Education
- high school industrial technology, mechanical drawing, carpentry, and math courses
- on-the-job training or apprenticeships

Employment

Where Carpenters Are Employed

Special Trade Carpenters 20%
Heavy Construction 12%
General Building Carpenters 33%
Manufacturing firms Government agencies Schools Wholesale/Retail Store 35%

Source: Occupational Outlook Handbook

---

interNET CONNECTION

Career Data For the latest information about a career in carpentry, visit:

www.geomconcepts.com

Are Golden Triangles Expensive?

Materials

 tracing paper

 straightedge

 protractor

 compass

Ratios of a Special Triangle

You may have heard of the golden rectangle, whose sides have a special ratio called the **golden ratio.** The golden ratio is approximately 1.618 to 1 or about 1.618. Artists and architects often use the golden rectangle because it is pleasing to the eye. A look at the Parthenon in Greece, the Taj Mahal in India, or the Lincoln Memorial in Washington, D.C., will reveal uses of the golden rectangle in architecture.

The Taj Mahal (golden rectangle)

Do you think there are golden triangles? How would they be constructed? Let's find out.

Investigate

1. Construct a golden triangle.

   a. Trace the regular pentagon below and use a photocopier to enlarge it. Draw $\overline{AC}$ and $\overline{AD}$.

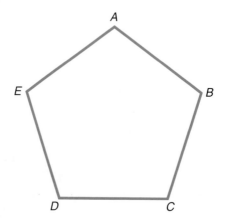

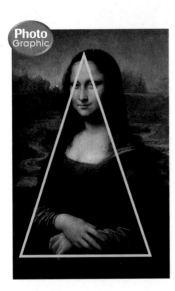

Mona Lisa (golden triangle)

b. Use a protractor to find the measures of the angles of △ADC. Classify △ADC.

c. Find the ratio of *AD* to *DC*. How does this ratio compare to the golden ratio?

Artists often use golden triangles to draw your eye toward the face of the subject. Notice how the folded arms and head of the *Mona Lisa* form a triangle.

2. Use your pentagon and triangle to construct another golden triangle. Follow the steps below.

a. Using a compass and straightedge, bisect ∠ADC. Label the point where the angle bisector intersects AC point *F*.

b. What are the measures of the angles in △DCF? Classify △DCF.

c. Find the ratio of *DC* to *FC*. How does this ratio compare to the golden ratio?

d. What conclusions can you draw from this activity?

Extending the Investigation

In this extension, you will investigate the golden ratio, golden triangles, and golden rectangles.

- Start with a regular pentagon. Draw at least four golden triangles, each one smaller than the previous one. Label the triangles and show the golden ratio in each triangle. You may want to use different colors to outline the different triangles.

- Are the golden triangles you drew within the pentagon similar? Explain.

Presenting Your Conclusions

Here are some ideas to help you present your conclusions to the class.

- Make a bulletin board to display your golden triangles.
- Research the golden ratio. Write a brief paper describing five examples where the golden ratio has been used.
- Research the golden rectangle. Make a poster demonstrating how to construct a golden rectangle.

 Investigation For more information on the golden ratio, visit: www.geomconcepts.com

Proportional Parts and Parallel Lines

What You'll Learn

You'll learn to identify and use the relationships between parallel lines and proportional parts.

Why It's Important

Real Estate Builders can use parallel lines and proportional parts to determine the length of a side of a building site. *See Exercise 24.*

The artistic concept of perspective combines proportion and the properties of parallel lines.

In the figure, points A and B lie on the horizon. $\overleftrightarrow{AC}$ and $\overleftrightarrow{BD}$ represent two lines extending from the horizon to the foreground. The two lines also form transversals for the parallel lines m, n, and p.

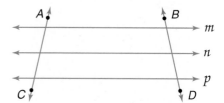

Along the transversals $\overleftrightarrow{AC}$ and $\overleftrightarrow{BD}$, the parallel lines cut segments of different lengths. Is there a relationship between the lengths of the segments?

The following activity investigates transversals that cross three parallel lines.

Hands-On Geometry

Materials: lined paper   ruler

Step 1   Draw a dark line over the line at the top of your lined paper. Count down four lines and draw a dark line over that line. Count down six more lines and draw a dark line over that line.

Step 2   Draw three different transversals that cross each of the parallel lines you drew. Label the points of intersection as shown.

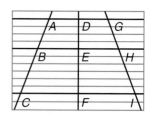

Try These

1. Measure $\overline{AB}$, $\overline{BC}$, $\overline{AC}$, $\overline{DE}$, $\overline{EF}$, $\overline{DF}$, $\overline{GH}$, $\overline{HI}$, and $\overline{GI}$.

2. Calculate each set of ratios. Determine whether the ratios in each set are equivalent to each other.

   a. $\dfrac{AB}{BC}$, $\dfrac{DE}{EF}$, $\dfrac{GH}{HI}$

   b. $\dfrac{AB}{AC}$, $\dfrac{DE}{DF}$, $\dfrac{GH}{GI}$

   c. $\dfrac{BC}{AC}$, $\dfrac{EF}{DF}$, $\dfrac{HI}{GI}$

3. Do the parallel lines divide the transversals proportionally?

The previous activity suggests Theorem 9–8.

Theorem 9–8	**Words:**	If three or more parallel lines intersect two transversals, the lines divide the transversals proportionally.
	Model:	
	Symbols:	If $\ell \parallel m \parallel n$, then $\frac{AB}{BC} = \frac{DE}{EF}$, $\frac{AB}{AC} = \frac{DE}{DF}$, and $\frac{BC}{AC} = \frac{EF}{DF}$.

Example ➊ **Complete the proportion $\frac{MN}{MP} = \frac{?}{RT}$.**

Since $\overleftrightarrow{MR} \parallel \overleftrightarrow{NS} \parallel \overleftrightarrow{PT}$, the transversals are divided proportionally. Therefore,
$\frac{MN}{MP} = \frac{RS}{RT}$.

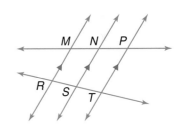

Your Turn

a. Use the figure above to complete the proportion $\frac{RS}{ST} = \frac{MN}{?}$.

You can use proportions from parallel lines to solve for missing measures.

Example ➋
Algebra Link

In the figure, $a \parallel b \parallel c$. Find the value of x.

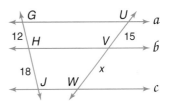

$\frac{GH}{HJ} = \frac{UV}{VW}$ *Theorem 9–8*

$\frac{12}{18} = \frac{15}{x}$ *GH = 12, HJ = 18, UV = 15, VW = x*

$12(x) = 18(15)$ *Cross Products*

$12x = 270$ *Simplify.*

$\frac{12x}{12} = \frac{270}{12}$ *Divide each side by 12.*

$x = \frac{45}{2}$ or $22\frac{1}{2}$ *Simplify.*

(continued on the next page)

Algebra Review
Solving One-Step
Equations, p. 722

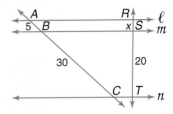

Your Turn

b. In the figure, $\ell \parallel m \parallel n$.
   Find the value of x.

Suppose three parallel lines intersect a transversal and divide the transversal into congruent segments. Refer to the figure below. The transversal on the left is divided into congruent segments. Is this true of the other two transversals? Find the values of x and y.

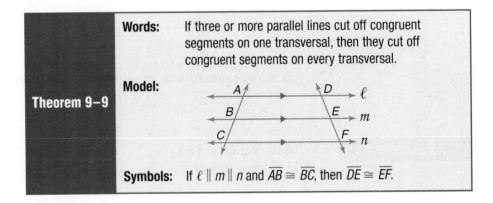

$$\frac{5}{5} = \frac{4}{x}$$ *Theorem 9–8* $$\frac{5}{5} = \frac{y}{6}$$

$5(x) = 5(4)$ *Cross Products* $5(6) = 5(y)$

$5x = 20$ *Simplify.* $30 = 5y$

$$\frac{5x}{5} = \frac{20}{5}$$ *Divide each side by 5.* $$\frac{30}{5} = \frac{5y}{5}$$

$x = 4$ *Simplify.* $6 = y$

These results suggest Theorem 9–9.

Theorem 9–9	**Words:**	If three or more parallel lines cut off congruent segments on one transversal, then they cut off congruent segments on every transversal.
	Model:	
	Symbols:	If $\ell \parallel m \parallel n$ and $\overline{AB} \cong \overline{BC}$, then $\overline{DE} \cong \overline{EF}$.

Check for Understanding

Communicating Mathematics

1. **Write** at least three proportions if $a \parallel b \parallel c$.

2. **Describe** how you would find GJ if $EF = 15$, $DF = 25$, and $GH = 12$.

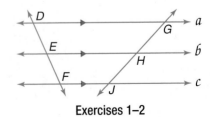

Exercises 1–2

3. **Writing Math** Draw a segment and label the endpoints A and B. Use the following steps to divide $\overline{AB}$ into three congruent segments.

a. Draw $\overrightarrow{AC}$ so that $\angle BAC$ is an acute angle.

b. With a compass, start at A and mark off three congruent segments on $\overrightarrow{AC}$. Label these points D, E, and F.

c. Draw $\overline{BF}$.

d. Construct lines through D and E that are parallel to $\overline{BF}$. These parallel lines will divide $\overline{AB}$ into three congruent segments.

e. Explain why this construction works.

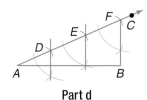

Parts a–c

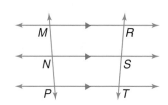

Part d

Look Back

Constructing Parallel Lines, Lesson 4–4

Guided Practice

Example 1

Complete each proportion.

4. $\dfrac{NP}{MP} = \dfrac{?}{RT}$

5. $\dfrac{ST}{?} = \dfrac{NP}{MN}$

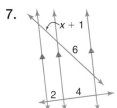

Example 2

Find the value of x.

6.

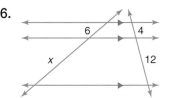

7.

8. Travel Alex is visiting the city of Melbourne, Australia. Elizabeth Street, Swanston Street, Russell Street, and Exhibition Street are parallel. If Alex wants to walk along La Trobe Street from Elizabeth Street to Exhibition Street, approximately how far will he walk? **Example 2**

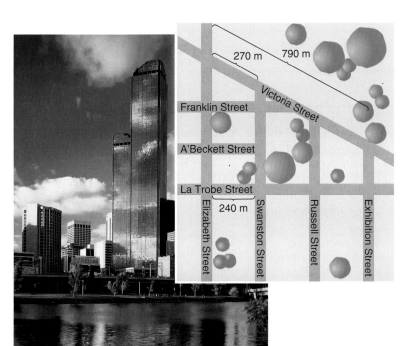

Melbourne, Australia

Exercises •

Practice

Homework Help

For Exercises	See Examples
9–14	1
15–24	2

Extra Practice
See page 743.

Complete each proportion.

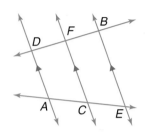

9. $\dfrac{DF}{FB} = \dfrac{?}{CE}$

10. $\dfrac{?}{DF} = \dfrac{CE}{AC}$

11. $\dfrac{AC}{?} = \dfrac{DF}{DB}$

12. $\dfrac{AC}{AE} = \dfrac{DF}{?}$

13. $\dfrac{?}{FB} = \dfrac{AE}{CE}$

14. $\dfrac{AE}{?} = \dfrac{DB}{DF}$

Find the value of x.

15.
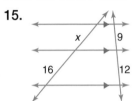

16.
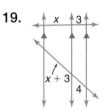

17.
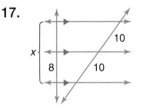

18.
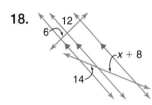

19.
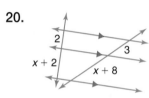

20.
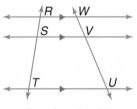

21. If $RT = 12$, $WV = 5$, and $WU = 13$, find RS.

22. If $RT = 20$, $ST = 15$, and $WU = 12$, find VU.

Exercises 21–22

Applications and Problem Solving

23. **City Planning** Numbered streets in Washington, D.C., run north and south. Lettered name streets run east and west. Other streets radiate out like spokes of a wheel and are named for states. The Metro subway system goes under some of the roads. Find the approximate distance between the two Metro stations indicated on the map by M.

24. Real Estate Hometown Builders are selling four building sites along Washburn River. If the total river frontage of the lots is 100 meters, find the river frontage for each lot.

25. Critical Thinking Exercise 3 on page 385 describes a construction that divides a segment into three congruent segments. Draw a line segment and describe how to divide it into three segments with the ratio 1:2:3. Then construct the divided segment.

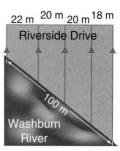

Exercise 24

Mixed Review

26. Triangle *CDE* is shown. Determine whether $\overline{DC} \parallel \overline{MN}$. *(Lesson 9–5)*

27. Use $\triangle XYZ$ to complete the proportion $\frac{YX}{YA} = \frac{YZ}{?}$. *(Lesson 9–4)*

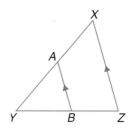

28. Is a triangle with measures 18, 24, and 30 a right triangle? *(Lesson 6–6)*

Standardized Test Practice
Ⓐ Ⓑ Ⓒ Ⓓ

29. Grid In $\overrightarrow{AC}$ and $\overrightarrow{AT}$ are opposite rays and $\overline{AR} \perp \overline{AS}$. If $m\angle CAR = 42$, find $m\angle SAT$. *(Lesson 3–7)*

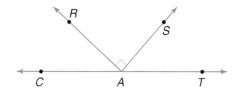

30. Short Response Draw and label a rectangle that has an area of 42 square centimeters. *(Lesson 1–6)*

Quiz 2 Lessons 9–3 through 9–6

Find the value of each variable. *(Lessons 9–3, 9–4, 9–5, and 9–6)*

1.
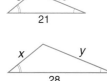

2.
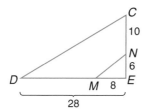

3.
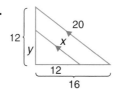

4.
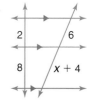

5. Carl is 6 feet tall and casts an 8-foot shadow. At the same time, a flagpole casts a 48-foot shadow. How tall is the flagpole? *(Lesson 9–3)*

If two triangles are similar, then the measures of their corresponding sides are proportional. Is there a relationship between the measures of the perimeters of the two triangles?

What You'll Learn

You'll learn to identify and use proportional relationships of similar triangles.

Why It's Important

Surveying Surveyors use scale factors to estimate distance. *See Exercise 12.*

Hands-On Geometry

Materials: grid paper

Step 1 On grid paper, draw right triangle *ABC* with legs 9 units and 12 units long as shown.

Step 2 On grid paper, draw right triangle *DEF* with legs 6 units and 8 units long as shown.

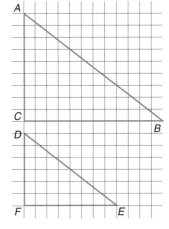

Look Back

Pythagorean Theorem: Lesson 6–6

Try These

1. Use the Pythagorean Theorem to find *AB* and *DE*.
2. Find the ratios $\frac{AB}{DE}$, $\frac{BC}{EF}$, and $\frac{CA}{FD}$.
3. Are the triangles similar? Explain.
4. Find the perimeters of $\triangle ABC$ and $\triangle DEF$.
5. Find the ratio $\frac{\text{perimeter of } \triangle ABC}{\text{perimeter of } \triangle DEF}$.
6. Compare the ratios $\frac{\text{perimeter of } \triangle ABC}{\text{perimeter of } \triangle DEF}$, $\frac{AB}{DE}$, $\frac{BC}{EF}$, and $\frac{CA}{FD}$. Describe your results.

The activity above suggests Theorem 9–10.

Theorem 9–10	**Words:**	If two triangles are similar, then the measures of the corresponding perimeters are proportional to the measures of the corresponding sides.
	Symbols:	If $\triangle ABC \sim \triangle DEF$, then $\frac{\text{perimeter of } \triangle ABC}{\text{perimeter of } \triangle DEF} = \frac{AB}{DE} = \frac{BC}{EF} = \frac{CA}{FD}$.

You can find the measures of all three sides of a triangle when you know the perimeter of the triangle and the measures of the sides of a similar triangle.

Example

Algebra Link

1 The perimeter of $\triangle RST$ is 9 units, and $\triangle MNP \sim \triangle RST$. Find the value of each variable.

$\dfrac{MN}{RS} = \dfrac{\text{perimeter of } \triangle MNP}{\text{perimeter of } \triangle RST}$ *Theorem 9–10*

$\dfrac{3}{x} = \dfrac{13.5}{9}$ *The perimeter of $\triangle MNP$ is $3 + 6 + 4.5$ or 13.5.*

$3(9) = x(13.5)$ *Cross Products*

$27 = 13.5x$ *Simplify.*

$\dfrac{27}{13.5} = \dfrac{13.5x}{13.5}$ *Divide each side by 13.5.*

$2 = x$ *Simplify.*

┌─ **Algebra Review** ─┐
Solving One-Step
Equations, p. 722
└─────────────────────┘

Because the triangles are similar, two other proportions can be written to find the value of the other two variables.

$\dfrac{MN}{RS} = \dfrac{NP}{ST}$ *Definition of Similar Polygons* $\dfrac{MN}{RS} = \dfrac{PM}{TR}$

$\dfrac{3}{2} = \dfrac{6}{y}$ *Substitution* $\dfrac{3}{2} = \dfrac{4.5}{z}$

$3(y) = 2(6)$ *Cross Products* $3(z) = 2(4.5)$

$3y = 12$ *Simplify.* $3z = 9$

$\dfrac{3y}{3} = \dfrac{12}{3}$ *Divide each side by 3.* $\dfrac{3z}{3} = \dfrac{9}{3}$

$y = 4$ *Simplify.* $z = 3$

Your Turn

a. The perimeter of $\triangle KQV$ is 72 units, and $\triangle GHJ \sim \triangle KQV$. Find the value of each variable.

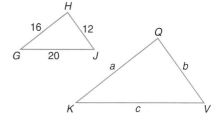

The ratio found by comparing the measures of corresponding sides of similar triangles is called the *constant of proportionality* or the **scale factor**.

If $\triangle ABC \sim \triangle DEF$, then $\dfrac{AB}{DE} = \dfrac{BC}{EF} = \dfrac{CA}{FD}$

or $\dfrac{3}{6} = \dfrac{7}{14} = \dfrac{5}{10}$. Each ratio is equivalent to $\dfrac{1}{2}$.

The scale factor of $\triangle ABC$ to $\triangle DEF$ is $\dfrac{1}{2}$.

The scale factor of $\triangle DEF$ to $\triangle ABC$ is $\dfrac{2}{1}$.

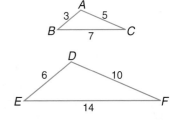

Determine the scale factor of △UVW to △XYZ.

$$\frac{UV}{XY} = \frac{10}{6} \text{ or } \frac{5}{3} \qquad \frac{VW}{YZ} = \frac{15}{9} \text{ or } \frac{5}{3}$$

$$\frac{WU}{ZX} = \frac{20}{12} \text{ or } \frac{5}{3}$$

The scale factor is $\frac{5}{3}$.

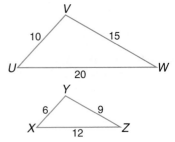

Your Turn

b. Determine the scale factor of △XYZ to △UVW.

In the figure, △ABC ~ △DEF. Suppose the scale factor of △ABC to △DEF is $\frac{1}{2}$.

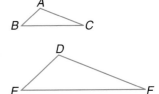

$$\frac{P \text{ of } \triangle ABC}{P \text{ of } \triangle DEF} = \frac{1}{2} \qquad \textit{P represents perimeter.}$$

$$\frac{P \text{ of } \triangle ABC}{\cancel{P \text{ of } \triangle DEF}} (\cancel{P \text{ of } \triangle DEF}) = \frac{1}{2}(P \text{ of } \triangle DEF) \quad \textit{Multiply each side by P of } \triangle DEF.$$

$$P \text{ of } \triangle ABC = \frac{1}{2}(P \text{ of } \triangle DEF) \quad \textit{Simplify.}$$

In general, if the scale factor of △ABC to △DEF is s,
P of △ABC = s(P of △DEF).

Suppose △SLC ~ △GFR and the scale factor of △SLC to △GFR is $\frac{2}{3}$. Find the perimeter of △GFR if the perimeter of △SLC is 14 inches.

$$\frac{\triangle SLC}{\triangle GFR} \qquad \frac{2}{3} = \frac{14}{x} \qquad \begin{array}{l} \textit{Perimeter of } \triangle SLC \\ \textit{Perimeter of } \triangle GFR \end{array}$$

$$2x = 42 \quad \textit{Find cross products.}$$

$$\frac{2x}{2} = \frac{42}{2} \quad \textit{Divide each side by 2.}$$

$$x = 21 \quad \textit{Simplify.}$$

The perimeter of △GFR is 21 inches.

Your Turn

c. Suppose △PQR ~ △XYZ and the scale factor of △PQR to △XYZ is $\frac{5}{6}$. Find the perimeter of △XYZ if the perimeter of △PQR is 25 centimeters.

Communicating Mathematics

1. **Confirm** that the ratio of the measures of the corresponding sides is the same as the ratio of the measures of the corresponding perimeters.

Vocabulary

scale factor

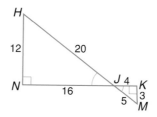

2. **Identify** the additional information needed to solve for x.

$\triangle TRS \sim \triangle PQO$
perimeter of $\triangle TRS = 20$
$QO = 15$

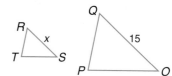

Guided Practice

⊙ **Getting Ready** Write each fraction in simplest form.

Sample: $\dfrac{18}{21}$ **Solution:** $\dfrac{18}{21} = \dfrac{18 \div 3}{21 \div 3}$ or $\dfrac{6}{7}$

3. $\dfrac{6}{42}$ 4. $\dfrac{10}{24}$ 5. $\dfrac{63}{18}$ 6. $\dfrac{91}{13}$

Example 1

For each pair of similar triangles, find the value of each variable.

7.

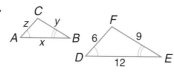

P of $\triangle ABC = 9$

8.

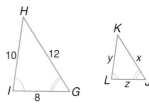

P of $\triangle JKL = 7.5$

Example 2

Determine the scale factor for each pair of similar triangles.

9. $\triangle MNO$ to $\triangle XYZ$

10. $\triangle XYZ$ to $\triangle MNO$

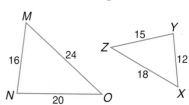

Example 3

11. Suppose $\triangle RST \sim \triangle UVW$ and the scale factor of $\triangle RST$ to $\triangle UVW$ is $\frac{3}{2}$. Find the perimeter of $\triangle UVW$ if the perimeter of $\triangle RST$ is 57 inches.

Example 2

12. Surveying Heather is using similar triangles to find the distance across Turtle Lake. What is the scale factor of $\triangle BCD$ to $\triangle ACE$?

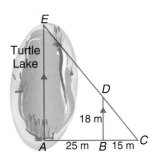

Exercises

Practice

For each pair of similar triangles, find the value of each variable.

13.

P of $\triangle DEF = 29$

14.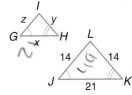

P of $\triangle GHI = 21$

15.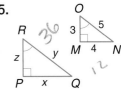

P of $\triangle PQR = 36$

Homework Help	
For Exercises	**See Examples**
13–18	1
19–24, 28	2
25–27	3
29	1, 2
Extra Practice	
See page 743.	

16.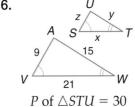

P of $\triangle STU = 30$

17.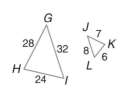

P of $\triangle BCD = 81$

18.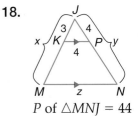

P of $\triangle MNJ = 44$

Determine the scale factor for each pair of similar triangles.

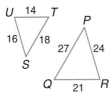

19. $\triangle STU$ to $\triangle PQR$
20. $\triangle PQR$ to $\triangle STU$

21. $\triangle GHI$ to $\triangle JKL$
22. $\triangle JKL$ to $\triangle GHI$

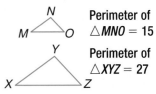

Perimeter of $\triangle MNO = 15$

Perimeter of $\triangle XYZ = 27$

23. $\triangle XYZ$ to $\triangle MNO$
24. $\triangle MNO$ to $\triangle XYZ$

25. The perimeter of $\triangle RST$ is 57 feet. If $\triangle RST \sim \triangle HKN$ and the scale factor of $\triangle RST$ to $\triangle HKN$ is $\frac{3}{2}$, find the perimeter of $\triangle HKN$.

26. Suppose $\triangle JKL \sim \triangle MVW$ and the scale factor of $\triangle JKL$ to $\triangle MVW$ is $\frac{4}{3}$. The lengths of the sides of $\triangle JLK$ are 12 meters, 10 meters, and 10 meters. Find the perimeter of $\triangle MVW$.

27. Suppose $\triangle ABC \sim \triangle DEF$ and the scale factor of $\triangle ABC$ to $\triangle DEF$ is $\frac{5}{3}$. Find the perimeter of $\triangle DEF$ if the perimeter of $\triangle ABC$ is 25 meters.

Applications and Problem Solving

28. **Drafting** In a blueprint of a house, 1 inch represents 3 feet. What is the scale factor of the blueprint to the actual house? (*Hint*: Change feet to inches.)

29. **Architecture** A bird's-eye view of the Pentagon reveals five similar pentagons. Each side of the outside pentagon is about 920 feet. Each side of the innermost pentagon is about 360 feet.

a. Find the scale factor of the outside pentagon to the innermost pentagon.

b. Find the perimeter of the outside pentagon.

c. Find the perimeter of the innermost pentagon.

d. Find the ratio of the perimeter of the outside pentagon to the perimeter of the innermost pentagon.

e. Tell how the ratio in part d compares to the scale factor of the pentagons.

30. **Critical Thinking** The perimeter of $\triangle RST$ is 40 feet. Find the perimeter of $\triangle XYZ$.

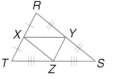

Mixed Review

31. In the figure shown, $\ell \parallel m \parallel n$. Find the value of x. (*Lesson 9–6*)

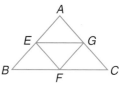

32. **Algebra** The midpoints of the sides of $\triangle ABC$ are E, F, and G. Find the measure of BC if $EG = 4b$. (*Lesson 9–5*)

33. *True* or *false*: The diagonals of a square bisect each other. (*Lesson 8–4*)

34. If $MN = 47$, $PQ = 63 - 4c$, and $MN \leq PQ$, what is the value of c? (*Lesson 7–1*)

Standardized Test Practice
Ⓐ Ⓑ Ⓒ Ⓓ

35. **Short Response** Suppose $AB = AD = 4$, $m\angle B = m\angle D = 65$, and $AC = 3.5$. Is $\triangle ABC \cong \triangle ADC$? (*Lesson 5–5*)

Exercise 35

36. **Multiple Choice** In $\triangle RST$, $m\angle R = 72$ and $m\angle S = 37$. What is $m\angle T$? (*Lesson 5–2*)

Ⓐ 19　　　Ⓑ 37　　　Ⓒ 71　　　Ⓓ 72

Understanding and Using the Vocabulary

*inter***NET**
CONNECTION **Review Activities**
For more review activities, visit:
www.geomconcepts.com

After completing this chapter, you should be able to define each term, property, or phrase and give an example or two of each.

cross products *(p. 351)*   means *(p. 351)*   scale drawing *(p. 358)*
extremes *(p. 351)*   polygon *(p. 356)*   scale factor *(p. 389)*
geometric mean *(p. 355)*   proportion *(p. 351)*   sides *(p. 356)*
golden ratio *(p. 380)*   ratio *(p. 350)*   similar polygons *(p. 356)*

Choose the correct term to complete each sentence.

1. Every proportion has two (similar figures, cross products).

2. A (proportion, ratio) is a comparison of two numbers by division.

3. The cross products are always equal in a (proportion, scale drawing).

4. In (proportions, similar figures), corresponding angles are congruent, and the measures of corresponding sides have equivalent ratios.

5. In the proportion $\frac{2}{5} = \frac{4}{10}$, the terms 2 and 10 are called the (extremes, means).

6. (Scale drawings, Proportions) are used to represent something that is too large or too small to be drawn at actual size.

7. A proportion has two cross products called the extremes and the (ratios, means).

8. The constant of proportionality is also called the (scale factor, scale drawing).

9. Knowing several measures of two (similar figures, scale drawings) may allow you to find the measures of missing parts.

10. The symbols a to b, $a{:}b$, and $\frac{a}{b}$, where $b \neq 0$ represent (ratios, cross products).

Skills and Concepts

Objectives and Examples	**Review Exercises**

• **Lesson 9–1** Use ratios and proportions to solve problems.

Solve $\frac{6}{45} = \frac{2}{3x}$.

$\frac{6}{45} = \frac{2}{3x}$

$6(3x) = 45(2)$ *Cross Products*

$18x = 90$ *Multiply.*

$\frac{18x}{18} = \frac{90}{18}$ *Divide each side by 18.*

$x = 5$ *Simplify.*

Write each ratio in simplest form.

11. $\frac{3}{9}$ 12. $\frac{45}{100}$ 13. $\frac{55}{22}$

Solve each proportion.

14. $\frac{3}{10} = \frac{9}{x}$ 15. $\frac{3}{2x} = \frac{12}{16}$

16. $\frac{84}{49} = \frac{12}{17 - x}$ 17. $\frac{16}{20} = \frac{x + 3}{10}$

www.geomconcepts.com/vocabulary_review

| **Objectives and Examples** | **Review Exercises** |

• **Lesson 9–2** Identify similar polygons.

Determine whether the polygons are similar.

Since $\frac{3}{9} = \frac{4}{12} = \frac{5}{15}$, the sides of the polygons are proportional. The corresponding angles are congruent. So, the polygons are similar.

18. Determine whether the polygons are similar. Justify your answer.

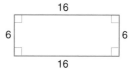

19. The polygons are similar. Find x and y.

 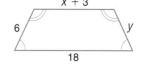

• **Lesson 9–3** Use AA, SSS, and SAS similarity tests for triangles.

Since $\angle A \cong \angle X$ and $\angle B \cong \angle Y$, the triangles are similar by AA Similarity.

Determine whether each pair of triangles is similar. If so, tell which similarity test is used.

20.   **21.**

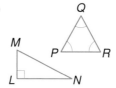

• **Lesson 9–4** Identify and use the relationships between proportional parts of triangles.

Since $\overline{BC} \parallel \overline{DE}$, $\triangle ABC \sim \triangle ADE$.

So, $\frac{AB}{AD} = \frac{BC}{DE}$.

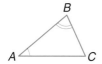

Complete each proportion.

22. $\dfrac{?}{PQ} = \dfrac{MN}{MP}$

23. $\dfrac{ML}{MQ} = \dfrac{?}{MP}$

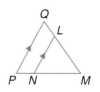

• **Lesson 9–5** Use proportions to determine whether lines are parallel to sides of triangles.

Determine whether $\overline{LM} \parallel \overline{JK}$.

$\dfrac{LJ}{NJ} \stackrel{?}{=} \dfrac{MK}{NK}$

$\dfrac{2}{8} \stackrel{?}{=} \dfrac{3}{12}$ *Substitute.*

$2(12) \stackrel{?}{=} 8(3)$ *Cross Products*

$24 = 24$ ✓

So, $\overline{LM} \parallel \overline{JK}$.

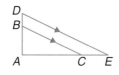

24. Determine whether $\overline{AB} \parallel \overline{CD}$.

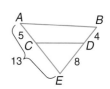

A, B, and C are the midpoints of the sides of △STU. Complete.

25. $\overline{AC} \parallel$ _____?_____

26. If $UT = 16$, then $AB =$ _____?_____

27. If $BC = 6$, then $SU =$ _____?_____

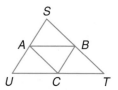

Mixed Problem Solving
See pages 758–765.

Objectives and Examples

Review Exercises

- **Lesson 9–6** Identify and use the relationships between parallel lines and proportional parts.

Since $\overleftrightarrow{AD} \parallel \overleftrightarrow{BE} \parallel \overleftrightarrow{CF}$,

$\dfrac{BC}{AC} = \dfrac{EF}{DF}$.

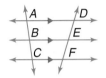

28. Complete the proportion $\dfrac{AB}{AC} = \dfrac{?}{DF}$ using the figure at the left.

29. Find the value of d if $x \parallel y \parallel z$.

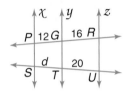

- **Lesson 9–7** Identify and use proportional relationships of similar triangles.

Determine the scale factor of $\triangle ABC$ to $\triangle XYZ$.

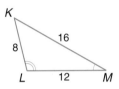

$\dfrac{AB}{XY} = \dfrac{AC}{XZ} = \dfrac{BC}{YZ} = \dfrac{1}{2}$

The scale factor of $\triangle ABC$ to $\triangle XYZ$ is $\dfrac{1}{2}$.

30. The triangles below are similar, and the perimeter of $\triangle NPQ$ is 27. Find the value of each variable.

31. Determine the scale factor of $\triangle DEF$ to $\triangle ABC$.

Applications and Problem Solving

32. Sailing The sail on John's boat is shaped like a right triangle. Its hypotenuse is 24 feet long, one leg is 16 feet long, and the other is about 18 feet long. The triangular sail on a model of the boat has a hypotenuse of 3 feet. If the two triangular sails are similar, how long are the legs of the model's sail? *(Lesson 9–3)*

33. Recreation The ends of the swing set at Parkdale Elementary School look like the letter A, as shown in the diagram. If the horizontal bar is parallel to the ground, how long is the horizontal bar? *(Lesson 9–5)*

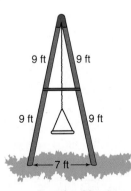

Exercise 33

1. **Name** three tests used to determine whether triangles are similar.

2. **Describe** how a scale drawing could be used by an architect.

Solve each proportion.

3. $\frac{x-2}{7} = \frac{20}{35}$

4. $\frac{5}{3} = \frac{x+7}{9}$

5. $\frac{27}{x} = \frac{36}{9}$

6. Determine if the polygons are similar. Justify your answer.

7. The polygons are similar. Find the values of x and z.

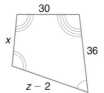

Determine whether each pair of triangles is similar. If so, tell which similarity test is used.

8.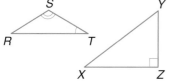

9.

10.

11. Find the values of x and y.

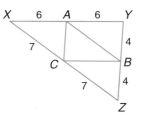

Complete each proportion.

12. $\frac{BF}{?} = \frac{BE}{BC}$

13. $\frac{BE}{BC} = \frac{?}{CD}$

Exercises 12–13

Complete each statement.

14. $AC = \underline{\quad ? \quad}$.

15. $\overline{BC} \parallel \underline{\quad ? \quad}$.

16. $\triangle ABC \sim \underline{\quad ? \quad}$.

17. List three proportions, given $x \parallel y \parallel z$.

18. Find the value of n.

Exercises 14–16

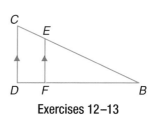

Exercises 17–18

19. A 42-foot tree casts a shadow of 63 feet. How long is the shadow of a 5-foot girl who is standing near it?

20. A puzzle consists of several triangles that are all similar. If the perimeter of $\triangle ABC$ is 39, find the value of each variable.

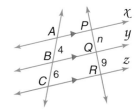

Exercise 20

Ratio and Proportion Problems

Standardized tests almost always include ratio and proportion problems. Remember that ratios can be written in several ways.

$$a \text{ to } b, \; a{:}b, \; \frac{a}{b}, \text{ or } a \div b, \text{ where } b \neq 0$$

> **Test-Taking Tip**
>
> A ratio often compares a part to a part. A fraction compares a part to a whole.

Example 1

A grocery store sells oranges at 3 for $1.29. How much do 10 oranges cost?

> **Hint** Write prices like 3 for $1.29 as a ratio.

Solution Write a proportion. Let one ratio represent the cost of the 3 oranges. Use x for the cost of 10 oranges. Let the second ratio represent the cost of the 10 oranges. Find the cross products. Solve the equation for x.

$$\begin{array}{l} \textit{cost of 3 oranges} \rightarrow \\ \textit{number of oranges} \rightarrow \end{array} \frac{1.29}{3} = \frac{x}{10} \begin{array}{l} \leftarrow \textit{cost of 10 oranges} \\ \leftarrow \textit{number of oranges} \end{array}$$

$$1.29(10) = 3(x)$$
$$12.9 = 3x$$
$$\frac{12.9}{3} = \frac{3x}{3}$$
$$4.3 = x$$

The cost of 10 oranges is $4.30.

Example 2

A bakery uses a special flour mixture that contains corn, wheat, and rye in the ratio of 3:5:2. If a bag of the mixture contains 5 pounds of rye, how many pounds of wheat does it contain?

(A) 2   (B) 5   (C) 7.5   (D) 10   (E) 12.5

> **Hint** Be on the lookout for extra information that is not needed to solve the problem.

Solution Read the question carefully. It contains a ratio of three quantities. Notice that the amount of corn is *not* part of the question. So you can ignore the part of the ratio that involves corn.

The ratio of wheat to rye is 5:2. The amount of rye is 5 pounds. Create a proportion. Let x represent the amount of wheat. Find the cross products. Solve the equation for x.

$$\frac{5}{2} = \frac{x}{5} \quad \begin{array}{l} \leftarrow \textit{wheat} \\ \leftarrow \textit{rye} \end{array}$$
$$5(5) = 2(x) \qquad \textit{Cross products}$$
$$25 = 2x \qquad \textit{Multiply.}$$
$$\frac{25}{2} = \frac{2x}{2} \qquad \textit{Divide each side by 2.}$$
$$12.5 = x \qquad \textit{Simplify.}$$

The bag contains 12.5 pounds of wheat. The answer is E.

After you work each problem, record your answer on the answer sheet provided or on a sheet of paper.

Multiple Choice

1. On a map, the distance from Springfield to Ames is 6 inches. The map scale is $\frac{1}{2}$ inch = 20 miles. How many miles is it from Springfield to Ames? *(Lesson 9–2)*
   - **A** 6.67 mi
   - **B** 60 mi
   - **C** 120 mi
   - **D** 240 mi

2. Which ordered pair represents the *y*-intercept of line *MN*? *(Lesson 4–6)*

   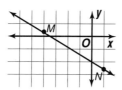

   - **A** (0, −2)
   - **B** (−2, 0)
   - **C** (0, −4)
   - **D** (−4, 0)

3. If 2 packages contain a total of 12 doughnuts, how many doughnuts are there in 5 packages? *(Algebra Review)*
   - **A** 60
   - **B** 36
   - **C** 30
   - **D** 24

4. Nathan earns $24,000 in salary and 8% commission on his sales. If he needs a total annual income of at least $30,000, how much does he need to sell? *(Percent Review)*
   - **A** at least $480
   - **B** at least $6000
   - **C** at least $26,400
   - **D** at least $75,000

5. Point *B*(4, 3) is the midpoint of line segment *AC*. If point *A* has coordinates (0, 1), what are the coordinates of point *C*? *(Lesson 2–5)*
   - **A** (−4, −1)
   - **B** (4, 1)
   - **C** (4, 4)
   - **D** (8, 5)
   - **E** (8, 9)

6. The ratio of girls to boys in a science class is 4 to 3. If the class has a total of 35 students, how many more girls are there than boys? *(Algebra Review)*
   - **A** 1
   - **B** 5
   - **C** 7
   - **D** 15

7. Jessica served cheese, peanut butter, and cucumber sandwiches at a luncheon. She also served iced tea and lemonade. Each guest chose one sandwich and one drink. Of the possible combinations of sandwich and drink, how many included iced tea? *(Statistics Review)*
   - **A** 1
   - **B** 2
   - **C** 3
   - **D** 6

8. The average of five numbers is 20. If one of the numbers is 18, then what is the sum of the other four numbers? *(Statistics Review)*
   - **A** 2
   - **B** 20.5
   - **C** 82
   - **D** 90
   - **E** 100

Grid In

9. An average of 3 out of every 10 students are absent from school because of illness during flu season. If there are normally 600 students attending a school, about how many students can be expected to attend during flu season? *(Algebra Review)*

Extended Response

10. The chart shows the average height for males ages 8 to 18. *(Statistics Review)*

Average Height for Males	
Age (yr)	Height (cm)
8	124
9	130
10	135
11	140
12	145
13	152
14	161
15	167
16	172
17	174
18	178

Part A Graph the information on a coordinate plane.

Part B Describe the relationship between age and height.

CHAPTER

10 Polygons and Area

Why It's Important

Spiders Spiders use different kinds of silk for different purposes, such as constructing cocoons or egg sacs, spinning webs, and binding prey. Spider silk is about five times as strong as steel of the same weight.

Polygons are used every day in fields like architecture, art, and science. You will use a polygon to estimate the area of a spider's web in Lesson 10–3.

✔ Check Your Readiness

✔ **Lesson 5–2,**
pp. 193–197

Find the value of each variable.

1.

26°
38° x°

2.

x°
44° x°

3.

53° x°

✔ **Lesson 1–6,**
pp. 35–40

Find the perimeter and area of each rectangle.

4.
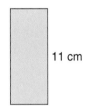
5 in.
12 in.

5.
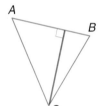
11 cm
4.4 cm

6.
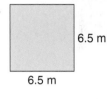
6.5 m
6.5 m

✔ **Lesson 6–2,**
pp. 234–239

For each triangle, tell whether the red segment or line is an
altitude, **a** ***perpendicular bisector, both,*** **or** ***neither.***

7.
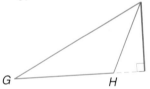
A
B
C

8.
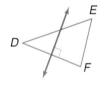
E
D
F

9.

I
G
H

FOLDABLES™
Study Organizer

Make this Foldable to help you organize your Chapter 10 notes. Begin with a sheet of plain $8\frac{1}{2}$" by 11" paper.

❶ **Fold** the short side in fourths.

❷ **Draw** lines along the folds and label each column *Prefix, Number of Sides, Polygon Names,* and *Figure.*

Reading and Writing Store the Foldable in a 3-ring binder. As you read and study the chapter, write definitions and examples of important terms in each column.

Recall that a *polygon* is a closed figure in a plane formed by segments called *sides*. A polygon is named by the number of its sides or angles. A triangle is a polygon with three sides. The prefix *tri-* means three. Prefixes are also used to name other polygons.

Prefix	Number of Sides	Name of Polygon
tri-	3	triangle
quadri-	4	quadrilateral
penta-	5	pentagon
hexa-	6	hexagon
hepta-	7	heptagon
octa-	8	octagon
nona-	9	nonagon
deca-	10	decagon

When you studied quadrilaterals in Lesson 8–1, you learned several terms that can be applied to all polygons.

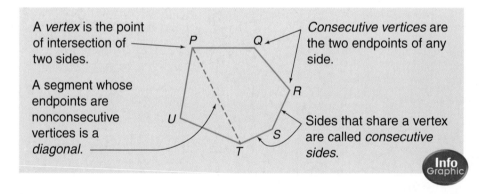

A *vertex* is the point of intersection of two sides.

A segment whose endpoints are nonconsecutive vertices is a *diagonal*.

Consecutive vertices are the two endpoints of any side.

Sides that share a vertex are called *consecutive sides*.

An *equilateral* polygon has all sides congruent, and an *equiangular* polygon has all angles congruent. A **regular polygon** is both equilateral and equiangular.

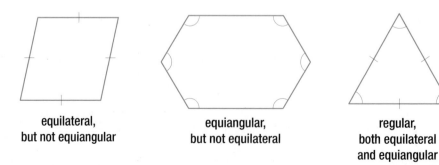

equilateral, but not equiangular

equiangular, but not equilateral

regular, both equilateral and equiangular

Examples

Science Link

Scientists are able to study the tops of forests from huge inflatable rafts.

1 **A. Identify polygon *ABCDEF* by its sides.**

The polygon has six sides. It is a hexagon.

B. Determine whether the polygon, as viewed from this angle, appears to be *regular* or *not regular*. If not regular, explain why.

The sides do not appear to be the same length, and the angles do not appear to have the same measure. The polygon is not regular.

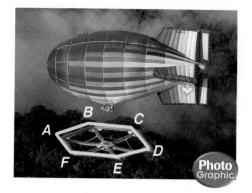

Photo Graphic

2 **Name two consecutive vertices of hexagon *ABCDEF*.**

B and *C* are consecutive vertices since they are the endpoints of $\overline{BC}$. Other pairs of consecutive vertices are listed below.

C, D   D, E   E, F   F, A   A, B

Your Turn

a. Identify polygon *LMNOPQRS* by its sides.

b. Determine whether the polygon appears to be *regular* or *not regular*. If not regular, explain why.

c. Name two nonconsecutive sides.

You can use the properties of regular polygons to find the perimeter.

Example

3 **Find the perimeter of a regular octagon whose sides are 7.6 centimeters long.**

$$\underbrace{\text{perimeter of regular polygon}}_{P} = \underbrace{\text{number of sides}}_{8} \times \underbrace{\text{length of each side}}_{7.6}$$

$$P = 60.8$$

The perimeter is 60.8 centimeters.

Your Turn

d. Find the perimeter of a regular decagon whose sides are 12 feet long.

A polygon can also be classified as convex or concave.

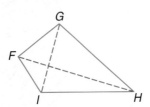

convex

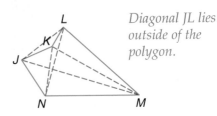

concave

Diagonal JL lies outside of the polygon.

If all of the diagonals lie in the interior of the figure, then the polygon is **convex**.

If any part of a diagonal lies outside of the figure, then the polygon is **concave**.

Example

4 **Classify polygon *STUVW* as *convex* or *concave*.**

When all the diagonals are drawn, no points lie outside of the polygon. So, *STUVW* is convex.

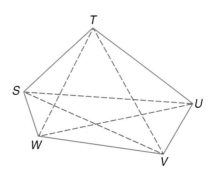

Your Turn

e. Classify the polygon at the right as *convex* or *concave*.

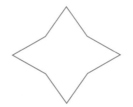

Check for Understanding

Communicating Mathematics

1. **Draw** a concave quadrilateral. Explain why it is concave.

2. **Determine** whether each figure is a polygon. Write *yes* or *no*. If no, explain why not.

a.   b.

3. **Writing Math** Find five words in the dictionary, each beginning with a different prefix listed in the table on page 402. Define each word.

Identify each polygon by its sides. Then determine whether it appears to be *regular* or *not regular*. If not regular, explain why.

4.

5.

Example 2

Name each part of pentagon *PENTA*.

6. all pairs of nonconsecutive vertices

7. any three consecutive sides

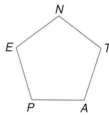

Example 3

8. Find the perimeter of a regular heptagon whose sides are 8.1 meters long.

Exercises 6–7

Example 4

Classify each polygon as *convex* or *concave*.

9.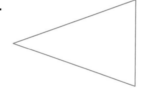

10.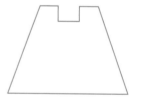

11. **Sewing**   Refer to the collar at the right.

Example 1

   a. Identify polygon *PQRSTUV* by its sides.

Example 4

   b. Classify the polygon as *convex* or *concave*.

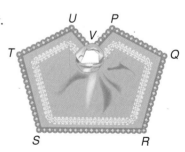

Exercises

Identify each polygon by its sides. Then determine whether it appears to be *regular* or *not regular*. If not regular, explain why.

12.

13.

14.

<table>
<tr><td colspan="2" align="center">**Homework Help**</td></tr>
<tr><td>For Exercises</td><td>See Examples</td></tr>
<tr><td>15–17</td><td>1</td></tr>
<tr><td>18–22</td><td>2</td></tr>
<tr><td>23–28</td><td>4</td></tr>
<tr><td>29–32</td><td>3</td></tr>
</table>

Extra Practice
See page 743.

Identify each polygon by its sides. Then determine whether it appears to be *regular* or *not regular*. If not regular, explain why.

15.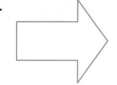

16.

17.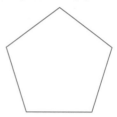

Name each part of octagon *MNOPQRST*.

18. two consecutive vertices

19. two diagonals

20. all nonconsecutive sides of $\overline{PQ}$

21. any three consecutive sides

22. any five consecutive vertices

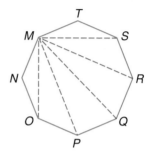

Classify each polygon as *convex* or *concave*.

23.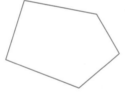

24.

25.

26.

27.

28.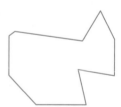

Find the perimeter of each regular polygon with the given side lengths.

29. pentagon, 20 in.

30. triangle, 16 km

31. nonagon, 3.8 mm

32. If the perimeter of a regular hexagon is 336 yards, what is the length of each side in yards? in feet?

33. Draw a convex pentagon. Label the vertices and name all of the diagonals.

34. Baseball The following excerpt is from the official rules of baseball.

Home base is a five-sided slab of whitened rubber. It is a 17-inch square with two of the corners removed so that one edge is 17 inches long, two consecutive sides are $8\frac{1}{2}$ inches long, and the remaining two sides are 12 inches long and set at an angle to make a point.

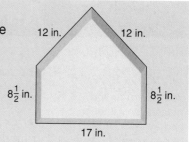

12 in.   12 in.

$8\frac{1}{2}$ in. $8\frac{1}{2}$ in.

17 in.

Explain why this statement is incorrect.

35. Chemistry Organic compounds are named using the same prefixes as polygons. Study the first two compounds below. Use what you know about polygons to name the last two compounds.

cyclopentene   cyclohexene

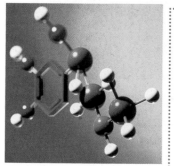

An organic compound model

36. Critical Thinking Use a straightedge to draw the following figures.
a. convex pentagon with two perpendicular sides
b. concave hexagon with three consecutive congruent sides

37. $\triangle XYZ$ is similar to $\triangle PQR$. Determine the scale factor for $\triangle XYZ$ to $\triangle PQR$.   *(Lesson 9–7)*

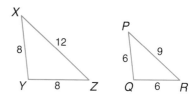

X
8   12
Y   8   Z

P
6   9
Q   6   R

38. Find the value of n.
(Lesson 9–6)

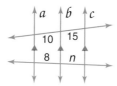

a   b   c
10   15
8   n

39. Determine whether $\overline{MJ} \parallel \overline{LK}$.
(Lesson 9–5)

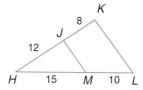

K
8
J
12
H   15   M   10   L

40. Write the ratio 2 yards to 2 feet in simplest form.   *(Lesson 9–1)*

41. Determine whether a triangle with side lengths 10 inches, 11 inches, and 15 inches is a right triangle.   *(Lesson 6–6)*

42. Short Response The letters at the right are written backward. What transformation did Donald use in writing his name?   *(Lesson 5–3)*

43. Multiple Choice If y varies directly as x and $y = 10.5$ when $x = 7$, find y when $x = 12$.   *(Algebra Review)*

Ⓐ 1.5   Ⓑ 8   Ⓒ 15.5   Ⓓ 18

10-2 Diagonals and Angle Measure

What You'll Learn

You'll learn to find measures of interior and exterior angles of polygons.

Why It's Important

Landscaping
Landscapers use angle measures of polygons in constructing garden borders.
See Example 1.

The constellation of stars at the right is called the *Scorpion*. The stars form pentagon *SCORP*. $\overline{SO}$ and $\overline{SR}$ are diagonals from vertex S and they divide the pentagon into triangles.

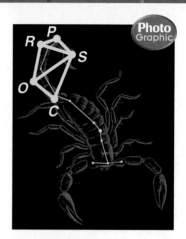

There is an important relationship between the number of sides of a convex polygon and the number of triangles formed by drawing the diagonals from one vertex. The hands-on activity explains this relationship.

Hands-On Geometry

Materials: straightedge

Step 1 Draw a convex quadrilateral.

Step 2 Choose one vertex and draw all possible diagonals from that vertex.

Step 3 How many triangles are formed?

Step 4 Make a table like the one below.

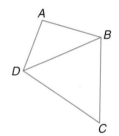

Recall that the sum of the measures of the angles of a triangle is 180.

Convex Polygon	Number of Sides	Number of Diagonals from One Vertex	Number of Triangles	Sum of Interior Angles
quadrilateral	4	1	2	2(180) = 360
pentagon				
hexagon				
heptagon				
n-gon	*n*			

Try These

1. Draw a pentagon, a hexagon, and a heptagon. Use the figures to complete all but the last row of the table.

2. A polygon with *n* sides is an *n-gon*. Determine the number of diagonals that can be drawn from one vertex and enter it in the table.

3. Determine the number of triangles that are formed in an *n*-gon by drawing the diagonals from one vertex. Enter it in the table.

4. What is the sum of the measures of the interior angles for a convex polygon with *n* sides? Write your answer in the table.

In the activity, you discovered that two triangles are formed in a quadrilateral when the diagonal is drawn from one vertex. So, the sum of measures of the interior angles is 2 × 180 or 360. You extended this pattern to other convex polygons and found the sum of interior angles of a polygon with *n* sides. The results are stated in the following theorem.

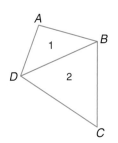

Theorem 10–1	If a convex polygon has *n* sides, then the sum of the measures of its interior angles is (*n* − 2)180.

You can use Theorem 10–1 to find the sum of the interior angles of any polygon or to find the measure of one interior angle of a regular polygon.

Examples

Landscaping Link

① **Landscapers often need to know interior angle measures of polygons in order to correctly cut wooden borders for garden beds. The border at the right is a regular hexagon. Find the sum of measures of the interior angles.**

sum of measures of interior angles = (*n* − 2)180   *Theorem 10–1*
= (6 − 2)180   *Replace n with 6.*
=   4 · 180   *Subtract.*
= 720   *Multiply.*

The sum of measures of the interior angles of a hexagon is 720.

② **Find the measure of one interior angle in the figure in Example 1.**

All interior angles of a regular polygon have the same measure. Divide the sum of measures by the number of angles.

measure of one interior angle = $\frac{720}{6}$ ←*sum of interior angle measures*
←*number of interior angles*

= 120   *Simplify.*

One interior angle of a regular hexagon has a measure of 120.

Your Turn

Refer to the regular polygon at the right.

a. Find the sum of measures of the interior angles.

b. Find the measure of one interior angle.

 Personal Tutor at geomconcepts.com

In Lesson 7–2, you identified exterior angles of triangles. Likewise, you can extend the sides of any convex polygon to form exterior angles. If you add the measures of the exterior angles in the hexagon at the right, you find that the sum is 360.

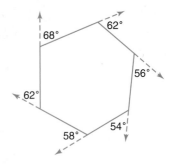

The figure above suggests a method for finding the sum of the measures of the exterior angles of a convex polygon. When you extend n sides of a polygon, n linear pairs of angles are formed. The sum of the angle measures in each linear pair is 180.

Algebra Review
Evaluating
Expressions, p. 718

$$\underset{\text{exterior angles}}{\text{sum of measures of}} = \underset{\text{of linear pairs}}{\underbrace{\text{sum of measures}}} - \underset{\text{of interior angles}}{\underbrace{\text{sum of measures}}}$$

$$= \quad n \cdot 180 \quad - \quad 180(n - 2)$$

$$= \quad 180n \quad - \quad 180n + 360$$

$$= \quad 360$$

So, the sum of the exterior angle measures is 360 for any convex polygon.

Theorem 10–2	In any convex polygon, the sum of the measures of the exterior angles, one at each vertex, is 360.

You can use Theorem 10–2 to find the measure of one exterior angle of a regular polygon.

Example
3

Find the measure of one exterior angle of a regular heptagon.

By Theorem 10–2, the sum of the measures of the exterior angles is 360. Since all exterior angles of a regular polygon have the same measure, divide this measure by the number of exterior angles, one at each vertex.

measure of one exterior angle $= \dfrac{360}{7}$ ←*sum of exterior angle measures*
←*number of exterior angles*

≈ 51 *Simplify.*

The measure of one exterior angle of a regular heptagon is about 51.

Your Turn

c. Find the measure of one exterior angle of a regular quadrilateral.

Check for Understanding

Communicating Mathematics

1. **Explain** how to find the interior angle measure of an *n*-sided regular polygon.

2. **Find a counterexample** to the following statement.
An exterior angle measure of any convex polygon can be found by dividing 360 by the number of interior angles.

3.  As part of a class assignment, Janelle was searching for polygons that are used in everyday life. She found the company logo at the right and reasoned, "Since this pentagon is divided into five triangles, the sum of the measures of the interior angles is 5(180), or 900." Is she correct? Explain why or why not.

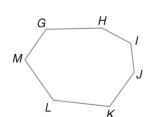

Guided Practice
Example 1

4. Find the sum of the measures of the interior angles of polygon *GHIJKLM*.

Example 2

5. Find the measure of one interior angle of a regular quadrilateral.

Example 3

6. What is the measure of one exterior angle of a regular triangle?

Example 1

7. **Soap** Soap bubbles form tiny polygons, as shown in the photograph at the right. Find the sum of the interior angle measures of polygon *ABCDEF*.

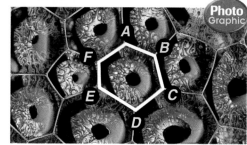

Exercises

Practice

Find the sum of the measures of the interior angles in each figure.

8.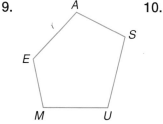

9.

10.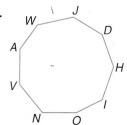

Homework Help	
For Exercises	See Examples
8–10	1
11–13	1, 2
14–17	2, 3
Extra Practice	
See page 744.	

Find the measure of one interior angle and one exterior angle of each regular polygon. If necessary, round to the nearest degree.

11. pentagon

12. heptagon

13. decagon

14. The sum of the measures of five exterior angles of a hexagon is 284. What is the measure of the sixth angle?

15. The measures of seven interior angles of an octagon are 142, 140, 125, 156, 133, 160, and 134. Find the measure of the eighth interior angle.

16. The measures of the exterior angles of a quadrilateral are x, $2x$, $3x$, and $4x$. Find x and the measure of each exterior angle of the quadrilateral.

17. The measure of an exterior angle of a regular octagon is $x + 10$. Find x and the measure of each exterior angle of the octagon.

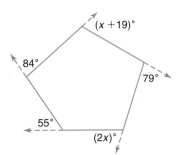

Applications and Problem Solving

18. **Algebra**   Find the value of x in the figure at the right.

Exercise 18

19. **Internet**   Some Web sites of real-estate companies offer "virtual tours" of houses that are for sale.

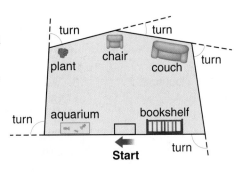

a. How many turns are made as you view the living room at the right?

b. If you view the entire room, how many degrees do you turn?

c. What is the sum of the measures of the interior angles of this room?

20. **Critical Thinking**   The sum of the measures of the interior angles of a convex polygon is 1800. Find the number of sides of the polygon.

Mixed Review

21. **Home Design**   Refer to Exercise 19. Identify the polygon formed by the walls of the room. Does it appear to be *regular* or *not regular*? *(Lesson 10–1)*

22. Find the value of x.   *(Lesson 9–4)*

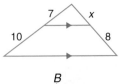

Find the measure of each angle.
(Lesson 7–2)

23. $\angle A$

24. $\angle B$

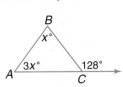

25. **Multiple Choice**   In any triangle, the point of intersection of the ____?____ is the same distance from all three vertices.   *(Lesson 6–2)*

Ⓐ   perpendicular bisectors

Ⓑ   altitudes

Ⓒ   medians

Ⓓ   angle bisectors

 www.geomconcepts.com/self_check_quiz

10-3 Areas of Polygons

What You'll Learn
You'll learn to estimate the areas of polygons.

Why It's Important
Home Improvement
Painters need to know the areas of decks before they can quote a price for refinishing. *See Exercise 24.*

Any polygon and its interior are called a **polygonal region**. In Lesson 1–6, you found the areas of rectangles.

Postulate 10–1 **Area Postulate**	For any polygon and a given unit of measure, there is a unique number *A* called the measure of the area of the polygon.

Area can be used to describe, compare, and contrast polygons. The two polygons below are congruent. How do the areas of these polygons compare?

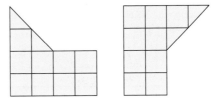

This suggests the following postulate.

Postulate 10–2	Congruent polygons have equal areas.

Reading Geometry

In this text, *area of a polygon* is used to mean *area of the polygonal region.*

The figures above are examples of **composite figures**. They are each made from a rectangle and a triangle that have been placed together. You can use what you know about the pieces to gain information about the figure made from them.

You can find the area of any polygon by dividing the original region into smaller and simpler polygonal regions, like squares, rectangles, and triangles. The area of the original polygonal region can then be found by adding the areas of the smaller polygons.

Postulate 10–3 **Area Addition** **Postulate**	**Words:**	The area of a given polygon equals the sum of the areas of the nonoverlapping polygons that form the given polygon.
	Model: 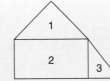	**Symbols:** $A_{total} = A_1 + A_2 + A_3$

You can use Postulates 10–2 and 10–3 to find the areas of various polygons.

Example ❶

Find the area of the polygon at the right. Each square represents 1 square centimeter.

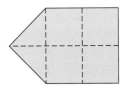

Since the area of each ☐ represents 1 square centimeter, the area of each ◺ represents 0.5 square centimeter.

$A = 5(1) + 4(0.5)$ *There are 5* ☐ *and 4* ◺.
$A = \;5\; + \;2$ *Multiply.*
$A = 7$ *Add.*

The area of the region is 7 square centimeters, or 7 cm².

Algebra Review
Evaluating
Expressions, p. 718

Your Turn

a. Find the area of the polygon at the right. Each square represents 1 square inch.

Irregular figures are not polygons, and they cannot be made from combinations of polygons. However, you can use combinations of polygons to approximate the areas of irregular figures.

Example ❷

Geography Link

If Lake Superior were drained, the resulting land area would be twice that of the Netherlands. Estimate the area of Lake Superior if each square represents 3350 square miles.

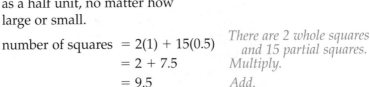

One way to estimate the area is to count each square as one unit and each partial square as a half unit, no matter how large or small.

number of squares $= 2(1) + 15(0.5)$ *There are 2 whole squares and 15 partial squares.*
$= 2 + 7.5$ *Multiply.*
$= 9.5$ *Add.*

Area $\approx 9.5 \times 3350$ *Each square represents 3350 square miles.*
$\approx 31{,}825$

The area of Lake Superior is about 31,825 square miles.

 www.geomconcepts.com/extra_examples

b. Estimate the area of the polygon at the right. Each square unit represents 1 acre.

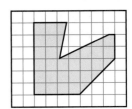

In the following activity, you will investigate the relationship between the area of a polygon drawn on dot paper and the number of dots on the figure.

Hands-On Geometry

Materials: ⬚ rectangular dot paper   ▱ straightedge

Step 1 On a piece of dot paper, draw polygons that go through 3 dots, 4 dots, 5 dots, and 6 dots, having no dots in the interiors, as shown at the right.

Step 2 Copy the table below. Find the areas of the figures at the right and write your answers in the appropriate places in the table.

Number of Dots on Figure	3	4	5	6	7	8	9	10
Area of Polygon (square units)								

Try These

1. Draw polygons that go through 7, 8, 9, and 10 dots, having no dots in the interiors. Then complete the table.

2. Predict the area of a figure whose sides go through 20 dots. Verify your answer by drawing the polygon.

3. Suppose there are n dots on a figure. Choose the correct relationship that exists between the number of dots n and the area of the figure A.

  a. $A = \frac{n}{2} + 1$ **b.** $A = \frac{n}{2}$ **c.** $A = \frac{n}{2} - 1$

Check for Understanding

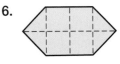
Communicating Mathematics

1. **Write** in your own words the Area Addition Postulate.

2. **Draw** a polygon with the same area as, but not congruent to the figure at the right. Use dot paper.

3. Carla says that if the area of a polygon is doubled, the perimeter also doubles. Kevin argues that this is not always the case. Who is correct? Why? Draw some figures to support your answer.

Guided Practice

Example 1

Find the area of each polygon in square units.

4.

5.

6.

Example 2

7. Estimate the area of the polygon in square units.

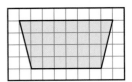

A natural geyser

8. **Geography** Two-thirds of all the geysers in the world are in Yellowstone National Park. Estimate the area of the park if each square represents 136 square miles. Use the map at the right. **Example 2**

Exercises

Practice

Find the area of each polygon in square units.

9.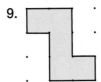

10.

11.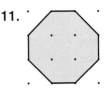

Homework Help	
For Exercises	**See Examples**
9–17	1
18–20, 23–24	2
Extra Practice	
See page 744.	

12.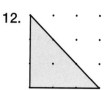

13.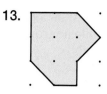

14.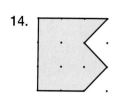

15.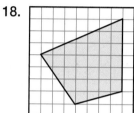

16.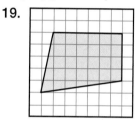

17.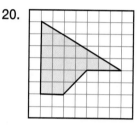

Estimate the area of each polygon in square units.

18. **19.** **20.**

21. Sketch two polygons that both have a perimeter of 12 units, but that have different areas.

22. Sketch a hexagon with an area of 16 square units.

Applications and Problem Solving

23. Spiders It takes about an hour for a spider to weave a web, and most spiders make a new web every single day. Estimate the area of the web at the right. Each square represents 1 square inch.

24. Home Improvement The Nakashis are having their wooden deck stained. The company doing the work charges $13.50 per square yard for staining.

  a. Each square on the grid represents 1 square yard. Estimate the area of the deck to the nearest square yard.

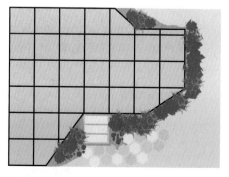

  b. About how much will it cost to stain the deck?

25. Critical Thinking   Use 3-by-3 arrays on square dot paper to draw all possible noncongruent convex pentagons. Determine the area of each pentagon.

Mixed Review

26. Tile Making. A floor tile is to be made in the shape of a regular hexagon. What is the measure of each interior angle?   *(Lesson 10–2)*

27. Find the perimeter of a regular hexagon whose sides are 3.5 feet long. *(Lesson 10–1)*

28. Find the values of x and y. *(Lesson 9–3)*

29. If $m\angle 1 = 3x$ and $m\angle 2 = x - 2$, find $m\angle 1$ and $m\angle 2$. *(Lesson 3–7)*

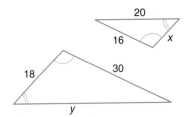

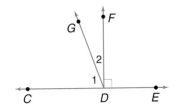

30. Multiple Choice   The top ten scores on Mr. Yunker's science test were 98, 96, 95, 100, 93, 95, 95, 96, 94, and 97. What is the median of this set of data?   *(Statistics Review)*

   Ⓐ 95.9   Ⓑ 95.5   Ⓒ 95.0   Ⓓ 94.9

Quiz 1   Lessons 10–1 through 10–3

Identify each polygon by its sides. Then classify the polygon as *convex* or *concave*.   *(Lesson 10–1)*

1.

2.

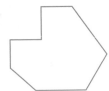

A *dodecagon* is a polygon with 12 sides. Find the measure of each angle of a regular dodecagon.   *(Lesson 10–2)*

3. one interior angle

4. one exterior angle

5. Swimming   Mineku needs to know the area of her swimming pool so she can order a cover. Each square represents 16 square feet. What is the area?   *(Lesson 10–3)*

www.geomconcepts.com/self_check_quiz

10-4 Areas of Triangles and Trapezoids

What You'll Learn
You'll learn to find the areas of triangles and trapezoids.

Why It's Important
Home Heating
Engineers design whole-house fans based on the interior areas of homes.
See Exercise 22.

Look at the rectangle below. Its area is bh square units. The diagonal divides the rectangle into two congruent triangles. The area of each triangle is half the area of the rectangle, or $\frac{1}{2}bh$ square units. This result is true of all triangles and is formally stated in Theorem 10–3.

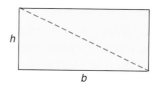

Look Back

Altitude:
Lesson 6–2

Theorem 10–3 **Area of a** **Triangle**	**Words:** If a triangle has an area of A square units, a base of b units, and a corresponding altitude of h units, then $A = \frac{1}{2}bh$.
	Model: 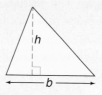 **Symbols:** $A = \frac{1}{2}bh$

Examples

1 Find the area of each triangle.

$A = \frac{1}{2}bh$ *Theorem 10–3*

$\quad = \frac{1}{2}(19)(14)$ *Replace b with 19 and h with 14.*

$\quad = 133$ *Simplify.*

The area is 133 square centimeters.

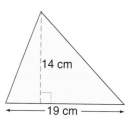

14 cm

19 cm

Algebra Review
Operations with Fractions, p. 721

2 $A = \frac{1}{2}bh$ *Theorem 10–3*

$\quad = \frac{1}{2}\left(11\frac{1}{2}\right)(7)$ *Replace b with $11\frac{1}{2}$ and h with 7.*

$\quad = \frac{1}{2}\left(\frac{23}{2}\right)(7)$ $11\frac{1}{2} = \frac{23}{2}$

$\quad = \frac{161}{4}$ or $40\frac{1}{4}$ *Simplify.*

The area is $40\frac{1}{4}$ square feet.

In a right triangle, a leg is also the altitude of the triangle.

7 ft

$11\frac{1}{2}$ ft

Find the area of △WXY.

$A = \frac{1}{2}bh$ *Theorem 10–3*

$= \frac{1}{2}(8)(4.2)$ *Replace b with 8*
                  and h with 4.2.

$= 16.8$ *Simplify.*

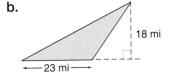

Note that $\overline{ZY}$ is not part of the triangle.

The area of △WXY is 16.8 square meters.

Your Turn   **Find the area of each triangle.**

a.
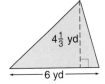
$4\frac{1}{3}$ yd

6 yd

b.
18 mi

23 mi

c.
12 mm

10.3 mm

 Personal Tutor at **geomconcepts.com**

Look Back

Trapezoids:
Lesson 8–5

You can find the area of a trapezoid in a similar way. The *altitude* of a trapezoid *h* is a segment perpendicular to each base.

h

The following activity leads you to find the area of a trapezoid.

Hands-On Geometry

Materials: grid paper   straightedge   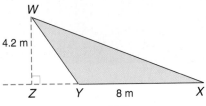 scissors

Step 1   Draw and label trapezoid *ABCD* on grid paper. The bases and altitude of the trapezoid can have any measure you choose. Draw the altitude and label it *h*. Label the bases b_1 and b_2.

Step 2   Draw trapezoid *FGHI* so that it is congruent to *ABCD*. Label the bases b_1, b_2, and *h*.

Step 3   Cut out both trapezoids. Arrange them so that two of the congruent legs are adjacent, forming a parallelogram.

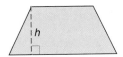

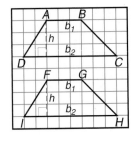

Try These

1. Find the length of the base of the parallelogram in terms of b_1 and b_2.

2. Recall that the formula for the area of a parallelogram is $A = bh$. Find the area of your parallelogram in terms of b_1, b_2, and *h*.

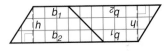 www.geomconcepts.com/extra_examples

3. How does the area of one trapezoid compare to the area of the parallelogram?
4. Write the formula for the area of a trapezoid using b_1, b_2, and h.

The results of this activity suggest the formula for the area of a trapezoid.

Theorem 10–4
Area of a Trapezoid

Words: If a trapezoid has an area of A square units, bases of b_1 and b_2 units, and an altitude of h units, then $A = \frac{1}{2}h(b_1 + b_2)$.

Model:

b_1

h

b_2

Symbols: $A = \frac{1}{2}h(b_1 + b_2)$

Example 4

Automobile Link

It costs $25.80 per square foot to replace a car window. How much would it cost to replace the window in the "microcar" at the right?

$\frac{2}{3}$ ft

1 ft

$1\frac{1}{6}$ ft

Explore You know the cost per square foot. You need to find the number of square feet of the window.

Plan The window is a trapezoid. Use Theorem 10–4 to find its area. Then multiply the number of square feet by $25.80 to find the total cost.

Solve $A = \frac{1}{2}h(b_1 + b_2)$ *Theorem 10–4*

$= \frac{1}{2}(1)\left(1\frac{1}{6} + \frac{2}{3}\right)$ *Replace h with 1, b_1 with $1\frac{1}{6}$, and b_2 with $\frac{2}{3}$.*

$= \frac{1}{2}(1)\left(\frac{7}{6} + \frac{4}{6}\right)$ *Rewrite as fractions with the same denominator.*

$= \frac{1}{2}(1)\left(\frac{11}{6}\right)$ or $\frac{11}{12}$ *Simplify.*

The area of the window is $\frac{11}{12}$ square foot. The cost to replace the window is $\frac{11}{12} \times \$25.80$, or $23.65.

(continued on the next page)

Examine   Check your answer by estimating.

$$A = \frac{1}{2}(1)(1 + 1)$$ *Round b_1 to 1 and b_2 to 1.*

$$= \frac{1}{2}(2) \text{ or } 1$$ The area is about 1 square foot.

Round $25.80 to $26. Then the total cost is $1 \times \$26$, or about $26. This is close to $23.65, so the solution seems reasonable.

Your Turn

d. Find the area of the trapezoid.

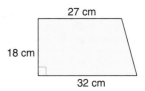

27 cm

18 cm

32 cm

Check for Understanding

1. Make a conjecture about how the area of a trapezoid changes if the lengths of its bases and altitude are doubled.

2. Use Theorem 10–3 to explain why the triangles below have equal areas.

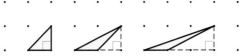

3. Writing Math The figure at the right is an isosceles trapezoid separated into four right triangles. On rectangular dot paper, draw three isosceles trapezoids. Separate them into the polygons below by drawing segments. Make each new vertex a dot on the trapezoid.

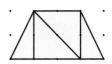

   a. 3 isosceles triangles
   b. 2 congruent trapezoids
   c. 5 polygonal regions (name the regions)

Guided Practice

⊕ **Getting Ready** **Evaluate each expression.**

Sample: $\frac{1}{2}(6)(9 + 4)$ **Solution:** $\frac{1}{2}(6)(9 + 4) = 3(13)$ or 39

4. $\frac{1}{2}(12)(7)$ **5.** $\frac{1}{2}(26 + 20)$ **6.** $\frac{1}{2}(18)(17 + 13)$

Find the area of each triangle or trapezoid.

7.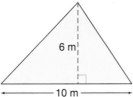
6 m
10 m

8.
3 in.
1.8 in.

9.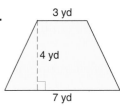
3 yd
4 yd
7 yd

10. **School** The Pep Club is making felt banners for basketball games. Each banner is shaped like an isosceles triangle with a base $\frac{2}{3}$ foot long and a height of 1 foot. **Example 1**

a. How much felt is needed to make 90 banners, assuming that there is no waste?

b. If felt costs \$1.15 per square foot, how much will it cost to make the banners?

Making banners

Exercises

Practice

Find the area of each triangle or trapezoid.

11.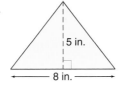
5 in.
8 in.

12. 24 m
21 m
20 m

13. 4 km
6 km

14.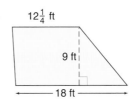
$15\frac{1}{3}$ cm
18 cm

15.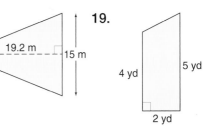
20 in.
18 in.
38 in.

16.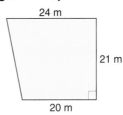
13 mm
13 mm
24.8 mm

17.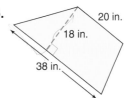
$12\frac{1}{4}$ ft
9 ft
18 ft

18.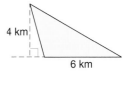
19.2 m
15 m

19.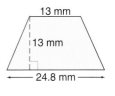
4 yd
5 yd
2 yd

20. Find the area of a trapezoid whose altitude measures 4 inches and whose bases are $5\frac{1}{3}$ inches and 9 inches long.

21. If the area of the triangle is 261 square meters, find the value of x.

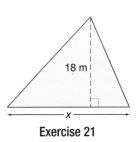
18 m
x
Exercise 21

22. Home Heating Whole-house fans are designed based on the interior square footage of living space in a home. Find the total area of the room at the right.

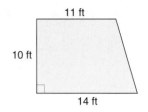

11 ft

10 ft

14 ft

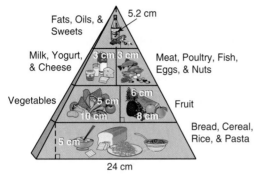

Fats, Oils, & Sweets   5.2 cm

Milk, Yogurt, & Cheese   3 cm 3 cm   Meat, Poultry, Fish, Eggs, & Nuts

Vegetables   6 cm

5 cm   Fruit

10 cm   8 cm

Bread, Cereal, Rice, & Pasta

5 cm

24 cm

Source: United States Department of Agriculture

23. Health The Food Guide Pyramid outlines foods you should eat for a healthy diet. One face of the pyramid is a triangle displaying the food groups. Find the area used for each food group below.

a. fats, oils, and sweets

b. fruit

c. bread, cereal, rice, and pasta

24. Construction It costs $1.59 per square yard to seal an asphalt parking area. How much will it cost to seal the parking lot surface at the right if all of the sections are 10 yards deep?

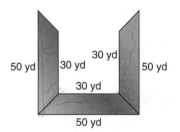

50 yd   30 yd   30 yd   50 yd

30 yd

50 yd

25. Critical Thinking Show how to separate isosceles trapezoid *RSTU* into four congruent trapezoidal regions.

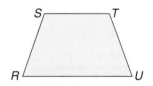

S   T

R   U

26. Estimate the area of the polygon at the right in square units. *(Lesson 10–3)*

27. Find the sum of the measures of the interior angles in the figure at the right. *(Lesson 10–2)*

Exercises 26–27

Find the length of the median in each trapezoid. *(Lesson 8–5)*

28.

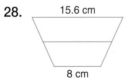

15.6 cm

8 cm

29.

22 ft

27 ft

30. Short Response A carpenter is building is building a triangular display case with a vertical support piece as shown at the right. Describe three ways to guarantee that shelves $\overline{AB}$ and $\overline{CD}$ are parallel. *(Lesson 10–4)*

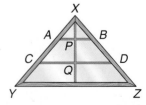

X

A   B

P

C   D

Q

Y   Z

www.geomconcepts.com/self_check_quiz

10-5 Areas of Regular Polygons

What You'll Learn
You'll learn to find the areas of regular polygons.

Why It's Important
Architecture For centuries, architects have used regular polygons in designing buildings. *See Exercises 15 and 17.*

Every regular polygon has a **center**, a point in the interior that is equidistant from all the vertices. A segment drawn from the center that is perpendicular to a side of the regular polygon is called an **apothem** (AP-ə-them). In any regular polygon, all apothems are congruent.

The following activity investigates the area of a regular pentagon by using its apothem.

Hands-On Geometry
Construction

Materials: compass straightedge

Step 1 Copy regular pentagon *PENTA* and its center, *O*.

Step 2 Draw the apothem from *O* to side $\overline{AT}$ by constructing the perpendicular bisector of $\overline{AT}$. Label the apothem measure *a*. Label the measure of $\overline{AT}$, *s*.

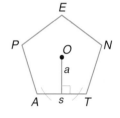

Step 3 Use a straightedge to draw $\overline{OA}$ and $\overline{OT}$.

Step 4 What measure in △*AOT* represents the base of the triangle? What measure represents the height?

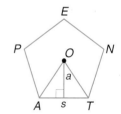

Step 5 Find the area of △*AOT* in terms of *s* and *a*.

Step 6 Draw $\overline{ON}$, $\overline{OE}$, and $\overline{OP}$. What is true of the five small triangles formed?

Step 7 How do the areas of the triangles compare?

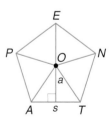

Try These

1. The area of pentagon *PENTA* can be found by adding the areas of the five triangles that make up the pentagonal region.

$$A = \frac{1}{2}sa + \frac{1}{2}sa + \frac{1}{2}sa + \frac{1}{2}sa + \frac{1}{2}sa$$
$$= \frac{1}{2}(sa + sa + sa + sa + sa) \text{ or } \frac{1}{2}(5sa)$$

What measure does 5*s* represent?

2. Rewrite the formula for the area of a pentagon using *P* for perimeter.

The results of this activity would be the same for any regular polygon.

Theorem 10–5 Area of a Regular Polygon	Words:	If a regular polygon has an area of A square units, an apothem of a units, and a perimeter of P units, then $A = \frac{1}{2}aP$.
	Model:	
	Symbols:	$A = \frac{1}{2}aP$

Real World

Example ──①

Game Link

The game at the right has a hexagon-shaped board. Find its area.

First find the perimeter of the hexagon.

$P = 6s$ *All sides of a regular hexagon are congruent.*

  $= 6(9)$ or 54   *Replace s with 9.*

Now you can find the area.

$A = \frac{1}{2}aP$ *Theorem 10–5*

  $= \frac{1}{2}(7.8)(54)$ or 210.6   *Replace a with 7.8 and P with 54.*

The area of the game board is 210.6 square inches.

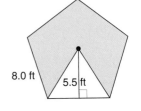

9 in.   7.8 in.

Photo Graphic

Your Turn

a. Each of the tiles in the game is also a regular hexagon. Find the area of one of the tiles if the sides are each 0.9 inch long and each apothem is 0.8 inch long.

Knowing how to find the area of a regular polygon is useful in finding other areas.

Example ──②

Find the area of the shaded region in the regular polygon at the right.

Explore   You need to find the area of the entire pentagon minus the area of the unshaded triangle.

Plan   Use Theorem 10–5 to find the area of the pentagon. Then find the area of the unshaded triangle and subtract.

8.0 ft   5.5 ft

  www.geomconcepts.com/extra_examples

Solve   **Area of Pentagon**

$P = 5s$ *All sides of a regular pentagon are congruent.*

$\quad = 5(8)$ or 40 *Replace s with 8.*

$A = \frac{1}{2}aP$ *Theorem 10–5*

$\quad = \frac{1}{2}(5.5)(40)$ *Replace a with 5.5 and P with 40.*

$\quad = 110 \text{ ft}^2$ *Simplify.*

Area of Triangle

$A = \frac{1}{2}bh$ *Theorem 10–3*

$\quad = \frac{1}{2}(8)(5.5)$ *Replace b with 8 and h with 5.5.*

$\quad = 22 \text{ ft}^2$ *Simplify.*

To find the area of the region, subtract the areas:
$110 - 22 = 88$. The area of the shaded region is 88 square feet.

Examine   The pentagon can be divided into five congruent triangles, as you discovered in the Hands-On activity. If each triangle is 22 square feet, then the area of the four shaded triangles is 22×4, or 88 square feet. The answer checks.

 Your Turn

b. Find the area of the shaded region in the regular polygon at the right.

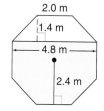

2.0 m
1.4 m
4.8 m
2.4 m

 Personal Tutor at geomconcepts.com

Significant digits represent the *precision* of a measurement. In the measurement 30.1 meters, there are three significant digits. If a 0 does not fall between two significant digits and is only a placeholder for locating the decimal point, it is not a significant digit. For example, the measure 73,000 has only two significant digits. In the figure above, the measures are stated with two significant digits. If all the measures were rounded to the nearest meter, then there would only be one significant digit. *How would this affect the answer?*

You can use a graphing calculator to calculate with significant digits.

**Graphing
Calculator Tutorial**
See pp. 782–785.

Graphing Calculator Exploration

The SIG-FIG CALCULATOR feature of the Science Tools App can be used to add, subtract, multiply, divide, or raise a number to a power and display the result using the correct number of significant digits.

Step 1 To open a Science Tools session, press APPS , select SciTools and press ENTER . Press any key to continue. Then press ENTER to select SIG-FIG CALCULATOR.

Step 2 Multiply 9.6 by 6 to find the perimeter of the hexagon. Notice that the calculator displays the number of significant digits in each number at the right.

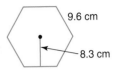
9.6 cm

8.3 cm

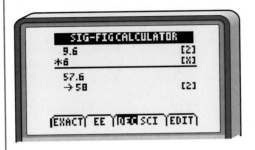

The 6 should be entered as an exact value because this is not a measured value. Exact values do not restrict the number of significant digits in the answer. To enter an exact value, enter the value and then press Y= . Press ENTER to calculate.

Try These

Find the area of the hexagon above using significant digits.

1. How does the answer differ from the answer calculated without using significant digits?

2. What value(s) should be entered as exact values when finding the area using significant digits? Why are the value(s) exact?

3. How does rounding or using significant figures at intermediate steps of a computation affect the final result?

4. Why do you think scientists use significant digits in calculations?

Check for Understanding

Communicating Mathematics

1. **Write** a sentence that describes the relationship between the number of significant digits in measures of a regular polygon and the number of significant digits in the measure of its area.

Vocabulary

center
apothem
significant digits

2. **Determine** whether the following statement is *true* or *false*. Explain.

   If the lengths of the sides of a regular polygon are doubled, then its area is also doubled.

3. **Describe** how to locate the center of an equilateral triangle, a square, a regular pentagon, and a regular hexagon.

Guided Practice
Examples 1 & 2

4. Find the area of the regular polygon below.

5.4 cm

5.2 cm

5. Find the area of the shaded region in the regular polygon below.

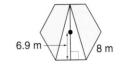
6.9 m   8 m

Example 1

6. **Traffic Signs**   A stop sign is a regular octagon whose sides are each 10 inches long and whose apothems are each 12 inches long. Find the area of a stop sign.

Exercises

Practice

Find the area of each regular polygon.

7.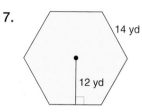
14 yd
12 yd

8.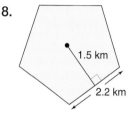
1.5 km
2.2 km

9.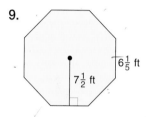
$6\frac{1}{5}$ ft
$7\frac{1}{2}$ ft

Find the area of the shaded region in each regular polygon.

10.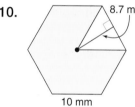
8.7 mm
10 mm

11.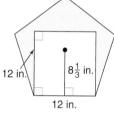
12 in.
$8\frac{1}{3}$ in.
12 in.

12.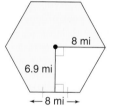
8 mi
6.9 mi
← 8 mi →

13. A regular nonagon has a perimeter of 45 inches and its apothems are each $6\frac{9}{10}$ inches long.

   a. Find the area.

   b. Round the length of an apothem to the nearest inch and find the area. How does it compare to the original area?

14. The area of a regular octagon is 392.4 square meters, and an apothem is 10.9 meters long.

   a. Find the perimeter.

   b. Find the length of one side.

Applications and Problem Solving

15. **Architecture**   Find the approximate area of Fort Jefferson in Dry Tortugas National Park, Florida, if each side is 460 feet long and an apothem is about 398 feet long.

center
a
s

Fort Jefferson

16. **Botany**   The petals in the flower form a polygon that is approximately a regular pentagon with an apothem 3.4 centimeters long.

5 cm

Photo Graphic

   a. Find the approximate area of the pentagon.

   b. There are five triangles in the pentagon that are not part of the flower. Assume that they have equal areas and have a height of 1.5 centimeters. Find the total area of the triangles.

   c. What is the approximate area of the flower?

17. **Architecture**   The Castel del Monte in Apulia, Italy, was built in the 13th century. The outer shape and the inner courtyard are both regular octagons.

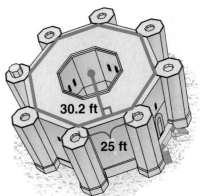
30.2 ft
25 ft

   a. Find the total area of the castle, including the courtyard.

   b. Find the area of the courtyard if the octagon has an apothem of 14.5 feet and sides of 12 feet.

   c. What is the area of the inside of the castle, not including the courtyard?

18. **Critical Thinking**   The regular polygons below all have the same perimeter. Are their areas equal? Explain.

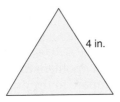
4 in.

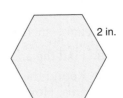
2 in.

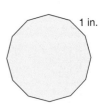
1 in.

Mixed Review

19. Find the area of a triangle whose altitude measures 8 inches and whose base is $5\frac{1}{2}$ inches long.   *(Lesson 10–4)*

20. Sketch an octagon with an area of 16 square units.   *(Lesson 10–3)*

Find the coordinates of the midpoint of each segment.   *(Lesson 2–5)*

21. $\overline{XY}$, with endpoints $X(4, -5)$ and $Y(-2, 1)$

22. $\overline{AB}$, with endpoints $A(-2, 6)$ and $B(8, 3)$

23. **Multiple Choice**   What is the solution of $45 \geq 7t + 10 \geq 3$?
   (Algebra Review)

   Ⓐ $-5 \geq t \geq 1$
   Ⓑ $5 \geq t \geq -1$
   Ⓒ $5 \leq t \leq -1$
   Ⓓ $-1 \geq t \geq 5$

www.geomconcepts.com/self_check_quiz

HVAC Technician

Do you like to put things together, or to figure out why something is not working and try to fix it? Then you might enjoy working as an HVAC (heating, ventilation, and air conditioning) technician.

Besides performing maintenance on heating and cooling systems, technicians also install new systems. To find the size of an efficient heating system needed for a home, the technician must estimate the amount of heat lost through windows, walls, and other surfaces that are exposed to the outside temperatures.

Use the formula $L = kDA$ to find the heat loss. In the formula, L is the heat loss in Btu (British thermal units) per hour, D is the difference between the outside and inside temperatures, A is the area of the surface in square feet, and k is the insulation rating of the surface.

HVAC Technician

1. Find the heat loss per hour for a 6 foot by 5 foot single-pane glass window ($k = 1.13$) when the outside temperature is 32°F and the desired inside temperature is 68°F. Write your answer in Btus.

2. How much more heat is lost through the window in Exercise 1 than would be lost through a surface with an insulation rating of 1.0 under the same conditions?

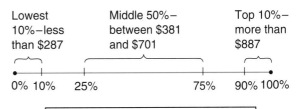

FAST FACTS About HVAC Technicians

Working Conditions

- usually work at different sites each day
- may work outside in cold or hot weather or inside in uncomfortable conditions
- usually work a 40-hour week, but overtime is often required during peak seasons

Education

- courses in applied math, mechanical drawing, applied physics and chemistry, electronics, blueprint reading, and computer applications
- some knowledge of plumbing or electrical work is also helpful

Earnings

Weekly Earnings

| Lowest 10%—less than $287 | Middle 50%— between $381 and $701 | Top 10%— more than $887 |

0% 10%   25%   75%   90% 100%

Median earnings— $536 per week

Source: *Occupational Outlook Handbook*

*inter*NET
CONNECTION
www.geomconcepts.com

Career Data For the latest information on careers in heating and air-conditioning, visit:

How About That Pythagoras!

Ratios of Perimeters and Areas of Similar Polygons

Materials

 ruler

 compass

 protractor

Squares may be different sizes, as in the tile at the right, but since they have the same shape, they are *similar polygons*. Likewise, all regular pentagons are similar polygons, all regular hexagons are similar polygons, and so on.

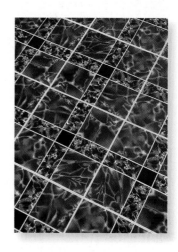

Look Back

Similar Polygons, Lesson 9–2

You have learned that for similar polygons, the ratio of the perimeters equals the ratio of the measures of the corresponding sides. Does this same proportionality hold true for the ratios of the areas and the measures of the corresponding sides, or for the ratios of corresponding perimeters and areas? Let's find out.

Investigate

1. The figure at the right is usually used to verify the Pythagorean Theorem. We can also use it to investigate relationships among side lengths, perimeters, and areas of regular polygons.

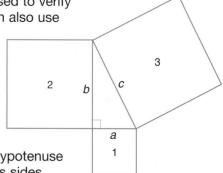

   a. Draw a right triangle. Make sure that its legs are no more than one-third the dimensions of your paper.

   b. Construct a square using the hypotenuse of the right triangle as one of its sides. Then construct squares on the two legs of the triangle. Label your drawing as shown.

c. Find and record the length of sides *a*, *b*, and *c*. Find and record the areas and perimeters of Squares 1, 2, and 3.

2. Use a spreadsheet or a calculator to write and compare each pair of ratios.

   a. $\frac{a}{b}$ and $\frac{\text{area of Square 1}}{\text{area of Square 2}}$

   b. $\frac{b}{c}$ and $\frac{\text{area of Square 2}}{\text{area of Square 3}}$

   c. $\frac{\text{perimeter of Square 1}}{\text{perimeter of Square 2}}$ and $\frac{\text{area of Square 1}}{\text{area of Square 2}}$

   d. $\frac{\text{perimeter of Square 2}}{\text{perimeter of Square 3}}$ and $\frac{\text{area of Square 2}}{\text{area of Square 3}}$

3. Use your results from Exercise 2 to solve each problem.

   a. The ratio of side lengths of two squares is $\frac{5}{4}$ or 5:4. What is the ratio of their corresponding areas? Explain how you know.

   b. The ratio of areas of two squares is $\frac{16}{30}$ or 16:30. What is the ratio of their corresponding perimeters? Explain.

Extending the Investigation

In this extension, you will determine whether there are relationships between ratios of perimeters and areas of other regular polygons.

- Use paper and construction tools or geometry drawing software to construct the following polygons on the sides of a right triangle.
  a. equilateral triangles   **b.** regular pentagons   **c.** regular hexagons

- Investigate the relationship between the ratios of side lengths of the right triangles and the ratios of areas of the corresponding polygons.

- For each set of polygons that you drew, investigate the ratios between areas and perimeters of corresponding polygons.

Presenting Your Conclusions

Here are some ideas to help you present your conclusions to the class.

- Make a booklet presenting your findings in this investigation.

- Research Pythagoras. Write a report about his mathematical achievements. Be sure to include at least two other ideas for which he is given credit.

 Investigation   For more information on Pythagoras, visit: www.geomconcepts.com

Snowflakes have puzzled scientists for decades. A curious fact is that all the branches of a snowflake grow at the same time in all six directions, preserving the **symmetry**. You can draw a line down the middle of any snowflake, and each half will be a mirror image of the other half. When this happens, a figure is said to have **line symmetry**, and the line is called a **line of symmetry**.

vertical line of symmetry

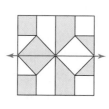

horizontal line of symmetry

diagonal line of symmetry

no line of symmetry

One way to determine whether a figure has line symmetry is to fold it in half along a line. If the two sides match exactly when the figure is folded, then the figure has line symmetry.

Example

① **Find all lines of symmetry for rhombus *ABCD*.**

Fold along possible lines of symmetry to see if the sides match.

$\overleftrightarrow{AC}$ is a line of symmetry.

$\overleftrightarrow{BD}$ is a line of symmetry.

not a line of symmetry

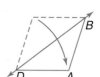

not a line of symmetry

Rhombus *ABCD* has two lines of symmetry, $\overleftrightarrow{AC}$ and $\overleftrightarrow{BD}$.

a. Draw all lines of symmetry for rectangle *RECT*.

Refer to the first figure below. You can turn this figure 90°, 180°, or 270° about point *C* and get the exact same figure. *Figures can be turned clockwise or counterclockwise.*

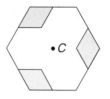   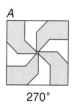

90° 180° 270°

Any figure that can be turned or rotated less than 360° about a fixed point so that the figure looks exactly as it does in its original position is said to have **rotational** or **turn symmetry**.

Example ❷ **Which of the figures below have rotational symmetry?**

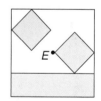

The first figure can be turned 120° and 240° about *C* and still look like the original. The second figure can be turned 180° about *D* and still look like the original. The third figure must be rotated 360° about *E* before it looks like the original. Therefore, the first and second figures have rotational symmetry, but the last figure does not.

Your Turn

Determine whether each figure has rotational symmetry. Write *yes* or *no*.

b.

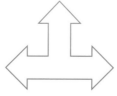

c.

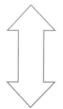

Check for Understanding

Communicating Mathematics

1. **Draw** a square with 8-inch sides on a sheet of notebook paper. Cut out the figure. How many different ways can you fold the square so that the fold line is a line of symmetry?

2. **Draw** a polygon that has line symmetry but not rotational symmetry. Then describe how you could change the figure so that it has rotational symmetry.

3. Writing Math Draw a polygon that has exactly three lines of symmetry. Draw the lines of symmetry.

Vocabulary

symmetry
line symmetry
line of symmetry
rotational symmetry
turn symmetry

Guided Practice

Getting Ready Is each letter symmetric? Write *yes* or *no*.

Sample 1: A

Solution 1: A Yes; the left and right halves are congruent, so the letter is symmetric.

Sample 2: J

Solution 2: J No; the left and right halves are not congruent, and the top and bottom halves are not congruent. The letter is *not* symmetric.

4. **B**   5. **N**   6. **Y**   7. **P**

Example 1

Determine whether each figure has line symmetry. If it does, copy the figure, and draw all lines of symmetry. If not, write *no.*

8.

9.

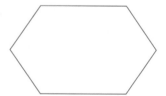

Example 2

Determine whether each figure has rotational symmetry. Write *yes* or *no.*

10.

11.

12. **Biology** In Central America, some starfish have as many as 50 arms. Does the top starfish at the right have *line symmetry*, *rotational symmetry*, *neither*, or *both*?

Exercises • • • • • • • • • • • • • • • • •

Practice

Determine whether each figure has line symmetry. If it does, copy the figure, and draw all lines of symmetry. If not, write *no*.

13.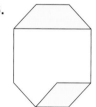

14.

15.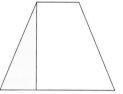

16.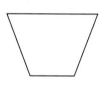

17.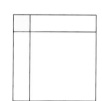

18.

Determine whether each figure has rotational symmetry. Write *yes* or *no*.

19.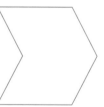

20.

21.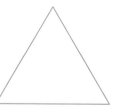

22.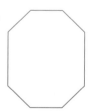

23.

24.

25. What kinds of triangles have line symmetry?

26. How many lines of symmetry does a regular hexagon have?

Applications and
Problem Solving

27. **Advertising**   All of the bank logos below have rotational symmetry. If you turn each logo a total of 360° about a fixed center, how many times does the rotated figure match the original? (*Hint:* Do not count the original figure at 360°.)

a.    b.    c.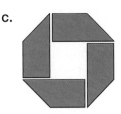

28. **Entertainment**   The design at the right was generated by a toy that produces symmetric designs. Does the design have *line symmetry, rotational symmetry, neither,* or *both*?

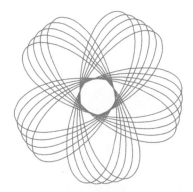

29. **Critical Thinking**   The figure at the right is the *Rainbow Star Mandala* from a Chinese temple. Suppose the dark blue shapes that are formed by the white lines inside the circle are cut out and placed in a pile. If you draw one piece at random, what is the probability that you have drawn a quadrilateral? (*Hint*: Use lines of symmetry and rotational symmetry to simplify the problem.)

The probability of an event is the ratio of the number of favorable outcomes to the total number of possible outcomes.

Mixed Review

Find the area of each polygon.   *(Lessons 10–4 and 10–5)*

30.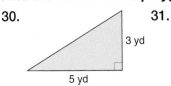
3 yd
5 yd

31.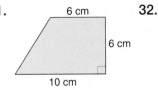
6 cm
6 cm
10 cm

32.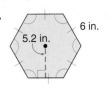
6 in.
5.2 in.

33. Determine if the pair of polygons is similar. Justify your answer. *(Lesson 9–2)*

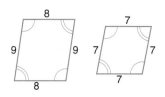

34. Short Response Which kind of quadrilateral is indicated by the following statement? The diagonals are congruent, perpendicular, and bisect each other. *(Lesson 8–5)*

35. Multiple Choice Find the missing measure in the figure at the right. *(Lesson 8–1)*

   Ⓐ 98   Ⓑ 100   Ⓒ 90   Ⓓ 99

Quiz 2 — Lessons 10–4 through 10–6

Find the area of each triangle or trapezoid. *(Lesson 10–4)*

1.
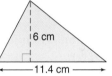
6 cm
11.4 cm

2.
$5\frac{1}{2}$ ft
3 ft
3 ft

3.
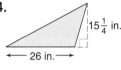
40.2 km
34.7 km
52 km

4.
$15\frac{1}{4}$ in.
26 in.

Find the area of each regular polygon. *(Lesson 10–5)*

5.
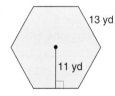
13 yd
11 yd

6.
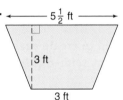
2 m
1.4 m

7.

16.8 cm
16.2 cm

8.

7 in.
$8\frac{1}{2}$ in.

9. Sports Refer to the soccer field at the right. *(Lesson 10–6)*

  a. Does the field have line symmetry? If so, how many lines of symmetry could be drawn?

  b. Does the field have rotational symmetry?

10. Draw a figure that has rotational symmetry when it is turned 180° only. *(Lesson 10–6)*

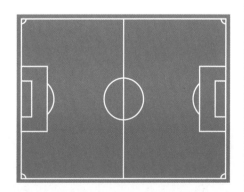

What You'll Learn

You'll learn to identify tessellations and create them by using transformations.

Why It's Important

Construction
Bricklayers use tessellations in building patios and walkways.
See Exercise 16.

The hexagon-tiled "floor" at the right formed about 100,000 years ago from molten lava.

Tiled patterns created by repeating figures to fill a plane without gaps or overlaps are called **tessellations**. Tessellations can be formed by translating, reflecting, or rotating polygons. When only one type of regular polygon is used to form a pattern, the pattern is called a **regular tessellation**. If two or more regular polygons are used in the same order at every vertex, it is called a **semi-regular tessellation**.

Giant's Causeway,
Northern Ireland

Examples

What types of transformations can be used to create these tessellations?

Identify the figures used to create each tessellation. Then identify the tessellation as *regular*, *semi-regular*, or *neither*.

1

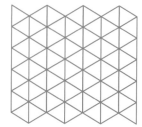

Only equilateral triangles are used. The tessellation is regular.

2

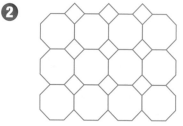

Squares and regular octagons are used in the same order at every vertex. The tessellation is semi-regular.

Your Turn

a.

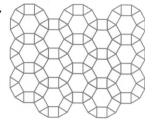

b.

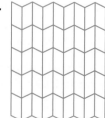

You can create tessellations easily using dot paper.

Example **3** **Use isometric dot paper to create a tessellation using regular hexagons and triangles.**

Here is one way to create the tessellation.

 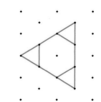 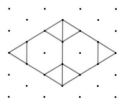

Draw a hexagon.

Add triangles so that there is no space between the polygons.

Draw another hexagon and more triangles. Continue the pattern.

COncepts in MOtion
Animation
geomconcepts.com

Your Turn

c. Use dot paper to create a tessellation using rhombi.

Check for Understanding

Communicating Mathematics

1. Find examples of tessellations in nature, in magazines, or on the Internet.

   a. Tell whether the tessellations are *regular*, *semi-regular*, or *neither*.

   b. Explain which transformations can be used to create the tessellations.

2. Predict the sum of the measures of the angles that meet at a common vertex in a tessellation, such as the tile pattern shown at the right. Explain how you made your prediction.

3. Explain why it is less expensive to make hexagonal pencils than round pencils.

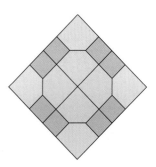

Exercise 2

4. Identify the polygons used to create the tessellation at the right. Then identify the tessellation as *regular*, *semi-regular*, or *neither*.

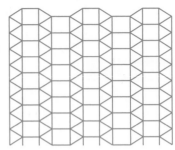

Example 2

5. Use rectangular dot paper to create a tessellation using two different-sized squares.

Example 1

6. **Biology**   Hard plates called *scutes* (scoots) cover the shell of a turtle. Each species has its own scute pattern, which gets larger as the turtle grows. Identify the polygons found in the tessellation on this turtle's shell.

Exercises

Practice

Identify the figures used to create each tessellation. Then identify the tessellation as *regular*, *semi-regular*, or *neither*.

7.

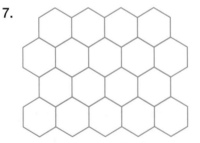

8.

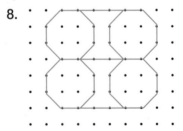

Homework Help	
For Exercises	See Examples
7–10	1
11–15	2
Extra Practice	
See page 745.	

9.

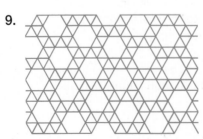

10.

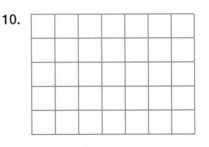

Use dot paper to create a tessellation using the given polygons.

11. trapezoids

12. large and small equilateral triangles

13. octagons and equilateral triangles

14. Refer to Exercise 4. Create a different tessellation using the same polygons.

15. Describe the tessellation used in the quilt block as *regular*, *semi-regular*, or *neither*.

Exercise 15

Applications and Problem Solving

16. **Construction** The stones or bricks in a patio are laid so that there is no space between them. One example is shown at the right.

    a. Design two different brick or stone tiling patterns that could be used in a patio.

    b. Describe the transformation or transformations you used to make the pattern.

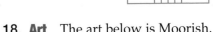

17. **Design** In the image below, the artist used arrows to create a tessellation. Use symbols, letters, or numbers to design your own tessellation.

George Abe, *Creating Direction*

18. **Art** The art below is Moorish.

    a. Describe the polygons in the art and explain how transformations can be used to make the design.

    b. Discuss the different types of symmetry in the design.

Alhambra Palace, Spain

19. **Technology** The graphic shows where people believed they would get most of their news by the year 2000. The survey included only people born since the 1971 invention of the computer chip.

    a. Use tracing paper and choose one, two, or three of the polygons in the graph to create a tessellation.

    b. Discuss the use of convex and concave polygons in your tessellation.

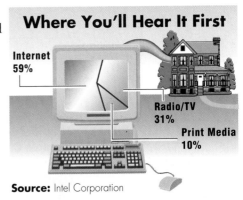

Where You'll Hear It First

Internet 59%

Radio/TV 31%

Print Media 10%

Source: Intel Corporation

20. Critical Thinking

    a. Copy and complete the table to show the measure of a vertex angle for each regular polygon listed.

Number of Sides	3	4	5	6	7	8	9
Vertex Angle Measure							

    b. In a rotation, how many degrees are in a full turn?

    c. What is the sum of the measures of the angles that meet at a vertex of a tessellation?

    d. Which angle measures in the table are factors of 360?

    e. Which regular polygons have those vertex angle measures?

    f. Write a conclusion based on your discoveries.

Mixed Review

21. Draw a regular octagon like the one shown. *(Lesson 10–6)*

    a. Draw all lines of symmetry.

    b. Does the octagon have rotational symmetry? If so, draw the fixed point about which the octagon rotates.

22. Painting Mrs. Davis is preparing to paint her house. If a gallon of paint covers 450 square feet, how many gallons of paint does she need to cover the side of the house shown at the right? *(Lesson 10–3)*

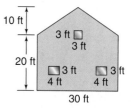

Determine whether each quadrilateral is a parallelogram. Write *yes* or *no*. If *yes*, give a reason for your answer. *(Lesson 8–3)*

23.

24.

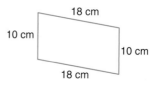

25. Short Response If the sides of △*CWO* have measures of $6y - 3$, $2y + 17$, and 70, write an inequality to represent the possible range of values for y. *(Lesson 7–4)*

26. Multiple Choice Find the value of x in the isosceles triangle shown at the right. *(Lesson 5–2)*

    Ⓐ 18   Ⓑ 22

    Ⓒ 23.1   Ⓓ 28.7

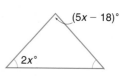

www.geomconcepts.com/self_check_quiz

Graphic Artist

If you enjoy drawing, or creating art on a computer, you may be interested in a career as a graphic artist. Graphic artists use a variety of print, electronic, and film media to create art that meets a client's needs. An artist may create a design or a logo by making a tessellation. The following steps show how to create a tessellation using a rotation.

Step 1 Draw an equilateral triangle. Then draw another triangle inside the right side of the triangle as shown below.

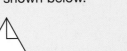

Step 2 Rotate the small triangle to the left side as shown below.

Step 3 Rotate the entire figure to create a tessellation of equilateral triangles. Use alternating colors to best show the tessellation.

Make a tessellation for each translation shown.

1.

2.

FAST FACTS About Graphic Artists

Working Conditions

- work in art and design studios located in office buildings or in their own studios
- odors from glues, paint, ink, or other materials may be present
- generally work a standard 40-hour week, with some overtime

Education

- bachelor's degree or other post-secondary training in art or design
- appropriate talent and skill, displayed in an artist's portfolio

Employment

Where Graphic Artists Work

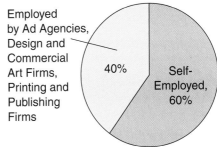

Employed by Ad Agencies, Design and Commercial Art Firms, Printing and Publishing Firms — 40%

Self-Employed, 60%

---

*inter*NET
CONNECTION   **Career Data** For the latest information on careers in graphic arts, visit:
www.geomconcepts.com

Understanding and Using the Vocabulary

After completing this chapter, you should be able to define each term, property, or phrase and give an example or two of each.

altitude *(p. 420)*
apothem *(p. 425)*
center *(p. 425)*
composite figure *(p. 413)*
concave *(p. 404)*
convex *(p. 404)*

irregular figure *(p. 414)*
line of symmetry *(p. 434)*
line symmetry *(p. 434)*
polygonal region *(p. 413)*
regular polygon *(p. 402)*
regular tessellation *(p. 440)*

rotational symmetry *(p. 435)*
semi-regular tessellation *(p. 440)*
significant digits *(p. 428)*
symmetry *(p. 434)*
tessellation *(p. 440)*
turn symmetry *(p. 435)*

Choose the term or terms from the list above that best complete each statement.

1. If all the diagonals of a polygon lie in its interior, then the polygon is ___?___ .

2. A(n) ___?___ is both equilateral and equiangular.

3. A segment perpendicular to the lines containing the bases of a trapezoid is a(n) ___?___ .

4. A segment drawn from the center perpendicular to a side of a regular polygon is a(n) ___?___ .

5. A pattern formed by repeating figures to fill a plane without gaps or overlaps is a(n) ___?___ .

6. Any polygon and its interior are called a(n) ___?___ .

7. A figure has ___?___ when a line drawn through the figure makes each half a mirror image of the other.

8. The ___?___ of a regular polygon is a point in the interior equidistant from all the vertices.

9. In a(n) ___?___ , only one kind of regular polygon is used to form the pattern.

10. A figure that can be turned less than 360° about a fixed point and look exactly as it does in its original position has ___?___ .

Skills and Concepts

Objectives and Examples	**Review Exercises**

• **Lesson 10–1** Name polygons according to the number of sides and angles.

Identify polygon *PQRST* and determine whether it appears to be *regular* or *not regular.*

Polygon *PQRST* has five sides, so it is a pentagon. It appears to be regular.

Identify each polygon by its sides. Then determine whether it appears to be *regular* or *not regular*.

11.   12.

Classify each polygon as *convex* or *concave*.

13.   14.

Objectives and Examples

Review Exercises

• **Lesson 10–2** Find measures of interior and exterior angles of polygons.

sum of measures of interior
angles $= (n - 2)180$
$\qquad = 8 \cdot 180$ or 1440

$\begin{array}{l} \text{measure of one} \\ \text{interior angle} \end{array} = \frac{1440}{10}$ or 144

Find the sum of the measures of the interior angles in each figure.

15.   **16.**

Find the measure of one interior angle and one exterior angle of each regular polygon.

17. octagon   **18.** nonagon

• **Lesson 10–3** Estimate the areas of polygons.

Find the area of the polygon.

$A \approx 4(1) + 3(0.5)$
$\quad \approx 5.5$ square units

19. Find the area of the polygon in square units.

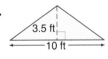

20. Estimate the area of the polygon in square units.

• **Lesson 10–4** Find the areas of triangles and trapezoids.

Area of a Triangle
$A = \frac{1}{2}bh$

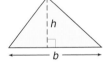

Area of a Trapezoid
$A = \frac{1}{2}h(b_1 + b_2)$

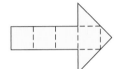

Find the area of each triangle or trapezoid.

21.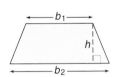

4 cm
6 cm
11 cm

22.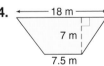

3.5 ft
10 ft

23.

30 in.
41 in.

24.

18 m
7 m
7.5 m

• **Lesson 10–5** Find the areas of regular polygons.

Find the area of the regular polygon.

$A = \frac{1}{2}aP$ *Area formula*
$A = \frac{1}{2}(12)(70)$ $P = 7(10)$ or 70 m
$A = 420$ square meters

10 m
12 m

Find the area of each regular polygon.

25.

8.7 cm
6 cm

26.

18 yd
23 yd

Mixed Problem Solving
See pages 758–765.

Objectives and Examples

- **Lesson 10–6** Identify figures with line symmetry and rotational symmetry.

Find all of the lines of symmetry for triangle *JKL*.

Triangle *JKL* has three lines of symmetry.

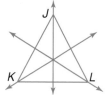

Review Exercises

Determine whether each figure has *line symmetry*, *rotational symmetry*, *both*, or *neither*.

27.

28.

29.

30.

- **Lesson 10–7** Identify tessellations and create them by using transformations.

Identify the figures used to create the tessellation. Then identify the tessellation as *regular*, *semi-regular*, or *neither*.

Regular hexagons and equilateral triangles are used to create the tessellation. It is semi-regular.

31. Identify the figures used to create the tessellation. Then identify the tessellation as *regular*, *semi-regular*, or *neither*.

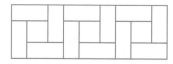

Use isometric or rectangular dot paper to create a tessellation using the given polygons.

32. isosceles triangles and pentagons

33. small and large squares and rectangles

Applications and Problem Solving

34. **Woodworking** A craftsman is making a wooden frame to replace the one on an octagonal antique mirror. Determine the measure of each interior angle of the frame if its shape is a regular octagon. *(Lesson 10–2)*

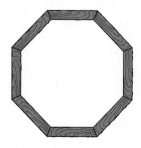

35. **Construction** The Deck Builders company has several designs for decks. Find the area of the deck at the right. *(Lesson 10–4)*

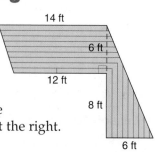

14 ft
6 ft
12 ft
8 ft
6 ft

36. **Architecture** The plans for a new high-rise office tower show that the shape of the building will be a regular hexagon with each side measuring 350 feet. Find the area of a floor if each apothem is 303 feet long. *(Lesson 10–5)*

1. **Describe** how a polygon is named.

2. **Compare and contrast** convex and concave polygons.

Identify each polygon by its sides. Then determine if it appears to be *regular* or *not regular*. If not regular, explain why.

3.

4.

Find the measure of one interior angle and one exterior angle of each regular polygon.

5. octagon

6. pentagon

7. nonagon

8. Find the area of the polygon in square units.

9. Estimate the area of the polygon in square units.

Find the area of each triangle or trapezoid.

10.
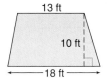
13 ft
10 ft
18 ft

11.

19 cm
27 cm

12.
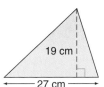
45 in.
37 in.

Find the area of each regular polygon.

13.

6 m
6.2 m

14.
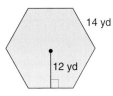
14 yd
12 yd

15.
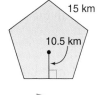
15 km
10.5 km

16. Copy the figure and draw any lines of symmetry.

17. Determine whether the figure has rotational symmetry.

18. Define *tessellation*. Describe one example of a tessellation in your school.

19. **Flooring** Ms. Lopez would like to have the wood floor in her kitchen refinished. An outline of the area is shown at the right. What is the area to be refinished?

20. **Maintenance** The base of a fountain at a shopping mall is in the shape of a regular hexagon, with each side 12 feet long and each apothem 10 feet long. The base of the fountain is to be repainted. What is the area of the base of the fountain?

Exercises 16–17

16 ft
14 ft
15.5 ft

Triangle and Quadrilateral Problems

Standardized tests always include geometry problems. You'll need to know the properties of geometric shapes like triangles and quadrilaterals.

Be sure you understand these concepts.

triangle	equilateral	isosceles	quadrilateral
rectangle	parallelogram	similar	congruent

> **Test-Taking Tip**
>
> Since a quadrilateral can be separated into two triangles, its interior angles must measure twice those of a triangle.
>
> 2(180) = 360

Example 1

Triangle PQR is isosceles, and side PQ is congruent to side QR. The measure of $\angle QRS$ is 110. What is the measure of $\angle PQR$?

  A 40

  B 55

  C 70

  D 110

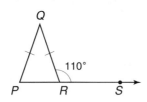

> **Hint** Look for words like *isosceles* that give you information about the figure.

Solution Start with the given information:

$$m\angle QRS = 110$$

$\angle QRS$ and $\angle PRQ$ are a linear pair, so they are supplementary.

$$m\angle PRQ + 110 = 180$$

So, $m\angle PRQ = 70$.

$\triangle PQR$ is isosceles with $\overline{PQ} \cong \overline{QR}$. Therefore, the two angles opposite these sides are congruent.

$$m\angle PRQ = m\angle RPQ = 70$$

Since the sum of the measures of the interior angles of a triangle is 180, $m\angle PQR + 70 + 70 = 180$. So, $m\angle PQR = 40$. The answer is A.

Example 2

Which statement must be true about $x + y$?

  A The sum is less than 90.

  B The sum is exactly 90.

  C The sum is greater than 90.

  D No relationship can be determined.

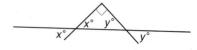

> **Hint** Use marks on the figure and relationships that are not marked to give you information about the sum.

Solution A right angle is marked on the figure; the degree measure of this angle is 90. The angle with degree measure x is one of a pair of vertical angles, and vertical angles have the same measure. The angle with degree measure y is also one of a pair of vertical angles. So the angles of the triangle measure 90, x, and y.

The sum of the measures of all angles in a triangle is 180. So $90 + x + y = 180$. So, $x + y = 90$, and the answer is B.

After you work each problem, record your answer on the answer sheet provided or on a sheet of paper.

Multiple Choice

1. Karen is making a larger sail for her model boat. Use the diagram to find the length of the base of the new sail. *(Lesson 9–4)*

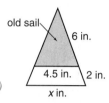

old sail

6 in.

4.5 in. 2 in.

x in.

   Ⓐ 6.5 in. Ⓑ 6 in. Ⓒ 8 in. Ⓓ 9.5 in.

2. The cost of a taxi is $3 plus $0.75 for each mile traveled. If a taxi fare is $7.50, which equation could be used to find *m*, the number of miles traveled? *(Algebra Review)*

   Ⓐ $(3 + 0.75)m = 7.50$
   Ⓑ $3 + 75m = 7.50$
   Ⓒ $3 + 0.75m = 7.50$
   Ⓓ $7.50 + 3 = 0.75m$

3. *ABCD* is a parallelogram. What must be the coordinates of point *C*? *(Lesson 8–2)*

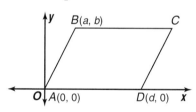

   Ⓐ (x, y) Ⓑ $(d + a, y)$ Ⓒ $(d − a, b)$
   Ⓓ $(d + x, b)$ Ⓔ $(d + a, b)$

4. Terry is making fertilizer for his garden. The directions call for 4 tablespoons in 1 gallon of water. Terry accidentally puts 5 tablespoons of fertilizer in the watering can. How many gallons of water should he use to keep the correct proportion of fertilizer to water? *(Algebra Review)*

   Ⓐ 0.8 Ⓑ 1 Ⓒ 1.25 Ⓓ 2

5. Solve $2(n + 5) − 6 = 3n + 9$. *(Algebra Review)*

   Ⓐ 10 Ⓑ −5 Ⓒ 13 Ⓓ no solution

6. A weather forecaster states that the probability of rain today is 40%. What are the odds that it will *not* rain today? *(Statistics Review)*

   Ⓐ 5:3 Ⓑ 2:3 Ⓒ 2:5 Ⓓ 3:2

7. Water flows through a hose at a rate of 5 gallons per minute. How many hours will it take to fill a 2400-gallon tank? *(Algebra Review)*

   Ⓐ 3 Ⓑ 5.5 Ⓒ 7.5 Ⓓ 8

8. Which expression is not equivalent to the others? *(Algebra Review)*

   Ⓐ $(x + 1)^2 − x^2$
   Ⓑ $(x + 1 + x)(x + 1 − x)$
   Ⓒ $2x + 1$
   Ⓓ $(x − 1)^2 + x^2$

Grid In

9. The measure of the base of a triangle is 13 and the other two sides are congruent. If the measures are integers, find the measure of the shortest possible side. *(Lesson 7–4)*

Extended Response

10. A Little League team of 24 children, along with 7 adults, is attending a minor league baseball game. The team raised $210, and the adults paid $100 to cover their expenses. *(Algebra Review)*

Part A There are 3 seats in the first row, 5 seats in the second row, 7 seats in the third row, and 9 seats in the fourth row. If this pattern continues, which row can they all sit in nearest the field? Show your work.

Part B The tickets cost $4 for children and $6 for adults. The adults all order hot dogs at $1.75 each and drinks at $2.25 each. Write an equation to find *S*, the amount of money left to spend.

What You'll Learn

Key Ideas

- Identify and use parts of circles. *(Lesson 11–1)*

- Identify major arcs, minor arcs, and semicircles and find the measures of arcs and central angles. *(Lesson 11–2)*

- Identify and use the relationships among arcs, chords, and diameters. *(Lesson 11–3)*

- Inscribe regular polygons in circles and explore lengths of chords. *(Lesson 11–4)*

- Solve problems involving circumferences, areas, and sectors of circles. *(Lessons 11–5 and 11–6)*

Key Vocabulary

center *(p. 454)*

circle *(p. 454)*

circumference *(p. 478)*

diameter *(p. 454)*

radius *(p. 454)*

Why It's Important

Navigation Latitude and longitude and the system for finding your position at any time were developed in order to allow safe travel by ships at sea. Horizontal circles called *latitude* lines indicate distance north or south of the equator. Meridians, or circles that pass through both poles, determine *longitude*.

Circles are used in sports, biology, carpentry, and other fields. You will investigate the measures of angles between the meridians in Lesson 11–2.

✔ Check Your Readiness

Refer to the line for Exercises 1–4.

1. If $LM = 8$ and $LP = 22$, find MP.
2. If $NP = 6$ and $PQ = 10$, find NQ.
3. If $LN = 16$ and $LM = 8$, find MN.
4. If $PR = 12$ and $LP = 22$, find LR.

Refer to the figure at the right.

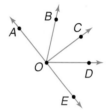

5. If $m\angle AOB = 55$ and $m\angle BOC = 32$, find $m\angle AOC$.
6. Find $m\angle DOB$ if $m\angle AOB = 55$ and $m\angle AOD = 130$.
7. If $m\angle EOD = 50$ and $m\angle COE = 93$, find $m\angle DOC$.

If c is the hypotenuse, find each missing measure. Round to the nearest tenth, if necessary.

8. $a = 9, b = 12, c = ?$
9. $a = ?, b = 15, c = 17$
10. $a = 40, b = ?, c = 41$
11. $a = 8, b = 14, c = ?$

Make this Foldable to help you organize your Chapter 11 notes. Begin with seven sheets of $8\frac{1}{2}$" by 11" paper.

❶ **Draw** and cut a circle from each sheet. Use a small plate or CD to outline the circles.

❷ **Staple** the circles together to form a booklet.

❸ **Label** the front of the booklet *Circles*. Label the six inside pages with the lesson titles.

Circles

Reading and Writing As you read and study the chapter, use each page to write definitions and theorems and draw models. Also include any questions that need clarifying.

What You'll Learn

You'll learn to identify and use parts of circles.

Why It's Important

Sports Many types of sports involve circles. *See Exercise 30.*

A circle is a special type of a geometric figure. All points on a **circle** are the same distance from a center point.

The measures of $\overline{OA}$ and $\overline{OB}$ are the same; that is, $OA = OB$.

Definition of a Circle	**Words:** A circle is the set of all points in a plane that are a given distance from a given point in the plane, called the **center** of the circle.
	Model: 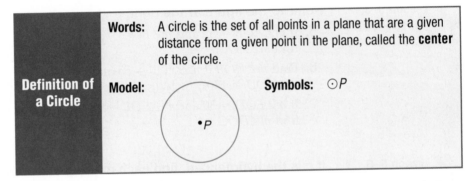 **Symbols:** $\odot P$

Note that a circle is named by its center. The circle above is named circle P.

Reading Geometry

The plural of *radius* is *radii*, which is pronounced RAY-dee-eye.

There are three kinds of segments related to circles. A **radius** is a segment whose endpoints are the center of the circle and a point on the circle. A **chord** is a segment whose endpoints are on the circle. A **diameter** is a chord that contains the center.

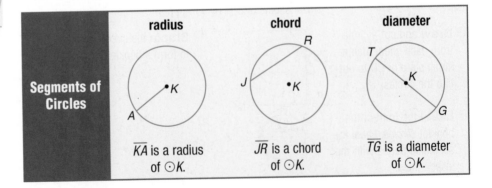

Segments of Circles

radius	chord	diameter
$\overline{KA}$ is a radius of $\odot K$.	$\overline{JR}$ is a chord of $\odot K$.	$\overline{TG}$ is a diameter of $\odot K$.

From the figures, you can note that the diameter is a special type of chord that passes through the center.

Use ⊙Q to determine whether each statement is *true* or *false*.

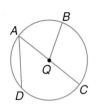

1 $\overline{AD}$ is a diameter of ⊙Q.

False; $\overline{AD}$ does not go through the center Q. Thus, $\overline{AD}$ is not a diameter.

2 $\overline{BQ}$ is a radius of ⊙Q.

True; the endpoints of $\overline{BQ}$ are the center Q and a point on the circle B. Thus, $\overline{BQ}$ is a radius.

Your Turn

a. $\overline{AC}$ is a chord of ⊙Q.   **b.** $\overline{AD}$ is a radius of ⊙Q.

Suppose the radius $\overline{PQ}$ of ⊙P is 5 centimeters long. So, the radius $\overline{PT}$ is also 5 centimeters. Then the diameter $\overline{QT}$ is 5 + 5 or 10 centimeters long. Notice that the diameter is twice as long as the radius. This leads to the next two theorems.

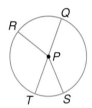

Theorem	Words	Symbols
11–1	All radii of a circle are congruent.	$\overline{PR} \cong \overline{PQ} \cong \overline{PS} \cong \overline{PT}$
11–2	The measure of the diameter *d* of a circle is twice the measure of the radius *r* of the circle.	$d = 2r$ or $\frac{1}{2}d = r$

Example

3 In ⊙T, $\overline{CD}$ is a diameter. If CD = 42, find TC.

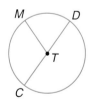

$\overline{TC}$ is a radius of ⊙T.

$d = 2r$ *Definition of radius*

$CD = 2(TC)$ *Replace d with CD and r with TC.*

$42 = 2(TC)$ *Replace CD with 42.*

$\frac{42}{2} = \frac{2(TC)}{2}$ *Divide each side by 2.*

$21 = TC$ *Simplify.*

Your Turn

c. In ⊙T, $\overline{TM}$ is a radius. If TM = 15.5, find CD.

Find the measure of radius $\overline{PC}$ if
$PC = 2x$ and $AB = 6x - 28$.

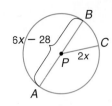

Algebra Review
Solving Multi-Step
Equations, p. 723

$$d = 2r$$
$$AB = 2(PC) \qquad \textit{Replace d with AB and r with PC.}$$
$$6x - 28 = 2(2x) \qquad \textit{Replace AB with } 6x - 28 \textit{ and PC with } 2x.$$
$$6x - 28 = 4x \qquad \textit{Simplify.}$$
$$6x - 28 - 6x = 4x - 6x \qquad \textit{Subtract 6x from each side.}$$
$$-28 = -2x \qquad \textit{Simplify.}$$
$$\frac{-28}{-2} = \frac{-2x}{-2} \qquad \textit{Divide each side by } -2.$$
$$14 = x \qquad \textit{Simplify.}$$

$PC = 2(14)$ or 28

Because all circles have the same shape, any two circles are similar. However, two circles are congruent if and only if their radii are congruent. Two circles are **concentric** if they meet the following three requirements.

- They lie in the same plane.
- They have the same center.
- They have radii of different lengths.

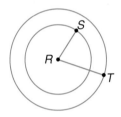

Circle R with radius $\overline{RS}$ and circle R with radius $\overline{RT}$ are concentric circles.

Check for Understanding

Communicating Mathematics

1. **Explain** why there are more than two radii in every circle. How many radii are there?

2. **Describe** how to find the measure of the radius if you know the measure of the diameter.

3. Jason says that every diameter of a circle is a chord. Amelia says that every chord of a circle is a diameter. Who is correct, and why?

Vocabulary
circle
center
radius
chord
diameter
concentric

Guided Practice

⊕ **Getting Ready** If r is the measure of the radius and d is the measure of the diameter, find each measure.

Sample: $d = 9.04$, $r = $ ___?___   **Solution:** $r = \dfrac{9.04}{2}$ or 4.52

4. $r = 3.8$, $d = $ ___?___   5. $d = 3\frac{1}{2}$, $r = $ ___?___   6. $r = \frac{x}{2}$, $d = $ ___?___

Examples 1 & 2

Use ⊙F to determine whether each statement is *true* or *false*.

7. $\overline{FD}$ is a radius of ⊙F.
8. $\overline{AB}$ is a diameter of ⊙F.
9. $\overline{CE}$ is a chord of ⊙F.

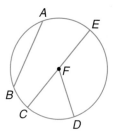

Example 3

Use ⊙F to complete the following.

10. If $CE = 15.2$, find FD.
11. If $FE = 19$, find CE.

Exercises 7–11

Example 4

12. **Algebra** Find the value of x in ⊙K.

Exercises

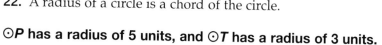

Use ⊙R to determine whether each statement is *true* or *false*.

Practice

13. $\overline{HB}$ is a radius of ⊙R.
14. $HD = 2(RD)$
15. $\overline{CG}$ is a diameter of ⊙R.
16. $\overline{BE}$ is a diameter of ⊙R.
17. $\overline{RE}$ is a chord of ⊙R.
18. $\overline{RC}$ is a radius of ⊙R.
19. $\overline{AF}$ is a chord of ⊙R.
20. $RH = RG$
21. A circle has exactly two radii.
22. A radius of a circle is a chord of the circle.

Homework Help	
For Exercises	See Examples
13–21	1, 2
23–29	3, 4
Extra Practice	
See page 746.	

⊙P has a radius of 5 units, and ⊙T has a radius of 3 units.

23. If $QR = 1$, find RT.
24. If $QR = 1$, find PQ.
25. If $QR = 1$, find AB.
26. If $AR = 2x$, find AP in terms of x.
27. If $TB = 2x$, find QB in terms of x.

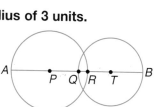

Applications and Problem Solving

28. **Music** Most music compact discs (CDs) have three concentric circles. The first circle forms the hole in the CD and has a diameter of 2.5 centimeters. The second circle forms an inner ring, on which no data are stored. It has a diameter of 4 centimeters. The third circle forms the disc itself. It has a diameter of 12 centimeters. What are the radii of the three circles?

29. **Algebra** In $\odot S$, $VK = 3x - 9$ and $JW = 2x + 15$. Find the measure of a radius of $\odot S$.

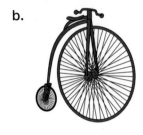

30. **Sports** Identify the types of circles shown below as *congruent*, *concentric*, or *similar*.

a.

archery target

b.

old-style bicycle

c.

Olympic rings

31. **Critical Thinking** Give a reason for each statement to show that a diameter is the longest chord of a circle. Assume that T is the center.
a. $QT + TR > QR$
b. $QT = AT$ and $TR = TB$
c. $AT + TB > QR$
d. $AB > QR$

Mixed Review

32. Identify the figures used to create the tessellation. Then identify the tessellation as *regular*, *semi-regular*, or *neither*. *(Lesson 10–7)*

33. Does the figure have line symmetry? rotational symmetry? *(Lesson 10–6)*

Exercises 32–33

Use the number line below to find each measure. *(Lesson 2–1)*

34. FD 35. EF 36. CG

37. **Extended Response** The caveman is using inductive reasoning to make a conjecture. *(Lesson 1–1)*
a. List the steps he uses to make his conjecture.
b. State the conjecture.

"Water boils down to nothing. . . snow boils down to nothing. . . ice boils down to nothing. . . everything boils down to nothing."

Reproduced by permission of Ed Fisher; ©1966 Saturday Review, Inc.

38. **Multiple Choice** Simplify $\dfrac{\sqrt{6} \cdot \sqrt{8}}{\sqrt{3}}$.
(Algebra Review)

Ⓐ $\sqrt{4}$ Ⓑ 4
Ⓒ 16 Ⓓ 48

www.geomconcepts.com/self_check_quiz

Woodworker

Frequently, a woodworker who makes furniture will need to cut a circular table top from a square piece of wood stock, as shown at the right. In order to cut the table top, the woodworker first needs to locate the center of the wood. Then a radius can be found and a circle can be drawn.

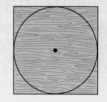

One method for finding the center is shown below.

Step 1 Make sure that the piece of stock is square.

Step 2 Use a ruler and pencil to draw the diagonals of the square.

Step 3 Mark the point at which the diagonals meet.

Step 4 Use a thumbtack, pencil, and string to sketch the circle. The radius of the circle will be one-half the length of a side of the square.

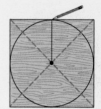

1. Sketch the largest circle that will fit inside a square 2 inches on a side.

2. Sketch the largest circle that will fit inside an equilateral triangle $1\frac{1}{2}$ inches on a side.

3. The table top shown will have a mahogany inlay in the shape of a circle. The inlay will be placed in the center of the rectangle and will have a diameter that is one-half the width of the rectangle. Copy the rectangle. Then sketch the circle described.

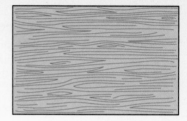

FAST FACTS About Woodworkers

Working Conditions

- production woodworkers usually work in large plants on mass-produced items
- precision woodworkers usually work in small shops on one-of-a-kind items
- woodworkers often wear eye and ear protection for safety

Education

- most are trained on the job; may require two or more years of training
- high school diploma desired
- ability to pay attention to detail a must

Earnings

Weekly Earnings for Cabinetmakers and Bench Carpenters

Lowest 10%—less than $290

Middle 50%—between $348 and $549

Top 10%—more than $688

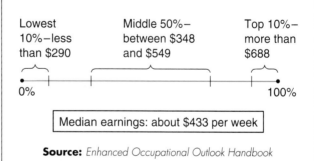

Median earnings: about $433 per week

Source: *Enhanced Occupational Outlook Handbook*

*inter*NET
CONNECTION
Career Update For the latest information about a career as a woodworker, visit:
www.geomconcepts.com

A Locus Is Not a Grasshopper!

COncepts in MOtion
Animation
geomconcepts.com

Materials

 cardboard

 string

 colored labeling dots

 two metersticks

 masking tape

 scissors

Loci

A **locus** is the set of all points that satisfy a given condition or conditions. The plural of locus is **loci** (LOW-sigh). You have already seen some sets of points that form loci. Let's investigate this idea further.

Investigate

1. Given a point *A*, what is the locus of points in a plane that are two feet from point *A*?

   a. Cut a square piece of cardboard with side length of 3 inches. Cut a piece of string about 30 inches in length. Poke a small hole in the center of the cardboard. Label the small hole *A*. Thread the string through the hole. Tie a knot in the string and cut it so that it measures two feet from point *A* to the end of the string.

   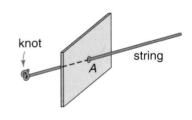

   b. Tape the cardboard square securely to the floor with the knot under the cardboard. Pull the string to its full length along the floor. Mark the floor at the end of the string with a colored dot. The dot will represent a point that is two feet from point A.

   c. Pick the string up and extend it in another direction. Mark the floor at the end of the string with another colored dot. Repeat this process until you have about 20 dots on the floor.

   d. Imagine placing more and more dots on the floor. Describe this set of points.

   e. Describe the locus of points in a plane at a given distance from a fixed point on the plane.

COncepts
in MOtion
Interactive Lab
geomconcepts.com

2. Given two parallel lines, what is the locus of points in a plane that are equidistant from the two parallel lines?

   a. Tape two metersticks to the floor so that they are parallel and at least one foot apart. Make sure that the ends of the metersticks are even.

   b. Cut a piece of string the length of the distance between the metersticks. To be accurate, stretch the string between the same markings on the sticks, say between 10 centimeters on each stick.

   c. Find and mark the midpoint of the piece of string. Lay the string between the same markings on the two sticks. Place a colored dot on the floor at the midpoint of the string. Pick up the string and mark about 10 additional points with colored dots.

   d. Imagine placing more and more dots. Describe this set of points.

   e. Describe the locus of points in a plane that are equidistant from the two parallel lines.

Extending the Investigation

1. Identify the locus of points that satisfies each condition.
   a. all points in a plane that are 2 feet from a given line
   b. all points in a plane that are equidistant from two given points
   c. all points in a plane that are equidistant from the sides of a given angle
   d. all points in a plane that are equidistant from two intersecting lines

2. A **compound locus** is the intersection of loci that satisfies two or more conditions. Identify the compound locus of points that satisfies each set of conditions.
   a. all points in a plane that are equidistant from the sides of a given 90° angle and 4 feet from the vertex of the angle
   b. all points in a plane that are equidistant from two given points and 10 centimeters from the line containing the two points
   c. all points on a coordinate plane that are 2 units from $P(1, 3)$ and 1 unit from $Q(1, 6)$

3. If the phrase "in a plane" was deleted from the two problems in the Investigation, what would the locus be for those two problems? Provide a sketch.

Presenting Your Conclusions

Here are some ideas to help you present your conclusions to the class.

• Make a poster with scale drawings and descriptions of the methods and materials you used to find the solution to each locus problem.

• Make a three-dimensional display using marbles for points, straws to show distance, and dowels for line segments or sides of angles.

interNET
CONNECTION **Investigation** For more information on loci, visit: www.geomconcepts.com

What You'll Learn
You'll learn to identify major arcs, minor arcs, and semicircles and find the measures of arcs and central angles.

Why It's Important
Food When some pizzas are sliced, central angles are formed.
See Exercise 12.

A **central angle** is formed when the two sides of an angle meet at the center of a circle. Each side intersects a point on the circle, dividing it into **arcs** that are curved lines.

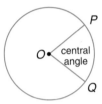

There are three types of arcs. A **minor arc** is part of the circle in the interior of the central angle with measure less than 180°. A **major arc** is part of the circle in the exterior of the central angle. **Semicircles** are congruent arcs whose endpoints lie on a diameter of the circle.

Arcs are named by their endpoints. Besides the length of an arc, the measure of an arc is also related to the corresponding central angle.

COncepts in MOtion
Animation
geomconcepts.com

Reading Geometry
We will use the *measure* of an arc to mean the degree measure of the arc.

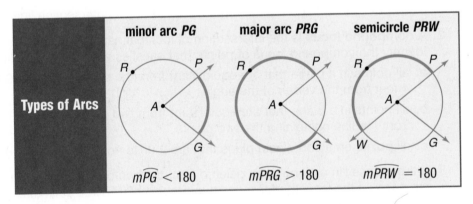

Note that for ⊙*A*, two letters are used to name the minor arc, but three letters are used to name the major arc and semicircle. These letters for naming arcs help us trace the set of points in the arc. In this way, there is no confusion about which arc is being considered.

Depending on the central angle, each type of arc is measured in the following way.

<table>
<tr>
<td rowspan="3">Definition of Arc Measure</td>
<td>1. The degree measure of a minor arc is the degree measure of its central angle.</td>
</tr>
<tr>
<td>2. The degree measure of a major arc is 360 minus the degree measure of its central angle.</td>
</tr>
<tr>
<td>3. The degree measure of a semicircle is 180.</td>
</tr>
</table>

Example ① In $\odot R$, $\overline{KN}$ is a diameter. Find $m\widehat{ON}$, $m\angle NRT$, $m\widehat{OTK}$ and $m\widehat{NTK}$.

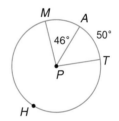

$m\widehat{ON} = m\angle ORN$ *Measure of minor arc*
      $= 42$ *Substitution*

$m\angle NRT = m\widehat{NT}$ *Measure of central angle*
      $= 89$ *Substitution*

Note that the sum of the measures of the central angles of $\odot R$ is 360.

$m\widehat{OTK} = 360 - m\angle ORK$ *Measure of major arc*
        $= 360 - \quad 138$ *Substitution*
        $= 222$ *Subtract.*

$m\widehat{NTK} = 180$ *Measure of semicircle*

Your Turn

a. In $\odot P$, find $m\widehat{AM}$, $m\angle APT$, and $m\widehat{THM}$.

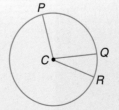

In $\odot P$ above, $\widehat{AM}$ and $\widehat{AT}$ are examples of **adjacent arcs**. Adjacent arcs have exactly one point in common. For $\widehat{AM}$ and $\widehat{AT}$, the common point is A. The measures of adjacent arcs can also be added.

<table>
<tr>
<td rowspan="3">Postulate 11–1
Arc Addition Postulate</td>
<td>Words: The sum of the measures of two adjacent arcs is the measure of the arc formed by the adjacent arcs.</td>
</tr>
<tr>
<td>Model:</td>
</tr>
<tr>
<td>Symbols:
If Q is a point on $\widehat{PR}$, then $m\widehat{PQ} + m\widehat{QR} = m\widehat{PQR}$.</td>
</tr>
</table>

In ⊙P, $\overline{RT}$ is a diameter. Find $m\overset{\frown}{RS}$, $m\overset{\frown}{ST}$, $m\overset{\frown}{STR}$, and $m\overset{\frown}{QS}$.

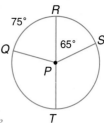

$m\overset{\frown}{RS} = m\angle RPS$ *Measure of minor arc*
     $= 65$ *Substitution*

$m\overset{\frown}{RS} + m\overset{\frown}{ST} = m\overset{\frown}{RST}$ *Arc addition postulate*
      $m\overset{\frown}{ST} = m\overset{\frown}{RST} - m\overset{\frown}{RS}$ *Subtract $m\overset{\frown}{RS}$ from each side.*
           $= 180 - 65$ or 115 *Substitution*

$m\overset{\frown}{STR} = 360 - m\angle RPS$ *Measure of major arc*
      $= 360 - 65$ or 295 *Substitution*

$m\overset{\frown}{QS} = m\overset{\frown}{QR} + m\overset{\frown}{RS}$ *Arc addition postulate*
     $= 75 + 65$ or 140 *Substitution*

Your Turn

b. In ⊙P, find $m\overset{\frown}{QT}$.   **c.** In ⊙P, find $m\overset{\frown}{STQ}$.

Suppose there are two concentric circles with $\angle ASD$ forming two minor arcs, $\overset{\frown}{BC}$ and $\overset{\frown}{AD}$. Are the two arcs congruent?

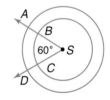

$m\overset{\frown}{BC} = m\angle BSC$ or 60
$m\overset{\frown}{AD} = m\angle ASD$ or 60

Although $\overset{\frown}{BC}$ and $\overset{\frown}{AD}$ each measure 60, they are not congruent. The arcs are in circles with different radii, so they have different lengths. However, in a circle, or in congruent circles, two arcs are congruent if they have the same measure.

Theorem 11–3	**Words:** In a circle or in congruent circles, two minor arcs are congruent if and only if their corresponding central angles are congruent. **Model:** 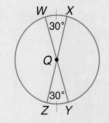 **Symbols:** $\overset{\frown}{WX} \cong \overset{\frown}{YZ}$ if and only if $m\angle WQX = m\angle YQZ$.

3 In ⊙M, $\overline{WS}$ and $\overline{RT}$ are diameters, $m\angle WMT = 125$, and $m\widehat{RK} = 14$.
Find $m\widehat{RS}$.

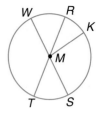

$\angle WMT \cong \angle RMS$	*Vertical angles are congruent.*
$m\angle WMT = m\angle RMS$	*Definition of congruent angles*
$m\widehat{WT} = m\widehat{RS}$	*Theorem 11–3*
$125 = m\widehat{RS}$	*Substitution*

Your Turn

d. Find $m\widehat{KS}$.   **e.** Find $m\widehat{ST}$.

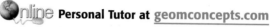 **Personal Tutor at** geomconcepts.com

Check for Understanding

Communicating Mathematics

1. Refer to ⊙R with diameter $\overline{PH}$.

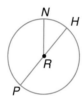

Vocabulary

central angle
arcs
minor arc
major arc
semicircle
adjacent arcs

a. Explain how to find $m\widehat{PNH}$.
b. Determine whether $\widehat{PNH} \cong \widehat{PHN}$. Explain.
c. Explain how to find $m\widehat{HPN}$ if $m\angle NRH = 35$.
d. Explain why diameter $\overline{PH}$ creates two arcs that measure 180.

2. Compare and contrast minor arcs and major arcs.

3. **You Decide?** Marisela says that two arcs can have the same measure and still not be congruent. Dexter says that if two arcs have the same measure, then they are congruent. Who is correct, and why?

Guided Practice

Getting Ready Determine whether each arc is a minor arc, major arc, or semicircle of ⊙M.

Sample: $\widehat{DAB}$
Solution: $m\widehat{DAB} > 180$, so $\widehat{DAB}$ is a major arc.

4. $\widehat{BDE}$
5. $\widehat{ECA}$
6. $\widehat{AB}$

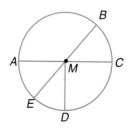

Find each measure in ⊙P if m∠APB = 30 and $\overline{AC}$ and $\overline{BD}$ are diameters.

Example 1
Example 2
Example 3

7. $m\widehat{AB}$

9. $m\widehat{BAC}$

11. $m\widehat{AD}$

8. $m\widehat{ACB}$

10. $m\widehat{BC}$

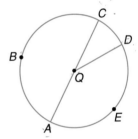

Example 1

12. **Food** Rosati's Pizza cuts their pizzas along four diameters, which separate each pizza into eight congruent pieces. What is the measure of the central angle of each piece?

Exercises •

Practice

Find each measure in ⊙P if m∠WPX = 28, $m\widehat{YZ}$ = 38, and $\overline{WZ}$ and $\overline{XV}$ are diameters.

13. $m\angle ZPY$

15. $m\widehat{VZ}$

17. $m\widehat{VWX}$

19. $m\widehat{WYZ}$

21. $m\angle XPY$

23. $m\widehat{XWY}$

14. $m\widehat{XZ}$

16. $m\angle VPZ$

18. $m\widehat{ZVW}$

20. $m\widehat{ZXW}$

22. $m\widehat{XY}$

24. $m\widehat{WZX}$

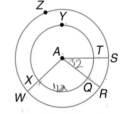

Homework Help	
For Exercises	See Examples
13–24, 38	3
25–37	1, 2
Extra Practice	
See page 746.	

In ⊙Q, $\overline{AC}$ is a diameter and m∠CQD = 40. Determine whether each statement is *true* or *false*.

25. $m\widehat{CBD} = 140$

26. $m\angle CQD = m\widehat{CD}$

27. $\angle AQD$ is a central angle.

28. $m\widehat{AD} = 320$

29. $m\widehat{ACD} = 140$

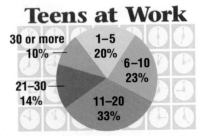

A is the center of two circles with radii $\overline{AQ}$ and $\overline{AR}$. If m∠SAR = 32 and $m\widehat{XQ}$ = 112, find each measure.

30. $m\widehat{SR}$

32. $m\widehat{WR}$

34. $m\widehat{TYX}$

31. $m\angle RAW$

33. $m\widehat{TQ}$

35. $m\widehat{SZW}$

Applications and Problem Solving

36. **Employment** Twenty-two percent of all teens ages 12 through 17 work either full- or part-time. The circle graph shows the number of hours they work per week. Find the measure of each central angle.

Teens at Work

30 or more
10%

1–5
20%

6–10
23%

21–30
14%

11–20
33%

Source: *ICRs TeenEXCEL* survey for Merrill Lynch

37. **Geography** Earth has 24 time zones, each of which is centered on a line called a *meridian*.

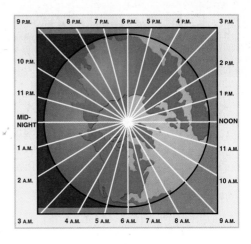

   a. What is the measure of the arc between 6 P.M. and 5 P.M.?

   b. What is the measure of the minor arc between 6 P.M. and 4 A.M.?

38. **Critical Thinking** In $\odot B$, $\overset{\frown}{PR} \cong \overset{\frown}{QS}$. Show that $\overset{\frown}{PQ} \cong \overset{\frown}{RS}$. Give a reason for each step of your argument.

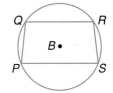

Mixed Review

39. **Basketball** Basketball rims are 18 inches in diameter. What is the radius of a rim? *(Lesson 11–1)*

40. Create your own tessellation using squares and triangles. *(Lesson 10–7)*

41. Solve $\frac{3x - 5}{4} = \frac{x}{2}$. *(Lesson 9–1)*

42. Use a straightedge to draw a quadrilateral that has exactly one diagonal in its interior. *(Lesson 8–1)*

Standardized Test Practice
Ⓐ Ⓑ Ⓒ Ⓓ

43. **Grid In** The brace shown at the right is used to keep a shelf perpendicular to the wall. If $m\angle AHM = 40$, find $m\angle HAT$. *(Lesson 7–2)*

44. **Short Response** Explain how you could use translations to draw a cube. *(Lesson 5–3)*

Exercise 43

Quiz 1   Lessons 11–1 and 11–2

Use $\odot P$ to determine whether each statement is *true* or *false*. $\overline{NL}$ and $\overline{MK}$ are diameters of $\odot P$. *(Lessons 11–1 & 11–2)*

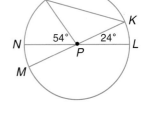

  1. $\overline{JP}$ is a radius.
  2. $\overline{JK}$ is a radius.
  3. $\overline{NP}$ is a chord.
  4. $NL = 2(NP)$
  5. $m\overset{\frown}{JM} = 54$
  6. $m\overset{\frown}{KL} = 336$
  7. $m\overset{\frown}{NM} = 24$
  8. $m\overset{\frown}{ML} = 126$
  9. $m\angle JPK = 102$

10. Time The hands of a clock form central angles. What is the approximate measure of the central angle at 6:00? at 12:05? at 6:05? *(Lesson 11–2)*

What You'll Learn

You'll learn to identify and use the relationships among arcs, chords, and diameters.

Why It's Important

Entertainment
Technical advisors ensure mathematical accuracy in movies. *See Exercise 24.*

In circle P below, each chord joins two points on a circle. Between the two points, an arc forms along the circle. By Theorem 11–3, $\overparen{AD}$ and $\overparen{BC}$ are congruent because their corresponding central angles are vertical angles, and therefore congruent ($\angle APD \cong \angle BPC$).

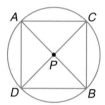

Can we show that $\overline{AD} \cong \overline{BC}$? You can prove that the two triangles $\triangle APD$ and $\triangle CPB$ are congruent by SAS. Therefore, $\overline{AD}$ and $\overline{BC}$ are congruent.

The following theorem describes the relationship between two congruent minor arcs and their corresponding chords.

Theorem 11–4	**Words:** In a circle or in congruent circles, two minor arcs are congruent if and only if their corresponding chords are congruent. **Model:** 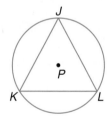 **Symbols:** $\overparen{AD} \cong \overparen{BC}$ if and only if $\overline{AD} \cong \overline{BC}$.

Example

1 The vertices of equilateral triangle *JKL* are located on $\odot P$. Identify all congruent minor arcs.

Since $\triangle JKL$ is equilateral, $\overline{JK} \cong \overline{KL} \cong \overline{JL}$. By Theorem 11–4, we know that $\overparen{JK} \cong \overparen{KL} \cong \overparen{JL}$.

Your Turn

a. The vertices of isosceles triangle *XYZ* are located on $\odot R$. If $\overline{XY} \cong \overline{YZ}$, identify all congruent arcs.

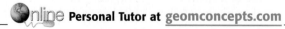 **Personal Tutor at** geomconcepts.com

You can use paper folding to find a special relationship between a diameter and a chord of a circle.

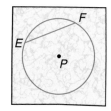
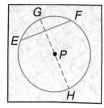
This activity suggests the following theorem.

Reading Geometry

When a minor arc and a chord share the same endpoints, we call the arc *the arc of the chord*.

Theorem 11–5	**Words:**	In a circle, a diameter bisects a chord and its arc if and only if it is perpendicular to the chord.
	Model:	**Symbols:** $\overline{AR} \cong \overline{BR}$ and $\overset{\frown}{AD} \cong \overset{\frown}{BD}$ if and only if $\overline{CD} \perp \overline{AB}$.

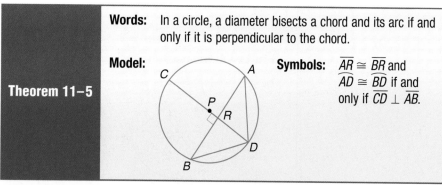

Like an angle, an arc can be bisected.

2 In ⊙P, if $\overline{PM} \perp \overline{AT}$, $PT = 10$, and $PM = 8$, find AT.

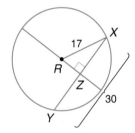

$\angle PMT$ is a right angle.	*Def. of perpendicular*
$\triangle PMT$ is a right triangle.	*Def. of right triangle*
$(MT)^2 + (PM)^2 = (PT)^2$	*Pythagorean Theorem*
$(MT)^2 + 8^2 = 10^2$	*Replace PM with 8 and PT with 10.*
$(MT)^2 + 64 = 100$	$8^2 = 64; 10^2 = 100$
$(MT)^2 + 64 - 64 = 100 - 64$	*Subtract 64 from each side.*
$(MT)^2 = 36$	*Simplify.*
$\sqrt{(MT)^2} = \sqrt{36}$	*Take the square root of each side.*
$MT = 6$	*Simplify.*

Look Back

Pythagorean Theorem: Lesson 6–6

By Theorem 11–5, $\overline{PM}$ bisects $\overline{AT}$. Therefore, $AT = 2(MT)$. So, $AT = 2(6)$ or 12.

3 In ⊙R, $XY = 30$, $RX = 17$, and $\overline{RZ} \perp \overline{XY}$. Find the distance from R to $\overline{XY}$.

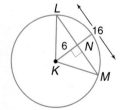

The measure of the distance from R to $\overline{XY}$ is RZ. Since $\overline{RZ} \perp \overline{XY}$, $\overline{RZ}$ bisects $\overline{XY}$, by Theorem 11–5. Thus, $XZ = \frac{1}{2}(30)$ or 15.

Recall that the distance from a point to a segment is measured on the perpendicular segment drawn from the point to the segment.

For right triangle RZX, the following equation can be written.

$(RZ)^2 + (XZ)^2 = (RX)^2$	*Pythagorean Theorem*
$(RZ)^2 + 15^2 = 17^2$	*Replace XZ with 15 and RX with 17.*
$(RZ)^2 + 225 = 289$	$15^2 = 225; 17^2 = 289$
$(RZ)^2 + 225 - 225 = 289 - 225$	*Subtract 225 from each side.*
$(RZ)^2 = 64$	*Simplify.*
$\sqrt{(RZ)^2} = \sqrt{64}$	*Take the square root of each side.*
$RZ = 8$	*Simplify.*

The distance from R to $\overline{XY}$, or RZ, is 8 units.

Your Turn

Find each measure in each ⊙K.

b. AB

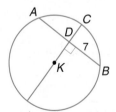

c. KM

In ⊙Q, $\overset{\frown}{KL} \cong \overset{\frown}{LM}$. If $CK = 2x + 3$ and $CM = 4x$, find x.

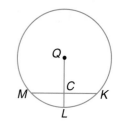

Since $\overset{\frown}{KL} \cong \overset{\frown}{LM}$, $\overline{QL}$ bisects $\overset{\frown}{KM}$. So, by Theorem 11–5, $\overline{QL}$ also bisects $\overline{KM}$. Thus, $\overline{CM} \cong \overline{CK}$.

$$CM = CK$$
$$4x = 2x + 3 \qquad \textit{Replace CM with 4x and CK with 2x + 3.}$$
$$4x - 2x = 2x + 3 - 2x \quad \textit{Subtract 2x from each side.}$$
$$2x = 3 \qquad \textit{Simplify.}$$
$$\frac{2x}{2} = \frac{3}{2} \qquad \textit{Divide each side by 2.}$$
$$x = \frac{3}{2} \qquad \textit{Simplify.}$$

Your Turn

d. Suppose $CK = 5x + 9$ and $CM = 6x - 13$. Find x.

Check for Understanding

Communicating Mathematics

1. **Complete** each statement.
   a. In the same circle, if two chords are ___?___ , then their arcs are congruent.

   b. If a diameter of a circle bisects a chord of the circle, then it is ___?___ to the chord and bisects its ___?___ .

   c. In a circle, if two ___?___ are congruent, then their chords are congruent.

2. Refer to ⊙J.
   a. **Explain** why △PAT is isosceles.
   b. **Explain** why $\overset{\frown}{AG} \cong \overset{\frown}{TG}$.

Guided Practice

Example 1

Use ⊙W to complete each statement.

3. $\overline{RS} \cong$ ___?___

4. $\overset{\frown}{ST} \cong$ ___?___

Example 2

Use ⊙D to find each measure.

5. DG

6. FH

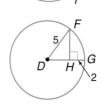

Example 3

7. In $\odot J$, radius $\overline{JL}$ and chord $\overline{MN}$ have lengths of 10 centimeters. Find the distance from J to $\overline{MN}$. Round to the nearest hundredth.

Example 4

8. Algebra In $\odot J$, $\widehat{KM} \cong \widehat{KN}$, $KM = 2x + 9$, and $KN = 5x$. Find x, KM, and KN.

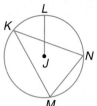

Exercises 7–8

Exercises

Practice

Homework Help	
For Exercises	**See Examples**
9–14	1
15–20	2
21–23	3, 4
Extra Practice	
See page 746.	

Use $\odot P$ to complete each statement.

9. If $\overline{CD} \cong \overline{DE}$, then $\widehat{CD} \cong$ ___?___ .

10. If $\widehat{CD} \cong \widehat{DE}$, then $\triangle PCD \cong \triangle$ ___?___ .

11. If $\overline{DP} \perp \overline{AB}$, then $\widehat{AD} \cong$ ___?___ .

12. If $\widehat{AE} \cong \widehat{BC}$, then $\overline{AE} \cong$ ___?___ .

13. If $\overline{AB} \perp \overline{CF}$, then $\overline{FG} \cong$ ___?___ .

14. If $\widehat{AE} \cong \widehat{BC}$, then $\overline{AC} \cong$ ___?___ .

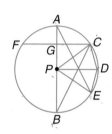

Use $\odot Q$, where $\overline{QE} \perp \overline{TN}$, to complete each statement.

15. If $QT = 8$, then $QN =$ ___?___ .

16. If $TE = 6$, then $TN =$ ___?___ .

17. If $TN = 82$, then $ET =$ ___?___ .

18. If $QE = 3$ and $EN = 4$, then $QN =$ ___?___ .

19. If $QN = 13$ and $EN = 12$, then $QE =$ ___?___ .

20. If $TN = 16$ and $QE = 6$, then $QN =$ ___?___ .

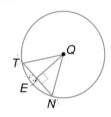

21. In $\odot M$, $RS = 24$ and $MW = 5$. Find MT.

22. In $\odot A$, $\overline{AL} \perp \overline{JK}$, $AM = 5$, and $AL = 3$. Find JK.

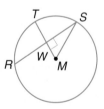

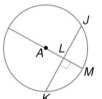

Applications and Problem Solving

23. Algebra In $\odot O$, $\overline{MN} \cong \overline{PQ}$, $MN = 7x + 13$, and $PQ = 10x - 8$. Find PS.

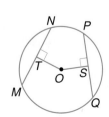

24. Entertainment In a recent movie, teacher Justin McCleod uses the following method to show his student how to find the center of a circle.

> (A man) wants to erect a pole in the center of his circle. But how does he find that center?. . . Draw a circle *ABC*. Draw within it any straight line *AB*. Now bisect (line) *AB* at *D* and draw a straight line *DC* at right angles to (line) *AB*. . . . Okay, (draw) any other straight line. . . *AC*. . . . Bisect (segment) *AC* (with a perpendicular line) and you get the center of your circle.

   a. Draw a figure that matches this description. Assume that "circle *ABC*" means a circle that goes through the points *A*, *B*, and *C*.

   b. Explain why this works.

25. Critical Thinking In ⊙*A*, $\overline{EB} \perp \overline{GD}$, and $\overline{CF}$ and $\overline{EB}$ are diameters.

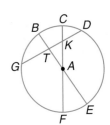

   a. Determine whether *AK* > *AT* or *AT* > *AK*. Why?

   b. Name the midpoint of $\overline{DG}$.

   c. Name an arc that is congruent to $\overset{\frown}{GE}$.

   d. If *K* is the midpoint of $\overline{TD}$, is *C* necessarily the midpoint of $\overline{BD}$?

Mixed Review

26. Architecture The Robinsons' front door has a semicircular window that is divided into four congruent sections, as shown at the right. What is the measure of each arc? *(Lesson 11–2)*

In the figure, the diameter of the circle is 26 centimeters. *(Lesson 11–1)*

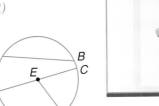

27. Name two chords.

28. Name three radii.

29. Find *ED*.

30. Grid In △*DFG* is isosceles with vertex angle *F*. Find *m∠G*. *(Lesson 6–4)*

$(7x - 32)°$ $(6x - 18)°$

31. Multiple Choice Eighty-six percent of U.S. adults said that they regularly participated in arts activities in a recent year. Their top five favorite activities are shown in the table below. Which type of graph would best illustrate this information? *(Statistics Review)*

   Ⓐ bar graph
   Ⓑ circle graph
   Ⓒ line graph
   Ⓓ pictograph

Activity	Percent
photography	44%
weaving/needlepoint/handwork	36%
painting/drawing	33%
dancing	30%
musical instrument	28%

Source: National Assembly of Local Arts Agencies, American Council for the Arts

What You'll Learn
You'll learn to inscribe regular polygons in circles and explore the relationship between the length of a chord and its distance from the center of the circle.

Why It's Important
Carpentry
Carpenters use inscribed polygons when they cut beams from logs.
See Exercise 6.

Concepts in MOtion
Animation
geomconcepts.com

For everyday meals, the Williams family's square kitchen table seats four people. On special occasions, however, the sides can be raised to change the square table to a circular one that seats six. When the table's top is open, its circular top is said to be **circumscribed** about the square. We also say that the square is **inscribed** in the circle.

Definition of Inscribed Polygon	A polygon is inscribed in a circle if and only if every vertex of the polygon lies on the circle.

Some regular polygons can be constructed by inscribing them in circles. In Lesson 1–5, you learned to construct a regular hexagon in this way. The following example demonstrates how to construct another regular polygon.

Example

1 **Construct a regular quadrilateral.**

- Construct ⊙P and draw a diameter $\overline{AC}$.
- Construct the perpendicular bisector of $\overline{AC}$, extending the line to intersect ⊙P at points B and D.
- Connect the consecutive points in order to form square $ABCD$.

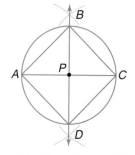

The construction of a hexagon can be used to discover another property of chords.

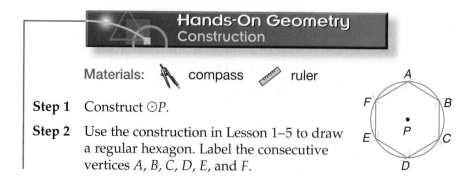

Hands-On Geometry
Construction

Materials: compass ruler

Step 1 Construct ⊙P.

Step 2 Use the construction in Lesson 1–5 to draw a regular hexagon. Label the consecutive vertices A, B, C, D, E, and F.

Step 3 Construct a perpendicular line from the center to each chord.

Step 4 Measure the distance from the center to each chord.

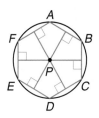

Try These

1. What is true about the distance from the center of $\odot P$ to each chord?
2. From the construction, what is true of $\overline{AB}$, $\overline{BC}$, $\overline{CD}$, $\overline{DE}$, $\overline{EF}$ and $\overline{FA}$?
3. **Make a conjecture** about the relationship between the measure of the chords and the distance from the chords to the center.

This activity suggests the following theorem.

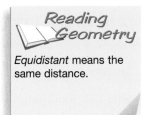

Reading Geometry

Equidistant means the same distance.

Theorem 11–6

Words:	In a circle or in congruent circles, two chords are congruent if and only if they are equidistant from the center.

Model:

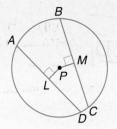

Symbols: $\overline{AD} \cong \overline{BC}$ if and only if $\overline{LP} \cong \overline{PM}$.

Example ──2

Algebra Link

In $\odot A$, $PR = 2x + 5$ and $QR = 3x - 27$. Find x.

The figure shows that $\overline{PR}$ and $\overline{QR}$ are equidistant from the center of the circle. From Theorem 11–6, we can conclude that $\overline{PR} \cong \overline{QR}$.

$$
\begin{array}{ll}
PR = QR & \text{\textit{Definition of congruent segments}} \\
2x + 5 = 3x - 27 & \text{\textit{Substitution}} \\
2x + 5 - 2x = 3x - 27 - 2x & \text{\textit{Subtract 2x from each side.}} \\
5 = x - 27 & \text{\textit{Simplify.}} \\
5 + 27 = x - 27 + 27 & \text{\textit{Add 27 to each side.}} \\
32 = x & \text{\textit{Simplify.}}
\end{array}
$$

Your Turn

In $\odot O$, O is the midpoint of $\overline{AB}$. If $CR = -3x + 56$ and $ST = 4x$, find x.

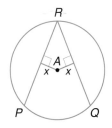

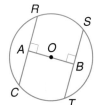

Check for Understanding

**Communicating
Mathematics**

1. **Look for a pattern** in the inscribed polygons. What would a polygon with 200 sides look like?

| equilateral triangle | regular pentagon | regular hexagon | regular octagon |

2. **Writing Math** Compare and contrast the meanings of the terms *circumscribed* and *inscribed*.

Guided Practice

Example 1

Example 2

3. Use the construction of a hexagon to construct an equilateral triangle. Explain each step.

4. In ⊙*T*, *CD* = 19. Find *AB*.

5. In ⊙*R*, if *AB* = 2*x* − 7 and *CD* = 5*x* − 22, find *x*.

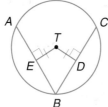

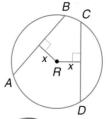

6. **Carpentry** The strongest rectangular beam that can be cut from a circular log is one whose width is 1.15 times the radius of the log. What is the width of the strongest beam that can be cut from a log 6 inches in diameter?

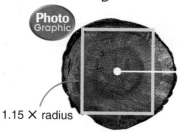

1.15 × radius

Exercises

Practice

Use a compass and straightedge to inscribe each polygon in a circle. Explain each step.

7. regular octagon

8. regular dodecagon (12 sides)

Homework Help	
For Exercises	See Examples
7, 8, 18	1
9–15, 17	2
Extra Practice	
See page 747.	

Use ⊙*P* to find *x*.

9. *AB* = 2*x* − 4, *CD* = *x* + 3

10. *AB* = 3*x* + 2, *CD* = 4*x* − 1

11. *AB* = 6*x* + 7, *CD* = 8*x* − 13

12. *AB* = 3(*x* + 2), *CD* = 12

13. *AB* = 2(*x* + 1), *CD* = 8*x* − 22

14. *AB* = 4(2*x* − 1), *CD* = 10(*x* − 3)

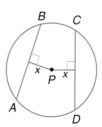

15. Square *MATH* is inscribed in ⊙*P* with a radius of 12 centimeters.

   a. Find *m*∠*HPT*.

   b. Find *TH*.

   c. What kind of triangle is △*PTH*?

   d. Find the distance from *P* to $\overline{HT}$.

   e. Are $\overline{AT}$ and $\overline{MA}$ equidistant from *P*?

Draw a figure and then solve each problem.

16. A regular hexagon is inscribed in a circle with a radius of 18 inches. Find the length of each side of the hexagon.

17. In ⊙*K*, chord $\overline{AT}$ is 7 units long, and chord $\overline{CR}$ is 3 units long. Which chord is closer to the center of ⊙*K*?

Applications and Problem Solving

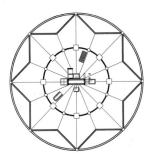

18. **Architecture** In 1457, the Italian architect Antonio Filarete designed a star-shaped city called Sforzinda. The plan for the city was constructed by inscribing two polygons within a circle. Which two polygons were used?

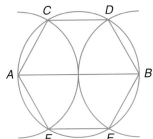

19. **Mechanical Drawing** Aisha Turner is drawing a plan for a hexagonal patio at the home of a client. She uses the method at the left to construct the hexagon. What is the one distance Ms. Turner needs to know in order to make the hexagon the correct size?

20. **Critical Thinking** To *truncate* means to change one shape into another by altering the corners. So, an octagon is a truncation of a square. Write a paragraph about the relationship between dodecagons and hexagons.

Mixed Review

21. Is it possible to determine whether $\overset{\frown}{AT} \cong \overset{\frown}{TB}$? *(Lesson 11–3)*

22. Find $m\overset{\frown}{QTS}$ if *m*∠*QRS* = 50 and $\overline{ST}$ is a diameter. *(Lesson 11–2)*

23. What is the sum of the measures of the interior angles of a dodecagon (12 sides)? *(Lesson 10–2)*

Exercises 21–22

Standardized Test Practice

Ⓐ Ⓑ Ⓒ Ⓓ

24. **Grid In** Janine's kite is a quadrilateral, as shown at the right. What is the perimeter in inches of the interior quadrilateral, assuming the quadrilaterals are similar? *(Lesson 9–7)*

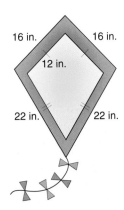

25. **Multiple Choice** On a blueprint, 1 inch represents 10 feet. Find the actual length of a room that is $2\frac{1}{4}$ inches long on the blueprint. *(Lesson 9–2)*

   Ⓐ 20 ft   Ⓑ $22\frac{1}{2}$ ft

   Ⓒ $20\frac{1}{4}$ ft   Ⓓ 25 ft

What You'll Learn
You'll learn to solve problems involving circumferences of circles.

Why It's Important
Law Enforcement
Police officers use circumference to measure skid marks at accident sites. *See Exercise 9.*

A brand of in-line skates advertises "80-mm clear wheels." The description "80-mm" refers to the diameter of the skates' wheels. As the wheels of an in-line skate complete one revolution, the distance traveled is the same as the circumference of the wheel.

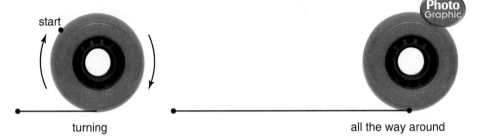

start

turning

all the way around

Just as the perimeter of a polygon is the distance around the polygon, the **circumference** of a circle is the distance around the circle. We can use a graphing calculator to find a relationship between the circumference and the diameter of a circle.

Graphing Calculator Tutorial
See pp. 782–785.

Graphing Calculator Exploration

Step 1 Use the Circle tool on the F2 menu to draw a circle.

Step 2 Use the Line tool on F2 to draw a line through the center of the circle.

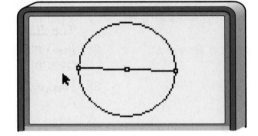

Step 3 Use the Segment tool on F2 to draw the segment connecting the two points at which the line intersects the circle.

Step 4 Use the Hide/Show tool on F5 to hide the line.

Try These

1. Use the Distance & Length tool on F5 to find the circumference of your circle and the length of the diameter.
2. Use the Calculate tool on F5 to find the ratio of the circumference to the diameter.
3. Use the drag key and the arrow keys to make your circle larger. What is the result?

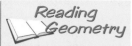
In this activity, the ratio of the circumference C of a circle to its diameter d appears to be a number slightly greater than 3, regardless of the size of the circle. The ratio of the circumference of a circle to its diameter is always fixed and equals an irrational number called **pi** or π. Thus, $\frac{C}{d} = \pi$ or $C = \pi d$. Since $d = 2r$, the relationship can also be written as $C = 2\pi r$.

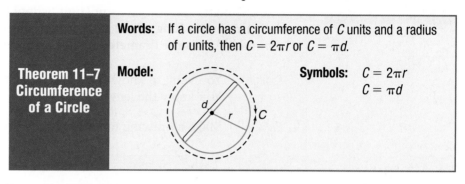

**Theorem 11–7
Circumference of a Circle**

Words: If a circle has a circumference of C units and a radius of r units, then $C = 2\pi r$ or $C = \pi d$.

Model:

Symbols: $C = 2\pi r$
$C = \pi d$

To the nearest hundredth, the irrational number π is approximated by 3.14. The *exact* circumference is a multiple of π.

Examples

1 **Find the circumference of ⊙O to the nearest tenth.**

$C = 2\pi r$ *Theorem 11–7*
$C = 2\pi(3)$ *Replace r with 3.*
$C = 6\pi$ *This is the exact circumference.*

3 cm

O

To estimate the circumference, use a calculator.

Enter: 6 ⊠ 2nd [π] ENTER *18.84955592*

The circumference is about 18.8 centimeters.

2 **The diameters of a penny, nickel, and quarter are 19.05 millimeters, 21.21 millimeters, and 24.26 millimeters, respectively. Find the circumference of each coin to the nearest millimeter.**

Penny	**Nickel**	**Quarter**
$C = \pi d$	$C = \pi d$	$C = \pi d$
$C = \pi(19.05)$	$C = \pi(21.21)$	$C = \pi(24.26)$
$C \approx 59.84734005$	$C \approx 66.63318018$	$C \approx 76.21503778$

The circumferences are about 60 millimeters, 67 millimeters, and 76 millimeters, respectively.

Your Turn

a. The circumference of a half dollar is about 96 millimeters. Find the diameter of the coin to the nearest tenth.

b. A circular flower garden has a circumference of 20 feet. Find the radius of the garden to the nearest hundredth.

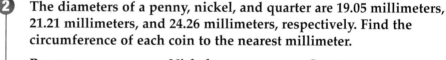 **Personal Tutor at geomconcepts.com**

You can use the formula for circumference to solve problems involving figures that have circular arcs.

Example ③

Sports Link

The 400-meter track at Jackson High School has two straightaways, each 100 meters long, and 2 semicircular ends, each 100 meters around. What is the diameter of each semicircle?

Explore   You want to know the diameter of each semicircle. You know the length of the semicircular ends.

Plan   Make a drawing to represent the problem.

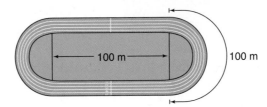

Solve   The two ends together form an entire circle. Since the circumference of a circle is given by $C = \pi d$, the length S of a semicircular end can be represented by $S = \frac{\pi d}{2}$.

$$S = \frac{\pi d}{2} \qquad \textit{Original formula}$$

$$100 = \frac{\pi d}{2} \qquad \textit{Replace S with 100.}$$

$$2 \cdot 100 = 2 \cdot \frac{\pi d}{2} \qquad \textit{Multiply each side by 2.}$$

$$200 = \pi d \qquad \textit{Simplify.}$$

$$\frac{200}{\pi} = \frac{\pi d}{\pi} \qquad \textit{Divide each side by } \pi.$$

$$63.66197724 \approx d \qquad \textit{Use a calculator.}$$

The diameter of each semicircle is about 64 meters.

Examine   Replace d with 64 in the formula to check the solution.

Enter: 64 ⨯ 2nd [π] ÷ 2 ENTER   *100.5309649* ✓

Check for Understanding

Communicating Mathematics

1. **Explain** why $C = 2\pi r$ and $C = \pi d$ are equivalent formulas.

2. **Determine** the exact circumference of the 80-mm diameter wheels described in the lesson.

Vocabulary
circumference
pi (π)

Guided Practice

Complete each chart. Round to the nearest tenth.

Examples 1 & 2

	r	d	C
3.		8 m	
5.			84.8 ft
7.	$5\frac{3}{4}$ yd		

	r	d	C
4.	2.4 km		
6.			32 cm
8.		3 in.	

Example 3

9. **Law Enforcement** Police officers use a *trundle wheel* to measure skid marks when investigating accidents. The diameter of the wheel is 24 centimeters. What is the distance measured when the wheel makes one complete revolution? Round to the nearest tenth.

Exercises

Practice

Find the circumference of each object to the nearest tenth.

10. dime

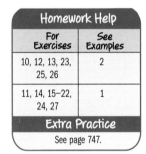

$d = 17.91$ mm

11. top of a can

$r = 3\frac{1}{4}$ in.

12. bicycle tire

$d = 1.7$ m

Homework Help	
For Exercises	**See Examples**
10, 12, 13, 23, 25, 26	2
11, 14, 15–22, 24, 27	1
Extra Practice	
See page 747.	

Find the circumference of each circle described to the nearest tenth.

13. $d = 4$ mm

14. $r = 6\frac{1}{2}$ ft

15. $r = 17$ yd

Find the radius of the circle to the nearest tenth for each circumference given.

16. 47.1 cm

17. 6.3 in.

18. 18 km

19. The circumference of the top of a tree stump is 8 feet. Find its radius.

20. If the radius of a circle is tripled, how does the circumference change?

Find the circumference of each circle to the nearest hundredth.

21.

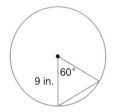

9 in. 60°

22.

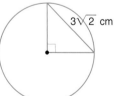

$3\sqrt{2}$ cm

23.

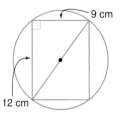

9 cm
12 cm

Applications and Problem Solving

24. **Electronics** Auto speakers are available with 2-inch and $2\frac{5}{8}$-inch radii. What are the circumferences of the two types of speakers to the nearest tenth?

25. **Bicycling** If the wheels of a bicycle have 24-inch diameters, about how many feet will the bicycle travel when the front wheel makes 200 revolutions?

26. **Gardening** A circular flower bed has a diameter of 5 meters. Different colors of tulip bulbs are to be planted in five equally-spaced concentric circles in the bed. The bulbs will be planted 20 centimeters apart.

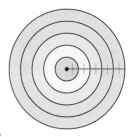

   a. What is the diameter of each circle?

   b. What is the circumference of each circle?

   c. How many bulbs will be needed for each circle?

27. **Critical Thinking** Arcs have degree measure, and they also have length. The length of an arc of a circle is a fractional part of the circumference. Find the length of the arc shown.

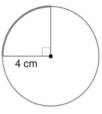

4 cm

Mixed Review

28. **Paper Folding** Draw a circle with a 4-inch radius and cut it out. Fold the circle in half. Fold it in half again twice, creasing the edges. Unfold the circle and draw a chord between each pair of adjacent endpoints created by the folds. What figure have you just drawn? *(Lesson 11–4)*

29. **Food** The grill on Mr. Williams' barbecue is circular with a diameter of 54 centimeters. The horizontal wires are supported by two wires that are 12 centimeters apart, as shown in the figure at the right. If the grill is symmetrical and the wires are evenly spaced, what is the length of each support wire? Round to the nearest hundredth. *(Lesson 11–3)*

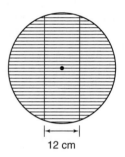

12 cm

30. What is the area of a trapezoid with bases of lengths 36 centimeters and 27 centimeters and a height of 18 centimeters? *(Lesson 10–4)*

Standardized Test Practice

(A) (B) (C) (D)

31. **Grid In** The measures of four interior angles of a pentagon are 110, 114, 99, and 107. Find the measure of the fifth interior angle. *(Lesson 10–2)*

32. **Multiple Choice** What is the value of x for the pair of congruent triangles? *(Lesson 5–4)*

   (A) 45   (B) 23

   (C) 11   (D) 6

$5x + 17$ $4x - 6$ $3x + 5$ 46 $3x + 39$ 46

Quiz 2   Lessons 11–3 through 11–5

1. Use $\odot K$ to complete the statement $\overarc{PQ} \cong$ ___?___ . *(Lesson 11–3)*

2. In $\odot K$, find x if $PQ = 3x - 5$ and $QR = 2x + 4$. *(Lesson 11–4)*

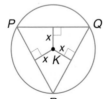

3. In $\odot Q$, $QC = 13$ and $QT = 5$. Find CB. *(Lesson 11–3)*

4. Find the circumference of $\odot Q$ to the nearest tenth. *(Lesson 11–5)*

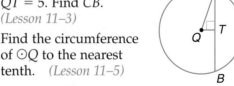

5. Find the radius of a circle whose circumference is 144.5 units. *(Lesson 11–5)*

www.geomconcepts.com/self_check_quiz

11-6 Area of a Circle

What You'll Learn
You'll learn to solve problems involving areas and sectors of circles.

Why It's Important
Biology Biologists use sectors to construct circle graphs.
See Exercise 26.

The space enclosed inside a circle is its area. By slicing a circle into equal pie-shaped pieces as shown below, you can rearrange the pieces into an approximate rectangle. Note that the length along the top and bottom of this rectangle equals the circumference of the circle, $2\pi r$. So, each "length" of this approximate rectangle is half the circumference, or πr.

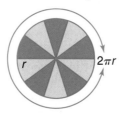
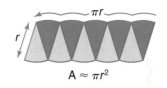

The "width" of this approximate rectangle is the radius r of the circle. Recall that the area of a rectangle is the product of its length and width. Therefore, the area of this approximate rectangle is $(\pi r)r$, or πr^2.

	Words:	If a circle has an area of A square units and a radius of r units, then $A = \pi r^2$.
Theorem 11–8 **Area of a Circle**	**Model:**	**Symbols:** $A = \pi r^2$

Example

1 Find the area of $\odot P$ to the nearest hundredth.

6.3 cm

$$A = \pi r^2 \qquad \textit{Theorem 11–8}$$
$$= \pi(6.3)^2 \quad \textit{Replace r with 6.3.}$$
$$= 39.69\pi \quad \textit{This is the exact area.}$$

To estimate the area, use a calculator.

Enter: 39.69 [×] [2nd] [π] [ENTER] *124.6898124*

The area is about 124.69 square centimeters.

Reading Geometry
Recall that area is always expressed in square units.

Your Turn

a. Find the area of $\odot C$ to the nearest hundredth if $d = 5$ inches.

 Online Personal Tutor at geomconcepts.com

You can use Theorem 11–8 to find the area of a circle if you know the circumference of the circle.

Example ➋ If ⊙A has a circumference of 10π inches, find the area of the circle to the nearest hundredth.

Use the circumference formula to find r.

$C = 2\pi r$ *Theorem 11–7*

$10\pi = 2\pi r$ *Replace C with 10π.*

$\dfrac{10\pi}{2\pi} = \dfrac{2\pi r}{2\pi}$ *Divide each side by 2π.*

$5 = r$ *Simplify.*

Now find the area of the circle.

$A = \pi r^2$ *Theorem 11–8*

$= \pi(5)^2$ *Replace r with 5.*

$= 25\pi$ $5^2 = 25$

≈ 78.54 *Use a calculator.*

To the nearest hundredth, the area is 78.54 square inches.

Your Turn

b. Find the area of the circle whose circumference is 6.28 meters. Round to the nearest hundredth.

You can use the area of a circle to solve problems involving probability.

Example ➌

Probability Link

To win a dart game at a carnival, the dart must land in the red section of the square board. What is the probability that a dart thrown onto the square at random will land in the red section? Assume that all darts thrown will land on the dartboard.

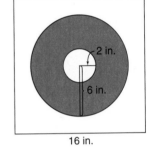

2 in.

6 in.

16 in.

To find the probability of landing in the red section, first subtract the area of the white circle from the area of the large circle.

area of red section = *area of large circle* − *area of white circle*

$$A = \pi r^2 - \pi r^2$$
$$= \pi(6^2) - \pi(2^2)$$
$$= 36\pi - 4\pi \text{ or } 32\pi$$

Use a calculator.   32 ☒ 2nd [π] ☰ *100.5309649*

The area of the red section is about 101 square inches. The area of the board is 16^2 or 256 square inches. So, find the probability as follows.

$P(\text{landing in the red section}) = \dfrac{\text{area of the red section}}{\text{area of the board}}$ or about $\dfrac{101}{256}$

The probability of landing in the red section is about $\dfrac{101}{256}$ or 0.395.

In Example 3, we used **theoretical probability**. The solution was based on the formulas for the areas of a circle and a square. This is different from **experimental probability**, in which the probability is calculated by repeating some action. To find the experimental probability for this situation, you would have to throw darts at a board like the one described above many times and record how many times the red section was hit and how many times it was not.

www.geomconcepts.com/extra_examples

Consider the circle at the right. The radius of
⊙C is 14 centimeters, and central angle ACB has
a measure of 90. The shaded region is called a
sector of the circle. A sector of a circle is a region
bounded by a central angle and its corresponding
arc. The sector shown is a 90° sector.

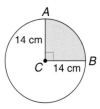

Since the sum of the measures of the central angles of a circle is 360, the
arc of the sector in ⊙C represents $\frac{90}{360}$ or $\frac{1}{4}$ of the circle. Therefore, the area
of the sector is $\frac{1}{4}$ the area of the circle.

Area of ⊙C = πr^2

$\qquad = \pi(14)^2$

$\qquad = 196\pi$ cm^2

Area of sector bounded
by $\angle ACB = \frac{1}{4}$(area of ⊙C)

$\qquad = \frac{1}{4}(196\pi)$

$\qquad = 49\pi$ cm^2

Theorem 11–9 **Area of a Sector** **of a Circle**	If a sector of a circle has an area of A square units, a central angle measurement of N degrees, and a radius of r units, then $A = \frac{N}{360}(\pi r^2)$.

Example ④ **Find the area of the shaded region in
⊙P to the nearest hundredth.**

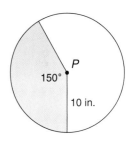

$A = \frac{N}{360}(\pi r^2)$ *Theorem 11–9*

$\quad = \frac{150}{360}[\pi(10)^2]$ *Substitution*

$\quad = \frac{150}{360}(100\pi)$ *$10^2 = 100$*

Enter: 150 ÷ 360 × 100 × 2nd [π] ENTER *130.8996939*

The area of the shaded region in ⊙P is 130.90 square inches.

Your Turn

c. Find the area of a 72° sector if the radius of the circle is $7\frac{1}{3}$ feet.
Round to the nearest hundredth.

Check for Understanding

**Communicating
Mathematics**

1. **Show** that Theorem 11–9 verifies the area
of the sector for ⊙C at the top of this page.

2. Writing Math Write a convincing argument
about whether the circumference and area of
a circle could have the same numeric value.

Vocabulary
sector experimental probability theoretical probability

Complete the chart. Round to the nearest hundredth.

	r	d	C	A
3.	2.75 m			
4.		14.14 cm		
5.			21.77 in.	

Example 4

6. In a circle with a radius of 9 inches, find the area of the sector whose central angle measures 90. Round to the nearest hundredth.

Example 3

7. **Probability** Assume that all darts thrown will will land on the dartboard at the right. Find the probability that a randomly-thrown dart will land in the red region.

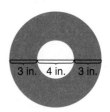

3 in. 4 in. 3 in.

Exercises • • • • • • • • • • • • • • • • • •

Find the area of each circle described to the nearest hundredth.

8. $r = 4$ mm

9. $r = 2\frac{1}{2}$ ft

10. $r = 6.7$ cm

11. $d = 15$ mi

12. $d = 1\frac{2}{3}$ in.

13. $d = 4.29$ m

14. $C = 81.68$ cm

15. $C = 37.07$ m

16. $C = 14\frac{3}{4}$ ft

17. Find the area of a circle whose diameter is 18 centimeters. Round to the nearest hundredth.

18. What is the radius to the nearest hundredth of a circle whose area is 719 square feet?

In a circle with radius of 6 centimeters, find the area of a sector whose central angle has the following measure.

19. 20

20. 90

21. 120

22. Find the area to the nearest hundredth of a 10° sector in a circle with diameter 12 centimeters.

23. The area of a 60° sector of a circle is 31.41 square meters. Find the radius of the circle.

Assume that all darts thrown will land on a dartboard. Find the probability that a randomly-thrown dart will land in the red region. Round to the nearest hundredth.

24.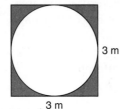

3 m

3 m

25.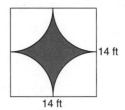

14 ft

14 ft

Homework Help	
For Exercises	See Examples
8–18	1, 2
19–23, 26	4
24, 28	3
25	3, 4
Extra Practice	
See page 747.	

26. Biology About 1.5 million species of animals have been named thus far. The circle graph shows how the various groups of named animals compare. If the sector representing insects is a 240° sector and the radius is $\frac{3}{4}$ inch, what is the area of that sector?

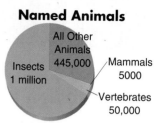

Named Animals

All Other Animals 445,000

Insects 1 million

Mammals 5000

Vertebrates 50,000

Source: *National Geographic World*

27. Cooking When Julio bakes a pie, he likes to put foil around the edges of the crust to keep it from getting too brown. He starts with a 12-inch square of foil, folds it in fourths, and tears out a sector with a radius of 4 inches. Then he places it over the pie. What is the area of the remaining piece of foil to the nearest hundredth?

28. Critical Thinking Refer to the circle at the right.

   a. Find the area of the shaded region to the nearest hundredth if r = 2, 3, 4, 5, 6, and 8 inches.

   b. What is the relationship between the area of the shaded region and the area of the large circle?

   c. Probability Suppose this figure represents a dartboard. What is the probability that a randomly-thrown dart will land in the yellow region?

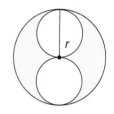

29. Animals Taylor is building a circular dog pen for her new puppy. If the diameter of the pen will be 12 meters, about how many meters of fencing will Taylor need to purchase? *(Lesson 11–5)*

30. In $\odot P$, if $JK = 3x - 4$ and $LM = 2x + 9$, find x. *(Lesson 11–4)*

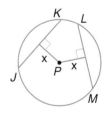

Exercise 30

31. Construction Jonathan Werner is building a a deck shaped like a regular octagon with an apothem 7.5 feet long and sides each 6.2 feet long. If wood for the deck floor costs $1.75 per square foot, how much will it cost Mr. Werner to install the deck floor? *(Lesson 10–5)*

32. Short Response Use the figure to complete the following statement. *(Lesson 9–6)*

$$\frac{AC}{BC} = \frac{?}{ED}$$

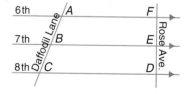

33. Multiple Choice Which angle forms a linear pair with $\angle SVT$? *(Lesson 3–4)*

   Ⓐ $\angle RVS$ Ⓑ $\angle PVQ$

   Ⓒ $\angle RVQ$ Ⓓ $\angle PVT$

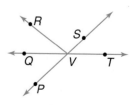

Understanding and Using the Vocabulary

Review Activities

For more review activities, visit:
www.geomconcepts.com

After completing this chapter, you should be able to define each term, property, or phrase and give an example or two of each.

adjacent arcs *(p. 463)*	circumscribed *(p. 474)*	major arc *(p. 462)*
arcs *(p. 462)*	concentric *(p. 456)*	minor arc *(p. 462)*
center *(p. 454)*	diameter *(p. 454)*	pi *(p. 479)*
central angle *(p. 462)*	experimental probability *(p. 484)*	radius *(p. 454)*
chord *(p. 454)*	inscribed *(p. 474)*	sector *(p. 485)*
circle *(p. 454)*	loci *(p. 460)*	semicircle *(p. 462)*
circumference *(p. 478)*	locus *(p. 460)*	theoretical probability *(p. 484)*

State whether each sentence is *true* or *false*. If false, replace the underlined word(s) to make a true statement.

1. A <u>central angle</u> is an angle whose vertex is at the center of the circle and whose sides intersect the circle.
2. A chord that contains the center of a circle is called a <u>radius</u>.
3. The <u>diameter</u> of a circle is the distance around the circle.
4. <u>Pi</u> is the ratio of the circumference of a circle to its diameter.
5. A <u>sector</u> is a segment whose endpoints are the center of the circle and a point on the circle.
6. A <u>chord</u> is a segment whose endpoints are on the circle.
7. A <u>semicircle</u> is an arc whose endpoints are on the diameter of a circle.
8. A sector of a circle is a region bounded by a central angle and its corresponding <u>center</u>.
9. A <u>major arc</u> is the part of the circle in the exterior of the central angle.
10. Two circles are <u>inscribed</u> if they lie in the same plane, have the same center, and have radii of different lengths.

Skills and Concepts

Objectives and Examples	**Review Exercises**
• **Lesson 11–1** Identify and use parts of circles.	**Refer to the circle at the left to complete each statement in Exercises 11–13.**
$\overline{LM}$ is a radius of $\odot M$. If $LM = 16$, find JK. 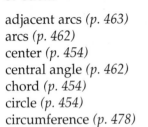	**11.** $\overline{KN}$ is a ___?___ of $\odot M$. **12.** A diameter of $\odot M$ is ___?___ . **13.** $\overline{JM}$, $\overline{KM}$, and $\overline{LM}$ are ___?___ of $\odot M$.
$d = 2r$ $JK = 2(LM)$ *Substitution* $JK = 2(16)$ or 32 *Substitution*	**14.** Find the measure of diameter $\overline{AC}$ if $BP = x$ and $AC = 5x - 6$.

 www.geomconcepts.com/vocabulary_review

Objectives and Examples

- **Lesson 11–2** Identify major arcs, minor arcs, and semicircles and find the measures of arcs and central angles.

In $\odot S$, find $m\widehat{UV}$ and $m\widehat{TUV}$.

$m\widehat{UV} = m\angle USV$
$\quad = 65$

$m\widehat{TUV} = 360 - m\angle VST$
$\quad = 360 - 115 \text{ or } 245$

Review Exercises

Find each measure in $\odot T$ if $\overline{PQ}$ is a diameter.

15. $m\widehat{PQS}$
16. $m\angle QTS$
17. $m\widehat{PS}$

B is the center of two circles with radii $\overline{BC}$ and $\overline{BD}$. If $m\angle DBJ = 113$ and $\overline{KL}$ and $\overline{JM}$ are diameters, find each measure.

18. $m\widehat{CK}$ 19. $m\widehat{DMJ}$
20. $m\widehat{CL}$

- **Lesson 11–3** Identify and use the relationships among arcs, chords, and diameters.

In $\odot Q$, if $\overline{QU} \perp \overline{RT}$ and $RT = 18$, find RS.

$RT = 2(RS)$
$18 = 2(RS)$ *Substitution*
$\dfrac{18}{2} = \dfrac{2(RS)}{2}$ *Divide each side by 2.*
$9 = RS$

Use $\odot X$ to complete each statement.

21. If $AC = 12$, then $CD = $ ___?___ .
22. If $DX = 18$ and $AC = 48$, then $CX = $ ___?___ .

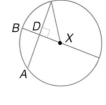

23. In $\odot L$, $\overline{MQ} \cong \overline{NQ}$. If $MN = 16$ and $LN = 10$, find LQ.

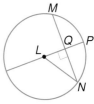

- **Lesson 11–4** Inscribe regular polygons in circles and explore the relationship between the length of a chord and its distance from the center of the circle.

In $\odot P$, $RT = 52$. Find XY.

$RT = XZ$ *Theorem 11–6*
$RT = 2(XY)$ *Theorem 11–5*
$52 = 2(XY)$ *Substitution*
$\dfrac{52}{2} = \dfrac{2(XY)}{2}$ *Divide each side by 2.*
$26 = XY$

24. In $\odot C$, $TU = 12x - 7$ and $TV = 3x + 20$. Find x.

Use $\odot C$ to determine whether each statement is *true* or *false*.

25. $\overline{AC} \cong \overline{BC}$
26. $\overline{AU} \cong \overline{BV}$
27. $\angle ATB \cong \angle ACB$
28. $\overline{BT} \cong \overline{BV}$

Mixed Problem Solving
See pages 758-765.

Objectives and Examples

- **Lesson 11–5** Solve problems involving circumferences of circles.

Find the circumference of ⊙Q to the nearest hundredth.

5 m
Q

$C = 2\pi r$ *Theorem 11–7*
$C = 2\pi(5)$ *Replace r with 5.*
$C = 10\pi$ or about 31.42

The circumference is about 31.42 meters.

- **Lesson 11–6** Solve problems involving areas and sectors of circles.

Find the area of ⊙T to the nearest tenth.

12.9 yd
T

$A = \pi r^2$
$A = \pi(12.9)^2$
$A = 166.41\pi$ or about 522.8

The area is about 522.8 square yards.

Review Exercises

Find the circumference of each circle described to the nearest tenth.

29. $d = 32$ ft
30. $d = 7$ km
31. $r = 22$ in.

Find the radius of the circle to the nearest tenth for each circumference given.

32. 46 yd
33. 17.3 cm
34. 325 m

Find the area of each circle described to the nearest hundredth.

35. $d = 50$ in.
36. $C = 33.3$ m
37. $r = 15.6$ ft

38. Find the area of the shaded region in ⊙M to the nearest hundredth.

M
12 yd
124°

Applications and Problem Solving

39. Cycling Suppose a bicycle wheel has 30 spokes evenly spaced and numbered consecutively from 1 through 30. Find the measure of the central angle formed by spokes 1 and 14. *(Lesson 11–2)*

40. Swimming Swimmers often use kickboards when they want to concentrate on their kicking. Find the width of the kickboard shown below. *(Lesson 11–5)*

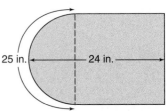
25 in.   24 in.

41. Food At a school function, a fruit pizza was served that had a diameter of 14 inches and was cut into 10 equal-sized wedges. By the end of the evening, 7 consecutive wedges had been eaten. Find the area of the remaining pizza to the nearest square inch. *(Lesson 11–6)*

1. **Define** *inscribed polygon* and give an example of one in everyday life.

2. **Explain** how the perimeter of a polygon and the circumference of a circle are related.

Use ⊙L to complete each statement.

3. $\overline{MP}$ is a ___?___ of ⊙L.

4. If $MN = 16$, then $LN = $ ___?___ .

5. All radii of a circle are ___?___ .

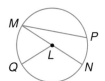

Exercises 3–8

Find each measure in ⊙L if m∠NLQ = 79.

6. $m\widehat{NQ}$

7. $m\widehat{MNQ}$

8. $m\angle MLQ$

Use ⊙W to complete each statement.

9. If $XZ = 18$, then $YZ = $ ___?___ .

10. If $WY = 12$ and $XY = 16$, then $XW = $ ___?___ .

11. If $XZ = 16$ and $WY = 6$, then $WZ = $ ___?___ .

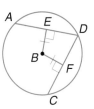

Exercise 12

12. In ⊙B, $CD = 62$. Find AE.

Find the circumference of each circle described to the nearest tenth.

13. $r = 14.3$ in.

14. $d = 33$ m

15. $r = 27$ ft

Find the area of each circle to the nearest hundredth.

16.

7.6 yd

17.

53 cm

18.

1.4 km

19. **Algebra** In ⊙S, $JL = 12x + 2$ and $MP = 3x + 20$. Find x.

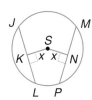

20. **Gardening** Eloy is preparing to plant flowers in his circular garden. The garden has a diameter of 4 meters and is divided into six equal portions. If he intends to fill the shaded portions of the garden with marigolds, what is the area that will remain for other types of flowers?

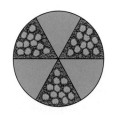

Function Problems

Standardized tests include many function problems. You'll look for patterns and extend sequences, interpret various representations of functions, create graphs, and write equations.

You'll need to understand these concepts.

equation of a line   function   graph of a line
negative reciprocal   sequence   slope
table of values

Example 1

The graph of a line is shown at the right. What is the equation of the line?

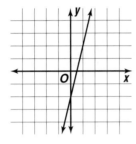

A $y = -\frac{1}{4}x - 2$ **B** $y = \frac{1}{4}x - 2$

C $y = 4x - 2$ **D** $y = -4x - 2$

Hint   Look carefully at the answer choices.

Solution   Locate two points on the graph. The y-intercept is at $(0, -2)$. Another point is at $(1, 2)$. Find the slope of the line, using these two points. Notice that the x-coordinate increases by 1 and the y-coordinate increases by 4. Therefore, the slope is $\frac{4}{1}$ or 4. Choose the equation with a slope of 4. The answer is C.

Check your answer by replacing x and y with the coordinates of the point $(0, -2)$ and see if the statement is true. Do the same for $(1, 2)$.

$y = 4x - 2$ $y = 4x - 2$

$-2 \stackrel{?}{=} 4(0) - 2$ $2 \stackrel{?}{=} 4(1) - 2$

$-2 = -2$ ✓ $2 = 2$ ✓

Example 2

If $x \otimes y = \frac{1}{x-y}$, what is the value of $\frac{1}{2} \otimes \frac{1}{3}$?

A   6   **B** $\frac{6}{5}$ **C** $\frac{1}{6}$ **D** -1 **E** -6

Hint   The SAT contains unique function problems. These problems use special symbols, like $\oplus$, #, or $\otimes$, to represent a function. For example, if $x \# y = x + 2y$, then what is $2 \# 5$? Just use the 2 in place of x and the 5 in place of y. So, $2 \# 5 = 2 + 2(5)$ or 12.

Solution   Substitute $\frac{1}{2}$ for x and $\frac{1}{3}$ for y in the expression. Then simplify the fraction.

$$\frac{1}{2} \otimes \frac{1}{3} = \frac{1}{\frac{1}{2} - \frac{1}{3}} \qquad \textit{Substitution}$$

$$= \frac{1}{\frac{3}{6} - \frac{2}{6}} \qquad \textit{The LCD is 6.}$$

$$= \frac{1}{\frac{1}{6}} \qquad \textit{Subtraction}$$

$$= 6 \qquad 1 \div \frac{1}{6} = 1 \times 6$$

The answer is A.

After you work each problem, record your answer on the answer sheet provided or on a sheet of paper.

Multiple Choice

1. The table shows how many sit-ups Luis can do after a number of weeks. How many sit-ups will he be able to do after 7 weeks? *(Algebra Review)*

Week	Number of Sit-ups
1	4
2	10
3	16
4	22

Ⓐ 28
Ⓑ 34
Ⓒ 40
Ⓓ 42

2. In the triangle at the right, what is $m\angle A$? *(Lesson 5–2)*

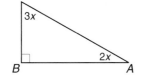

Ⓐ 9   Ⓑ 18
Ⓒ 36   Ⓓ 54

3. Which equation best describes the graph? *(Lesson 4–6)*

Ⓐ $y - 1 = 3(x - 4)$
Ⓑ $y - 4 = 3(x - 1)$
Ⓒ $y = -3x + 1$
Ⓓ $y = \frac{1}{3}x + 1$

4. For all integers Z, suppose $\triangle{Z} = Z^2$ (if Z is odd), and $\triangle{Z} = \sqrt{Z}$ (if Z is even). What is the value of $\triangle{36} + \triangle{9}$? *(Algebra Review)*

Ⓐ 9   Ⓑ 15   Ⓒ 39
Ⓓ 45   Ⓔ 87

5. A plumber charges $75 for the first 30 minutes of each house call plus $2 for each additional minute. She charged Mr. Adams $113. For how long did the plumber work? *(Algebra Review)*

Ⓐ 38 min   Ⓑ 44 min   Ⓒ 49 min
Ⓓ 59 min   Ⓔ 64 min

6. Which equation represents the relationship shown in the table? *(Algebra Review)*

Ⓐ $y = 3x - 2$
Ⓑ $y = 2x + 1$
Ⓒ $y = 2x^2 - x$
Ⓓ $y = x^2 - 2$

x	1	2	3	4
y	1	4	7	10

7. What is $m\angle PQR$? *(Lesson 7–2)*

Ⓐ 40°
Ⓑ 55°
Ⓒ 70°
Ⓓ 110°

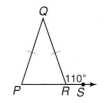

8. If 10% of x is 20% of 100, what is x? *(Percent Review)*

Ⓐ 200   Ⓑ 100   Ⓒ 20   Ⓓ 5

Grid In

9. What number should come next in this sequence:
$\frac{17}{4}, \frac{31}{8}, \frac{7}{2}, \frac{25}{8}, \dots$? *(Algebra Review)*

Extended Response

10.

Dry Pavement Stopping Distance	
Speed (mph)	**Distance (ft)**
55	289
60	332
65	378
70	426

Part A Graph the ordered pairs (speed, distance) on a coordinate plane. *(Algebra Review)*

Part B Describe how stopping distance is related to speed.

CHAPTER 12

Surface Area and Volume

What You'll Learn

Key Ideas

- Identify solid figures.
  (Lesson 12–1)
- Find the lateral areas, surface areas, and volumes of prisms, cylinders, regular pyramids, and cones.
  (Lessons 12–2 to 12–5)
- Find the surface areas and volumes of spheres.
  (Lesson 12–6)
- Identify and use the relationships between similar solid figures.
  (Lesson 12–7)

Key Vocabulary

cone *(p. 497)*

cylinder *(p. 497)*

prism *(p. 497)*

pyramid *(p. 497)*

sphere *(p. 528)*

Why It's Important

Horticulture A greenhouse is an enclosed glass house used for growing plants in regulated temperatures, humidity, and ventilation. Greenhouses range from small rooms to large heated buildings covering many acres. Millions of dollars worth of plant products are raised in greenhouses each year.

Surface area and volume are important concepts used in fields like architecture, automotive design, and interior design. You will find the surface area of a prism to determine the amount of glass for a greenhouse in Lesson 12–2.

✓ Check Your Readiness

Find the area and perimeter of each rectangle described.

1. $\ell = 12$ m, $w = 4$ m

2. $\ell = 23$ in., $w = 8$ in.

3. $\ell = 13$ ft, $w = 40$ ft

4. $\ell = 6.2$ cm, $w = 1.8$ cm

Find the area of each triangle or trapezoid.

5.
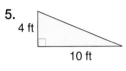
4 ft
10 ft

6.

12 cm
5.6 cm

7.
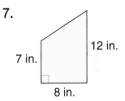
12 in.
7 in.
8 in.

Find the area of each regular polygon.

8.

5.5 cm
3.8 cm

9.

13 m
15 m

10.

8 in.
9.2 in.

Find the circumference and area of each circle described to the nearest hundredth.

11. $r = 4$ m

12. $d = 10$ in.

13. $d = 6.4$ cm

14. $r = 8\frac{1}{2}$ ft

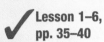

FOLDABLES™
Study Organizer

Make this Foldable to help you organize your Chapter 12 notes. Begin with a sheet of 11" by 17" paper.

❶ Fold in thirds lengthwise.

❷ Open and fold a 2" tab along the short side. Then fold the rest in fifths.

❸ Draw lines along the folds and label as shown.

Reading and Writing As you read and study the chapter, use each page to write definitions and theorems and draw models. Also include any questions that need clarifying.

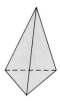

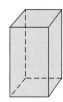

What You'll Learn
You'll learn to identify solid figures.

Why It's Important
Construction
Architects use various solid shapes to design buildings.
See Example 2.

Study the figures below. How are these figures alike? How do they differ?

All of the above figures are examples of **solid figures** or *solids*. In geometry, solids enclose a part of space. Solids with flat surfaces that are polygons are called **polyhedrons** or *polyhedra*. Which of the above figures are polyhedrons?

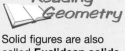

Reading Geometry

Solid figures are also called **Euclidean solids**.

Info Graphic

Parts of a Polyhedron

The 2-dimensional surfaces called **faces** are polygons and their interiors. △*ABC*, △*ADC*, △*ABD*, and △*BCD* are the faces of the polyhedron.

Two faces intersect in a segment called an **edge**. $\overline{CD}$, $\overline{CA}$, $\overline{CB}$, $\overline{AB}$, $\overline{AD}$, and $\overline{BD}$ are the edges of the polyhedron.

Three or more edges intersect at a point called a *vertex*. *A*, *B*, *C*, and *D* are the vertices of the polyhedron.

Example 1

Name the faces, edges, and vertices of the polyhedron.

The faces are quadrilaterals *EFKJ*, *FGLK*, *GHML*, *HINM*, and *IEJN* and pentagons *EFGHI* and *JKLMN*. The edges are $\overline{EJ}$, $\overline{FK}$, $\overline{GL}$, $\overline{HM}$, $\overline{IN}$, $\overline{EF}$, $\overline{FG}$, $\overline{GH}$, $\overline{HI}$, $\overline{IE}$, $\overline{JK}$, $\overline{KL}$, $\overline{LM}$, $\overline{MN}$, and $\overline{NJ}$.
The vertices are *E*, *F*, *G*, *H*, *I*, *J*, *K*, *L*, *M*, and *N*.

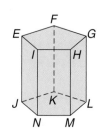

Your Turn

a. Name the faces, edges, and vertices of the polyhedron.

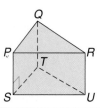

Prisms and **pyramids** are two types of polyhedrons.

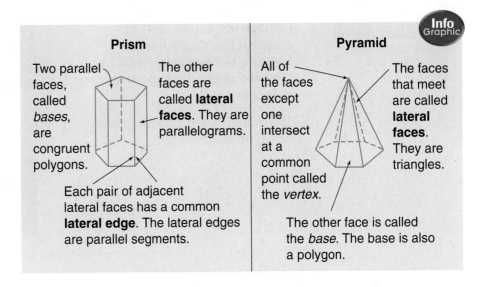

Prism

Two parallel faces, called *bases*, are congruent polygons.

The other faces are called **lateral faces**. They are parallelograms.

Each pair of adjacent lateral faces has a common **lateral edge**. The lateral edges are parallel segments.

Pyramid

All of the faces except one intersect at a common point called the *vertex*.

The faces that meet are called **lateral faces**. They are triangles.

The other face is called the *base*. The base is also a polygon.

Both prisms and pyramids are classified by the shape of their bases.

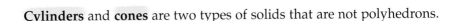

triangular prism   rectangular prism   hexagonal prism   rectangular pyramid   pentagonal pyramid

A **cube** is a special rectangular prism in which all of the faces are squares. A cube is one of the **Platonic solids,** which include all regular polyhedrons.

A **tetrahedron** is another name for a triangular pyramid. All of its faces are triangles.

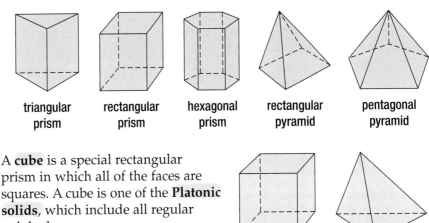

cube   tetrahedron

Cylinders and **cones** are two types of solids that are not polyhedrons.

Reading Geometry

In this text, circular cylinders and circular cones will be referred to as cylinders and cones.

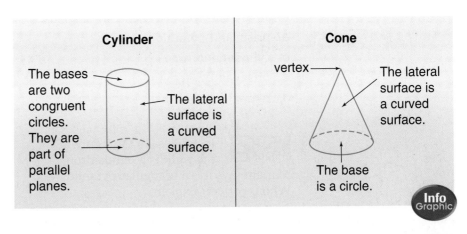

Cylinder

The bases are two congruent circles. They are part of parallel planes.

The lateral surface is a curved surface.

Cone

vertex

The lateral surface is a curved surface.

The base is a circle.

Example

Architecture Link

② Eero Saarinen designed the chapel at the Massachusetts Institute of Technology (M.I.T.). Describe the basic shape of the chapel as a solid.

The chapel has two circular bases. It resembles a cylinder.

A **composite solid** is a solid that is made by combining two or more solids. The solid shown in the figure is a combination of a cylinder and a cone.

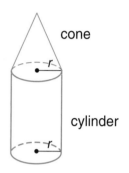

cone

cylinder

Check for Understanding

Communicating Mathematics

1. **Draw** an example of each solid.
   a. cylinder
   b. pentagonal prism
   c. cone
   d. rectangular pyramid

2. **Compare and contrast** each pair of solids.
   a. prism and pyramid
   b. cylinder and cone
   c. prism and cylinder
   d. pyramid and cone

> **Vocabulary**
>
> solid figures
> polyhedrons
> faces
> edge
> prisms
> pyramids
> lateral faces
> lateral edge
> cube
> tetrahedron
> cylinders
> cones
> composite solid
> Platonic solid

3. **You Decide?** Jasmine says the figure at the right is a tetrahedron. Mariam says it is a triangular pyramid. Who is correct? Explain.

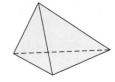

⊖ Getting Ready Name the polygons that form the lateral faces of each polyhedron.

Sample 1: pentagonal pyramid
Solution: triangles

Sample 2: triangular prism
Solution: rectangles

4. triangular pyramid 5. cube 6. hexagonal prism
7. rectangular pyramid 8. pentagonal prism 9. tetrahedron

Example 1

10. Name the faces, edges, and vertices of the polyhedron.

Example 2

Describe the basic shape of each item as a solid figure.

11.

12.

Example 2

13. **Art** David Smith creates geometric sculptures. What geometric solids did Mr. Smith use for this sculpture?

David Smith, *Cubi IX*

Exercises ● ● ● ● ● ● ● ● ● ● ● ● ● ● ● ● ●

Practice

Name the faces, edges, and vertices of each polyhedron.

14.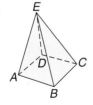

15.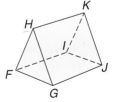

16.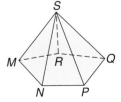

Homework Help	
For Exercises	See Examples
14–16	1
17–22, 30, 31	2
Extra Practice	
See page 748.	

Describe the basic shape of each item as a solid.

17.

18.

19.

Describe the basic shape of each item as a solid.

20.

21.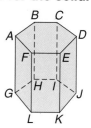

22.

Determine whether each statement is *true* or *false* for the solid.

23. The figure is a pyramid.
24. The figure is a polyhedron.
25. Hexagon *ABCDEF* is a base.
26. Hexagon *GHIJKL* is a lateral face.
27. There are six lateral faces.
28. The figure has 12 lateral edges.
29. *BHIC* is a lateral face.

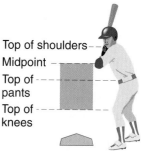

Applications and Problem Solving

30. **Sports**   In baseball, the ball is in the strike zone if the following criteria are met.

    - It is over the home plate. (Home plate is a pentagon placed on the ground.)
    - It is between the top of the batter's knees and the midpoint between the batter's shoulders and the top of his pants.

    Describe the strike zone in terms of a geometric solid.

Strike Zone

The Flatiron Building

31. **History**   In 1902, the Flatiron Building, New York's first skyscraper, was built. The base of the building is a polygon having three sides. Describe the basic shape of the building as a geometric solid.

32. **Orthographic Drawings**   An *orthographic drawing* shows the top, the front, and the right-side views of a solid figure.

    a. Use the orthographic drawing to sketch the solid it represents.

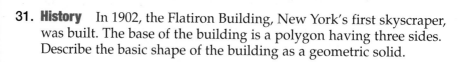

    top   front   right

    b. Select a solid object in your home and draw the top, the front, and the right-side views.

33. Critical Thinking The Swiss mathematician Leonhard Euler (1707–1783) was the first to discover the relationship among the number of vertices V, edges E, and faces F of a polyhedron.

a. Copy and complete the table.

Solid	Number of Vertices (V)	Number of Faces (F)	Number of Edges (E)
triangular prism	6	5	9
triangular pyramid			
rectangular prism			
rectangular pyramid			
pentagonal prism			

b. Look for a pattern, and write a formula using V, F, and E to show the relationship discovered by Euler.

c. If a prism has 10 faces and 24 edges, how many vertices does it have? Use the formula you found in part b.

Mixed Review

34. Recreation The area of a circular pool is approximately 707 square feet. The owner wishes to purchase a new cover for the pool. What is the diameter of the cover? *(Lesson 11–6)*

35. Transportation Suppose the wheels of a car have 29-inch diameters. How many full revolutions will each front wheel make when the car travels 1 mile? *(Lesson 11–5)*

36. In $\odot K$, find $m\widehat{MN}$, $m\widehat{MWP}$, and $m\widehat{NWP}$. *(Lesson 11–2)*

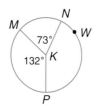

Determine whether each figure has line symmetry. If it does, copy the figure, and draw all lines of symmetry. If not, write *no*. *(Lesson 10–6)*

37.   38.   39.

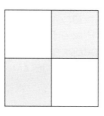      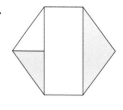

40. Multiple Choice What is the area of the polygonal region in square units? *(Lesson 10–3)*

Ⓐ 6 square units   Ⓑ 8 square units

Ⓒ 10 square units   Ⓓ 12 square units

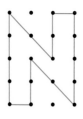

Take a Slice

Materials

 modeling clay

 dental floss

Cross Sections of Solids

What would happen if you sliced through the center of a basketball? The shape you would see is a circle. If you sliced through the ball away from the center, you would see a smaller circle. Let's find out what happens when you slice through solid figures.

Investigate

1. Use modeling clay to investigate slicing a right circular cylinder.

   a. Roll a piece of modeling clay on your desk to form a thick tube. Then use dental floss to cut off the ends. Make your cuts perpendicular to the sides of the tube. You have created a cylinder.

   b. Use dental floss to slice through the cylinder horizontally as shown at the right. Place the cut surface on a piece of paper and trace around it. What shape do you get?

   c. Remake the cylinder. Use dental floss to slice through the cylinder on an angle as shown at the right. Place the cut surface on a piece of paper and trace around it. What shape do you get?

   d. Remake the cylinder. Use dental floss to slice through the cylinder vertically as shown at the right. Place the cut surface on a piece of paper and trace around it. What shape do you get?

2. Use modeling clay to investigate slicing a cone.

   a. Use modeling clay to form a cone.

   b. Use dental floss to slice through the cone as shown in each diagram. Place the cut surface on a piece of paper and trace around it. Identify the shape of each cross section. Remember to remake your cone after each slice.

horizontal slice

angled slice

vertical slice

Extending the Investigation

In this extension, you will use modeling clay to identify the cross sections of solid figures. Make a horizontal slice, an angled slice, and a vertical slice of each solid figure.

1. rectangular prism or cube
2. triangular prism
3. square pyramid
4. a solid of your choice

Presenting Your Conclusions

Here are some ideas to help you present your conclusions to the class.

- Make a poster with drawings of your solids and the tracings of the shapes of the cross sections.
- Write a paragraph about each solid and its various cross sections.
- Consider a solid that you have not studied in this Investigation. Draw and make the solid. Identify the cross sections that would result from this solid.

*inter*NET
CONNECTION
Investigation   For more information on cross sections of solids, visit: www.geomconcepts.com

You'll learn to find the lateral areas and surface areas of prisms and cylinders.

Why It's Important
Home Improvement
Plumbers use lateral area to determine the amount of insulation needed to cover the sides of a water heater.
See Example 4.

The heights of the two decks of cards are the same, but the shapes are different. One is oblique and the other is right.

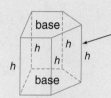

In a **right prism**, a lateral edge is also an altitude.

An *altitude* of a prism is a segment perpendicular to the two planes that contain the bases. *The length of an altitude is also called the height of the prism. Its measure is represented by h.*

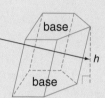

In an **oblique prism**, a lateral edge is *not* an altitude.

Info Graphic

The **lateral area** of a solid figure is the sum of the areas of its lateral faces. The **surface area** of a solid figure is the sum of the areas of all its surfaces.

The activity shows how the formula for the lateral area and surface area are derived.

Theorem 12–1 Lateral Area of a Prism	**Words:** If a prism has a lateral area of L square units and a height of h units and each base has a perimeter of P units, then $L = Ph$.
	Model: 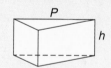 **Symbols:** $L = Ph$

Reading Geometry

In this text, you can assume that all prisms are right unless noted otherwise.

To find the surface area of a prism, the areas of the two bases must be included. Remember that the two bases are congruent.

Theorem 12–2 Surface Area of a Prism	**Words:** If a prism has a surface area of S square units and a height of h units and each base has a perimeter of P units and an area of B square units, then $S = Ph + 2B$.
	Model: **Symbols:** $S = Ph + 2B$

The formula for surface area can also be written in terms of lateral area, $S = L + 2B$.

1 Find the lateral area and the surface area of the rectangular prism.

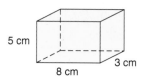

5 cm · 8 cm · 3 cm

First, find the perimeter of the base, P. Then, find the area of the base, B.

Perimeter of Base	**Area of Base**
$P = 2\ell + 2w$	$B = \ell w$
$= 2(8) + 2(3)$	$= 8(3)$ or 24
$= 16 + 6$ or 22	

Algebra Review
Evaluating Expressions, p. 718

Use this information to find the lateral area and the surface area.

$L = Ph$	$S = L + 2B$
$= 22(5)$	$= 110 + 2(24)$
$= 110$	$= 110 + 48$ or 158

The lateral area is 110 square centimeters, and the surface area is 158 square centimeters.

2 Find the lateral area and the surface area of the triangular prism.

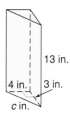

13 in. · 4 in. · 3 in. · c in.

First, use the Pythagorean Theorem to find c, the measure of the hypotenuse. Use the value of c to find the perimeter of the base. Then, find the area.

	Perimeter of Base	**Area of Base**
$c^2 = 3^2 + 4^2$	$P = 4 + 3 + c$	$B = \frac{1}{2}bh$
$c^2 = 9 + 16$	$= 4 + 3 + 5$	$= \frac{1}{2}(4)(3)$
$c^2 = 25$	$= 12$	$= 6$
$c = \sqrt{25}$ or 5		

Look Back

Pythagorean Theorem: Lesson 6–6

Use this information to find the lateral area and surface area.

$L = Ph$	$S = L + 2B$
$= 12(13)$	$= 156 + 2(6)$
$= 156$	$= 156 + 12$ or 168

The lateral area is 156 square inches, and the surface area is 168 square inches.

Your Turn

Find the lateral area and the surface area of each prism.

a.

8 cm · 8 cm · 8 cm

b.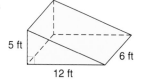

5 ft · 6 ft · 12 ft

You can use the CellSheet™ App on a TI-83 Plus/TI-84 Plus graphing calculator to investigate how changing a dimension of a rectangular prism changes the surface area of the prism.

Graphing Calculator Tutorial

See pp. 782–785.

Graphing Calculator Exploration

To open a CellSheet session, press APPS, select CelSheet and press ENTER. Press any key twice to continue. A blank spreadsheet is opened and cell A1 is selected.

Step 1 Enter the labels LENGTH, WIDTH, HEIGHT, and SA in cells A1, B1, C1, and D1, respectively. To enter text press 2nd ALPHA ["], then press each letter.

S01	A	B	C
1	LENGTH	WIDTH	HEIGH
2	8	5	3
3	16	5	3
4	8	10	3
5	8	5	6
6	24	15	9

C6: 9   [Menu]

Step 2 Enter the dimensions of a rectangular prism in cells A2, B2, and C2.

Step 3 Enter the formula for the surface area of a prism in cell D2. To enter the formula, press STO▶ (2 × ALPHA [A] 2 + 2 × ALPHA [B] 2) × ALPHA [C] 2 + 2 × ALPHA [A] 2 +2 × ALPHA [B] 2 ENTER. The formula uses the information in cells A2, B2, and C2 to automatically calculate the surface area.

Try These

1. Use CellSheet to find the surface area of a prism. Then double one dimension of the prism. How is the surface area affected?

2. How is the surface area of a prism affected if all of the dimensions are doubled?

3. How do you think the surface area of a prism is affected if all of the dimensions are tripled? Use CellSheet to verify your conjecture.

Reading Geometry

In this text, you can assume that all cylinders are right circular cylinders unless noted otherwise.

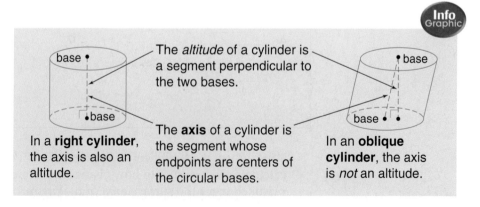

The *altitude* of a cylinder is a segment perpendicular to the two bases.

base

base

In a **right cylinder**, the axis is also an altitude.

The **axis** of a cylinder is the segment whose endpoints are centers of the circular bases.

In an **oblique cylinder**, the axis is *not* an altitude.

base

base

Info Graphic

The lateral area of a cylinder is the area of the curved surface. If a cylinder were cut across the lateral side and unfolded, it would resemble a rectangle. Its **net** is shown at the right. A net is a two-dimensional pattern that folds to form a solid. The width of the rectangle is the height h of the cylinder. The length of the rectangle is the distance around the circular base, or the circumference, $2\pi r$. Since $\ell = 2\pi r$ and $w = h$, $L = \ell w$ becomes $L = (2\pi r)h$.

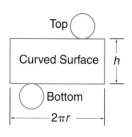

Theorem 12–3 **Lateral Area** **of a Cylinder**	**Words:**	If a cylinder has a lateral area of L square units and a height of h units and the bases have radii of r units, then $L = 2\pi rh$.
	Model:	**Symbols:** $L = 2\pi rh$

The surface area of a cylinder is still found by using $S = L + 2B$. However, L can be replaced with $2\pi rh$, and B can be replaced with πr^2.

Theorem 12–4 **Surface Area** **of a Cylinder**	**Words:**	If a cylinder has a surface area of S square units and a height of h units and the bases have radii of r units, then $S = 2\pi rh + 2\pi r^2$.
	Model:	**Symbols:** $S = 2\pi rh + 2\pi r^2$

COncepts in MOtion
Interactive Lab
geomconcepts.com

Example ③ Find the lateral area and surface area of the cylinder to the nearest hundredth.

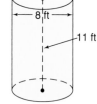

Lateral Area	Surface Area
$L = 2\pi rh$ $\quad d = 8,$	$S = 2\pi rh + 2\pi r^2$
$= 2\pi(4)(11)$ $\quad r = 4$	$= 2\pi(4)(11) + 2\pi(4)^2$
≈ 276.46	≈ 376.99

To the nearest hundredth, the lateral area is about 276.46 square feet, and the surface area is about 376.99 square feet.

Your Turn

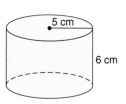

c. Find the lateral area and the surface area of the cylinder to the nearest hundredth.

 Personal Tutor at geomconcepts.com

Example ④ **4**

Plumbing Link

Find the amount of insulation needed to cover the sides of the 50-gallon hot water heater to the nearest hundredth.

The hot water heater is in the shape of a cylinder. Since the insulation covers only the sides of the hot water heater, find the lateral area. The diameter is 21 inches. So, the radius is 10.5 inches.

21 in.

55 in.

$L = 2\pi rh$ *Theorem 12–3*

   $= 2\pi(10.5)(55)$ *Replace r with 10.5 and h with 55.*

   ≈ 3628.54 *Use a calculator.*

The amount of insulation needed is 3628.54 square inches.

Check for Understanding

Communicating Mathematics

1. **Explain** the difference between lateral area and surface area.

2. **Draw** an oblique cylinder and a right cylinder. Write a sentence or two explaining the difference between the two kinds of cylinders.

Vocabulary
right prism
oblique prism
lateral area
surface area
net
right cylinder
axis
oblique cylinder

Guided Practice

Examples 1–3

Find the lateral area and the surface area for each solid. Round to the nearest hundredth, if necessary.

3.

6 cm

3 cm   2 cm

4.

11 in.

6 in.   8 in.

5.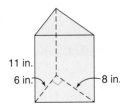

15 ft

6 ft

Examples 3 & 4

6. **Manufacturing** A soup can has a height of 10 centimeters and a diameter of 6.5 centimeters.

   a. Find the amount of paper needed to cover the can.

   b. Find the amount of steel needed to make the can.

Exercises

• • • •   • • • • • • • • • • • • • •

Practice

Find the lateral area and the surface area for each solid. Round to the nearest hundredth, if necessary.

7.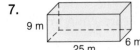

9 m

25 m   6 m

8.

32 in.   5 in.

9.

3 m

3 m

3 m

Homework Help

For Exercises	See Examples
7–19, 16, 19	1
10–12	2
13–15, 17	3, 4
18	1, 2

Extra Practice
See page 748.

10.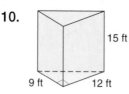

11.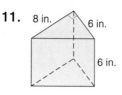

12.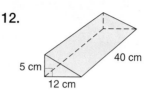

13.

14.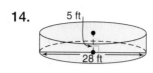

15.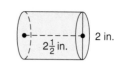

16. Draw a rectangular prism that is 4 centimeters by 5 centimeters by 8 centimeters. Find the surface area of the prism.

17. A cylinder has a diameter of 32 feet and a height of 20 feet.
  a. Find the lateral area of the cylinder.
  b. Find the surface area of the cylinder.

Applications and Problem Solving

18. Architecture Find the amount of glass needed to build the greenhouse.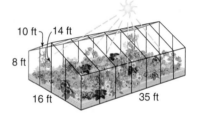

19. Painting A rectangular room is 12 feet by 21 feet. The walls are 8 feet tall. Paint is sold in one-gallon containers. If a gallon of paint covers 450 square feet, how many gallons of paint should Poloma buy to paint the walls of the room?

20. Critical Thinking Identify the solid by its net shown at the right.

Mixed Review

21. Draw an example of a pentagonal pyramid. *(Lesson 12–1)*

22. Find the area of a 30° sector of a circle if the radius of the circle is 36 meters. Round to the nearest hundredth. *(Lesson 11–6)*

23. Nature What is the circumference of a bird's nest if its diameter is 7 inches? *(Lesson 11–5)*

Standardized Test Practice

Ⓐ Ⓑ Ⓒ Ⓓ

24. Grid In Find the perimeter in centimeters of a regular nonagon whose sides are 3.9 centimeters long. *(Lesson 10–1)*

25. Multiple Choice Candace bought a cordless screwdriver on sale for $22.50. The regular price was $30. What was the percent of discount? *(Percent Review)*

  Ⓐ 15%   Ⓑ 20%   Ⓒ 25%   Ⓓ 30%

Volumes of Prisms and Cylinders

What You'll Learn

You'll learn to find the volumes of prisms and cylinders.

Why It's Important

Automotive Design
Engineers calculate the volume of cylinders to determine the displacement of an engine.
See Example 4.

The amount of water a fish tank can hold, the amount of grain a silo can hold, or the amount of concrete needed for a patio floor are all examples of volume.

Volume measures the space contained within a solid. Volume is measured in cubic units. The cube below has a volume of 1 cubic centimeter or 1 cm^3. Each of its sides is 1 centimeter long.

1 cm
1 cm
1 cm

Hands-On Geometry

Materials: 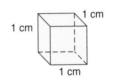 cubes/blocks

Step 1 Make a prism like the one shown at the right.

Step 2 Make at least three different rectangular prisms.

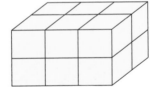

Try These

1. Assume that the edge of each cube represents 1 unit. Then, the area of each surface of each cube is 1 square unit, and the volume of each cube is 1 cubic unit. Copy and complete the table for your prisms.

Prism	Area of Base	Height	Volume
1	6	2	12
2			
3			
4			

2. Describe how the area of the base and the height of a prism are related to its volume.

3. The corner view of the prism is shown. Draw the top, front, and side views of the prism with one cube removed.

The activity above suggests the following postulate.

<table>
<tr>
<td rowspan="3">**Theorem 12–5**
Volume of a
Prism</td>
<td>**Words:**</td>
<td colspan="2">If a prism has a volume of V cubic units, a base with an area of B square units, and a height of h units, then $V = Bh$.</td>
</tr>
<tr>
<td>**Model:**</td>
<td></td>
<td>**Symbols:** $V = Bh$</td>
</tr>
</table>

Examples

1 **Find the volume of the triangular prism.**

The area of the triangular base can be calculated using the measures of the two legs.

Therefore, $B = \frac{1}{2}(4)(4)$ or 8.

$V = Bh$ *Theorem 12–5*

   $= 8(12)$ *Replace B with 8 and h with 12.*

   $= 96$ The volume of the triangular prism is 96 cubic feet.

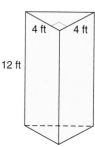
4 ft   4 ft
12 ft

Look Back

Area of Regular
Polygons:
Lesson 10–5

2 **The base of the prism is a regular pentagon with sides of 4 centimeters and an apothem of 2.75 centimeters. Find the volume of the prism.**

The perimeter of the base is 5(4) or 20.

$B = \frac{1}{2}Pa$ $V = Bh$

   $= \frac{1}{2}(20)(2.75)$ $= 27.5(9)$

   $= 27.5$ $= 247.5$

The volume of the pentagonal prism is 247.5 cubic centimeters.

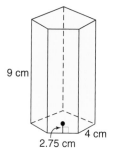
9 cm
2.75 cm   4 cm

Your Turn   **Find the volume of each prism.**

a.

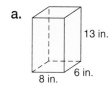
13 in.
8 in.   6 in.

b.

12.5 m
4 m   6.5 m

A stack of coins is shaped like a cylinder. The volume of a cylinder is given by the same general formula as for a prism. The volume is the product of the cylinder's base and height. The base of this cylinder is a circle with area πr^2, and its height is h.

$V = Bh$
   $= (\pi r^2)h$ or $\pi r^2 h$

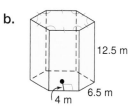
Photo Graphic
h

<table>
<tr>
<td rowspan="3">Theorem 12-6
Volume of a
Cylinder</td>
<td>Words:</td>
<td colspan="2">If a cylinder has a volume of V cubic units, a radius of r units, and a height of h units, then $V = \pi r^2 h$.</td>
</tr>
<tr>
<td>Model:</td>
<td></td>
<td>Symbols: $V = \pi r^2 h$</td>
</tr>
</table>

Example ❸

Find the volume of the cylinder to the nearest hundredth.

$V = \pi r^2 h$ *Theorem 12-6*

$ = \pi(8^2)(12.5)$ *Replace r with 8 and h with 12.5.*

$ \approx 2513.27$ *Use a calculator.*

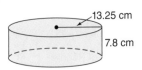

The volume of the cylinder is 2513.27 cubic centimeters.

Your Turn

c. Find the volume of the cylinder to the nearest hundredth.

Real World

Example ❹

Automotive Link

In an automotive engine, the piston moves up and down in a cylinder. The volume of space through which the piston moves is called the *displacement*. A certain automobile has a stroke of 3.54 inches and a bore of 3.23 inches. The bore is the diameter of the cylinder.

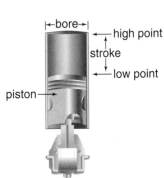

A. **Find the displacement of each cylinder.**

The diameter d of the cylinder is 3.23 inches. So, the radius is 1.615 inches.

$V = \pi r^2 h$ *Theorem 12-6*

$ = \pi(1.615)^2(3.54)$ *Replace r with 1.615 and h with 3.54.*

$ \approx 29$ *Use a calculator.*

The displacement of each cylinder is about 29 cubic inches.

B. **Find the total displacement of the car's four cylinders.**

Since the car has four cylinders, the total displacement is about 4×29 or about 116 cubic inches.

 Personal Tutor at geomconcepts.com

Check for Understanding

Communicating Mathematics

1. **Draw** two right rectangular prisms with a volume of 24 cubic inches, but with different dimensions.

2. **Compare and contrast** surface area and volume. Be sure to include the type of unit used for each measurement.

3. **You Decide** A fish tank is 18 inches by 12 inches by 10 inches. Rebecca says that if she were to double the dimensions of the fish tank, she would need twice as much water to fill the tank. Caitlin disagrees. Who is correct? Explain.

Guided Practice

⏱ Getting Ready Find the area of each figure to nearest hundredth.

Sample: rectangle: length, 4.9 inches; width, 5 inches
Solution: $A = \ell w$
$A = 4.9 \times 5$ or 24.5 The area is 24.5 square inches.

4. triangle: base, $3\frac{3}{4}$ feet; height, 4 feet
5. circle: diameter, 7 meters
6. regular pentagon: side, 6 yd; apothem, 4.1 yd

Examples 1–3

Find the volume of each solid. Round to the nearest hundredth, if necessary.

7.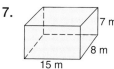
7 m
8 m
15 m

8. 4 ft 10 ft
15 ft

9. 36 mm
6 mm
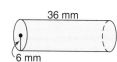

Example 4

10. **Environmental Engineering** A classroom is 30 feet long, 24 feet wide, and 10 feet high. If each person in the room needs 300 cubic feet of air, find the maximum capacity of the room.

Exercises

Practice

Find the volume of each solid. Round to the nearest hundredth, if necessary.

11.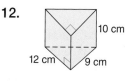
3 m
4 m 4 m

12. 10 cm
12 cm 9 cm

13. 18 ft
16 ft
11 ft
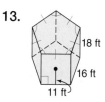

Homework Help

For Exercises	See Examples
11–13, 17, 19–21, 25	1, 2
14–16, 18, 22, 23	3, 4
24	1–3

Extra Practice

See page 748.

Find the volume of each solid. Round to the nearest hundredth, if necessary.

14.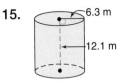
2 in.
$3\frac{1}{2}$ in.

15.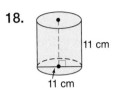
6.3 m
12.1 m

16.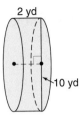
2 yd
10 yd

17.
9 ft
2 ft   4 ft

18.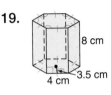
11 cm
11 cm

19.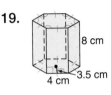
8 cm
4 cm   3.5 cm

20. What is the volume of a cube that has a 5-inch edge?

21. Draw a rectangular prism that is 4 centimeters by 8 centimeters by 3 centimeters. Find the volume of the prism.

22. A cylinder has a base diameter of 14 inches and a height of 18 inches. What is the volume of the cylinder?

Applications and Problem Solving

23. **Weather**   Rain enters the rain gauge through a funnel-shaped top. It is measured in the cylindrical collector. Find the volume of the cylindrical collector of the rain gauge.

20 in.
2 in.

24. **Packaging**   Salt is usually packaged in cylindrical boxes. In 1976, the Leslie Salt Company in Newark, California, tried to package salt in rectangular boxes to save space on supermarket shelves.

    a. Compare the volumes of the two salt boxes.

    b. Which design is better for storing, stacking, and shipping? Explain your reasoning.

8.4 cm
SALT
13.5 cm

SALT
15 cm
5 cm
10 cm

25. **Critical Thinking**   The areas of the faces of a rectangular prism are 12 square inches, 32 square inches, and 24 square inches. The lengths of the edges are represented by whole numbers. Find the volume of the prism. Explain how you solved the problem.

26. Find the lateral area and the surface area of the cylinder to the nearest hundredth. *(Lesson 12–2)*

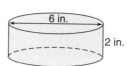

6 in.

2 in.

27. Describe the basic shape of a package of computer disks as a geometric solid. *(Lesson 12–1)*

28. Algebra In $\odot H$, $\overarc{DE} \cong \overarc{FG}$. If $DE = 3x$ and $FG = 4x - 6$, what is the value of x? *(Lesson 11–3)*

29. Short Response Find the area of a regular octagon whose perimeter is 49.6 feet and whose apothem is 7.5 feet long. *(Lesson 10–5)*

30. Multiple Choice Which ordered pair is a solution to $3x + 4y \geq 17$? *(Algebra Review)*

   (A) $(1, 1)$ (B) $(-3, 7)$ (C) $(4, 0)$ (D) $(2, -3)$

Quiz 1   Lessons 12–1 through 12–3

Identify each solid. *(Lesson 12–1)*

1. **2.** **3.** **4.**

Find the lateral area and the surface area for each solid. Round to the nearest hundredth, if necessary. *(Lesson 12–2)*

5. 5 m   8 m   12 m

6. 3 ft   4 ft   8 ft

7. |←16 in.→| 7 in.

8. 6 cm   18 cm

9. Find the volume of a cylinder with a base diameter of 24 meters and a height of 28 meters. *(Lesson 12–3)*

10. Construction How many cubic yards of concrete will be needed for a driveway that is 40 feet long, 18 feet wide, and 4 inches deep? Round the answer to the nearest tenth. (*Hint:* $27 \text{ ft}^3 = 1 \text{ yd}^3$) *(Lesson 12–3)*

What You'll Learn

You'll learn to find the lateral areas and surface areas of regular pyramids and cones.

Why It's Important

Building Architects can determine the area of the outside walls of a building in the shape of a pyramid.
See Example 2.

Just as there are right and oblique prisms and cylinders, there are also right and oblique pyramids and cones.

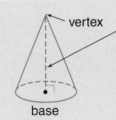

Info Graphic

vertex

The *altitude* of a pyramid or cone is the segment from the vertex perpendicular to the plane containing the base.

vertex →

base

In a **right pyramid** or **right cone**, the altitude is perpendicular to the base at its center.

In an **oblique pyramid** or **oblique cone**, the altitude is perpendicular to the base at a point other than its center.

The Great American Pyramid is an arena in Memphis, Tennessee. It is an example of a regular pyramid.

Reading Geometry

In this text, you can assume that all pyramids are right unless noted otherwise.

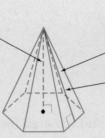

A pyramid is a **regular pyramid** if and only if it is a right pyramid and its base is a regular polygon.

The lateral faces of a regular pyramid form congruent isosceles triangles.

The height of each lateral face is called the **slant height** of the pyramid. The measure of the slant height is represented by ℓ.

Info Graphic

Consider the regular pentagonal pyramid below. Its lateral area L can be found by adding the areas of all its congruent triangular faces. Because the base is a regular polygon, the length of each edge along the base has equal measure s.

Solid

Net

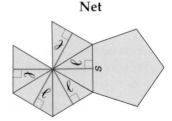

$$L = \frac{1}{2}s\ell + \frac{1}{2}s\ell + \frac{1}{2}s\ell + \frac{1}{2}s\ell + \frac{1}{2}s\ell$$

$$= \frac{1}{2}(s + s + s + s + s)\ell \qquad \textit{Distributive Property}$$

$$= \frac{1}{2}P\ell \qquad\qquad P = s + s + s + s + s$$

Theorem 12–7 Lateral Area of a Regular Pyramid	**Words:** If a regular pyramid has a lateral area of L square units, a base with a perimeter of P units, and a slant height of ℓ units, then $L = \frac{1}{2}P\ell$.
	Model: **Symbols:** $L = \frac{1}{2}P\ell$

With prisms and cylinders, the formula for the surface area is $S = L + 2B$. Since a pyramid has only one base, the formula for its surface area is $S = L + B$, where $L = \frac{1}{2}P\ell$.

Theorem 12–8 Surface Area of a Regular Pyramid	**Words:** If a regular pyramid has a surface area of S square units, a slant height of ℓ units, and a base with perimeter of P units and area of B square units, then $S = \frac{1}{2}P\ell + B$.
	Model: **Symbols:** $S = \frac{1}{2}P\ell + B$

The formula for surface area can also be written in terms of lateral area, $S = L + B$.

Examples

1 **Find the lateral area and the surface area of the regular hexagonal pyramid.**

First, find the perimeter and area of the hexagon. For a regular hexagon, the perimeter is 6 times the length of one side. The area is one-half the perimeter times the apothem.

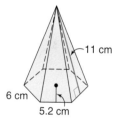

$P = 6s$ $\qquad\qquad B = \frac{1}{2}Pa$

$\quad = 6(6)$ or 36 $\qquad = \frac{1}{2}(36)(5.2)$ or 93.6

(continued on the next page)

Now use this information to find the lateral area and surface area.

$L = \frac{1}{2}P\ell$ $S = L + B$

$\quad = \frac{1}{2}(36)(11)$ $= 198 + 93.6$

$\quad = 198$ $= 291.6$

The lateral area is 198 square centimeters, and the surface area is 291.6 square centimeters.

Your Turn

Find the lateral area and the surface area of each regular pyramid.

a.

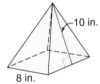

10 in.

8 in.

b.

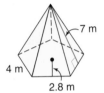

7 m

4 m

2.8 m

Architecture Link

Real World

2 **The Great American Pyramid in Memphis, Tennessee, is a regular pyramid with a square base. Each side of the base is about 544 feet long. The slant height is about 420 feet. Find the area of the outside walls of this structure.**

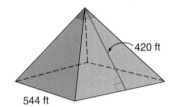

420 ft

544 ft

The area of the outside walls is the lateral area of the pyramid. The perimeter of the base is 4(544) or 2176 feet.

$L = \frac{1}{2}P\ell$ *Theorem 12–7*

$\quad = \frac{1}{2}(2176)(420)$ *Replace P with 2176 and ℓ with 420.*

$\quad = 456{,}960$

The area of the outside walls of the Great American Pyramid is about 456,960 square feet.

The slant height of a cone is the length of any segment whose endpoints are the vertex of the cone and a point on the circle that forms the base.

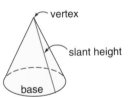

vertex

slant height

base

The formulas for finding the lateral area and surface area of a cone are similar to those for a regular pyramid. However, since the base is a circle, the perimeter becomes the circumference, and the area of the base is πr^2 square units.

<table>
<tr><td rowspan="2">**Theorem 12–9**
Lateral Area
of a Cone</td><td>**Words:**</td><td>If a cone has a lateral area of L square units, a slant height of ℓ units, and a base with a radius of r units, then $L = \frac{1}{2} \cdot 2\pi r \cdot \ell$ or $\pi r \ell$.</td></tr>
<tr><td>**Model:**</td><td>   **Symbols:** $L = \pi r \ell$</td></tr>
</table>

To find the surface area of a cone, add its lateral area and the area of its base.

<table>
<tr><td rowspan="2">**Theorem 12–10**
Surface Area
of a Cone</td><td>**Words:**</td><td>If a cone has a surface area of S square units, a slant height of ℓ units, and a base with a radius of r units, then $S = \pi r \ell + \pi r^2$.</td></tr>
<tr><td>**Symbols:**</td><td>$S = \pi r \ell + \pi r^2$</td></tr>
</table>

Example

3 **Find the lateral area and the surface area of the cone to the nearest hundredth.**

Since the diameter of the base is 10 feet, the radius is 5 feet. Use the Pythagorean Theorem to find ℓ.

$\ell^2 = 5^2 + 12^2$ *Use the Pythagorean Theorem.*
$\ell^2 = 25 + 144$ *$5^2 = 25$; $12^2 = 144$*
$\ell^2 = 169$ *Add 25 and 144.*
$\ell = \sqrt{169}$ or 13   *Take the square root of each side.*

Use this value to find the lateral area and the surface area.

$L = \pi r \ell$	$S = \pi r \ell + \pi r^2$ *Theorems 12–9 and 12–10*
$= \pi (5)(13)$	$\approx 204.20 + \pi(5^2)$ *Substitution*
≈ 204.20	≈ 282.74 *Use a calculator.*

The lateral area is 204.20 square feet, and the surface area is 282.74 square feet.

Your Turn

Find the lateral area and the surface area of each cone. Round to the nearest hundredth.

c.

9 m

4 m

d.

8 in.

12 in.

 Personal Tutor at geomconcepts.com

Communicating Mathematics

1. **Explain** the difference between the slant height and the altitude of a regular pyramid.

2. **Draw** a cone in which the altitude is not perpendicular to the base at its center. What is the name of this solid?

3. *Writing Math* Write a paragraph comparing the methods for finding the surface areas of prisms, cylinders, pyramids, and cones.

Vocabulary
right pyramid
right cone
oblique pyramid
oblique cone
regular pyramid
slant height

Guided Practice
Examples 1 & 3

4. Find the lateral area and the surface area of the regular pentagonal pyramid.

5. Find the lateral area and the surface area of the cone. Round to the nearest hundredth.

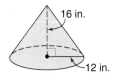

Example 3

6. **Manufacturing** The Party Palace makes cone-shaped party hats out of cardboard. If the diameter of the hat is $6\frac{1}{2}$ inches and the slant height is 7 inches, find the amount of cardboard needed for each hat.

Exercises

Practice

Find the lateral area and the surface area of each regular pyramid.

7.

8.

9.

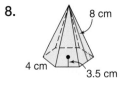

Homework Help	
For Exercises	See Examples
7–9, 13, 15	1, 2
10–12, 14, 17	3
Extra Practice	
See page 749.	

Find the lateral area and the surface area of each cone. Round to the nearest hundredth.

10.

11.

12.

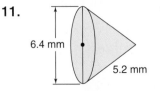

13. A regular pyramid has an altitude of 4 feet. The base is a square with sides 6 feet long. What is the surface area of the pyramid?

14. Determine which cone has the greater lateral area.

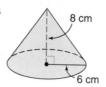

15. Tourism A tropical resort maintains cabanas on the beach for their guests. Find the amount of canvas needed to cover one cabana.

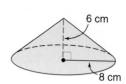

16. Aircraft Design An important design factor of aircraft is known as *wing loading*. Wing loading is the total weight of the aircraft and its load on take-off divided by the surface area of its wings.

  a. Determine the wing loading for each aircraft to the nearest hundredth.

Aircraft	Maximum Takeoff Weight (lb)	Surface Area of Wings (ft²)
Wright brothers' plane	750	532
Concorde	408,000	3856
Nighthawk	52,500	913
Tomcat	70,280	565

Source: *Aircraft of the World*

  b. What do the resulting wing loading numbers mean?

17. Critical Thinking A **truncated solid** is the remaining part of a solid after one or more vertices have been cut off.

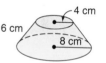

  a. Find the surface area of the truncated cone to the nearest hundredth. (*Hint*: Sketch the entire cone.)

  b. A lampshade resembles a truncated cone without a solid top or bottom. Find the lateral area of a lampshade in your home.

18. Construction The town of West Mountfort recently built a new cylindrical water tower. If the tower is 275 feet tall and has a diameter of 87 feet, how many cubic feet of water can the tank hold? Round the answer to the nearest cubic foot. (*Lesson 12–3*)

19. Draw and label a rectangular prism with a length of 3 centimeters, a width of 2 centimeters, and a height of 4 centimeters. Then find the surface area of the prism. (*Lesson 12–2*)

20. Find the area of a trapezoid whose height measures 8 inches and whose bases are 7 inches and 12 inches. (*Lesson 10–4*)

21. Short Response Given $L(2, 1)$, $M(4, 3)$, $P(0, 2)$, and $Q(3, -1)$, determine if $\overline{LM}$ and $\overline{PQ}$ are *parallel*, *perpendicular*, or *neither*. (*Lesson 4–5*)

22. Multiple Choice Find the value of x in the figure. (*Lesson 3–6*)

  Ⓐ 3   Ⓑ 5
  Ⓒ 7   Ⓓ 9

What You'll Learn

You'll learn to find the volumes of pyramids and cones.

Why It's Important

Entertainment
Theater managers can determine the amount of popcorn in a container.
See Example 3.

The pyramid and prism at the right have the same base and height. The cone and cylinder have the same base and height. If you consider these figures in terms of volume, you can see that the volume of the pyramid is less than the volume of the prism. Likewise, the volume of the cone is less than the volume of the cylinder.

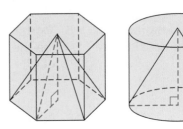

Let's investigate the relationship between the volumes of a prism and a pyramid with the same base and height.

Hands-On Geometry

Materials: card stock / ruler / compass
scissors / tape / rice

Step 1 Draw nets for a cube and a pyramid on card stock using a ruler and a compass.

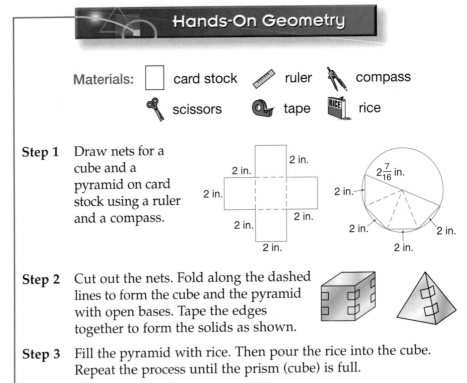

Step 2 Cut out the nets. Fold along the dashed lines to form the cube and the pyramid with open bases. Tape the edges together to form the solids as shown.

Step 3 Fill the pyramid with rice. Then pour the rice into the cube. Repeat the process until the prism (cube) is full.

Try These

1. Set the cube and pyramid on your desk with the base of each figure on the desk. Compare the heights of the cube and pyramid.
2. Place the base of the pyramid on top of the cube. Compare their bases.
3. How many times did you fill the pyramid in order to fill the cube?
4. In general, if a pyramid and a prism have the same height and base, how would you expect their volumes to compare?

This activity suggests Theorem 12–11. Recall that Bh is the volume of a cube with height h and area of its base B.

| Theorem 12–11 Volume of a Pyramid | **Words:** | If a pyramid has a volume of V cubic units and a height of h units and the area of the base is B square units, then $V = \frac{1}{3}Bh$. |
| | **Model:** | | **Symbols:** $V = \frac{1}{3}Bh$ |

Example ❶

Find the volume of the rectangular pyramid to the nearest hundredth.

In a rectangle, $A = \ell w$. Therefore, $B = 8(5)$ or 40.

$V = \frac{1}{3}Bh$ *Theorem 12–11*

$ = \frac{1}{3}(40)(10)$ *Replace B with 40 and h with 10.*

$ \approx 133.33$ *Simplify.*

The volume of the pyramid is 133.33 cubic centimeters.

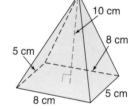

Your Turn

Find the volume of each pyramid. Round to the nearest hundredth.

a.

b.

COncepts in MOtion

Interactive Lab

geomconcepts.com

The relationship between the volumes of a cone and a cylinder is similar to the relationship between the volumes of a pyramid and a prism. The volume of a cone is $\frac{1}{3}$ the volume of cylinder with the same base and height. The volume of a cylinder is $\pi r^2 h$, so the volume of a cone is $\frac{1}{3}\pi r^2 h$.

| Theorem 12–12 Volume of a Cone | **Words:** | If a cone has a volume of V cubic units, a radius of r units, and a height of h units, then $V = \frac{1}{3}\pi r^2 h$. |
| | **Model:** | | **Symbols:** $V = \frac{1}{3}\pi r^2 h$ |

Find the volume of the cone to the nearest hundredth.

The triangle formed by the height, radius, and slant height is a right triangle. So, you can use the Pythagorean Theorem to find the measure of the height h.

$a^2 + b^2 = c^2$	*Pythagorean Theorem*
$h^2 + 9^2 = 15^2$	*Replace a with h, b with 9, and c with 15.*
$h^2 + 81 = 225$	$9^2 = 81; 15^2 = 225$
$h^2 + 81 - 81 = 225 - 81$	*Subtract 81 from each side.*
$h^2 = 144$	*Simplify.*
$h = \sqrt{144}$ or 12	*Take the square root of each side.*

Use this value of h to find the volume of the cone.

$V = \frac{1}{3}\pi r^2 h$	*Theorem 12–12*
$= \frac{1}{3}\pi(9)^2(12)$	*Replace r with 9 and h with 12.*
≈ 1017.88	*Use a calculator.*

The volume of the cone is 1017.88 cubic inches.

Your Turn

Find the volume of each cone to the nearest hundredth.

c.
13 ft
5 ft

d.
4 m
5 m

nline **Personal Tutor at geomconcepts.com**

Real World

Fernando Soto is the manager of a theater. He can purchase one of two containers to hold a small order of popcorn. How much more does the box hold than the cone?

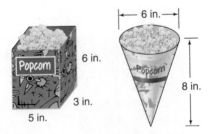

Popcorn
6 in.
3 in.
5 in.

6 in.
8 in.
Popcorn

First, find the volume of each container.

<table>
<tr><td>

Volume of Prism

$V = Bh \qquad$ *Theorem 12–5*

$\quad = 15(6) \quad$ *B = 5(3) or 15*

$\quad = 90 \qquad$ *Simplify.*

</td><td>

Volume of Cone

$V = \frac{1}{3}\pi r^2 h \qquad$ *Theorem 12–12*

$\quad = \frac{1}{3}\pi(3)^2(8) \quad$ *If d = 6, r = 3.*

$\quad \approx 75.40 \qquad$ *Use a calculator.*

</td></tr>
</table>

The volume of the box is 90 cubic inches. The volume of the cone is about 75.40 cubic inches. Thus, the box holds 90 − 75.40 or 14.6 cubic inches more than the cone.

Check for Understanding

Communicating Mathematics

1. **Compare and contrast** the formulas for the volume of a prism and the volume of a pyramid.

2. **Explain** why both $V = \frac{1}{3}Bh$ and $V = \frac{1}{3}\pi r^2 h$ can be used to find the volume of a cone.

3. Darnell believes that doubling the radius of a cone increases the volume more than doubling the height. Nicole disagrees. Who is correct? Explain.

Guided Practice
Examples 1 & 2

Find the volume of each solid. Round to the nearest hundredth, if necessary.

4.
7 in.
5 in. 9 in.

5.
6 ft
5 ft 4 ft

6.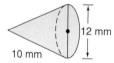
12 mm
10 mm

Example 3

7. **Architecture** The Muttart Conservatory in Edmonton, Alberta consists of four greenhouses in the shape of pyramids. Each of the two largest pyramids has a height of about 24 meters and a base with an area of about 625 square meters. Find the total volume of the two pyramids.

The Muttart Conservatory

Exercises

Practice

Find the volume of each solid. Round to the nearest hundredth, if necessary.

8.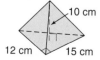
10 cm
12 cm 15 cm

9.
5 cm
2 cm

10.
1.0 m
2.6 m

Find the volume of each solid. Round to the nearest hundredth, if necessary.

Homework Help

For Exercises	See Examples
8, 11–14, 16, 17, 20	1
9, 10, 15 18, 19, 21	2

Extra Practice

See page 749.

11.
7 in.

8 in.   8 in.

12.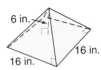
6 in.

16 in.

16 in.

13.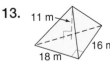
11 m

18 m   16 m

14.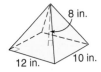
8 in.

12 in.   10 in.

15.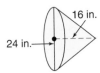
16 in.

24 in.

16.
6.9 cm   11 cm

8 cm

17. A pyramid has a height of 5 centimeters and a base with area of 18 square centimeters. What is its volume?

18. A cone has a height of 10 meters and a base with a radius of 3 meters. Find the volume of the cone.

19. The diameter of the base of a cone is 18 feet. The height of the cone is 12 feet. What is the volume of the cone?

20. The base of a pyramid is a triangle with a base of 12 inches and a height of 8 inches. The height of the pyramid is 10 inches. Find the volume of the pyramid.

Applications and Problem Solving

21. Geology A stalactite in Endless Caverns in Virginia is shaped like a cone. It is 4 feet tall and has a diameter at the roof of $1\frac{1}{2}$ feet. Find the volume of the stalactite.

22. History The Great Pyramid at Giza was built about 2500 B.C. It is a square pyramid.

  a. Originally the Great Pyramid was about 481 feet tall. Each side of the base was about 755 feet long. What was the original volume of the pyramid?

  b. Today the Great Pyramid is about 450 feet tall. Each side of the base is still about 755 feet long. What is the current volume of the pyramid?

  c. What is the difference between the volume of the original pyramid and the current pyramid?

  d. What was the yearly average (mean) loss in the volume of the pyramid from 2500 B.C. to 2000 A.D.?

23. Critical Thinking A cone and a cylinder have the same volume, and their radii have the same measure. What is true about these two solids?

Stalactite formation

24. Find the lateral area of the cone to the nearest hundredth. *(Lesson 12–4)*

25. Find the volume of the triangular prism. *(Lesson 12–3)*

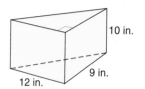

26. Construction   In a blueprint, 1 inch represents an actual length of 13 feet. If the dimensions of a kitchen on the blueprint are 1.25 inches by 0.75 inch, what are the actual dimensions? *(Lesson 9–2)*

Standardized Test Practice
Ⓐ Ⓑ Ⓒ Ⓓ

27. Grid In   In □*RSTU*, find *RW* if *RT* = 47.4 units. *(Lesson 8–2)*

28. Short Response   In △*ABC*, $\overline{AT}$, $\overline{BR}$, and $\overline{CS}$ are medians. What is the measure of $\overline{XR}$ if *BR* = 18? *(Lesson 6–1)*

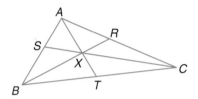

Quiz 2   Lessons 12–4 and 12–5

Find the lateral area and the surface area of each solid. Round to the nearest hundredth, if necessary. *(Lesson 12–4)*

1.

6 in.
5 in.

2.

12 cm
10 cm   8.6 cm

3.

←—14 m—→
9 m

4.

24 ft
10 ft

5. Architecture   The base of the glass pyramid that serves as the entrance to the Louvre in Paris is a square with sides measuring 115 feet. The slant height of the pyramid is about 92 feet. To the nearest square foot, find the area of the glass covering the outside of the pyramid. *(Lesson 12–4)*

Find the volume of each solid. Round to the nearest hundredth, if necessary. *(Lesson 12–5)*

6.

15 cm
13 cm

7.

6 in.
7 in.   9 in.

8.

2 ft
5 ft

9.

16 m
10 m

10. History   Monk's Mound in Illinois is an earthen pyramid built around A.D. 600. It is 30.5 meters high. The base of this pyramid is a 216.6-meter by 329.4-meter rectangle. Find the volume of the soil used to build this mound to the nearest cubic meter. *(Lesson 12–5)*

What You'll Learn

You'll learn to find the surface areas and volumes of spheres.

Why It's Important

Sports

Manufacturers need to determine the amount of leather needed to cover a baseball. *See Exercise 6.*

A circle is the set of all points in a plane that are a fixed distance from a point in the plane, called the center. Suppose you were not limited to a plane. From a center, all points in space at a fixed distance form a hollow shell called a **sphere**. The hollow shell of the sphere is its surface.

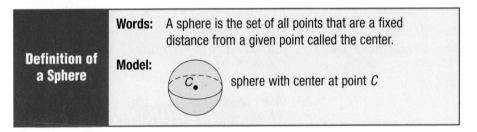

| **Definition of a Sphere** | **Words:** | A sphere is the set of all points that are a fixed distance from a given point called the center. |
| | **Model:** | sphere with center at point *C* |

A sphere has many properties like those of a circle.

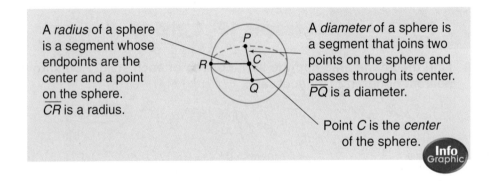

A *radius* of a sphere is a segment whose endpoints are the center and a point on the sphere. $\overline{CR}$ is a radius.

A *diameter* of a sphere is a segment that joins two points on the sphere and passes through its center. $\overline{PQ}$ is a diameter.

Point *C* is the *center* of the sphere.

Info Graphic

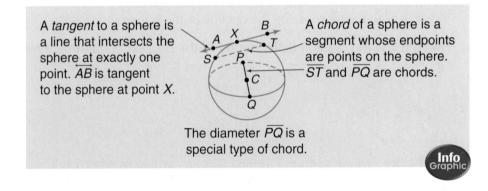

A *tangent* to a sphere is a line that intersects the sphere at exactly one point. $\overleftrightarrow{AB}$ is tangent to the sphere at point *X*.

A *chord* of a sphere is a segment whose endpoints are points on the sphere. $\overline{ST}$ and $\overline{PQ}$ are chords.

The diameter $\overline{PQ}$ is a special type of chord.

Info Graphic

The formulas for finding the surface area and the volume of a sphere are given on the next page.

Theorem	Words	Model and Symbols
12–13 **Surface Area** **of a Sphere**	If a sphere has a surface area of S square units and a radius of r units, then $S = 4\pi r^2$.	 $S = 4\pi r^2$
12–14 **Volume of** **a Sphere**	If a sphere has a volume of V cubic units and a radius of r units, then $V = \frac{4}{3}\pi r^3$.	$V = \frac{4}{3}\pi r^3$

Example

1 **Find the surface area and volume of the sphere.**

Since the diameter is 36 meters, the radius is 18 meters.

Surface Area

$S = 4\pi r^2$ *Theorem 12–13*

$= 4\pi(18)^2$ *Replace r with 18.*

≈ 4071.50 *Use a calculator.*

Volume

$V = \frac{4}{3}\pi r^3$ *Theorem 12–14*

$= \frac{4}{3}\pi(18)^3$ *Replace r with 18.*

$\approx 24{,}429.02$ *Use a calculator.*

The surface area is about 4071.50 square meters. The volume is about 24,429.02 cubic meters.

Your Turn

Find the surface area and volume of each sphere. Round to the nearest hundredth.

a.

b.

COncepts
in MOtion
Animation
geomconcepts.com

 Personal Tutor at geomconcepts.com

When a plane intersects a sphere so that congruent halves are formed, each half is called a hemisphere.

Example **2**

Aerospace Link

The large external tank attached
to the space shuttle at the time of
launch contains the propellants
for takeoff. It holds three tanks,
including the liquid hydrogen
tank. If the ends of the liquid
hydrogen tank are hemispheres,
find the volume of this tank to
the nearest hundredth.

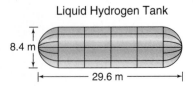

Liquid Hydrogen Tank

8.4 m

29.6 m

Explore   You know the length and the diameter of the tank. You need
to find the volume of the tank.

Plan   Think of the tank as
two hemispheres and a
cylinder. The radius of
each hemisphere and
the cylinder is $\frac{1}{2}(8.4)$ or
4.2 meters. Draw a
diagram showing the
measurements, and
calculate the sum of the volumes of each shape. You can
think of the two hemispheres as one sphere.

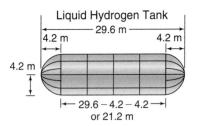

Liquid Hydrogen Tank

29.6 m

4.2 m   4.2 m

4.2 m

|← 29.6 − 4.2 − 4.2 →|
or 21.2 m

Solve   **Volume of Sphere**   **Volume of Cylinder**

$V = \frac{4}{3}\pi r^3$ $V = \pi r^2 h$

$\quad = \frac{4}{3}\pi(4.2)^3$ $\quad = \pi(4.2)^2(21.2)$

$\quad \approx 310.34$ $\quad \approx 1174.86$

$310.34 + 1174.86 = 1485.2$

The volume of the liquid
hydrogen tank is about
1485.2 cubic meters.

Examine   Find the volume of a
cylindrical tank with a radius
of 4.2 meters and a height
of 29.6 meters. Its volume
should be slightly more than
the liquid hydrogen tank.

$V = \pi r^2 h$

$\quad = \pi(4.2)^2(29.6)$

$\quad \approx 1640.36$

The answer seems reasonable.

Check for Understanding

Communicating Mathematics

1. **Compare and contrast** circles and spheres.

2. **Draw and label** a sphere with center at point P and chord $\overline{MN}$ that is not a diameter.

3. Writing Math Describe real-world examples of five different solids. Write the formula used to find the volume of each type of solid.

Guided Practice

Example 1

Find the surface area and volume of each sphere. Round to the nearest hundredth.

4.

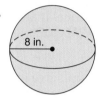

8 in.

5.

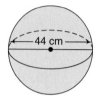

44 cm

Example 2

6. **Sports** Find the amount of leather needed to cover an official major league baseball if its diameter is 7.4 centimeters.

Exercises

Practice

Find the surface area and volume of each sphere. Round to the nearest hundredth.

Homework Help	
For Exercises	**See Examples**
7–18, 20, 21	1
19	2
Extra Practice	
See page 749.	

7.

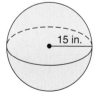

15 in.

8.

5 ft

9.

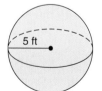

8 m

10.

2.15 cm

11.

22 cm

12.

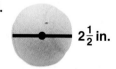

$2\frac{1}{2}$ in.

13. Find the surface area of a sphere with a diameter of 24 meters. Round to the nearest hundredth.

14. Find the volume of a sphere with a radius of 10 inches. Round to the nearest hundredth.

15. What is the volume of a sphere with a radius of 7 feet? Express the answer in cubic yards. Round to the nearest hundredth.

16. What is the surface area of a sphere with a diameter of 18 centimeters? Express the answer in square meters. Round to the nearest hundredth.

Applications and Problem Solving

17. **Food** An ice cream cone is 10 centimeters deep and has a diameter of 4 centimeters. A spherical scoop of ice cream that is 4 centimeters in diameter rests on top of the cone. If all the ice cream melts into the cone, will the cone overflow? Explain.

18. **Housing** The Algonquin people live in northern Canada, Greenland, Alaska, and eastern Siberia. Their traditional homes are igloos that resemble hemispheres.

   a. Find the square footage of the living area of the igloo to the nearest hundredth.

   b. Find the volume of the living area of the igloo to the nearest hundredth.

19. **Aerospace** The liquid oxygen tank in the external tank of the space shuttle resembles a combination of a hemisphere, a cylinder, and a cone. Find the volume of the liquid oxygen tank to the nearest hundredth.

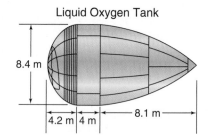

20. **Sales** The Boy Scouts of America recently made a giant popcorn ball to promote their popcorn sales. Scouts and other helpers added to the popcorn ball as it traveled around the country.

   a. What was the volume of the popcorn ball when it had a diameter of 6 feet? Round to the nearest hundredth.

   b. Suppose the diameter of the popcorn ball was enlarged to 9 feet. What would be its volume to the nearest hundredth?

   c. What was the volume of the popcorn needed to increase the popcorn ball from a diameter of 6 feet to a diameter of 9 feet?

21. Science The diameter of Earth is about 7900 miles.

    a. Find the surface area of Earth to the nearest hundred square miles.

    b. Find the volume of Earth to the nearest hundred cubic miles.

    c. Most of Earth's atmosphere is less than 50 miles above the surface. Find the volume of Earth's atmosphere to the nearest hundred cubic miles.

22. Critical Thinking A plane slices a sphere as shown. If the radius of the sphere is 10 centimeters, find the area of the circle formed by the intersection of the sphere and the plane to the nearest hundredth.

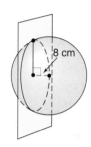

8 cm

Mixed Review

23. Cooking A cake-decorating bag is in the shape of a cone. To the nearest hundredth, how much frosting will fit into a cake-decorating bag that has a diameter of 7 inches and a height of 12 inches? *(Lesson 12–5)*

7 in.

12 in.

24. A square pyramid has a base with sides 3.6 meters long and with a slant height of 6.5 meters. *(Lesson 12–4)*

    a. Draw a figure to represent this situation.

    b. Find the surface area of the pyramid to the nearest hundredth.

Identify the figures used to create each tessellation. Then identify the tessellation as *regular*, *semi-regular*, or *neither*. *(Lesson 10–7)*

25.

26.

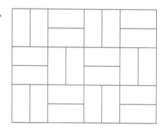

27. Short Response Determine the scale factor of △MNP to △RST.

(Lesson 9–7)

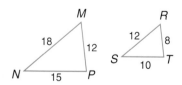

28. Multiple Choice Which of the following represents the distance between the points with coordinates $(a, 0)$ and $(0, b)$? *(Lesson 6–7)*

    Ⓐ $\sqrt{2a + 2b}$ Ⓑ $\sqrt{a^2 + b^2}$ Ⓒ $\sqrt{a + b}$ Ⓓ $\sqrt{a^2 - b^2}$

What You'll Learn

You'll learn to identify and use the relationships between similar solid figures.

Why It's Important

Automotive Design
Car designers make models that are similar to proposed new cars. *See Exercise 6.*

Similar solids are solids that have the same shape but are not necessarily the same size. Just as with similar polygons, corresponding linear measures have equivalent ratios. For example, two triangles are similar if the ratio of their corresponding sides are equal. In other words, the ratio of their corresponding sides are proportional.

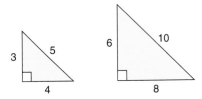

$$\frac{8}{4} = \frac{6}{3} = \frac{10}{5}$$

The two cones shown below are similar because the ratios of the diameters to the bases are proportional, that is, $\frac{6}{12} = \frac{4}{8}$.

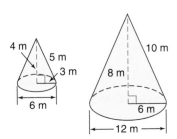

Recall that the ratio of measures is called a *scale factor*.

Characteristics of Similar Solids

Words:	For similar solids, the corresponding lengths are proportional, and the corresponding faces are similar.
Model:	

Symbols: $\dfrac{AB}{EF} = \dfrac{BC}{FG} = \dfrac{CA}{GE} = \dfrac{AD}{EJ} = \dfrac{BD}{FJ} = \dfrac{CD}{GJ} = \dfrac{h_1}{h_2} = \dfrac{\ell_1}{\ell_2};$

$\triangle ABC \sim \triangle EFG, \triangle ABD \sim \triangle EFJ,$
$\triangle BCD \sim \triangle FGJ, \triangle ACD \sim \triangle EGJ$

Determine whether each pair of solids is similar.

1

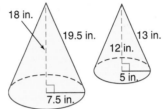

2

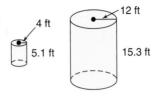
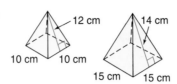

$$\frac{10}{15} \stackrel{?}{=} \frac{12}{14}$$

$$10(14) \stackrel{?}{=} 15(12)$$

$$140 \neq 180$$

The pyramids are not similar.

$$\frac{4}{12} \stackrel{?}{=} \frac{5.1}{15.3}$$

$$4(15.3) \stackrel{?}{=} 12(5.1)$$

$$61.2 = 61.2 \;\checkmark$$

The cylinders are similar.

Your Turn

a.

b.

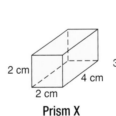
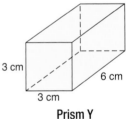

Online Personal Tutor at **geomconcepts.com**

Study the two similar prisms at the right. The scale factor of prism X to prism Y is $\frac{2}{3}$. What is the ratio of the surface area of prism X to the surface area of prism Y?

Prism X

Prism Y

Surface Area of Prism X	Surface Area of Prism Y
$S = Ph + 2B$	$S = Ph + 2B$
$\;\;= 12(2) + 2(8)$	$\;\;= 18(3) + 2(18)$
$\;\;= 40$	$\;\;= 90$

The ratio of the surface area of prism X to the surface area of prism Y is $\frac{40}{90}$ or $\frac{4}{9}$. Notice that $\frac{4}{9} = \frac{2^2}{3^2}$.

Now, compare the ratio of the volume of prism X to the volume of prism Y.

Volume of Prism X	Volume of Prism Y
$V = Bh$	$V = Bh$
$\;\;= 8(2)$	$\;\;= 18(3)$
$\;\;= 16$	$\;\;= 54$

The ratio of the volume of prism X to the volume of prism Y is $\frac{16}{54}$ or $\frac{8}{27}$. Notice that $\frac{8}{27} = \frac{2^3}{3^3}$.

The relationships between prism X and prism Y suggest Theorem 12–15.

Theorem 12–15

Words: If two solids are similar with a scale factor of $a{:}b$, then the surface areas have a ratio of $a^2{:}b^2$ and the volumes have a ratio of $a^3{:}b^3$.

Model:

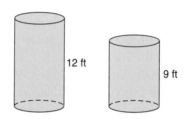

Solid A   Solid B

Symbols: scale factor of solid A to solid B $= \dfrac{a}{b}$

$\dfrac{\text{surface area of solid A}}{\text{surface area of solid B}} = \dfrac{a^2}{b^2}$

$\dfrac{\text{volume of solid A}}{\text{volume of solid B}} = \dfrac{a^3}{b^3}$

Example ③ For the similar cylinders, find the scale factor of the cylinder on the left to the cylinder on the right. Then find the ratios of the surface areas and the volumes.

12 ft   9 ft

The scale factor is $\frac{12}{9}$ or $\frac{4}{3}$.

The ratio of the surface areas is $\frac{4^2}{3^2}$ or $\frac{16}{9}$.

The ratio of the volumes is $\frac{4^3}{3^3}$ or $\frac{64}{27}$.

Your Turn

For each pair of similar solids, find the scale factor of the solid on the left to the solid on the right. Then find the ratios of the surface areas and the volumes.

c.

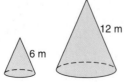

6 m   12 m

d.

6 cm   15 cm

Example **4**

Social Studies Link

Mrs. Gomez's social studies class is using cardboard to build a scale model of the Great Pyramid of Khufu in Egypt. The surface area of this pyramid is about 1,496,510 square feet. If the scale factor of the model to the original is 1:100, how much cardboard will the class need to make the model?

Let S represent the surface area of the model.

$$\frac{\text{surface area of model}}{\text{surface area of Great Pyramid}} = \frac{1^2}{100^2} \qquad \textit{Theorem 12–15}$$

$$\frac{S}{1,496,510} = \frac{1}{10,000} \qquad \textit{Substitution}$$

$$S(10,000) = 1,496,510(1) \qquad \textit{Cross Products}$$

$$10,000S = 1,496,510 \qquad \textit{Multiply.}$$

$$\frac{10,000S}{10,000} = \frac{1,496,510}{10,000} \qquad \textit{Divide each side by 10,000.}$$

$$S = 149.651 \qquad \textit{Simplify.}$$

The class will need about 150 square feet of cardboard.

Great Pyramid of Khufu

Check for Understanding

Communicating Mathematics

1. **Explain** the meaning of similar solids. Can two solids which have the same size and shape be similar? Explain.

2. **Draw** two spheres such that the ratio of their volumes is 1:64.

> **Vocabulary**
> similar solids

Guided Practice

Examples 1 & 2

Determine whether each pair of solids is similar.

3.
9 ft
15 ft
6 ft
9 ft

4.
9 m
3 m
12 m
4 m

Example 3

5. For the similar cones, find the scale factor of the cone on the left to the cone on the right. Then find the ratios of the surface areas and the volumes.

24 in.
16 in.
2:3

Example 4

6. **Automotive Design** Car designers often build clay models of the concept car they are creating.

   a. If the 30-inch model represents a 15-foot car, what is the scale factor of the model to the actual car? (*Hint*: Change feet to inches.)

   b. What is the ratio of the surface areas of the model to the actual car?

Practice

Determine whether each pair of solids is similar.

7.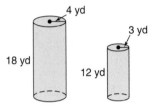

8.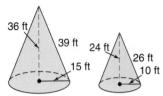

9.

10.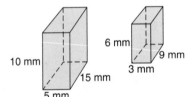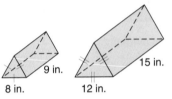

11.

12.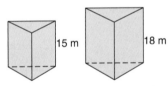

For each pair of similar solids, find the scale factor of the solid on the left to the solid on the right. Then find the ratios of the surface areas and the volumes.

13.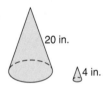

14.

15.

16.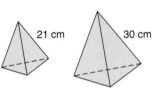

17. The dimensions of a prism are doubled.
   a. How does the surface area change?
   b. How does the volume change?

18. The ratio of the heights of two similar prisms is 5:3.
   a. Find the ratio of their surface areas.
   b. Find the ratio of their volumes.

19. The ratio of the surface areas of two similar cones is 9:16.
   a. What is the scale factor of the cones?
   b. What is the ratio of the volumes of the cones?

20. The ratio of the volumes of two similar pyramids is 27:1000.
   a. Find the scale factor of the pyramids.
   b. Find the ratio of the surface areas of the pyramids.

21. Baking   In 1989, a very large pecan pie was created for the Pecan Festival in Okmulgee, Oklahoma. The pie was 40 feet in diameter. If the pie was similar to a normal pie with an 8-inch diameter, find the ratio of volume of the large pie to the volume of the normal pie.

22. Miniatures   The Carole & Barry Kaye Museum of Miniatures in Los Angeles displays a tiny desk made from a thousand pieces of wood.
   a. If the 3-inch tall desk represents a real desk that is 30 inches tall, what is the scale factor of the miniature to the real desk?
   b. What is the ratio of the surface areas of the miniature to the real desk?
   c. What is the ratio of the volumes of the miniature to the real desk?

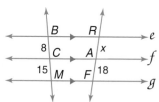

23. Critical Thinking   Explain why all cubes are similar to each other. Name another type of solid that is always similar to others in the category.

24. Sports   What is the surface area and volume of a racquetball if its diameter is 2.25 inches? Round to the nearest hundredth. *(Lesson 12–6)*

25. Find the volume of the rectangular pyramid to the nearest hundredth. *(Lesson 12–5)*

9 ft
7 ft
4 ft

26. Algebra   Find the measure of radius $\overline{JG}$ if $JG = 3x$ and $EF = 4x + 3$.   *(Lesson 11–1)*

E
G 3x J
4x + 3
F

27. Grid In   Find the measure of one exterior angle of a regular hexagon. *(Lesson 10–2)*

28. Multiple Choice   In the figure at the right, $e \parallel f \parallel g$. Find the value of x. *(Lesson 9–6)*

Ⓐ $8\frac{1}{4}$ Ⓑ $8\frac{2}{3}$

Ⓒ $9\frac{1}{2}$ Ⓓ $9\frac{3}{5}$

B   R   e
8 C   A x   f
15 M   F 18   g

Understanding and Using the Vocabulary

After completing this chapter, you should be able to define each term, property, or phrase and give an example or two of each.

*inter*NET
CONNECTION **Review Activities**
For more review activities, visit:
www.geomconcepts.com

axis *(p. 506)*	oblique cone *(p. 516)*	right prism *(p. 504)*
composite solid *(p. 498)*	oblique cylinder *(p. 506)*	right pyramid *(p. 516)*
cone *(p. 497)*	oblique prism *(p. 504)*	similar solids *(p. 534)*
cube *(p. 497)*	oblique pyramid *(p. 516)*	slant height *(p. 516)*
cylinder *(p. 497)*	Platonic solid *(p. 497)*	solid figures *(p. 496)*
edge *(p. 496)*	polyhedron *(p. 496)*	sphere *(p. 528)*
face *(p. 496)*	prism *(p. 496)*	surface area *(p. 504)*
lateral area *(p. 504)*	pyramid *(p. 496)*	tetrahedron *(p. 497)*
lateral edge *(p. 497)*	regular pyramid *(p. 516)*	truncated solid *(p. 521)*
lateral face *(p. 497)*	right cone *(p. 516)*	volume *(p. 510)*
net *(p. 507)*	right cylinder *(p. 506)*	

Choose the letter of the term that best matches each phrase.

1. the sum of the areas of a solid's surfaces
2. a special rectangular prism in which all of the faces are squares
3. the measurement of the space occupied by a solid region
4. the height of each lateral face of a regular pyramid
5. solids that have the same shape, but not necessarily the same size
6. solid with flat surfaces that are polygons
7. figure that encloses a part of space
8. another name for a special kind of triangular pyramid
9. the set of all points that are a given distance from the center
10. intersection of two faces of a polyhedron

a. polyhedron
b. volume
c. cube
d. surface area
e. slant height
f. sphere
g. edge
h. tetrahedron
i. similar solids
j. solid

Skills and Concepts

Objectives and Examples	Review Exercises

• **Lesson 12–1** Identify solid figures.

The polyhedron has 6 faces, 12 edges, and 8 vertices. It is a rectangular prism.

11. Name the faces, edges, and vertices of the polyhedron at the left.

Refer to the figure at the left to determine whether each statement is *true* or *false*.

12. *EFGH* is a lateral face.
13. *ABCD* and *EFGH* are bases.
14. *CG* is a lateral edge.

 www.geomconcepts.com/vocabulary_review

Objectives and Examples	Review Exercises

• Lesson 12–2 Find the lateral areas and surface areas of prisms and cylinders.

Prism
$L = Ph$
$S = Ph + 2B$

Cylinder
$L = 2\pi rh$
$S = 2\pi rh + 2\pi r^2$

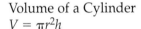

Find the lateral area and the surface area for each solid. Round to the nearest hundredth, if necessary.

15.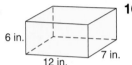
6 in. 12 in. 7 in.

16.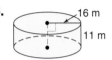
16 m, 11 m

• Lesson 12–3 Find the volumes of prisms and cylinders.

Volume of a Prism
$V = Bh$

Volume of a Cylinder
$V = \pi r^2 h$

Find the volume of each solid described. Round to the nearest hundredth, if necessary.

17. rectangular prism that is 5 in. by 9 in. by 9 in.

18. cylinder with a base diameter of 6 cm and a height of 5 cm

19. triangular prism with a base area of 4 m² and a height of 8 m

---

• Lesson 12–4 Find the lateral areas and surface areas of regular pyramids and cones.

Regular Pyramid
$L = \frac{1}{2}P\ell$
$S = \frac{1}{2}P\ell + B$

Cone
$L = \pi r\ell$
$S = \pi r\ell + \pi r^2$

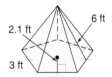

Find the lateral area and the surface area for each solid. Round to the nearest hundredth, if necessary.

20.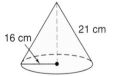
2.1 ft, 6 ft, 3 ft

21. 16 cm, 21 cm

• Lesson 12–5 Find the volumes of pyramids and cones.

Volume of a Pyramid
$V = \frac{1}{3}Bh$

Volume of a Cone
$V = \frac{1}{3}\pi r^2 h$

Find the volume of each solid. Round to the nearest hundredth, if necessary.

22.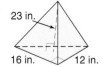
23 in., 16 in., 12 in.

23.
4 m, 3.5 m

Mixed Problem Solving
See pages 758-765.

Objectives and Examples

Review Exercises

- **Lesson 12–6** Find the surface areas and volumes of spheres.

Surface Area of a Sphere
$S = 4\pi r^2$

Volume of a Sphere
$V = \frac{4}{3}\pi r^3$

Find the surface area and volume of each sphere. Round to the nearest hundredth.

24.

6.5 yd

25.

7 cm

- **Lesson 12–7** Identify and use the relationships between similar solid figures.

The pyramids are similar.

$\frac{4}{6} = \frac{8}{12}$

$4(12) = 6(8)$

$48 = 48$

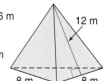

6 m

12 m

4 m 4 m

8 m 8 m

The scale factor of the pyramid on the left to the pyramid on the right is $\frac{1}{2}$. The ratio of the surface area is $\frac{1^2}{2^2}$ or $\frac{1}{4}$. The ratio of the volumes is $\frac{1^3}{2^3}$ or $\frac{1}{8}$.

Determine whether each pair of solids is similar.

26.

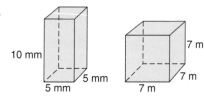

10 mm   5 mm   5 mm   7 m   7 m   7 m

27.

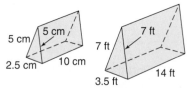

5 cm   5 cm   7 ft   7 ft
2.5 cm   10 cm   3.5 ft   14 ft

28. The ratio of the heights of two similar right cylinders is 3:4. Find the ratio of their surface areas and the ratio of their volumes.

Applications and Problem Solving

29. **Hobbies** Jeanette's aquarium is a regular hexagonal prism. Find the volume of water the aquarium holds when it is completely full to the nearest cubic inch. *(Lesson 12–3)*

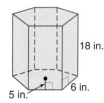

18 in.

5 in. 6 in.

30. **Landscaping** A truckload of fill dirt is dumped in front of a newly-built home. The pile of dirt is cone-shaped. It has a height of 7 feet and a diameter of 15 feet. Find the volume of the dirt to the nearest hundredth. *(Lesson 12–5)*

31. **Astronomy** Find the surface area and volume of the moon if its diameter is approximately 2160 miles. *(Lesson 12–6)*

1. **Compare and contrast** surface area and volume.
2. **Define** the term *sphere*, and name three common items that are shaped like spheres.

Determine whether each statement is *true* or *false* for the geometric solid.

3. The figure has 10 edges.
4. The figure is a polyhedron.
5. The figure is a circular cone.
6. The figure is a tetrahedron.

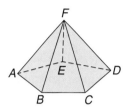

Find the lateral area and the surface area of each geometric solid. Round to the nearest hundredth, if necessary.

7.

4 yd
3 yd

8.

5.3 ft
2 ft

9.
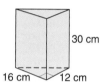
30 cm
16 cm   12 cm

10.

11 m
18 m

11.
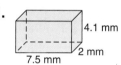
4.1 mm
7.5 mm   2 mm

12.
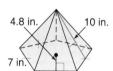
4.8 in.   10 in.
7 in.

Find the volume of each solid. Round to the nearest hundredth, if necessary.

13.
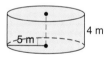
4 m
5 m

14.

7 in.
8 in.   9 in.

15.

15 ft
7 ft

16.

12 cm
5 cm   7 cm

17.
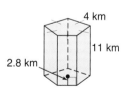
4 km
11 km
2.8 km

18.

8 m
4 m   3 m

19. **Recreation**   A beach ball has a diameter of 24 inches. Find the surface area and volume of the beach ball to the nearest hundredth.

20. **Storage**   ABC Lumber Company sells plans and materials for several storage sheds. The two designs shown have a similar shape, but differ in size.
   a. Find the scale factor of the shed on the left to the shed on the right.
   b. Find the ratios of the surface areas and the volumes.

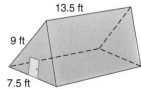
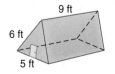
13.5 ft
9 ft   9 ft
6 ft
5 ft   7.5 ft

Angle, Line, and Arc Problems

Geometry problems on standardized tests often involve parallel lines and circles.

Review these concepts.

Angles: vertical angles, supplementary angles, complementary angles

Parallel lines: transversals, alternate interior angles

Circles: inscribed angles, central angles, arc length, tangent line

Example 1	**Example 2**

Example 1

Name each of the following in the figure below.

a. an arc

b. a sector

c. a chord

Hint Before taking a standardized test, it is a good idea to review vocabulary words.

Solution

a. *BF* is an arc.

b. *BCF* is a sector.

c. *AB* is a chord.

Example 2

In the figure below, line ℓ is parallel to line *m*. Line *n* intersects both ℓ and *m*, with angles 1, 2, 3, 4, 5, 6, 7, and 8 as shown. Which of the following lists includes all of the angles that are supplementary to ∠1?

Ⓐ angles 2, 4, 6, and 8

Ⓑ angles 3, 5, and 7

Ⓒ angles 2, 4, and 3

Ⓓ angles 5, 6, 7, and 8

Ⓔ angles 4, 3, 8, and 7

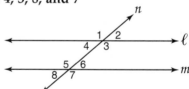

Hint Look for words like *supplementary*.

Solution Look carefully at the figure.

• Find ∠1. Notice that ∠1 and ∠2 form a linear pair, so ∠2 is supplementary to ∠1.

• Since ∠2 and ∠4 are vertical angles, they are equal in measure. So ∠4 is also supplementary to ∠1.

• Since ∠4 and ∠6 are alternate interior angles, they are equal. So ∠6 is supplementary to ∠1.

• And since ∠6 and ∠8 are vertical angles, ∠8 is supplementary to ∠1.

The angles supplementary to ∠1 are angles 2, 4, 6, and 8. The answer is A.

After you work each problem, record your answer on the answer sheet provided or on a sheet of paper.

Multiple Choice

1. The figures at the right are similar. Find the value of x. (Lesson 9–2)

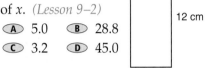

   (A) 5.0   (B) 28.8
   (C) 3.2   (D) 45.0

2. Which expression could be used to find the value of y? (Algebra Review)

   (A) $2x + 1$
   (B) $1 - 3x$
   (C) $3x - 1$
   (D) $3x + 1$

x	y
1	2
2	5
3	8
4	11

3. If line ℓ is parallel to line m in the figure below, what is the value of x? (Lesson 5–2)

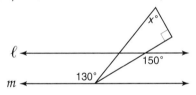

   (A) 20   (B) 50   (C) 70
   (D) 80   (E) 90

4. $5\frac{1}{3} - 6\frac{1}{4} = ?$ (Algebra Review)

   (A) $-\frac{11}{12}$ (B) $-\frac{1}{2}$ (C) $-\frac{2}{7}$
   (D) $\frac{1}{2}$ (E) $\frac{9}{12}$

5. Two number cubes are rolled. What is the probability that both number cubes will show a number less than 4? (Statistics Review)

   (A) $\frac{1}{4}$ (B) $\frac{1}{3}$ (C) $\frac{4}{9}$ (D) $\frac{1}{2}$

6. What is the slope of a line perpendicular to the line represented by the equation $3x - 6y = 12$? (Lesson 4–5)

   (A) -2 (B) $-\frac{1}{2}$ (C) $\frac{1}{3}$ (D) $\frac{1}{2}$

7. What is the height of a triangle with an area of 36 square centimeters and a base of 4 centimeters? (Lesson 10–4)

   (A) 9 cm   (B) 18 cm   (C) 36 cm   (D) 72 cm

8. Which statement about $\angle OAB$ is true? (Lesson 11–2)

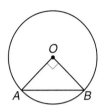

   (A) $\angle OAB$ measures more than 90°.
   (B) $\angle OAB$ measures less than 90° but more than 45°.
   (C) $\angle OAB$ measures exactly 45°.
   (D) $\angle OAB$ measures less than 45°.

Grid In

9. If ℓ_1 is parallel to ℓ_2 in the figure below, what is the value of y? (Lesson 5–2)

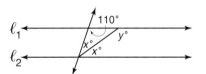

Extended Response

10. The height h, in feet, of a ball t seconds after being hit in the air from a height of 4 feet is $h = 4 + vt - 16t^2$, where v is the initial upward velocity. (Algebra Review)

   Part A   Make a table of values showing the height of a baseball hit with an initial upward velocity of 128 feet per second. Track the height for every half second for the first four seconds.

   Part B   How long will it take for the ball to reach a height of 256 feet on its way up?

13

Right Triangles and Trigonometry

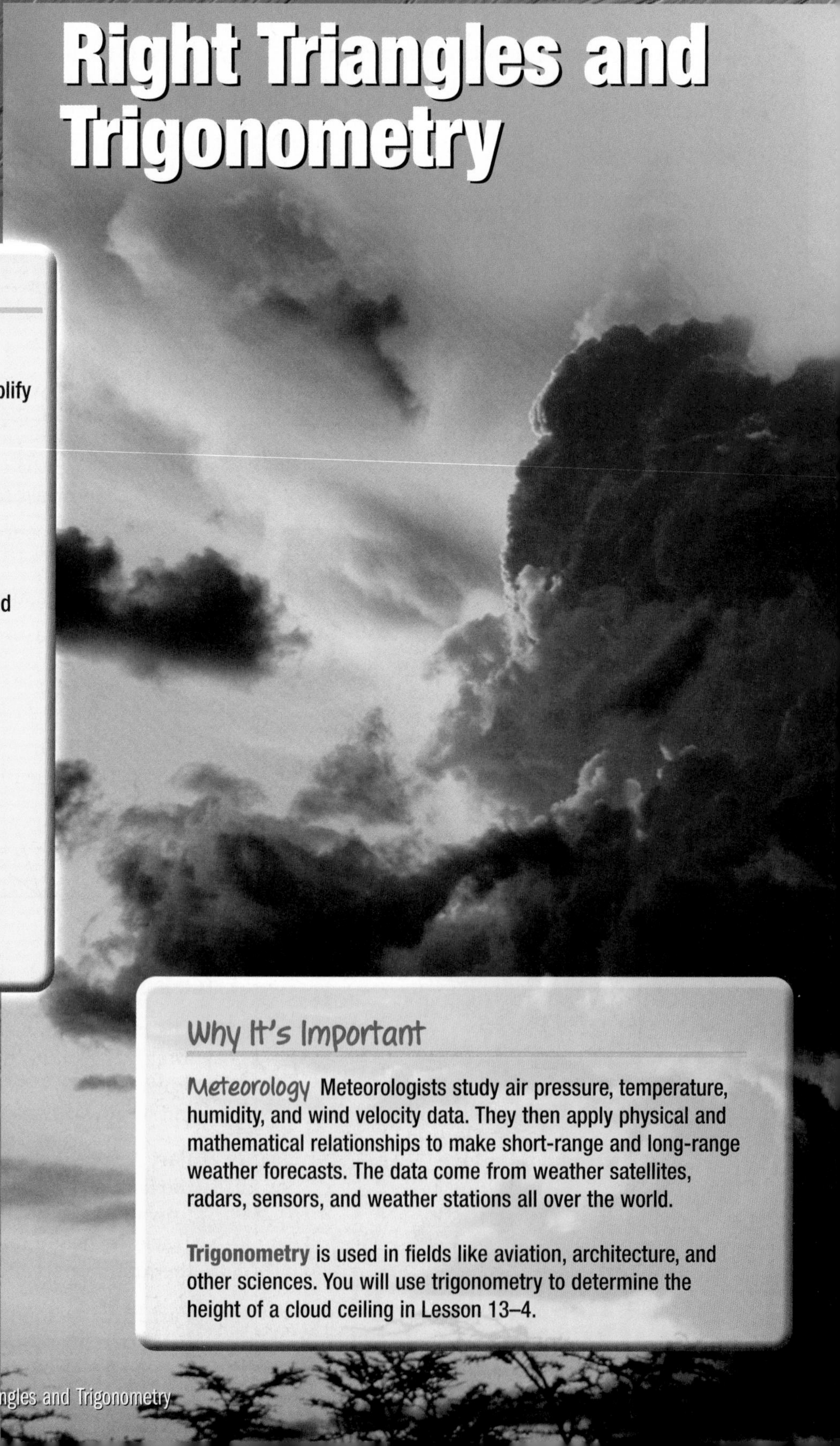

What You'll Learn

Key Ideas

- Multiply, divide, and simplify radical expressions. *(Lesson 13–1)*

- Use the properties of 45°–45°–90° and 30°–60°–90° triangles. *(Lessons 13–2 and 13–3)*

- Use the tangent, sine, and cosine ratios to solve problems. *(Lessons 13–4 and 13–5)*

Key Vocabulary

cosine *(p. 572)*

sine *(p. 572)*

tangent *(p. 564)*

radical expression *(p. 549)*

trigonometry *(p. 564)*

Why It's Important

Meteorology Meteorologists study air pressure, temperature, humidity, and wind velocity data. They then apply physical and mathematical relationships to make short-range and long-range weather forecasts. The data come from weather satellites, radars, sensors, and weather stations all over the world.

Trigonometry is used in fields like aviation, architecture, and other sciences. You will use trigonometry to determine the height of a cloud ceiling in Lesson 13–4.

 **Check Your Readiness**

If *c* is the hypotenuse, find each missing measure. Round to the nearest tenth, if necessary.

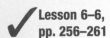 **Lesson 6–6,**
pp. 256–261

1. $a = 8, b = 14, c = ?$
2. $a = ?, b = 18, c = 26$
3. $a = 18, b = 18, c = ?$
4. $a = 6, b = ?, c = 12$
5. $a = ?, b = 22, c = 28$
6. $a = 31, b = 33, c = ?$

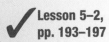 **Lesson 5–2,**
pp. 193–197

Find the value of each variable.

7.

8.

9.
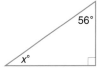

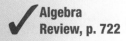 **Algebra**
Review, p. 722

Solve each equation. Check your solution.

10. $x = \dfrac{18}{24}$
11. $y = \dfrac{36}{24}$
12. $n = \dfrac{36}{8}$

13. $0.5 = \dfrac{x}{5}$
14. $0.75 = \dfrac{y}{24}$
15. $0.62 = \dfrac{n}{12}$

16. $0.25 = \dfrac{x}{5}$
17. $0.44 = \dfrac{y}{28}$
18. $0.16 = \dfrac{n}{44}$

FOLDABLES™
Study Organizer

Make this Foldable to help you organize your Chapter 13 notes. Begin with three sheets of notebook paper.

❶ **Stack** the sheets of paper with edges $\dfrac{1}{4}$ inch apart.

❷ **Fold** up the bottom edges. All the tabs should be the same size.

❸ **Crease** and staple along the fold.

❹ **Turn** and label the tabs with the lesson titles.

Reading and Writing As you read and study the chapter, use each page to write main ideas, theorems, and examples for each lesson.

What You'll Learn
You'll learn to multiply, divide, and simplify radical expressions.

Why It's Important
Aviation Pilots use a formula with a radical expression to determine the distance to the horizon.
See Example 9.

Squaring a number means multiplying a number by itself. The area of a square is found by squaring the measure of a side.

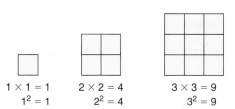

$1 \times 1 = 1$ $2 \times 2 = 4$ $3 \times 3 = 9$ $4 \times 4 = 16$
$1^2 = 1$ $2^2 = 4$ $3^2 = 9$ $4^2 = 16$

The numbers 1, 4, 9, and 16 are called **perfect squares** because $1 = 1^2$, $4 = 2^2$, $9 = 3^2$, and $16 = 4^2$. Note that perfect squares are products of two equal factors.

The inverse (opposite) of squaring is finding a **square root**. To find a square root of 16, find two equal factors whose product is 16. The symbol $\sqrt{}$, called a **radical sign**, is used to indicate the positive square root.

$$\sqrt{16} = 4 \text{ because } 4^2 = 16.$$

Read the symbol $\sqrt{16}$ as the square root of 16.

Examples

Simplify each expression.

1 $\sqrt{49}$

$\sqrt{49} = 7$ because $7^2 = 49$.

2 $\sqrt{64}$

$\sqrt{64} = 8$ because $8^2 = 64$.

Your Turn

a. $\sqrt{25}$ b. $\sqrt{144}$

There are many squares that have area measures that are *not* perfect squares. For example, the center square has an area of 12 square units.

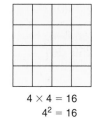

3 units | 9 units² | ? | 12 units² | 4 units | 16 units²
3 units | ? | 4 units

Look Back

Irrational Number: Lesson 2–1

However, there are no whole number values for $\sqrt{12}$. It is an irrational number. You can use a calculator to find an approximate value for $\sqrt{12}$.

2nd [$\sqrt{}$] 12 ENTER 3.464101615

A **radical expression** is an expression that contains a square root. The number under the radical sign is called a **radicand.** To simplify a radical expression, make sure that the radicand has no perfect square factors other than 1.

To simplify $\sqrt{12}$, use prime factorization to find a perfect square that is a factor of 12. The tree diagram shows two ways to find the prime factorization. The prime factorization of 12 is $2 \times 2 \times 3$.

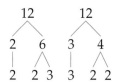

To complete the simplification of $\sqrt{12}$, use the following property.

Product Property of Square Roots	**Words:**	The square root of a product is equal to the product of each square root.
	Symbols:	$\sqrt{ab} = \sqrt{a} \cdot \sqrt{b}$ $a \geq 0, b \geq 0$
	Numbers:	$\sqrt{6} = \sqrt{2} \cdot \sqrt{3}$

Example 3

Simplify $\sqrt{12}$.

$\sqrt{12} = \sqrt{2 \cdot 2 \cdot 3}$ *The prime factorization of 12 is $2 \times 2 \times 3$.*

$= \sqrt{2 \cdot 2} \cdot \sqrt{3}$ *Use the Product Property of Square Roots to group any factors that occur in pairs.*

$= 2 \cdot \sqrt{3}$ $\sqrt{2 \cdot 2} = \sqrt{4}$ *or* 2

$= 2\sqrt{3}$ $2\sqrt{3}$ *means 2 times* $\sqrt{3}$.

Technology Tip

Use a calculator to find the value of $2\sqrt{3}$. Compare it to the value of $\sqrt{12}$.

Your Turn

c. $\sqrt{8}$ d. $\sqrt{75}$ e. $\sqrt{20}$

You can also use the Product Property to multiply square roots.

Example 4

Simplify $\sqrt{3} \cdot \sqrt{6}$.

$\sqrt{3} \cdot \sqrt{6} = \sqrt{3 \cdot 6}$ *Product Property of Square Roots*

$= \sqrt{3 \cdot 3 \cdot 2}$ *Replace 6 with $3 \cdot 2$.*

$= \sqrt{3 \cdot 3} \cdot \sqrt{2}$ *Product Property of Square Roots*

$= 3 \cdot \sqrt{2}$ or $3\sqrt{2}$ $\sqrt{3 \cdot 3} = 3$

Your Turn

f. $\sqrt{5} \cdot \sqrt{10}$ g. $\sqrt{3} \cdot \sqrt{15}$ h. $\sqrt{3} \cdot \sqrt{7}$

You can divide square roots and simplify radical expressions that involve fractions by using the following property.

Quotient Property of Square Roots	**Words:**	The square root of a quotient is equal to the quotient of each square root.
	Symbols:	$\sqrt{\dfrac{a}{b}} = \dfrac{\sqrt{a}}{\sqrt{b}}$ $a \geq 0, b > 0$
	Numbers:	$\sqrt{\dfrac{5}{2}} = \dfrac{\sqrt{5}}{\sqrt{2}}$

Examples

Simplify each expression.

5 $\dfrac{\sqrt{16}}{\sqrt{8}}$

$\dfrac{\sqrt{16}}{\sqrt{8}} = \sqrt{\dfrac{16}{8}}$ *Quotient Property*

$= \sqrt{2}$ *Simplify.*

6 $\sqrt{\dfrac{9}{4}}$

$\sqrt{\dfrac{9}{4}} = \dfrac{\sqrt{9}}{\sqrt{4}}$ *Quotient Property*

$= \dfrac{3}{2}$ *Simplify.*

i. $\dfrac{\sqrt{81}}{\sqrt{100}}$

j. $\sqrt{\dfrac{49}{64}}$

Reading Geometry

The process of simplifying a fraction with a radical in the denominator is called *rationalizing the denominator.*

To simplify a radical expression with fractions, make sure that it does not have a radical in the denominator. Remember that a fraction can be changed to an equivalent fraction by multiplying it by another fraction that is equivalent to 1. To simplify $\dfrac{\sqrt{3}}{\sqrt{5}}$, multiply it by $\dfrac{\sqrt{5}}{\sqrt{5}}$ because $\sqrt{5} \cdot \sqrt{5} = \sqrt{25}$ or 5, and the new denominator will not have a radical sign.

Example

7 Simplify $\dfrac{\sqrt{3}}{\sqrt{5}}$.

$\dfrac{\sqrt{3}}{\sqrt{5}} = \dfrac{\sqrt{3}}{\sqrt{5}} \cdot \dfrac{\sqrt{5}}{\sqrt{5}}$ $\dfrac{\sqrt{5}}{\sqrt{5}} = 1$

$= \dfrac{\sqrt{3 \cdot 5}}{\sqrt{5 \cdot 5}}$ *Product Property of Square Roots*

$= \dfrac{\sqrt{15}}{5}$ $\sqrt{5 \cdot 5} = 5$

Example —⑧ Simplify $\dfrac{2}{\sqrt{3}}$.

$$\dfrac{2}{\sqrt{3}} = \dfrac{2}{\sqrt{3}} \cdot \dfrac{\sqrt{3}}{\sqrt{3}} \qquad \dfrac{\sqrt{3}}{\sqrt{3}} = 1$$

$$= \dfrac{2 \cdot \sqrt{3}}{\sqrt{3} \cdot \sqrt{3}} \qquad \textit{Product Property of Square Roots}$$

$$= \dfrac{2\sqrt{3}}{\sqrt{9}} \qquad \sqrt{3} \cdot \sqrt{3} = \sqrt{9}$$

$$= \dfrac{2\sqrt{3}}{3} \qquad 2\sqrt{3} \textit{ is in simplest form, and } \sqrt{9} = 3.$$

Your Turn

k. $\dfrac{\sqrt{7}}{\sqrt{2}}$

l. $\dfrac{4}{\sqrt{3}}$

A radical expression is said to be in **simplest form** when the following conditions are met.

Rules for Simplifying Radical Expressions	**1.** There are no perfect square factors other than 1 in the radicand.
	2. The radicand is not a fraction.
	3. The denominator does not contain a radical expression.

When working with square roots of numbers that are not perfect squares, the radical form is used to show an exact number. However, decimal approximations are often used when applying radical expressions in real-life situations.

Real World

Example —⑨

Aviation Link

Pilots use the formula $d = 1.5\sqrt{h}$ to find the distance in miles that an observer can see under ideal conditions. In the formula, d is the distance in miles, and h is the height in feet of the plane. If an observer on a plane is flying at a height of 2000 feet, how far can he or she see? Round to the nearest mile.

$d = 1.5\sqrt{h}$

$\quad = 1.5 \cdot \sqrt{2000} \qquad \textit{Replace h with 2000.}$

1.5 [2nd] [√] 2000 [ENTER] *67.08203932*

To the nearest mile, the observer can see a distance of about 67 miles.

Communicating Mathematics

1. **Write** the symbol for the square root of 100.

2. **Find** the next three perfect squares after 16.

3. **You Decide?** Talisa says that $\sqrt{30}$ is in simplest form. Robbie says $\sqrt{30}$ is *not* in simplest form. Who is correct? Explain.

Guided Practice

⊕ Getting Ready Simplify each expression.

Sample: $\sqrt{5} \cdot \sqrt{5}$	**Solution:** 5

4. $\sqrt{7} \cdot \sqrt{7}$ 5. $\sqrt{2} \cdot \sqrt{2}$ 6. $\sqrt{11} \cdot \sqrt{11}$

Simplify each expression.

Examples 1 & 2   7. $\sqrt{36}$ 8. $\sqrt{81}$

Examples 3 & 4   9. $\sqrt{27}$ 10. $\sqrt{2} \cdot \sqrt{10}$

Examples 5 & 6   11. $\dfrac{\sqrt{45}}{\sqrt{15}}$ 12. $\sqrt{\dfrac{25}{36}}$

Examples 7 & 8   13. $\dfrac{\sqrt{7}}{\sqrt{3}}$ 14. $\dfrac{1}{\sqrt{5}}$

Example 9   15. **Buildings** The Empire State Building is about 1500 feet tall. From that height, how far can an observer see? Use the formula in Example 9 and round to the nearest mile.

Exercises •

Practice

Simplify each expression.

16. $\sqrt{100}$ 17. $\sqrt{121}$ 18. $\sqrt{28}$ 19. $\sqrt{32}$

20. $\sqrt{50}$ 21. $\sqrt{48}$ 22. $\sqrt{45}$ 23. $\sqrt{200}$

24. $\sqrt{2} \cdot \sqrt{6}$ 25. $\sqrt{5} \cdot \sqrt{15}$ 26. $\sqrt{3} \cdot \sqrt{5}$ 27. $\sqrt{8} \cdot \sqrt{9}$

28. $\dfrac{\sqrt{9}}{\sqrt{16}}$ 29. $\dfrac{\sqrt{30}}{\sqrt{5}}$ 30. $\dfrac{\sqrt{21}}{\sqrt{7}}$ 31. $\sqrt{\dfrac{16}{81}}$ 32. $\dfrac{\sqrt{16}}{\sqrt{4}}$

33. $\dfrac{\sqrt{3}}{\sqrt{2}}$ 34. $\dfrac{\sqrt{2}}{\sqrt{3}}$ 35. $\dfrac{4}{\sqrt{7}}$ 36. $\sqrt{\dfrac{5}{10}}$ 37. $\dfrac{1}{\sqrt{8}}$

Homework Help	
For Exercises	**See Examples**
16–23, 38, 41, 42	1–3
24–27, 39, 40, 44	4
28–37	5–8
43	9
Extra Practice	
See page 750.	

38. What is the square root of 400?

39. Multiply $\sqrt{19}$ and $\sqrt{2}$.

40. Write $\sqrt{14} \cdot \sqrt{2}$ in simplest form.

Applications and Problem Solving

41. Comics Help the character from *Shoe* find the square root of 225.

42. Measurement The area of a square is 50 square meters. In simplest form, find the length of one of its sides.

43. Fire Fighting The velocity of water discharged from a nozzle is given by the formula $V = 12.14\sqrt{P}$, where V is the velocity in feet per second and P is the pressure at the nozzle in pounds per square inch. Find the velocity of water if the nozzle pressure is 64 pounds per square inch.

44. Critical Thinking You know that $\sqrt{5} \cdot \sqrt{5} = 5$ and $\sqrt{2} \cdot \sqrt{2} = 2$. Find the value of $\sqrt{n} \cdot \sqrt{n}$ if $n \geq 0$. Explain your reasoning.

Mixed Review

45. Food The world's largest cherry pie was made by the Oliver Rotary Club of Oliver, British Columbia, Canada. It measured 20 feet in diameter and was completed on July 14, 1990. Most pies are 8 inches in diameter. If the largest pie was similar to a standard pie, what is the ratio of the volume of the larger pie to the volume of the smaller pie? *(Lesson 12–7)*

46. What is the volume of a sphere with a radius of 7.3 yards? *(Lesson 12–6)*

Refer to the figure for Exercises 47–49. *(Lesson 7–3)*

47. Find the shortest segment in the figure. (The figure is not drawn to scale.)

48. Which segment is longer, $\overline{AC}$ or $\overline{CE}$? Explain.

49. Find $m\angle CBD$. Does this show that $\overline{DC} \cong \overline{AB}$?

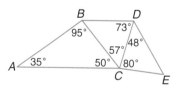

If $\triangle PRQ \cong \triangle YXZ$, $m\angle P = 63$, $m\angle Q = 57$, $XY = 10$, and $YZ = 11$, find each measure. *(Lesson 5–4)*

50. $m\angle R$

51. PQ

52. Multiple Choice A basketball player made 9 out of 40 free throws last season. What percent of the free throws did she make? *(Percent Review)*

　Ⓐ 0.225%　　Ⓑ 4.4%　　Ⓒ 22.5%　　Ⓓ 44.4%

13-2 45°-45°-90° Triangles

If you're a baseball fan, you know that home plate, first base, second base, and third base form the baseball "diamond." But geometrically, a baseball diamond is actually a square.

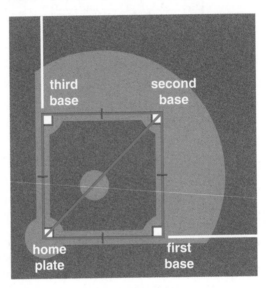

The line segment from home plate to second base is a diagonal of the square. The diagonal of a square separates the square into two **45°-45°-90° triangles**.

Hands-On Geometry

Materials: ruler   protractor

Step 1 Draw a square with sides 4 centimeters long. Label its vertices *A*, *B*, *C*, and *D*.

Step 2 Draw the diagonal $\overline{AC}$.

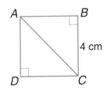

Try These

1. Use a protractor to measure ∠*CAB* and ∠*ACB*.
2. Use the Pythagorean Theorem to find *AC*. Write your answer in simplest form.
3. Repeat Exercise 2 for squares with sides 6 centimeters long and 8 centimeters long.
4. **Make a conjecture** about the length of the diagonal of a square with sides 7 inches long.

The results you discovered in the activity lead to Theorem 13–1.

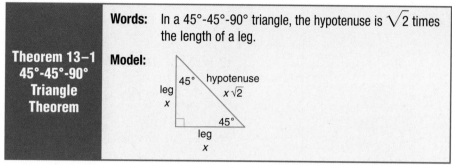

Theorem 13–1
45°-45°-90°
Triangle
Theorem

Words: In a 45°-45°-90° triangle, the hypotenuse is $\sqrt{2}$ times the length of a leg.

Model:

45°
leg
x
hypotenuse
$x\sqrt{2}$
45°
leg
x

A 45°-45°-90° triangle is also called an isosceles right triangle.

If you know the leg length of a 45°-45°-90° triangle, you can use the above theorem to find the hypotenuse length.

Real World

Example ❶

Baseball Link

An official baseball diamond is a square with sides 90 feet long. How far is it from home plate to second base? Round your answer to the nearest tenth.

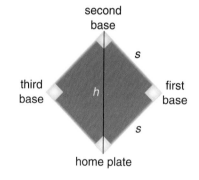

second base

third base

first base

s

h

s

home plate

Explore   The sides of a baseball diamond are 90 feet long. You need to find the distance from home plate to second base.

Plan   The triangle formed by first base, second base, and home plate is a 45°-45°-90° triangle. The distance h from home plate to second base is the length of the hypotenuse of the triangle. Let s represent the length of the legs.

Solve $h = s\sqrt{2}$ *The hypotenuse is $\sqrt{2}$ times the length of a leg.*

   $h = 90\sqrt{2}$ *Replace s with 90.*

   90 [2nd] [$\sqrt{}$] 2 [ENTER] *127.27922*

   To the nearest tenth, the distance from home plate to second base is 127.3 feet.

Examine   The value of $\sqrt{2}$ is about 1.5. So, the distance from home plate to second base is about 90 · 1.5 or 135. The answer seems reasonable.

Example — **2**

If △*PQR* is an isosceles right triangle and the measure of the hypotenuse is 12, find *s*. Write the answer in simplest form.

$h = s\sqrt{2}$ *The hypotenuse is $\sqrt{2}$ times the length of a leg.*

$12 = s\sqrt{2}$ *Replace h with 12.*

$\dfrac{12}{\sqrt{2}} = \dfrac{s\sqrt{2}}{\sqrt{2}}$ *Divide each side by $\sqrt{2}$.*

$\dfrac{12}{\sqrt{2}} = s$ *Simplify.*

$\dfrac{12}{\sqrt{2}} \cdot \dfrac{\sqrt{2}}{\sqrt{2}} = s$ *Simplify the radical expression.*

$\dfrac{12\sqrt{2}}{\sqrt{4}} = s$ *Product property of square roots*

$\dfrac{12\sqrt{2}}{2} = s$ *Simplify.*

Therefore, $s = \dfrac{12\sqrt{2}}{2}$ or $6\sqrt{2}$.

Your Turn

△*ABC* is an isosceles right triangle. Find *s* for each value of *h*.

a. 4   **b.** 5   **c.** $3\sqrt{2}$

Check for Understanding

Communicating Mathematics

1. **Draw and label** a 45°-45°-90° triangle in which the sides are 3 centimeters, 3 centimeters, and about 4.2 centimeters.

2. **Explain** two different methods for finding the length of the hypotenuse of a 45°-45°-90° triangle.

3. 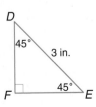 Jamie says that the length of a leg of △*DEF* is $3\sqrt{2}$ inches. Kyung says the length of a leg is $\dfrac{3\sqrt{2}}{2}$ inches. Who is correct?

Explain your reasoning.

Find the missing measures. Write all radicals in simplest form.

4.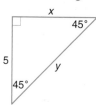

5.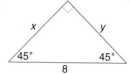

Example 1

6. **Maps**   On the map of Milwaukee, Wisconsin, the shape formed by Route 190, Route 57, and Route 145 closely resembles an isosceles right triangle. If the distances on Routes 190 and 57 are each 2.8 miles, find the distance on Route 145 between Route 190 and Route 57. Round to the nearest tenth.

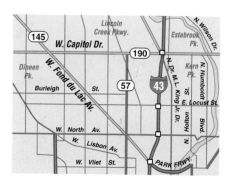

Exercises

• • • • • • • • • • • • • • • • •

Practice

Find the missing measures. Write all radicals in simplest form.

Homework Help	
For Exercises	**See Examples**
7, 8, 11, 14, 15, 17	1
9, 10, 12, 13, 16	2
Extra Practice	
See page 750.	

7.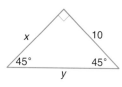

8.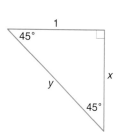

9.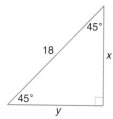

10.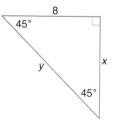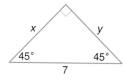

11.

12.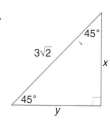

13. The length of the hypotenuse of an isosceles right triangle is $6\sqrt{2}$ feet. Find the length of a leg.

Applications and Problem Solving

14. **Measurement**   The length of one side of a square is 12 meters. To the nearest tenth, find the length of a diagonal of the square.

15. **Machine Technology**   A square bolt 2 centimeters on each side is to be cut from round stock. To the nearest tenth, what diameter stock is needed?

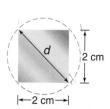

Exercise 15

16. **Logging**   In order to make flat boards from a log, a miller first trims off the four sides to make a square beam. Then the beam is cut into flat boards. If the diameter of the original log was 15 inches, find the maximum width of the boards. Round your answer to the nearest tenth.

  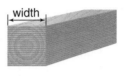

17. **Critical Thinking**   Use the figure at the right to find each measure.

   a. u **b.** v

   c. w **d.** x

   e. y **f.** z

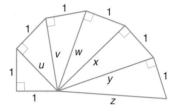

Mixed Review

Simplify each expression.   *(Lesson 13–1)*

18. $\sqrt{20}$ 19. $\sqrt{5} \cdot \sqrt{15}$ 20. $\dfrac{\sqrt{2}}{\sqrt{3}}$

21. **Manufacturing**   The Purely Sweet Candy Company has just released its latest candy. It is a chocolate sphere filled with a candy surprise. If the sphere is 6 centimeters in diameter, what is the minimum amount of foil paper it will take to wrap the sphere?   *(Lesson 12–6)*

Standardized Test Practice
Ⓐ Ⓑ Ⓒ Ⓓ

22. **Grid In**   In a circle with a radius of 7 inches, a chord is 5 inches from the center of the circle. To the nearest tenth of an inch, what is the length of the chord?   *(Lesson 11–3)*

23. **Short Response**   Draw a semi-regular tessellation. *(Lesson 10–7)*

Quiz 1   Lessons 13–1 and 13–2

Simplify each expression.   *(Lesson 13–1)*

1. $\sqrt{12}$ 2. $\sqrt{6} \cdot \sqrt{3}$ 3. $\dfrac{6}{\sqrt{2}}$

4. The measure of a leg of an isosceles right triangle is 15. Find the measure of the hypotenuse in simplest form.   *(Lesson 13–2)*

5. **Measurement**   The length of a diagonal of a square is $4\sqrt{2}$ inches.   *(Lesson 13–2)*
   a. Find the length of a side of the square.
   b. Find the perimeter of the square.

13-3 30°-60°-90° Triangles

What You'll Learn

You'll learn to use the properties of 30°-60°-90° triangles.

Why It's Important

Architecture
Architects use 30°-60°-90° triangles to design certain kinds of arches.
See Example 4.

An equilateral triangle has three equal sides and three equal angles. Because the sum of the measures of the angles in a triangle is 180°, the measure of each angle in an equilateral triangle is 60°. If you draw a median from vertex A to side $\overline{BC}$, the median bisects the angle A. The median of an equilateral triangle separates it into two **30°-60°-90° triangles**.

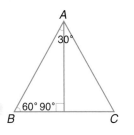

The activity explores some of the special properties of 30°-60°-90° triangles.

Hands-On Geometry

Materials: compass protractor ruler

Step 1 Construct an equilateral triangle with sides 2 in. long. Label its vertices A, B, and C.

Step 2 Find the midpoint of $\overline{AB}$ and label it D. Draw $\overline{CD}$, a median.

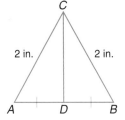

Try These

1. Use a protractor to measure $\angle ACD$, $\angle A$, and $\angle CDA$.

2. Use a ruler to measure $\overline{AD}$.

3. Copy and complete the table below. Use the Pythagorean Theorem to find CD. Write your answers in simplest form.

AC	AD	CD
2 in.		
4 in.		
3 in.		

4. Suppose the length of a side of an equilateral triangle is 10 inches. What values would you expect for AC (hypotenuse), AD (shorter leg), and CD (longer leg)?

The results you discovered in the activity lead to Theorem 13–2.

Theorem 13–2
30°-60°-90°
Triangle
Theorem

Words: In a 30°-60°-90° triangle, the hypotenuse is twice the length of the shorter leg, and the longer leg is $\sqrt{3}$ times the length of the shorter leg.

Model:

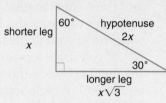

Examples ──① In $\triangle ABC$, $b = 7$. Find a and c. Write in simplest form.

Since b is opposite the 30° angle, b is the measure of the shorter leg. Therefore, a is the measure of the longer leg, and c is the measure of the hypotenuse.

Find c.
$c = 2b$ *The hypotenuse is twice the shorter leg.*
$c = 2(7)$ *Replace b with 7.*
$c = 14$ *Multiply.*

Find a.
$a = b\sqrt{3}$ *The longer leg is $\sqrt{3}$ times the length of the shorter leg.*
$a = 7\sqrt{3}$ *Replace b with 7.*

② In $\triangle ABC$, $c = 18$. Find a and b. Write in simplest form.

Since b is opposite the 30° angle, b is the measure of the shorter leg.

Find b.
$c = 2b$ *The hypotenuse is twice the shorter leg.*
$18 = 2b$ *Replace c with 18.*
$9 = b$ *Divide each side by 2.*

Now, find a.
$a = b\sqrt{3}$ *The longer leg is $\sqrt{3}$ times the shorter leg.*
$a = 9\sqrt{3}$ *Replace b with 9.*

a. Refer to $\triangle ABC$ above. If $b = 8$, find a and c.
b. Refer to $\triangle ABC$ above. If $c = 10$, find a and b.

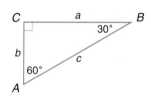

Example ③

In △*DEF*, *DE* = 12. Find *EF* and *DF*.
Write in simplest form.

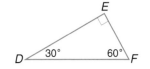

First, find *EF*.

$DE = (EF)\sqrt{3}$ *The longer leg is $\sqrt{3}$ times the shorter leg.*

$12 = (EF)\sqrt{3}$ *Replace DE with 12.*

$\dfrac{12}{\sqrt{3}} = \dfrac{(EF)\sqrt{3}}{\sqrt{3}}$ *Divide each side by $\sqrt{3}$.*

$EF = \dfrac{12}{\sqrt{3}} \cdot \dfrac{\sqrt{3}}{\sqrt{3}}$ $\dfrac{\sqrt{3}}{\sqrt{3}} = 1$ │   Now, find *DF*.

$EF = \dfrac{12\sqrt{3}}{3}$ $\sqrt{3} \cdot \sqrt{3} = 3$ │ $DF = 2(EF)$ *The hypotenuse is twice the shorter leg.*

$EF = 4\sqrt{3}$ *Simplify.*   │ $= 2(4\sqrt{3})$ *Replace EF with $4\sqrt{3}$.*

  │ $= (2 \cdot 4)\sqrt{3}$ *Associative Property*

  │ $= 8\sqrt{3}$ *Multiply.*

Your Turn

c. Refer to △*DEF* above. If *DE* = 8, find *EF* and *DF*.

●nline **Personal Tutor at geomconcepts.com**

You can use the properties of a 30°-60°-90° triangle to solve real-world problems that involve equilateral triangles.

Example ④
Architecture Link

The Gothic arch is based on an equilateral triangle. Find the height of the arch to the nearest tenth.

The height *h* separates the triangle into two 30°-60°-90° triangles. The hypotenuse is 12 feet, and the side opposite the 30° angle is 6 feet. The height of the arch is the length of the side opposite the 60° angle.

$h = 6\sqrt{3}$ *Theorem 13–2*

6 [2nd] [√] 3 [ENTER] *10.39230485*

Math Data
●nline **Update**
For more
information on Gothic
architecture, visit:
geomconcepts.com

To the nearest tenth, the height of the arch is 10.4 feet.

Communicating Mathematics

1. **Draw and label** a 30°-60°-90° triangle in which the sides are 5 inches, 10 inches, and $5\sqrt{3}$ inches.

2. **Writing Math** Compare and contrast the 30°-60°-90° Triangle Theorem and the 45°-45°-90° Triangle Theorem.

Guided Practice

Examples 1–3

Find the missing measures. Write all radicals in simplest form.

3.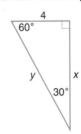

4.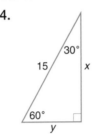

5.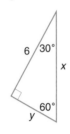

Example 4

6. **Design** The hexagons in the stained-glass window are made of equilateral triangles. If the length of a side of a triangle is 14 centimeters, what is the height of the triangle? Round to the nearest tenth.

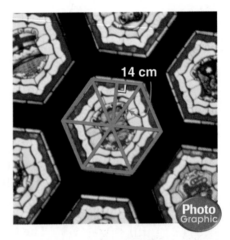

14 cm

Photo Graphic

Find the missing measures. Write all radicals in simplest form.

Practice

Homework Help	
For Exercises	See Examples
7, 9–11, 13–17	1, 3
8, 12	2, 4
Extra Practice	
See page 751.	

7.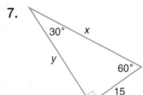

8.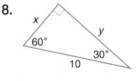

9.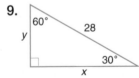

10.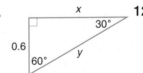

11.

12.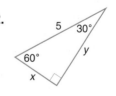

13.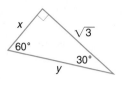

14.

15.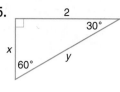

16. The length of the shorter leg of a 30°-60°-90° triangle is 24 meters. Find the length of the hypotenuse.

Applications and Problem Solving

17. Measurement At the same time that the sun's rays make a 60° angle with the ground, the shadow cast by a flagpole is 24 feet. To the nearest foot, find the height of the flagpole.

18. Measurement The length of one side of an equilateral triangle is 10 meters.
   a. Find the length of an altitude.
   b. Find the area of the triangle.

Exercise 17

19. Critical Thinking A regular hexagon is made up of six congruent equilateral triangles. Find the area of a regular hexagon whose perimeter is 24 feet.

Mixed Review

The measure of one side of the triangle is given. Find the missing measures to complete the table. *(Lesson 13–2)*

	a	b	c
20.	18		
21.		$5\sqrt{2}$	

22. Electricity The current that can be generated in a circuit is given by the formula $I = \sqrt{\dfrac{P}{R}}$, where I is the current in amperes, P is the power in watts, and R is the resistance in ohms. Find the current when $P = 25$ watts and $R = 400$ ohms. *(Lesson 13–1)*

Standardized Test Practice
Ⓐ Ⓑ Ⓒ Ⓓ

23. Grid In Find the distance between $A(0, 0)$ and $B(3, 4)$. *(Lesson 6–7)*

24. Multiple Choice Line ℓ has a slope of $\dfrac{1}{2}$ and a y-intercept of -7. Which of the following is the equation of ℓ written in slope-intercept form? *(Algebra Review)*

Ⓐ $y = \dfrac{1}{2}x - 7$ Ⓑ $y = \dfrac{1}{2}x + 7$

Ⓒ $2x + y = 7$ Ⓓ $x - 2y = 14$

What You'll Learn
You'll learn to use the tangent ratio to solve problems.

Why It's Important
Surveying Surveyors use the tangent ratio to find distances that cannot be measured directly.
See Example 2.

In Pittsburgh, Pennsylvania, the Duquesne Incline transports passengers from the river valley up to Mount Washington. The Incline has a 403-foot rise and a 685-foot run. What angle is made by the track and the run? *This problem will be solved in Example 4.*

Photo Graphic

403 ft

685 ft

The Duquesne Incline

Trigonometry is the study of the properties of triangles. The word *trigonometry* comes from two Greek words meaning *angle measurement*. A **trigonometric ratio** is a ratio of the measures of two sides of a right triangle. Using trigonometric ratios, you can find the unknown measures of angles and sides of right triangles.

One of the most common trigonometric ratios is the **tangent** ratio.

Reading Geometry
The symbol tan *A* is read *the tangent of angle A.*

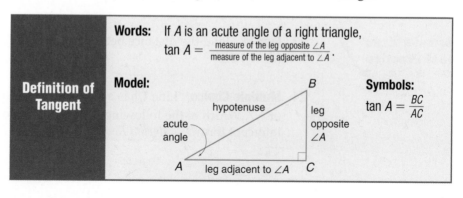

Definition of Tangent	**Words:** If *A* is an acute angle of a right triangle, $\tan A = \dfrac{\text{measure of the leg opposite } \angle A}{\text{measure of the leg adjacent to } \angle A}$.

Model:

hypotenuse

acute angle

leg opposite ∠A

A leg adjacent to ∠A *C*

B

Symbols:
$\tan A = \dfrac{BC}{AC}$

Example ➊ **Find tan A and tan B.**

$\tan A = \dfrac{BC}{AC}$ *opposite* / *adjacent*

    $= \dfrac{9}{40}$ or 0.225   *Replace BC with 9 and AC with 40.*

$\tan B = \dfrac{AC}{BC}$ *AC is the leg opposite $\angle B$.*
                   BC is the leg adjacent to $\angle B$.

    $= \dfrac{40}{9}$ or 4.4444   *Replace AC with 40 and BC with 9.*

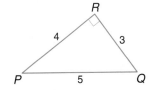

Reading Geometry

When trigonometric ratios are expressed as decimals, they are usually rounded to four decimal places.

Your Turn

a. Find tan P and tan Q.

The tangent of an angle in a right triangle depends on the measure of the angle, not on the size of the triangle. So, an expression like tan 70° has a unique value. You can use the TAN function on your calculator to find the value of tan 70°, or you can find the value in a trigonometric table.

If you know the measure of an angle and one leg, you can use the tangent ratio to find distances that cannot be measured directly.

Example ➋

Surveying Link

A surveyor standing at the edge of a canyon made the measurements shown at the right. Find the distance across the canyon from D to F.

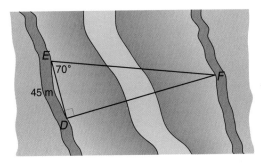

$\tan 70° = \dfrac{DF}{DE}$ *opposite* / *adjacent*

$\tan 70° = \dfrac{DF}{45}$ *Replace DE with 45.*

$45 \cdot \tan 70° = \cancel{45} \cdot \dfrac{DF}{\cancel{45}}$ *Multiply each side by 45.*

$45 \cdot \tan 70° = DF$ *Use a calculator. Be sure it is in degree mode.*

$45 \boxed{\times} \boxed{\text{TAN}} 70 \boxed{\text{ENTER}}$ *123.6364839*

Therefore, the distance across the canyon is about 123.6 meters.

Some applications of trigonometry use an angle of elevation or angle of depression. In the photo below, the angle made by the line of sight from the boat and a horizontal line is called an **angle of elevation**. The angle made by the line of sight from the parasail and a horizontal line is called an **angle of depression**.

You can use parallel lines and alternate interior angles to show that the angle of elevation and the angle of depression are congruent.

Look Back

Alternate Interior
Angles: Lesson 4–2

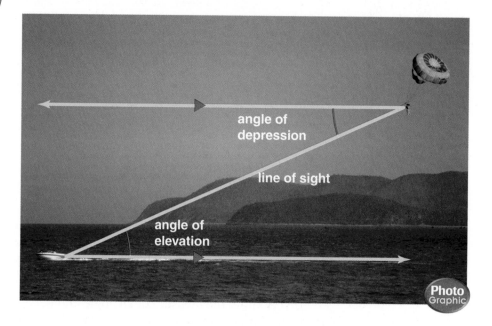

angle of
depression

line of sight

angle of
elevation

Photo
Graphic

Real World

Example ❸

Forestry Link

A ranger standing 100 feet from a tree sights the top of the tree at a 40° angle of elevation. Find the height of the tree.

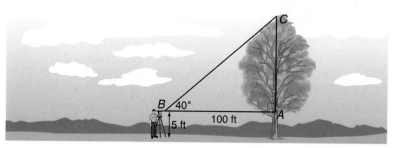

$$\tan 40° = \frac{AC}{AB} \qquad \frac{opposite}{adjacent}$$

$$\tan 40° = \frac{AC}{100} \qquad \textit{Replace AB with 100.}$$

$$100 \cdot \tan 40° = 100 \cdot \frac{AC}{100} \qquad \textit{Multiply each side by 100.}$$

$$100 \cdot \tan 40° = AC \qquad \textit{Use a calculator.}$$

100 $\boxed{\times}$ $\boxed{\text{TAN}}$ 40 $\boxed{\text{ENTER}}$ *83.90996312*

The height of the ranger is 5 feet. Therefore, the height of the tree is about 83.9 + 5 or 88.9 feet.

You can use the TAN^{-1} function on your calculator to find the measure of an acute angle of a right triangle when you know the measures of the legs. The TAN^{-1} function is called the *inverse tangent*. *The inverse tangent is also called the* **arctangent**.

Example 4

Real World

Engineering Link

Refer to the beginning of the lesson. Find the angle between the track and the run of the Duquesne Incline to the nearest tenth.

$$\tan A = \frac{BC}{AB} \quad \frac{opposite}{adjacent}$$

$$= \frac{403}{685} \quad \textit{Replace BC with 403 and AB with 685.}$$

Now, use a calculator to find the measure of $\angle A$, an angle whose tangent ratio is $\frac{403}{685}$. $m\angle A = tan^{-1}\left(\frac{403}{685}\right)$

[2nd] [TAN^{-1}] 403 [÷] 685 [ENTER] *30.46920015*

To the nearest tenth, the measure of $\angle A$ is about 30.5°.

Your Turn

b. The access ramp to a parking lot has a rise of 10 feet and a run of 48 feet. To the nearest degree, find the angle between the ramp and the run.

Check for Understanding

Communicating Mathematics

1. **Write** a definition of *tangent*.

2. **Draw** right triangle *LMN* in which $\angle N$ is an acute angle. Label the leg opposite $\angle N$ and the leg adjacent to $\angle N$.

> **Vocabulary**
> trigonometry
> trigonometric ratio
> tangent
> angle of elevation
> angle of depression

Guided Practice
Example 1

Find each tangent. Round to four decimal places, if necessary.

3. tan *J*

4. tan *K*

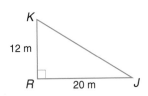

Find each missing measure. Round to the nearest tenth.

5.

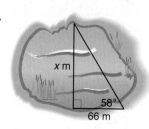

6.

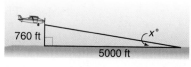

Example 4

7. **Travel** The distance from a boat to a bridge is 200 meters. A person aboard measures the angle of elevation to the bridge as 12°. To the nearest tenth, how far above the water is the bridge?

Exercises

· · · · · ● · · · · · · · · · · · · · · · · ·

Practice

Homework Help	
For Exercises	See Examples
8–11	1
12, 14, 17, 20	2, 3
13, 15, 16, 21	4
Extra Practice	
See page 751.	

Find each tangent. Round to four decimal places, if necessary.

8. tan *A*   9. tan *B*

10. tan *E*   11. tan *Z*

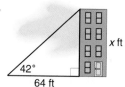

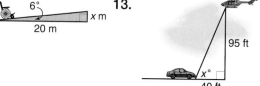

Find each missing measure. Round to the nearest tenth.

12.

13.

14.

15.

16.

17.

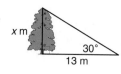

18. If the leg adjacent to a 29° angle in a right triangle is 9 feet long, what is the measure of the other leg to the nearest tenth?

Applications and Problem Solving

19. **Meteorology** A searchlight located 200 meters from a weather office is shined directly overhead. If the angle of elevation to the spot of light on the clouds is 35°, what is the altitude of the cloud ceiling? Round to the nearest tenth.

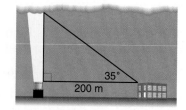

20. Farming A pile of corn makes an angle of 27.5° with the ground. If the distance from the center of the pile to the outside edge is 25 feet, how high is the pile of corn?

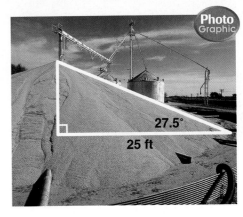

21. Critical Thinking In a right triangle, the tangent of one of the acute angles is 1. How are the measures of the two legs related?

Mixed Review

22. Find the missing measures. Write all radicals in simplest form. *(Lesson 13–3)*

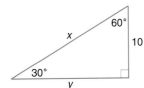

23. Sports Many younger children like to play a game similar to baseball called tee-ball. Instead of trying to hit a ball thrown by a pitcher, the batter hits the ball off a tee. To accommodate younger children, the bases are only 40 feet apart. Find the distance between home plate and second base. *(Lesson 13–2)*

Determine whether it is possible for a trapezoid to have the following conditions. Write *yes* or *no*. If *yes*, draw the trapezoid. *(Lesson 8–5)*

24. congruent diagonals

25. three obtuse angles

Standardized Test Practice
Ⓐ Ⓑ Ⓒ Ⓓ

26. Multiple Choice Which is *not* a name for this angle? *(Lesson 3–1)*

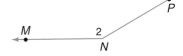

Ⓐ ∠MNP Ⓑ ∠2
Ⓒ ∠NPM Ⓓ ∠PNM

Quiz 2   Lessons 13–3 and 13–4

Find each missing measure. Write radicals in simplest form. Round decimals to the nearest tenth. *(Lessons 13–3 & 13–4)*

1.
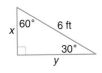

2.

3.

4.

5. Transportation The steepest grade of any standard railway system in the world is in France between Chedde and Servoz. The track rises 1 foot for every 11 feet of run. Find the measure of the angle formed by the track and the run. *(Lesson 13–4)*

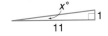

Chapter 13

Investigation

Materials

 protractor

 index card

 straws

 paper clips

 string

 tape

Indirect Measurement Using a Hypsometer

The General Sherman giant sequoia in Sequoia National Park, California, has been measured at 275 feet tall. But it's unlikely that someone actually measured the height with a tape measure. It's more likely that the height was calculated using trigonometry.

A **hypsometer** is an instrument that measures angles of elevation. You can use a hypsometer and what you know about the tangent ratio to find the heights of objects that are difficult to measure directly. *A hypsometer is sometimes called a clinometer.*

Investigate

1. Make a hypsometer by following these steps.

   a. Tape a protractor on an index card so that both zero points align with the edge of the card. Mark the center of the protractor on the card and label it *C*. Then mark and label every 10° on the card like the one shown below.

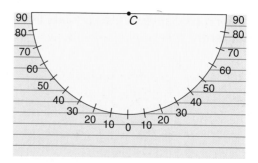

   b. Tie a piece of string to a large paper clip. Attach the other end of the string to the index card at *C*.

   c. Tape a straw to the edge of the index card that contains *C*.

2. To use the hypsometer, look through the straw at an object like the top of your classroom door. Ask another student to read the angle from the scale. This is the angle of elevation.

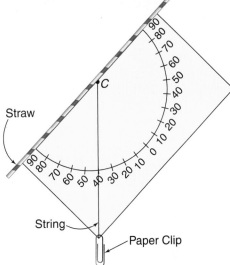

Straw

String

Paper Clip

3. Use your hypsometer to find the angle of elevation to the top of a tree or flagpole on your school property. Also, measure the distance from the hypsometer to the ground and from your foot to the base of the tree.

4. Make a sketch that you can use to find the height of the tree. Use your measurements from Exercise 3.

5. Explain how you can use trigonometry to find the height of the tree. Then find the height of the tree.

Extending the Investigation

In this project, you will use your hypsometer to find the height of three objects on your school property. Here are some suggestions.

- tree
- flagpole
- basketball hoop
- goal post
- school building

Presenting Your Conclusions

Here are some ideas to help you present your conclusions to the class.

- Make scale drawings that show the angle of elevation, distance to the ground, and distance from the object.
- Research *indirect measurement*. Write a paragraph that explains how indirect measurement was used in ancient times.

 Investigation For more information on indirect measurement,
visit: www.geomconcepts.com

13-5 Sine and Cosine Ratios

In Lesson 13–4, you learned about the tangent ratio. In a right triangle, the tangent ratio $\frac{BC}{AC}$ is related to the angle A.

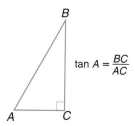

$$\tan A = \frac{BC}{AC}$$

Two other useful trigonometric ratios are the **sine** ratio and the **cosine** ratio. Both sine and cosine relate an angle measure to the ratio of the measures of a triangle's leg to its hypotenuse.

Definition of Sine and Cosine

Words: If A is an acute angle of a right triangle,

$$\sin A = \frac{\text{measure of the leg opposite } \angle A}{\text{measure of the hypotenuse}}, \text{ and}$$

$$\cos A = \frac{\text{measure of the leg adjacent to } \angle A}{\text{measure of the hypotenuse}}.$$

Model:

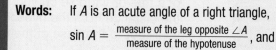

Symbols: $\sin A = \frac{BC}{AB}$ $\cos A = \frac{AC}{AB}$

Example

1 Find sin A, cos A, sin B, and cos B.

$\sin A = \frac{BC}{AB}$ *opposite / hypotenuse* $\sin B = \frac{AC}{AB}$

$\quad = \frac{4}{5}$ or 0.8 $\quad = \frac{3}{5}$ or 0.6

$\cos A = \frac{AC}{AB}$ *adjacent / hypotenuse* $\cos B = \frac{BC}{AB}$

$\quad = \frac{3}{5}$ or 0.6 $\quad = \frac{4}{5}$ or 0.8

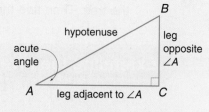

Your Turn   Find each value.

a. sin P **b.** sin R

c. cos P **d.** cos R

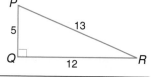

Example **2**

Recreation Link

The Aerial Ski Run in Snowbird, Utah, is 8395 feet long and, on average, has a 20° angle of elevation. What is the vertical drop? Round to the nearest tenth.

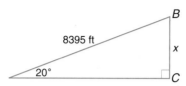

You know the length of the hypotenuse and the measure of ∠A. You need to find the measure of the leg opposite ∠A. Use the sine ratio.

Graphing Calculator Tutorial

See pp. 782–785.

$$\sin A = \frac{BC}{AB} \qquad \frac{opposite}{hypotenuse}$$

$$\sin 20° = \frac{x}{8395} \qquad \text{Replace } AB \text{ with 8395,} \\ BC \text{ with } x, \text{ and } A \text{ with 20°.}$$

$$8395 \cdot \sin 20° = 8395 \cdot \frac{x}{8395} \qquad \text{Multiply each side by 8395.}$$

$$8395 \cdot \sin 20° = x \qquad \text{Simplify.}$$

8395 $\boxed{\times}$ $\boxed{\text{SIN}}$ 20 $\boxed{\text{ENTER}}$ *2871.259103*

Therefore, the vertical drop in the Aerial Ski Run is about 2871.3 feet.

 Personal Tutor at geomconcepts.com

You can use the SIN^{-1} or COS^{-1} function on your calculator to find the measure of an acute angle of a right triangle when you know the measures of a leg and the measure of the hypotenuse. The SIN^{-1} function is called the *inverse sine*, and the COS^{-1} function is called the *inverse cosine*. *The inverse sine and cosine are also called the **arcsine** and **arccosine**, respectively.*

Example **3**

Find the measure of ∠H to the nearest degree.

You know the lengths of the side adjacent to ∠H and the hypotenuse. You can use the cosine ratio.

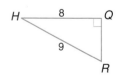

$$\cos H = \frac{HQ}{HR} \qquad \frac{adjacent}{hypotenuse}$$

$$\cos H = \frac{8}{9} \qquad \text{Replace } HQ \text{ with 8 and } HR \text{ with 9.}$$

Now, use a calculator to find the measure of ∠H, an angle whose tangent ratio is $\frac{8}{9}$. $m∠H = \cos^{-1}\left(\frac{8}{9}\right)$

$\boxed{\text{2nd}}$ [COS^{-1}] 8 $\boxed{\div}$ 9 $\boxed{\text{ENTER}}$ *27.26604445*

To the nearest degree, the measure of ∠H is about 27°.

e. Find the measure of $\angle R$ to the nearest degree if $HQ = 35$ and $HR = 37$.

There are many relationships that can be proven among trigonometric functions. Such relationships are called **trigonometric identities.** In the following theorem, the tangent function is defined as the ratio $\dfrac{\sin x}{\cos x}$. As long as $\cos x \neq 0$, this relationship is true and defines the tangent function. You can use a TI–92 to show that Theorem 13–3 is a trigonometric identity.

Theorem 13–3	If x is the measure of an acute angle of a right triangle, then $\dfrac{\sin x}{\cos x} = \tan x.$

 Graphing Calculator Exploration

To find $\sin 37°$, type $\boxed{\text{SIN}}$ 37 $\boxed{)}$ $\boxed{\text{ENTER}}$. To four decimal places, the result is 0.6018. *Make sure your calculator is in degree mode.*

Try These

1. Copy and complete the table below. Round to four decimal places, if necessary.

x	0°	15°	30°	45°	60°	75°
$\dfrac{\sin x}{\cos x}$						
$\tan x$						

2. Compare your results in rows 2 and 3.

3. What happens when you find both results for $x = 90°$? Why?

4. You can also use graphing to verify Theorem 13–3. Change the viewing window by entering $\boxed{\text{WINDOW}}$ -90 $\boxed{\text{ENTER}}$ 720 $\boxed{\text{ENTER}}$ 90 $\boxed{\text{ENTER}}$ -4 $\boxed{\text{ENTER}}$ 4 $\boxed{\text{ENTER}}$ 1 $\boxed{\text{ENTER}}$. Then enter $\boxed{\text{Y=}}$ $\boxed{\text{SIN}}$ $\boxed{\text{X,T,}\theta\text{,}n}$ $\boxed{)}$ $\boxed{\div}$ $\boxed{\text{COS}}$ $\boxed{\text{X,T,}\theta\text{,}n}$ $\boxed{)}$ $\boxed{\text{ENTER}}$ $\boxed{\text{GRAPH}}$. Repeat with $y = \tan x$. What do you notice about the graphs?

Check for Understanding

Communicating Mathematics

1. **Compare and contrast** the sine and cosine ratios.

Vocabulary

sine
cosine
trigonometric identities

2. **Draw** a right triangle *DEF* for which $\sin D = \frac{4}{5}$, $\cos D = \frac{3}{5}$, and $\tan D = \frac{4}{3}$.

3. **Writing Math** Some **O**ld **H**orse—**C**aught **A** **H**orse—**T**aking **O**ats **A**way is a helpful mnemonic device for remembering the trigonometric ratios. S, C, and T represent sine, cosine, and tangent, respectively, while O, H, and A represent opposite, hypotenuse, and adjacent, respectively. Make up your own mnemonic device for remembering the ratios.

Guided Practice

> **Getting Ready** Identify each segment in the figure below.
>
> | **Sample:** leg opposite ∠S | **Solution:** $\overline{RT}$ |

4. leg adjacent to ∠S
5. leg opposite ∠R
6. leg adjacent to ∠R

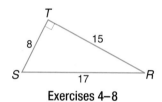

Exercises 4–8

Example 1 **Find each sine or cosine. Round to four decimal places, if necessary.**

7. $\sin R$

8. $\cos R$

Examples 2 & 3 **Find each missing measure. Round to the nearest tenth.**

9.

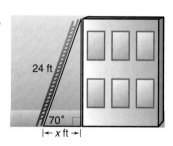

24 ft

70°

|← x ft →|

10.

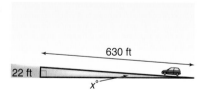

630 ft

22 ft

$x°$

Example 3

11. **Recreation** Sierra is flying a kite. She has let out 55 feet of string. If the angle of elevation is 35° and the hand holding the string is 6 feet from the ground, what is the altitude of the kite? Round to the nearest tenth.

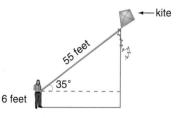

kite

55 feet

35°

6 feet

Practice

Homework Help	
For Exercises	See Examples
12–15, 22–29, 35	1
16, 18, 19	3
17, 20, 21, 31–34	2
Extra Practice	
See page 751.	

Find each sine or cosine. Round to four decimal places, if necessary.

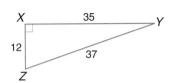

12. $\sin Q$

13. $\sin R$

14. $\cos Z$

15. $\cos Y$

Find each measure. Round to the nearest tenth.

16.

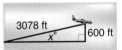

17.

18.

19.

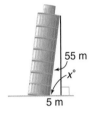

20.

21.

Use the 30°-60°-90° and 45°-45°-90° triangles to find each value. Round to four decimal places, if necessary.

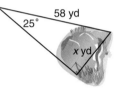

22. $\sin 30°$ **23.** $\sin 60°$

24. $\sin 45°$ **25.** $\cos 30°$

26. $\cos 60°$ **27.** $\cos 45°$

28. $\tan 30°$ **29.** $\tan 60°$

30. $\tan 45°$

31. If the hypotenuse of a right triangle is 5 feet and m$\angle A = 68$, find the measure of the leg adjacent to $\angle A$.

32. In a right triangle, the hypotenuse is 24 centimeters and m$\angle D = 16$. What is the measure of a leg opposite $\angle D$?

Applications and Problem Solving

33. Safety To guard against a fall, a ladder should make an angle of 75° or less with the ground. What is the maximum height that a 20-foot ladder can reach safely?

34. Engineering According to the Parking Standards in Santa Clarita, California, an access ramp to a parking lot cannot have a slope exceeding 11°. Suppose a parking lot is 10 feet above the road. If the length of the ramp is 60 feet, does this access ramp meet the requirements of the code? Explain your reasoning.

35. **Critical Thinking**   Verify each step in parts a through e. Then solve parts f and g.

Theorem 13–4	If x is the measure of an acute angle of a right triangle, then $\sin^2 x + \cos^2 x = 1$.

a. $\sin P = \dfrac{p}{q}$ and $\cos P = \dfrac{r}{q}$

b. $\sin^2 P = \dfrac{p^2}{q^2}$ and $\cos^2 P = \dfrac{r^2}{q^2}$

c. $\sin^2 P + \cos^2 P = \dfrac{p^2}{q^2} + \dfrac{r^2}{q^2}$ or $\dfrac{p^2 + r^2}{q^2}$

d. $p^2 + r^2 = q^2$

e. $\sin^2 P + \cos^2 P = \dfrac{q^2}{q^2}$ or 1

f. Find $\sin x$ if $\cos x = \dfrac{3}{5}$.

g. Find $\cos x$ if $\sin x = \dfrac{5}{13}$.

Mixed Review

The heights of several tourist attractions are given in the table. Find the angle of elevation from a point 100 feet from the base of each attraction to its top. *(Lesson 13–4)*

36. Chief Crazy Horse Statue

37. Washington Monument

38. Empire State Building

Heights of Tourist Attractions
Empire State Building, New York City . . . . .1250 ft
Gateway to the West Arch, St. Louis . . . . .630 ft
Chief Crazy Horse Statue, South Dakota . . . . .563 ft
Washington Monument, Washington, D.C. . . . . .555 ft
Statue of Liberty, New York City . . . . .305 ft

Chief Crazy Horse Statue, South Dakota

39. What is the area of the square? *(Lesson 13–2)*

8 cm

Exercise 39

40. **Grid In**   If the radius of a circle is 15 inches and the diameter is $(3x + 7)$ inches, what is the value of x? Round to the nearest hundredth. *(Lesson 11–1)*

41. **Multiple Choice** $\triangle PIG \sim \triangle COW$. If $PI = 6$, $IG = 4$, $CO = x + 3$, and $OW = x$, find the value of x.   *(Lesson 9–2)*

   Ⓐ 1.5   Ⓑ 6   Ⓒ 7.5   Ⓓ 9

Understanding and Using the Vocabulary

After completing this chapter, you should be able to define each term, property, or phrase and give an example or two of each.

*inter*NET CONNECTION **Review Activities**
For more review activities, visit:
www.geomconcepts.com

30°-60°-90° triangle *(p. 559)*
45°-45°-90° triangle *(p. 554)*
angle of depression *(p. 566)*
angle of elevation *(p. 566)*
cosine *(p. 572)*
hypsometer *(p. 570)*

perfect square *(p. 548)*
radical expression *(p. 549)*
radical sign *(p. 548)*
radicand *(p. 549)*
simplest form *(p. 551)*
sine *(p. 572)*

square root *(p. 548)*
tangent *(p. 564)*
trigonometric identity *(p. 574)*
trigonometric ratio *(p. 564)*
trigonometry *(p. 564)*

Choose the correct term to complete each sentence.

1. A (perfect square, trigonometric ratio) is a ratio of the measures of two sides of a right triangle.

2. The symbol used to indicate a square root is called a (radical sign, tangent).

3. (Square roots, Trigonometry) can be used to find the measures of unknown angles as well as unknown sides of a right triangle.

4. The number 25 is an example of a (perfect square, radical expression).

5. The opposite of squaring is finding a(n) (angle of elevation, square root).

6. Sine, tangent, and (radical sign, cosine) are three common trigonometric ratios.

7. Parallel lines and alternate interior angles can be used to show that the angle of elevation and the (tangent, angle of depression) are congruent.

8. A simplified definition of (tangent, sine) is opposite over adjacent.

9. The (angle of depression, angle of elevation) is below the line of sight.

10. Cosine and (sine, tangent) are both trigonometric ratios that use the hypotenuse.

Skills and Concepts

Objectives and Examples	Review Exercises

• **Lesson 13–1**   Multiply, divide, and simplify radical expressions.

Simplify $\sqrt{6} \cdot \sqrt{10}$.
$$\sqrt{6} \cdot \sqrt{10} = \sqrt{6 \cdot 10} \qquad \textit{Product Property}$$
$$= \sqrt{2 \cdot 3 \cdot 2 \cdot 5}$$
$$= \sqrt{2 \cdot 2 \cdot 3 \cdot 5}$$
$$= \sqrt{2 \cdot 2} \cdot \sqrt{3 \cdot 5} \qquad \textit{Product Property}$$
$$= 2\sqrt{15} \qquad \sqrt{2 \cdot 2} = 2$$

Simplify each expression.

11. $\sqrt{36}$ 12. $\sqrt{56}$

13. $\dfrac{\sqrt{25}}{\sqrt{10}}$ 14. $\dfrac{2}{\sqrt{6}}$

15. $\sqrt{6} \cdot \sqrt{8}$ 16. $\sqrt{3} \cdot \sqrt{18}$

      www.geomconcepts.com/vocabulary_review

Objectives and Examples

- **Lesson 13–2**   Use the properties of 45°-45°-90° triangles.

$\triangle LMN$ is an isosceles right triangle. Find the length of the hypotenuse.

$h = s\sqrt{2}$

$h = 7\sqrt{2}$

The length of the hypotenuse is $7\sqrt{2}$ feet.

Review Exercises

Find the missing measures. Write all radicals in simplest form.

17.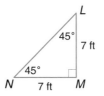

18.

19.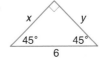

20.

- **Lesson 13–3**   Use the properties of 30°-60°-90° triangles.

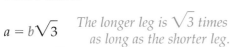

In the triangle, $b = 15$. Find a and c.

$a = b\sqrt{3}$ *The longer leg is $\sqrt{3}$ times as long as the shorter leg.*

$a = 15\sqrt{3}$ *Replace b with 15.*

$c = 2b$ *The hypotenuse is twice the shorter leg.*

$c = 2(15)$ *Replace b with 15.*

$c = 30$ *Simplify.*

Find the missing measures. Write all radicals in simplest form.

21.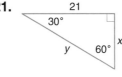

22.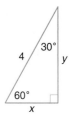

23.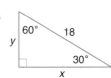

24.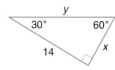

- **Lesson 13–4**   Use the tangent ratio to solve problems.

Find tan P rounded to four decimal places.

$\tan P = \dfrac{QR}{PR}$ $\dfrac{opposite}{adjacent}$

$\tan P = \dfrac{5}{12}$ *Replace QR with 5 and PR with 12.*

$\tan P = 0.4167$ *Use a calculator.*

Find each tangent. Round to four decimal places, if necessary.

25. tan W

26. tan Z

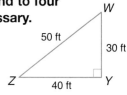

27. Find the missing measure. Round to the nearest tenth.

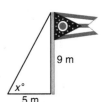

Mixed Problem Solving
See pages 758-765.

Objectives and Examples

- **Lesson 13–5** Use the sine and cosine ratios to solve problems.

Find sin D
and cos D.

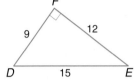

$\sin D = \dfrac{EF}{DE}$ $\dfrac{opposite}{hypotenuse}$

    $= \dfrac{12}{15}$ or 0.8   *Substitute.*

$\cos D = \dfrac{DF}{DE}$ $\dfrac{adjacent}{hypotenuse}$

    $= \dfrac{9}{15}$ or 0.6   *Substitute.*

Review Exercises

Find each sine or cosine. Round to four decimal places, if necessary.

28. sin A

29. cos B

30. cos A

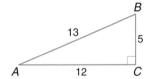

Find each missing measure. Round to the nearest tenth.

31.

32.

Applications and Problem Solving

33. **Arts and Crafts** Alyssa has a square piece of construction paper with a perimeter of 68 inches. Suppose she cuts the paper diagonally to form two congruent triangles. To the nearest inch, what is the sum of the perimeters of the two triangles? *(Lesson 13–2)*

34. **Surveying** A forest ranger sights a tree through a surveying instrument. The angle of elevation to the top of the tree is 27°. The instrument is 4 feet above the ground. The surveyor is 100 feet from the base of the tree. To the nearest foot, how tall is the tree? *(Lesson 13–4)*

35. **Construction** Mr. Boone is building a wooden ramp to allow people who use wheelchairs easier access to the public library. The ramp must be 2 feet tall. Find the angle of elevation if the ramp begins 24 feet away from the library. Round to the nearest tenth. *(Lesson 13–4)*

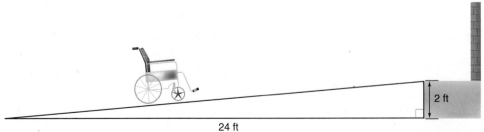

1. **Define** the term perfect square and list all perfect squares less than 100.

2. **Compare and contrast** angles of elevation and angles of depression.

Simplify each expression.

3. $\dfrac{\sqrt{3}}{\sqrt{7}}$

4. $\sqrt{44}$

5. $\sqrt{6} \cdot \sqrt{3}$

6. $\sqrt{\dfrac{16}{3}}$

Find the missing measures. Write all radicals in simplest form.

7.
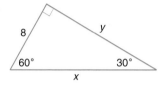

8.

9.
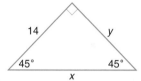

10.
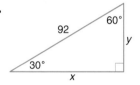

11.

12.
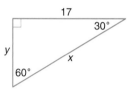

Find each trigonometric ratio. Round to four decimal places, if necessary.

13. $\sin Q$

14. $\tan P$

15. $\cos Q$

Find each missing measure. Round to the nearest tenth.

16.
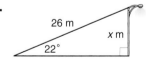

17.
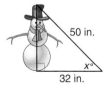

18.
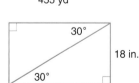

19. **Pets** Vincent's rectangular hamster cage is 18 inches wide. He would like to divide the cage into two triangular areas to separate his two hamsters. How long must the divider be in order to completely separate the two areas?

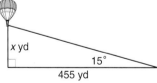

Exercise 19

20. **Transportation** A train travels 5000 meters along a track whose angle of elevation has a measurement of 3°. How much did the train rise during this distance? Round to the nearest tenth.

Perimeter, Circumference, and Area Problems

Standardized test problems often ask you to calculate the perimeter, circumference, or area of geometric shapes. You need to apply formulas for triangles, quadrilaterals, and circles.

Be sure you understand the following concepts.

area	base	circumference
diameter	height	perimeter
radius		

Example 1

Maxine's family is replacing a window with one in the shape shown in the drawing. If the top is a semicircle, what is the area, to the nearest tenth, of the glass needed for the window?

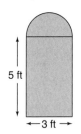

5 ft

←3 ft→

Hint Break a complicated problem into smaller parts and solve each part.

Solution Find the area of the rectangle and the area of the semicircle. Then add them.

The formula for the area of a rectangle is $A = \ell w$. The area of the rectangle is (5)(3) or 15 square feet.

The area of the semicircle is one-half the area of the circle. Find the area of the circle.

$A = \pi r^2$ *The diameter is 3 feet, so*
$A = (\pi)(1.5)^2$ *the radius is 1.5 feet.*
$A \approx 7.1$

The area of the circle is about 7.1 square feet.

Now find the area of the window.
$A = (\text{area of rectangle}) + \frac{1}{2}(\text{area of circle})$
$A \approx 15 + \frac{1}{2}(7.1)$
$A \approx 18.55$

The answer is about 18.6 square feet.

Example 2

What is the area of parallelogram *ABCD*?

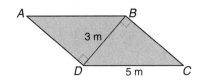

A 12 m² B 15 m² C 18 m² D 20 m²

E It cannot be determined from the information given.

Hint Look for triangles inside quadrilaterals.

Solution Recall that the formula for the area of a parallelogram is $A = bh$. You know that the height is 3 meters, but its base is *not* 5 meters, because $\overline{BD}$ is *not* perpendicular to $\overline{DC}$.

However, $\overline{BD}$ is perpendicular to $\overline{AD}$. Therefore, $\triangle ABD$ is a right triangle. The hypotenuse of the triangle is 5 meters long, and one side is 3 meters long. Note that this is a 3-4-5 right triangle. So, the base is 4 meters long.

Use the formula for the area of a parallelogram.
$A = bh$ *Area formula*
$A = (4)(3)$ *Substitution*
$A = 12$ *Simplify.*

The area of the parallelogram is 12 square meters.

The answer is A.

After you work each problem, record your answer on the answer sheet provided or on a sheet of paper.

Multiple Choice

1. If you double the length and the width of a rectangle, how does its perimeter change? *(Lesson 9–2)*

   A It increases by $1\frac{1}{2}$.

   B It doubles.

   C It quadruples.

   D It does not change.

2. Bacteria are growing at a rate of 1.3×10^5 per half hour. There are 5×10^5 bacteria. How many are there after 2 hours? *(Algebra Review)*

   A 2.26×10^5 **B** 5.2×10^5

   C 6.3×10^5 **D** 1.02×10^6

3. S is the set of all n such that $0 < n < 100$ and $\sqrt{n}$ is an integer. What is the median of the members of set S? *(Statistics Review)*

   A 5   **B** 5.5   **C** 25

   D 50   **E** 99

4. For all integers $n \neq 1$, let $<n> = \frac{n+1}{n-1}$. Which is greatest? *(Algebra Review)*

   A $<0>$ **B** $<2>$ **C** $<3>$

   D $<4>$ **E** $<5>$

5. What is the value of $\dfrac{2 \times 4}{36 \div 2 - 5 \times 2}$? *(Algebra Review)*

   A -9.9 **B** 1 **C** $\frac{4}{11}$ **D** $-\frac{1}{3}$

6. If the perimeter of rectangle $ABCD$ is p and $x = \frac{2}{3}y$, what is y? *(Lesson 9–2)*

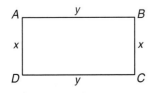

Note: Figure not drawn to scale.

   A $\frac{p}{10}$ **B** $\frac{3p}{10}$ **C** $\frac{p}{3}$ **D** $\frac{2p}{5}$ **E** $\frac{3p}{5}$

7. In the figure, $\overline{AC} \parallel \overline{ED}$. If $BD = 3$, what is BE? *(Lesson 4–3)*

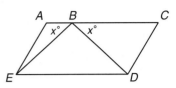

   A 3   **B** 4

   C 5   **D** $3\sqrt{3}$

   E It cannot be determined from the information given.

8. What is the difference between the mean of Set B and the median of Set A? *(Statistics Review)*

   Set A: $\{2, -1, 7, -4, 11, 3\}$

   Set B: $\{12, 5, -3, 4, 7, -7\}$

   A -0.5 **B** 0

   C 0.5 **D** 1

Grid In

9. For a July 4th celebration, members of the school band wrap red, white, and blue ribbons around a circular bandstand. Its radius is 25 feet. If each colored ribbon is used once around the bandstand, about how many feet of ribbon are needed to encircle it? *(Lesson 11–5)*

Extended Response

10. Mr. Huang has a rectangular garden that measures 15 meters by 20 meters. He wants to build a concrete walk of the same width around the garden. His budget for the project allows him to buy enough concrete to cover an area of 74 m^2. *(Lesson 1–6)*

   Part A   Draw a diagram of the walk.

   Part B   How wide can he build the walk?

CHAPTER 14

Circle Relationships

What You'll Learn

Key Ideas

- Identify and use properties of inscribed angles and tangents to circles. *(Lessons 14–1 and 14–2)*

- Find measures of arcs and angles formed by secants and tangents. *(Lessons 14–3 and 14–4)*

- Find measures of chords, secants, and tangents. *(Lesson 14–5)*

- Write equations of circles. *(Lesson 14–6)*

Key Vocabulary

inscribed angle *(p. 586)*

intercepted arc *(p. 586)*

secant segment *(p. 600)*

tangent *(p. 592)*

Why It's Important

Astronomy Planets and stars were observed by many ancient civilizations. Later scientists like Isaac Newton developed mathematical theories to further their study of astronomy. Modern observatories like Kitt Peak National Observatory in Arizona use optical and radio telescopes to continue to expand the understanding of our universe.

Circle relationships can be applied in many scientific fields. You will use circles to investigate two galaxies in Lesson 14–2.

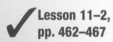

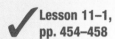

Study these lessons to improve your skills.

✓ Check Your Readiness

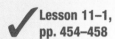**Lesson 11–1, pp. 454–458**

Use ⊙Q to determine whether each statement is *true* or *false*.

1. $\overline{WU}$ is a radius of ⊙Q.

2. $\overline{VT}$ is a diameter of ⊙Q.

3. $\overline{RS}$ is a chord of ⊙Q.

4. $\overline{QT}$ is congruent to $\overline{QU}$.

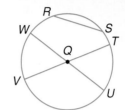

Lesson 11–2, pp. 462–467

Find each measure in ⊙O if $m\angle AOB = 36$, $m\widehat{BC} = 24$, and $\overline{AD}$ and $\overline{BE}$ are diameters.

5. $m\angle EOD$ 6. $m\widehat{AD}$

7. $m\angle COD$ 8. $m\widehat{BD}$

9. $m\widehat{CBE}$ 10. $m\angle DOB$

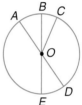

Lesson 6–6, pp. 256–261

If c is the hypotenuse, find each missing measure. Round to the nearest tenth, if necessary.

11. $a = 8, b = 15, c = ?$ 12. $a = 12, b = ?, c = 19$

13. $a = 10, b = 10, c = ?$ 14. $a = ?, b = 7.5, c = 16.8$

Study Organizer

Make this Foldable to help you organize your Chapter 14 notes. Begin with four sheets of $8\frac{1}{2}$" by 11" paper.

❶ **Fold** each sheet in half along the width.

❷ **Open** and fold the bottom edge up to form a pocket. Glue the sides.

❸ **Repeat** Steps 1 and 2 three times and glue all four pieces together.

❹ **Label** each pocket with a lesson title. Use the last two for vocabulary. Place an index card in each pocket.

Circle Relationships

Reading and Writing As you read and study the chapter, you can write main ideas, examples of theorems, and postulates on the index cards.

What You'll Learn
You'll learn to identify and use properties of inscribed angles.

Why It's Important
Architecture
Inscribed angles are important in the overall symmetry of many ancient structures.
See Exercise 8.

Recall that a polygon can be inscribed in a circle. An angle can also be inscribed in a circle. An **inscribed angle** is an angle whose vertex is on the circle and whose sides contain chords of the circle. Angle *JKL* is an inscribed angle.

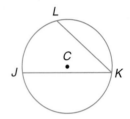

Notice that *K*, the vertex of ∠*JKL*, lies on ⊙*C*. The sides of ∠*JKL* contain chords *LK* and *JK*. Therefore, ∠*JKL* is an inscribed angle. Each side of the inscribed angle intersects the circle at a point. The two points *J* and *L* form an arc. We say that ∠*JKL* intercepts $\widehat{JL}$, or that $\widehat{JL}$ is the **intercepted arc** of ∠*JKL*.

Definition of Inscribed Angle	**Words:** An angle is inscribed if and only if its vertex lies on the circle and its sides contain chords of the circle.
	Model: 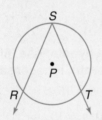 **Symbols:** ∠*RST* is inscribed in ⊙*P*.

Look Back
Central Angles, Lesson 11–2

Example ❶ Determine whether ∠*APB* is an inscribed angle. Name the intercepted arc for the angle.

Point *P*, the vertex of ∠*APB*, is not on ⊙*P*. So, ∠*APB* is not an inscribed angle. The intercepted arc of ∠*APB* is $\widehat{AB}$.

Your Turn Determine whether each angle is an inscribed angle. Name the intercepted arc for the angle.

a. ∠*CTL*

b. ∠*QRS*

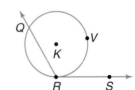

You can find the measure of an inscribed angle if you know the measure of its intercepted arc. This is stated in the following theorem.

Theorem 14–1	**Words:** The degree measure of an inscribed angle equals one-half the degree measure of its intercepted arc.
	Model: 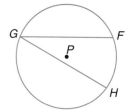 **Symbols:** $m\angle PQR = \frac{1}{2}m\widehat{PR}$

You can use Theorem 14–1 to find the measure of an inscribed angle or the measure of its intercepted arc if one of the measures is known.

Examples

2 If $m\widehat{FH} = 58$, find $m\angle FGH$.

$m\angle FGH = \frac{1}{2}(m\widehat{FH})$ *Theorem 14–1*

$m\angle FGH = \frac{1}{2}(58)$ *Replace $m\widehat{FH}$ with 58.*

$m\angle FGH = 29$ *Multiply.*

Game Link

Real World

3 In the game shown at the right, $\triangle WPZ$ is equilateral. Find $m\widehat{WZ}$.

$m\angle WPZ = \frac{1}{2}(m\widehat{WZ})$ *Theorem 14–1*

$60 \quad = \frac{1}{2}(m\widehat{WZ})$ *Replace $m\angle WPZ$ with 60.*

$2 \cdot 60 = 2 \cdot \frac{1}{2}(m\widehat{WZ})$ *Multiply each side by 2.*

$120 = m\widehat{WZ}$ *Simplify.*

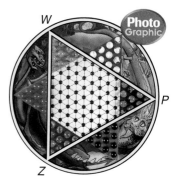

Chinese Checkers

Your Turn

c. If $m\widehat{JK} = 80$, find $m\angle JMK$.

d. If $m\angle MKS = 56$, find $m\widehat{MS}$.

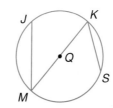

In ⊙B, if the measure of $\overset{\frown}{NO}$ is 74, what is the measure of inscribed angle *NCO*? What is the measure of inscribed angle *NDO*? Notice that both of the inscribed angles intercept the same arc, $\overset{\frown}{NO}$. This relationship is stated in Theorem 14–2.

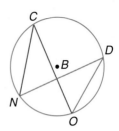

Theorem 14–2	**Words:** If inscribed angles intercept the same arc or congruent arcs, then the angles are congruent.
	Model: 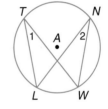 **Symbols:** $\angle 1 \cong \angle 2$

Example
4

Algebra Link

In ⊙A, $m\angle 1 = 2x$ and $m\angle 2 = x + 14$. Find the value of x.

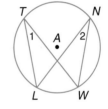

$\angle 1$ and $\angle 2$ both intercept $\overset{\frown}{LW}$.

$\angle 1 \cong \angle 2$	*Theorem 14–2*
$m\angle 1 = m\angle 2$	*Definition of congruent angles*
$2x = x + 14$	*Replace $m\angle 1$ with $2x$ and $m\angle 2$ with $x + 14$.*
$2x - x = x + 14 - x$	*Subtract x from each side.*
$x = 14$	*Simplify.*

— **Algebra Review** —

Solving Equations with the Variable on Both Sides, p. 724

Your Turn

e. In ⊙J, $m\angle 3 = 3x$ and $m\angle 4 = 2x + 9$. Find the value of x.

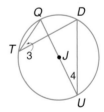

Suppose $\angle MTD$ is inscribed in ⊙C and intercepts semicircle $\overset{\frown}{MYD}$. Since $m\overset{\frown}{MYD} = 180$, $m\angle MTD = \frac{1}{2} \cdot 180$ or 90. Therefore, $\angle MTD$ is a right angle. This relationship is stated in Theorem 14–3.

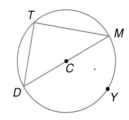

<table>
<tr><td rowspan="3">Theorem 14–3</td><td>Words: If an inscribed angle of a circle intercepts a semicircle, then the angle is a right angle.</td></tr>
<tr><td>Model:
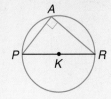</td><td>Symbols: $m\angle PAR = 90$</td></tr>
</table>

 Example

Algebra Link

⑤ In ⊙T, $\overline{CS}$ is a diameter. Find the value of x.

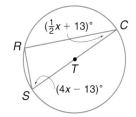

Inscribed angle CRS intercepts semicircle $\overset{\frown}{CS}$. By Theorem 14–3, $\angle CRS$ is a right angle. Therefore, $\triangle CRS$ is a right triangle and $\angle C$ and $\angle S$ are complementary.

$$m\angle C + m\angle S = 90 \qquad \textit{Definition of complementary angles}$$

$$\left(\tfrac{1}{2}x + 13\right) + (4x - 13) = 90 \qquad \textit{Substitution}$$

$$\tfrac{9}{2}x = 90 \qquad \textit{Combine like terms.}$$

$$\left(\tfrac{2}{9}\right)\tfrac{9}{2}x = \left(\tfrac{2}{9}\right)90 \qquad \textit{Multiply each side by } \tfrac{2}{9}.$$

$$x = 20 \qquad \textit{Simplify.}$$

 Your Turn

f. In ⊙K, $\overline{GH}$ is a diameter and $m\angle GNH = 4x - 14$. Find the value of x.

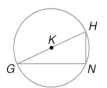

🌐nline **Personal Tutor at** geomconcepts.com

Check for Understanding

Communicating Mathematics

1. **Describe** an intercepted arc of a circle. State how its measure relates to the measure of an inscribed angle that intercepts it.

2. **Draw** inscribed angle QLS in ⊙T that has a measure of 100. Include all labels.

Guided Practice

Example 1

3. Determine whether $\angle WLS$ is an inscribed angle. Name the intercepted arc for the angle.

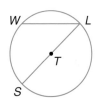

Find each measure.

4. $m\angle ABC$

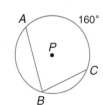

5. $m\widehat{PT}$

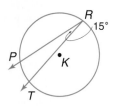

In each circle, find the value of x.

6.

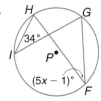

7.

Example 2

8. **Architecture** Refer to $\odot C$ in the application at the beginning of the lesson. If $m\widehat{JL} = 84$, find $m\angle JKL$.

Exercises

Determine whether each angle is an inscribed angle. Name the intercepted arc for the angle.

Homework Help	
For Exercises	**See Examples**
9–11	1
12–17, 26, 27	2, 3
18–21, 24	4
22, 23	5
Extra Practice	
See page 752.	

9. $\angle DEF$

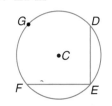

10. $\angle NZQ$

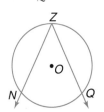

11. $\angle JTS$

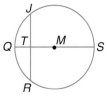

Find each measure.

12. $m\angle HKI$

13. $m\angle IKJ$

14. $m\widehat{IK}$

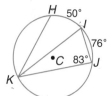

15. $m\widehat{XW}$

16. $m\angle TXV$

17. $m\widehat{VW}$

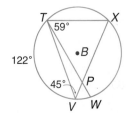

In each circle, find the value of x.

18.

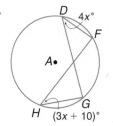

19.

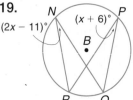

20.

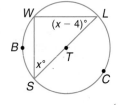

21.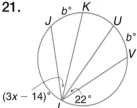

22.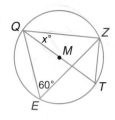

23.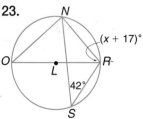
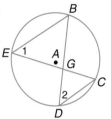

24. In ⊙A, $m\angle 1 = 13x - 9$ and $m\angle 2 = 27x - 65$.
  a. Find the value of x.
  b. Find $m\angle 1$ and $m\angle 2$.
  c. If $m\angle BGE = 92$, find $m\angle ECD$.

**Applications and
Problem Solving**

25. Literature Is Dante's suggestion in the quote at the right always possible? Explain why or why not.

> Or draw a triangle inside a semicircle
> That would have no right angle.
>
> —Dante, *The Divine Comedy*

26. History The symbol at the right appears throughout the Visitor Center in Texas' Washington-on-the-Brazos State Historical Park. If $\widehat{DH} \cong \widehat{HG} \cong \widehat{GF} \cong \widehat{FE} \cong \widehat{ED}$, find $m\angle HEG$.

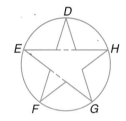

27. Critical Thinking Quadrilateral *MATH* is inscribed in ⊙R. Show that the opposite angles of the quadrilateral are supplementary.

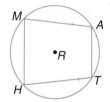

Mixed Review

28. Use △HJK to find cos H. Round to four decimal places. *(Lesson 13–5)*

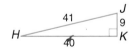

29. A right cylinder has a base radius of 4 centimeters and a height of 22 centimeters. Find the lateral area of the cylinder to the nearest hundredth. *(Lesson 12–2)*

30. Find the area of a 20° sector in a circle with diameter 15 inches. Round to the nearest hundredth. *(Lesson 11–6)*

**Standardized
Test Practice**
Ⓐ Ⓑ Ⓒ Ⓓ

31. Grid In Students are using a slide projector to magnify insects' wings. The ratio of actual length to projected length is 1:25. If the projected length of a wing is 8.14 centimeters, what is the actual length in centimeters? Round to the nearest hundredth. *(Lesson 9–1)*

32. Multiple Choice Solve $\sqrt{2q + 7} = 19$. *(Algebra Review)*
  Ⓐ 36   Ⓑ 177   Ⓒ 184   Ⓓ 736

14-2 Tangents to a Circle

What You'll Learn
You'll learn to identify and apply properties of tangents to circles.

Why It's Important
Astronomy Scientists use tangents to calculate distances between stars.
See Example 2.

A **tangent** is a line that intersects a circle in exactly one point. Also, by definition, a line segment or ray can be tangent to a circle if it is a part of a line that is tangent to the circle. Using tangents, you can find more properties of circles.

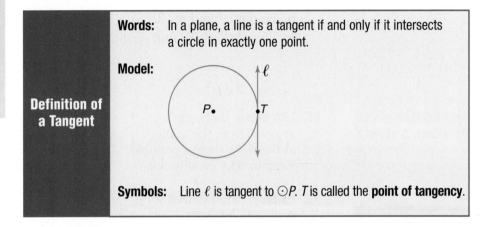

Definition of a Tangent	**Words:** In a plane, a line is a tangent if and only if it intersects a circle in exactly one point.
	Model:
	Symbols: Line ℓ is tangent to $\odot P$. T is called the **point of tangency**.

Two special properties of tangency are stated in the theorems below.

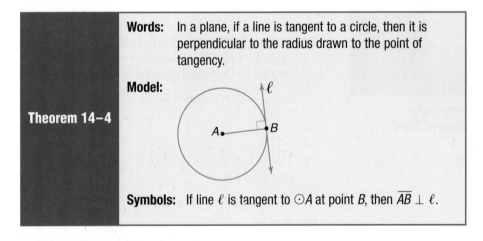

Theorem 14–4	**Words:** In a plane, if a line is tangent to a circle, then it is perpendicular to the radius drawn to the point of tangency.
	Model:
	Symbols: If line ℓ is tangent to $\odot A$ at point B, then $\overline{AB} \perp \ell$.

The converse of Theorem 14–4 is also true.

Theorem 14–5	**Words:** In a plane, if a line is perpendicular to a radius of a circle at its endpoint on the circle, then the line is a tangent.
	Symbols: If $\overline{AB} \perp \ell$, then ℓ is tangent to $\odot A$ at point B.

$\overline{TD}$ is tangent to $\odot K$ at T. Find KD.

From Theorem 14–4, $\overline{KT} \perp \overline{TD}$. Thus, $\angle KTD$ is a right angle, and $\triangle KTD$ is a right triangle.

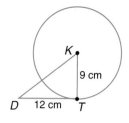

$(KD)^2 = (KT)^2 + (TD)^2$	*Pythagorean Theorem*
$(KD)^2 = 9^2 + 12^2$	*Replace KT with 9 and TD with 12.*
$(KD)^2 = 81 + 144$	*Square 9 and 12.*
$\sqrt{(KD)^2} = \sqrt{225}$	*Take the square root of each side.*
$KD = 15$	*Simplify.*

Your Turn

a. $\overrightarrow{QR}$ is tangent to $\odot P$ at R. Find RQ.

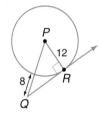

In the following activity, you'll find a relationship between two tangents that are drawn from a point outside a circle.

Hands-On Geometry
Paper Folding

Materials: compass patty paper straightedge

Step 1 Use a compass to draw a circle on patty paper.

Step 2 Draw a point outside the circle.

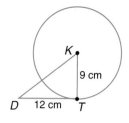

Step 3 Carefully fold the paper so that a tangent is formed from the point to one side of the circle. Use a straightedge to draw the segment. Mark your point of tangency.

Step 4 Repeat Step 3 for a tangent line that intersects the tangent line in Step 3.

Try These

1. Fold the paper so that one tangent covers the other. Compare their lengths.

2. **Make a conjecture** about the relationship between two tangents drawn from a point outside a circle.

The results of the activity suggest the following theorem.

Theorem 14–6

Words: If two segments from the same exterior point are tangent to a circle, then they are congruent.

Model:

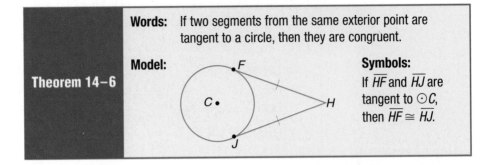

Symbols:
If $\overline{HF}$ and $\overline{HJ}$ are tangent to $\odot C$, then $\overline{HF} \cong \overline{HJ}$.

Example 2

Astronomy Link

The ring of stars in the photograph appeared after the small blue galaxy on the right S crashed through the large galaxy on the left G. The two galaxies are 168 thousand light-years apart (GS = 168 thousand light-years), and $\odot G$ has a radius of 75 thousand light-years. Find ST and SL if they are tangent to $\odot G$.

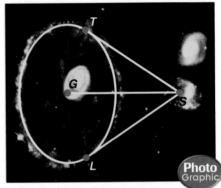

Cartwheel Galaxy

Explore From Theorem 14–6, $\overline{ST} \cong \overline{SL}$, so we only need to find the measure of one of the segments.

Plan By Theorem 14–4, $\overline{GT} \perp \overline{ST}$. Thus, $\angle GTS$ is a right angle and $\triangle GTS$ is a right triangle. We can use the Pythagorean Theorem to find ST.

Solve

$$(GS)^2 = (GT)^2 + (ST)^2 \qquad \textit{Pythagorean Theorem}$$
$$168^2 = 75^2 + (ST)^2 \qquad \textit{Substitution}$$
$$28{,}224 = 5625 + (ST)^2 \qquad \textit{Square 168 and 75.}$$
$$28{,}224 - 5625 = 5625 + (ST)^2 - 5625 \qquad \textit{Subtract 5625 from each side.}$$
$$22{,}599 = (ST)^2$$
$$\sqrt{22{,}599} = \sqrt{(ST)^2} \qquad \textit{Take the square root of each side.}$$
$$150.33 \approx ST \qquad \textit{Simplify.}$$

Examine Check your answer by substituting into the original equation.
$$(GS)^2 = (GT)^2 + (ST)^2$$
$$168^2 \stackrel{?}{=} 75^2 + 150.33^2$$
$$28{,}224 \approx 28{,}224.11 \quad \checkmark \qquad \text{The answer checks.}$$

If you round your final answer to the nearest tenth, the measure of $\overline{ST}$ is about 150.3 thousand light-years. By Theorem 14–6, the measure of $\overline{SL}$ is also about 150.3 thousand light-years.

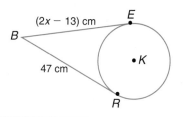

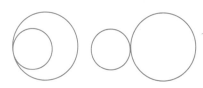

Your Turn

b. $\overline{BE}$ and $\overline{BR}$ are tangent to $\odot K$. Find the value of x.

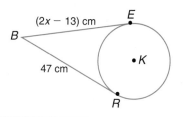

Two circles can be tangent to each other. If two circles are tangent and one circle is inside the other, the circles are **internally tangent.** If two circles are tangent and neither circle is inside the other, the circles are **externally tangent.**

Check for Understanding

Communicating Mathematics

1. **Determine** how many tangents can be drawn to a circle from a single point outside the circle. Explain why these tangents must be congruent.

2. **Explain** why $\overrightarrow{CD}$ is tangent to $\odot P$, but $\overrightarrow{CA}$ is not tangent to $\odot P$.

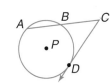

Guided Practice

⊕ Getting Ready — **Evaluate each expression. Round to the nearest tenth.**

Sample: $\sqrt{16^2 - 9^2}$ **Solution:** $\sqrt{16^2 - 9^2} = \sqrt{256 - 81}$
$$= \sqrt{175} \approx 13.2$$

3. $\sqrt{441 - 20^2}$ **4.** $\sqrt{7^2 + 10^2}$ **5.** $\sqrt{19^2 - 12^2}$

Examples 1 & 2

6. $\overline{JT}$ is tangent to $\odot S$ at T. Find SJ to the nearest tenth.

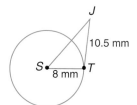

7. $\overline{QA}$ and $\overline{QB}$ are tangent to $\odot O$. Find QB.

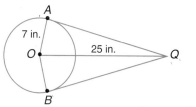

Example 2

8. **Music** The figure at the right shows a compact disc (CD) packaged in a square case.

a. Obtain a CD case and measure to the nearest centimeter from the corner of the disc case to each point of tangency, such as $\overline{AB}$ and $\overline{AC}$.

b. Which theorem is verified by your measures?

Practice

Find each measure. If necessary, round to the nearest tenth. Assume segments that appear to be tangent are tangent.

9. *CE*

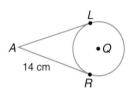

10. *HJ*

11. *m∠PTS*

12. *AL*

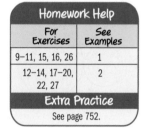

13. *AC*

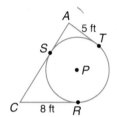

14. *BD*

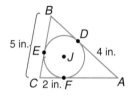

In the figure, $\overline{GC}$ and $\overline{GK}$ are both tangent to ⊙P. Find each measure.

15. *m∠PCG*

16. *m∠CGP*

17. *CG*

18. *GK*

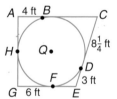

19. Find the perimeter of quadrilateral *AGEC*. Explain how you found the missing measures.

$\overline{BI}$ **and** $\overline{BC}$ **are tangent to ⊙P.**

20. If $BI = 3x - 6$ and $BC = 9$, find the value of x.

21. If $m\angle PIC = x$ and $m\angle CIB = 2x + 3$, find the value of x.

Supply a reason to support each statement.

22. $\overline{BI} \cong \overline{BC}$

23. $\overline{PI} \cong \overline{PC}$

24. $\overline{PB} \cong \overline{PB}$

25. $\triangle PIB \cong \triangle PCB$

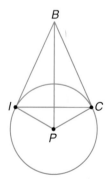

Exercises 20–25

Applications and Problem Solving

26. **Science** The science experiment at the right demonstrates zero gravity. When the frame is dropped, the pin rises to pop the balloon. If the pin is 2 centimeters long, find *x*, the distance the pin must rise to pop the balloon. Round to the nearest tenth.

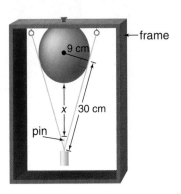

27. **Algebra** Regular pentagon *PENTA* is *circumscribed* about ⊙*K*. This means that each side of the pentagon is tangent to the circle.

    a. If *NT* = 12*x* − 30 and *ER* = 2*x* + 9, find *GP*.

    b. Why is the point of tangency the midpoint of each side?

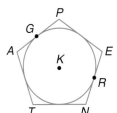

28. **Critical Thinking** How many tangents intersect both circles, each at a single point? Make drawings to show your answers.

    a.   **b.**    **c.**

Mixed Review

29. In ⊙*N*, find *m∠TUV*.
    (Lesson 14–1)

30. **Building** A ladder leaning against the side of a house forms a 72° angle with the ground. If the foot of the ladder is 6 feet from the house, find the height that the top of the ladder reaches. Round to the nearest tenth. *(Lesson 13–4)*

31. **Recreation** How far is the kite off the ground? Round to the nearest tenth.
    (Lesson 13–3)

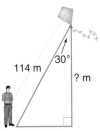

Standardized Test Practice
Ⓐ Ⓑ Ⓒ Ⓓ

32. **Grid In** The plans for Ms. Wathen's new sunroom call for a window in the shape of a regular octagon. What is the measure of one interior angle of the window? *(Lesson 10–2)*

33. **Multiple Choice** In parallelogram *RSTV*, *RS* = 4*p* + 9, *m∠V* = 75, and *TV* = 45. What is the value of *p*? *(Lesson 8–2)*

    Ⓐ 45   Ⓑ 13.5   Ⓒ 9   Ⓓ 7

Investigation

The Ins and Outs of Polygons

Materials

 ruler

 compass

 protractor

Areas of Inscribed and Circumscribed Polygons

Circles and polygons are paired together everywhere. You can find them in art, advertising, and jewelry designs. How do you think the area of a circle compares to the area of a regular polygon inscribed in it, or to the area of a regular polygon circumscribed about it? Let's find out.

Investigate

1. Use construction tools to draw a circle with a radius of 2 centimeters. Label the circle *O*.

2. Follow these steps to inscribe an equilateral triangle in ⊙*O*.

    a. Draw radius $\overline{OA}$ as shown. Find the area of the circle to the nearest tenth.

    b. Since there are three sides in a triangle, the measure of a central angle is 360 ÷ 3, or 120. Draw a 120° angle with side $\overline{OA}$ and vertex *O*. Label point *B* on the circle as shown.

    c. Using $\overline{OB}$ as one side of an angle, draw a second 120° angle as shown at the right. Label point *C*.

    d. Connect points *A*, *B*, and *C*. Equilateral triangle *ABC* is inscribed in ⊙*O*.

    e. Use a ruler to find the measures of one height and base of △*ABC*. Then find and record its area to the nearest tenth.

Look Back

Constructing
Perpendicular Line
Segments,
Lesson 3–7

3. Now circumscribe an equilateral triangle about ⊙O by constructing a line tangent to ⊙O at A, B, and C.

4. Find and record the area of the circumscribed triangle to the nearest tenth.

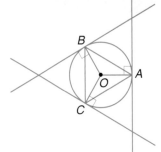

Extending the Investigation

In this extension, you will compare the areas of regular inscribed and circumscribed polygons to the area of a circle.

- Make a table like the one below. Record your triangle information in the first row.

Regular Polygon	Area of Circle (cm^2)	Area of Inscribed Polygon (cm^2)	Area of Circumscribed Polygon (cm^2)	(Area of Inscribed Polygon) ÷ (Area of Circumscribed Polygon)
triangle	12.6	5.2	20.7	
square				
pentagon				
hexagon				
octagon				

- Use a compass to draw four circles congruent to ⊙O. Record their areas in the table.

- Follow Steps 2 and 3 in the Investigation to inscribe and circumscribe each regular polygon listed in the table.

- Find and record the area of each inscribed and circumscribed polygon. *Refer to Lesson 10–5 to review areas of regular polygons.*

- Find the ratios of inscribed polygon area to circumscribed polygon area. Record the results in the last column of the table. What do you notice?

- **Make a conjecture** about the area of inscribed polygons compared to the area of the circle they inscribe.

- **Make a conjecture** about the area of circumscribed polygons compared to the area of the circle they circumscribe.

Presenting Your Conclusions

Here are some ideas to help you present your conclusions to the class.

- Make a poster displaying your table and the drawings of your circles and polygons.

- Summarize your findings about the areas of inscribed and circumscribed polygons.

Investigation For more information on inscribed and circumscribed polygons, visit: www.geomconcepts.com

What You'll Learn
You'll learn to find measures of arcs and angles formed by secants.

Why It's Important
Marketing
Understanding secant angles can be helpful in locating the source of data on a map. *See Exercise 24.*

A circular saw has a flat guide to help cut accurately. The edge of the guide represents a **secant segment** to the circular blade of the saw. A line segment or ray can be a secant of a circle if the line containing the segment or ray is a secant of the circle.

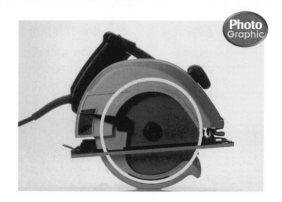

Photo Graphic

Theorem 14–7	**Words:** A line or line segment is a secant to a circle if and only if it intersects the circle in two points.
	Model:
	Symbols: $\overleftrightarrow{CD}$ is a secant of $\odot P$. Chord CD is a secant segment.

When two secants intersect, the angles formed are called **secant angles**. There are three possible cases.

Case 1 Vertex On the Circle	**Case 2** Vertex Inside the Circle	**Case 3** Vertex Outside the Circle
Secant angle CAB intercepts $\widehat{BC}$ and is an inscribed angle.	Secant angle DHG intercepts $\widehat{DG}$, and its vertical angle intercepts $\widehat{EF}$.	Secant angle JQL intercepts $\widehat{JL}$ and $\widehat{PK}$.

When a secant angle is inscribed, as in Case 1, recall that its measure is one-half the measure of the intercepted arc. The following theorems state the formulas for Cases 2 and 3.

	Words:	If a secant angle has its vertex inside a circle, then its degree measure is one-half the sum of the degree measures of the arcs intercepted by the angle and its vertical angle.
Theorem 14–8	**Model:**	**Symbols:** $m\angle 1 = \frac{1}{2}(m\widehat{MA} + m\widehat{HT})$

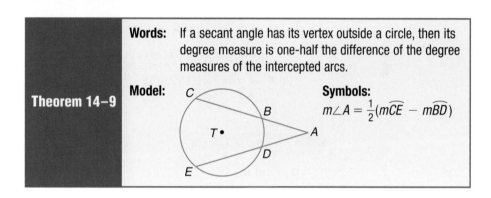

	Words:	If a secant angle has its vertex outside a circle, then its degree measure is one-half the difference of the degree measures of the intercepted arcs.
Theorem 14–9	**Model:**	**Symbols:** $m\angle A = \frac{1}{2}(m\widehat{CE} - m\widehat{BD})$

You can use these theorems to find the measures of arcs and angles formed by secants.

Example **1** **Find** $m\angle WSK$.

The vertex of $\angle WSK$ is inside $\odot T$.
Apply Theorem 14–8.

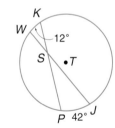

$m\angle WSK = \frac{1}{2}(m\widehat{WK} + m\widehat{PJ})$ *Theorem 14–8*

$m\angle WSK = \frac{1}{2}(12 + 42)$ *Replace $m\widehat{WK}$ with 12 and $m\widehat{PJ}$ with 42.*

$m\angle WSK = \frac{1}{2}(54)$ or 27   *Simplify.*

You also could have used this method to find $m\angle PSJ$.

Your Turn

a. Find $m\widehat{OT}$.

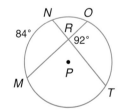

Example ❷

Art Link

Examine the objects in a student's painting at the right. Since they are difficult to identify, the painting is an example of *non-objective* art. If $m\angle T = 64$ and $m\overset{\frown}{NQ} = 19$, find $m\overset{\frown}{PR}$.

The vertex of $\angle T$ is outside the circle. Apply Theorem 14–9.

$$m\angle T = \tfrac{1}{2}(m\overset{\frown}{PR} - m\overset{\frown}{NQ}) \qquad \textit{Theorem 14–9}$$

$$64 = \tfrac{1}{2}(m\overset{\frown}{PR} - 19) \qquad \textit{Replace } m\angle T \textit{ with 64 and } m\overset{\frown}{NQ} \textit{ with 19.}$$

$$2 \cdot 64 = 2 \cdot \tfrac{1}{2}(m\overset{\frown}{PR} - 19) \qquad \textit{Multiply each side by 2.}$$

$$128 = m\overset{\frown}{PR} - 19 \qquad \textit{Simplify.}$$

$$128 + 19 = m\overset{\frown}{PR} - 19 + 19 \qquad \textit{Add 19 to each side.}$$

$$147 = m\overset{\frown}{PR} \qquad \textit{Simplify.}$$

Your Turn

b. Find $m\angle C$.

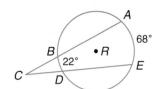

You can also use algebra to solve problems involving secant angles.

Example ❸

Algebra Link

Find $m\overset{\frown}{FG}$.

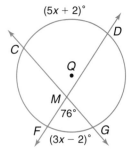

Explore   First, find the value of x. Then find $m\overset{\frown}{FG}$.

Plan   The vertex of $\angle FMG$ is inside $\odot Q$. Apply Theorem 14–8.

> **Algebra Review**
> Solving Multi-Step Equations, p. 723

Solve $m\angle FMG = \tfrac{1}{2}(m\overset{\frown}{CD} + m\overset{\frown}{FG})$ *Theorem 14–8*

$$76 = \tfrac{1}{2}(5x + 2 + 3x - 2) \qquad \textit{Substitution}$$

$$76 = \tfrac{1}{2}(8x) \qquad \textit{Simplify inside the parentheses.}$$

$$76 = 4x \qquad \textit{Simplify.}$$

$$\frac{76}{4} = \frac{4x}{4} \qquad \textit{Divide each side by 4.}$$

$$19 = x \qquad \textit{Simplify.}$$

The value of x is 19. Now substitute to find $m\widehat{FG}$.

$$m\widehat{FG} = 3x - 2 \qquad \textit{Substitution}$$
$$= 3(19) - 2 \text{ or } 55 \quad \textit{Replace x with 19.}$$

Examine Find $m\widehat{CD}$ and substitute into the original equation $m\angle FMG = \frac{1}{2}(m\widehat{CD} + m\widehat{FG})$. The solution checks.

Your Turn

c. Find the value of x. Then find $m\angle R$.

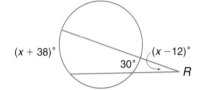

Check for Understanding

Communicating Mathematics

1. **Determine** the missing information needed for $\odot K$ if you want to use Theorem 14–9 to find $m\angle A$.

2. **Explain** how to find $m\angle A$ using only the given information.

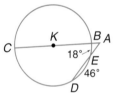

Vocabulary

secant segment
secant angles

Exercises 1–2

3. Writing Math The word *secant* comes from the Latin word *secare*. Use a dictionary to find the meaning of the word and explain why secant is used for a line that intersects a circle in exactly two points.

Guided Practice

Examples 1 & 2

Find each measure.

4. $m\angle 2$

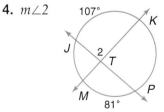

5. $m\widehat{LH}$

Example 3

In each circle, find the value of x. Then find the given measure.

6. $m\widehat{GR}$

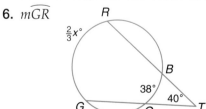

7. $m\angle MRO$

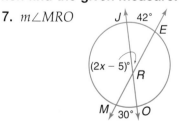

Example 1

8. **Food** A cook uses secant segments to cut a round pizza into rectangular pieces. If $\overline{PQ} \perp \overline{CL}$ and $m\widehat{QL} = 140$, find $m\widehat{PC}$.

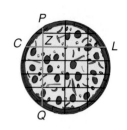

Exercises

Practice

Find each measure.

9. $m\widehat{GZ}$

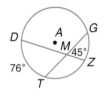

10. $m\angle 1$

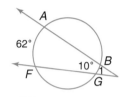

11. $m\angle Q$

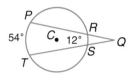

Homework Help	
For Exercises	**See Examples**
9, 12, 14, 15, 19, 20, 23, 24, 26	1, 3
10, 11, 13, 16–18, 25	2
21, 22	1–3
Extra Practice	
See page 752.	

12. $m\angle HLI$

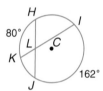

13. $m\widehat{AK}$

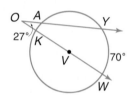

14. $m\widehat{LC}$

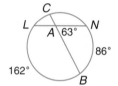

In each circle, find the value of x. Then find the given measure.

15. $m\widehat{SV}$

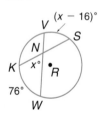

16. $m\angle M$

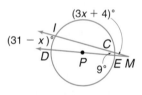

17. $m\widehat{HI}$

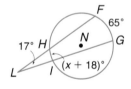

18. $m\widehat{RS}$

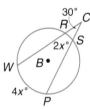

19. $m\angle LTQ$

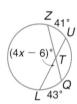

20. $m\widehat{JH}$

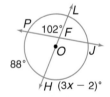

21. If $m\angle 4 = 38$ and $m\widehat{BC} = 38$, find $m\widehat{AE}$.

22. If $m\widehat{BAE} = 198$ and $m\widehat{CD} = 64$, find $m\angle 3$.

23. In a circle, chords AC and BD meet at P. If $m\angle CPB = 115$, $m\widehat{AB} = 6x + 16$, and $m\widehat{CD} = 3x - 12$. Find x, $m\widehat{AB}$, and $m\widehat{CD}$.

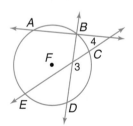

Exercises 21–22

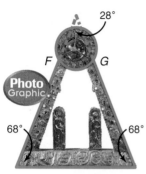

Exercise 25

24. Marketing The figure at the
right is a "one-mile" circle of
San Diego used for research
and marketing purposes.
What is $m\widehat{SD}$?

25. History The gold figurine
at the left was made by the
Germanic people in the 8th
century. Find $m\widehat{FG}$.

26. Critical Thinking In ⊙*P*,
$\overline{CR} \perp \overline{AT}$. Find $m\widehat{AC} + m\widehat{TR}$.

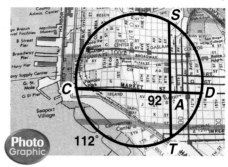

Mixed Review

27. $\overline{EF}$ and $\overline{EG}$ are tangent to ⊙*C*.
Find the value of *x*.   *(Lesson 14–2)*

28. A pyramid has a height of 12 millimeters and a base with area of
34 square millimeters. What is its volume?   *(Lesson 12–5)*

29. Find the circumference of a circle whose diameter is 26 meters. Round
to the nearest tenth.   *(Lesson 11–5)*

**Standardized
Test Practice**
Ⓐ Ⓑ Ⓒ Ⓓ

30. Short Response Find the area of a trapezoid whose height measures
8 centimeters and whose bases are 11 centimeters and 9 centimeters
long.   *(Lesson 10–4)*

31. Multiple Choice Find the value for *y* that verifies
that the figure is a parallelogram.   *(Lesson 8–3)*

Ⓐ 4   Ⓑ 12   Ⓒ 12.8   Ⓓ 14

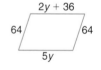

Quiz 1   Lessons 14–1 through 14–3

1. Determine whether ∠*NGP* is an inscribed angle.
Name the intercepted arc.   *(Lesson 14–1)*

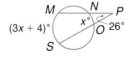

**Find each measure. Assume segments
that appear to be tangent are tangent.**
(Lesson 14–2)

2. *CH*

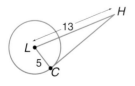

3. *AC*

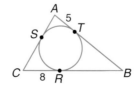

**In each circle, find the value of x. Then
find the given measure.**   *(Lesson 14–3)*

4. $m\widehat{GF}$

5. $m\widehat{MS}$

14–4 Secant-Tangent Angles

What You'll Learn

You'll learn to find measures of arcs and angles formed by secants and tangents.

Why It's Important

Archaeology
Scientists can learn a lot about an ancient civilization by using secant-tangent angles to find pottery measurements.
See Exercise 20.

When a secant and a tangent of a circle intersect, a **secant-tangent angle** is formed. This angle intercepts an arc on the circle. The measure of the arc is related to the measure of the secant-tangent angle.

There are two ways that secant-tangent angles are formed, as shown below.

Case 1 Vertex Outside the Circle	Case 2 Vertex On the Circle
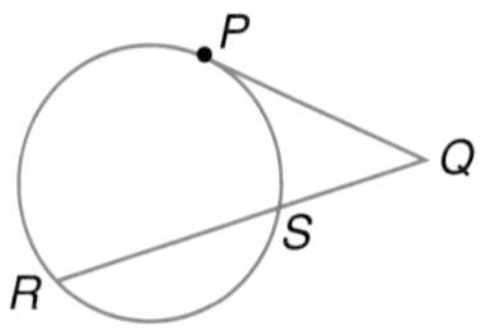	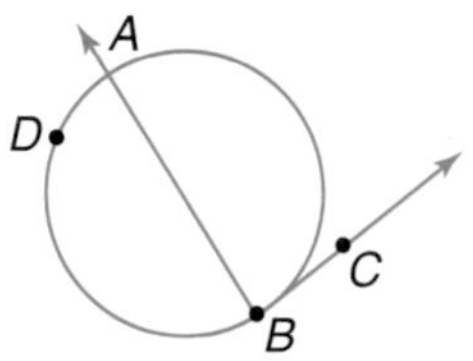
Secant-tangent angle *PQR* intercepts $\widehat{PR}$ and $\widehat{PS}$.	Secant-tangent angle *ABC* intercepts $\widehat{AB}$.

Notice that the vertex of a secant-tangent angle cannot lie inside the circle. This is because the tangent always lies outside the circle, except at the single point of contact.

The formulas for the measures of these angles are shown in Theorems 14–10 and 14–11.

Theorem	Words	Models and Symbols
14–10	If a secant-tangent angle has its vertex outside the circle, then its degree measure is one-half the difference of the degree measures of the intercepted arcs.	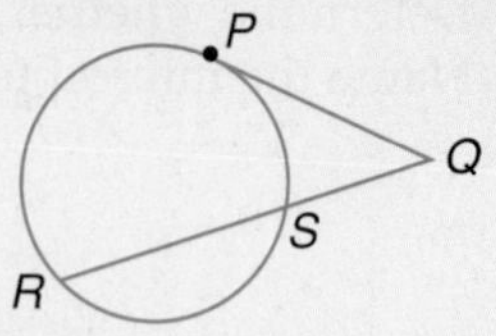 $m\angle PQR = \frac{1}{2}(m\widehat{PR} - m\widehat{PS})$
14–11	If a secant-tangent angle has its vertex on the circle, then its degree measure is one-half the degree measure of the intercepted arc.	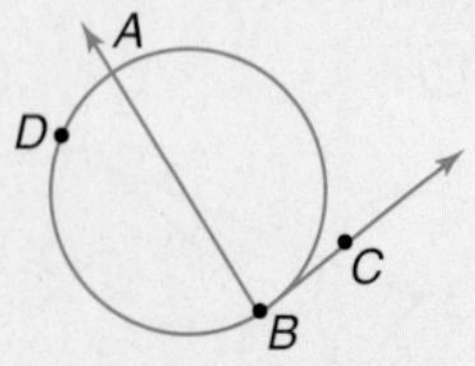 $m\angle ABC = \frac{1}{2}(m\widehat{AB})$

1 $\overline{CR}$ is tangent to $\odot T$ at C. If $m\widehat{CDN} = 200$, find $m\angle R$.

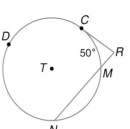

Vertex R of the secant-tangent angle is outside of $\odot T$. Apply Theorem 14–10.

Algebra Review
Evaluating Expressions, p. 718

$m\angle R = \frac{1}{2}(m\widehat{CDN} - m\widehat{CM})$ *Theorem 14–10*

$m\angle R = \frac{1}{2}(200 - 50)$ *Substitution*

$m\angle R = \frac{1}{2}(150)$ or 75   *Simplify.*

2 $\overrightarrow{BA}$ is tangent to $\odot P$ at B. Find $m\angle ABC$.

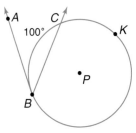

Vertex B of the secant-tangent angle is on $\odot P$. Apply Theorem 14–11.

$m\angle ABC = \frac{1}{2}(m\widehat{BC})$ *Theorem 14–11*

$m\angle ABC = \frac{1}{2}(100)$ or 50   *Substitution*

Your Turn

$\overline{AC}$ is tangent to $\odot P$ at C and $\overrightarrow{DE}$ is tangent to $\odot P$ at D.

a. Find $m\angle A$.

b. Find $m\angle BDE$.

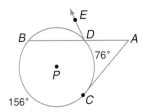

A **tangent-tangent angle** is formed by two tangents. The vertex of a tangent-tangent angle is always outside the circle.

Theorem 14–12	**Words:**	The degree measure of a tangent-tangent angle is one-half the difference of the degree measures of the intercepted arcs.
	Model:	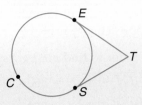
	Symbols:	$m\angle ETS = \frac{1}{2}(m\widehat{ECS} - m\widehat{ES})$

You can use a TI–83 Plus/TI–84 Plus calculator to verify the relationship stated in Theorem 14–12.

Graphing Calculator Tutorial
See pp. 782–785.

 Graphing Calculator Exploration

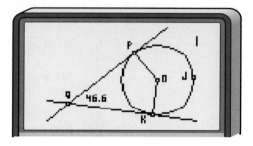

The calculator screen at the right shows an acute angle, $\angle Q$. To verify Theorem 14–12, you can measure $\angle Q$, find the measures of the intercepted arcs, and then perform the calculation.

Try These

1. Use the calculator to construct and label a figure like the one shown above. Then use the Angle tool on the F5 menu to measure $\angle Q$. What measure do you get?

2. How can you use the Angle tool on F5 to find $m\overarc{PK}$ and $m\overarc{PJK}$? Use the calculator to find these measures. What are the results?

3. Use the Calculate tool on F5 to find $\frac{1}{2}(m\overarc{PJK} - m\overarc{PK})$. How does the result compare with $m\angle Q$ from Exercise 1? Is your answer in agreement with Theorem 14–12?

4. Move point P along the circle so that Q moves farther away from the center of the circle. Describe how this affects the arc measures and the measure of $\angle Q$.

5. Suppose you change $\angle Q$ to an obtuse angle. Do the results from Exercises 1–3 change? Explain your answer.

You can use Theorem 14–12 to solve problems involving tangent-tangent angles.

Example ❸

Architecture Link

The Duomo,
Florence, Italy

In the 15th century, Brunelleschi, an Italian architect, used his knowledge of mathematics to create a revolutionary design for the dome of a cathedral in Florence. A close-up of one of the windows is shown at the right. Find $m\angle B$.

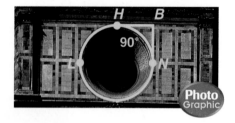

$\angle B$ is a tangent-tangent angle. Apply Theorem 14–12.

In order to find $m\angle B$, first find $m\overarc{HLN}$.

$m\overarc{HLN} + m\overarc{HN} = 360$ *The sum of the measures of a minor arc and its*
$m\overarc{HLN} + \quad 90 \quad = 360$ *major arc is 360.*

$m\overarc{HLN} = 270$ *Subtract 90 from each side.*

$$m\angle B = \tfrac{1}{2}(m\widehat{HLN} - m\widehat{HN}) \quad \text{\textit{Theorem 14–12}}$$

$$m\angle B = \tfrac{1}{2}(270 - 90) \qquad \text{\textit{Substitution}}$$

$$m\angle B = \tfrac{1}{2}(180) \text{ or } 90 \qquad \text{\textit{Simplify.}}$$

Your Turn

c. Find $m\angle A$.

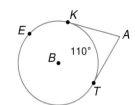

 Personal Tutor at geomconcepts.com

Check for Understanding

Communicating Mathematics

1. **Explain** how to find the measure of a tangent-tangent angle.

2. **Name** three secant-tangent angles in $\odot K$.

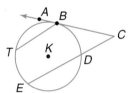

3. 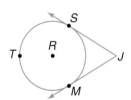 In $\odot R$, $\overrightarrow{JS}$ and $\overrightarrow{JM}$ are tangents. Maria says that if $m\angle J$ increases, $m\widehat{STM}$ increases. Is she correct? Make some drawings to support your conclusion.

Guided Practice

Find the measure of each angle. Assume segments that appear to be tangent are tangent.

Examples 1–3

4. $\angle 3$

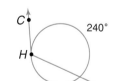

5. $\angle CHM$

6. $\angle Q$

Example 1

7. **Billiards** Refer to the application at the beginning of the lesson. If $x = 31$ and $y = 135$, find $m\angle 1$, the angle measure of the cue ball's spin.

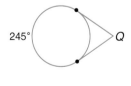

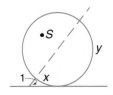

Practice

Find the measure of each angle. Assume segments that appear to be tangent are tangent.

8. ∠2

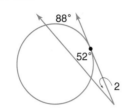

9. ∠1

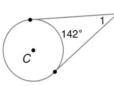

10. ∠*BAN*

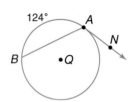

Homework Help	
For Exercises	See Examples
8, 13, 14, 17–20	1
9, 12, 16, 22	3
10, 11, 15, 21	2
Extra Practice	
See page 753.	

11. ∠*RAV*

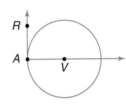

12. ∠*T*

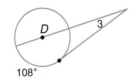

13. ∠*WNG*

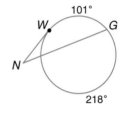

14. ∠3

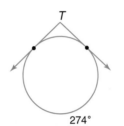

15. ∠4

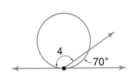

16. ∠*S*

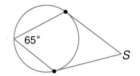

17. In ⊙*N*, find the value of *x*.

18. What is $m\widehat{PK}$?

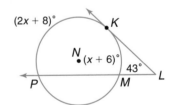

Exercises 17–18

Applications and Problem Solving

19. Algebra $\overline{IL}$ is a secant segment, and $\overline{LK}$ is tangent to ⊙*T*. Find $m\widehat{IJ}$ in terms of *x*. (*Hint*: First find $m\widehat{IK}$ in terms of *x*.)

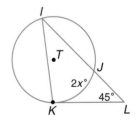

20. Mechanics In the piston and rod diagram at the right, the throw arm moves from position *A* to position *B*. Find $m\widehat{AB}$.

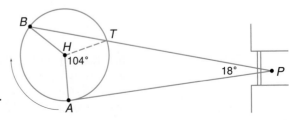

21. Archaeology The most commonly found artifact on an archaeological dig is a pottery shard. Many clues about a site and the group of people who lived there can be found by studying these shards. The piece at the right is from a round plate.

a. If $\overrightarrow{HD}$ is a tangent at H, and $m\angle SHD = 60$, find $m\widehat{SH}$.

b. Suppose an archaeologist uses a tape measure and finds that the distance along the outside edge of the shard is 8.3 centimeters. What was the circumference of the original plate? Explain how you know.

22. Critical Thinking $\overline{AB}$ and $\overline{BC}$ are tangent to $\odot K$.

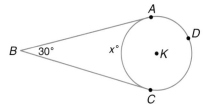

a. If x represents $m\widehat{AC}$, what is $m\widehat{ADC}$ in terms of x?

b. Find $m\widehat{AC}$.

c. Find $m\angle B + m\widehat{AC}$.

d. Is the sum of the measures of a tangent-tangent angle and the smaller intercepted arc always equal to the sum in part c? Explain.

Mixed Review

Find each measure.

23. $m\angle 3$ *(Lesson 14–3)*

24. FG and GE *(Lesson 13–2)*

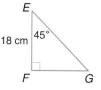

25. Museums A museum of miniatures in Los Angeles, California, has 2-inch violins that can actually be played. If the 2-inch model represents a 2-foot violin, what is the scale factor of the model to the actual violin? (*Hint*: Change feet to inches.) *(Lesson 12–7)*

Standardized Test Practice
Ⓐ Ⓑ Ⓒ Ⓓ

26. Short Response The perimeter of $\triangle QRS$ is 94 centimeters. If $\triangle QRS \sim \triangle CDH$ and the scale factor of $\triangle QRS$ to $\triangle CDH$ is $\frac{4}{3}$, find the perimeter of $\triangle CDH$. *(Lesson 9–7)*

27. Multiple Choice Find the solution to the system of equations. *(Algebra Review)*

$y = 3x + 5$
$5x + 3y = 43$

Ⓐ $(2, 11)$　　Ⓑ $(-11, 2)$　　Ⓒ $(-2, 11)$　　Ⓓ $(11, 2)$

In the circle at the right, chords AC and BD intersect at E. Notice the two pairs of segments that are formed by these intersecting chords.

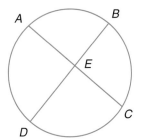

$\overline{AE}$ and $\overline{EC}$ are segments of $\overline{AC}$.

$\overline{BE}$ and $\overline{ED}$ are segments of $\overline{BD}$.

There exists a special relationship for the measures of the segments formed by intersecting chords. This relationship is stated in the following theorem.

Theorem 14–13	**Words:**	If two chords of a circle intersect, then the product of the measures of the segments of one chord equals the product of the measures of the segments of the other chord.
	Model:	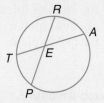
	Symbols:	$TE \cdot EA = RE \cdot EP$

Example

Algebra Link

1 In $\odot P$, find the value of x.

$LE \cdot EQ = JE \cdot EM$	*Theorem 14–13*
$x \cdot 6 = 3 \cdot 4$	*Substitution*
$6x = 12$	*Multiply.*
$\dfrac{6x}{6} = \dfrac{12}{6}$	*Divide each side by 6.*
$x = 2$	*Simplify.*

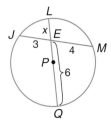

Algebra Review
Solving One-Step Equations, p. 722

Your Turn

a. In $\odot C$, find UW.

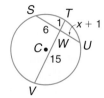

 Personal Tutor at geomconcepts.com

$\overline{RP}$ and $\overline{RT}$ are secant segments of $\odot A$. $\overline{RQ}$ and $\overline{RS}$ are the parts of the segments that lie outside the circle. They are called **external secant segments**.

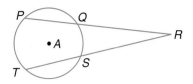

Definition of External Secant Segment	**Words:** A segment is an external secant segment if and only if it is the part of a secant segment that is outside a circle. **Model:** 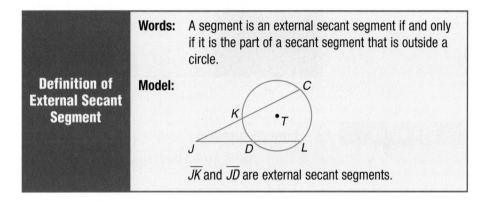 $\overline{JK}$ and $\overline{JD}$ are external secant segments.

A special relationship between secant segments and external secant segments is stated in the following theorem.

Theorem 14–14	**Words:** If two secant segments are drawn to a circle from an exterior point, then the product of the measures of one secant segment and its external secant segment equals the product of the measures of the other secant segment and its external secant segment. **Model:** 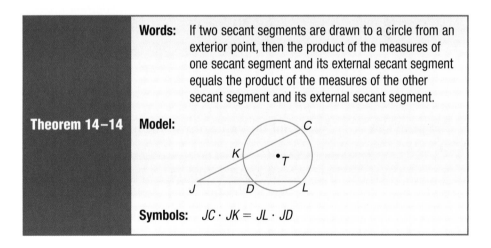 **Symbols:** $JC \cdot JK = JL \cdot JD$

In $\odot D$, a similar relationship exists if one segment is a secant and one is a tangent. $\overline{PA}$ is a tangent segment.

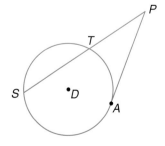

$$PA \cdot PA = PS \cdot PT$$
$$(PA)^2 = PS \cdot PT$$

This result is formally stated in the following theorem.

Words: If a tangent segment and a secant segment are drawn to a circle from an exterior point, then the square of the measure of the tangent segment equals the product of the measures of the secant segment and its external secant segment.

Model:

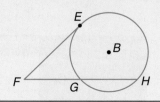

Symbols:
$(FE)^2 = FH \cdot FG$

Examples ② **Find _AV_ and _RV_.**

$$AC \cdot AB = AV \cdot AR \quad \text{\textit{Theorem 14–14}}$$
$$(3 + 9) \cdot 3 = AV \cdot 4 \quad \text{\textit{Substitution}}$$
$$12 \cdot 3 = AV \cdot 4 \quad \text{\textit{Add.}}$$
$$36 = 4(AV) \quad \text{\textit{Multiply.}}$$
$$\frac{36}{4} = \frac{4(AV)}{4} \quad \text{\textit{Divide each side by 4.}}$$
$$9 = AV \quad \text{\textit{Simplify.}}$$

$$AR + RV = AV \quad \text{\textit{Segment Addition Property}}$$
$$4 + RV = 9 \quad \text{\textit{Substitution}}$$
$$4 + RV - 4 = 9 - 4 \quad \text{\textit{Subtract 4 from each side.}}$$
$$RV = 5 \quad \text{\textit{Simplify.}}$$

Algebra Link ③ **Find the value of _x_ to the nearest tenth.**

$$(TU)^2 = TP \cdot TW \quad \text{\textit{Theorem 14–15}}$$
$$x^2 = (10 + 10) \cdot 10 \quad \text{\textit{Substitution}}$$
$$x^2 = 20 \cdot 10 \quad \text{\textit{Add.}}$$
$$x^2 = 200 \quad \text{\textit{Multiply.}}$$
$$\sqrt{x^2} = \sqrt{200} \quad \text{\textit{Take the square root of each side.}}$$
$$x \approx 14.1 \quad \text{\textit{Use a calculator.}}$$

Your Turn

b. Find the value of x to the nearest tenth.

c. Find MN to the nearest tenth.

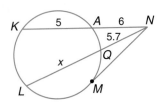

Communicating Mathematics

1. **Draw** and label a circle that fits the following description.

- Has center *K*.
- Contains secant segments *AM* and *AL*.
- Contains external secant segments *AP* and *AN*.
- $\overleftrightarrow{JM}$ is tangent to the circle at *M*.

2. **Complete** the steps below to prove Theorem 14–13. Refer to ⊙*R* shown at the right.

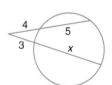

   a. $\angle BAE \cong \angle CDE$ and
   $\angle ABE \cong \angle DCE$ *Theorem*

   b. $\triangle ABE \sim \triangle$ ____   *AA Similarity Postulate*

   c. $\dfrac{AE}{DE} =$ ____   *Definition of Similar Polygons*

   d. $AE \cdot CE = DE \cdot BE$

3. 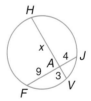 Leon wrote the equation $4 \cdot 5 = 3x$ to find the value of *x* in the figure at the right. Yoshica wrote the equation $9 \cdot 4 = (3 + x) \cdot 3$. Who wrote the correct equation? Explain.

Guided Practice
Example 1

4. Find the value of *x*.

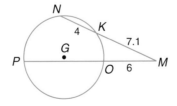

Examples 2 & 3

Find each measure. If necessary, round to the nearest tenth.

5. *OP*   6. *TR*

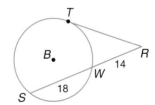

Example 1

7. Find *DE* to the nearest tenth.

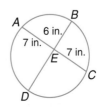

Practice

In each circle, find the value of *x*. If necessary, round to the nearest tenth.

Homework Help	
For Exercises	**See Examples**
8–10, 20	1
11, 14, 16, 21	2
12, 13, 15, 19	3
17, 18	1, 2
Extra Practice	
See page 753.	

8.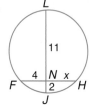

9.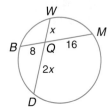

10.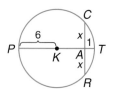

Find each measure. If necessary, round to the nearest tenth.

11. *AC*

12. *LP*

13. *KM*

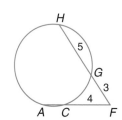

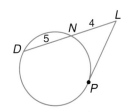

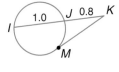

14. *QR*

15. *BE*

16. *NV*

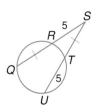

17. If $CH = 13$, $EH = 3.2$, and $DH = 6$, find *FH* to the nearest tenth.

18. If $AF = 7.5$, $FH = 7$, and $DH = 6$, find *BA* to the nearest tenth.

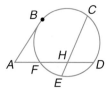

Applications and Problem Solving

19. **Space** The space shuttle *Discovery D* is 145 miles above Earth. The diameter of Earth is about 8000 miles. How far is its longest line of sight $\overline{DA}$ to Earth?

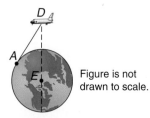

Figure is not drawn to scale.

20. **Native American Art** The traditional sun design appears in many phases of Hopi art and decoration. Find the length of $\overline{TJ}$.

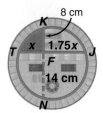

21. Critical Thinking Find the radius of ⊙N:

a. using the Pythagorean Theorem.

b. using Theorem 14–4. (*Hint*: Extend $\overline{TN}$ to the other side of ⊙N.)

c. Which method seems more efficient? Explain.

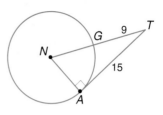

Mixed Review

22. In ⊙R, find the measure of ∠STN. *(Lesson 14–4)*

23. Simplify $\dfrac{\sqrt{8}}{\sqrt{36}}$. *(Lesson 13–1)*

24. In a circle, the measure of chord *JK* is 3, the measure of chord *LM* is 3, and $m\widehat{JK} = 35$. Find $m\widehat{LM}$. *(Lesson 11–3)*

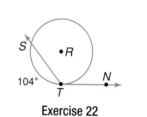

Exercise 22

25. Short Response Determine whether the face of the jaguar has *line symmetry*, *rotational symmetry*, *both*, or *neither*. *(Lesson 10–6)*

26. Short Response Sketch and label isosceles trapezoid *CDEF* and its median *ST*. *(Lesson 8–5)*

Exercise 25

Quiz 2 · Lessons 14–4 and 14–5

Find the measure of each angle.
(Lesson 14–4)

1. ∠C

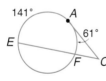

2. ∠3

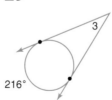

In each circle, find the value of x.
(Lesson 14–5)

3.

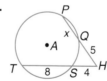

4.

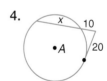

5. Astronomy A *planisphere* is a "flattened sphere" that shows the whole sky. The smaller circle inside the chart is the area of sky that is visible to the viewer. Find the value of *x*. *(Lesson 14–5)*

14-6 Equations of Circles

What You'll Learn

You'll learn to write equations of circles using the center and the radius.

Why It's Important

Meteorology
Equations of circles are important in helping meteorologists track storms shown on radar.
See Exercise 30.

In Lesson 4–6, you learned that the equation of a straight line is linear. In slope-intercept form, this equation is written as $y = mx + b$. A circle is not a straight line, so its equation is not linear. You can use the Distance Formula to find the equation of any circle.

Circle C has its center at $C(3, 2)$. It has a radius of 4 units. Let $P(x, y)$ represent any point on $\odot C$. Then d, the measure of the distance between P and C, must be equal to the radius, 4.

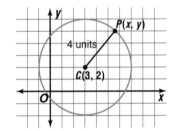

$$\sqrt{(x_2 - x_1)^2 + (y_2 - y_1)^2} = d \quad \textit{Distance Formula}$$

$$\sqrt{(x - 3)^2 + (y - 2)^2} = 4 \quad \textit{Replace } (x_1, y_1) \textit{ with } (3, 2) \textit{ and } (x_2, y_2) \textit{ with } (x, y).$$

$$\left(\sqrt{(x - 3)^2 + (y - 2)^2}\right)^2 = 4^2 \quad \textit{Square each side of the equation.}$$

$$(x - 3)^2 + (y - 2)^2 = 16 \quad \textit{Simplify.}$$

Therefore, the equation of the circle with center at $(3, 2)$ and a radius of 4 units is $(x - 3)^2 + (y - 2)^2 = 16$. This result is generalized in the equation of a circle given below.

Theorem 14–16 General Equation of a Circle	**Words:** The equation of a circle with center at (h, k) and a radius of r units is $(x - h)^2 + (y - k)^2 = r^2$.
	Model: 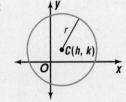

Example ① Write an equation of a circle with center $C(-1, 2)$ and a radius of 2 units.

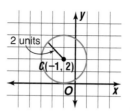

$(x - h)^2 + (y - k)^2 = r^2$ *General Equation of a Circle*
$[x - (-1)]^2 + (y - 2)^2 = 2^2$ *$(h, k) = (-1, 2), r = 2$*
$(x + 1)^2 + (y - 2)^2 = 4$ *Simplify.*

The equation for the circle is $(x + 1)^2 + (y - 2)^2 = 4$.

Your Turn

a. Write an equation of a circle with center at $(3, -2)$ and a diameter of 8 units.

nline Personal Tutor at geomconcepts.com

You can also use the equation of a circle to find the coordinates of its center and the measure of its radius.

Example ②

Geography Link

The lake in Crater Lake Park was formed thousands of years ago by the explosive collapse of Mt. Mazama. If the park entrance is at $(0, 0)$, then the equation of the circle representing the lake is $(x + 1)^2 + (y + 11)^2 = 9$. Find the coordinates of its center and the measure of its diameter. Each unit on the grid represents 2 miles.

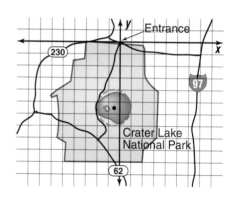

Rewrite the equation in the form $(x - h)^2 \quad + \quad (y - k)^2 \quad = r^2$.

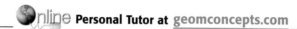

$$[(x - (-1)]^2 + [(y - (-11)]^2 = 3^2$$

Since $h = -1$, $k = -11$, and $r = 3$, the center of the circle is at $(-1, -11)$. Its radius is 3 miles, so its diameter is 6 miles.

Crater Lake, Oregon

Your Turn

b. Find the coordinates of the center and the measure of the radius of a circle whose equation is $x^2 + \left(y - \frac{3}{4}\right)^2 = \frac{25}{4}$.

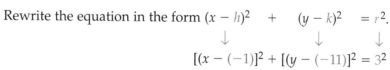

Check for Understanding

1. **Draw** a circle on a coordinate plane. Use a ruler to find its radius and write its general equation.

2. **Match** each graph below with one of the equations at the right.

   (1) $(x + 1)^2 + (y - 4)^2 = 5$
   (2) $(x - 1)^2 + (y + 4)^2 = 5$
   (3) $(x + 1)^2 + (y - 4)^2 = 25$
   (4) $(x - 1)^2 + (y + 4)^2 = 25$

a.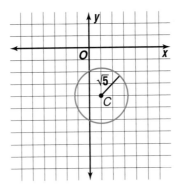

b.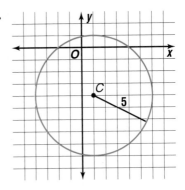

c.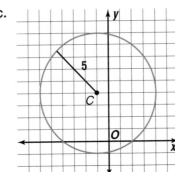

d.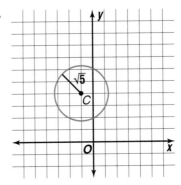

3. **Explain** how you could find the equation of a line that is tangent to the circle whose equation is $(x - 4)^2 + (y + 6)^2 = 9$.

4. **Writing Math** How could you find the equation of a circle if you are given the coordinates of the endpoints of a diameter? First, make a sketch of the problem and then list the information that you need and the steps you could use to find the equation.

Guided Practice

⊕ Getting Ready If *r* represents the radius and *d* represents the diameter, find each missing measure.

Sample: $d = \frac{1}{3}, r^2 = $ ___?___   **Solution:** $r^2 = \left(\frac{1}{2} \cdot \frac{1}{3}\right)^2$ or $\frac{1}{36}$

5. $r^2 = 169, d = $ ___?___

6. $d = 2\sqrt{18}, r^2 = $ ___?___

7. $d = \frac{2}{5}, r^2 = $ ___?___

8. $r^2 = \frac{16}{49}, d = $ ___?___

Example 1   Write an equation of a circle for each center and radius or diameter measure given.

9. $(1, -5), d = 8$

10. $(3, 4), r = \sqrt{2}$

Example 2   Find the coordinates of the center and the measure of the radius for each circle whose equation is given.

11. $(x - 7)^2 + (y + 5)^2 = 4$

12. $(x - 6)^2 + y^2 = 64$

Example 1   13. **Botany** Scientists can tell what years had droughts by studying the rings of bald cypress trees. If the radius of a tree in 1612 was 14.5 inches, write an equation that represents the cross section of the tree. Assume that the center is at $(0, 0)$.

Exercises ·

Practice   Write an equation of a circle for each center and radius or diameter measure given.

14. $(2, -11), r = 3$

15. $(-4, 2), d = 2$

16. $(0, 0), r = \sqrt{5}$

17. $(6, 0), r = \dfrac{2}{3}$

18. $(-1, -1), d = \dfrac{1}{4}$

19. $(-5, 9), d = 2\sqrt{20}$

Homework Help

For Exercises	See Examples
14–19, 28, 29, 31, 32	1
20–27, 30	2

Extra Practice
See page 753.

Find the coordinates of the center and the measure of the radius for each circle whose equation is given.

20. $(x - 9)^2 + (y - 10)^2 = 1$

21. $x^2 + (y + 5)^2 = 100$

22. $(x + 7)^2 + (y - 3)^2 = 25$

23. $\left(x + \dfrac{1}{2}\right)^2 + \left(y + \dfrac{1}{3}\right)^2 = \dfrac{16}{25}$

24. $(x - 19)^2 + y^2 = 20$

25. $(x - 24)^2 + (y + 8.1)^2 - 12 = 0$

Graph each equation on a coordinate plane.

26. $(x + 5)^2 + (y - 2)^2 = 4$

27. $x^2 + (y - 3)^2 = 16$

28. Write an equation of the circle that has a diameter of 12 units and its center at $(-4, -7)$.

29. Write an equation of the circle that has its center at $(5, -13)$ and is tangent to the y-axis.

30. Meteorology Often when a hurricane is expected, all people within a certain radius are evacuated. Circles around a radar image can be used to determine a safe radius. If an equation of the circle that represents the evacuated area is given by $(x + 42)^2 + (y - 11)^2 = 1024$, find the coordinates of the center and measure of the radius of the evacuated area. Units are in miles.

Math **Data**
nline **Update**
For the latest information on the percents of international internet users, visit:
geomconcepts.com

31. Technology Although English is the language used by more than half the Internet users, over 56 million people worldwide use a different language, as shown in the circle graph at the right. If the circle displaying the information has a center $C(0, -3)$ and a diameter of 7.4 units, write an equation of the circle.

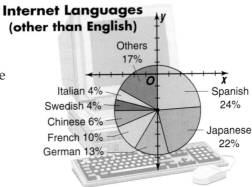

Internet Languages (other than English)

Others 17%
Italian 4%
Swedish 4%
Chinese 6%
French 10%
German 13%
Spanish 24%
Japanese 22%

Source: Euro-Marketing Associates

32. Critical Thinking The graphs of $x = 4$ and $y = -1$ are both tangent to a circle that has its center in the fourth quadrant and a diameter of 14 units. Write an equation of the circle.

Mixed Review

33. Find AB to the nearest tenth. *(Lesson 14–5)*

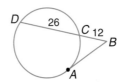

34. Toys Describe the basic shape of the toy as a geometric solid. *(Lesson 12–1)*

35. Find the area of a regular pentagon whose perimeter is 40 inches and whose apothems are each 5.5 inches long. *(Lesson 10–5)*

36. Short Response Find the values of x and y. *(Lesson 9–3)*

10
9
6
15
x
y

37. Multiple Choice Find the length of the diagonal of a rectangle whose length is 12 meters and whose width is 4 meters. *(Lesson 6–6)*

Ⓐ 48 m　　Ⓑ 160 m　　Ⓒ 6.9 m　　Ⓓ 12.6 m

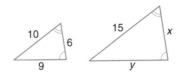

Math In the Workplace

Meteorologist

Do you enjoy watching storms? Have you ever wondered why certain areas of the country have more severe weather conditions such as hurricanes or tornadoes? If so, you may want to consider a career as a meteorologist. In addition to forecasting weather, meteorologists apply their research of Earth's atmosphere in areas of agriculture, air and sea transportation, and air-pollution control.

1. Suppose your home is located at (0, 0) on a coordinate plane. If the "eye of the storm," or the storm's center, is located 25 miles east and 12 miles south of you, what are the coordinates of the storm's center?

2. If the storm has a 7-mile radius, write an equation of the circle representing the storm.

3. Graph the equation of the circle in Exercise 2.

FAST FACTS About Meteorologists

Working Conditions

- may report from radio or television station studios
- must be able to work as part of a team
- those not involved in forecasting work regular hours, usually in offices
- may observe weather conditions and collect data from aircraft

Education

- high school math and physical science courses
- bachelor's degree in meteorology
- A master's or Ph.D. degree is required for research positions.

Employment

4 out of 10 meteorologists have federal government jobs.

Government Position	Tasks Performed
Beginning Meterologist	collect data, perform computations or analysis
Entry-Level Intern	learn about the Weather Service's forecasting equipment and procedures
Permanent Duty	handle more complex forecasting jobs

*inter*NET
CONNECTION
www.geomconcepts.com

Career Data For the latest information about a career as a meteorologist, visit:

Understanding and Using the Vocabulary

After completing this chapter, you should be able to define each term, property, or phrase and give an example or two of each.

*inter*NET
CONNECTION **Review Activities**
For more review activities, visit:
www.geomconcepts.com

external secant segment *(p. 613)*
externally tangent *(p. 595)*
inscribed angle *(p. 586)*
intercepted arc *(p. 586)*

internally tangent *(p. 595)*
point of tangency *(p. 592)*
secant angle *(p. 600)*
secant-tangent angle *(p. 606)*

secant segment *(p. 600)*
tangent *(p. 592)*
tangent-tangent angle *(p. 607)*

Choose the term or terms from the list above that best complete each statement.

1. When two secants intersect, the angles formed are called ___?___ .
2. The vertex of a(n) ___?___ is on the circle and its sides contain chords of the circle.
3. A tangent-tangent angle is formed by two ___?___ .
4. A tangent intersects a circle in exactly one point called the ___?___ .
5. The measure of an inscribed angle equals one-half the measure of its ___?___ .
6. A(n) ___?___ is the part of a secant segment that is outside a circle.
7. A(n) ___?___ is formed by a vertex outside the circle or by a vertex on the circle.
8. A ___?___ is a line segment that intersects a circle in exactly two points.
9. The measure of a(n) ___?___ is always one-half the difference of the measures of the intercepted arcs.
10. If a line is tangent to a circle, then it is perpendicular to the radius drawn to the ___?___ .

Skills and Concepts

Objectives and Examples	**Review Exercises**

• **Lesson 14–1** Identify and use properties of inscribed angles.

$m\angle ABC = \frac{1}{2}m\widehat{AC}$

$\angle 1 \cong \angle 2$

$m\angle LMN = 90$

Find each measure.

11. $m\widehat{XZ}$

12. $m\angle ABC$

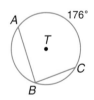

In each circle, find the value of x.

13. $(x + 2)°$ ... $3x°$

14.

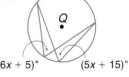
$(6x + 5)°$ ... $(5x + 15)°$

www.geomconcepts.com/vocabulary_review

Objectives and Examples

Review Exercises

- **Lesson 14-2** Identify and apply properties of tangents to circles.

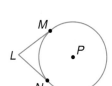

If line ℓ is tangent to ⊙C, then $\overline{CD} \perp \ell$.

If $\overline{CD} \perp \ell$, then ℓ must be tangent to ⊙C.

If $\overline{LM}$ and $\overline{LN}$ are tangent to ⊙P, then $\overline{LM} \cong \overline{LN}$.

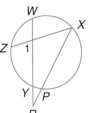

Find each measure. Assume segments that appear to be tangent are tangent.

15. *MN*   **16.** *BD*

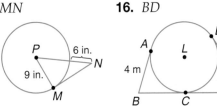

17. Find $m\angle RQS$ and *QS*.

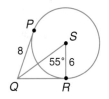

- **Lesson 14-3** Find measures of arcs and angles formed by secants.

$m\angle 1 = \frac{1}{2}(m\widehat{WX} + m\widehat{YZ})$

$m\angle R = \frac{1}{2}(m\widehat{WX} - m\widehat{YP})$

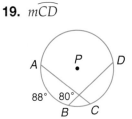

Find each measure.

18. $m\angle J$ **19.** $m\widehat{CD}$

20. Find the value of *x*. Then find $m\widehat{RS}$.

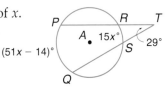

- **Lesson 14-4** Find measures of arcs and angles formed by secants and tangents.

$m\angle ABC = \frac{1}{2}(m\widehat{AC} - m\widehat{CD})$

$m\angle JEF = \frac{1}{2}(m\widehat{JGE})$

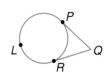

$m\angle PQR = \frac{1}{2}(m\widehat{PLR} - m\widehat{PR})$

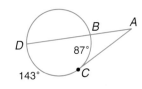

Find the measure of each angle. Assume segments that appear to be tangent are tangent.

21. $\angle CAD$ **22.** $\angle PQR$

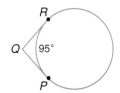

Mixed Problem Solving
See pages 758–765.

Objectives and Examples

• **Lesson 14–5** Find measures of chords, secants, and tangents.

$AP \cdot PD = BP \cdot PC$

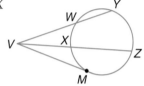

$VY \cdot VW = VZ \cdot VX$

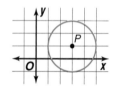

$(VM)^2 = VZ \cdot VX$

Review Exercises

In each circle, find the value of *x*. If necessary, round to the nearest tenth.

23.

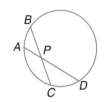

24.

Find each measure. If necessary, round to the nearest tenth.

25. *AB*

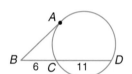

26. *ST*

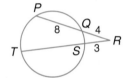

• **Lesson 14–6** Write equations of circles using the center and the radius.

Write the equation of a circle with center $P(3, 1)$ and a radius of 2 units.

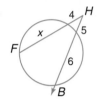

$(x - h)^2 + (y - k)^2 = r^2$ *General equation*
$(x - 3)^2 + (y - 1)^2 = 2^2$ $(h, k) = (3, 1); r = 2$

The equation is $(x - 3)^2 + (y - 1)^2 = 4$.

Write the equation of a circle for each center and radius or diameter measure given.

27. $(-3, 2), r = 5$
28. $(6, 1), r = 6$
29. $(5, -5), d = 4$

Find the coordinates of the center and the measure of the radius for each circle whose equation is given.

30. $(x + 2)^2 + (y + 3)^2 = 36$
31. $(x - 9)^2 + (y + 6)^2 = 16$
32. $(x - 5)^2 + (y - 7)^2 = 169$

Applications and Problem Solving

33. Lumber A lumber yard receives perfectly round logs of raw lumber for further processing. Determine the diameter of the log at the right. *(Lesson 14–1)*

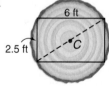

34. Algebra Find *x*. Then find $m\angle A$. *(Lesson 14–3)*

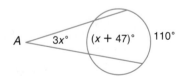

1. **Compare and contrast** a tangent to a circle and a secant of a circle.
2. **Draw** a circle with the equation $(x - 1)^2 + (y + 1)^2 = 4$.
3. **Define** the term *external secant segment*.

$\odot O$ is inscribed in $\triangle XYZ$, $m\overset{\frown}{AB} = 130$, $m\overset{\frown}{AC} = 100$, and $m\angle DOB = 50$. Find each measure.

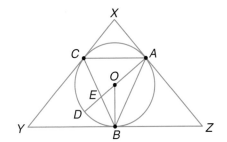

4. $m\angle YXZ$
5. $m\angle CAD$
6. $m\angle XZY$
7. $m\angle AEC$
8. $m\angle OBZ$
9. $m\angle ACB$

Find each measure. If necessary, round to the nearest tenth. Assume segments that appear to be tangent are tangent.

10. $m\overset{\frown}{QR}$

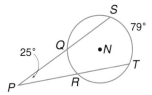

11. AE

12. $m\angle XYZ$

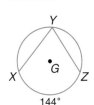

13. BD

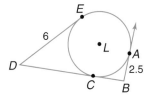

14. $m\overset{\frown}{JM}$

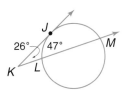

15. WX

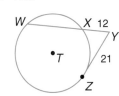

Find each value of x. Then find the given measure.

16. $m\overset{\frown}{JK}$

17. TZ

Write the equation of a circle for each center and radius or diameter measure given.

18. $(6, -1)$, $d = 12$
19. $(3, 7)$, $r = 1$

20. **Antiques** A round stained-glass window is divided into three sections, each a different color. In order to replace the damaged middle section, an artist must determine the exact measurements. Find the measure of $\angle A$.

Right Triangle and Trigonometry Problems

Many geometry problems on standardized tests involve right triangles and the Pythagorean Theorem.

The ACT also includes trigonometry problems. Memorize these ratios.

$$\sin \theta = \frac{\text{opposite}}{\text{hypotenuse}}, \cos \theta = \frac{\text{adjacent}}{\text{hypotenuse}}, \tan \theta = \frac{\text{opposite}}{\text{adjacent}}$$

Standardized tests often use the Greek letter θ (*theta*) for the measure of an angle.

Example 1

A 32-foot telephone pole is braced with a cable that runs from the top of the pole to a point 7 feet from the base. What is the length of the cable rounded to the nearest tenth?

(A) 31.2 ft   **(B)** 32.8 ft

(C) 34.3 ft   **(D)** 36.2 ft

Hint   If no diagram is given, draw one.

Solution   Draw a sketch and label the given information.

32 ft

7 ft

You can assume that the pole makes a right angle with the ground. In this right triangle, you know the lengths of the two sides. You need to find the length of the hypotenuse. Use the Pythagorean Theorem.

$c^2 = a^2 + b^2$ *Pythagorean Theorem*

$c^2 = 32^2 + 7^2$ *a = 32 and b = 7*

$c^2 = 1024 + 49$ *$32^2 = 1024$ and $7^2 = 49$*

$c^2 = 1073$ *Add.*

$c = \sqrt{1073}$ *Take the square root of each side.*

$c \approx 32.8$ *Use a calculator.*

To the nearest tenth, the hypotenuse is 32.8 feet. The answer is B.

Example 2

In the figure at the right, $\angle A$ is a right angle, $\overline{AB}$ is 3 units long, and $\overline{BC}$ is 5 units long. If the measure of $\angle C$ is θ, what is the value of $\cos \theta$?

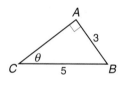

(A) $\frac{3}{5}$ **(B)** $\frac{3}{4}$ **(C)** $\frac{4}{5}$ **(D)** $\frac{5}{4}$ **(E)** $\frac{5}{3}$

Hint   In trigonometry problems, label the triangle with the words *opposite*, *adjacent*, and *hypotenuse*.

Solution

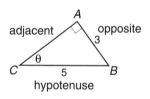

adjacent   opposite

hypotenuse

To find $\cos \theta$, you need to know the length of the adjacent side. Notice that the hypotenuse is 5 and one side is 3, so this is a 3-4-5 right triangle. The adjacent side is 4 units.

Use the ratio for $\cos \theta$.

$\cos \theta = \frac{\text{adjacent}}{\text{hypotenuse}}$ *Definition of cosine*

$= \frac{4}{5}$ *Substitution*

The answer is C.

After you work each problem, record your answer on the answer sheet provided or on a sheet of paper.

Multiple Choice

1. Fifteen percent of the coins in a piggy bank are nickels and 5% are dimes. If there are 220 coins in the bank, how many are not nickels or dimes? *(Percent Review)*

   Ⓐ 80   Ⓑ 176   Ⓒ 180   Ⓓ 187   Ⓔ 200

2. A bag contains 4 red, 10 blue, and 6 yellow balls. If three balls are removed at random and no ball is returned to the bag after removal, what is the probability that all three balls will be blue? *(Statistics Review)*

   Ⓐ $\frac{1}{2}$ Ⓑ $\frac{1}{8}$ Ⓒ $\frac{3}{20}$ Ⓓ $\frac{2}{19}$ Ⓔ $\frac{3}{8}$

3. Which point represents a number that could be the product of two negative numbers and a positive number greater than 1? *(Algebra Review)*

   Ⓐ P and Q Ⓑ P only
   Ⓒ R and S Ⓓ S only

4. What is the area of $\triangle ABC$ in terms of x? *(Lesson 13–5)*

   Ⓐ $10 \sin x$
   Ⓑ $40 \sin x$
   Ⓒ $80 \sin x$
   Ⓓ $40 \cos x$
   Ⓔ $80 \cos x$

5. Suppose $\triangle PQR$ is to have a right angle at Q and an area of 6 square units. Which could be coordinates of point R? *(Lesson 10–4)*

   Ⓐ $(2, 2)$ Ⓑ $(5, 8)$
   Ⓒ $(5, 2)$ Ⓓ $(2, 8)$

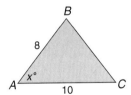

6. What is the diagonal distance across a rectangular yard that is 20 yd by 48 yd? *(Lesson 6–6)*

   Ⓐ 52 yd   Ⓑ 60 yd
   Ⓒ 68 yd   Ⓓ 72 yd

7. What was the original height of the tree? *(Lesson 6–6)*

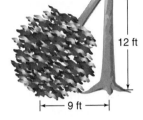

   Ⓐ 15 ft
   Ⓑ 20 ft
   Ⓒ 27 ft
   Ⓓ 28 ft

8. Points A, B, C, and D are on the square. $ABCD$ is a rectangle, but not a square. Find the perimeter of $ABCD$ if the distance from E to A is 1 and the distance from E to B is 1. *(Lesson 6–6)*

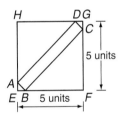

   Ⓐ 64 units   Ⓑ $10\sqrt{2}$ units
   Ⓒ 10 units   Ⓓ 8 units

Grid In

9. Segments AB and BD are perpendicular. Segments AB and CD bisect each other at x. If $AB = 8$ and $CD = 10$, what is BD? *(Lesson 2–3)*

Extended Response

10. The base of a ladder should be placed 1 foot from the wall for every 3 feet of length. *(Lesson 6–6)*

Part A How high can a 15-foot ladder safely reach? Draw a diagram.

Part B How long a ladder is needed to reach a window 24 feet above the ground?

15

Formalizing Proof

What You'll Learn

Key Ideas

- Find the truth values of simple and compound statements. *(Lesson 15–1)*

- Use the Law of Detachment and the Law of Syllogism in deductive reasoning. *(Lesson 15–2)*

- Use paragraph proofs, two-column proofs, and coordinate proofs to prove theorems. *(Lessons 15–3, 15–5, and 15–6)*

- Use properties of equality in algebraic and geometric proofs. *(Lesson 15–4)*

Key Vocabulary

Why It's Important

Engineering   Designers and engineers use Computer-Aided Design, or CAD, to prepare drawings and make specifications for products in fields such as architecture, medical equipment, and automotive design. An engineer can also use CAD to simulate an operation to test a design, rather than building a prototype.

Geometric proof allows you to determine relationships among figures. You will investigate a CAD design in Lesson 15–6.

✓ Check Your Readiness

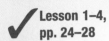

Lesson 1–4,
pp. 24–28

Identify the hypothesis and conclusion of each statement.

1. If it rains, then we will not have soccer practice.

2. If two segments are congruent, then they have the same measure.

3. I will advance to the semi-finals if I win this game.

4. All dogs are mammals.

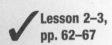

Lesson 2–3,
pp. 62–67

Determine whether each statement is *true* or *false*. Explain your reasoning.

5. If $AB = DE$, then $\overline{AB} \cong \overline{DE}$.

6. If $\overline{MN} \cong \overline{OP}$ and $\overline{MN} \cong \overline{RT}$, then $\overline{OP} \cong \overline{RT}$.

7. If $\overline{JK} \cong \overline{KL}$, then K is the midpoint of $\overline{JL}$.

8. If $\overline{XY} \cong \overline{WZ}$ and $\overline{WZ} \cong \overline{WY}$, then $\overline{XZ} \cong \overline{WY}$.

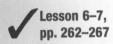

Lesson 6–7,
pp. 262–267

Find the distance between each pair of points. Round to the nearest tenth, if necessary.

9. $A(2, 1)$, $B(5, 5)$

10. $C(-3, -2)$, $D(2, 10)$

11. $E(0, 4)$, $F(-3, 1)$

12. $G(9, 9)$, $H(-1, 8)$

13. $I(-2, -4)$, $J(-6, -5)$

14. $K(12, 10)$, $(-5, -6)$

FOLDABLES™
Study Organizer

Make this Foldable to help you organize your Chapter 15 notes. Begin with four sheets of $8\frac{1}{2}$" by 11" grid paper.

❶ **Fold** each sheet in half along the width. Then cut along the crease.

❷ **Staple** the eight half-sheets together to form a booklet.

❸ **Cut** seven lines from the bottom of the top sheet, six lines from the second sheet, and so on.

❹ **Label** each tab with a lesson title. The last tab is for vocabulary.

Reading and Writing As you read and study the chapter, use the pages to write main ideas, theorems, and examples for each lesson.

What You'll Learn
You'll learn to find the truth values of simple and compound statements.

Why It's Important
Advertising
Advertisers use conditional statements to sell products.
See Exercise 35.

Every time you take a true-false test, you are using a building block of logic. Here's an example.

True or false:
Albany is the capital of New York.

A **statement** is any sentence that is either true or false, but not both. Therefore, every statement has a **truth value**, true (T) or false (F). The map shows that the statement above is true.

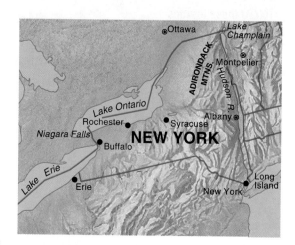

A convenient way to represent a statement is with a letter such as *p* or *q*.

p: Albany is the capital of New York.

Suppose you want to say that Albany is *not* the capital of New York. This statement is called *not p*.

not p: Albany is *not* the capital of New York.

The statement *not p* is the **negation** of *p*.

Definition of Negation	**Words:**	If a statement is represented by *p*, then *not p* is the negation of the statement.
	Symbols:	$\sim p$

Examples

Let *p* represent "It is raining" and *q* represent "$15 - 8 = 5$." Write the statements for each negation.

1 $\sim p$
   p: It is raining.
   $\sim p$: It is *not* raining.

2 $\sim q$
   q: $15 - 8 = 5$
   $\sim q$: $15 - 8 \neq 5$

Your Turn

Let *r* represent "Today is Monday" and *s* represent "$4 + 3 = 7$."
   a. $\sim r$ **b.** $\sim s$

There is a relationship between the truth value of a statement and its negation. If a statement is true, its negation is false. If a statement is false, its negation is true. The truth values can be organized in a **truth table** like the one shown below.

Negation	
p	$\sim p$
T	F
F	T

← If p is a true statement, then $\sim p$ is a false statement.
← If p is a false statement, then $\sim p$ is a true statement.

Any two statements can be joined to form a **compound statement**. Consider the following two statements.

p: I am taking geometry. q: I am taking Spanish.

The two statements can be joined by the word *and*.

p *and* q: I am taking geometry, *and* I am taking Spanish.

Definition of Conjunction	**Words:**	A **conjunction** is a compound statement formed by joining two statements with the word *and*.
	Symbols:	$p \wedge q$

The two statements can also be joined by the word *or*.

p *or* q: I am taking geometry, *or* I am taking Spanish.

Definition of Disjunction	**Words:**	A **disjunction** is a compound statement formed by joining two statements with the word *or*.
	Symbols:	$p \vee q$

A conjunction is true only when *both* of the statements are true. In this case, the conjunction is true only if you are taking both geometry and Spanish. The disjunction is true if you are taking either geometry or Spanish, or both. In this case, the disjunction is false only if you are taking neither geometry nor Spanish. This information is summarized in the truth tables below.

Conjunction		
p	q	$p \wedge q$
T	T	T
T	F	F
F	T	F
F	F	F

A conjunction is true only when both statements are true.

Disjunction		
p	q	$p \vee q$
T	T	T
T	F	T
F	T	T
F	F	F

A disjunction is false only when both statements are false.

Lesson 15–1 Logic and Truth Tables **633**

Let *p* represent "10 + 3 = 13", *q* represent "June has 31 days," and *r* represent "A triangle has three sides." Write the statement for each conjunction or disjunction. Then find the truth value.

3 *p* ∧ *q*

10 + 3 = 13 and June has 31 days.

p ∧ *q* is false because *p* is true and *q* is false.

4 *p* ∨ *r*

10 + 3 = 13 or a triangle has three sides.

p ∨ *r* is true because both *p* and *r* are true.

5 ~*q* ∧ *r*

June does not have 31 days and a triangle has three sides.

Because *q* is false, ~*q* is true. Therefore, ~*q* ∧ *r* is true because both ~*q* and *r* are true.

Your Turn

c. *q* ∧ *r*   **d.** *p* ∨ *q*   **e.** ~*p* ∨ *q*

You can use truth values for conjunctions and disjunctions to construct truth tables for more complex compound statements.

Example

6 **Construct a truth table for the conjunction *p* ∧ ~*q*.**

Step 1 Make columns with the headings *p*, *q*, ~*q*, and *p* ∧ ~*q*.

Step 2 List all of the possible combinations of truth values for *p* and *q*.

Step 3 Use the truth values for *q* to write the truth values for ~*q*.

Step 4 Use the truth values for *p* and ~*q* to write the truth values for *p* ∧ ~*q*.

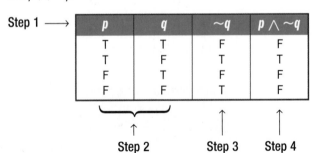

Step 1 ⟶

p	*q*	~*q*	*p* ∧ ~*q*
T	T	F	F
T	F	T	T
F	T	F	F
F	F	T	F

Step 2   Step 3   Step 4

Look Back

Conditional Statements, Lesson 1-4

Your Turn

f. Construct a truth table for the disjunction ~*p* ∨ *q*.

In this text, you have been using another compound statement that is formed by joining statements with *if. . . . then*. Recall that these statements are called *conditional statements*. Consider the following statements.

p: A figure is a rectangle.   *q*: The diagonals are congruent.

If p, then q: *If* a figure is a rectangle, *then* the diagonals are congruent.

When is a conditional statement true? If a figure is a rectangle and its diagonals are congruent, the statement is true. If the figure is a rectangle, but its diagonals are *not* congruent, the statement is false.

If the figure is *not* a rectangle, it is not possible to tell whether the diagonals are congruent. In this case, we will consider the conditional to be true.

A truth table for conditional statements is shown at the right.

Conditional		
p	*q*	*p* → *q*
T	T	T
T	F	F
F	T	T
F	F	T

A conditional is false only when p is true and q is false.

In Chapter 1, you learned about the *converse* of a conditional. The converse is formed by exchanging the hypothesis and the conclusion.

Conditional: *If* a figure is a rectangle, *then* the diagonals are congruent.

Converse: *If* the diagonals are congruent, *then* the figure is a rectangle.

Using symbols, if *p* → *q* is a conditional, *q* → *p* is its converse.

Example ⑦ Construct a truth table for the converse *q* → *p*.

Converse		
p	*q*	*q* → *p*
T	T	T
T	F	T
F	T	F
F	F	T

← *The converse of a conditional is false when p is false and q is true.*

Your Turn

g. The **inverse** of a conditional is formed by negating both *p* and *q*. So, if *p* → *q* is the conditional, ~*p* → ~*q* is its inverse. Construct a truth table for the inverse ~*p* → ~*q*.

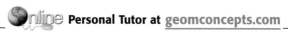 **Personal Tutor at geomconcepts.com**

Check for Understanding

Communicating Mathematics

1. **Explain** the difference between a conjunction and a disjunction.

2. **Write** a compound sentence that meets each set of conditions.
   a. a true conjunction
   b. a false disjunction
   c. a true conditional

Guided Practice

> ⏱ **Getting Ready** Tell whether each statement is *true* or *false*.

> **Sample:** Abraham Lincoln was a president of the United States.
> **Solution:** This statement is *true*.

3. $5 + 6 = 14$
4. France is a country in South America.
5. 0.5 is a rational number.

Examples 1 & 2

Let *p* represent "$5 + 8 = 13$" and *q* represent "Mark Twain is a famous author." Write the statements for each negation.

6. $\sim p$ 7. $\sim q$

Examples 3–5

Let *r* represent "A square has congruent sides," *s* represent "A scalene triangle has congruent sides," and *t* represent "A parallelogram has parallel sides." Write a statement for each conjunction or disjunction. Then find the truth value.

8. $r \wedge s$ 9. $r \vee t$ 10. $\sim r \vee s$

Examples 6 & 7

Construct a truth table for each compound statement.

11. $p \vee \sim q$ 12. $\sim p \to q$

Example 7

13. **Advertising** A cat food company's slogan is *If you love your cat, feed her Tasty Bits*. Let *p* represent "you love your cat" and *q* represent "feed her Tasty Bits." If *p* is false and *q* is true, find the truth value of $q \to p$.

Exercises

Practice

Use conditionals *p*, *q*, *r*, and *s* for Exercises 14–25.

p: Water freezes at 32°F.   *q*: Memorial Day is in July.
r: $20 \times 5 = 90$ *s*: A pentagon has five sides.

Write the statements for each negation.

14. $\sim p$ 15. $\sim q$ 16. $\sim r$ 17. $\sim s$

Homework Help

For Exercises	See Examples
14–17	1, 2
18–25	3–5
26–29, 32–34	6
30, 31, 35	7
36, 37	6, 7

Extra Practice

See page 754.

Applications and Problem Solving

Write a statement for each conjunction or disjunction. Then find the truth value.

18. $p \vee q$ **19.** $p \wedge q$ **20.** $q \vee r$ **21.** $p \wedge s$

22. $\sim p \vee r$ **23.** $\sim p \wedge \sim s$ **24.** $\sim q \wedge s$ **25.** $\sim q \vee \sim r$

Construct a truth table for each compound statement.

26. $\sim p \vee \sim q$ **27.** $\sim(p \vee q)$ **28.** $\sim p \wedge q$ **29.** $\sim p \wedge \sim q$

30. $p \rightarrow \sim q$ **31.** $\sim p \rightarrow q$ **32.** $\sim(p \vee \sim q)$ **33.** $\sim(\sim p \wedge q)$

34. Geography   Use the map on page 632 to determine whether each statement is true or false.

   a. Albany is *not* located on the Hudson River.

   b. Either Rochester or Syracuse is located on Lake Ontario.

   c. It is false that Buffalo is located on Lake Erie.

35. Advertising   *If you want clear skin, use Skin-So-Clear.*

   a. Write the converse of the conditional.

   b. What do you think the advertiser wants people to conclude about Skin-So-Clear?

   c. Is the conclusion in Exercise 35b valid? Explain.

36. Critical Thinking   The **contrapositive** of $p \rightarrow q$ is $\sim q \rightarrow \sim p$.

   a. Construct a truth table for the contrapositive $\sim q \rightarrow \sim p$.

   b. Two statements are **logically equivalent** if their truth tables are the same. Compare the truth tables for a conditional, converse, inverse, and contrapositive. Which of the statements is logically equivalent to a conditional?

37. Critical Thinking   The conjunction $(p \rightarrow q) \wedge (q \rightarrow p)$ is called a **biconditional**. For which values of p and q is a biconditional true?

Mixed Review

38. Write the equation of a circle with center $C(-2, 3)$ and a radius of 3 units.   *(Lesson 14–6)*

Find each value of x.   *(Lesson 14–5)*

39.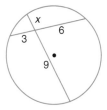

40.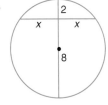

41.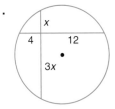

Find each ratio in △ABC.   *(Lesson 13–5)*

42. $\sin A$ **43.** $\sin B$

44. $\cos A$ **45.** $\cos B$

Exercises 42–45

Standardized Test Practice
Ⓐ Ⓑ Ⓒ Ⓓ

46. Short Response   Draw a diagram in which the angle of depression to an object is 30°.   *(Lesson 13–4)*

15-2 Deductive Reasoning

What You'll Learn

You'll learn to use the Law of Detachment and the Law of Syllogism in deductive reasoning.

Why It's Important

Literature Mystery writers use logical arguments.
See Exercise 25.

Look Back

Inductive Reasoning: Lesson 1–1

When you make a prediction based on a pattern in data, you are using inductive reasoning. Inductive reasoning is very useful for developing conjectures in mathematics. In this text, you have developed the foundation of geometric definitions, postulates, and theorems using inductive reasoning with the help of models. Here's an example.

Suppose you place six points on a circle and draw each segment that connects a pair of points. What is the greatest number of regions within the circle that are formed by the segments? Look for a pattern.

Model				
Points	2	3	4	5
Regions	2	4	8	16

Make a conjecture about the number of regions formed by six points. It seems that the number of regions increases by a power of 2. Based on the pattern, you can reason that 6 points should determine 32 regions. Now, test your conjecture by counting the regions in the figure below.

Remember that only one counterexample is needed to disprove a conjecture.

COncepts in MOtion

Interactive Lab
geomconcepts.com

The maximum number of regions is only 31. The counterexample shows that the conjecture based on inductive reasoning did not give the correct conclusion.

Even though patterns can help you make a conjecture, patterns alone do *not* guarantee that the conjecture will be true. In logic you can *prove* that a statement is true for all cases by using deductive reasoning. **Deductive reasoning** is the process of using facts, rules, definitions, or properties in a logical order.

Here's an example of a conclusion that is arrived at by deductive reasoning using a conditional statement.

Words	Symbols	Meaning
If Marita obeys the speed limit, then she will not get a speeding ticket.	$p \rightarrow q$	If p is true, then q is true.
Marita obeyed the speed limit.	p	p is true.
Therefore, Marita did not get a speeding ticket.	q	Therefore, q is true.

In deductive reasoning, if you arrive at the conclusion using correct reasoning, then the conclusion is valid. The rule that allows us to reach a conclusion from conditional statements in the example is called the **Law of Detachment**.

Law of Detachment	If $p \rightarrow q$ is a true conditional and p is true, then q is true.

Examples

Use the Law of Detachment to determine a conclusion that follows from statements (1) and (2). If a valid conclusion does not follow, write *no valid conclusion*.

 (1) If $\overline{AD} \parallel \overline{CB}$ and $\overline{AD} \cong \overline{CB}$, then ABCD is a parallelogram.

(2) $\overline{AD} \parallel \overline{CB}$ and $\overline{AD} \cong \overline{CB}$.

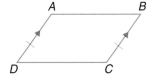

Let p and q represent the parts of the statements.
p: $\overline{AD} \parallel \overline{CB}$ and $\overline{AD} \cong \overline{CB}$
q: ABCD is a parallelogram
Statement (1) indicates that $p \rightarrow q$ is true, and statement (2) indicates that p is true. So, q is true. Therefore, ABCD is a parallelogram.

 (1) If a figure is a square, it has four right angles.

(2) A figure has four right angles.

p: a figure is a square
q: a figure has four right angles
Statement (1) is true, but statement (2) indicates that q is true. It does not provide information about p. Therefore, there is no valid conclusion.

(continued on the next page)

a. Use the Law of Detachment to determine a conclusion that follows from statements (1) and (2).

(1) In a plane, if two lines are cut by a transversal so that a pair of alternate interior angles is congruent, then the lines are parallel.

(2) Lines ℓ and m are cut by transversal t and $\angle 1 \cong \angle 2$.

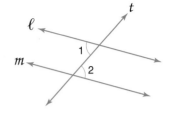

Look Back

Transitive Property:
Lesson 2–2

Another rule of logic is the **Law of Syllogism**. This rule is similar to the Transitive Property of Equality.

Law of Syllogism	If $p \rightarrow q$ and $q \rightarrow r$ are true conditionals, then $p \rightarrow r$ is also true.

Example — ③

Use the Law of Syllogism to determine a conclusion that follows from statements (1) and (2).

(1) If $\angle R \cong \angle M$, then $\triangle RAM$ is an isosceles triangle.

(2) If $\triangle RAM$ is an isosceles triangle, then $\overline{AR} \cong \overline{AM}$.

Let p, q, and r represent the parts of the statements.
p: $\angle R \cong \angle M$
q: $\triangle RAM$ is an isosceles triangle
r: $\overline{AR} \cong \overline{AM}$
Use the Law of Syllogism to conclude $p \rightarrow r$.
Therefore, if $\angle R \cong \angle M$, then $\overline{AR} \cong \overline{AM}$.

b. (1) If a triangle is a right triangle, the sum of the measures of the acute angles is 90.

(2) If the sum of the measures of two angles is 90, then the angles are complementary.

 Personal Tutor at geomconcepts.com

Check for Understanding

1. **Explain** the difference between inductive and deductive reasoning.

2. **Write** your own example to illustrate the correct use of the Law of Syllogism.

Vocabulary
deductive reasoning
Law of Detachment
Law of Syllogism

3. 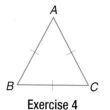 Joel and Candace found the conclusion to this conditional using the Law of Detachment.

> If crocuses are blooming, it must be spring.
> Crocuses are not blooming.

Joel said the conclusion is *It must not be spring.* Candace said there is no valid conclusion. Who is correct? Explain your reasoning.

Guided Practice

Use the Law of Detachment to determine a conclusion that follows from statements (1) and (2). If a valid conclusion does not follow, write *no valid conclusion*.

Examples 1 & 2

4. (1) If a triangle is equilateral, then the measure of each angle is 60.

    (2) $\triangle ABC$ is an equilateral triangle.

5. (1) If Amanda is taller than Teresa, then Amanda is at least 6 feet tall.

    (2) Amanda is older than Teresa.

Exercise 4

Example 3

Use the Law of Syllogism to determine a conclusion that follows from statements (1) and (2). If a valid conclusion does not follow, write *no valid conclusion*.

6. (1) If I work part-time, I will save money.
    (2) If I save money, I can buy a computer.

7. (1) If two angles are vertical angles, then they are congruent.
    (2) If two angles are congruent, then their supplements are congruent.

Examples 1 & 2

8. Media The following statement is part of a message frequently played by radio stations across the country.

> *If this had been an actual emergency, the attention signal you just heard would have been followed by official information, news, or instruction.*

Suppose there were an actual emergency. What would you expect to happen?

Exercises • • • • • • • • • • • • • • • • • •

Practice

Use the Law of Detachment to determine a conclusion that follows from statements (1) and (2). If a valid conclusion does not follow, write *no valid conclusion*.

9. (1) If I lose my textbook, I will fail my math test.
    (2) I did not lose my textbook.

10. (1) If x is an integer, then x is a real number.
    (2) x is an integer.

Homework Help

For Exercises	See Examples
9–14, 25	1, 2
15–20	3

Extra Practice
See page 754.

Use the Law of Detachment to determine a conclusion that follows from statements (1) and (2). If a valid conclusion does not follow, write *no valid conclusion*.

11. (1) If two odd numbers are added, their sum is an even number.
    (2) 5 and 3 are added.

12. (1) If three sides of one triangle are congruent to three corresponding sides of another triangle, then the triangles are congruent.
    (2) In $\triangle ABC$ and $\triangle DEF$, $\overline{AB} \cong \overline{DE}$, $\overline{BC} \cong \overline{EF}$, and $\overline{CA} \cong \overline{FD}$.

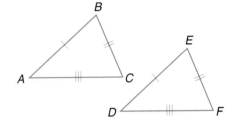

13. (1) If the measure of an angle is less than 90, it is an acute angle.
    (2) $m\angle B = 45$.

14. (1) If a figure is a rectangle, then its opposite sides are congruent.
    (2) $\overline{AB} \cong \overline{DC}$ and $\overline{AD} \cong \overline{BC}$.

Use the Law of Syllogism to determine a conclusion that follows from statements (1) and (2). If a valid conclusion does not follow, write *no valid conclusion*.

15. (1) If a parallelogram has four congruent sides, it is a rhombus.
    (2) If a figure is a rhombus, then the diagonals are perpendicular.

16. (1) If it is sunny tomorrow, I'll go swimming.
    (2) If I go swimming, I'll miss the baseball game.

17. (1) If Morgan studies hard, she'll get a good grade on her test.
    (2) If Morgan studies hard, she'll miss her favorite television show.

18. (1) If M is the midpoint of $\overline{AB}$, then $AM = MB$.
    (2) If the measures of two segments are equal, then they are congruent.

19. (1) All integers are rational numbers.
    (2) All integers are real numbers.

20. (1) All cheerleaders are athletes.
    (2) All athletes can eat at the training table at lunch.

Determine whether each situation is an example of inductive or deductive reasoning.

21. Lessie's little sister found a nest of strange eggs near the beach. The first five eggs hatched into lizards. She concluded that all of the eggs were lizard eggs.

22. Carla has had a quiz in science every Friday for the last two months. She concludes that she will have a quiz this Friday.

23. Vincent's geometry teacher told his classes at the beginning of the year that there would be a quiz every Friday. Vincent concluded that he will have a quiz this Friday.

24. A number is divisible by 4 if its last two digits make a number that is divisible by 4. Dena concluded that 624 is divisible by 4.

25. Literature Sherlock Holmes was a master of deductive reasoning. Consider this argument from *The Hound of the Baskervilles*.

> If the initials C.C.H. mean Charing Cross Hospital, then the owner is a physician. The initials C.C.H. mean Charing Cross Hospital.

What conclusion can you draw from this argument?

26. Logic There are three women—Alicia, Brianne, and Charlita—each of whom has two occupations from the following: doctor, engineer, teacher, painter, writer, and lawyer. No two have the same occupation.

- The doctor had lunch with the teacher.
- The teacher and writer went to the movies with Alicia.
- The painter is related to the engineer.
- Brianne lives next door to the writer.
- The doctor hired the painter to do a job.
- Charlita beat Brianne and the painter at tennis.

Which two occupations does each woman have?

Sherlock Holmes with
Dr. Watson in *The Hound of
the Baskervilles*

27. Critical Thinking In addition to being the author of *Alice in Wonderland*, Lewis Carroll also wrote a book called *Symbolic Logic*. What conclusion can you draw from the following argument that is adapted from his book on logic?

> Babies are illogical.
> Nobody is despised who can manage a crocodile.
> Illogical people are despised.

Let *p* represent "Dogs are mammals," *q* represent "Snakes are reptiles," and *r* represent "Birds are insects." Write a statement for each compound sentence. Then find the truth value. *(Lesson 15–1)*

28. $p \land \sim r$ **29.** $p \lor q$ **30.** $q \rightarrow r$

Find the coordinates of the center and measure of the radius of each circle whose equation is given. *(Lesson 14–6)*

31. $(x - 3)^2 + (y - 5)^2 = 1$ **32.** $(x + 5)^2 + y^2 = 49$

33. Grid In A small kitchen garden is shaped like a 45°-45°-90° triangle. If the legs of the triangle each measure 8 feet, find the length of the hypotenuse to the nearest tenth of a foot. *(Lesson 13–2)*

34. Multiple Choice Suppose a triangle has two sides measuring 12 units and 15 units. If the third side has a length of *x* units, which inequality must be true? *(Lesson 7–4)*

Ⓐ $4 < x < 26$ Ⓑ $4 < x < 29$ Ⓒ $3 < x < 27$ Ⓓ $2 < x < 27$

In mathematics, proofs are used to show that a conjecture is valid. A **proof** is a logical argument in which each statement you make is backed up by a reason that is accepted as true. Throughout this text, some informal proofs have been presented, and you have been preparing to write proofs. In the next few lessons, you will learn to write them.

One type of proof is a **paragraph proof**. In this kind of proof, you write your statements and reasons in paragraph form. The following is a paragraph proof of a theorem you studied in Lesson 5–2.

Conjecture: If $\triangle PQR$ is an equiangular triangle, then the measure of each angle is 60.

Paragraph Proof

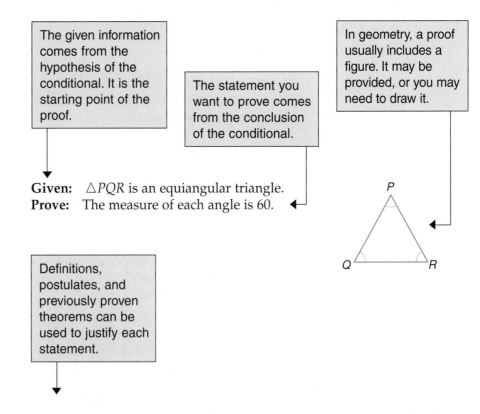

The given information comes from the hypothesis of the conditional. It is the starting point of the proof.

The statement you want to prove comes from the conclusion of the conditional.

In geometry, a proof usually includes a figure. It may be provided, or you may need to draw it.

Given: $\triangle PQR$ is an equiangular triangle.
Prove: The measure of each angle is 60.

Definitions, postulates, and previously proven theorems can be used to justify each statement.

You know that $\triangle PQR$ is an equilangular triangle. All of the angles of an equiangular triangle are congruent. The Angle Sum Theorem states that the sum of the measures of the angles of a triangle is 180. Since all of the angles have equal measure, the measure of each angle is $180 \div 3$ or 60. Therefore, the measure of each angle of an equiangular triangle is 60.

Before you begin to write a paragraph proof, you should make a plan. One problem-solving strategy that you might use is *work backward*. Start with what you want to prove, and work backward step-by-step until you can decide on a plan for completing the proof.

Examples

Write a paragraph proof for each conjecture.

1 In $\triangle ABC$, if $\overline{AB} \cong \overline{CB}$ and D is the midpoint of $\overline{AC}$, then $\angle 1 \cong \angle 2$.

Given: $\overline{AB} \cong \overline{CB}$
$\quad\quad\quad$ D is the midpoint of $\overline{AC}$.

Prove: $\angle 1 \cong \angle 2$

Plan: $\angle 1 \cong \angle 2$ if they are corresponding parts of congruent triangles. Try to prove $\triangle ABD \cong \triangle CBD$ by SSS, SAS, ASA, or AAS.

Look Back
Isosceles Triangles:
Lesson 6–4,
SAS: Lesson 5–5

You know that $\overline{AB} \cong \overline{CB}$. If two sides of a triangle are congruent, then the angles opposite those sides are congruent. So, $\angle BAD \cong \angle BCD$. Also, $\overline{DA} \cong \overline{DC}$ because D is the midpoint of $\overline{AC}$. Since $\overline{AB} \cong \overline{CB}$, $\angle BAD \cong \angle BCD$, and $\overline{DA} \cong \overline{DC}$, the triangles are congruent by SAS. Therefore, $\angle 1 \cong \angle 2$ because corresponding parts of congruent triangles are congruent (*CPCTC*).

2 If $p \parallel q$, then $\angle 1$ is supplementary to $\angle 4$.

Given: $p \parallel q$

Prove: $\angle 1$ is supplementary to $\angle 4$

Plan: $\angle 1$ is supplementary to $\angle 4$ if $m\angle 1 + m\angle 4 = 180$. Use corresponding angles and linear pairs to show that the angles are supplementary.

Look Back
Corresponding Angles:
Lesson 4–3,
Supplementary Angles:
Lesson 3–5

You know that $p \parallel q$. If two parallel lines are cut by a transversal, their corresponding angles are congruent. So, $\angle 1 \cong \angle 3$. Also, $\angle 3$ and $\angle 4$ are supplementary because they are a linear pair. Since $m\angle 3 + m\angle 4 = 180$ and $m\angle 1 = m\angle 3$, $m\angle 1 + m\angle 4 = 180$ by substitution. Therefore, $\angle 1$ and $\angle 4$ are supplementary.

Your Turn

If $\overline{PM} \parallel \overline{RN}$ and $\overline{PM} \cong \overline{RN}$, then $\triangle MPT \cong \triangle RNT$.

Plan: Use alternate interior angles to show $\angle P \cong \angle N$ or $\angle M \cong \angle R$.

 Personal Tutor at geomconcepts.com

Communicating Mathematics

1. **List** three things that can be used to justify a statement in a paragraph proof.

2. **Explain** how deductive reasoning is used in paragraph proofs.

Guided Practice

Example 1

Write a paragraph proof for each conjecture.

3. If T bisects $\overline{PN}$ and $\overline{RM}$, then $\angle M \cong \angle R$.
   Plan: Use a triangle congruence postulate.

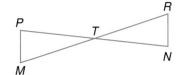

Example 2

4. If p and q are cut by transversal t, and $\angle 1$ is supplementary to $\angle 2$, then $p \parallel q$.

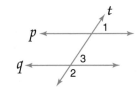

Example 2

5. **Carpentry**   A carpenter is building a flight of stairs. The tops of the steps are parallel to the floor, and the bottom of the stringer makes a 25° angle with the floor. Prove that the top of the steps makes a 25° angle with the top of the stringer.

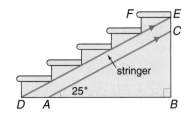

Exercises

Practice

Write a paragraph proof for each conjecture.

6. If $\angle D \cong \angle T$ and M is the midpoint of $\overline{DT}$, then $\triangle DEM \cong \triangle TEM$.
   Plan: Use a triangle congruence postulate.

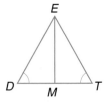

Homework Help	
For Exercises	See Examples
6–10, 12–14	1
7	2
Extra Practice	
See page 754.	

7. If $\overline{MQ} \parallel \overline{NP}$ and $m\angle 4 = m\angle 3$, then $m\angle 1 = m\angle 5$.
   Plan: Use corresponding angles and alternate interior angles.

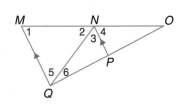

8. If $\angle 3 \cong \angle 4$, then $\overline{MA} \cong \overline{MC}$.

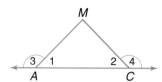

9. If $\triangle GMK$ is an isosceles triangle with vertex $\angle GMK$ and $\angle 1 \cong \angle 6$, then $\triangle GMH \cong \triangle KMJ$.

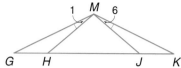

10. If $\overline{PH}$ bisects $\angle YHX$ and $\overline{HP} \perp \overline{YX}$, then $\triangle YHX$ is an isosceles triangle.

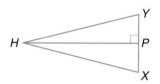

11. If $\angle 5 \cong \angle 6$ and $\overline{FR} \cong \overline{GS}$, then $\angle 4 \cong \angle 3$.

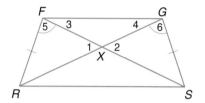

Draw and label a figure for each conjecture. Then write a paragraph proof.

12. In quadrilateral $EFGH$, if $\overline{EF} \cong \overline{GH}$ and $\overline{EH} \cong \overline{GF}$, then $\triangle EFH \cong \triangle GHF$.

13. If an angle bisector of a triangle is also an altitude, then the triangle is isosceles.

14. The medians drawn to the congruent sides of an isosceles triangle are congruent.

Applications and Problem Solving

15. Law When a prosecuting attorney presents a closing argument in a trial, he or she gives a summary of the trial. Explain how the closing argument is like a paragraph proof.

An attorney presenting a summary to the jury

16. Sewing Abby needs to divide a rectangular piece of fabric into three strips, each having the same width. The width of the fabric is 10.5 inches. Instead of dividing 10.5 by 3, Abby angles her ruler as shown in the figure, divides 12 by 3, and makes marks at 4 inches and 8 inches. Explain why this method divides the fabric into three strips having the same width.

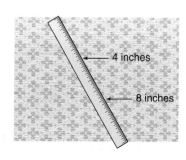

17. **Critical Thinking** What conclusion can you draw about the sum of $m\angle 1$ and $m\angle 4$ if $m\angle 1 = m\angle 2$ and $m\angle 3 = m\angle 4$?

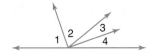

Mixed Review

18. If two lines are perpendicular, then they form four right angles. Lines ℓ and m are perpendicular. What conclusion can you derive from these statements? *(Lesson 15–2)*

Construct a truth table for each compound statement. *(Lesson 15–1)*

19. $\sim p \wedge \sim q$
20. $\sim q \rightarrow \sim p$
21. $p \vee \sim q$

Standardized Test Practice
Ⓐ Ⓑ Ⓒ Ⓓ

22. **Short Response** Simplify $\sqrt{8} \cdot \sqrt{9}$. *(Lesson 13–1)*

23. **Multiple Choice** The median of a trapezoid is 8 meters. The height of the trapezoid is 4 meters. How is the area of this trapezoid changed when the median is doubled? *(Lesson 10–4)*
   Ⓐ The area is halved.
   Ⓑ The area is not changed.
   Ⓒ The area is doubled.
   Ⓓ The area is tripled.

Quiz 1   Lessons 15–1 through 15–3

Suppose p is a true statement and q is a false statement. Find the truth value of each compound statement. *(Lesson 15–1)*

1. $p \rightarrow q$
2. $p \vee q$

Use the Law of Detachment or the Law of Syllogism to determine a conclusion that follows from statements (1) and (2). If a valid conclusion does not follow, write *no valid conclusion*. *(Lesson 15–2)*

3. (1) If school is in session, then it is not Saturday.
   (2) It is not Saturday.

4. (1) If a parallelogram has four right angles, it is a rectangle.
   (2) If a figure is a rectangle, its diagonals are congruent.

5. If $\triangle CAN$ is an isosceles triangle with vertex $\angle N$ and $\overline{CA} \parallel \overline{BE}$, write a paragraph proof that shows $\triangle NEB$ is also an isosceles triangle. *(Lesson 15–3)*

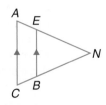

15-4 Preparing for Two-Column Proofs

What You'll Learn
You'll learn to use properties of equality in algebraic and geometric proofs.

Why It's Important
Science Scientists use properties of equality when they solve formulas for a specific variable.
See Example 2.

When you solve an equation, you are using a deductive argument. Each step can be justified by an algebraic property.

$$3(y + 2) = 12 \qquad \textit{Given}$$
$$3y + 6 = 12 \qquad \textit{Distributive Property}$$
$$3y = 6 \qquad \textit{Subtraction Property of Equality}$$
$$y = 2 \qquad \textit{Division Property of Equality}$$

Notice that the column on the left is a step-by-step process that leads to a solution. The column on the right contains the reasons for each statement.

In geometry, you can use a similar format to prove theorems. A **two-column proof** is a deductive argument with statements and reasons organized in two columns. A two-column proof and a paragraph proof contain the same information. They are just organized differently. The following is an example of a two-column proof. You may want to compare it to the paragraph proof on page 644.

Conjecture: If $\triangle PQR$ is an equiangular triangle, the measure of each angle is 60.

Two-Column Proof

Given: $\triangle PQR$ is an equiangular triangle.

Prove: The measure of each angle is 60.

Proof:

There is a reason for each statement.

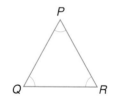

Statements	Reasons
1. $\triangle PQR$ is an equiangular triangle.	1. Given
2. $\angle P \cong \angle Q \cong \angle R$	2. Definition of equiangular triangle
3. $m\angle P = m\angle Q = m\angle R$	3. Definition of congruent angles
4. $m\angle P + m\angle Q + m\angle R = 180$	4. Angle Sum Theorem
5. $m\angle P + m\angle P + m\angle P = 180$	5. Substitution Property of Equality
6. $3(m\angle P) = 180$	6. Combining like terms
7. $m\angle P = 60$	7. Division Property of Equality
8. The measure of each angle of $\triangle PQR$ is 60.	8. Substitution Property of Equality

The first statement(s) contains the given information.

The last statement is what you want to prove.

Notice that algebraic properties are used as reasons in the proof of a geometric theorem. Algebraic properties can be used because segment measures and angle measures are real numbers, so they obey the algebraic properties.

Example ① **Justify the steps for the proof of the conditional.**
If AC = BD, then AB = CD.

Given: $AC = BD$

Prove: $AB = CD$

Remember that AC, BD, AB, and CD represent real numbers.

Proof:

Statements	Reasons
1. $AC = BD$	1. ___?___
2. $AB + BC = AC$ $BC + CD = BD$	2. ___?___
3. $AB + BC = BC + CD$	3. ___?___ *Hint: Use statements 1 and 2.*
4. $BC = BC$	4. ___?___
5. $AB = CD$	5. ___?___

Look Back

Properties of Equality: Lesson 2–2

Reason 1: Given
Reason 2: Segment Addition Postulate
Reason 3: Substitution Property of Equality
Reason 4: Reflexive Property of Equality
Reason 5: Subtraction Property of Equality

Your Turn

a. Justify the steps for the proof of the conditional.
   If m∠1 = m∠2, then m∠PXR = m∠SXQ.

   Given: $m\angle 1 = m\angle 2$

   Prove: $m\angle PXR = m\angle SXQ$

   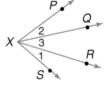

   Proof:

Statements	Reasons
1. $m\angle 1 = m\angle 2$	1. ___?___
2. $m\angle 3 = m\angle 3$	2. ___?___
3. $m\angle 1 + m\angle 3 =$ $m\angle 2 + m\angle 3$	3. ___?___
4. $m\angle PXR = m\angle 2 + m\angle 3$ $m\angle SXQ = m\angle 1 + m\angle 3$	4. ___?___
5. $m\angle PXR = m\angle SXQ$	5. ___?___

 Personal Tutor at geomconcepts.com

www.geomconcepts.com/extra_examples

Example ❷
Science Link

The formula $d = rt$ describes the relationship between distance, speed, and time. In the formula, d is the distance, r is the speed, and t is the time. Show that if $d = rt$, then $r = \frac{d}{t}$.

Given: $d = rt$

Prove: $r = \frac{d}{t}$

Proof:

Statements	Reasons
1. $d = rt$	1. Given
2. $\frac{d}{t} = \frac{rt}{t}$	2. Division Property of Equality
3. $\frac{d}{t} = r$	3. Substitution Property of Equality
4. $r = \frac{d}{t}$	4. Symmetric Property of Equality

Check for Understanding

Communicating Mathematics

1. **List** the parts of a two-column proof.

2. **Explain** why algebraic properties can be used in geometric proofs.

3. **Writing Math** Compare and contrast paragraph proofs and two-column proofs.

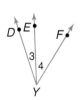

> **Vocabulary**
> two-column proof

Guided Practice
Example 1

4. Copy and complete the proof.
   If $m\angle AXC = m\angle DYF$ and $m\angle 1 = m\angle 3$, then $m\angle 2 = m\angle 4$.

   Given: $m\angle AXC = m\angle DYF$ and
   $m\angle 1 = m\angle 3$

   Prove: $m\angle 2 = m\angle 4$

   Proof:

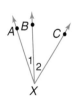

Statements	Reasons
a. $m\angle AXC = m\angle DYF$ $m\angle 1 = m\angle 3$	a. ___?___
b. $m\angle AXC = m\angle 1 + m\angle 2$ $m\angle DYF = m\angle 3 + m\angle 4$	b. ___?___
c. $m\angle 1 + m\angle 2 = m\angle 3 + m\angle 4$	c. ___?___
d. $m\angle 3 + m\angle 2 = m\angle 3 + m\angle 4$	d. ___?___
e. $m\angle 2 = m\angle 4$	e. ___?___

Example 2

5. **Algebra** Solve the equation $-2x + 5 = -13$ by using a two-column proof.

Practice

Homework Help	
For Exercises	See Examples
6, 8	1
7, 9	2
Extra Practice	
See page 755.	

Copy and complete each proof.

6. If $AC = DF$ and $AB = DE$, then $BC = EF$.

Given: $AC = DF$ and $AB = DE$

Prove: $BC = EF$

Proof:

Statements	Reasons
a. $AC = DF$	**a.** ___?___
b. $AC = AB + BC$ $DF = DE + EF$	**b.** ___?___
c. $AB + BC = DE + EF$	**c.** ___?___
d. $AB = DE$	**d.** ___?___
e. $BC = EF$	**e.** ___?___

7. If $\frac{5x}{3} = 15$, then $x = 9$.

Given: $\frac{5x}{3} = 15$

Prove: $x = 9$

Proof:

Statements	Reasons
a. $\frac{5x}{3} = 15$	**a.** ___?___
b. $5x = 45$	**b.** ___?___
c. $x = 9$	**c.** ___?___

8. If $m\angle TUV = 90$, $m\angle XWV = 90$, and $m\angle 1 = m\angle 3$, then $m\angle 2 = m\angle 4$.

Given: $m\angle TUV = 90$, $m\angle XWV = 90$, and $m\angle 1 = m\angle 3$

Prove: $m\angle 2 = m\angle 4$

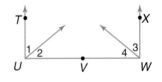

Proof:

Statements	Reasons
a. ___?___	**a.** Given
b. $m\angle TUV = m\angle XWV$	**b.** ___?___
c. $m\angle TUV = m\angle 1 + m\angle 2$ $m\angle XWV = m\angle 3 + m\angle 4$	**c.** ___?___
d. ___?___	**d.** Substitution Property, =
e. $m\angle 1 + m\angle 2 = m\angle 1 + m\angle 4$	**e.** ___?___
f. ___?___	**f.** Subtraction Property, =

9. **Physics** The mass, force, and acceleration of a motorcycle and its rider are related by the formula $F = ma$, where F is the force, m is the mass, and a is the acceleration. Show that if $F = ma$, then $m = \frac{F}{a}$.

10. **Critical Thinking** There are ten boys lined up in gym class. They are arranged in order from the shortest to the tallest. Max is taller than Nate, Nate is taller than Rey, and Rey is taller than Ted. Brian is taller than Rey, but shorter than Nate. Mike is standing between Sal and Chet. Chet is shorter than Max but taller than Mike. Van is standing between Max and Omar. Omar is standing next to Chet. There are seven boys standing between Van and Ted. Name the ten boys in order from shortest to tallest.

Mixed Review

11. Write a paragraph proof for the conjecture. *(Lesson 15–3)*
If $\overline{VT}$ and $\overline{RU}$ intersect at S, $\overline{VR} \perp \overline{RS}$, $\overline{UT} \perp \overline{SU}$, and $\overline{RS} \cong \overline{US}$, then $\overline{VR} \cong \overline{TU}$.

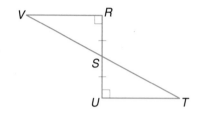

12. **Biology** Use the Law of Syllogism to determine a conclusion that follows from statements (1) and (2). If a valid conclusion does not follow, write *no valid conclusion.* *(Lesson 15–2)*
   (1) Sponges belong to the phylum porifera.
   (2) Sponges are animals.

Solve each equation. *(Lesson 9–1)*

13. $\frac{120}{b} = \frac{24}{60}$

14. $\frac{n}{2} = \frac{0.7}{0.4}$

15. $\frac{18}{x + 1} = \frac{9}{4}$

Math Data
nline Update
For the latest information on population trends, visit:
geomconcepts.com

16. **Geography** The graph shows how the center of population in the United States has changed since 1790. Half of the population lives north of this point and half lives south of it; half lives west of the point and half lives east of it. Use inductive reasoning to predict the position of the center of population in 2010. *(Lesson 1–1)*

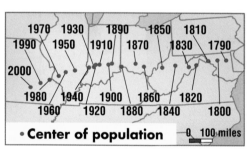

• **Center of population**   0   100 miles

Source: Statistical Abstract of the United States, 2003

Standardized Test Practice
Ⓐ Ⓑ Ⓒ Ⓓ

17. **Multiple Choice** If x and y are positive integers and $x < y$, then $x - y$— *(Algebra Review)*
   Ⓐ is positive.
   Ⓑ is negative.
   Ⓒ equals zero.
   Ⓓ cannot be determined.

Before a computer programmer writes a program, a flowchart is developed to organize the main steps of the program. A flowchart helps identify what the computer program needs to do at each step.

You can use the steps in the flowchart below to organize your thoughts before you begin to write a two-column proof.

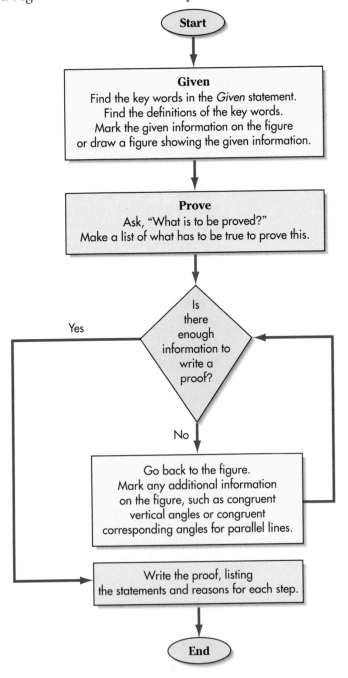

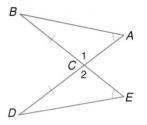

1 Write a two-column proof for the conjecture.
If $\overline{BC} \cong \overline{DC}$ and $\angle A \cong \angle E$, then $\overline{AB} \cong \overline{ED}$.

Given: $\overline{BC} \cong \overline{DC}$
$\angle A \cong \angle E$

Prove: $\overline{AB} \cong \overline{ED}$

Explore   You know that $\overline{BC} \cong \overline{DC}$ and $\angle A \cong \angle E$. Even though it is not mentioned in the given statement, $\angle 1 \cong \angle 2$ because they are vertical angles. Mark this information on the figure. You want to prove that $\overline{AB} \cong \overline{ED}$.

Plan $\overline{AB} \cong \overline{ED}$ if they are corresponding parts of congruent triangles. You can use the AAS Theorem to show that $\triangle BAC \cong \triangle DEC$.

Solve

Statements	Reasons
1. $\overline{BC} \cong \overline{DC}$	1. Given
2. $\angle A \cong \angle E$	2. Given
3. $\angle 1 \cong \angle 2$	3. Vertical angles are congruent.
4. $\triangle BAC \cong \triangle DEC$	4. AAS Theorem
5. $\overline{AB} \cong \overline{ED}$	5. CPCTC

Look Back

AAS Theorem: Lesson 5–6

Examine   Check your proof to be sure you haven't used any information that is not given or derived from definitions, postulates, or previously proved theorems. Never assume information that is not given.

Your Turn

a. If $\overline{AM} \parallel \overline{CR}$ and B is the midpoint of $\overline{AR}$, then $\overline{AM} \cong \overline{RC}$.

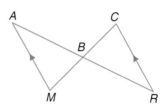

 Personal Tutor at geomconcepts.com

Sometimes the figure you are given contains triangles that overlap. If so, try to visualize them as two separate triangles. You may want to redraw them so they are separate.

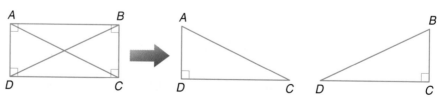

2 Write a two-column proof.

Given: *ABCD* is a rectangle with diagonals $\overline{AC}$ and $\overline{BD}$.

Prove: $\overline{AC} \cong \overline{BD}$

Plan: Show that $\overline{AC}$ and $\overline{BD}$ are corresponding parts of congruent triangles. Redraw the figure as two separate triangles.

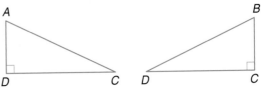

Statements	Reasons
1. *ABCD* is a rectangle with diagonals $\overline{AC}$ and $\overline{BD}$.	**1.** Given
2. $\overline{DC} \cong \overline{DC}$	**2.** Congruence of segments is reflexive.
3. $\overline{AD} \cong \overline{BC}$	**3.** Definition of rectangle
4. $\angle ADC$ and $\angle BCD$ are right angles.	**4.** Definition of rectangle
5. $\triangle ADC$ and $\triangle BCD$ are right triangles.	**5.** Definition of right triangle
6. $\triangle ADC \cong \triangle BCD$	**6.** LL Theorem
7. $\overline{AC} \cong \overline{BD}$	**7.** CPCTC

Look Back

LL Theorem:
Lesson 6–5

Your Turn

b. Given: *JKLM* is a square with diagonals $\overline{KM}$ and $\overline{LJ}$.

Prove: $\overline{KM} \cong \overline{LJ}$

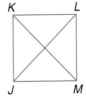

COncepts in MOtion
Interactive Lab
geomconcepts.com

Check for Understanding

Communicating Mathematics

1. Suppose you are given that $\overline{RS} \parallel \overline{WT}$ and $\overline{RW} \cong \overline{ST}$. Can you use $\angle 1 \cong \angle 4$ as a statement in a proof? Explain your reasoning.

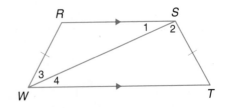

2. Writing Math Write a paragraph in which you explain the process of writing a two-column proof.

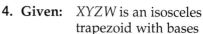

Guided Practice

Examples 1 & 2

Write a two-column proof.

3. **Given:** $\overline{EF} \cong \overline{GH}$
   $\overline{EH} \cong \overline{GF}$

   Prove: $\triangle EFH \cong \triangle GHF$

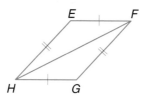

4. **Given:** $XYZW$ is an isosceles trapezoid with bases $\overline{XY}$ and $\overline{WZ}$.

   Prove: $\angle 1 \cong \angle 2$

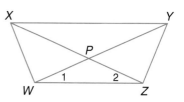

Example 1

5. **Kites** You can make a simple kite using paper, two sticks, and some string. The sticks meet so that $\overline{AC} \perp \overline{BD}$ and $\overline{AF} \cong \overline{CF}$. Prove that $\overline{AB} \cong \overline{CB}$ and $\overline{CD} \cong \overline{AD}$.

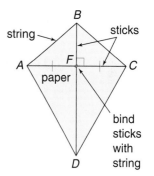

Exercises

• • • • • • • • • • • • • • • • •

Practice

Homework Help	
For Exercises	See Examples
6–15	1, 2
Extra Practice	
See page 755.	

Write a two-column proof.

6. **Given:** $\angle A \cong \angle E$
   $\angle 1 \cong \angle 2$
   $\overline{AC}$ bisects $\overline{BD}$.

   Prove: $\overline{AB} \cong \overline{ED}$

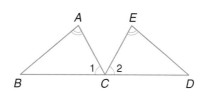

7. **Given:** $\overline{AB} \cong \overline{CD}$
   $\angle 1 \cong \angle 2$

   Prove: $\overline{AD} \cong \overline{CB}$

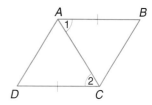

8. **Given:** $PRSV$ is a parallelogram.
   $\overline{PT} \perp \overline{SV}$
   $\overline{QS} \perp \overline{PR}$

   Prove: $\triangle PTV \cong \triangle SQR$

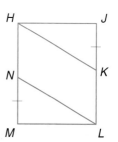

9. **Given:** $HJLM$ is a rectangle.
   $\overline{KJ} \cong \overline{NM}$

   Prove: $\overline{HK} \cong \overline{LN}$

Write a two-column proof.

10. Given: $\angle 1 \cong \angle 4$
$\overline{NA} \cong \overline{TC}$

Prove: $\angle 3 \cong \angle 2$

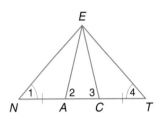

11. Given: $\overline{CD}$ is a diameter of $\odot E$.
$\overline{CD} \perp \overline{AB}$

Prove: $\overline{AF} \cong \overline{BF}$

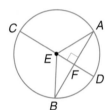

Draw and label a figure for each conjecture. Then write a two-column proof.

12. The measure of an exterior angle of a triangle is equal to the sum of the measures of its two remote interior angles.

13. The diagonals of an isosceles trapezoid are congruent.

14. The median from the vertex angle of an isosceles triangle is the perpendicular bisector of the base.

Applications and Problem Solving

15. Construction Before laying the foundation of a rectangular house, the construction supervisor sets the corner points so that $\overline{AB} \cong \overline{DC}$ and $\overline{AD} \cong \overline{BC}$. In order to guarantee that the corners are right angles, the supervisor measures both diagonals to be sure they are congruent.

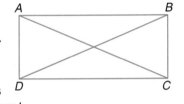

a. Write the conjecture that the supervisor is using.

b. Write a two-column proof for the conjecture.

16. Critical Thinking Consider the following proof of the conjecture *A diagonal of a parallelogram bisects opposite angles.*

Given: $\square MATH$ with diagonal $\overline{MT}$.

Prove: $\overline{MT}$ bisects $\angle AMH$ and $\angle ATH$.

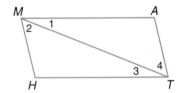

Statements	Reasons
a. $\square MATH$ is a parallelogram.	**a.** Given
b. $\overline{MH} \cong \overline{AT}, \overline{MA} \cong \overline{HT}$	**b.** Definition of parallelogram
c. $\overline{MT} \cong \overline{MT}$	**c.** Congruence of segments is reflexive.
d. $\triangle MHT \cong \triangle MAT$	**d.** SSS
e. $\angle 1 \cong \angle 2, \angle 3 \cong \angle 4$	**e.** CPCTC
f. $\overline{MT}$ bisects $\angle AMH$ and $\overline{MT}$ bisects $\angle ATH$.	**f.** Definition of angle bisector

Is this proof correct? Explain your reasoning.

17. Algebra Solve $-4x + 5 = -15$ using a two-column proof.
(Lesson 15–4)

18. Use a paragraph proof to prove the following conjecture.
If two sides of a quadrilateral are parallel and congruent, the quadrilateral is a parallelogram. *(Lesson 15–3)*

Find each measure. *(Lesson 14–1)*

19. $m\angle ADB$

20. $m\angle BDC$

21. $m\widehat{AD}$

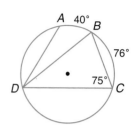

22. Short Response Draw a polygon with the same area as, but not congruent to, the figure at the right. *(Lesson 10–3)*

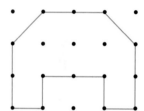

Quiz 2 — Lessons 15–4 and 15–5

Copy and complete the proof of the conditional.
If m∠ABC = m∠DEF and m∠1 = m∠4, then m∠2 = m∠3. *(Lesson 15–4)*

Given: $m\angle ABC = m\angle DEF$
$m\angle 1 = m\angle 4$

Prove: $m\angle 2 = m\angle 3$

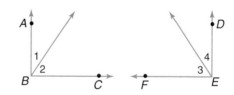

Statements	Reasons
1. $m\angle ABC = m\angle DEF$ $m\angle 1 = m\angle 4$	**1.** ___?___
2. $m\angle ABC = m\angle 1 + m\angle 2$ $m\angle DEF = m\angle 3 + m\angle 4$	**2.** ___?___
3. $m\angle 1 + m\angle 2 = m\angle 3 + m\angle 4$	**3.** ___?___
4. $m\angle 2 = m\angle 3$	**4.** ___?___

5. Write a two-column proof. *(Lesson 15–5)*

Given: T is the midpoint of $\overline{BQ}$.
$\angle B$ and $\angle Q$ are right angles.
$\angle 1 \cong \angle 2$

Prove: $\overline{AT} \cong \overline{PT}$

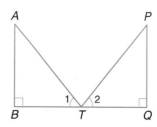

15-6 Coordinate Proofs

What You'll Learn
You'll learn to use coordinate proofs to prove theorems.

Why It's Important
Computer-Aided Design A coordinate system is used to plot points on a CAD drawing.
See Exercise 23.

Some recent movies have been made entirely with computer graphics. Animators first make wireframe drawings and assign numbers to control points in the drawings based on a coordinate system. The animator then makes objects move by using commands that move the control points.

Important relationships in geometry can also be demonstrated using a coordinate system.

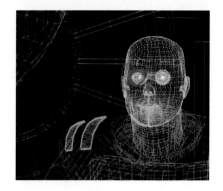

Hands-On Geometry

Materials: grid paper straightedge

Step 1: Draw and label the first quadrant of a rectangular coordinate system.

Step 2: Graph the vertices of $\triangle ABC$ and $\triangle RST$ at $A(2, 6)$, $B(5, 5)$, $C(3, 3)$, and $R(9, 5)$, $S(8, 2)$, $T(6, 4)$. Draw the triangles.

Look Back

Distance Formula:
Lesson 6–7

Try These

1. Use the Distance Formula to find the length of each side of each triangle.
2. Based on your calculations, what conclusion can you make about the two triangles?

In this activity, you used a coordinate plane and the Distance Formula to show that two triangles are congruent. You can also use a coordinate plane to prove many theorems in geometry. A proof that uses figures on a coordinate plane is called a **coordinate proof**.

One of the most important steps in planning a coordinate proof is deciding how to place the figure on a coordinate plane.

COncepts in MOtion
Animation
geomconcepts.com

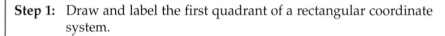

Guidelines for Placing Figures on a Coordinate Plane	1. Use the origin as a vertex or center.
	2. Place at least one side of a polygon on an axis.
	3. Keep the figure within the first quadrant, if possible.
	4. Use coordinates that make computations as simple as possible.

1 Position and label a square with sides a units long on a coordinate plane.

- Use the origin as a vertex.
- Place one side on the x-axis and one side on the y-axis.
- Label the vertices A, B, C, and D.
- B is on the x-axis. So, its y-coordinate is 0, and its x-coordinate is a.
- D is on the y-axis. So, its x-coordinate is 0, and its y-coordinate is a.
- Since the sides of a square are congruent, the x-coordinate of C is a, and the y-coordinate of C is a.

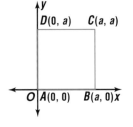

2 Position and label a parallelogram with base a units long and a height of b units on a coordinate plane.

- Use the origin as a vertex.
- Place the base along the x-axis.
- Label the vertices E, F, G, and H.
- Since F is on the x-axis, its y-coordinate is 0, and its x-coordinate is a.
- Since the height of the parallelogram is b units, the y-coordinate of H and G is b.
- Let the x-coordinate of H be c. Therefore, the x-coordinate of G is $c + a$.

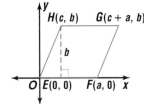

Your Turn

Position and label a right triangle with legs a units long and b units long on a coordinate plane.

In coordinate proofs, you use algebraic tools like variables, equations, and formulas.

- First, draw the figure on a coordinate plane.
- Next, use all of the known properties of the figure to assign coordinates to the vertices. Use as few letters as possible.
- Finally, use variables, equations, and formulas to show the relationships among segments in the figure. The most common formulas are shown below. In the formulas, (x_1, y_1) and (x_2, y_2) are coordinates of the endpoints of a segment.

Midpoint Formula: $\left(\dfrac{x_1 + x_2}{2}, \dfrac{y_1 + y_2}{2} \right)$

Distance Formula: $d = \sqrt{(x_2 - x_1)^2 + (y_2 - y_1)^2}$

Slope Formula: $m = \dfrac{y_2 - y_1}{x_2 - x_1}$

Write a coordinate proof for each conjecture.

❸ The diagonals of a rectangle are congruent.

Given: rectangle $QRST$ with diagonals $\overline{QS}$ and $\overline{RT}$

Prove: $\overline{QS} \cong \overline{RT}$

Plan: Use the Distance Formula to find the length of each diagonal.

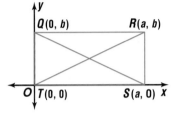

Place rectangle $QRST$ with width a and height b on a coordinate plane and label the coordinates as shown. Use the Distance Formula to find QS and RT.

First, find QS.

$$d = \sqrt{(x_2 - x_1)^2 + (y_2 - y_1)^2} \quad \textit{Distance Formula}$$
$$QS = \sqrt{(a - 0)^2 + (0 - b)^2} \quad (x_1, y_1) = (0, b), (x_2, y_2) = (a, 0)$$
$$QS = \sqrt{a^2 + b^2} \quad \textit{Simplify.}$$

Find RT.

$$d = \sqrt{(x_2 - x_1)^2 + (y_2 - y_1)^2} \quad \textit{Distance Formula}$$
$$RT = \sqrt{(0 - a)^2 + (0 - b)^2} \quad (x_1, y_1) = (a, b), (x_2, y_2) = (0, 0)$$
$$RT = \sqrt{a^2 + b^2} \quad \textit{Simplify.}$$

The measures of the diagonals are equal. Therefore, $\overline{QS} \cong \overline{RT}$.

❹ The diagonals of a parallelogram bisect each other.

Given: parallelogram $HGFE$ with diagonals $\overline{HF}$ and $\overline{GE}$

Prove: $\overline{HF}$ and $\overline{GE}$ bisect each other.

Plan: Use the Midpoint Formula to find the coordinates of the midpoint of each diagonal. If the coordinates are the same, the midpoint of each diagonal is the same point, and the diagonals bisect each other.

Look Back

Midpoint Formula: Lesson 2–5

Place parallelogram $HGFE$ on a coordinate plane and label the coordinates as shown. Use the Midpoint Formula to find the coordinates of the midpoint of each diagonal.

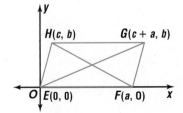

First, find the midpoint of $\overline{HF}$.

$$\left(\frac{x_1 + x_2}{2}, \frac{y_1 + y_2}{2}\right) = \left(\frac{c + a}{2}, \frac{b + 0}{2}\right) \quad \textit{Use the Midpoint Formula.}$$

$$= \left(\frac{c + a}{2}, \frac{b}{2}\right) \quad \textit{Simplify.}$$

Find the midpoint of $\overline{GE}$.

$$\left(\frac{x_1 + x_2}{2}, \frac{y_1 + y_2}{2}\right) = \left(\frac{c + a + 0}{2}, \frac{b + 0}{2}\right) \quad \textit{Use the Midpoint Formula.}$$

$$= \left(\frac{c + a}{2}, \frac{b}{2}\right) \quad \textit{Simplify.}$$

The midpoints of the diagonals have the same coordinates. Therefore, they name the same point, and the diagonals bisect each other.

Online Personal Tutor at geomconcepts.com

Check for Understanding

Communicating Mathematics

1. **Explain** how you should position a figure on the coordinate plane if it is to be used in a coordinate proof.

2. Michael placed a right triangle on a coordinate plane with an acute angle at the origin. Latisha placed the right angle at the origin. Whose placement is best? Explain your reasoning.

Guided Practice

Getting Ready If the coordinates of the endpoints of a segment are given, find the coordinates of the midpoint.

Sample: $(a, 0), (0, a)$ **Solution:** $\left(\frac{a + 0}{2}, \frac{0 + a}{2}\right) = \left(\frac{a}{2}, \frac{a}{2}\right)$

3. $(6, 3), (2, -5)$ **4.** $(a, b), (0, 0)$ **5.** $(2e, 0), (0, 2f)$

Examples 1 & 2

Position and label each figure on a coordinate plane.

6. a rectangle with length a units and width b units

7. a parallelogram with base m units and height n units

Examples 3 & 4

Write a coordinate proof for each conjecture.

8. The diagonals of a square are congruent.

9. The diagonals of a rectangle bisect each other.

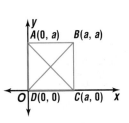

Exercises • • • • • • • • • • • • • • • •

Practice

Homework Help	
For Exercises	See Examples
10–15	1, 2
16–22	3, 4
23, 24	3
Extra Practice	
See page 755.	

Position and label each figure on a coordinate plane.

10. a square with sides r units long

11. a rectangle with base x units and width y units

12. an isosceles right triangle with legs b units long

13. a right triangle with legs c and d units long

14. an isosceles triangle with base b units long and height h units long

15. a parallelogram with base r units long and height t units long

Write a coordinate proof for each conjecture.

16. The diagonals of a square are perpendicular.

17. The midpoint of the hypotenuse of a right triangle is equidistant from each of the vertices.

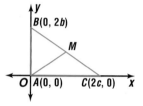

18. The line segments joining the midpoints of the sides of a rectangle form a rhombus.

19. The medians to the legs of an isosceles triangle are congruent.

20. The measure of the median to the hypotenuse of a right triangle is one-half the measure of the hypotenuse. (*Hint:* Use the figure from Exercise 17.)

21. The diagonals of an isosceles trapezoid are congruent.

22. The line segments joining the midpoints of the sides of any quadrilateral form a parallelogram.

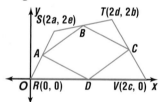

Applications and Problem Solving

A CAD Designer •········

23. Computer-Aided Design
CAD systems produce accurate drawings because the user can input exact points, such as the ends of line segments. In CAD, the *digitizing tablet* and its *puck* act as a keyboard and mouse. The figure at the right shows an outline of the foundation of a house. Prove that *PQRS* is a rectangle.

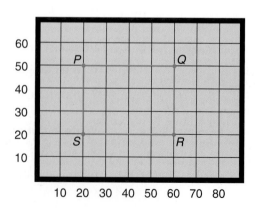

24. **Critical Thinking** Point A has coordinates $(0, 0)$, and B has coordinates (a, b). Find the coordinates of point C so $\triangle ABC$ is a right triangle.

Mixed Review

25. Write a two-column proof. *(Lesson 15–5)*

   Given: $ACDE$ is a rectangle.
   $ABCE$ is a parallelogram.

   Prove: $\triangle ABD$ is isosceles.

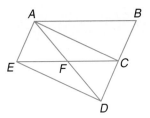

26. Copy and complete the proof. *(Lesson 15–4)*

   Given: $\angle 1$ and $\angle 3$ are supplementary.
   $\angle 2$ and $\angle 3$ are supplementary.

   Prove: $\angle 1 \cong \angle 2$

Statements	Reasons
a. $\angle 1$ and $\angle 3$ are supplementary. $\angle 2$ and $\angle 3$ are supplementary.	a. ___?___
b. $m\angle 1 + m\angle 3 = 180$ $m\angle 2 + m\angle 3 = 180$	b. ___?___
c. $m\angle 1 + m\angle 3 = m\angle 2 + m\angle 3$	c. ___?___
d. $m\angle 1 = m\angle 2$	d. ___?___
e. $\angle 1 \cong \angle 2$	e. ___?___

27. **Manufacturing** Many baking pans are given a special coating to make food stick less to the surface. A rectangular cake pan is 9 inches by 13 inches and 2 inches deep. What is the area of the surface to be coated? *(Lesson 12–2)*

Find the value of each variable. *(Lesson 5–2)*

28.

29.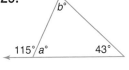

30.

Standardized Test Practice
Ⓐ Ⓑ Ⓒ Ⓓ

31. **Grid In** Use the pattern in the table to find the unit digit for 7^{41}. *(Lesson 1–1)*

Power	7^1	7^2	7^3	7^4	7^5	7^6	7^7	7^8	7^9
Unit Digit	7	9	3	1	7	9	3	1	7

32. **Multiple Choice** Lawana's math test scores are 90, 85, 78, 92, and 99. What must she score on the next math test so that her average is exactly 90? *(Statistics Review)*

   Ⓐ 90   Ⓑ 91   Ⓒ 95   Ⓓ 96

DON'T TOUCH THE POISON IVY

CONcepts in MOtion
Interactive Lab
geomconcepts.com

Indirect Proofs

If you've ever felt the itch of a rash from poison ivy, you are very careful about what plants you touch. You probably also know that poison ivy leaves are grouped in threes. What about the plant shown below at the left? How could you prove that the plant is *not* poison ivy?

You can use a technique called **indirect reasoning**. The following steps summarize the process of indirect reasoning. Use these steps when writing an **indirect proof**. *Indirect proofs are also called proofs by contradiction.*

Step 1: Assume that the *opposite* of what you want to prove is true.

Step 2: Show that this assumption leads to a contradiction of the hypothesis or some other fact. Therefore, the assumption must be false.

Step 3: Finally, state that what you want to prove is true.

Indirect Proof

Given: You have a picture of a plant.
Prove: The plant is *not* poison ivy.
Assume: The plant *is* poison ivy.

Proof: If the plant is poison ivy, its leaves would be in groups of three. However, the picture shows that the leaves are *not* in groups of three. This is a contradiction. Therefore the assumption *The plant is poison ivy* is false. So the statement *The plant is not poison ivy* is true.

Investigate

1. State the assumption you would make to start an indirect proof of each statement.

 a. Kiley ate the pizza.

 b. The defendant is guilty.

 c. Lines ℓ and *m* intersect at point *X*.

d. If two lines are cut by a transversal and the alternate interior angles are congruent, then the lines are parallel.

e. Angle B is not a right angle.

f. $\overline{XY} \cong \overline{AB}$

g. If a number is odd, its square is odd.

h. $m\angle 1 < m\angle 2$

2. Fill in the blanks with a word, symbol, or phrase to complete an indirect proof of this conjecture.

A triangle has no more than one right angle.

Given: $\triangle ABC$

Prove: $\triangle ABC$ has no more than one right angle.

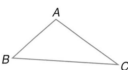

Assume: _____

Assume that $\angle A$ and $\angle B$ are both right angles. So, $m\angle A =$ _____ and $m\angle B =$ _____. According to the Angle Sum Theorem, $m\angle A + m\angle B + m\angle C =$ _____. By substitution, _____ + _____ + $m\angle C = 180$. Therefore, $m\angle C =$ _____. This is a contradiction because _____. Therefore, the assumption $\triangle ABC$ *has more than one right angle* is _____. The statement _____ is true.

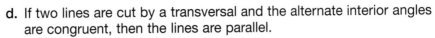

Extending the Investigation

In this extension, you will write an indirect proof and compare indirect proofs to other forms of proof. Here are some suggestions.

- Write an indirect proof of the following statement.
  A quadrilateral has no more than three acute angles.
- Work with a partner and choose a theorem from Chapters 4–8. Write a paragraph proof, a two-column proof, a coordinate proof, or an indirect proof of your theorem.

Presenting Your Conclusions

Here are some ideas to help you present your conclusions to the class.

- Put your theorem on poster board and explain the method you used to prove it.
- Make a notebook that contains your proofs.
- Make a bulletin board that compares the different kinds of proof.

 Investigation   For more information on proofs, visit: www.geomconcepts.com

Understanding and Using the Vocabulary

After completing this chapter, you should be able to define each term, property, or phrase and give an example or two of each.

*inter***NET**
CONNECTION **Review Activities**
For more review activities, visit:
www.geomconcepts.com

Geometry

coordinate proof *(p. 660)*
indirect proof *(p. 666)*
paragraph proof *(p. 644)*
proof *(p. 644)*
proof by contradiction *(p. 666)*
two-column proof *(p. 649)*

Logic

compound statement *(p. 633)*
conjunction *(p. 633)*
contrapositive *(p. 637)*
deductive reasoning *(p. 638)*
disjunction *(p. 633)*
indirect reasoning *(p. 666)*
inverse *(p. 635)*

Law of Detachment *(p. 639)*
Law of Syllogism *(p. 640)*
logically equivalent *(p. 637)*
negation *(p. 632)*
statement *(p. 632)*
truth table *(p. 633)*
truth value *(p. 632)*

Choose the letter of the term that best matches each phrase.

1. a rule similar to the Transitive Property of Equality
2. the process of using facts, rules, definitions, or properties in a logical order
3. a proof that uses figures on a coordinate plane
4. a proof containing statements and reasons that are organized with numbered steps and reasons
5. a logical argument in which each statement made is backed up by a reason that is accepted as true
6. $p \vee q$
7. if $p \rightarrow q$ is a true conditional and p is true, then q is true
8. $p \wedge q$
9. where a vertex or center of the figure should be placed in a coordinate proof
10. a table that lists all the truth values of a statement

a. proof
b. Law of Detachment
c. deductive reasoning
d. truth table
e. Law of Syllogism
f. disjunction
g. two-column proof
h. origin
i. coordinate proof
j. conjunction

Skills and Concepts

Objectives and Examples

- **Lesson 15–1** Find the truth values of simple and compound statements.

p	q	$p \vee q$	$p \wedge q$	$p \rightarrow q$
T	T	T	T	T
T	F	T	F	F
F	T	T	F	T
F	F	F	F	T

Review Exercises

Let p represent a true statement and q represent a false statement. Find the truth value of each compound statement.

11. $p \vee q$ 12. $p \wedge q$
13. $p \rightarrow q$ 14. $p \wedge \sim q$
15. $\sim p \vee q$ 16. $p \rightarrow \sim q$

17. *True* or *false*: $3 + 4 = 7$ and $2 + 5 = 8$.

 www.geomconcepts.com/vocabulary_review

Objectives and Examples

Review Exercises

- **Lesson 15–2** Use the Law of Detachment and the Law of Syllogism in deductive reasoning.

Law of Detachment: If $p \rightarrow q$ is a true conditional and p is true, then q is true.

Law of Syllogism: If $p \rightarrow q$ and $q \rightarrow r$ are true conditionals, then $p \rightarrow r$ is also true.

Determine a conclusion that follows from statements (1) and (2). If a valid conclusion does not follow, write *no valid conclusion*.

18. (1) If $x < 0$, then x is a negative number.
(2) $x < 0$

19. (1) Sean is on a field trip.
(2) All art students are on a field trip.

20. (1) If today is Tuesday, Katie has basketball practice.
(2) If Katie has basketball practice, she will eat dinner at 7:00.

- **Lesson 15–3** Use paragraph proofs to prove theorems.

Given: $r \parallel s$

Prove: $\angle 4 \cong \angle 8$

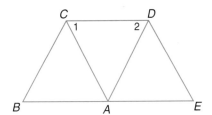

You know that $r \parallel s$. If two parallel lines are cut by a transversal, their corresponding angles are congruent. So, $\angle 6 \cong \angle 8$. Also, $\angle 6 \cong \angle 4$ because vertical angles are congruent. Since $\angle 6 \cong \angle 8$ and $\angle 6 \cong \angle 4$, then $\angle 4 \cong \angle 8$ by substitution.

21. If $m\angle BCD = m\angle EDC$, $\overline{AC}$ bisects $\angle BCD$, and $\overline{AD}$ bisects $\angle EDC$, write a paragraph proof that shows $\triangle ACD$ is isosceles.

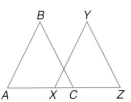

- **Lesson 15–4** Use properties of equality in algebraic and geometric proofs.

Given: $AX = ZC$

Prove: $AC = ZX$

Statements	Reasons
1. $AX = ZC$	1. Given
2. $XC = XC$	2. Reflexive, =
3. $AX + XC = ZC + XC$	3. Addition, =
4. $AC = ZX$	4. Substitution, =

22. Copy and complete the proof.

Given: $SA = BP$
$AT = QB$

Prove: $ST = QP$

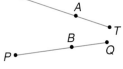

Statements	Reasons
a. $SA = BP$, $AT = QB$	a. ?
b. $SA + AT = BP + QB$	b. ?
c. $ST = SA + AT$ $QP = BP + QB$	c. ?
d. $ST = QP$	d. ?

Mixed Problem Solving
See pages 758–765.

Objectives and Examples

Review Exercises

• **Lesson 15–5** Use two-column proofs to prove theorems.

Given: $\angle 1 \cong \angle 2$

Prove: $\angle 1 \cong \angle 3$

Statements	Reasons
1. $\angle 1 \cong \angle 2$	1. Given
2. $\angle 2 \cong \angle 3$	2. Vertical angles are $\cong$.
3. $\angle 1 \cong \angle 3$	3. Transitive Prop, $\cong$

Write a two-column proof.

23. **Given:**
$m\angle AEC = m\angle DEB$
Prove:
$m\angle AEB = m\angle DEC$

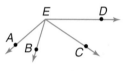

24. **Given:**
$m\angle 1 = m\angle 3$,
$m\angle 2 = m\angle 4$
Prove:
$m\angle MHT = m\angle MAT$

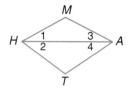

• **Lesson 15–6** Use coordinate proofs to prove theorems.

Guidelines for placing figures on a coordinate plane:

1. Use the origin as a vertex or center.
2. Place at least one side of a polygon on an axis.
3. Keep the figure within the first quadrant if possible.
4. Use coordinates that make computations as simple as possible.

Position and label each figure on a coordinate plane.

25. an isosceles right triangle with legs a units long
26. a parallelogram with base b units long and height h units

Write a coordinate proof for each conjecture.

27. The diagonals of a square bisect each other.
28. The opposite sides of a parallelogram are congruent.

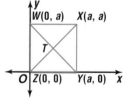

Exercise 27

Applications and Problem Solving

29. **School** Use the Law of Detachment to determine a conclusion that follows from statements (1) and (2).
   (1) If Julia scores at least 90 on the math final exam, she will earn an A for the semester.
   (2) Julia scored 93 on the math final exam.
   (Lesson 15–2)

30. **Maps** If Oak Street is parallel to Center Street, write a paragraph proof to show that $\angle 1 \cong \angle 4$. *(Lesson 15–3)*

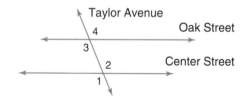

1. **List** three types of proofs and describe each type.

2. **Compare and contrast** a conjunction and a disjunction.

Construct a truth table for each compound statement.

3. $\sim p \vee q$

4. $\sim(p \wedge \sim q)$

Use the Law of Detachment or the Law of Syllogism to determine a conclusion that follows from statements (1) and (2). If a valid conclusion does not follow, write *no valid conclusion*.

5. (1) Central Middle School's mascot is a polar bear.
   (2) Dan is on the baseball team at Central Middle School.

6. (1) If I watch television, I will waste time.
   (2) If I waste time, I will not be able to complete my homework.

Write a paragraph proof for each conjecture.

7. If $\overline{QS}$ bisects $\angle PQR$ and $\overline{RS} \parallel \overline{QP}$, then $\angle 2 \cong \angle 3$.

8. If $\angle 3$ is complementary to $\angle 1$ and $\angle 4$ is complementary to $\angle 2$, then $\angle 3 \cong \angle 4$.

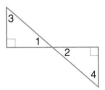

Copy and complete the proof of the conditional.
If $m\angle ABE = m\angle CBE$ and $m\angle 1 = m\angle 2$, then $m\angle 3 = m\angle 4$.

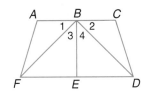

Statements	Reasons
9. $m\angle ABE = m\angle CBE,\ m\angle 1 = m\angle 2$	9. ___?___
10. $m\angle ABE - m\angle 1 = m\angle CBE - m\angle 2$	10. ___?___
11. $m\angle 3 = m\angle 4$	11. ___?___

Write a two-column proof.

12. **Given:** $\overrightarrow{SR}$ bisects $\angle QSP$,
    $\angle 1$ is complementary to $\angle 2$,
    $\angle 4$ is complementary to $\angle 3$.
    Prove: $\angle 1 \cong \angle 4$

13. **Given:** $\overline{XZ}$ bisects $\angle WXY$, $\overline{XZ} \perp \overline{WY}$
    Prove: $\angle W \cong \angle Y$

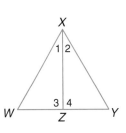

14. Position and label a rectangle with base b units long and width w units long on a coordinate plane.

15. Write a coordinate proof for this conjecture.
    The segment joining the midpoints of two legs of an isosceles triangle is half the length of the base.

16. **Algebra** Write a two-column proof to show that if $3x + 5 = 23$, then $x = 6$.

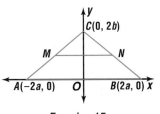

Exercise 15

Solid Figure Problems

State proficiency tests often include geometry problems with 3-dimensional shapes. The ACT and SAT may contain just one or two problems with solid figures. Formulas for surface area and volume are usually provided in the test itself, but you'll save time if you understand and memorize these formulas.

> **Test-Taking Tip**
>
> The volume of a right prism and a right cylinder is the area of the base times the height.

You'll need to know the following concepts.

cone　　cylinder　　prism　　pyramid　　surface area　　volume

Example 1

The right circular cone at the right has radius of 5 centimeters and height of 15 centimeters. A cross section of the cone is parallel to and 6 centimeters above the base of the cone. What is the area of this cross section, to the nearest square centimeter?

15 cm

6 cm

5 cm

  Ⓐ 13   Ⓑ 19   Ⓒ 28   Ⓓ 50

> **Hint** Look carefully for similar triangles and right triangles.

Solution   There are two right triangles. These triangles are similar by AA Similarity, since the radii of the cross section and the base are parallel.

Find the length of the radius. Use a proportion of the sides of the two similar triangles. The height of the larger triangle is 15. The height of the smaller triangle is $15 - 6$ or 9.

$$\frac{5}{15} = \frac{x}{9} \quad \rightarrow \quad 15x = 45 \quad \rightarrow \quad x = 3$$

The radius is 3 centimeters.

Now calculate the area.

$$A = \pi r^2 \quad \rightarrow \quad A = \pi(3)^2 \quad \rightarrow \quad A \approx 28$$

The answer is C.

Example 2

A rectangular swimming pool has a volume of 16,500 cubic feet, a depth of 10 feet, and a length of 75 feet. What is the width of the pool, in feet?

  Ⓐ 22   Ⓑ 26   Ⓒ 32
  Ⓓ 110   Ⓔ 1650

> **Hint**   Draw a figure to help you understand the problem.

Solution

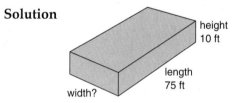

height
10 ft

length
75 ft

width?

Use the formula for the volume of a rectangular prism. Use the given information for volume, length, and height. Solve for width.

$$V = Bh$$

$$V = \ell w h$$

$$16{,}500 = 75 \times w \times 10$$

$$16{,}500 = 750w$$

$$22 = w$$

The answer is A.

After you work each problem, record your answer on the answer sheet provided or on a sheet of paper.

Multiple Choice

1. About how much paper is needed for the label on this can? *(Lesson 12–2)*

   3 in.
   4 in.

   Ⓐ 6 in^2　Ⓑ 12 in^2
   Ⓒ 18 in^2　Ⓓ 38 in^2

2. Which is equivalent to $\sqrt{72}$? *(Algebra Review)*

   Ⓐ $2\sqrt{6}$　Ⓑ $6\sqrt{2}$　Ⓒ 12　Ⓓ 36

3. If $\tan \theta = \frac{4}{3}$, what is $\sin \theta$? *(Lesson 13–5)*

   Ⓐ $\frac{3}{4}$　Ⓑ $\frac{4}{5}$　Ⓒ $\frac{5}{4}$　Ⓓ $\frac{5}{3}$　Ⓔ $\frac{7}{3}$

4. A rectangular garden is surrounded by a 60-foot fence. One side of the garden is 6 feet longer than the other side. Which equation could be used to find s, the shorter side of the garden? *(Lesson 1–6)*

   Ⓐ $60 = 8s + s$　Ⓑ $4s = 60 + 12$
   Ⓒ $60 = s(s + 6)$　Ⓓ $60 = 2(s - 6) + 2s$
   Ⓔ $60 = 2(s + 6) + 2s$

5. A plane intersects the square pyramid so that the smaller pyramid formed has a height of 6 meters and a slant height of 7.5 meters. What is the area of the shaded cross section? *(Lesson 12–4)*

   Ⓐ 20.25 m^2
   Ⓑ 36 m^2
   Ⓒ 56.25 m^2
   Ⓓ 81 m^2

   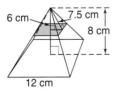
   6 cm　7.5 cm
   8 cm
   12 cm

6. Ashley subscribes to four magazines that cost $12.90, $16.00, $18.00, and $21.90 per year. If she makes a down payment of one-half the total amount and pays the rest in four equal payments, how much is each payment? *(Algebra Review)*

   Ⓐ $8.60　Ⓑ $9.20　Ⓒ $9.45
   Ⓓ $17.20　Ⓔ $34.40

7. Which equation could you use to find the length of the missing side? *(Lessson 6–6)*

   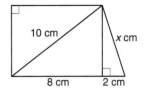
   10 cm　x cm
   8 cm　2 cm

   Ⓐ $6^2 + 2^2 = x^2$　Ⓑ $6^2 + 10^2 = x^2$
   Ⓒ $8 + 2 = x$　Ⓓ $6 + 2 = x$

8. A rectangle with area 4 square units has sides of length r units and s units. Which expression shows the perimeter of the rectangle in terms of s? *(Lesson 1–6)*

   Ⓐ $4s$ units
   Ⓑ $2r + 2s$ units
   Ⓒ $\frac{4}{r}$ units
   Ⓓ $\frac{8}{s} + 2s$ units

Grid In

9. Segment AD bisects $\angle BAC$, and segment CD bisects $\angle BCA$. What is the measure of $\angle B$? *(Lesson 3–3)*

   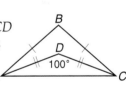
   B　D　100°
   A　C

Extended Response

10. Adding heat to ice causes it to change from a solid to a liquid. The figures below represent two pieces of ice. *(Lesson 12–2)*

    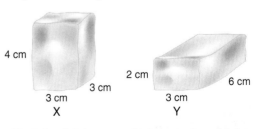
    4 cm
    3 cm　3 cm
    X
    2 cm　6 cm
    3 cm
    Y

    Part A　Make a conjecture as to which piece of ice will melt faster. Explain.

    Part B　Defend your reasoning in a short paragraph.

More Coordinate Graphing and Transformations

What You'll Learn

Key Ideas

- Solve systems of equations by graphing, substitution, or elimination. *(Lessons 16–1 and 16–2)*

- Investigate and draw translations, reflections, rotations, and dilations on a coordinate plane. *(Lessons 16–3 to 16–6)*

Key Vocabulary

dilation *(p. 703)*

reflection *(p. 692)*

rotation *(p. 697)*

translation *(p. 687)*

system of equations *(p. 676)*

Why It's Important

Architecture The Parthenon in Athens, Greece is one of the greatest monuments of antiquity. Made completely from marble, it took ten years to build and was finished in 438 B.C. The famous frieze that decorated the inside walls was 524 feet long and contained 115 plaques.

Coordinate graphing and transformations allow for exact study of geometric figures. You will use transformations to create a repeating frieze in Lesson 16–3.

✔ Check Your Readiness

Graph each equation using the slope and *y*-intercept.

1. $y = 2x - 3$ **2.** $y = -x + 2$ **3.** $y = -1$

4. $2x + 3y = 6$ **5.** $3x + 4y = 12$ **6.** $5x - y = 2$

Determine whether each figure has line symmetry. If it does, copy the figure, and draw all lines of symmetry. If not, write *no*.

7.    **8.**    **9.**

Determine whether each figure has rotational symmetry. Write *yes* or *no*.

10.   **11.**   **12.**

Determine whether each pair of polygons is similar. Justify your answer.

13.

9 cm, 10 cm, 4 cm, 5 cm

14.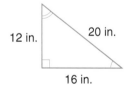

3 in., 5 in., 4 in., 12 in., 20 in., 16 in.

FOLDABLES™
Study Organizer

Make this Foldable to help you organize your Chapter 16 notes. Begin with six sheets of grid paper and an $8\frac{1}{2}$" by 11" poster board.

❶ **Staple** the six sheets of grid paper onto the poster board.

❷ **Label** each sheet with a lesson title.

Lesson 16-1

Reading and Writing As you read and study the chapter, use the pages to write the main ideas and graph examples of each lesson.

16–1 Solving Systems of Equations by Graphing

What You'll Learn
You'll learn to solve systems of equations by graphing.

Why It's Important
Business Business analysts can find the break-even point in producing and selling a product by solving systems of equations. *See Exercise 25.*

A set of two or more equations is called a **system of equations**. The solution of a system of equations is the intersection point of the graphs of these equations. The ordered pair for this point satisfies all equations in the system.

To solve a system of equations, first graph each equation in the system. To graph the equations, you can find ordered pairs, use the slope and y-intercept, or use a graphing calculator. *Choose the method that is easiest for you to use. The results will be the same for all methods.*

When solving a system of two linear equations in two variables, there are three possibilities.

- The lines intersect, so the point of intersection is the solution.

- The lines are parallel, so there is no point of intersection and, therefore, no solution.

- The lines coincide (both graphs are the same), so there is an infinite number of solutions.

Example Solve the system of equations by graphing.

$$y = x + 4$$
$$y = -2x + 1$$

In this example, we find ordered pairs by choosing values for x and finding the corresponding values of y.

$y = x + 4$			
x	$x + 4$	y	(x, y)
0	4	4	(0, 4)
1	5	5	(1, 5)

$y = -2x + 1$			
x	$-2x + 1$	y	(x, y)
−1	3	3	(−1, 3)
0	1	1	(0, 1)

Graph the ordered pairs and draw the graphs of the equations. The graphs intersect at the point whose coordinates are $(-1, 3)$. Therefore, the solution of the system of equations is $(-1, 3)$.

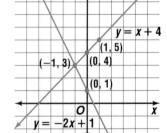

Your Turn

a. $y = -x + 1$
$y = x - 5$

When the graphs of the equations are parallel lines, a system of equations has no solution.

Example — **②** Solve the system of equations by graphing.

$$y = 3x$$
$$2y = 6x - 8$$

In this example, we use the slope and y-intercept to graph each equation. Write the second equation in slope-intercept form.

Look Back

Graphing Equations Using the Slope and y-intercept:
Lesson 4–6

$2y = 6x - 8$ *Original equation*

$\dfrac{2y}{2} = \dfrac{6x - 8}{2}$ *Divide each side by 2.*

$y = 3x - 4$ *Simplify.*

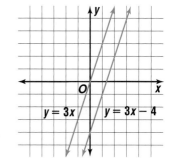

Equation	Slope	y-intercept
$y = 3x$	3	0
$y = 3x - 4$	3	−4

Note that the slope of each line is 3. So, the graphs are parallel and do not intersect. Therefore, there is no solution to the system of equations.

Your Turn

b. $y = 2x - 3$
     $y - 2x = 4$

You can use systems of equations to solve problems.

Example — **③**

Gardening Link

Toshiro is making a vegetable garden. He wants the length to be twice the width, and he has 24 feet of fencing to put around the garden. If w represents the width of the garden and ℓ represents the length, solve the system of equations below to find the dimensions of Toshiro's garden.

$$2w + 2\ell = 24$$
$$\ell = 2w$$

Solve the first equation for ℓ.

$2w + 2\ell = 24$ *The perimeter is 24 feet.*

$2w + 2\ell - 2w = 24 - 2w$ *Subtract $2w$ from each side.*

$2\ell = 24 - 2w$ *Simplify.*

$\dfrac{2\ell}{2} = \dfrac{24 - 2w}{2}$ *Divide each side by 2.*

$\ell = 12 - w$ *Simplify.*

(continued on the next page)

Use a TI–83/84 Plus graphing calculator to graph the equations $\ell = 12 - w$ and $\ell = 2w$ and to find the coordinates of the intersection point. *Note that these equations can be written as $y = 12 - x$ and $y = 2x$ and then graphed.*

Use an appropriate viewing window such as $[-5, 15]$ by $[-5, 15]$.

Enter: $\boxed{Y=}$ 12 $\boxed{-}$
$\boxed{X,T,\theta,n}$ $\boxed{ENTER}$ 2
$\boxed{X,T,\theta,n}$ $\boxed{GRAPH}$

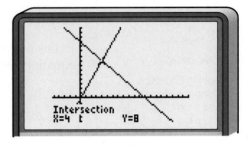

Next, use the Intersection tool on the CALC menu to find the coordinates of the point of intersection.

The solution is (4, 8). Since $w = 4$ and $\ell = 8$, the width of the garden will be 4 feet, and the length will be 8 feet.

Check your answer by examining the original problem.

Is the length of the garden twice the width? ✓
Does the garden have a perimeter of 24 feet? ✓

The solution checks.

Check for Understanding

Communicating Mathematics

1. **Draw** a system of equations whose solution is $P(5, 1)$. Explain why the ordered pair for P is the solution.

Vocabulary

system of equations

2. **State** the solution of the system of equations represented by each pair of lines.
   a. a and d
   b. b and d
   c. a and c
   d. b and c

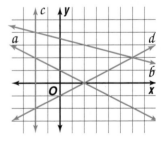

3. **a.** Writing Math Graph the system of equations.
   $y = x + 5$
   $2y = 2x + 10$
   b. List five solutions.
   c. Make a conjecture about systems of equations in part a.

Guided Practice

⊕ **Getting Ready** Write each equation in slope-intercept form.

Sample: $5x - y = 6$ **Solution:** $5x - y = 6$
$-y = 6 - 5x$
$y = 5x - 6$

4. $-x + y = -11$ **5.** $2y + 4x = 7$ **6.** $3x = -2y + 15$

Examples 1 & 2

Solve each system of equations by graphing.

7. $y = -x + 6$ **8.** $y = x - 1$ **9.** $3 - y = 2$
$y = x - 2$ $x + y = 11$ $4y = 16$

Example 3

10. Pets Marieta has 50 feet of fencing to make a dog pen. She wants the length of the pen to be 5 feet longer than its width. The system of equations below represents this problem.

$2w + 2\ell = 50$
$\ell = w + 5$

a. Graph the system of equations.

b. Determine the dimensions of the pen.

c. Explain how the dimensions of the pen are related to the ordered pair solution.

Exercises • • • • • • • • • • • • • • • • •

Practice

Solve each system of equations by graphing.

11. $y = -3x + 6$ **12.** $x + y = 4$ **13.** $x + y = 6$
$y = x - 6$ $x + y = 2$ $y = x - 2$

14. $3x - 2y = 10$ **15.** $x + 2y = 12$ **16.** $y = \frac{1}{2}x - 4$
$x + y = 0$ $y = 2$ $y = -x + 5$

17. $2x + y = -7$ **18.** $y = 2x + 1$ **19.** $x + 2y = -9$
$\frac{1}{3}x - y = -7$ $x + 2y = 7$ $x - y = 6$

Homework Help	
For Exercises	See Examples
11, 16, 24, 25	1, 2
12–15, 17–23, 26	3
Extra Practice	
See page 756.	

State the letter of the ordered pair that is a solution of both equations.

20. $3x = 15$ **a.** $(5, 0)$ **b.** $(0, 5)$ **c.** $(5, 1)$ **d.** $(5, 8)$
$2x - y = 9$

21. $2x - 5y = -1$ **a.** $(0, 5)$ **b.** $(2, 1)$ **c.** $(-0.5, 0)$ **d.** $(-2, -1)$
$x + 2y = 4$

22. Find the solution of the system $y = x - 2$ and $x + 2y = -10$ by graphing.

23. The graphs of $x + 2y = 6$, $3x - y = 4$, $x + 5y = -4$, and $-3x + 4y = 12$ intersect to form a quadrilateral.

a. Graph the system of equations.

b. Find the coordinates of the vertices of the quadrilateral.

24. **Games**   In 1998, high school sophomore Whitney Braunstein of Columbus, Ohio, created the board game *Get-a-Pet*, in which players circle the board trying to collect pets. The equation $y = 80$ represents the number of points needed to buy one pet. The equation $y = 20 + 20x$ represents the number of points a player can collect by walking the neighbor's dog once and by mowing the lawn x times.

   a. Solve the system of equations by graphing.

   b. What does this solution mean?

25. **Business**   The number of products that a company must sell so their cost equals income is called the *break-even point*. For the Gadget Company, the equation $y = 3x + 150$ represents the weekly cost of producing x gadgets. The equation that represents the income from selling the gadgets is $y = 4x$.

   a. Graph the system of equations.

   b. Find the break-even point and explain what this point means.

26. **Critical Thinking**   Graph $x + y = 3$ and $x - y = 4$. Describe the relationship between the two lines. Explain your reasoning.

27. Position and label a rectangle with length s units and width t units on a coordinate plane.   *(Lesson 15–6)*

Find each measure.

28. $m\angle B$ *(Lesson 14–4)*

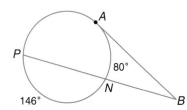

29. $m\widehat{JK}$ *(Lesson 14–3)*

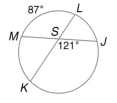

30. **Finances**   A recent survey asked nearly 200,000 students in grades 6–12 the question, "Who should pay for your movies, CDs, etc.?" The results are shown in the graph at the right. If $m\angle HTG = 68$ and $m\angle GTF = 25$, find $m\widehat{HJF}$.   *(Lesson 11–2)*

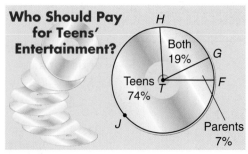

Source: *USA Weekend*

31. **Multiple Choice**   Factor $4c^2 - 25d^2$.   *(Algebra Review)*

   Ⓐ $(2c - 5d)(2c + 5d)$

   Ⓑ $(5d + 2c)(5d + 2c)$

   Ⓒ $(4c + 25d)(4c - 25d)$

   Ⓓ $(2d - 5c)(2d + 5c)$

16-2 Solving Systems of Equations by Using Algebra

What You'll Learn

You'll learn to solve systems of equations by using the substitution or elimination method.

Why It's Important

Consumer Choices
You can use systems of equations to compare phone rates. *See Exercise 30.*

In the previous lesson, you learned to solve systems of equations by graphing. You can also solve systems of equations by using algebra. One algebraic method is **substitution**. The substitution method for solving a system of equations is illustrated in Example 1.

❶ Example

Use substitution to solve the system of equations.

$2x + y = 5$
$3x - 2y = 4$

Step 1 Solve the first equation for y since the coefficient of y is 1.

$$2x + y = 5$$
$$2x + y - 2x = 5 - 2x \quad \text{Subtract 2x from each side.}$$
$$y = 5 - 2x$$

Step 2 In the solution of the system, y must have the same value in both equations. So, substitute $5 - 2x$ for y in the second equation. Then solve for x.

$$3x - 2y = 4 \qquad \text{Original equation}$$
$$3x - 2(5 - 2x) = 4 \qquad \text{Replace y with } 5 - 2x.$$
$$3x - 10 + 4x = 4 \qquad \text{Distributive property}$$
$$7x - 10 = 4 \qquad \text{Add like terms.}$$
$$7x - 10 + 10 = 4 + 10 \qquad \text{Add 10 to each side.}$$
$$7x = 14 \qquad \text{Simplify.}$$
$$\frac{7x}{7} = \frac{14}{7} \qquad \text{Divide each side by 7.}$$
$$x = 2 \qquad \text{Simplify.}$$

Step 3 Substitute 2 for x in the first equation and solve for y.

$$2x + y = 5 \qquad \text{Original equation}$$
$$2(2) + y = 5 \qquad \text{Replace x with 2.}$$
$$4 + y = 5 \qquad \text{Multiply.}$$
$$4 + y - 4 = 5 - 4 \qquad \text{Subtract 4 from each side.}$$
$$y = 1 \qquad \text{Simplify.}$$

Check your solution by substituting in both equations or by graphing.

The solution to this system of equations is (2, 1).

Your Turn Use substitution to solve each system of equations.

a. $x = y + 1$
$x + y = 8$

b. $3x + y = -6$
$-2x + 3y = 4$

Another algebraic method for solving systems of equations is called **elimination**. You can eliminate one of the variables by adding or subtracting the equations.

Example ─ **2** | Use elimination to solve the system of equations.

$x + y = 9$
$-x + 2y = -1$

$$\begin{aligned} x + y &= 9 \\ (+) \ -x + 2y &= -1 \\ \hline 0 + 3y &= 8 \end{aligned}$$ *Add the equations to eliminate the x terms.*

$$\frac{3y}{3} = \frac{8}{3}$$ *Divide each side by 3.*

$$y = \frac{8}{3}$$ *Simplify.*

The value of y in the solution is $\frac{8}{3}$.

Now substitute in either equation to find the value of x. *Choose the equation that is easier for you to solve.*

$$x + y = 9$$ *Original equation*

$$x + \frac{8}{3} = 9$$ *Replace y with $\frac{8}{3}$.*

$$x + \frac{8}{3} - \frac{8}{3} = 9 - \frac{8}{3}$$ *Subtract $\frac{8}{3}$ from each side.*

$$x = \frac{27}{3} - \frac{8}{3} \text{ or } \frac{19}{3}$$ *Simplify.*

The value of x in the solution is $\frac{19}{3}$.

The solution to the system of equations is $\left(\frac{19}{3}, \frac{8}{3}\right)$.

Your Turn | Use elimination to solve each system of equations.

c. $7x - 2y = 20$
$5x - 2y = 16$

d. $x + y = 10$
$x - y = 5$

When neither the x nor y terms can easily be eliminated by addition or subtraction, you can multiply one or both of the equations by some number.

  www.geomconcepts.com/extra_examples

Example
Business Link
③

The Helping Hearts is a nonprofit company started by students in Spokane, Washington. The company donated money to charities from the sale of beeswax candles made by the students. Suppose small candles cost $0.60 each to make and package and large candles cost $0.90. The cost of making x small candles and y large candles is $81. A total of 115 candles were made.

Solve the system of equations to find the number of small and large candles that were made.

$0.60x + 0.90y = 81$ *x represents the number of small candles, and*
$x + y = 115$ *y represents the number of large candles.*

Explore   Neither of the variables can be eliminated by simply adding or subtracting the equations.

Plan   If you multiply the second equation by 0.60 and subtract, the x variables will be eliminated.

Solve
$$0.60x + 0.90y = 81$$
$$x + y = 115 \quad \boxed{\text{Multiply by 0.60.}} \rightarrow$$

$$0.60x + 0.90y = 81$$
$$(-)\ 0.60x + 0.60y = 69$$
$$\overline{\quad 0 + 0.30y = 12}$$
$$\frac{0.30y}{0.30} = \frac{12}{0.30}$$
$$y = 40$$

Now substitute 40 for y in either of the original equations and find the value of x.

$x + y = 115$	*Original equation*
$x + 40 = 115$	*Replace y with 40.*
$x + 40 - 40 = 115 - 40$	*Subtract 40 from each side.*
$x = 75$	*Simplify.*

The solution of this system is (75, 40). Therefore, the students made 75 small candles and 40 large candles.

Examine   Check the answer by looking at the original problem.

Are there 115 candles?   *$75 + 40 = 115$* ✓

Is the total cost $81?   *$0.60(75) + 0.90(40) = \$81$* ✓

The answer checks.

Your Turn   Use elimination to solve each system of equations.

e. $3x + 5y = 12$
     $4x - y = -7$

f. $6x - 2y = 11$
     $-9x + 5y = -17$

 Personal Tutor at geomconcepts.com

Communicating Mathematics

1. **Explain** how you know that $(3, -3)$ is a solution of the system $x - 3y = 12$ and $2x + y = 3$.

2. **Write** a system of equations that has $(0, 1)$ as a solution.

3. Writing Math Describe the difference between the substitution and elimination methods. Explain when it is better to use each method.

Guided Practice

⏱ **Getting Ready** Find the value of *y* for each given value of *x*.

Sample: $4x - 5y = 12$; $x = 2y$ **Solution:** $4x - 5y = 12$

$$4(2y) - 5y = 12$$
$$3y = 12 \text{ or } y = 4$$

4. $2x + 7y = 2$;
    $x = 8$

5. $9y = 11 - x$;
    $x = y + 3$

6. $2x - 6y = 4$;
    $x = 2y - 1$

Example 1

Use substitution to solve each system of equations.

7. $x + 2y = 5$
    $2x + y = 7$

8. $y = x - 1$
    $3x - 6y = 16$

Examples 2 & 3

Use elimination to solve each system of equations.

9. $5x - 4y = 10$
    $x + 4y = 2$

10. $x - 3y = 3$
     $2x + 9y = 11$

11. **Business** *Food From the 'Hood* is a company in Los Angeles in which students sell their own vegetables and salad dressings. Suppose in one week they sell 250 bottles of creamy Italian and garlic herb dressings. This can be represented by the equation $x + y = 250$. The creamy Italian x is \$3 a bottle, and the garlic herb y is \$2.40 a bottle. If they earn \$668.40 from the sales of these two dressings, this can be represented by the equation $3x + 2.4y = 668.4$. **Example 1**

a. Use substitution to solve the system of equations.

b. How many bottles of each type of salad dressing did they sell?

Exercises • • • • • • • • • • • • • • • • • • •

Practice

Use substitution to solve each system of equations.

12. $y = 4$
     $x + y = 9$

13. $x = 1 - 4y$
     $3x + 2y = 23$

14. $9x + y = 20$
     $4x + 3y = 14$

15. $3x + 4y = -7$
     $2x + y = -3$

16. $2x = 5$
     $x + y = 7$

17. $x + 2y = 4$
     $\frac{3}{4}x + \frac{1}{2}y = 2$

Homework Help

For Exercises	See Examples
12–17, 28, 29	1
18–23, 27, 31	2, 3

Extra Practice
See page 756.

Use elimination to solve each system of equations.

18. $x + y = 2$
$x - y = 6$

19. $x - y = 6$
$x + y = 7$

20. $2x - y = 32$
$2x + y = 60$

21. $y - x = 2$
$3y - 8x = 9$

22. $2x + 5y = 13$
$4x - 3y = -13$

23. $3x - 2y = 15$
$2x - 5y = -1$

State whether *substitution* or *elimination* would be better to solve each system of equations. Explain your reasoning. Then solve the system.

24. $y = 7 - x$
$2x - y = 8$

25. $2x - 5y = -2$
$x = y - 7$

26. $3x + 5y = -16$
$2x - 2y = 0$

27. Solve the system of equations by using elimination. Round to the nearest hundredth.

$3x + 14y = 6$
$2x + 3y = 5$

28. What is the value of x in the solution of this system of equations?

$x - y = 16$
$\frac{1}{2}x + \frac{1}{2}y = 37$

Applications and Problem Solving

29. Transportation Josh is 3 miles from home riding his bike averaging 7 miles per hour. The miles traveled y can be represented by the equation $y = 3 + 7x$, where x represents the time in hours. His mother uses her car to catch up to him because he forgot his water bottle. If she averages 25 miles per hour, the equation $y = 25x$ represents her distance traveled.

   a. Explain how you can find the time it takes Josh's mother to catch up to him. (Assume that the speeds are constant.)

   b. How many minutes does it take Josh's mother to catch up to him?

   c. Explain the meaning of the y value in the solution.

30. Cellular Phones Tonisha is looking for the best cellular phone rate. Discount Cellular's plan costs $7.95 a month plus 36¢ per minute. Maland Communications advertises their Basic Plan for $9.95 a month plus 29¢ per minute.

   a. If x represents the number of minutes used in a month and y represents the total monthly cost, write an equation for each cellular phone plan.

   b. Solve the system of equations. Round to the nearest tenth.

   c. Explain what this solution represents.

   d. Which plan should Tonisha choose if she plans on talking 30 minutes per month? Explain how you determined your answer.

31. **Critical Thinking** If you use elimination to solve each system of equations, what do you get when you add or subtract the equations? Describe what this means in terms of the solution.

    a. $5x - 2y = 4$

         $10x - 4y = 8$

    b. $2x + y = 15$

         $4x + 2y = -3$

Mixed Review

32. Solve the system of equations by graphing. *(Lesson 16–1)*

    $y = 4x - 3$

    $x + 2y = 12$

33. Refer to the figures at the right. Write the *Given* statements and the *Prove* statement of a two-column proof showing that angle S is congruent to angle V. *(Lesson 15–3)*

34. $\overline{GI}$ is a diameter of $\odot B$, and $\angle GHI$ is inscribed in $\odot B$. What is the measure of $\angle GHI$? *(Lesson 14–1)*

35. **Sports** Find the volume of the tennis ball can. Round to the nearest hundredth. *(Lesson 12–3)*

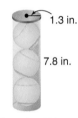

1.3 in.

7.8 in.

Exercise 35

36. **Short Response** Determine whether the following statement is *true* or *false*. Every rectangle is a parallelogram. *(Lesson 8–4)*

37. **Short Response** Draw two intersecting lines and name a pair of supplementary angles formed by them. *(Lesson 3–5)*

Quiz 1   Lessons 16–1 and 16–2

Solve each system of equations by graphing. *(Lesson 16–1)*

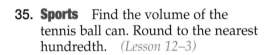

1. $y = \frac{1}{2}x + 3$

    $x = 4$

2. $-x + y = 6$

    $-3x + 3y = 3$

Use substitution or elimination to solve each system of equations. *(Lesson 16–2)*

3. $x = 5 - y$

    $3y = 3x + 1$

4. $2x + 3y = 11$

    $4x - 7y = 35$

5. Sales Used Music sells CDs for $9 and cassette tapes for $4. Suppose x represents the number of CDs sold and y represents the number of cassette tapes sold. Then $x + y = 38$ describes the total number sold in one afternoon and $9x + 4y = 297$ describes the total sales during that time. *(Lesson 16–2)*

    a. Solve the system of equations.

    b. How many more CDs than cassette tapes were sold?

16-3 Translations

What You'll Learn

You'll learn to investigate and draw translations on a coordinate plane.

Why It's Important

Animation Animators use translations to make their drawings appear to move on the screen.
See Exercises 1 and 2 on page 691.

The map below shows that the center of a hurricane has moved from 30°N latitude, 75°W longitude to 32°N latitude, 78°W longitude.

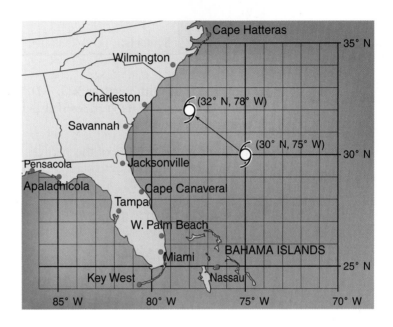

Tracking a hurricane by its latitude and longitude is like the translation of a geometric figure on a coordinate plane. In Lesson 5–3, you learned that a **translation** is a figure that is moved from one position to another without turning. In the figure below, $\triangle ABC$ is translated 3 units right and 5 units up. The image is $\triangle A'B'C'$.

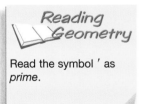

Reading Geometry

Read the symbol ' as *prime*.

preimage   *image*

$A(-1, 1)$ +3 → $A'(2, 6)$
         +5 ↑

$B(5, 4)$ +3 → $B'(8, 9)$
         +5 ↑

$C(4, 1)$ +3 → $C'(7, 6)$
         +5 ↑

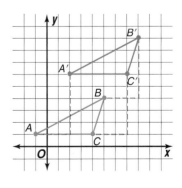

This translation can be written as the ordered pair (3, 5). To find the image of any point of $\triangle ABC$, add 3 to the x-coordinate of the ordered pair and add 5 to the y-coordinate.

$$(x, y) \rightarrow (x + 3, y + 5)$$

Definition of Translation	**Words:**	A translation is the sliding of a figure from one position to another.
	Model:	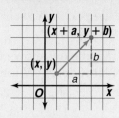
	Symbols:	(x, y) translated by (a, b) is $(x + a, y + b)$.

Example ❶ Graph $\triangle XYZ$ with vertices $X(3, -2)$, $Y(-1, 0)$, and $Z(4, 1)$. Then find the coordinates of its vertices if it is translated by $(-2, 3)$. Graph the translation image.

To find the coordinates of the vertices of $\triangle X'Y'Z'$, add -2 to each x-coordinate and add 3 to each y-coordinate of $\triangle XYZ$: $(x - 2, y + 3)$.

Algebra Review
Operations with Integers, p. 719

$X(3, -2) + (-2, 3) \rightarrow$
$X'(3 + (-2), (-2) + 3)$ or $X'(1, 1)$

$Y(-1, 0) + (-2, 3) \rightarrow$
$Y'(-1 + (-2), 0 + 3)$ or $Y'(-3, 3)$

$Z(4, 1) + (-2, 3) \rightarrow$
$Z'(4 + (-2), 1 + 3)$ or $Z'(2, 4)$

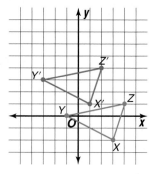

The coordinates of the vertices of $\triangle X'Y'Z'$ are $X'(1, 1)$, $Y'(-3, 3)$, and $Z'(2, 4)$.

Your Turn

Graph $\triangle LMN$ with vertices $L(0, 3)$, $M(4, 2)$, and $N(-3, -1)$. Then find the coordinates of its vertices if it is translated by $(-2, -2)$. Graph the translation image.

🌐nline **Personal Tutor at** geomconcepts.com

Check for Understanding

Communicating Mathematics

1. Write a sentence to describe a figure that is translated by $(-2, -4)$.

   Vocabulary
   translation

2. **Compare** the lengths of corresponding sides in $\triangle ABC$ and $\triangle A'B'C'$ on the next page. What can you conclude about the relationship between $\triangle ABC$ and its image? Explain.

www.geomconcepts.com/self_check_quiz

3. 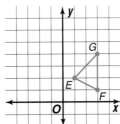 Mikasi thinks the figure at the right shows the translation (3, −1). Nicole thinks it shows the translation (3, 1). Who is correct? Explain your reasoning.

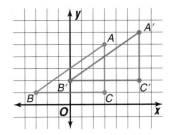

Guided Practice
Example 1

4. Find the coordinates of the vertices of △*XYZ* if it is translated by (2, 1). Then graph the translation image.

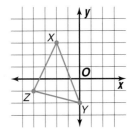

Example 1

5. Graph △*RST* with vertices *R*(5, 2), *S*(−2, 4), and *T*(−1, 1). Then find the coordinates of its vertices if it is translated by (1, −4). Graph the translation image.

6. Engineering In 1870, the Cape Hatteras Lighthouse was built 1600 feet from the ocean. By July 1999, the shore was just a few feet away. Engineers placed the lighthouse on tracks and moved it inland to protect it from soil erosion. Suppose three of the vertices at its base were *L*(16, 7), *M*(22, 12), and *N*(30, 7). If the lighthouse was moved 540 feet up the shoreline and 1506 feet inland, the translation can be given by (540, −1506). Find the coordinates of *L*′, *M*′, and *N*′ after the move.

Cape Hatteras
Lighthouse,
North Carolina

Exercises • • • • • • • • • • • • • • • • • •

Practice

Find the coordinates of the vertices of each figure after the given translation. Then graph the translation image.

7. (−2, 2)   **8.** (3, 0)   **9.** (−4, −1)

Homework Help	
For Exercises	See Examples
7–14	1
Extra Practice	
See page 756.	

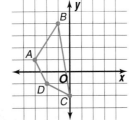

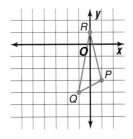

Graph each figure. Then find the coordinates of the vertices after the given translation and graph the translation image.

	Figure	Vertices	Translated By:
10.	△JKL	J(4, 0), K(2, −1), L(0, 1)	(0, −4)
11.	△XYZ	X(−5, −2), Y(−2, 7), Z(1, −6)	(6, 2)
12.	square QRST	Q(2, 1), R(4, 3), S(2, 5), T(0, 3)	(−3, 5)

13. Quadrilateral *TUVW* has vertices *T*(8, 1), *U*(0, −7), *V*(−10, −3), and *W*(−5, 2). Suppose you translate the figure 3 units right and 2 units down. What are the coordinates of its vertices *T′*, *U′*, *V′*, and *W′*? Graph the translation image.

14. Triangle *BCD* has vertices *B*(2, 4), *C*(6, 1), and *D*(0, 1). Suppose △*BCD* is translated along the *y*-axis until *D′* has coordinates (0, −8).

    a. Describe this translation using an ordered pair.

    b. Find the coordinates of *B′* and *C′*.

Applications and Problem Solving

15. Art A *frieze* is a pattern running across the upper part of a wall. Make a frieze by drawing a basic pattern on grid paper and then translating it as many times as you can.

16. Music When music is *transposed* to a different key, each note is moved the same distance up or down the musical scale. The music below shows "Deck the Halls" in two different keys.

Key of F

Key of G

Copy the music at the left and translate it into the key of G. The first measure is done for you.

17. Critical Thinking Suppose a triangle is translated by (3, −2) and then the image is translated by (−3, 2). Without graphing, what is the final position of the figure? Explain your reasoning.

Mixed Review

18. Solve the system of equations $y = 4x − 5$ and $2x + 7y = 10$. *(Lesson 16–2)*

19. Write the equation of a circle that has a diameter of 9 units and its center at (6, −21). *(Lesson 14–6)*

Find the missing measures. Write all radicals in simplest form.

20. *(Lesson 13–3)*

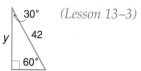

21. *(Lesson 13–2)*

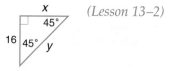

Standardized Test Practice
Ⓐ Ⓑ Ⓒ Ⓓ

22. Multiple Choice Find the surface area of a bowling ball if its diameter is 8.5 inches. Round to the nearest hundredth. *(Lesson 12–6)*

    Ⓐ 106.81 in²   **Ⓑ** 226.98 in²   **Ⓒ** 229.98 in²   **Ⓓ** 907.92 in²

Math In the Workplace

Animator

If you loved being scared by Godzilla or by the dinosaurs in *Jurassic Park*, thank an animator. Animators create sequences of motion-based art for the motion picture and television industry. Most animators use computer animation programs.

Computer animation results from thousands of changes that occur on a coordinate plane. Let's start with one change. This graphing calculator program draws a triangle and translates it a certain distance horizontally and vertically.

```
PROGRAM:MAPPING              L2(3))
:Disp "ENTER VERTICES"       :Line(L1(1), L2(1), L1(3),
 For(N, 1, 3)                 L2(3))
:Input "X", X                :Pause
:X→L1(N)                     :Input "HORIZONTAL MOVE", H
:Input "Y", Y                :Input "VERTICAL MOVE", V
:Y→L2(N)                     :Line(L1(1)+H, L2(1)+V,
:End                          L1(2)+H, L2(2)+V)
:ClrDraw                     :Line(L1(2)+H, L2(2)+V,
:ZStandard                    L1(3)+H, L2(3)+V)
:Line(L1(1), L2(1), L1(2),   :Line(L1(1)+H, L2(1)+V,
 L2(2))                       L1(3)+H, L2(3)+V)
:Line(L1(2), L2(2), L1(3),   :Stop
```

1. Use the program to draw $\triangle ABC$ and its translated image.
   a. $A(2, 3)$, $B(5, 9)$, $C(0, 4)$; horizontal move: 2; vertical move: 1
   b. $A(-4, 3)$, $B(1, 7)$, $C(3, -2)$; horizontal move: 4; vertical move: -3
   c. $A(3, 6)$, $B(5, 2)$, $C(-2, 8)$; horizontal move: -4; vertical move: -5
2. Modify the program so that the first triangle is erased before the image is drawn.

FAST FACTS About Animators

Working Conditions
- work as part of a team
- use computers for extended periods of time

Education
- internship or college degree in graphic art, computer-aided design, or visual communications
- Knowledge and training in computer techniques are critical.

Employment

Animators who stay at least 10 years in the field — 20%

Animators who pursue other careers — 80%

*inter*NET CONNECTION **Career Data** For the latest information on a career as an animator, visit:

www.geomconcepts.com

What You'll Learn
You'll learn to investigate and draw reflections on a coordinate plane.

Why It's Important
Printing The use of reflections is important in some types of printing.
See Exercise 15.

The photograph at the right shows the reflection of an ancient Roman bridge in the Tiber River. Every point on the bridge has a corresponding point on the water. In mathematics, this type of one-to-one correspondence is also called a **reflection**.

In the following activity, you'll investigate reflections over coordinate axes.

Tiber River, Rome

Hands-On Geometry

Materials: grid paper   tracing paper

 straightedge

Step 1 On a coordinate graph, use a straightedge to draw a quadrilateral with vertices $A(1, 0)$, $B(2, 3)$, $C(4, 1)$, and $D(3, -3)$.

Step 2 Fold a piece of tracing paper twice to create coordinate axes. Unfold the paper and label the x- and y-axes.

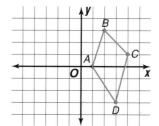

Step 3 Place the tracing paper on top of the coordinate graph, lining up the axes on both pieces of paper. Trace quadrilateral $ABCD$.

Step 4 Turn over the tracing paper so that the figure is flipped over the y-axis.

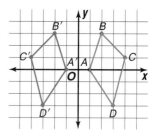

Try These

1. Name the coordinates for the reflected quadrilateral $A'B'C'D'$ from Step 4.

2. Repeat Step 4, but this time flip $ABCD$ over the x-axis. Name the coordinates for this reflected quadrilateral $A''B''C''D''$.

3. Compare the coordinates of the original quadrilateral with the coordinates of each reflection. What do you notice?

The results of this activity are stated in the following definition.

<table>
<tr><td rowspan="2">Definition of Reflection</td><td>Words: A reflection flips a figure over a line.</td></tr>
<tr><td>
Model:

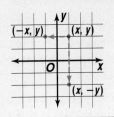

Symbols:
reflection over x-axis:
$(x, y) \rightarrow (x, -y)$

reflection over y-axis:
$(x, y) \rightarrow (-x, y)$
</td></tr>
</table>

Examples ①

Graph $\triangle PQR$ with vertices $P(-5, 3)$, $Q(-4, -1)$, and $R(-2, 2)$. Then find the coordinates of its vertices if it is reflected over the x-axis and graph its reflection image.

To find the coordinates of the vertices of $\triangle P'Q'R'$, use the definition of reflection over the x-axis: $(x, y) \rightarrow (x, -y)$.

preimage   *image*

$P(-5, 3) \quad \rightarrow P'(-5, -3)$

$Q(-4, -1) \rightarrow Q'(-4, 1)$

$R(-2, 2) \quad \rightarrow R'(-2, -2)$

The vertices of $\triangle P'Q'R'$ are $P'(-5, -3)$, $Q'(-4, 1)$, and $R'(-2, -2)$.

Art Link

Real World

②

A *dyrnak gul* is a symmetrical motif found in oriental carpets. Reflect the design over the y-axis below to find the coordinates of J', K', L', and M'. Then graph the reflection points.

To find the coordinates of J', K', and L', use the definition of reflection over the y-axis: $(x, y) \rightarrow (-x, y)$.

preimage   *image*

$J(-3, 2) \quad \rightarrow J'(3, 2)$
$K(-5, 0) \quad \rightarrow K'(5, 0)$
$L(-8, 0) \quad \rightarrow L'(8, 0)$
$M(-3, -2) \rightarrow M'(3, -2)$

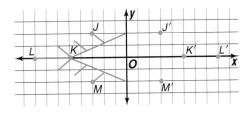

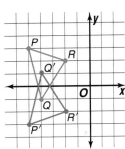

Dyrnak Gul

Your Turn

Graph each figure. Then find the coordinates of the vertices after a reflection over the given axis and graph the reflection image.

a. $\triangle HIJ$: $H(3, 1)$, $I(4, 4)$, $J(-2, 3)$, x-axis
b. trapezoid $LMNP$: $L(-3, -3)$, $M(-3, 2)$, $N(-1, 2)$, $P(1, -3)$, y-axis

 nline Personal Tutor at geomconcepts.com

Check for Understanding

Communicating Mathematics

1. **Explain** whether the translated image of a figure can ever be the same as its reflected image. Show an example to support your answer.

> **Vocabulary**
> reflection

2. Suppose you reflect a figure over the *x*-axis and then reflect that image over the *y*-axis. Is this double reflection the same as a translation? Explain why or why not.

3. **You Decide?** Manuel says that if a figure with parallel sides is reflected over the *x*-axis, the reflected figure will also have parallel sides. Natalie says that such a reflected figure may not always have parallel sides. Who is correct? Explain why.

Guided Practice
Example 1

4. Graph $\triangle ABC$ with vertices $A(-4, -4)$, $B(0, 2)$, and $C(1, -3)$. Then find the coordinates of its vertices if it is reflected over the *x*-axis and graph the reflection image.

Example 2

5. Find the coordinates of the vertices of quadrilateral *HIJK* if it is reflected over the *y*-axis. Then graph the reflection image.

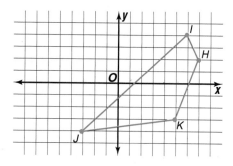

Example 2

6. **Architecture** To preserve the symmetry of his house, George Washington had the second window from the left, upstairs, painted on. If this fake window has coordinates $A(-22, -0.5)$, $B(-18, -0.5)$, $C(-18, -4)$, and $D(-22, -4)$, then what are the coordinates of its reflection over the *y*-axis, window $A'B'C'D'$?

Look Back

Symmetry: Lesson 10–6

George Washington's home, Mount Vernon, Virginia

Practice

Find the coordinates of the vertices of each figure after a reflection over the given axis. Then graph the reflection image.

Homework Help	
For Exercises	See Examples
7, 10, 12, 14, 16	2
8, 9, 11, 13, 15, 17	1
Extra Practice	
See page 757.	

7. y-axis

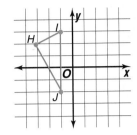

8. x-axis

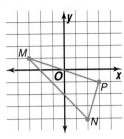

9. x-axis

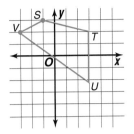

Graph each figure. Then find the coordinates of the vertices after a reflection over the given axis and graph the reflection image.

	Figure	Vertices	Reflected Over:
10.	$\triangle EFG$	$E(-1, 2)$, $F(2, 4)$, $G(2, -4)$	y-axis
11.	$\triangle PQR$	$P(1, 2)$, $Q(4, 4)$, $R(2, -3)$	x-axis
12.	quadrilateral $VWXY$	$V(0, -1)$, $W(1, 1)$, $X(4, -1)$, $Y(1, -6)$	y-axis

13. Suppose $\triangle CDE$ is reflected over the line $x = 2$.

   a. Find the coordinates of the vertices after the reflection.

   b. Graph the reflection image.

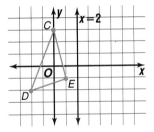

14. The combination of a reflection and a translation is called a *glide reflection*. An example of this is a set of footprints.

   a. State the steps for finding the coordinates of each footprint, given points $A(7, -2)$ and $B(13, 2)$.

   b. Find the coordinates for point C.

Applications and Problem Solving

15. Printing In *lithographic printing,* a printed image is a reflection of an inked surface. In the letter A shown at the right, suppose points $H(-5, -7)$, $J(0, -4)$, and $K(7, -12)$ lie on the inked surface. Name the coordinates of the corresponding points on the printed image if the inked surface is reflected over the x-axis.

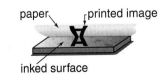

paper   printed image

inked surface

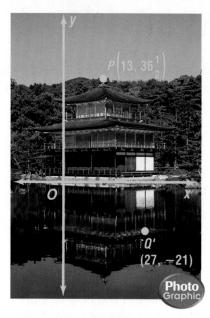

P$\left(13, 36\frac{1}{2}\right)$

O

x

Q'
(27, −21)

Photo Graphic

16. Nature The *pagoda*, or Far Eastern tower, shown at the left is reflected in the lake.

   a. Suppose point $P\left(13, 36\frac{1}{2}\right)$ lies on the pagoda. Name the coordinates of this point reflected in the water.

   b. If the reflected point $Q'(27, -21)$ appears in the water, what are the coordinates of the original point Q on the pagoda?

17. Critical Thinking Triangle *CDE* is the preimage of a double reflection over line *j* and then line *k*. Copy the figure at the right. Label the vertices of the first image C', D', and E'. Then label the vertices of the second image C'', D'', and E''.

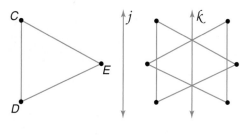

Mixed Review

18. Graph $\triangle JKL$ with vertices $J(-4, 4)$, $K(-1, 1)$, and $L(-3, -2)$. Then find the coordinates of its vertices if it is translated by $(3, -2)$. Graph the translation image. *(Lesson 16–3)*

19. Logic Write a conclusion that follows from statements (1) and (2). If a valid conclusion does not follow, write *no valid conclusion*. *(Lesson 15–1)*

   (1) If two angles are complementary to the same angle, then they are congruent.

   (2) $m\angle A + m\angle R = 90$ and $m\angle D + m\angle R = 90$

20. Find sin A. Round to four decimal places. *(Lesson 13–5)*

Standardized Test Practice
Ⓐ Ⓑ Ⓒ Ⓓ

21. Short Response Simplify $\sqrt{3} \cdot \sqrt{18}$. *(Lesson 13–1)*

22. Multiple Choice What is the value of x? *(Lesson 6–4)*

   Ⓐ 62

   Ⓑ 7

   Ⓒ 56

   Ⓓ 110

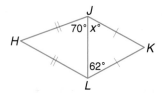

16-5 Rotations

What You'll Learn

You'll learn to investigate and draw rotations on a coordinate plane.

Why It's Important

Art Rotations are often used to create patterns for stained glass windows. *See Exercise 15.*

A trapeze artist swings on the trapeze in a circular motion. This type of movement around a fixed point is called a **turn** or a **rotation**. The fixed point is called the **center of rotation**. This point may be in the center of an object, as in a spinner, or outside an object, as with the swinging trapeze artist whose center of rotation is at the top of the trapeze.

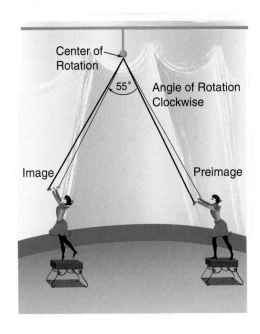

Center of Rotation

55° Angle of Rotation Clockwise

Image   Preimage

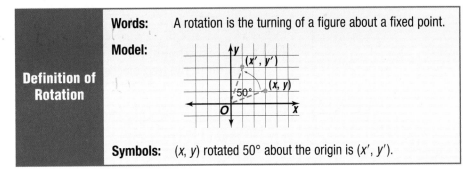

Definition of Rotation	**Words:**	A rotation is the turning of a figure about a fixed point.
	Model:	(x', y') (x, y) 50°
	Symbols:	(x, y) rotated 50° about the origin is (x', y').

Rotations can be either clockwise or counterclockwise. As with translations and reflections, there is a one-to-one correspondence between the preimage and the image, and the resulting image after a rotation is congruent to the original figure.

In each case below, $\overline{JK}$ is rotated 60° counterclockwise about the origin.

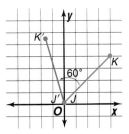

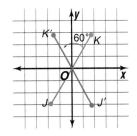

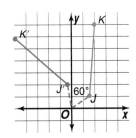

You can use tracing paper to rotate figures whose centers of rotation are on the figure or outside of the figure.

Real World

Example ①	**The pattern in the Hungarian needlework at the left is formed by 120°-rotations. Rotate quadrilateral *RSTU* 120° clockwise about point *O* by tracing the figure.**
Art Link	

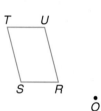

Step 1 Draw a segment from point *O* to *R*.

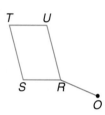

Step 2 Use a protractor to draw an angle 120° clockwise about point *O*. Draw segment *OR'* congruent to $\overline{OR}$.

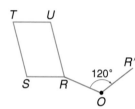

Step 3 Trace the figure on a sheet of tracing paper. Label the corresponding vertices *R'*, *S'*, *T'*, and *U'*.

Step 4 Place a straight pin through the two pieces of paper at point *O*. Rotate the top paper clockwise, keeping point *O* in the same position, until the figure is rotated 120°.

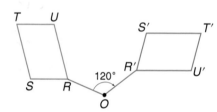

R'S'T'U' is the rotation image of *RSTU*.

Bird Motif,
Northern Hungary

Your Turn

Rotate each figure about point *C* by tracing the figure. Use the given angle of rotation.

a. 110° clockwise

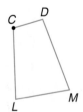

b. 60° counterclockwise

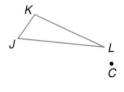

www.geomconcepts.com/extra_examples

You can also draw rotations on the coordinate plane without using tracing paper.

Example ——② **Find the coordinates of the vertices of △LNH if it is rotated 90° counterclockwise about the origin. Graph the rotation image.**

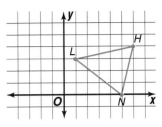

Step 1 Draw a segment from the origin to point L.

Step 2 Use a protractor to draw ∠LOL′ so that its measure is 90 and OL = OL′.

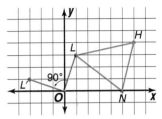

Step 3 Draw a segment from the origin to point N.

Step 4 Use a protractor to draw ∠NON′ so that its measure is 90 and ON = ON′.

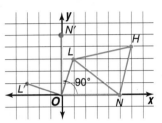

Step 5 Draw a segment from the origin to point H.

Step 6 Use a protractor to draw ∠HOH′ so that its measure is 90 and OH = OH′.

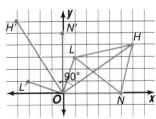

Step 7 Connect the points to form △L′N′H′.

The vertices of △L′N′H′ are L′(−3, 1), N′(0, 5), and H′(−4, 6).

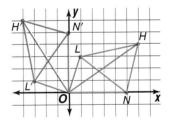

Your Turn

Graph each figure. Then find the coordinates of the vertices after the given rotation about the origin and graph the rotation image.

c. segment LM with vertices L(−2, 4) and M(0, 6) rotated 90° clockwise

d. △STR with vertices S(0, 0), T(0, −4), and R(3, −2), rotated 180° counterclockwise

You can use a TI–83/84 Plus calculator to rotate figures about a selected point.

Graphing Calculator Tutorial
See pp. 782–785.

Graphing Calculator Exploration

Step 1   Construct a triangle by using the Triangle tool on the ⬚F2⬚ menu. Label this triangle *ABC*.

Step 2   Draw the point about which you want to rotate the triangle. Label it Point *P*.

Step 3   Use the segment tool on ⬚F2⬚ to draw two segments with a common endpoint. Then use the Angle Measure tool on ⬚F5⬚ to find the measure of the angle. This will be your angle of rotation.

Step 4   Use the Rotation tool on ⬚F4⬚ to rotate the triangle about the point using the angle. Label the corresponding vertices of the rotated image as *X*, *Y*, and *Z*.

Step 5   Draw segments from *P* to the corresponding vertices *B* and *Y*. Use the Display tool on ⬚F5⬚ to make these segments stand out.

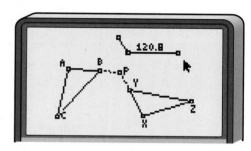

Try These

1. Use the angle tool on ⬚F6⬚ to measure ∠*BPY* and ∠*CPZ*.
2. What is the relation of the angle measures from Exercise 1 to the angle measure you used for the rotation?
3. Describe what happens to the angle measures if you drag point *P* to a different location.
4. Change the angle measure for the rotation from 120° to 110°. What happens to the image of the original triangle and the angle measures?

Check for Understanding

Communicating Mathematics

1. **Study** Example 2. What can you conclude about the coordinates of the image and preimage in a 90° rotation?

2. **Describe** two techniques for finding the rotation of a figure about a fixed point.

3. ***You Decide*** Diem rotated quadrilateral *HIJK* 305° in a clockwise direction about the origin. Lakesha said she could have rotated the quadrilateral 55° in a counterclockwise direction and found the same image. Is Lakesha correct? Explain why or why not.

> **Vocabulary**
> turn
> rotation
> center of rotation

Example 1

Rotate each figure about point _P_ by tracing the figure. Use the given angle of rotation.

4. 130° counterclockwise

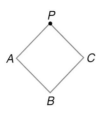

5. 85° clockwise

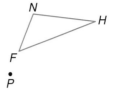

Example 2

6. Graph △*QRS* with vertices *Q*(1, 1), *R*(4, −3), and *S*(1, −3). Then find the coordinates of its vertices if the figure is rotated clockwise 180° about the origin. Graph the rotation image.

Example 1

7. **Art** Rotate *HIJK* 120° clockwise about *H* and 120° counterclockwise about *H* to complete the Islamic mosaic pattern.

Exercises

· · · · · ● · · · · · · · · · · · · · ·

Practice

Rotate each figure about point _K_ by tracing the figure. Use the given angle of rotation.

8. 90° clockwise

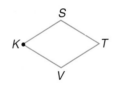

9. 120° counterclockwise

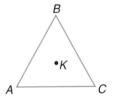

Homework Help	
For Exercises	**See Examples**
8, 10–12, 14	1
9, 13, 17	2
Extra Practice	
See page 757.	

10. 60° clockwise

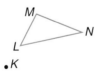

11. 180° clockwise

Find the coordinates of the vertices of each figure after the given rotation about the origin. Then graph the rotation image.

12. 90° clockwise

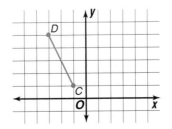

13. 180° counterclockwise

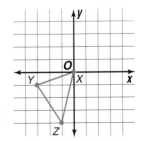

14. Segment *QR* has endpoints *Q*(−4, 3) and *R*(0, 1). Find the coordinates of the vertices of the segment if it is rotated 90° clockwise about the origin.

Applications and Problem Solving

15. Art In the stained glass window at the right, how are rotations used to create the pattern? Use tracing paper and label the figures to show your answer.

16. Art The design at the left is by Canadian scientist Francois Brisse. Describe the transformations he could have used to create the design.

Notre Dame Cathedral, Paris

17. Critical Thinking Triangle *ABC* has been rotated in a counterclockwise direction about point *D*. Find the angle of rotation.

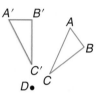

Mixed Review

18. Triangle *QRS* has vertices *Q*(−2, 6), *R*(1, 1), and *S*(4, −3). Find the coordinates of its vertices if it is reflected over the *x*-axis. *(Lesson 16–4)*

19. Solve the system of equations 2*x* − *y* = 8 and 6*x* − *y* = 7. *(Lesson 16–2)*

20. Draw a figure and write a two-column proof to show that opposite angles of a rhombus are congruent. *(Lesson 15–4)*

Standardized Test Practice
Ⓐ Ⓑ Ⓒ Ⓓ

21. Short Response If $\overline{ML}$ and $\overline{MN}$ are tangent to ⊙*E* at *L* and *N* respectively, then $\overline{ML} \cong$ ___?___ . *(Lesson 14–2)*

22. Multiple Choice A purse has an original price of $45. If the purse is on sale for 25% off, what is the sale price? *(Percent Review)*

Ⓐ $11.25   Ⓑ $56.25   Ⓒ $20.00   Ⓓ $33.75

Quiz 2   Lessons 16–3 through 16–5

▶ **Graph each figure. Then find the coordinates of the vertices after the given transformation and graph the image.**

Figure	Vertices	Transformation
1. △*HIJ*	*H*(−2, 1), *I*(2, 3), *J*(0, 0)	translated by (2, 4) *(Lesson 16–3)*
2. △*EFG*	*E*(2, 0), *F*(−1, −1), *G*(1, 3)	translated by (−3, 1) *(Lesson 16–3)*
3. △*ABC*	*A*(−4, −2), *B*(−1, −4), *C*(2, −2)	reflected over *x*-axis *(Lesson 16–4)*
4. quadrilateral *QRST*	*Q*(1, 0), *R*(2, −3), *S*(0, −3), *T*(−3, −1)	reflected over *y*-axis *(Lesson 16–4)*

5. Design Describe how rotation was used to create the design at the right and list the angles of rotation that were used. *(Lesson 16–5)*

702   Chapter 16 More Coordinate Graphing and Transformations

www.geomconcepts.com/self_check_quiz

What You'll Learn
You'll learn to investigate and draw dilations on a coordinate plane.

Why It's Important
Publishing Artists must understand dilations when sizing art for textbooks.
See Exercise 11.

Alexis went to the camera shop to have a picture enlarged from a 4 × 6 to an 8 × 12. This transformation is called a **dilation**.

In previous lessons, we learned that in translations, reflections, and rotations, the image and preimage are congruent. Dilations are different because they alter the size of an image, but not its shape. The ratio of a dilated image to its preimage is called the *scale factor*.

COncepts in MOtion
Animation
geomconcepts.com

Definition of Dilation	**Words:** A dilation reduces or enlarges a figure by a scale factor k.
	Model: 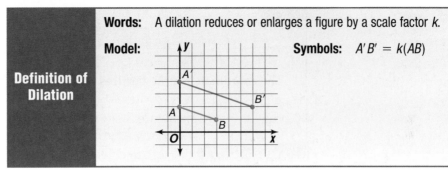 **Symbols:** $A'B' = k(AB)$

A figure is enlarged if the scale factor is greater than 1, and reduced if the scale factor is between 0 and 1. Note this in the following examples.

Real World Example 1
Animal Link

$\overline{CM}$ **with endpoints** $C(4, 2)$ **and** $M(2, 6)$ **represents the length of a baby dolphin. If the mother dolphin is one and a half times the size of her baby, find the coordinates of the dilation image of** $\overline{CM}$ **with a scale factor of 1.5, and graph its dilation image.**

Since $k > 1$, this is an enlargement. To find the dilation image, multiply each coordinate in the ordered pairs by 1.5.

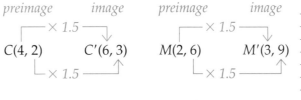

The coordinates of the endpoints of the dilation image are $C'(6, 3)$ and $M'(3, 9)$.

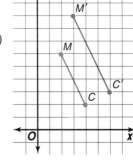

Graph △QRS with vertices Q(−2, 6) and R(8, 0), and S(6, 5). Then find the coordinates of the dilation image with a scale factor of $\frac{1}{2}$ and graph its dilation image.

Since $k < 1$, this is a reduction. To find the dilation image, multiply each coordinate in the ordered pairs by $\frac{1}{2}$.

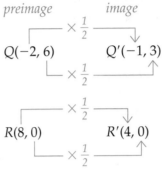

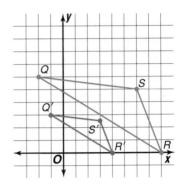

preimage   *image*

$$Q(-2, 6) \xrightarrow{\times \frac{1}{2}} Q'(-1, 3)$$

$$R(8, 0) \xrightarrow{\times \frac{1}{2}} R'(4, 0)$$

$$S(6, 5) \xrightarrow{\times \frac{1}{2}} S'\left(3, \frac{5}{2}\right)$$

The coordinates of the vertices of the dilation image are $Q'(-1, 3)$, $R'(4, 0)$, and $S'\left(3, \frac{5}{2}\right)$.

Your Turn

Graph each figure. Then find the coordinates of the dilation image for the given scale factor k, and graph the dilation image.

a. △GHI with vertices G(0, −2), H(1, 3), and I(4, 1); $k = 2$

b. $\overline{LN}$ with endpoints L(8, 8) and N(4, 5); $k = \frac{1}{4}$

☁nline Personal Tutor at geomconcepts.com

Check for Understanding

Communicating Mathematics

1. **Explain** how you can determine whether a dilation is a reduction or an enlargement.

2. **Compare and contrast** the difference between a dilation and the other transformations that you have studied.

3. **Write** about dilations that you can find in everyday life. Explain whether they are reductions or enlargements and estimate what the scale factors might be.

Guided Practice

⊕ **Getting Ready** **Find each product.**

| **Sample:** $(2, -1) \times 7$ | **Solution:** $(2 \times 7, -1 \times 7)$ or $(14, -7)$ |

4. $(0, 7) \times 4$ **5.** $(-8, 2) \times \frac{1}{4}$ **6.** $(10, -6.5) \times 0.5$

Examples 1 & 2

Find the coordinates of the dilation image for the given scale factor k, and graph the dilation image.

7. 2 **8.** $\frac{1}{3}$

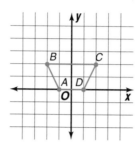 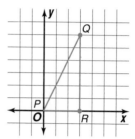

Graph each figure. Then find the coordinates of the dilation image for the given scale factor k, and graph the dilation image.

Example 1 **9.** $\overline{FG}$ with endpoints $F(-2, 1)$ and $G(1, -2)$; $k = 3$

Example 2 **10.** $\triangle STW$ with vertices $S(4, 8)$, $T(10, 0)$, and $W(0, -4)$; $k = \frac{1}{4}$

Example 2 **11. Publishing** When an artist receives art specifications for a textbook, the artist must size the art so that it fits the space on the page. In order for the triangle at the right to fit the given space, the artist must make it $\frac{3}{4}$ its original size.

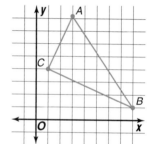

  a. What will be the coordinates of the vertices of the new image?

  b. Graph the new image.

Exercises

• • • • • • • • • • • • • • • • • • • •

Practice

Find the coordinates of the dilation image for the given scale factor k, and graph the dilation image.

Homework Help	
For Exercises	See Examples
12, 15, 17, 20, 23, 26	1
13, 14, 16, 18, 19, 21	2
Extra Practice	
See page 757.	

12. 2 **13.** $\frac{1}{3}$ **14.** $\frac{1}{4}$

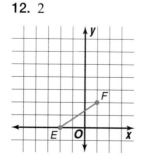 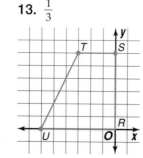 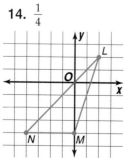

Find the coordinates of the dilation image for the given scale factor _k_, and graph the dilation image.

15. 3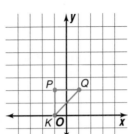

16. $\frac{1}{2}$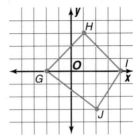

17. 4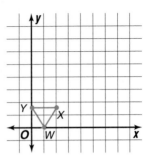

Graph each figure. Then find the coordinates of the dilation image for the given scale factor _k_, and graph the dilation image.

18. $\overline{ST}$ with endpoints $S(-2, 4)$ and $T(4, 0)$; $k = \frac{1}{2}$

19. $\triangle ABC$ with vertices $A(2, 0)$, $B(0, -6)$, and $C(-4, -4)$; $k = \frac{1}{4}$

20. quadrilateral $NPQR$ with vertices $N(1, -2)$, $P(1, 0)$, $Q(2, 2)$, and $R(3, 0)$; $k = 2$

21. Graph quadrilateral $JKLM$ with vertices $J(0, 0)$, $K(5, 3)$, $L(7, -2)$, and $M(4, -4)$. Then find the coordinates of the dilation image for the scale factor $\frac{3}{4}$, and graph the dilation image.

22. Triangle $J'L'M'$ is the dilation image of $\triangle JLM$. Find the scale factor.

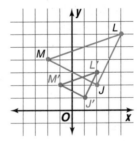

23. Technology Mr. Hernandez wants to project a 2×2-inch slide onto a wall to create an image 64 inches $\times$ 64 inches. If the slide projector makes the image twice as large each yard that it is moved away from the wall, how far away from the wall should Mr. Hernandez place the projector?

24. Photography Refer to the application at the beginning of the lesson. Can a 5×7 picture be directly enlarged to an 11×14 poster? If so, state the scale factor. If not, explain why.

25. Art What type of transformations did artist Norman Rockwell use in his painting at the right? List the transformations and explain how each one was used.

Norman Rockwell,
Triple Self-Portrait

26. Critical Thinking Suppose rectangle *HIJK* is dilated with a scale factor of 2.

a. Graph the dilation image.

b. Compare the perimeter of the dilation image with the perimeter of the preimage.

c. Compare the area of the dilation image with the area of the preimage.

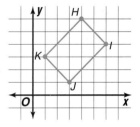

Mixed Review

27. Rotate $\triangle JLN$ 35° counterclockwise about point *K* by tracing the figure. *(Lesson 16–5)*

28. Solve the system of equations by graphing. *(Lesson 16–1)*
$y = 2x$
$\frac{1}{2}x - y = 3$

29. Find *GM*. Round to the nearest tenth. *(Lesson 14–5)*

30. Find tan *B*. Round to four decimal places. *(Lesson 13–4)*

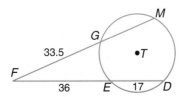

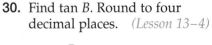

31. Multiple Choice The length of the diagonal of a square is $15\sqrt{2}$ feet. Find the length of a side. *(Lesson 13–2)*

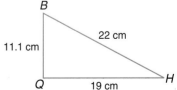

Ⓐ 7.5 ft   Ⓑ 10.6 ft   Ⓒ 30 ft   Ⓓ 15 ft

Artists Do Math, Don't They?

Composition of Transformations

Materials

 straightedge

 compass

 protractor

 grid paper

You can find many examples of transformations in art—repetitions of congruent or similar figures. You can also find **composition of transformations**. This is when artists use more than one transformation, like in the figure at the right. Can you name two transformations that the artist used?

Investigate

1. Use a pencil, straightedge, and grid paper to explore composition of transformations.

   a. Draw and label the *x*-axis and *y*-axis on the grid paper. Use a straightedge to draw a triangle with vertices *A*(3, 1), *B*(5, 2), and *C*(4, 6).

   b. Draw the reflection of △*ABC* over the *x*-axis and label it △*A'B'C'*.

   c. Draw the reflection of △*A'B'C'* over the *y*-axis and label it △*A"B"C"*. Record the coordinates of the vertices of △*A"B"C"* in a table.

   d. Rotate △*ABC* 180° clockwise about the origin. Record the coordinates of the vertices in a table. How do these coordinates compare to the coordinates in part c? What can you conclude?

2. Repeat Exercise 1, this time reflecting the triangle over the *x*-axis, and then over the line $y = -7$. Is there a way to get to this image by performing just one transformation? Explain.

3. **Fractals** Mathematician Benoit Mandelbrot used the term *fractal* to describe things in nature that are irregular in shape, such as clouds, coastlines, and trees. Fractals are self-similar; that is, the smaller details of the shape have the same characteristics as the original form. Describe how transformations and compositions of transformations could be used to draw a picture of a fractal.

Extending the Investigation

In this extension, you will create your own artwork using composition of transformations.

- Use paper and construction tools or geometry software to construct any figure with three or more vertices on a coordinate grid.
- Make a table and record the coordinates for each vertex of the figure.
- Perform two transformations on the figure. This can be any combination of translations, reflections, rotations, and dilations. After each transformation, record the coordinates of the vertices in your table.
- Perform the two transformations again, and record the coordinates of the vertices.
- Repeat this procedure several times.
- Are any of your combinations the same as a single rotation? Explain.

Presenting Your Conclusions

Here are some ideas to help you present your conclusions to the class.

- Make a poster displaying your table and your final artwork.
- Write a paper summarizing how you used multiple transformations to create your artwork.

 Investigation For more information on transformations and art, **visit:** www.geomconcepts.com

Understanding and Using the Vocabulary

After completing this chapter, you should be able to define each term, property, or phrase and give an example or two of each.

 Review Activities
For more review activities, visit:
www.geomconcepts.com

Geometry

center of rotation *(p. 697)*
composition of transformations *(p. 708)*
dilation *(p. 703)*
reflection *(p. 692)*

rotation *(p. 697)*
translation *(p. 687)*
turn *(p. 697)*

Algebra

elimination *(p. 682)*
substitution *(p. 681)*
system of equations *(p. 676)*

State whether each sentence is *true* or *false*. If false, replace the underlined word(s) to make a true statement.

1. A <u>dilation</u> alters the size of a figure but does not change its shape.
2. Substitution and elimination are methods for solving <u>translations</u>.
3. A <u>reflection</u> is the turning of a figure about a fixed point.
4. A figure is reduced in a dilation if the scale factor is between <u>0 and 1</u>.
5. In a <u>reflection</u>, a figure is moved from one position to another without turning.
6. Systems of equations can be solved algebraically by <u>elimination</u>.
7. The fixed point about which a figure is rotated is called the <u>center of rotation</u>.
8. A <u>rotation</u> flips a figure over a line.
9. It is better to use <u>substitution</u> to solve a system of equations when one of the equations is already solved for a variable.
10. Another name for a rotation is a <u>reflection</u>.

Skills and Concepts

Objectives and Examples

• **Lesson 16–1** Solve systems of equations by graphing.

Solve the system of equations by graphing.
$y = 3x + 1$
$y = x - 1$

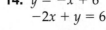

The solution is $(-1, -2)$.

Review Exercises

Solve each system of equations by graphing.

11. $x + y = 4$
    $y = 2x + 1$
12. $y = x - 5$
    $y = 5x + 7$
13. $3y = 9$
    $15 - y = 9$
14. $y = -x + 6$
    $-2x + y = 6$

 www.geomconcepts.com/vocabulary_review

Objectives and Examples

- **Lesson 16–2** Solve systems of equations by using the substitution or elimination method.

$$x + 2y = 4 \text{ and } 3x - 2y = 4$$

Substitution:
1. Solve the first equation for x.
2. Substitute the result into the second equation and solve for y.
3. Substitute the value of y into the first equation and solve for x.

Elimination:
1. Add the equations to eliminate the y terms.
2. Solve the resulting equation for x.
3. Substitute the value of x into either original equation and solve for y.

Review Exercises

State whether *substitution* or *elimination* would be better to solve each system of equations. Explain your reasoning. Then solve the system.

15. $5x + y = 11$
$x - y = 7$

16. $-2x + y = 40$
$3x + 7y = 195$

17. $7x - y = -10$
$2x + 2y = 52$

18. $-x + 6y = 8$
$x - y = 2$

19. $4x + 12y = -4$
$-x - 5y = 3$

- **Lesson 16–3** Investigate and draw translations on a coordinate plane.

Find the coordinates of the vertices of $\triangle PQR$ if it is translated by $(3, 3)$.

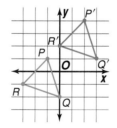

$P(-1, 1) + (3, 3) \rightarrow P'(-1 + 3, 1 + 3)$ or $P'(2, 4)$
$Q(0, -2) + (3, 3) \rightarrow Q'(0 + 3, -2 + 3)$ or $Q'(3, 1)$
$R(-3, -1) + (3, 3) \rightarrow R'(-3 + 3, -1 + 3)$ or $R'(0, 2)$

Graph each figure. Then find the coordinates of the vertices after the given translation and graph its translation image.

20. quadrilateral $ABCD$ with vertices $A(1, 1)$, $B(1, 5)$, $C(7, 5)$, and $D(7, 1)$ translated by $(-2, -3)$

21. triangle LMN with vertices $L(-3, 1)$, $M(0, 3)$, and $N(1, -1)$ translated by $(1, 1)$

- **Lesson 16–4** Investigate and draw reflections on a coordinate plane.

Find the coordinates of the vertices of $\triangle XYZ$ if it is reflected over the y-axis.
$X(-3, 1) \rightarrow X'(3, 1)$
$Y(-2, 3) \rightarrow Y'(2, 3)$
$Z(-1, 0) \rightarrow Z'(1, 0)$

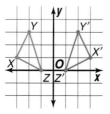

Graph each figure. Then find the coordinates of the vertices after a reflection over the given axis and graph the reflection image.

22. triangle DEF with vertices $D(0, 3)$, $E(3, 3)$, and $F(1, 1)$ reflected over the x-axis

23. quadrilateral $STUV$ with vertices $S(0, 2)$, $T(4, 1)$, $U(2, -1)$, and $V(-1, -2)$ reflected over the y-axis

Objectives and Examples

Review Exercises

• **Lesson 16–5** Investigate and draw rotations on a coordinate plane.

Find the coordinates of the vertices of $\triangle ABC$ if it is rotated 90° clockwise about the origin.

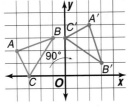

The coordinates of $\triangle A'B'C'$ are $A'(2, 4)$, $B'(3, 1)$, and $C'(0, 3)$.

24. Rotate quadrilateral $WXYZ$ 45° counterclockwise about point W by tracing.

25. Find the coordinates of $\triangle STR$ if the figure is rotated 90° clockwise about the origin. Then graph the rotation image.

• **Lesson 16–6** Investigate and draw dilations on a coordinate plane.

Find the coordinates of the dilation image of $\triangle STU$ with a scale factor of 2.

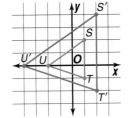

Multiply each coordinate by 2.

$S(1, 2) \longrightarrow S'(2, 4)$

$T(1, -1) \longrightarrow T'(2, -2)$

$U(-2, 0) \longrightarrow U'(-4, 0)$

Graph each figure. Then find the coordinates of the dilation image for the given scale factor k, and graph the dilation image.

26. $\triangle DEF$ with vertices $D(0, 0)$, $E(2, -4)$, and $F(-2, -2)$; $k = \frac{1}{2}$

27. quadrilateral $QRST$ with vertices $Q(0, 1)$, $R(0, 0)$, $S(-1, -1)$, and $T(-2, 1)$; $k = 3$

28. quadrilateral $ABCD$ with vertices $A(-2, 2)$, $B(2, 2)$, $C(2, -1)$, and $D(-2, -1)$; $k = 2$

Applications and Problem Solving

29. **Fund-raiser** The Central Middle School chorale sold hot dogs for $2 and cookies for $1 to raise funds for a new piano. If x represents the number of hot dogs sold and y represents the number of cookies sold, then $x + y = 220$ describes the total number of items sold and $2x + y = 368$ describes the money they made. How many hot dogs and cookies were sold? *(Lesson 16–2)*

30. **School Spirit** Gina is designing a large banner for an after-school pep rally. If the scale factor of her design to the actual banner is $\frac{1}{8}$ and the dimensions of the design are 2 feet by 3 feet, what will be the dimensions of the completed banner? *(Lesson 16–6)*

1. **List** the four types of transformations in this chapter and give a brief description and an example of each.

2. **Describe** three methods for solving a system of equations.

Solve each system of equations by graphing.

3. $x + 4y = 0$
$y = -x - 3$

4. $y = -\frac{1}{2}x - 2$
$x - 2y = 12$

5. $y = 5x + 1$
$x + y = 7$

Solve each system of equations by substitution or elimination.

6. $-x + y = 2$
$3x - 2y = 5$

7. $x - 4y = 1$
$3x + 4y = 7$

8. $2x + 2y = -18$
$6x - y = 2$

Graph each figure. Then find the coordinates of the vertices after the given transformation and graph the image.

Figure	Vertices	Transformation
9. $\triangle PQR$	$P(-2, 0)$, $Q(0, -1)$, $R(-3, -3)$	reflected over the y-axis
10. quadrilateral $HIJK$	$H(-2, 4)$, $I(-1, 2)$, $J(-2, 1)$, $K(-4, 2)$	rotated 90° clockwise about the origin
11. $\triangle ABC$	$A(-2, 3)$, $B(-1, 2)$, $C(-4, 1)$	translated by $(5, -2)$
12. square $WXYZ$	$W(1, 2)$, $X(3, 2)$, $Y(3, 0)$, $Z(1, 0)$	reflected over the x-axis
13. $\triangle EFG$	$E(-1, 0)$, $F(0, -3)$, $G(-3, -1)$	rotated 90° counterclockwise about the origin

14. Find the coordinates of the dilation image for a scale factor of 2. Then graph the dilation image.

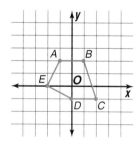

15. **Landscaping** Mr. Collins is landscaping his yard. He plans to move a storage shed from the side of his house to the corner of the yard. Suppose the vertices at the corners of the base of the shed are $S(32, 35)$, $H(42, 35)$, $E(42, 27)$, and $D(32, 27)$. If the shed is moved 15 feet across the yard and 81 feet back, the translation can be given by $(15, 81)$. Find the coordinates of the vertices after the shed is moved.

16. **Office Equipment** Most copy machines can reduce and enlarge images. It is necessary to reduce a rectangular image from 16 cm by 20 cm to 12 cm by 15 cm. What scale factor should be used?

Systems of Equations and Polynomial Problems

Standardized tests often include problems on systems of equations. You usually can add or subtract the equations to solve them.

The ACT and SAT also contain several problems that ask you to simplify rational and polynomial expressions.

Test-Taking Tip
Memorize these polynomial relationships. $a^2 - b^2 = (a + b)(a - b)$ $a^2 + 2ab + b^2 = (a + b)^2$ $a^2 - 2ab + b^2 = (a - b)^2$

Example 1

At the school store, Zoe spent $13.05 for 2 pens and 3 notebooks, and Kenji spent $9.94 for 5 pens and 1 notebook. Which system of equations would allow you to determine the cost of each pen and each notebook?

A $2x + 3y = 13.05$
$5x + y = 9.94$

B $3x + 2y = 13.05$
$5x + y = 9.94$

C $2x + 3y = 9.94$
$5x + y = 13.05$

D $2x + 3y = 9.94$
$5x + y = 13.05$

Hint Write the equations and then look to see which answer choice matches your equations.

Solution Since the answer choices use the variables x and y, let x represent the cost of a pen, and let y represent the cost of a notebook.

Translate Zoe's purchase into an equation.

cost of 2 pens	plus	cost of 3 notebooks	equals	total cost
$2x$	$+$	$3y$	$=$	13.05

Translate Kenji's purchase into an equation.

cost of 5 pens	plus	cost of 1 notebook	equals	total cost
$5x$	$+$	y	$=$	9.94

Compare your equations to the choices. The answer is A.

Example 2

If $4x + 2y = 24$ and $\frac{7y}{2x} = 7$, then $x = ?$

Hint Simplify equations before solving.

Solution There are two equations and two variables, so this is a system of equations. First simplify the equations, if possible. Start with the first equation. Divide each side by 2.

$$4x + 2y = 24 \boxed{\div 2} \quad 2x + y = 12$$

Now solve the second equation for y.

$\frac{7y}{2x} = 7$	*Original equation*
$2x \cdot \frac{7y}{2x} = 2x \cdot 7$	*Multiply each side by $2x$.*
$7y = 14x$	*Simplify.*
$\frac{7y}{7} = \frac{14x}{7}$	*Divide each side by 7.*
$y = 2x$	*Simplify.*

You need to find the value of x. Substitute $2x$ for y in the first equation.

$2x + y = 12$	*Original equation*
$2x + 2x = 12$	*Replace y with $2x$.*
$4x = 12$	*Add like terms.*
$\frac{4x}{4} = \frac{12}{4}$	*Divide each side by 4.*
$x = 3$	*Simplify.*

The answer is 3.

Preparing for Standardized Tests
For test-taking strategies and more practice,
see pages 766–781.

After you work each problem, record your answer on the answer sheet provided or on a sheet of paper.

Multiple Choice

1. Rachel has $100 in her savings account and deposits an additional $25 per week. Nina has $360 in her account and is saving $5 per week. After how many weeks will the girls have the same amount? *(Algebra Review)*

Ⓐ 10 Ⓑ 11 Ⓒ 12 Ⓓ 13

2. If $\triangle WXY$ is translated 5 units right and 3 units up to become $\triangle W'X'Y'$, where is Y'? *(Lesson 16–3)*

Ⓐ (6, 3) Ⓑ (7, 3)
Ⓒ (7, 4) Ⓓ (3, 7)

3. For all $y \neq 3$, $\dfrac{y^2 - 9}{3y - 9} = ?$ *(Algebra Review)*

Ⓐ y Ⓑ $y + 1$ Ⓒ $\dfrac{y}{3}$ Ⓓ $\dfrac{y + 3}{3}$

4. What is the mode of the ages? *(Statistics Review)*

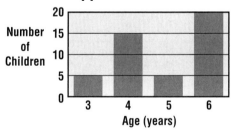

Puppet Show Attendance

Ⓐ 3 Ⓑ 4 Ⓒ 5 Ⓓ 6

5. If $\square ABCD$ is reflected over the y-axis to become $\square A'B'C'D'$, what are the coordinates of C'? *(Lesson 16–4)*

Ⓐ (1, −1)
Ⓑ (−1, −1)
Ⓒ (1, 1)
Ⓓ (−1, 1)

6. $(10x^4 - x^2 + 2x - 8) - (3x^4 + 3x^3 + 2x + 8) =$ *(Algebra Review)*

Ⓐ $7x^4 - 3x^3 - x^2 - 16$
Ⓑ $7x^4 - 4x^2 - 16$
Ⓒ $7x^4 + 3x^3 - x^2 + 4x$
Ⓓ $7x^4 + 2x^2 + 4x$
Ⓔ $13x^4 - 3x^3 + x^2 + 4x$

7. A two-digit number is 7 times its unit digit. If 18 is added to the number, its digits are reversed. Find the number. *(Algebra Review)*

Ⓐ 24 Ⓑ 30 Ⓒ 32 Ⓓ 35

8. What is the area of the triangle in terms of the radius r of the circle? *(Lesson 14–1)*

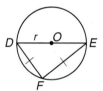

Note: Figure not drawn to scale.

Ⓐ $\frac{1}{2}r^2$ units^2

Ⓑ r^2 units^2

Ⓒ $\frac{\pi r^2}{2}$ units^2

Ⓓ $2r^2$ units^2

Grid In

9. Grid-In The average of x and y is 100, and the ratio of x to y is 3 to 2. What is the value of $x - y$? *(Statistics Review)*

Extended Response

10. A student group is taking a field trip on a bus that holds at most 40 people. Student tickets cost $4, and chaperone tickets cost $7. The group has $196. *(Algebra Review)*

Part A Write a system of two inequalities to find the number of students s and number of chaperones c that can go on the trip.

Part B Graph the inequalities in the first quadrant. Label the region that is the solution. Give one example of a solution.

Student Handbook

Algebra Review

Evaluating Expressions

When evaluating expressions, use the order of operations.

Order of Operations	1. Simplify the expressions inside grouping symbols, such as parentheses and brackets, and as indicated by fraction bars.
	2. Evaluate all powers.
	3. Do all multiplications and divisions from left to right.
	4. Do all additions and subtractions from left to right.

Examples

1 Evaluate $4(1 + 5)^2 \div 8$.

$$
\begin{aligned}
4(1 + 5)^2 \div 8 &= 4(6)^2 \div 8 && \text{Add 1 and 5.} \\
&= 4(36) \div 8 && 6^2 = 36 \\
&= 144 \div 8 && \text{Multiply 4 and 36.} \\
&= 18 && \text{Divide 144 by 8.}
\end{aligned}
$$

2 Evaluate $\dfrac{a^2 - b^2}{2 + c}$ if $a = 6$, $b = 4$, and $c = \dfrac{1}{2}$.

$$
\begin{aligned}
\frac{a^2 - b^2}{2 + c} &= \frac{6^2 - 4^2}{2 + \frac{1}{2}} && \text{Replace } a \text{ with 6, } b \text{ with 4, and } c \text{ with } \tfrac{1}{2}. \\
&= \frac{36 - 16}{2 + \frac{1}{2}} && \text{Evaluate all powers.} \\
&= \frac{20}{2\frac{1}{2}} && \text{Evaluate the numerator and denominator separately.} \\
&= 8 && \text{Divide 20 by } 2\tfrac{1}{2}.
\end{aligned}
$$

Evaluate each expression.

1. $3 + 12 + 9$

2. $3 \cdot 5 \cdot 12$

3. $7^2 - 6(2)$

4. $9^2 - 4(9) + 6$

5. $5(4)^2 \div 20 + 9^2$

6. $6(15^2 - 24) - 6(8)$

7. $\dfrac{13 + 12}{5}$

8. $\dfrac{5^3 + 19}{16 - 2(4)}$

9. $\dfrac{3^2 - 2^3}{7 + 2(4)}$

Evaluate each expression if $a = 2$, $b = 7$, $c = 10$, and $d = 14$.

10. $a + c$

11. bd

12. $b(d - a) + 6$

13. $d^3 - b^3$

14. $\dfrac{c}{5a}$

15. $\dfrac{c - 3a}{2d}$

16. $\dfrac{(c + a)^2}{b + 2}$

17. $\dfrac{d^2 - c^2}{(d - c)^2}$

18. $\dfrac{4\left(b^2 + \frac{a}{2}\right)}{2c} + d$

Operations with Integers

- The sum of two positive integers is positive.
- The sum of two negative integers is negative.
- The sign of the sum of a positive integer and a negative integer matches the integer with the greater absolute value.
- To subtract an integer, add its opposite.

Examples ❶ **Solve $y = 7 + (-4)$.**

$y = 7 + (-4)$ $|7| > |-4|$, *so the sum is positive.*
The difference of 7 and 4 is 3, so $y = 3$.

❷ **Solve $-4 - (-2) = t$.**

$-4 - (-2) = t$ *To subtract –2, add 2.*
$-4 + 2 = t$ $|-4| > |2|$, *so the sum is negative.*
The difference of 4 and 2 is 2, so $t = -2$.

- The product and quotient of two positive integers is positive.
- The product and quotient of two negative integers is positive.
- The product and quotient of a positive integer and a negative integer is negative.

Examples ❸ **Solve $d = (-7)(-2)$.**

$d = (-7)(-2)$ *The product is positive.*
$d = 14$

❹ **Solve $\frac{56}{-7} = z$.**

$\frac{56}{-7} = z$ *The quotient is negative.*
$-8 = z$

Solve each equation.

1. $-5 + (-8) = x$
2. $-4 + 9 = s$
3. $v = -5 + 5$
4. $6 + 6 = a$
5. $47 + (-29) = y$
6. $d = 82 + (-14) + (-35)$
7. $5 - (-6) = p$
8. $c = -9 - 3$
9. $90 - 43 = g$
10. $-23 - 45 = z$
11. $28 - (-14) = k$
12. $w = 3 - 9$
13. $h = (-8)(-4)$
14. $(-3)5 = j$
15. $b = 8(-9)$
16. $\ell = 8(6)$
17. $(-7)(-2) = n$
18. $(-1)(45)(-45) = t$
19. $\frac{-64}{8} = u$
20. $m = \frac{-24}{-6}$
21. $\frac{42}{6} = f$
22. $r = \frac{72}{-8}$
23. $\frac{992}{-32} = q$
24. $a = \frac{189}{9}$

Operations with Decimals

- When adding or subtracting decimals, align the decimal points. You may want to insert zeros to help align the columns. Then add or subtract.

Examples **1** Solve $57.5 + 7.94 = m$.

$$
\begin{array}{r}
57.50 \\
+\ 7.94 \\
\hline
65.44
\end{array}
$$ *Annex zeros to align the columns.*

2 Solve $8 - 3.49 = n$.

$$
\begin{array}{r}
8.00 \\
-\ 3.49 \\
\hline
4.51
\end{array}
$$

- When multiplying decimals, count the number of decimal places in each number. Then find the sum of these two numbers. The product should have the same number of decimal places as this sum.
- When dividing decimals, move the decimal point in the divisor to the right. Then move the decimal point in the dividend the same number of places. Align the decimal point in the quotient with the decimal point in the dividend.

Examples **3** Solve $2.1(0.59) = w$.

$$
\begin{array}{r}
2.1 \\
\times\ 0.59 \\
\hline
189 \\
105 \\
\hline
1.239
\end{array}
$$ *1 decimal place*
2 decimal places

3 decimal places

4 Solve $m = 15.54 \div 2.1$.

$$
\begin{array}{r}
7.4 \\
2.1\overline{)15.54} \\
14\ 7 \\
\hline
8\ 4 \\
8\ 4 \\
\hline
0
\end{array}
$$ *Move each decimal point 1 place.*

Solve each equation.

1. $14.75 + 0.18 = k$
2. $y = -12 + (-9.6)$
3. $c = 9.8 + (-2.5)$

4. $-12.5 + 20.13 = w$
5. $-0.47 + 0.62 = h$
6. $0.2 + 6.51 + 2.03 = a$

7. $12.01 - 0.83 = s$
8. $66.4 - 5.28 = d$
9. $-0.17 - (-14.6) = g$

10. $1.2 - 6.73 = j$
11. $-4.23 - 2.47 = \ell$
12. $m = 10 - 13.46$

13. $b = 108(0.9)$
14. $r = -67(5.89)$
15. $-4.07(-1.95) = q$

16. $627(-0.14) = n$
17. $p = 59.8(100.23)$
18. $t = 1.21(0.47)(-9.3)$

19. $-6.25 \div 5 = x$
20. $4.72 \div 0.8 = v$
21. $-7.02 \div (-1.08) = f$

22. $u = \dfrac{-81.4}{-37}$
23. $z = \dfrac{-15.54}{2.1}$
24. $a = \dfrac{9374.4}{100.8}$

Operations with Fractions

- To add or subtract fractions with like denominators, add or subtract the numerators.
- To add or subtract fractions with unlike denominators, find the least common denominator (LCD), rewrite each fraction with the LCD, and add or subtract the numerators.

Examples ❶ Solve $y = \frac{6}{7} + \frac{3}{7}$.

$$
\begin{array}{r}
\frac{6}{7} \\
+ \frac{3}{7} \\
\hline
\frac{9}{7} = 1\frac{2}{7}
\end{array}
$$

❷ Solve $7 - 1\frac{4}{5} = a$.

$$
\begin{array}{r}
7 \quad \rightarrow \quad 6\frac{5}{5} \\
- 1\frac{4}{5} \quad \rightarrow \quad - 1\frac{4}{5} \\
\hline
5\frac{1}{5}
\end{array}
$$

- To multiply fractions, multiply the numerators and multiply the denominators. Then simplify as necessary.
- To divide fractions, multiply by the reciprocal of the second fraction.

Examples ❸ Solve $1\frac{2}{3}\left(3\frac{5}{8}\right) = w$.

$$1\frac{2}{3} = \frac{1 \cdot 3 + 2}{3} \text{ or } \frac{5}{3}$$

$$3\frac{5}{8} = \frac{3 \cdot 8 + 5}{8} \text{ or } \frac{29}{8}$$

$$\frac{5}{3} \cdot \frac{29}{8} = w$$

$$\frac{145}{24} = w \quad \textit{Multiply the numerators and the denominators.}$$

❹ Solve $n = \frac{3}{5} \div \frac{6}{7}$.

$$n = \frac{3}{5} \div \frac{6}{7}$$

$$= \frac{3}{5} \cdot \frac{7}{6} \quad \textit{Multiply by the reciprocal.}$$

$$= \frac{21}{30} \quad \textit{Multiply the numerators and the denominators.}$$

$$= \frac{7}{10} \quad \textit{The LCD is 3.}$$

Solve each equation.

1. $p = \frac{7}{12} + \frac{4}{12}$

2. $n = \frac{3}{16} + \frac{7}{12}$

3. $f = -\frac{3}{5} + \left(-3\frac{1}{4}\right)$

4. $b = \frac{5}{7} - \frac{3}{7}$

5. $q = \frac{1}{12} - \left(-\frac{7}{12}\right)$

6. $a = 1\frac{1}{2} - \left(\frac{3}{4}\right)$

7. $h = 5\frac{11}{20} + 4\frac{7}{12}$

8. $y = 9\frac{2}{7} - 5\frac{5}{6}$

9. $j = 7\frac{5}{6} + \left(-8\frac{7}{8}\right)$

10. $t = \frac{3}{2}\left(-\frac{4}{9}\right)$

11. $c = \frac{10}{33}\left(4\frac{2}{5}\right)$

12. $g = 4\frac{1}{4}\left(2\frac{1}{3}\right)$

13. $s = -\frac{1}{2} \div \frac{1}{3}$

14. $v = -\frac{11}{7} \div 1\frac{2}{7}$

15. $k = -6\frac{1}{7} \div \frac{4}{21}$

16. $m = -7\frac{3}{8}\left(-9\frac{1}{2}\right)$

17. $d = 3\frac{3}{4} \div 3\frac{4}{7}$

18. $\ell = -10\frac{1}{5} \div 5\frac{2}{5}$

Solving One-Step Equations

To solve equations involving subtraction or addition, add the same number to or subtract the same number from each side of the equation.

Examples   **1** Solve $k + 18 = -9$.

$$k + 18 = -9$$
$$k + 18 - 18 = -9 - 18 \quad \text{\textit{Subtract 18 from each side.}}$$
$$k = -27$$

2 Solve $c - 21 = 40$.

$$c - 21 = 40$$
$$c - 21 + 21 = 40 + 21 \quad \text{\textit{Add 21 to each side.}}$$
$$c = 61$$

To solve equations involving division or multiplication, multiply or divide each side of the equation by the same number.

Examples   **3** Solve $-7t = -98$.

$$-7t = -98$$
$$\frac{-7t}{-7} = \frac{-98}{-7} \quad \text{\textit{Divide each side by} } -7.$$
$$t = 14$$

4 Solve $\frac{y}{6} = -3$.

$$\frac{y}{6} = -3$$
$$6\left(\frac{y}{6}\right) = 6(-3) \quad \text{\textit{Multiply each side by 6.}}$$
$$y = -18$$

Solve each equation. Check your solution.

1. $k - 17 = 40$

2. $g - 11 = -15$

3. $-40 - s = -9$

4. $-34 = -5 - r$

5. $-6 = a + (-7)$

6. $z + (-9) = 7$

7. $15 = n + 18$

8. $-17 = c + 4$

9. $x + 5 = 2$

10. $v - (-12) = 10$

11. $q - (-6) = 2$

12. $d - (-15) = -12$

13. $81 = -9m$

14. $2f = -100$

15. $7\ell = -49$

16. $-3h = -51$

17. $-41t = -1476$

18. $-1815 = -33u$

19. $5 = \frac{p}{-9}$

20. $\frac{b}{-8} = -4$

21. $\frac{1}{4}j = -16$

22. $\frac{v}{8} = -8$

23. $-\frac{5}{2}y = 15$

24. $\frac{w}{-21} = -14$

Solving Multi-Step Equations

When solving some equations, you must perform more than one operation on both sides. First, determine what operations have been done to the variable. Then undo these operations in the reverse order.

Examples

1 Solve $5x + 3 = 23$.

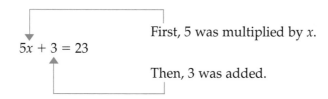

$5x + 3 = 23$ First, 5 was multiplied by x.

Then, 3 was added.

To solve, undo the operations in reverse order.

$5x + 3 - 3 = 23 - 3$ *Because 3 was added, subtract 3 from each side.*

$5x = 20$

$\dfrac{5x}{5} = \dfrac{20}{5}$ *Because 5 was multiplied, divide each side by 5.*

$x = 4$

2 Solve $10 = 7 - \dfrac{r}{2}$.

$10 - 7 = 7 - \dfrac{r}{2} - 7$ *Because 7 was added, subtract 7 from each side.*

$3 = -\dfrac{r}{2}$

$-2(3) = -2\left(-\dfrac{r}{2}\right)$ *Because $-\dfrac{r}{2}$ means r divided by -2, multiply each side by -2.*

$-6 = r$

Solve each equation. Check your solution.

1. $2x + 3 = 11$

2. $7f - 2 = -9$

3. $3y - 5 = -8$

4. $5n - 2 = 8$

5. $14g - 8 = 34$

6. $5t + 16 = 51$

7. $\dfrac{m}{2} - 4 = 9$

8. $\dfrac{k}{8} + 9 = -3$

9. $\dfrac{3}{4}c + 1 = 10$

10. $\dfrac{z}{-5} + 3 = -13$

11. $\dfrac{7u}{8} - 4 = 10$

12. $8 + \dfrac{3r}{12} = 13$

13. $-8(a - 20) = -96$

14. $75 = 5(-4 + 2w)$

15. $4(9q + 3) = -6$

16. $4(e + 7) = 2(9)$

17. $5(3 - \ell) - 7 = 8$

18. $6(p + 3) - 2(p - 1) = 16$

19. $\dfrac{j - 8}{-6} = 7$

20. $6 = \dfrac{h - 12}{14}$

21. $-4 = \dfrac{7d + 1}{-8}$

22. $\dfrac{4b + 8}{-2} = 10$

23. $\dfrac{v + 12}{-4} = 5$

24. $-14 = \dfrac{s + 12}{-6}$

Solving Equations with the Variable on Both Sides

When an equation has the variable on both sides, the goal is to write equivalent equations until the variable is alone on one side.

Examples

❶ Solve $4a - 25 = 6a + 51$.

$$4a - 25 = 6a + 51$$
$$4a - 25 - 6a = 6a + 51 - 6a \quad \textit{Subtract 6a from each side.}$$
$$-2a - 25 = 51$$
$$-2a - 25 + 25 = 51 + 25 \quad \textit{Add 25 to each side.}$$
$$-2a = 76$$
$$\frac{-2a}{-2} = \frac{76}{-2} \quad \textit{Divide each side by } -2.$$
$$a = -38$$

❷ Solve $8p - 5(p + 3) = 3(7p + 1)$.

$$8p - 5(p + 3) = 3(7p + 1)$$
$$8p - 5p - 15 = 21p + 3 \quad \textit{Use the Distributive Property.}$$
$$3p - 15 = 21p + 3 \quad \textit{Combine like terms.}$$
$$3p - 15 - 3p = 21p + 3 - 3p \quad \textit{Subtract 3p from each side.}$$
$$-15 = 18p + 3$$
$$-15 - 3 = 18p + 3 - 3 \quad \textit{Subtract 3 from each side.}$$
$$-18 = 18p$$
$$\frac{-18}{18} = \frac{18p}{18} \quad \textit{Divide each side by 18.}$$
$$-1 = p$$

Solve each equation. Check your solution.

1. $z = 5z - 28$

2. $-3b = 96 + b$

3. $2f = 3f + 2$

4. $2w + 3 = 5w$

5. $6n - 42 = 4n$

6. $5y = 2y - 12$

7. $21 - j = -87 - 2j$

8. $-5 - 8v = -7v + 21$

9. $6r - 12 = 2r + 36$

10. $4x - 9 = 7x + 12$

11. $6a - 14 = 9a - 5$

12. $8n - 13 = 13 - 8n$

13. $3d + 20 = -7 - 6d$

14. $25c + 17 = 5c - 143$

15. $-45m + 68 = 84m - 61$

16. $4(2k - 1) = -10(k - 5)$

17. $-8(8 + 9g) = 7(-2 - 11g)$

18. $1 - 3b = 2b - 3$

19. $6 + 17h = -7 - 9h$

20. $4p - 25 = 6p - 50$

21. $2(s - 3) + 5 = 4(s - 1)$

22. $-3(\ell - 8) - 5 = 9(\ell + 2) + 1$

23. $2(q - 8) + 7 = 6(q + 2) - 3q - 19$

Solving Inequalities

Inequalities are sentences that compare two quantities that are not equal.
The symbols below are used in inequalities.

Symbols	Words
$<$	less than
$>$	greater than
$\leq$	less than or equal to
$\geq$	greater than or equal to
$\neq$	not equal to

Inequalities usually have more than one solution.

Examples **1** **Solve $-13u > 143$.**

$-13u > 143$

$\dfrac{-13u}{-13} < \dfrac{143}{-13}$ *Divide each side by -13. Because you are dividing by a negative number, reverse the direction of the inequality.*

$u < -11$

2 **Solve $2x + 7 \leq 13$.**

$2x + 7 \leq 13$

$2x + 7 - 7 \leq 13 - 7$ *Subtract 7 from each side.*

$2x \leq 6$

$\dfrac{2x}{2} \leq \dfrac{6}{2}$ *Divide each side by 2.*

$x \leq 3$

To graph the solution on a number line, draw a bullet at 3.
Then draw an arrow to show all numbers less than or equal to 3.

Solve each inequality. Graph the solution on a number line.

1. $m + 3 < 8$

2. $c + 7 \geq 15$

3. $j + 5 < -8$

4. $a - 5 \leq -2$

5. $d - 14 \geq -9$

6. $x - 4 < 10$

7. $6w > 18$

8. $-7f \leq 63$

9. $-29z < -29$

10. $\dfrac{y}{2} \geq 3$

11. $\dfrac{g}{4} < -6$

12. $-\dfrac{3}{4}n \leq 12$

13. $2t - 1 > 9$

14. $-4\ell - 7 \geq 13$

15. $-1 - 2h < -15$

16. $4(k - 3) \leq 8$

17. $7(2 - v) \leq 5$

18. $5(b + 2) > b - 3(6)$

Lesson 1–1 *(Pages 4–9)* **Find the next three terms of each sequence.**

1. 2, 4, 6, . . .

2. 10, 7, 4, . . .

3. 97, 86, 75, . . .

4. 30, 31, 34, 39, 46, . . .

5. 4, 2, −2, −8, . . .

6. 1, 4, 9, 16, . . .

Draw the next figure in each pattern.

7.

8.

Lesson 1–2 *(Pages 12–17)* **Use the figure to name examples of each term.**

1. a line

2. a ray *not* containing *A*

3. a segment

4. three collinear points

5. a point *not* on $\overleftrightarrow{AD}$

Determine whether each model suggests a point, a line, a ray, a segment or a plane.

6. grain of salt

7. ceiling tile

8. hand of a clock

Draw and label a figure for each situation described.

9. three noncollinear points

10. plane *CAT*

Lesson 1–3 *(Pages 18–23)* **Name all the different lines that can be drawn through each set of points.**

1. *K*   *L*   *M*

2. *W*   *X*   *Z*   *Y*

Name the intersection of each pair of lines.

3. $\overleftrightarrow{XY}$ and $\overleftrightarrow{YZ}$

4. $\overleftrightarrow{HJ}$ and $\overrightarrow{HK}$

Name all the planes that are represented in each figure.

5.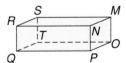

6.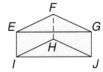

Determine whether each statement is *true* or *false*.

7. Two distinct planes intersect in a line.

8. Three points determine a line.

9. Three noncollinear points determine a plane.

10. Two lines can intersect in a point.

Lesson 1-4 *(Pages 24–28)* Identify the hypothesis and the conclusion of each statement.

1. If a road is 5280 feet long, then it is a mile long.
2. We will play baseball if it is not raining.
3. If I am hungry, then I will eat.

Write two other forms of each statement.

4. An equilateral triangle has three congruent sides.
5. Any purebred dog is not mixed with another type of dog.
6. A quadrilateral has exactly four sides.

Write the converse of each statement.

7. If the race is 5 kilometers, then it is about 3.1 miles.
8. Broccoli is a vegetable.

Lesson 1-5 *(Pages 29–34)* Match the term with the definition that best describes it.

1. compass
2. construction
3. straightedge
4. midpoint
5. optical illusion

a. a point in the middle of a segment
b. geometry tool used for drawing circles and arcs
c. a misleading image
d. object used to draw a straight line
e. a special drawing created using compass and straightedge

Use a straightedge or compass to determine which segment is longer, $\overline{AB}$ or $\overline{CD}$.

6.

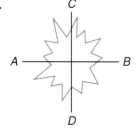

7.

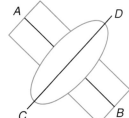

Lesson 1-6 *(Pages 35–41)* Find the perimeter and area of each rectangle.

1.

8 cm
2 cm

2.

6 ft
6 ft

3. $\ell = 4$ in., $w = 8$ in.
4. $\ell = 12$ m, $w = 4.2$ m

Find the area of each parallelogram.

5.

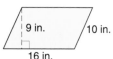
9 in.
10 in.
16 in.

6.
10 m
8 m
6.5 m

7. $b = 15$ mi, $h = 8$ mi
8. $b = 4$ cm, $h = 11$ cm

Lesson 2–1 *(Pages 50–55)* **For each situation, write a real number with ten digits to the right of the decimal point.**

1. a rational number between 0 and 1 with a 2-digit repeating pattern

2. a rational number between 5 and 5.4 with a 4-digit repeating pattern

3. an irrational number between 2.5 and 3

4. an irrational number less than –5

Use the number line to find each measure.

5. *AE*　　　　　　　**6.** *GH*　　　　　　　**7.** *EC*　　　　　　　**8.** *BF*

Lesson 2–2 *(Pages 56–61)* **Three segment measures are given. The three points named are collinear. Determine which point is between the other two.**

1. $XY = 25$, $YZ = 22$, $XZ = 47$　　　　　　**2.** $XY = 25$, $YZ = 22$, $XZ = 3$

Refer to the line for Exercises 3–6.

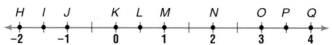

3. If $RS = 8$ and $SV = 22$, find RV.　　　　**4.** If $QT = 24$ and $QV = 40$, find TV.

5. If $SU = 11.2$ and $UW = 12.9$, find SW.　　**6.** If $QR = 5$, $RT = 8$, and $TV = 12$, find QV.

Find the length of each segment in centimeters and in inches.

7. ————————————————　　　　　**8.** ————————————————

9. ——————————————————————————

Lesson 2–3 *(Pages 62–67)* **Use the number line to determine whether each statement is *true* or *false*. Explain your reasoning.**

1. $\overline{JK}$ is congruent to $\overline{MN}$.　　　　　　**2.** $\overline{JM}$ is congruent to $\overline{JH}$.

3. $\overline{HI}$ is congruent to $\overline{PQ}$.　　　　　　　**4.** K is the midpoint of $\overline{LM}$.

5. If $\overline{JK} \cong \overline{OQ}$, then $JK = OQ$.

6. If $\overline{HI} \cong \overline{IJ}$, $\overline{LM} \cong \overline{KL}$, and $\overline{IJ} \cong \overline{LM}$, then $\overline{HI} \cong \overline{KL}$.

Determine whether each statement is *true* or *false*. Explain your reasoning.

7. If $\overline{MN} \cong \overline{RS}$, then $MN = RS$.　　　　**8.** If $\overline{WX} \cong \overline{YZ}$, then $\overline{YX} \cong \overline{WZ}$.

9. If $\overline{GH} \cong \overline{HI}$, then H is the midpoint of $\overline{GI}$.

10. The point at which a line bisects a segment is called its midpoint.

11. A point, ray, line, segment, and plane can bisect a segment.

Lesson 2-4 *(Pages 68–73)* **Draw and label a coordinate plane on a piece of grid paper. Then graph and label each point.**

1. $A(2, -3)$ **2.** $B(0, 4)$ **3.** $C(-4, -2)$ **4.** $D(5, 1)$

Refer to the coordinate plane at the right.
Name the ordered pair for each point.

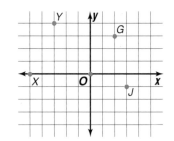

5. G

6. J

7. X

8. Y

9. origin

Lesson 2-5 *(Pages 76–81)* **Use the number line to find the coordinate of the midpoint of each segment.**

1. $\overline{UW}$ **2.** $\overline{WY}$ **3.** $\overline{UX}$ **4.** $\overline{YU}$

The coordinates of the endpoints of a segment are given. Find the coordinates of the midpoint of each segment.

5. $(0, 0); (4, -6)$ **6.** $(5, -8); (13, -2)$

7. $(-3, 2); (5, -3)$ **8.** $(g, h); (j, k)$

The coordinates of one endpoint A and the midpoint M of a segment are given. Find the coordinates of the other endpoint B.

9. $A(8, 10); M(4, 5)$ **10.** $A(-2, 7); M(3, 3)$

Lesson 3-1 *(Pages 90–95)* **Name each angle in four ways. Identify its vertex and its sides.**

1.

2.

3.

Name all the angles having X as their vertex.

4.

5.

6.

Tell whether each point is in the *interior*, *exterior*, or *on* the angle.

7.

8.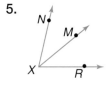

9.

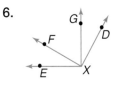

Lesson 3–2 *(Pages 96–101)* **Use a protractor to find the measure of each angle. Then classify each angle as** *acute*, *obtuse*, **or** *right*.

1. $m\angle AXB$

2. $m\angle CXE$

3. $m\angle DXA$

4. $m\angle EXD$

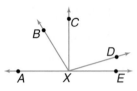

Use a protractor to draw an angle having each measurement. Then classify each angle as *acute*, *obtuse*, **or** *right*.

5. 40° angle

6. 120° angle

7. 90° angle

Given the measure of each angle, solve for *x*.

8. $m\angle B = 122$

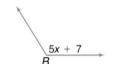

9. $m\angle A = 65$

Lesson 3–3 *(Pages 104–109)* **Refer to the figure at the right.**

1. If $m\angle KPJ = 32$ and $m\angle JPH = 58$, find $m\angle KPH$.

2. If $\angle HPM$ is a right angle and $m\angle LPM = 41$, find $m\angle LPH$.

3. Find $m\angle KPL$ if $m\angle KPJ = 28$, $m\angle JPH = 56$, and $m\angle HPL = 45$.

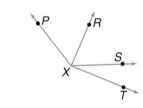

Refer to the figure at the right.

4. If $\overrightarrow{XR}$ bisects $\angle PXS$ and $m\angle PXS = 92$, find $m\angle PXR$.

5. If $m\angle RXS = 55$ and $\overrightarrow{XS}$ bisects $\angle RXT$, find $m\angle RXT$.

6. If $m\angle PXR = 23$, $m\angle RXS = 57$, and $m\angle SXT = 20$, find $m\angle PXT$.

7. If $m\angle PXR = 2x$, $m\angle RXS = 4x$, $m\angle SXT = 3x - 8$, and $m\angle PXT = 127$, find x.

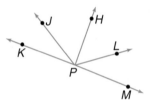

Lesson 3–4 *(Pages 110–114)* **Use the terms** *adjacent angles, linear pair,* **or** *neither* **to describe angles 1 and 2 in as many ways as possible.**

1.

2.

3.

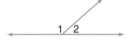

In the figure, $\overrightarrow{BF}$ **and** $\overrightarrow{BA}$ **and** $\overrightarrow{BC}$ **and** $\overrightarrow{BD}$ **are opposite rays.**

4. Which angle forms a linear pair with $\angle CBF$?

5. Name two angles adjacent to $\angle ABE$.

6. Name two angles that do not form a linear pair.

7. Do $\angle EBD$ and $\angle DBF$ form a linear pair? Explain.

Lesson 3–5 *(Pages 116–121)* **Refer to the figure at the right.**

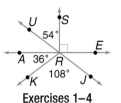

Exercises 1–4

1. Name a pair of adjacent complementary angles.
2. Name a pair of nonadjacent supplementary angles.
3. Find the measure of the angle that is supplementary to $\angle ARU$.
4. Find the measure of $\angle ERJ$.

5. Angles A and B are complementary. If $m\angle A$ is 5 times the value of $m\angle B$, find $m\angle B$.

6. Angles C and D are supplementary. If $m\angle C$ is 15 more than twice the value of $m\angle D$, find $m\angle C$.

7. Angles E and F form a linear pair. Find x if $m\angle E = 2x$ and $m\angle F = 4x + 18$.

8. Angles G and H are two adjacent angles that form a right angle. Find x if $m\angle G = 3x + 5$ and $m\angle H = 4x - 6$.

Lesson 3–6 *(Pages 122–127)* **Find the value of x in each figure.**

1.

2.

3.

4.

5.

6.

7. If $\angle 1 \cong \angle 2$, what is the measure of an angle that is complementary to $\angle 2$?

8. If $\angle 3$ is complementary to $\angle 4$ and $\angle 4$ is complementary to $\angle 5$, find $m\angle 3$ and $m\angle 5$.

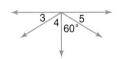

Lesson 3–7 *(Pages 128–133)* $\overline{HJ} \perp \overrightarrow{IO}$, $\overline{NP} \perp \overrightarrow{IO}$, and O is the midpoint of $\overline{NP}$. **Determine whether each of the following is *true* or *false*.**

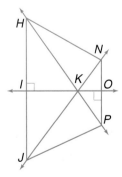

1. $\overline{HI} \perp \overleftrightarrow{KO}$
2. $\overline{NP} \perp \overline{HJ}$
3. $\angle HKN \cong \angle JKO$
4. $\angle HKJ \cong \angle NKP$
5. $\overline{NO} \cong \overline{OP}$
6. $\angle HIK + \angle NOK = 180$

7. $\angle JKO$ and $\angle JKI$ are supplementary angles.
8. $\angle POK$ and $\angle NOK$ are complementary angles.
9. Name four right angles.
10. If $\overline{HI} \perp \overleftrightarrow{IO}$ and $m\angle HIO = 4x + 10$, solve for x.

Lesson 4–1 *(Pages 142–147)* **Describe each pair of segments in the prism as** *parallel, skew,* **or** *intersecting.*

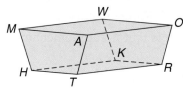

1. $\overline{MA}, \overline{WO}$
2. $\overline{WM}, \overline{HT}$
3. $\overline{OR}, \overline{AT}$
4. $\overline{AO}, \overline{TA}$
5. $\overline{WO}, \overline{AT}$

Name the parts of the figure shown.

6. all pairs of parallel planes
7. all segments skew to $\overline{MH}$
8. all segments parallel to $\overline{TR}$

Lesson 4–2 *(Pages 148–153)* **Identify each pair of angles as** *alternate interior, alternate exterior, consecutive interior,* **or** *vertical.*

1. $\angle 1$ and $\angle 14$
2. $\angle 10$ and $\angle 15$
3. $\angle 5$ and $\angle 2$
4. $\angle 6$ and $\angle 7$
5. $\angle 9$ and $\angle 16$

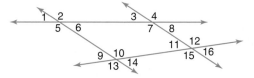

If $m\angle 1 = 112$, find the measure of each angle. Give a reason for each answer.

6. $\angle 2$ 7. $\angle 3$
8. $\angle 5$ 9. $\angle 7$

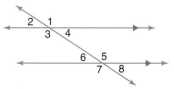

Lesson 4–3 *(Pages 156–161)* **In the figure, $x \parallel y$. Name all angles congruent to the given angle. Give a reason for each answer.**

1. $\angle 1$
2. $\angle 10$
3. $\angle 6$
4. $\angle 11$

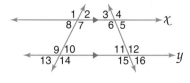

Find the measure of each numbered angle.

5.

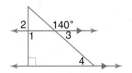

6.

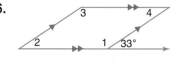

7. If $m\angle 1 = 3x + 10$ and $m\angle 2 = 2x + 5$, find x, $m\angle 1$ and $m\angle 2$.

Lesson 4–4 *(Pages 162–167)* Find x so that $a \parallel b$.

1.

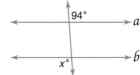

2.

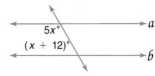

3.

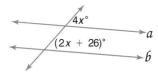

4.

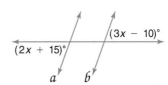

Name the pairs of parallel lines or segments.

5.

6.

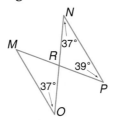

7.

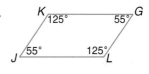

Lesson 4–5 *(Pages 168–173)* Find the slope of each line.

1.

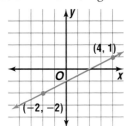

2.

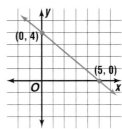

3. the line through points at $(2, -3)$ and $(4, 7)$

4. the line through points at $(5, 8)$ and $(0, -4)$

Given each set of points, determine if $\overleftrightarrow{AB}$ and $\overleftrightarrow{CD}$ are *parallel, perpendicular,* or *neither.*

5. $A(6, 8)$, $B(4, 0)$, $C(-3, 4)$, $D(-7, 5)$

6. $A(0, 8)$, $B(4, 0)$, $C(2, 1)$, $D(3, 3)$

7. $A(3, 6)$, $B(0, 4)$, $C(0, 5)$, $D(2, 2)$

Lesson 4–6 *(Pages 174–179)* Name the slope and y-intercept of the graph of each equation.

1. $y = 4x - 2$

2. $2x + 3y = 12$

3. $x = 4$

4. $4y = 3x + 8$

5. $y = 8$

6. $\frac{1}{2}y + 2x = 5$

Graph each equation using the slope and y-intercept.

7. $y = 3x - 2$

8. $4x - 3y = 6$

9. $\frac{1}{2}x + \frac{1}{4}y = -1$

Write an equation of the line satisfying the given conditions.

10. slope $= 5$, goes through the point at $(-2, 3)$

11. parallel to the graph of $y = 2x + 9$, passes through the point at $(4, 1)$

12. passes through the point at $(4, 2)$ and perpendicular to the graph of $y = -4x + 1$

Lesson 5–1 *(Pages 188–192)* **Classify each triangle by its angles and by its sides.**

1.

2.
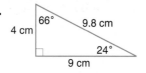

3.
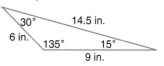

Make a sketch of each triangle. If it is not possible to sketch the figure, write *not possible*.

4. acute, equilateral

5. right, isosceles

6. obtuse, not scalene

7. right, obtuse

Lesson 5–2 *(Pages 193–197)* **Find the value of each variable.**

1.

2.

3.

4.

5.

6.
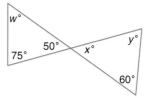

Find the measure of each angle in each triangle.

7.
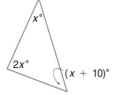

8.

9.
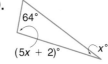

Lesson 5–3 *(Pages 198–202)* **Identify each motion as a *translation*, *reflection*, or *rotation*.**

1.

2.

3.

4.

In the figure at the right, $\triangle HIJ \rightarrow \triangle RST$.

5. Which angle corresponds to $\angle I$? $\angle T$?

6. Name the image of point P and point K.

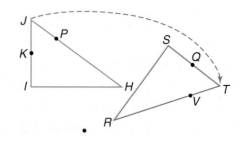

Lesson 5–4 *(Pages 203–207)* For each pair of congruent triangles, name the congruent angles and sides. Then draw the triangles, using arcs and slash marks to show the congruent angles and sides.

1. 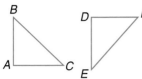 △BAC ≅ △FDE   2. 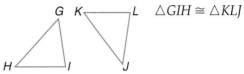 △GIH ≅ △KLJ

Complete each congruence statement.

3.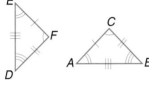

△EFD ≅ △ _____

4.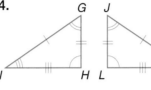

△IHG ≅ △ _____

5.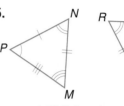

△PNM ≅ △ _____

6.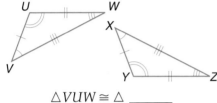

△VUW ≅ △ _____

7.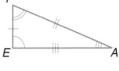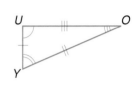

△IEA ≅ △ _____

Lesson 5–5 *(Pages 210–214)* Determine whether each pair of triangles is congruent. If so, write a congruence statement and explain why the triangles are congruent.

1. 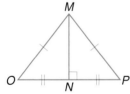   2. 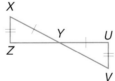   3.   4.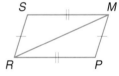

Determine whether the two triangles described are congruent by *SSS, SAS,* or *neither.*

5. $\angle A \cong \angle N, \overline{PM} \cong \overline{CA}, \overline{BA} \cong \overline{MN}$
6. $\overline{AB} \cong \overline{NM}, \overline{BC} \cong \overline{MP}, \overline{AC} \cong \overline{NP}$
7. $\angle B \cong \angle M, \overline{CB} \cong \overline{PM}, \overline{BA} \cong \overline{MN}$

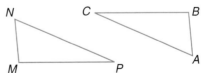

Lesson 5–6 *(Pages 215–219)* Name the additional congruent parts needed so that the triangles are congruent by the postulate or theorem indicated.

1. AAS 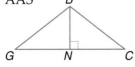   2. AAS 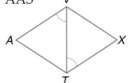   3. K   ASA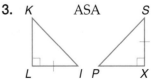

Lesson 6–1 *(Pages 228–233)* In △*EFG*, $\overline{EK}$, $\overline{FJ}$, and $\overline{GH}$ are medians.

1. Find *JX* if *FX* = 18.
2. If *EJ* = 6, find *EG*.
3. What is *HF* if *EF* = 14?
4. What is *EX* if *XK* = 6?
5. If *HG* = 15, find *HX*.

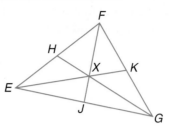

6. In △*ABC*, $\overline{AG}$, $\overline{BH}$, and $\overline{CI}$ are medians. If *AH* = 4*x* − 5, *HC* = 2*x* + 1, and *BI* = 3*x* − 1, what is *AI*?

7. In △*XYZ*, $\overline{XA}$, $\overline{YB}$, and $\overline{ZC}$ are medians. If *XC* = 7*x*, *YA* = 3*x* + 2, and *CY* = 5*x* + 8, what is *AZ*?

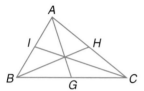

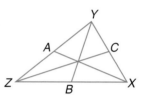

Lesson 6–2 *(Pages 234–239)* For each triangle, tell whether the red segment or line is an *altitude*, a *perpendicular bisector*, *both*, or *neither*.

1.

2.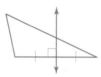

3.

4.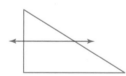

5.

6.

Lesson 6–3 *(Pages 240–243)* In △*LMN*, $\overline{LA}$ bisects ∠*MLN*, $\overline{MB}$ bisects ∠*NML*, and $\overline{NC}$ bisects ∠*LNM*.

1. If *m*∠*MLA* = 30, what is *m*∠*MLN*?
2. If *m*∠*LMN* = 70, what is *m*∠*LMB*?
3. Find *m*∠*MNC*, if *m*∠*LNC* = 25.
4. Find *m*∠*NMB*, if *m*∠*NML* = 88.

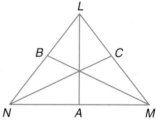

In △*UVW*, $\overline{VX}$ and $\overline{WZ}$ are angle bisectors.

5. If *m*∠*XVW* = 3*x* − 8 and *m*∠*XVU* = 2*x* + 10, find *x*.
6. If *m*∠*UWV* = 8*y* and *m*∠*UWZ* = 3*y* + 5, find *m*∠*VWZ*.
7. If *P* is equidistant from $\overline{UV}$ and $\overline{UW}$, what is $\overline{UP}$ called?

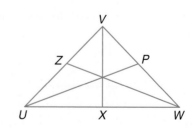

Lesson 6–4 *(Pages 246–250)* **For each triangle, find the values of the variables.**

1.

2.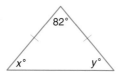

3.

4.

5.
4x − 10
2x + 8

6.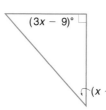
(3x − 9)°
(x + 27)°

7.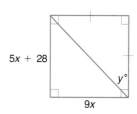
5x + 28
9x
y°

Lesson 6–5 *(Pages 251–255)* **Determine whether each pair of right triangles is congruent by LL, HA, LA, or HL. If it is not possible to prove that they are congruent, write *not possible*.**

1.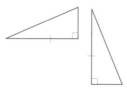

2.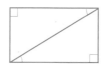

3.

Name the corresponding parts needed to prove the triangles congruent. Then complete the congruence statement and name the theorem used.

4.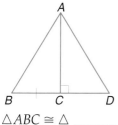
△ABC ≅ △ _____

5.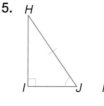
△HIJ ≅ △ _____

6.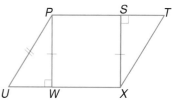
△PUW ≅ △ _____

Lesson 6–6 *(Pages 256–261)* **Find the missing measure in each right triangle. Round to the nearest tenth, if necessary.**

1.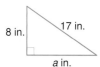
8 in. 17 in.
a in.

2.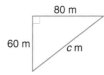
80 m
60 m
c m

3.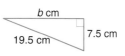
b cm
19.5 cm 7.5 cm

4.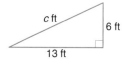
c ft 6 ft
13 ft

If *c* is the measure of the hypotenuse, find each missing measure. Round to the nearest tenth, if necessary.

5. $a = 3$, $b = 4$, $c = ?$

6. $a = 5$, $c = 13$, $b = ?$

7. $b = 10$, $c = 15$, $a = ?$

8. $a = \sqrt{2}$, $b = \sqrt{7}$, $c = ?$

The lengths of three sides of a triangle are given. Determine whether each triangle is a right triangle.

9. 25, 20, 15

10. 1.6, 3.0, 3.4

11. 5, 10, 14

12. 14, 48, 50

Lesson 6–7 *(Pages 262–267)* **Find the distance between each pair of points. Round to the nearest tenth, if necessary.**

1. $Y(0, 8)$, $Z(5, 0)$

2. $U(2, 3)$, $V(-2, 5)$

3. $W(5, -4)$, $X(-9, -2)$

4. Determine whether $\triangle DEF$ with vertices $D(0, 0)$, $E(6, 4)$, and $F(2, -5)$ is a scalene triangle. Explain.

5. Determine whether $\triangle MNO$ with vertices $M(6, 14)$, $N(1, 2)$, and $O(13, -3)$ is an isosceles triangle. Explain.

6. Is $\triangle QRS$ with vertices $Q(-4, -4)$, $R(0, 2)$, and $S(-7, -2)$ a right triangle? Explain.

Lesson 7–1 *(Pages 276–281)*

Exercises 1–8

Replace each ● with <, >, or = to make a true sentence.

1. GI ● KM

2. IP ● MW

3. PW ● GK

4. KW ● EK

Determine if each statement is *true* or *false*.

5. $MP \leq KW$

6. $GM \not\equiv KW$

7. $PI \neq PE$

8. $ME \geq GP$

Lines *AB*, *ED*, and *FG* intersect at *C*. Replace each ● with <, >, or = to make a true sentence.

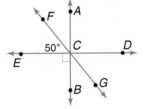

9. $m\angle ECF$ ● $m\angle FCA$

10. $m\angle ACD$ ● $m\angle DCB$

11. $m\angle BCG$ ● $m\angle DCB$

12. $m\angle FCD$ ● $m\angle FCB$

Lesson 7–2 *(Pages 282–287)* **Name the angles.**

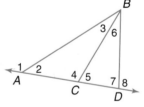

1. an exterior angle of $\triangle ABC$

2. an interior angle of $\triangle CBD$

3. a remote interior angle of $\triangle CBD$ with respect to $\angle ACB$.

Find the measure of each angle.

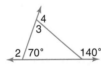

4. $\angle 2$

5. $\angle 3$

6. $\angle 4$

Find the value of x.

7.

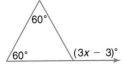

8.

9.

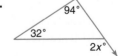

Lesson 7–3 *(Pages 290–295)* List the angles in order from least to greatest measure.

1.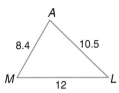

2.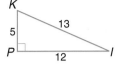

3.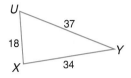

List the sides in order from least to greatest measure.

4.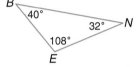

5.

6.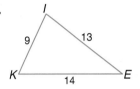

Identify the angle and side with the greatest measure.

7.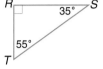

8.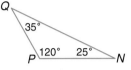

9.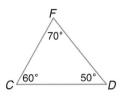

Lesson 7–4 *(Pages 296–301)* Determine if the three numbers can be measures of the sides of a triangle. Write *yes* or *no*. Explain.

1. 12, 11, 17

2. 5, 100, 100

3. 4.7, 9, 4.1

4. 2.3, 12, 12.2

If two sides of a triangle have the following measures, find the range of possible measures for the third side.

5. 12, 15

6. 4, 13

7. 21, 17

8. 20, 34

Lesson 8–1 *(Pages 310–315)* Refer to quadrilaterals *ABCD* and *EFGH*.

1. Name a side that is consecutive with $\overline{AD}$.

2. Name the diagonals of *ABCD*.

3. Name all pairs of nonconsecutive angles in quadrilateral *EFGH*.

4. Name the side opposite $\overline{EF}$.

5. Name the vertex that is opposite *H*.

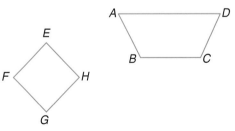

Find the missing measure(s) in each figure.

6.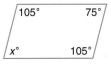

7.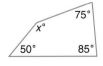

8.

Lesson 8–2 *(Pages 316–321)* Find each measure in parallelogram *ABCD*.

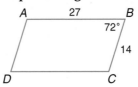

1. $m\angle C$

2. DC

3. $m\angle D$

4. AD

In parallelogram *KLMN*, *LO* = 12 and *OM* = 8. Find each measure.

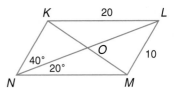

5. OK

6. NL

7. $m\angle KNM$

8. $m\angle NLK$

9. NM

10. NO

Lesson 8–3 *(Pages 322–326)* Determine whether each quadrilateral is a parallelogram. Write *yes* or *no*. If *yes*, give a reason for your answer.

1.

2.

3.

4.

5.

6.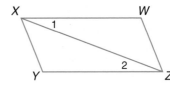

7. In quadrilateral *WXYZ*, $\overline{WX} \cong \overline{ZY}$ and $\overline{WX} \parallel \overline{ZY}$. Show that *WXYZ* is a parallelogram by providing a reason for each step.

   a. $\angle 1 \cong \angle 2$
   b. $\overline{ZX} \cong \overline{ZX}$
   c. $\triangle WXZ \cong \triangle YZX$
   d. $\overline{WZ} \cong \overline{XY}$
   e. *WXYZ* is a parallelogram.

Lesson 8–4 *(Pages 327–332)* Identify each parallelogram as a *rectangle, rhombus, square,* or *none of these.*

1.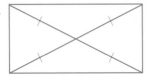

2.

3.

4.

5.

6.

Lesson 8–5 *(Pages 333–339)* For each trapezoid, name the bases, the legs, and the base angles.

1.

2.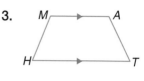

3.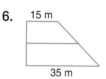

Find the length of the median in each trapezoid.

4.
27 mm
9 mm

5.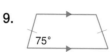
24 in.
48 in.

6. 15 m
35 m

Find the missing angle measures in each isosceles trapezoid.

7.
115°

8. 110°

9. 75°

Lesson 9–1 *(Pages 350–355)* Write each ratio in simplest form.

1. $\frac{4}{12}$

2. $\frac{15}{45}$

3. $\frac{49}{56}$

4. $\frac{81}{18}$

Solve each proportion.

5. $\frac{x}{5} = \frac{21}{35}$

6. $\frac{14}{3} = \frac{x+3}{6}$

7. $\frac{x-1}{7} = \frac{3}{6}$

8. $\frac{2}{3x+1} = \frac{1}{x}$

Lesson 9–2 *(Pages 356–361)* Determine whether each pair of polygons is similar. Justify your answer.

1.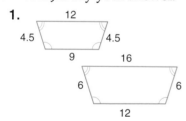
12
4.5
4.5
9
16
6
6
12

2.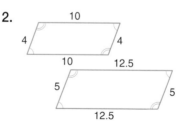
10
4
4
10
12.5
5
5
12.5

3.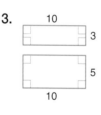
10
3
5
10

Each pair of polygons is similar. Find the values of x and y.

4.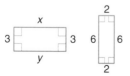
x
3
3
y
2
6
6
2

5.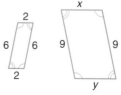
2
6
6
2
x
9
9
y

6.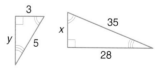
3
y
5
x
35
28

Determine whether each statement is *always*, *sometimes*, or *never* true.

7. Similar polygons have the same shape and same size.

8. In a similar polygon, corresponding sides are proportional.

9. If corresponding angles are congruent, then the figures are similar.

10. Figures with same size and different shapes are similar.

Lesson 9–3 *(Pages 362–367)* **Determine whether each pair of triangles is similar. If so, tell which similarity test is used and complete the statement.**

1.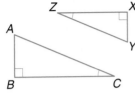

$\triangle ABC \sim \triangle$ __?__

2.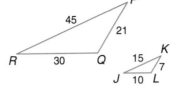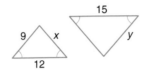

$\triangle JKL \sim \triangle$ __?__

3.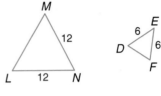

$\triangle LMN \sim \triangle$ __?__

Find the value of each variable.

4.

5.

6.

Lesson 9–4 *(Pages 368–373)* **Complete each proportion.**

1. $\dfrac{EF}{EG} = \dfrac{?}{EH}$

2. $\dfrac{EH}{EI} = \dfrac{GH}{?}$

3. $\dfrac{?}{FE} = \dfrac{HI}{IE}$

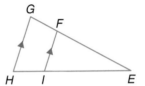

Find the value of each variable.

4.

5.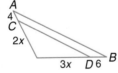

6.

7.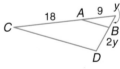

Lesson 9–5 *(Pages 374–379)* **In each figure, determine whether $\overline{AB} \parallel \overline{CD}$.**

1.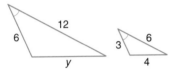

2.

3.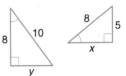

4.

X, *Y*, and *Z* are the midpoints of the sides of $\triangle LMN$. **Complete each statement.**

5. $\overline{XZ} \parallel$ _____

6. If $XY = 15$, then $LN =$ _____ .

7. If $m\angle MXY = 72$, then $m\angle MLN =$ _____ .

8. If $ML = 42$, then $YZ =$ _____ .

9. $\overline{ML} \parallel$ _____

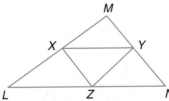

Lesson 9–6 (Pages 382–387) Complete each proportion.

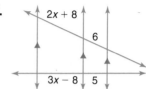

1. $\dfrac{DC}{CA} = \dfrac{FE}{?}$

2. $\dfrac{DA}{AC} = \dfrac{?}{BE}$

3. $\dfrac{?}{FB} = \dfrac{DC}{DA}$

4. $\dfrac{AC}{?} = \dfrac{BE}{BF}$

Find the value of x.

5.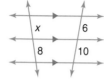

6.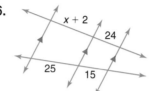

7.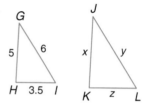

Lesson 9–7 (Pages 388–393) For each pair of similar triangles, find the value of each variable.

1.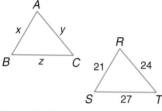

   P of $\triangle ABC = 24$

2.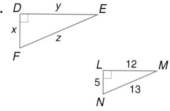

   P of $\triangle DEF = 90$

3.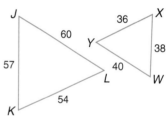

   P of $\triangle JKL = 58$

Determine the scale factor for each pair of similar triangles.

4.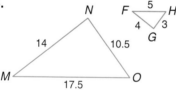

   $\triangle MNO$ to $\triangle FGH$

5.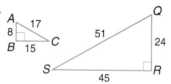

   $\triangle ABC$ to $\triangle QRS$

6.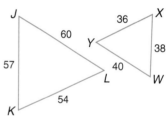

   $\triangle WXY$ to $\triangle JKL$

7. The perimeter of $\triangle EDF$ is 48 centimeters. If $\triangle EDF \sim \triangle NOP$ and the scale factor is $\frac{1}{2}$, find the perimeter of $\triangle NOP$.

8. The perimeter of $\triangle JKL$ is 30 inches. If $\triangle JKL \sim \triangle XVU$ and the scale factor is $\frac{5}{6}$, find the perimeter of $\triangle XVU$.

Lesson 10–1 (Pages 402–407) Identify each polygon by its sides. Then determine whether it appears to be *regular* or *not regular*. If not regular, explain why.

1.

2.

3.

4.

Classify each polygon as *convex* or *concave*.

5.

6.

7.

8.

Lesson 10–2 *(Pages 408–412)* **Find the sum of the measures of the interior angles in each figure.**

1.
2.
3.
4.

Find the measure of one interior angle and one exterior angle of each regular polygon.

5. octagon
6. quadrilateral
7. nonagon
8. 20-gon

9. The sum of the measures of six interior angles of a heptagon is 790. What is the measure of the seventh angle?

10. The sum of the measures of four exterior angles of a pentagon is 290. What is the fifth angle's measure?

Lesson 10–3 *(Pages 413–418)* **Find the area of each polygon in square units.**

1.
2.
3.
4.

Estimate the area of each polygon in square units.

5.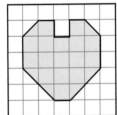
6.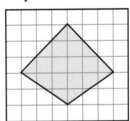
7.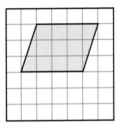

Lesson 10–4 *(Pages 419–424)* **Find the area of each triangle or trapezoid.**

1.
   5 in.   5 in.   4 in.   6 in.
2.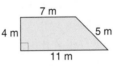
   7 m   4 m   5 m   11 m
3.
   5 ft   12 ft
4.
   9 mm   6 mm   6 mm

5.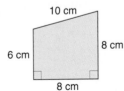
   10 cm   8 cm   6 cm   8 cm
6.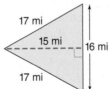
   17 mi   15 mi   16 mi   17 mi
7.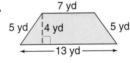
   7 yd   5 yd   4 yd   5 yd   13 yd
8.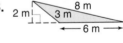
   2 m   3 m   8 m   6 m

9. The area of the triangle is 48 square inches. If the height is 8 inches, find the length of the base.

10. The area of the trapezoid is 108 square centimeters. If the sum of the bases is 27 centimeters, find the height.

Lesson 10–5 *(Pages 425–431)* Find the area of each regular polygon.

1.
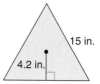
15 in.
4.2 in.

2.

4 m   5 m

3.

7.3 ft
6 ft

Find the area of the shaded region in each regular polygon.

4.
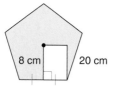
8 cm   20 cm

5.

5 yd
5 yd

6.
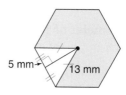
5 mm   13 mm

7. A regular decagon has an area of 210 square meters and a perimeter of 120 meters. Find the length of one side and the length of the apothem.

Lesson 10–6 *(Pages 434–439)* Determine whether each figure has line symmetry. If it does, copy the figure and draw all lines of symmetry. If not, write *no*.

1.

2.

3.

4.

Determine whether each figure has rotational symmetry. Write *yes* or *no*.

5.

6.

7.

8.

Lesson 10–7 *(Pages 440–445)* Identify the figures used to create each tessellation. Then identify the tessellation as *regular*, *semi-regular*, or *neither*.

1.

2.

3.

4.
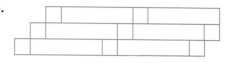

Use isometric or rectangular dot paper to create a tessellation using the given polygons.

5. large and small squares

6. hexagons

7. parallelograms and triangles

8. parallelograms

Lesson 11-1 *(Pages 454–459)* Use ⊙K to determine whether each statement is *true* or *false*.

1. $\overline{PN}$ is a diameter of ⊙K.
2. $ML = 2(MQ)$
3. $\overline{OP}$ is a chord of ⊙K.
4. $KO = MK$
5. A radius is a chord.
6. A diameter contains the center of the circle.

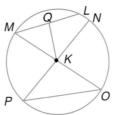

⊙C has a diameter of 12 units, and ⊙E has a diameter of 8 units.

7. If $AB = 1$, find AC.
8. If $AB = 1$, find BD.
9. If $AB = 1$, find CE.
10. If $AE = 3x$, find AD in terms of x.

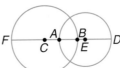

Lesson 11-2 *(Pages 462–467)* Find each measure in ⊙M if $m\angle JMK = 32$, $m\widehat{LN} = 58$, and $\overline{KO}$, $\overline{JN}$, and $\overline{PL}$ are diameters.

1. $m\angle PMJ$
2. $m\widehat{NO}$
3. $m\widehat{OP}$
4. $m\angle KML$
5. $m\angle NMO$
6. $m\widehat{JP}$

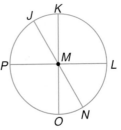

In ⊙C, $\overline{AD}$ is a diameter, and $m\angle ACB = 55$. Determine whether each statement is *true* or *false*.

7. $m\widehat{AB} = 135$
8. $m\angle BCD = m\widehat{BD}$
9. $\angle ACE$ is a central angle.
10. $m\widehat{AEB} = 320$

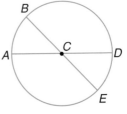

Lesson 11-3 *(Pages 468–473)* Use ⊙C to complete each statement.

1. If $\overline{LJ} \cong \overline{HJ}$, then $LJ =$ _____ .
2. If $\overline{LI} \perp \overline{HJ}$, then $\overline{HG} \cong$ _____ .
3. If $HG = GJ$, then $\triangle JCG \cong \triangle$ _____ .
4. If $CH = 12$, then $JC =$ _____ .
5. If $\overline{CH} \perp \overline{HJ}$, $HJ = 24$, and $CH = 13$, then $CG =$ _____ .

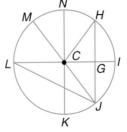

6. In ⊙V, $XY = 48$ and $XW = 50$. Find ZU and UV.

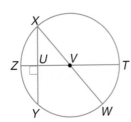

7. If $\overline{XY} \cong \overline{WU}$, $XY = 4x + 4$, $WU = 7x - 11$, and $VY = 8x - 10$, find x and VZ.

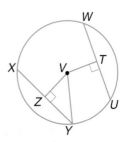

Lesson 11–4 *(Pages 474–477)*   Use ⊙C to find x.

1. $\overline{FC} \cong \overline{CG}$, $AB = 12$, and $DE = 7x - 9$
2. $\overline{AB} \cong \overline{DE}$, $FC = 3x - 1$ and $CG = 2x + 4$
3. $\overline{FC} \cong \overline{CG}$, $AB = 30$, and $DG = 4x + 7$
4. $\overline{AB} \cong \overline{DE}$, $FC = 4(x + 1)$, and $CG = 3(2x - 8)$

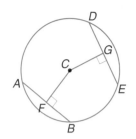

Lesson 11–5 *(Pages 478–482)*   Find the circumference of each object to the nearest tenth.

1. quarter, $d = 2.5$ cm

2. swimming pool, $r = 3.5$ ft

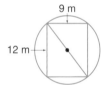

3. bass drum, $d = 3.0$ ft

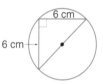

Find the circumference of each circle described to the nearest tenth.

4. $d = 6$ mm
5. $r = 4\frac{1}{4}$ yd
6. $r = 2.7$ mi

Find the radius of the circle to the nearest tenth for each circumference given.

7. 64.3 km
8. 18.9 in.
9. 126.8 cm

Find the circumference of each circle to the nearest hundredth.

10.    11.   12.

Lesson 11–6 *(Pages 483–487)*   Find the area of each circle described to the nearest hundredth.

1. $r = 8$ in.
2. $d = 10.6$ ft
3. $r = 6.3$ mm
4. $d = 26$ mi
5. $C = 427.8$ m
6. $C = 20\frac{1}{4}$ yd

In a circle with radius of 8 meters, find the area of a sector whose central angle has the following measure.

7. 45
8. 150
9. 270

Assume that all darts thrown will land on a dartboard. Find the probability that a randomly-thrown dart will land in the red region. Round to the nearest hundredth.

10.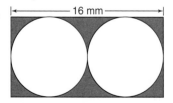

|← 16 mm →|

11.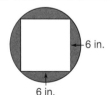

6 in.

6 in.

Lesson 12–1 *(Pages 496–501)* Name the faces, edges, and vertices of each polyhedron.

1.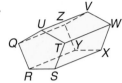

2.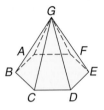

3.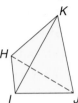

Describe the basic shape of each item as a solid.

4.

5.

6.

Lesson 12–2 *(Pages 504–509)* Find the lateral area and the surface area for each solid. Round each to the nearest hundredth, if necessary.

1.

2.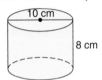

3.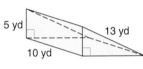

4.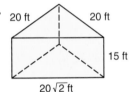

5.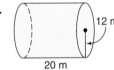

6.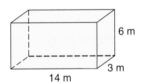

7. Draw a rectangular prism that is 3 inches by 6 inches by 9 inches. Find the surface area of the prism.

Lesson 12–3 *(Pages 510–515)* Find the volume of each solid. Round to the nearest hundredth, if necessary.

1.

2.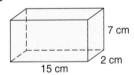

3.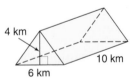

4.

5.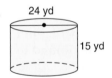

6.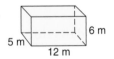

7. What is the volume of a cube that has a 7-centimeter edge?

8. Draw a cylinder that has a base diameter of 12 inches and a height of 9 inches. What is the volume of the cylinder?

Lesson 12–4 *(Pages 516–521)* **Find the lateral area and surface area for each solid. Round each to the nearest hundredth, if necessary.**

1.
5 cm
7 cm
7 cm

2.
5 m
6 m

3.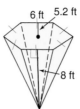
6 ft   5.2 ft
8 ft

4.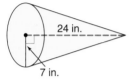
24 in.
7 in.

5.
16 in.
8 in.

6.
13 cm
10 cm
10 cm

7. A regular pyramid has a lateral area of 80 square centimeters. If the base is a square with length 4 centimeters, find the length of the slant height.

Lesson 12–5 *(Pages 522–527)* **Find the volume of each solid. Round to the nearest hundredth, if necessary.**

1.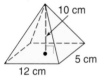
10 cm
5 cm
12 cm

2.
18 m
13 m

3.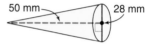
50 mm   28 mm

4.
17 ft
24 ft

5.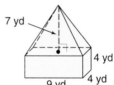
7 yd
4 yd
9 yd   4 yd

6.
9 in.
10 in.
12 in.

7. A pyramid has a volume of 729 cubic units. If the area of the base is 243 square units, what is the height of the pyramid?

Lesson 12–6 *(Pages 528–533)* **Find the surface area and volume of each sphere. Round each to the nearest hundredth.**

1.
5 in.

2.
22 cm

3.
8 ft

4.
3 in.

5.
3.8 in.

6.
4.2 in.

7. Find the surface area of a sphere with a diameter of 12 meters. Round your answer to the nearest hundredth.

Lesson 12–7 *(Pages 534–539)* **Determine whether each pair of solids is similar.**

1. 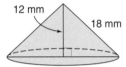 2 in. 2 in. 6 in. 3 in. 3 in. 10 in.

2. 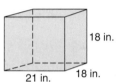 4 in. 8 in. 2.5 in. 5 in.

3. 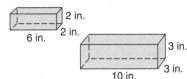 6 cm 4.5 cm 4 cm 3 cm

4. 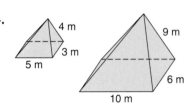 4 m 3 m 5 m 9 m 6 m 10 m

For each pair of similar solids, find the scale factor of the solid on the left to the solid on the right. Then find the ratios of the surface areas and the volumes.

5. 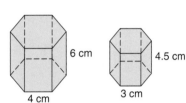 12 mm 18 mm 8 mm 12 mm

6. 12 in. 14 in. 12 in. 18 in. 21 in. 18 in.

7. The ratio of the lateral edges of two similar pyramids is 6:5.
   a. Find the ratio of their surface areas.
   b. Find the ratio of their volumes.

Lesson 13–1 *(Pages 548–553)* **Simplify each expression.**

1. $\sqrt{49}$ 2. $\sqrt{120}$ 3. $\sqrt{9} \cdot \sqrt{9}$ 4. $\sqrt{11} \cdot \sqrt{7}$ 5. $\sqrt{8} \cdot \sqrt{6}$

6. $\dfrac{\sqrt{16}}{\sqrt{25}}$ 7. $\dfrac{\sqrt{50}}{\sqrt{5}}$ 8. $\dfrac{\sqrt{10}}{\sqrt{3}}$ 9. $\sqrt{\dfrac{1}{8}}$ 10. $\dfrac{6}{\sqrt{18}}$

Lesson 13–2 *(Pages 554–558)* **Find the missing measures. Write all radicals in simplest form.**

1. x $14\sqrt{2}$ 45° y

2. y 8 45° 45° x

3. 12 x y

4. y x 45° 45° 10

5. y x $5\sqrt{6}$

6. x 45° 42 y 45°

7. The length of the hypotenuse of an isosceles right triangle is $10\sqrt{2}$ feet. Find the length of a leg.

8. The length of one leg in an isosceles right triangle is $15\sqrt{2}$ centimeters. What is the length of the hypotenuse?

9. The length of the hypotenuse of a 45°-45°-90° triangle is 36 units. Find the length of one leg in the triangle.

Lesson 13–3 *(Pages 559–563)* **Find the missing measures. Write all radicals in simplest form.**

1.

2.

3.

4.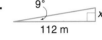

5.

6.

7. The measure of the length of the hypotenuse of a 30°-60°-90° triangle is 4. Find the measure of the length of the two legs.

8. The measure of the length of the shorter leg of a 30°-60°-90° triangle is $7\sqrt{2}$. Find the measure of the length of the longer leg and hypotenuse.

Lesson 13–4 *(Pages 564–569)* **Find each tangent. Round to four decimal places, if necessary.**

1. tan *A*

2. tan *V*

3. tan *B*

4. tan *W*

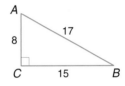

Find each missing measure. Round to the nearest tenth.

5.

6.

7.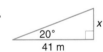

8.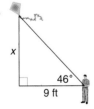

9.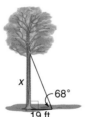

10.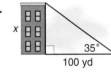

Lesson 13–5 *(Pages 572–577)* **Find each sine or cosine. Round to four decimal places, if necessary.**

1. sin *R*

2. sin *A*

3. cos *T*

4. cos *A*

Find each measure. Round to the nearest tenth.

5.

6.

7.

Use the 30°-60°-90° and 45°-45°-90° triangles to find each value. Round to four decimal places, if necessary.

8. cos 60°

9. sin 45°

10. sin 30°

11. cos 45°

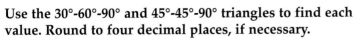

Lesson 14-1 *(Pages 586–591)* **Determine whether each angle is an inscribed angle. Name the intercepted arc for the angle.**

1. ∠BAD

2. ∠XYZ

3. ∠RST

In each circle, find the value of x.

4.

5.

6.

Lesson 14-2 *(Pages 592–597)* **Find each measure. If necessary, round to the nearest tenth. Assume segments that appear to be tangent are tangent.**

1. AQ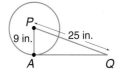

2. m∠FDE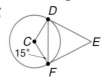

3. LM

4. JK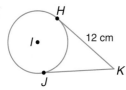

5. XU

6. NP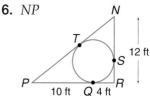

In the figure, $\overline{AB}$ and $\overline{AD}$ are both tangent to ⊙C. Find each measure. If necessary, round to the nearest tenth.

7. m∠DAC

8. m∠CBA

9. CA

10. AB

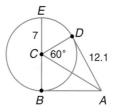

Lesson 14-3 *(Pages 600–605)* **Find each measure.**

1. m$\widehat{IJ}$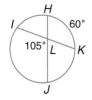

2. m$\widehat{FE}$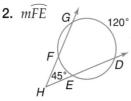

3. m∠SRT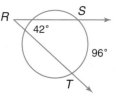

In each circle, find the value of x. Then find the given measure.

4. m$\widehat{KL}$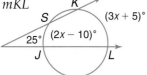

5. m$\widehat{AB}$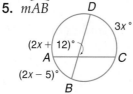

6. m$\widehat{PQ}$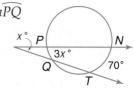

Lesson 14-4 *(Pages 606–611)*   Find the measure of each angle. Assume segments that appear to be tangent are tangent.

1. $\angle A$

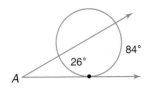

2. $\angle B$

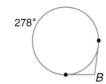

3. $\angle GAP$

4. $\angle 1$

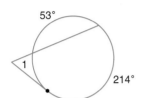

5. $\angle 2$

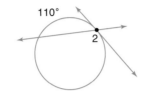

6. $\angle K$

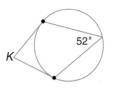

Lesson 14-5 *(Pages 612–617)*   In each circle, find the value of x. If necessary, round to the nearest tenth.

1.

2.

3.

Find each measure. If necessary, round to the nearest tenth.

4. BC

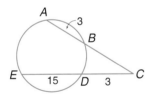

5. KJ

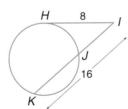

6. XY

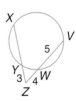

Lesson 14-6 *(Pages 618–623)*   Write an equation of a circle for each center and radius or diameter measure given.

1. $(3, -4), r = 5$

2. $(-1, 0), d = 28$

3. $(0, 0), r = \sqrt{7}$

4. $(-8, 3), d = \frac{2}{3}$

5. $(3, 10), r = \frac{1}{2}$

6. $(0, -4), d = 4\sqrt{6}$

Find the coordinates of the center and the measure of the radius for each circle whose equation is given.

7. $(x - 8)^2 + (y + 3)^2 = 49$

8. $x^2 + (y - 7)^2 = 1$

9. $(x + 12)^2 + (y - 11)^2 = 50$

10. $(x + 5)^2 + y^2 = \frac{25}{36}$

Graph each equation on a coordinate plane.

11. $(x - 2)^2 + (y + 1)^2 = 9$

12. $x^2 + (y - 5)^2 = 100$

Lesson 15–1 *(Pages 632–637)* Use conditionals *p*, *q*, *r*, and *s* for Exercises 1–8.

p: An octagon has eight sides.
r: 14 × 6 = 84

q: Labor Day is in May.
s: Puerto Rico is one of the fifty states.

Write the statements for each negation.

1. ~*p*

2. ~*q*

3. ~*r*

4. ~*s*

Write a statement for each conjunction.

5. *p* ∨ *q*

6. ~*q* ∧ *r*

7. ~*r* ∨ *s*

8. *p* ∧ ~*s*

Construct a truth table for each compound statement.

9. *p* ∨ ~*r*

10. ~(~*q* ∧ *s*)

11. *r* → *s*

Lesson 15–2 *(Pages 638–643)* Use the Law of Detachment to determine a conclusion that follows from statements (1) and (2). If a valid conclusion does not follow, write *no valid conclusion*.

1. (1) If two lines are parallel, then the lines do not intersect.
   (2) *k* ∥ *m*

2. (1) If alternate interior angles are congruent, then lines are parallel.
   (2) ∠*A* ≅ ∠*B*

Use the Law of Syllogism to determine a conclusion that follows from statements (1) and (2). If a valid conclusion does not follow, write *no valid conclusion*.

3. (1) If a triangle has three congruent sides, then it is an equilateral triangle.
   (2) If a triangle is equilateral, then it is equiangular.

4. (1) If two odd numbers are multiplied, their product is an odd number.
   (2) If the product of two numbers is odd, then the product is not divisible by 2.

Determine whether each situation is an example of inductive or deductive reasoning.

5. Jimmy's family eats chicken every Sunday for dinner. Today is Sunday. Jimmy concluded that he will have chicken tonight for dinner.

6. A number is divisible by 9 if the sum of the digits is divisible by 9. Dana concluded that 639 is divisible by 9.

Lesson 15–3 *(Pages 644–648)* Write a paragraph proof for each conjecture.

1. If $\overline{AB} \parallel \overline{CD}$ and $\overline{AC} \parallel \overline{BD}$, then △*ABC* ≅ △*DCB*.
   Plan: Use a triangle congruence postulate.

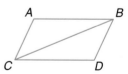

2. If *WX* = *YZ*, then *WY* = *XZ*.

3. If ∠1 ≅ ∠3 and ∠2 ≅ ∠4, then ∠*HIJ* ≅ ∠*KLM*.

4. If ∠1 ≅ ∠2 and *G* is the midpoint of $\overline{FH}$, then △*EFG* ≅ △*IHG*.

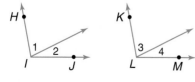

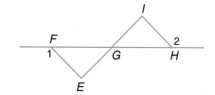

Lesson 15–4 *(Pages 649–653)* **Copy and complete each proof.**

1. If $\overrightarrow{RS}$ is the angle bisector of $\angle QRP$, then $\angle QRS = \frac{1}{2}(\angle QRP)$.

Given: $\overrightarrow{RS}$ is the angle bisector of $\angle QRP$.

Prove: $\angle QRS = \frac{1}{2}(\angle QRP)$

Proof:

Statements	Reasons
a. $\overrightarrow{RS}$ is the angle bisector of $\angle QRP$.	**a.** ___?___
b. $\angle QRP = \angle QRS + \angle SRP$	**b.** ___?___
c. $\angle QRS = \angle SRP$	**c.** ___?___
d. $\angle QRP = \angle QRS + \angle QRS = 2(\angle QRS)$	**d.** ___?___
e. $\frac{1}{2}(\angle QRP) = \angle QRS$	**e.** ___?___

2. If $\frac{2x+4}{3} = 5$, then $x = \frac{11}{2}$.

Given: $\frac{2x+4}{3} = 5$

Prove: $x = \frac{11}{2}$

Proof:

Statements	Reasons
a. $\frac{2x+4}{3} = 5$	**a.** ___?___
b. $2x + 4 = 15$	**b.** ___?___
c. $2x = 11$	**c.** ___?___
d. $x = \frac{11}{2}$	**d.** ___?___

Lesson 15–5 *(Pages 654–659)* **Write a two-column proof.**

1. Given: $\angle 1 \cong \angle 2$, $\angle 1 \cong \angle 3$
Prove: $\overline{AB} \parallel \overline{DE}$

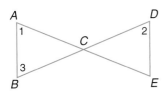

2. Given: $\overline{KM}$ is a perpendicular bisector of $\triangle JKL$.
Prove: $\triangle JKM \cong \triangle LKM$

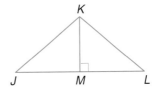

3. Given: $\angle 1 \cong \angle 2$, $\overline{QU} \perp \overline{US}$
Prove: $\overline{RT} \perp \overline{US}$

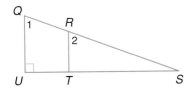

4. Given: X is the midpoint of $\overline{YZ}$ and $\overline{WV}$.
Prove: $\triangle WXY \cong \triangle VXZ$

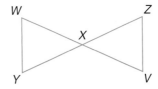

Lesson 15–6 *(Pages 660–665)* **Position and label each figure on a coordinate plane.**

1. a rectangle with length a units and width b units

2. a right triangle with legs y and z units long

3. a parallelogram with base b units and height h units

4. an isosceles triangle with base a units long and height h units

Lesson 16–1 *(Pages 676–680)* Solve each system of equations by graphing.

1. $5x - 2y = 11$
$y = x - 1$

2. $y + 1 = x$
$4x - y = 19$

3. $4x - 3y = 6$
$2x - 3y = 12$

4. $x - 2y = 0$
$y = 2x - 3$

5. $x + 4y = 10$
$x - \frac{1}{2}y = 1$

6. $2x + y = 4$
$y = 2x$

State the letter of the ordered pair that is a solution of both equations.

7. $4x = 36$
$y = 3x - 15$

  a. $(9, 0)$ **b.** $(5, 0)$ **c.** $(9, 12)$ **d.** $(-5, 9)$

8. $y = \frac{1}{3}x$
$x + 2y = -3$

  a. $(0, 0)$ **b.** $\left(-\frac{9}{5}, -\frac{3}{5}\right)$ **c.** $\left(\frac{1}{3}, -\frac{5}{3}\right)$ **d.** $\left(-\frac{3}{2}, -2\right)$

9. The graphs of $x + 2y = 6$, $y = 5$, and $x - 2y = -6$ intersect to form a triangle.

  a. Graph the system of equations.

  b. Find the coordinates of the vertices of the triangle.

Lesson 16–2 *(Pages 681–686)* Use substitution to solve each system of equations.

1. $y = 3x$
$x + 2y = -21$

2. $x + y = 6$
$x - y = 2$

3. $x + 2y = 5$
$y = x - 3$

Use elimination to solve each system of equations.

4. $x - y = 5$
$x + y = 25$

5. $9x + 7y = 4$
$6x - 3y = 18$

6. $x - 2y = 5$
$3x - 2y = 21$

State whether *substitution* or *elimination* would be better to solve each system of equations. Explain your reasoning. Then solve the system.

7. $x - 2y = 5$
$3x - 5y = 8$

8. $y = x - 1$
$x + y = 11$

9. $3x - 2y = 10$
$x + y = 0$

Lesson 16–3 *(Pages 687–691)* Find the coordinates of the vertices of each figure after the given translation. Then graph the translation image.

1. $(-1, 2)$ **2.** $(0, -3)$ **3.** $(2, 3)$

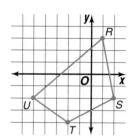 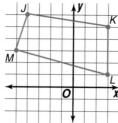 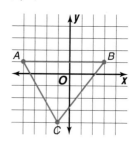

Graph each figure. Then find the coordinates of the vertices after the given translation and graph the translation image.

	Figure	Vertices	Translated By:
4.	$\triangle WIG$	$W(0, 3)$, $I(4, -5)$, $G(5, 2)$	$(3, 1)$
5.	$\triangle MOP$	$M(-2, -2)$, $O(-4, 0)$, $P(0, 3)$	$(4, -2)$
6.	square $DISH$	$D(3, 2)$, $I(-1, 2)$, $S(-1, 6)$, $H(3, 6)$	$(3, -4)$

Lesson 16–4 *(Pages 692–696)* Find the coordinates of the vertices of each figure after a reflection over the given axis. Then graph the reflection image.

1. *x*-axis

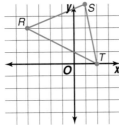

2. *y*-axis

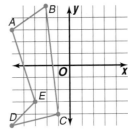

3. *y*-axis

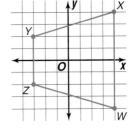

Graph each figure. Then find the coordinates of the vertices after a reflection over the given axis and graph the reflection image.

	Figure	Vertices	Reflected Over:
4.	$\triangle ABC$	$A(2, -3)$, $B(3, 4)$, $C(-1, 0)$	*y*-axis
5.	$\triangle DEF$	$D(5, 4)$, $E(1, 1)$, $F(0, -3)$	*x*-axis
6.	quadrilateral $QRST$	$Q(-3, 4)$, $R(2, 5)$, $S(3, -3)$, $T(-2, -4)$	*x*-axis

Lesson 16–5 *(Pages 697–702)* Rotate each figure about point *P* by tracing the figure. Use the given angle of rotation.

1. 90° clockwise

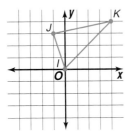

2. 60° counterclockwise

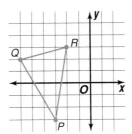

3. 120° clockwise

Find the coordinates of the vertices of each figure after the given rotation about the origin. Then graph the rotation image.

4. 180° counterclockwise

5. 90° clockwise

6. 60° clockwise

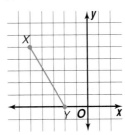

Lesson 16–6 *(Pages 703–707)* Find the coordinates of the dilation image for the given scale factor *k*, and graph the dilation image.

1. 2

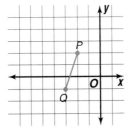

2. $\frac{1}{3}$

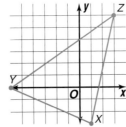

3. 3

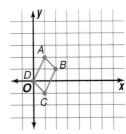

Mixed Problem Solving

Chapter 1 Reasoning in Geometry

1. **Mail** The graph shows the cost to mail a book using the media mail rate for a given number of pounds. Predict the cost for mailing a book that weighs 5 pounds. *(Lesson 1-1)*

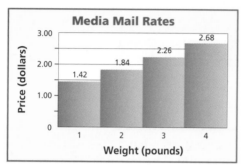

Media Mail Rates

Source: *Time Almanac*

2. **Building** Zach is building the birdhouse pictured. Name a point that is coplanar with points A and B. *(Lesson 1-2)*

For Exercises 3-5, use the prism. *(Lesson 1-3)*

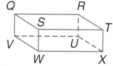

3. Name three planes that intersect.

4. Name two planes that do not intersect.

5. Name two lines that lie in the same plane.

For Exercises 6 and 7, consider this statement. *If two words begin with the same letter, then they form an alliteration.* *(Lesson 1-4)*

6. Identify the hypothesis and the conclusion of this statement.

7. Write the converse of the statement.

8. **Sewing** Juana wants to put binding around three fleece blankets. If each blanket is 1 yard by 2 yards and each package of binding is 2.5 feet long, how many packages of binding will she need? *(Lesson 1-6)*

Chapter 2 Segment Measure and Coordinate Graphing

1. **Geography** In Idaho, the highest point is Borah Peak with an elevation of 12,662 feet above sea level. The lowest point is the Snake River at 710 feet above sea level. What is the difference in their elevations? *(Lesson 2-1)*

2. **Art** Julia measures the length of a piece of paper to be 9 centimeters and the width to be 76 millimeters. Which measure is more precise? *(Lesson 2-2)*

3. C is the midpoint of $\overline{AB}$ and B is the midpoint of $\overline{AD}$. What is AC? *(Lesson 2-3)*

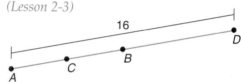

Geography **The table shows the area and highest point of each continent.** *(Lesson 2-4)*

Continent	Area (square miles)	Highest Point (feet)
North America	8,300,000	20,320
South America	6,800,000	22,834
Europe	8,800,000	18,510
Asia	12,000,000	29,035
Africa	11,500,000	19,340
Australia and Oceania	3,200,000	7310
Antarctica	5,400,000	16,864

Source: *The World Almanac For Kids*

4. If the x-coordinate of an ordered pair represents the area and the y-coordinate represents the highest point, write an ordered pair for each continent.

5. Graph the ordered pairs.

6. Look for patterns in the graph. Do larger continents have taller highest points?

7. **Design** Tevin is making a graphic design. Two points in the design are at $(-4, 5)$ and $(-5, 7)$. Find the midpoint of the segment joining these points. *(Lesson 2-5)*

Chapter 3 Angles

1. **Hiking** Enrique plans to hike the path shown below. Use a protractor to find the measure of the angle and classify it as *acute*, *obtuse*, or *right*. *(Lesson 3-2)*

For Exercises 2-4, use the diagram below.
(Lesson 3-3)

2. If $m\angle AEC = 93$ and $m\angle CED = 35$, find $m\angle AED$.

3. Find $m\angle AEB$ if $\overrightarrow{EB}$ bisects $\angle AEC$ and $m\angle AEC = 102$.

4. If $m\angle AED = 152$ and $m\angle CED = 34$, find $m\angle AEC$.

5. **Geography** List all of the adjacent angles and linear pairs that are formed by the states of Utah, Colorado, New Mexico, and Arizona. *(Lesson 3-4)*

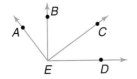

6. **Gardening** Janine is planting a rose garden in her yard. She wants to make a walking path at a diagonal through the garden. Find x. *(Lesson 3-5)*

7. $\angle 1$ is supplementary to $\angle 3$ and $\angle 2$ is supplementary to $\angle 3$. If $m\angle 2 = 129$, what are $m\angle 1$ and $m\angle 3$? *(Lesson 3-6)*

8. Which hallways of South High School appear to be perpendicular? *(Lesson 3-7)*

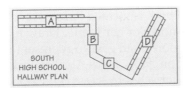

Chapter 4 Parallels

1. **Construction** Ebony built this bookcase for her room. Describe a pair of parts that are parallel and a pair that intersect. *(Lesson 4-1)*

For Exercises 2-7, use the following information. In the diagram below, $m\angle 2 = 95$, find the measure of each angle. *(Lesson 4-2)*

2. $\angle 1$
3. $\angle 3$
4. $\angle 4$
5. $\angle 5$
6. $\angle 6$
7. $\angle 7$

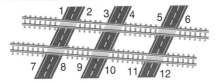

8. **Transportation** The train tracks pictured below cross three parallel streets. Make a conjecture about which angles are congruent. Explain your reasoning. *(Lesson 4-3)*

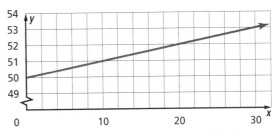

9. **Construction** What is the pitch, or slope, of this roof? *(Lesson 4-5)*

Car Rental The graph below shows the charges for renting a car when it is driven different numbers of miles. *(Lesson 4-6)*

10. What is the slope of the line and what does it represent?

11. What is the y-intercept of the line and what does it represent?

12. Write an equation of the line.

Chapter 5 Triangles and Congruence

1. Classify the blue triangles in the flag of the United Kingdom by their angles and by their sides. *(Lesson 5-1)*

2. **Astronomy** Pegasus is a constellation represented by a horse. The horse's head is three stars that form a triangle. If the angles have measures as shown in the figure, find the value of *x*. *(Lesson 5-2)*

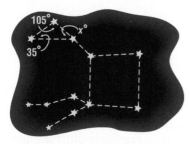

3. **Design** Aliyah painted this design as a border on her wall. Identify the type of transformation that she used. *(Lesson 5-3)*

4. **Construction** Sean built a deck in the shape of △*ABC* in the corner of his yard. He wants to build an identical deck in the opposite corner of his yard so that △*ABC* ≅ △*RST*. What will be the measure of ∠*R* in the second deck? *(Lesson 5-4)*

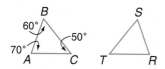

5. **Kites** A kite has two pairs of consecutive congruent sides. Determine which, if any, of the triangles in the kite are congruent. If so, write a congruence statement and explain why the triangles are congruent. If not, explain why not. *(Lesson 5-5)*

Chapter 6 More About Triangles

1. **Sailing** Is the seam marked $\overline{GH}$ on the sail an *altitude*, a *perpendicular bisector*, *both*, or *neither*? *(Lesson 6-2)*

2. **Construction** In plans for a deck, $\overline{WZ}$ is an angle bisector of ∠*XZY*. If *m*∠*XZY* = 120, find *m*∠2. *(Lesson 6-3)*

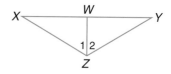

3. **Ice Cream** What type of triangle is the picture of the cone? If the measure of angle 2 is 75, what are the measures of the other two angles of the triangle? *(Lesson 6-4)*

4. **Sewing** Keely is creating a quilt design that has congruent triangles. She created a template for a triangle as shown below. What theorem can you use to prove that △*LMO* ≅ △*LMR*? Explain. *(Lesson 6-5)*

5. **Walking** Levi walked 6 blocks west then turned and walked 8 blocks north to get to the library from his house. How far was he from his house? *(Lesson 6-5)*

6. **Maps** The place Abril works is located 7 miles west and 3 miles south of her home. Her school is located 5 miles east and 6 miles north of her home. Draw a diagram on a coordinate grid to represent this situation. How far is Abril's workplace from her school? *(Lesson 6-7)*

Chapter 7 Triangle Inequalities

1. **Driving** Randa lives 14 miles from Jack. Use the Comparison Property to tell how this distance might compare to the distance from Randa's house to Sadie's. *(Lesson 7-1)*

2. **Art** Mia has drawn the shape below for an art project. If she shortens each of the sides by 3 centimeters, write an inequality comparing *AB* and *DC*. *(Lesson 7-1)*

3. **Camping** In the diagram below, find the measure of ∠3. *(Lesson 7-2)*

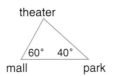

4. **Directions** If Jamel is at the mall, is he closer to the park or to the theater? *(Lesson 7-3)*

5. **Graphics** Huang wants the angle with the greatest measure to be in the lower left-hand corner of the graphic he is drawing. Which vertex should be in the lower-left hand corner? *(Lesson 7-3)*

6. **Gardening** Elyse is making a triangular garden that will be lined with rose bushes on two of the three sides. She has enough roses to cover 12 feet on one side and 10 feet on the other side. What are the possible lengths of the third side of the garden? *(Lesson 7-4)*

Chapter 8 Quadrilaterals

1. **Astronomy** Four of the stars of the Big Dipper constellation form a quadrilateral. Which of the stars labeled are vertices of the quadrilateral? *(Lesson 8-1)*

2. **Sewing** The quilt square shown is called the Lone Star pattern. Each of the pieces in the star forms a parallelogram. If the measure of one of the angles in one of the small parallelograms is 50°, what are the measures of the other 3 angles? *(Lesson 8-2)*

3. **Gardening** Hulleah is planting a garden. She has drawn the diagram pictured below and says that if she connects the vertices *W*, *X*, *Y*, and *Z* to form the border, she will have a parallelogram. Is she correct? Explain. *(Lesson 8-3)*

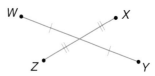

4. **Games** The diamond from a playing card is a quadrilateral. List all types of quadrilaterals that apply to the diamond shown. Explain your reasoning. *(Lesson 8-4)*

5. **Health** Keiran is making a poster of the food pyramid for his health class. What is the value of *x*? *(Lesson 8-5)*

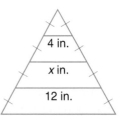

Chapter 9 Proportions and Similarity

1. **School** At West High, the ratio of male to female students in the ninth grade is 6:5. If there are 308 ninth graders, how many are female? *(Lesson 9-1)*

2. **Building** If 1 inch on Esteban's plans represents 6 feet in the house, what are the dimensions of a room that is 2.5 inches by 3 inches on the drawing? *(Lesson 9-2)*

3. **Entertainment** How far are the bumper cars from the Ferris wheel? *(Lesson 9-3)*

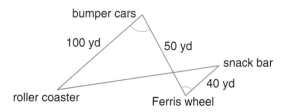

4. **Construction** Ian is building a swing set. How far apart should the posts be? *(Lesson 9-4)*

5. **Roads** The distance from the intersection of Broadway and Maple and the intersection of Broadway and Sunburst is 2 miles. If Maple, Washington, Olearia, and Sunburst run parallel, what is the distance between the intersection of Broadway and Washington and the intersection of Broadway and Olearia? *(Lesson 9-6)*

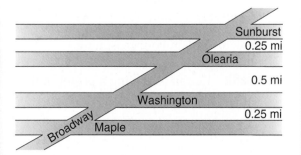

6. **Dollhouses** Megan's dollhouse is a miniature of her house, where 0.5 inch represents 4 feet. What is the scale factor of the dollhouse to the house? *(Lesson 9-7)*

Chapter 10 Polygons and Area

1. **Cars** The door of the car below forms a polygon. Identify the polygon by its sides and as convex or concave. *(Lesson 10-1)*

2. **Construction** A park has a hexagonal gazebo. What is the value of *x*? *(Lesson 10-2)*

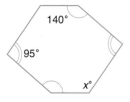

3. **Real Estate** Estimate the area of Tavon's backyard if each square represents 10 square feet. *(Lesson 10-3)*

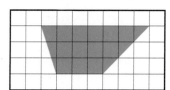

4. **Carpet** Carpet costs $3.49 per square foot. How much will it cost to carpet this stage? *(Lesson 10-4)*

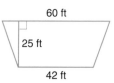

5. **Construction** Isaac is building a platform for a hot tub shaped like a regular hexagon. The apothem is 7 feet and each side is 9 feet. What is the area of the platform? *(Lesson 10-5)*

6. **Apples** Does the picture of the cross section of an apple have *line symmetry, rotational symmetry, neither,* or *both*? *(Lesson 10-6)*

7. **Quilting** Explain how transformations can be used to make the quilt square. *(Lesson 10-7)*

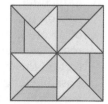

Chapter 11 Circles

1. **Olympics** A discus is circular with a metal rim and a weight in the center. The men's discus has a diameter of approximately 220 millimeters and the women's discus has a diameter of approximately 181 millimeters. What are the radii of the men's and women's discuses? *(Lesson 11-1)*

2. **Stadiums** The roof of the Astrodome is circular with 12 equally-spaced trusses as shown. *NM* is a diameter and *m∠MJL* = 30. What is *m∠NJL*? *(Lesson 11-2)*

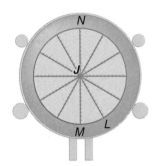

3. **Art** Abby makes glass paperweights in the shape of a sphere from which a flat surface is cut to make a base. If the complete sphere has a radius of 4 centimeters and the diameter of the flat base is 6 centimeters, what is the height of the paperweight? *(Lesson 11-3)*

4. **Archaeology** A mound built by ancient Indians in Ohio consists of a square inscribed in a circle. If the diameter of the circle is 120 feet, what is the distance from the center of the circle to one of the sides of the square? *(Lesson 11-4)*

5. **Astronomy** The diameter of the Moon is 2160 miles. If you could walk around the Moon at the equator, how far would you walk? Round to the nearest hundredth. *(Lesson 11-5)*

6. **Ponds** Maria has a round pond with a diameter of 8 feet. She wants to build a walkway around the pond that would have a width of 2 feet. What will be the area of the walkway? Round to the nearest hundredth. *(Lesson 11-6)*

Chapter 12 Surface Area and Volume

Baking Use the picture of the wedding cake.

1. What geometric solids are used to make the cake? *(Lesson 12-1)*

2. The radii of the three tiers of the cake are 4, 7, and 11 inches and each tier is 5 inches tall. How much area needs to be covered with icing? (*Hint:* The top of each tier is iced, but the bottom is not.) *(Lesson 12-2)*

3. **Storage** If Kyle has 6700 cubic inches of packing peanuts to store, how many boxes like the one below will he need? *(Lesson 12-3)*

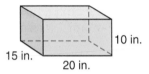

10 in.
15 in.
20 in.

4. **Gardening** How much glass was needed for the walls of the backyard greenhouse? *(Lesson 12-4)*

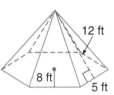

12 ft
8 ft
5 ft

5. **Candles** What is the volume of the candle? Round to the nearest hundredth. *(Lesson 12-5)*

6 in.
4 in.

6. **Astronomy** The diameter of the Sun is approximately 864,000 miles. What is the volume of the Sun? *(Lesson 12-6)*

7. **Decorating** Sierra drew a model footstool with a diameter of 3 inches and a height of 1 inch. The cushion has a diameter of 18 inches and a height of 6 inches, what is the ratio of the surface area of the model to the surface area of the actual cushion? *(Lesson 12-7)*

Chapter 13 Right Triangles and Trigonometry

1. **Furniture** The area of the base of a storage cube is 28 square inches. How long is one of the sides of the base? *(Lesson 13-1)*

2. **Decorating** Aquilah is decorating tables for a family reunion. The tables are square with sides 5 feet long. She wants to use ribbon to place across the diagonal of the table as shown below. What is the length of the ribbon that she will need for each table? *(Lesson 13-2)*

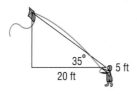

3. **Biking** Jaime built a bike ramp. How long is the ramp? *(Lesson 13-3)*

8 ft

30°

4. **Kites** Kari sights her kite at an angle of elevation of 35 degrees. Her eyes are 5 feet above the ground. If she is 20 feet from the point directly below the kite, how high above the ground is the kite? Round to the nearest foot. *(Lesson 13-4)*

35°
20 ft
5 ft

5. **Rivers** Melanie stands across a river from Jiro. She sights Jiro at a 65° angle when she is 15 feet downstream from him. How far is Melanie from Jiro? Round to the nearest hundredth. *(Lesson 13-5)*

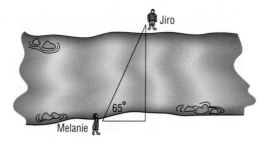

Jiro

65°

Melanie

Chapter 14 Circle Relationships

1. **Graphics** Hilary drew an isosceles triangle inscribed in a circle for a computer graphic. In the diagram $m\angle Q = m\angle R$ and $m\widehat{QR} = 70$. Find $m\angle Q$, $m\angle R$, and $m\angle S$. *(Lesson 14-1)*

S

Q   R

2. **Pizza** The first two cuts Deshawn made in a pizza made four slices as shown. What is the measure of $\angle 1$? *(Lesson 14-3)*

178°

1   48°

3. **Meteorology** A rainbow is really a full circle with a center at a point in the sky directly opposite the Sun. The position of a rainbow varies according to the viewer's position, but its angular size $\angle ABC$ is always 42°. If $m\widehat{CD} = 160$, find the measure of the visible part of the rainbow, $m\widehat{AC}$. *(Lesson 14-4)*

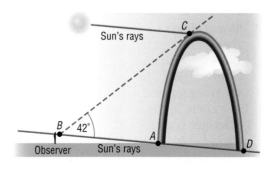

Sun's rays   C

B   42°
A
Observer   Sun's rays   D

4. **Astronomy** The planet Mercury has a diameter of 3032 miles. Write an equation that represents the cross section of Mercury at the center. Assume that the center is at (0, 0). *(Lesson 14-6)*

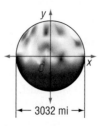

y

C   x

← 3032 mi →

Chapter 15 Formalizing Proof

1. **Advertising**   A university advertises *If you want a superb education, then attend Lincoln University.* Let p represent "if you want a superb education" and q represent "then attend Lincoln University". What is the converse of the conditional? *(Lesson 15-1)*

2. **Airlines**   If your luggage weighs over 50 pounds, you must pay $5 for each extra pound. Talisa's luggage weighs 51 pounds. What conclusion can you draw from the information? *(Lesson 15-2)*

3. **Proof**   Write a paragraph proof to show that if $\angle 4$ is supplementary to $\angle 2$, then $\angle 2 \cong \angle 3$. *(Lesson 15-3)*

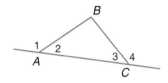

4. **Area**   Show that if $A = \frac{1}{2}bh$, then $h = \frac{2A}{B}$. *(Lesson 15-4)*

5. **Proof**   Write a two-column proof.
   Given: $\angle A \cong \angle D$; $\angle B \cong \angle E$
   Prove: $\angle C \cong \angle F$
   (Lesson 15-5)

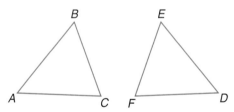

6. **Proof**   Write a coordinate proof.
   Given: R is the midpoint of $\overline{OP}$.
   S is the midpoint of $\overline{QP}$.
   Prove: $RS = \frac{1}{2}OQ$
   (Lesson 15-6)

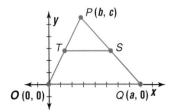

Chapter 16 More Coordinate Graphing and Transformations

1. **Farming**   Mr. Garcia wants to fence a pasture. He wants the length to be 3 times the width and he has 320 yards of fencing to use. If w represents the width of the pasture and ℓ represents the length. Write and solve a system of equations by graphing to find the dimensions of the pasture. *(Lesson 16-1)*

2. **Candy**   Kasa bought x pounds of chocolate candies that cost $7.50 per pound and y pounds of mint candies that cost $4.50 per pound. The cost of buying a total of 6 pounds was $33. Write and solve a system of equations to find the number of pounds of chocolate candies and the number of pounds of mint candies that she purchased. *(Lesson 16-2)*

3. **Storage**   Eli wants to move his shed further back in his yard and to the right. If the vertices of the shed were at $A(18, 24)$, $B(30, 24)$, $C(30, 38)$ and $D(18, 38)$ and he moves the shed 10 feet back and 5 feet to the right, the translation can be given by $(10, 5)$. Find the coordinates of A', B', C', and D' after the move. *(Lesson 16-3)*

Graphic Design   Malaya is using transformations to design a graphic for a T-shirt. Malaya places one triangle on the coordinate grid system as shown.

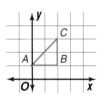

4. What will be the coordinates of the reflection over the y-axis?
   (Lesson 16-4)

5. What will be the vertices of $\triangle A''B''C''$ if it is created by rotating $\triangle ABC$ 180° counterclockwise about the origin.
   (Lesson 16-5)

6. **Yearbooks**   Luis is working on the yearbook. He placed a picture on a page so that the vertices are at $S(0, 0)$, $T(2, 0)$, $U(2, 3)$, and $V(0, 3)$. He wants to dilate the picture by a scale factor of 2. What will be the coordinates for the dilation image of the picture? *(Lesson 16-6)*

Becoming a Better Test-Taker

At some time in your life, you will have to take a standardized test. Sometimes this test may determine if you go on to the next grade or course, or even if you will graduate from high school. This section of your textbook is dedicated to making you a better test-taker.

TYPES OF TEST QUESTIONS   In the following pages, you will see examples of four types of questions commonly seen on standardized tests. A description of each type of question is shown in the table below.

Type of Question	Description	See Pages
multiple choice	Four or five possible answer choices are given from which you choose the best answer.	767–768
gridded response	You solve the problem. Then you enter the answer in a special grid and color in the corresponding circles.	769–772
short response	You solve the problem, showing your work and/or explaining your reasoning.	773–776
extended response	You solve a multi-part problem, showing your work and/or explaining your reasoning.	777–781

PRACTICE   After being introduced to each type of question, you can practice that type of question. Each set of practice questions is divided into five sections that represent the categories most commonly assessed on standardized tests.

- Number and Operations
- Algebra
- Geometry
- Measurement
- Data Analysis and Probability

USING A CALCULATOR   On some tests, you are permitted to use a calculator. You should check with your teacher to determine if calculator use is permitted on the test you will be taking, and, if so, what type of calculator can be used.

Test-Taking Tip

If you are allowed to use a calculator, make sure you are familiar with how it works so that you won't waste time trying to figure out the calculator when taking the test.

TEST-TAKING TIPS   In addition to the Test-Taking Tips like the one shown at the right, here are some additional thoughts that might help you.

- Get a good night's rest before the test. Cramming the night before does not improve your results.

- Budget your time when taking a test. Don't dwell on problems that you cannot solve. Just make sure to leave that question blank on your answer sheet.

- Watch for key words like NOT and EXCEPT. Also look for order words like LEAST, GREATEST, FIRST, and LAST.

Multiple-Choice Questions

Multiple-choice questions are the most common type of question on standardized tests. These questions are sometimes called *selected-response questions*. You are asked to choose the best answer from four or five possible answers.

To record a multiple-choice answer, you may be asked to shade in a bubble that is a circle or an oval or just to write the letter of your choice. Always make sure that your shading is dark enough and completely covers the bubble.

Sometimes a question does not provide you with a figure that represents the problem. Drawing a diagram may help you to solve the problem. Once you draw the diagram, you may be able to eliminate some of the possibilities by using your knowledge of mathematics. Another answer choice might be that the correct answer is not given.

Example

Strategy

Diagrams
Draw a diagram of the playground.

1 **A coordinate plane is superimposed on a map of a playground. Each side of each square represents 1 meter. The slide is located at (5, −7), and the climbing pole is located at (−1, 2). What is the distance between the slide and the pole?**

Ⓐ $\sqrt{15}$ m Ⓑ 6 m Ⓒ 9 m Ⓓ $9\sqrt{13}$ m Ⓔ none of these

Draw a diagram of the playground on a coordinate plane. Notice that the difference in the *x*-coordinates is 6 meters and the difference in the *y*-coordinates is 9 meters.

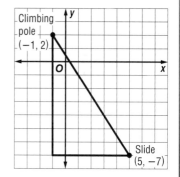

Since the two points are two vertices of a right triangle, the distance between the two points must be greater than either of these values. So we can eliminate Choices B and C.

Use the Distance Formula or the Pythagorean Theorem to find the distance between the slide and the climbing pole. Let's use the Pythagorean Theorem.

$a^2 + b^2 = c^2$	Pythagorean Theorem
$6^2 + 9^2 = c^2$	Substitution
$36 + 81 = c^2$	$6^2 = 36$ and $9^2 = 81$
$117 = c^2$	Add.
$3\sqrt{13} = c$	Take the square root of each side and simplify.

So, the distance between the slide and pole is $3\sqrt{13}$ meters. Since this is not listed as choice A, B, C, or D, the answer is Choice E.

If you are short on time, you can test each answer choice to find the correct answer. Sometimes you can make an educated guess about which answer choice to try first.

Multiple Choice Practice

Choose the best answer.

Number and Operations

1. Carmen designed a rectangular banner that was 5 feet by 8 feet for a local business. The owner of the business asked her to make a larger banner measuring 10 feet by 20 feet. What was the percent increase in size from the first banner to the second banner?

   Ⓐ 4%   Ⓑ 20%   Ⓒ 80%   Ⓓ 400%

2. A roller coaster casts a shadow 57 yards long. Next to the roller coaster is a 35-foot tree with a shadow that is 20 feet long at the same time of day. What is the height of the roller coaster to the nearest whole foot?

   Ⓐ 98 ft   Ⓑ 100 ft   Ⓒ 299 ft   Ⓓ 388 ft

Algebra

3. At Speedy Car Rental, it costs $32 per day to rent a car and then $0.08 per mile. If y is the total cost of renting the car and x is the number of miles, which equation describes the relation between x and y?

   Ⓐ $y = 32x + 0.08$ Ⓑ $y = 32x - 0.08$
   Ⓒ $y = 0.08x + 32$ Ⓓ $y = 0.08x - 32$

4. Eric plotted his house, school, and the library on a coordinate plane. Each side of each square represents one mile. What is the distance from his house to the library?

   Ⓐ $\sqrt{24}$ mi
   Ⓑ 5 mi
   Ⓒ $\sqrt{26}$ mi
   Ⓓ $\sqrt{29}$ mi

   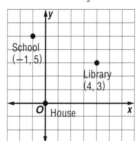

Geometry

5. The grounds outside of the Custer County Museum contain a garden shaped like a right triangle. One leg of the triangle measures 8 feet, and the area of the garden is 18 square feet. What is the length of the other leg?

   Ⓐ 2.25 in.   Ⓑ 4.5 in.   Ⓒ 13.5 in.
   Ⓓ 27 in.   Ⓔ 54 in.

Test-Taking Tip
Questions 2, 5 and 7
The units of measure given in the question may not be the same as those given in the answer choices. Check that your solution is in the proper unit.

6. The circumference of a circle is equal to the perimeter of a regular hexagon with sides that measure 22 inches. What is the length of the radius of the circle to the nearest inch? Use 3.14 for π.

   Ⓐ 7 in.   Ⓑ 14 in.   Ⓒ 121 in.
   Ⓓ 24 in.   Ⓔ 28 in.

Measurement

7. Eduardo is planning to install carpeting in a rectangular room that measures 12 feet 6 inches by 18 feet. How many square yards of carpet does he need for the project?

   Ⓐ 25 yd²   Ⓑ 50 yd²
   Ⓒ 225 yd²   Ⓓ 300 yd²

8. A cylinder has a diameter of 14 centimeters and a height of 30 centimeters. A cone has a radius of 15 centimeters and a height of 14 centimeters. Find the ratio of the volume of the cylinder to the volume of the cone.

   Ⓐ 3 to 1   Ⓑ 2 to 1
   Ⓒ 7 to 5   Ⓓ 7 to 10

Data Analysis and Probability

9. Refer to the table. Which statement is true about this set of data?

Country	Spending per Person
Japan	$8622
United States	$8098
Switzerland	$6827
Norway	$6563
Germany	$5841
Denmark	$5778

   Source: *Top 10 of Everything 2003*

   Ⓐ The median is less than the mean.
   Ⓑ The mean is less than the median.
   Ⓒ The range is 2844.
   Ⓓ A and C are true.
   Ⓔ B and C are true.

Gridded-Response Questions

Gridded-response questions are another type of question on standardized tests. These questions are sometimes called *student-produced response* or *grid-in*, because you must create the answer yourself, not just choose from four or five possible answers.

For gridded response, you must mark your answer on a grid printed on an answer sheet. The grid contains a row of four or five boxes at the top, two rows of ovals or circles with decimal and fraction symbols, and four or five columns of ovals, numbered 0–9. Since there is no negative symbol on the grid, answers are never negative. An example of a grid from an answer sheet is shown at the right.

How do you correctly fill in the grid?

Example ❶ **In the diagram, $\triangle MPT \sim \triangle RPN$. Find PR.**

What value do you need to find?

You need to find the value of x so that you can substitute it into the expression $3x + 3$ to find PR. Since the triangles are similar, write a proportion to solve for x.

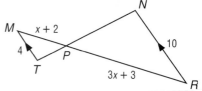

$$\frac{MT}{RN} = \frac{PM}{PR} \qquad \text{Definition of similar polygons}$$

$$\frac{4}{10} = \frac{x + 2}{3x + 3} \qquad \text{Substitution}$$

$$4(3x + 3) = 10(x + 2) \qquad \text{Cross products}$$

$$12x + 12 = 10x + 20 \qquad \text{Distributive Property}$$

$$2x = 8 \qquad \text{Subtract 12 and 10x from each side.}$$

$$x = 4 \qquad \text{Divide each side by 2.}$$

Now find PR.
$$PR = 3x + 3$$
$$= 3(4) + 3 \text{ or } 15$$

How do you fill in the grid for the answer?

- Write your answer in the answer boxes.

- Write only one digit or symbol in each answer box.

- Do not write any digits or symbols outside the answer boxes.

- You may write your answer with the first digit in the left answer box, or with the last digit in the right answer box. You may leave blank any boxes you do not need on the right or the left side of your answer.

- Fill in only one bubble for every answer box that you have written in. Be sure not to fill in a bubble under a blank answer box.

Many gridded-response questions result in an answer that is a fraction or a decimal. These values can also be filled in on the grid.

How do you grid decimals and fractions?

Example ❷ A triangle has a base of length 1 inch and a height of 1 inch. What is the area of the triangle in square inches?

Use the formula $A = \frac{1}{2}bh$ to find the area of the triangle.

$A = \frac{1}{2}bh$ Area of a triangle

$= \frac{1}{2}(1)(1)$ Substitution

$= \frac{1}{2}$ or 0.5   Simplify.

How do you grid the answer?

You can either grid the fraction or the decimal. Be sure to write the decimal point or fraction bar in the answer box. The following are acceptable answer responses.

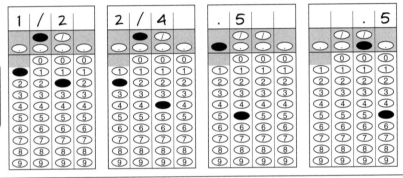

> Do not leave a blank answer box in the middle of an answer.

Sometimes an answer is an improper fraction. Never change the improper fraction to a mixed number. Instead, grid either the improper fraction or the equivalent decimal.

How do you grid mixed numbers?

Example ❸ The shaded region of the rectangular garden will contain roses. What is the ratio of the area of the garden to the area of the shaded region?

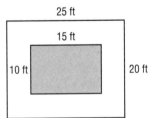

First, find the area of the garden.

$A = \ell w$

$= 25(20)$ or 500

Then find the area of the shaded region.

$A = \ell w$

$= 15(10)$ or 150

Write the ratio of the areas as a fraction.

$\frac{\text{area of garden}}{\text{area of shaded region}} = \frac{500}{150}$ or $\frac{10}{3}$

Leave the answer as the improper fraction $\frac{10}{3}$, as there is no way to correctly grid $3\frac{1}{3}$.

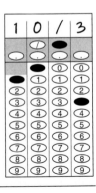

> **Strategy**
>
> **Formulas**
> If you are unsure of a formula, check the reference sheet.

Gridded-Response Practice

Solve each problem and complete the grid.

Number and Operations

1. A large rectangular meeting room is being planned for a community center. Before building the center, the planning board decides to increase the area of the original room by 40%. When the room is finally built, budget cuts force the second plan to be reduced in area by 25%. What is the ratio of the area of the room that is built to the area of the original room?

2. Greenville has a spherical tank for the city's water supply. Due to increasing population, they plan to build another spherical water tank with a radius twice that of the current tank. How many times as great will the volume of the new tank be as the volume of the current tank?

3. In Earth's history, the Precambrian period was about 4600 million years ago. If this number of years is written in scientific notation, what is the exponent for the power of 10?

4. A virus is a microorganism so small it must be viewed with an electron microscope. The largest shape of virus has a length of about 0.0003 millimeter. To the nearest whole number, how many viruses would fit end to end on the head of a pin measuring 1 millimeter?

Algebra

5. Kaia has a painting that is 10 inches by 14 inches. She wants to make her own frame that has an equal width on all sides. She wants the total area of the painting and frame to be 285 square inches. What will be the width of the frame in inches?

6. The diagram shows a triangle graphed on a coordinate plane. If $\overline{AB}$ is extended, what is the value of the y-intercept?

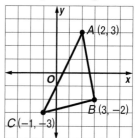

7. Tyree networks computers in homes and offices. In many cases, he needs to connect each computer to every other computer with a wire. The table shows the number of wires he needs to connect various numbers of computers. Use the table to determine how many wires are needed to connect 20 computers.

Computers	Wires	Computers	Wires
1	0	5	10
2	1	6	15
3	3	7	21
4	6	8	28

8. A line perpendicular to $9x - 10y = -10$ passes through $(-1, 4)$. Find the x-intercept of the line.

9. Find the positive solution of $6x^2 - 7x = 5$.

Geometry

10. The diagram shows $\triangle RST$ on the coordinate plane. The triangle is first rotated $90°$ counterclockwise about the origin and then reflected in the y-axis. What is the x-coordinate of the image of T after the two transformations?

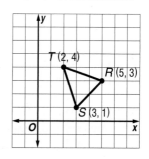

11. An octahedron is a solid with eight faces that are all equilateral triangles. How many edges does the octahedron have?

12. Find the measure of $\angle A$ to the nearest tenth of a degree.

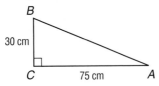

Measurement

13. The Pep Club plans to decorate some large garbage barrels for Spirit Week. They will cover only the sides of the barrels with decorated paper. How many square feet of paper will they need to cover 8 barrels like the one in the diagram? Use 3.14 for π. Round to the nearest square foot.

3 ft

14 in.

14. Kara makes decorative paperweights. One of her favorites is a hemisphere with a diameter of 4.5 centimeters. What is the surface area of the hemisphere including the bottom on which it rests? Use 3.14 for π. Round to the nearest tenth of a square centimeter.

4.5 cm

15. The record for the fastest land speed of a car traveling for one mile is approximately 763 miles per hour. The car was powered by two jet engines. What was the speed of the car in feet per second? Round to the nearest whole number.

16. On average, a B-777 aircraft uses 5335 gallons of fuel on a 2.5-hour flight. At this rate, how much fuel will be needed for a 45-minute flight? Round to the nearest gallon.

Data Analysis and Probability

17. The table shows the heights of the tallest buildings in Kansas City, Missouri. To the nearest tenth, what is the positive difference between the median and the mean of the data?

Name	Height (m)
One Kansas City Place	193
Town Pavilion	180
Hyatt Regency	154
Power and Light Building	147
City Hall	135
1201 Walnut	130

Source: skyscrapers.com

18. A long-distance telephone service charges 40 cents per call and 5 cents per minute. If a function model is written for the graph, what is the rate of change of the function?

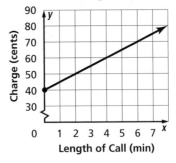

19. In a dart game, the dart must land within the innermost circle on the dartboard to win a prize. If a dart hits the board, what is the probability, as a percent, that it will hit the innermost circle?

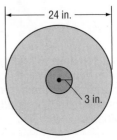

24 in.

3 in.

Short-Response Questions

Short-response questions require you to provide a solution to the problem, as well as any method, explanation, and/or justification you used to arrive at the solution. These are sometimes called *constructed-response, open-response, open-ended, free-response,* or *student-produced questions.* The following is a sample rubric, or scoring guide, for scoring short-response questions.

Credit	Score	Criteria
Full	2	Full credit: The answer is correct and a full explanation is provided that shows each step in arriving at the final answer.
Partial	1	Partial credit: There are two different ways to receive partial credit. • The answer is correct, but the explanation provided is incomplete or incorrect. • The answer is incorrect, but the explanation and method of solving the problem is correct.
None	0	No credit: Either an answer is not provided or the answer does not make sense.

On some standardized tests, no credit is given for a correct answer if your work is not shown.

Example ❶ Mr. Solberg wants to buy all the lawn fertilizer he will need for this season. His front yard is a rectangle measuring 55 feet by 32 feet. His back yard is a rectangle measuring 75 feet by 54 feet. Two sizes of fertilizer are available—one that covers 5000 square feet and another covering 15,000 square feet. He needs to apply the fertilizer four times during the season. How many bags of each size should he buy to have the least amount of waste?

Full Credit Solution

Find the area of each part of the lawn and multiply by 4 since the fertilizer is to be applied 4 times. Each portion of the lawn is a rectangle, so $A = lw$.

$$4[(55 \times 32) + (75 \times 54)] = 23,240 \text{ ft}^2$$

If Mr. Solberg buys 2 bags that cover 15,000 ft², he will have too much fertilizer. If he buys 1 large bag, he will still need to cover $23,240 - 15,000$ or 8240 ft².

Find how many small bags it takes to cover 82400 ft².

$$8240 \div 5000 = 1.648$$

Since he cannot buy a fraction of a bag, he will need to buy 2 of the bags that cover 5000 ft² each.

Mr. Solberg needs to buy 1 bag that covers 15,000 square feet and 2 bags that cover 5000 square feet each.

Strategy

Estimation
Use estimation to check your solution.

The steps, calculations, and reasoning are clearly stated.

The solution of the problem is clearly stated.

Partial Credit Solution

In this sample solution, the answer is correct. However, there is no justification for any of the calculations.

There is not an explanation of how 23,240 was obtained.

23,240

$$23{,}240 - 15{,}000 = 8240$$

$$8240 \div 5000 = 1.648$$

Mr. Solberg needs to buy 1 large bag and 2 small bags.

Partial Credit Solution

In this sample solution, the answer is incorrect. However, after the first statement, all of the calculations and reasoning are correct.

The first step of multiplying the area by 4 was left out.

First find the total number of square feet of lawn. Find the area of each part of the yard.

$$(55 \times 32) + (75 \times 54) = 5810 \text{ ft}^2$$

The area of the lawn is greater than 5000 ft^2, which is the amount covered by the smaller bag, but buying the bag that covers 15,000 ft^2 would result in too much waste.

$$5810 \div 5000 = 1.162$$

Therefore, Mr. Solberg will need to buy 2 of the smaller bags of fertilizer.

No Credit Solution

In this sample solution, the response is incorrect and incomplete.

The wrong operations are used, so the answer is incorrect. Also, there are no units of measure given with any of the calculations.

$$55 + 75 = 130$$
$$32 + 54 = 86$$
$$130 \times 86 \times 4 = 44{,}720$$
$$44{,}720 \div 15{,}000 = 2.98$$

Mr. Solberg will need 3 bags of fertilizer.

Short-Response Practice

Solve each problem. Show all your work.

Number and Operations

1. In 2000, approximately $191 billion in merchandise was sold by a popular retail chain store in the United States. The population at that time was 281,421,906. Estimate the average amount of money spent at this store by each person in the U.S.

2. At a theme park, three educational movies run continuously all day long. At 9 A.M., the three shows begin. One runs for 15 minutes, the second for 18 minutes, and the third for 25 minutes. At what time will the movies all begin at the same time again?

3. Ming found a sweater on sale for 20% off the original price. However, the store was offering a special promotion, where all sale items were discounted an additional 60%. What was the total percent discount for the sweater?

4. The serial number of a DVD player consists of three letters of the alphabet followed by five digits. The first two letters can be any letter, but the third letter cannot be O. The first digit cannot be zero. How many serial numbers are possible with this system?

Algebra

5. Solve and graph $2x - 9 \leq 5x + 4$.

6. Vance rents rafts for trips on the Jefferson River. You have to reserve the raft and provide a $15 deposit in advance. Then the charge is $7.50 per hour. Write an equation that can be used to find the charge for any amount of time, where y is the total charge in dollars and x is the amount of time in hours.

Test-Taking Tip Ⓐ Ⓑ Ⓒ Ⓓ

Question 4
Be sure to completely and carefully read the problem before beginning any calculations. If you read too quickly, you may miss a key piece of information.

7. Hector is working on the design for the container shown below that consists of a cylinder with a hemisphere on top. He has written the expression $\pi r^2 + 2\pi rh + 2\pi r^2$ to represent the surface area of any size container of this shape. Explain the meaning of each term of the expression.

8. Find all solutions of the equation $6x^2 + 13x = 5$.

9. In 2001, there were 2,148,630 farms in the U.S., while in 2003, there were 2,126,860 farms. Let x represent years since 2001 and y represent the total number of farms in the U.S. Suppose the number of farms continues to decrease at the same rate as from 2001 to 2003. Write an equation that models the number of farms for any year after 2001.

Geometry

10. Refer to the diagram. What is the measure of $\angle 1$?

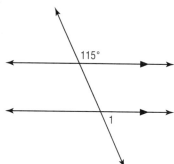

11. Quadrilateral $JKLM$ is to be reflected in the line $y = x$. What are the coordinates of the vertices of the image?

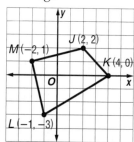

12. Write an equation in standard form for a circle that has a diameter with endpoints at $(-3, 2)$ and $(4, -5)$.

13. In the Columbia Village subdivision, an unusually shaped lot, shown below, will be used for a small park. Find the exact perimeter of the lot.

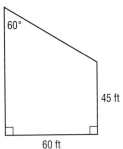

Measurement

14. The Astronomical Unit (AU) is the distance from Earth to the Sun. It is usually rounded to 93,000,000 miles. The star Alpha Centauri is 25,556,250 million miles from Earth. What is this distance in AU?

15. Linesse handpaints unique designs on shirts and sells them. It takes her about 4.5 hours to create a design. At this rate, how many shirts can she design if she works 22 days per month for an average of 6.5 hours per day?

16. The world's largest pancake was made in England in 1994. To the nearest cubic foot, what was the volume of the pancake?

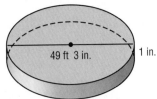

17. Find the ratio of the volume of the cylinder to the volume of the pyramid.

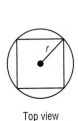

Top view

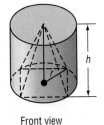

Front view

Data Analysis and Probability

18. The table shows the winning times for the Olympic men's 1000-meter speed skating event. Make a scatter plot of the data and describe the pattern in the data. Times are rounded to the nearest second.

Men's 1000-m Speed Skating Event		
Year	Country	Time(s)
1976	U.S.	79
1980	U.S.	75
1984	Canada	76
1988	USSR	73
1992	Germany	75
1994	U.S.	72
1998	Netherlands	71
2002	Netherlands	67

Source: *The World Almanac*

19. Bradley surveyed 70 people about their favorite spectator sport. If a person is chosen at random from the people surveyed, what is the probability that the person's favorite spectator sport is basketball?

Favorite Spectator Sport

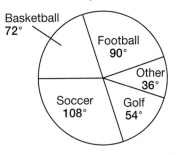

20. The graph shows the altitude of a small airplane. Write a function to model the graph. Explain what the model means in terms of the altitude of the airplane.

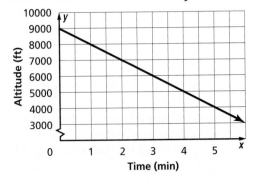

Extended-Response Questions

Extended-response questions are often called *open-ended* or *constructed-response questions*. Most extended-response questions have multiple parts. You must answer all parts to receive full credit.

Extended-response questions are similar to short-response questions in that you must show all of your work in solving the problem. A rubric is also used to determine whether you receive full, partial, or no credit. The following is a sample rubric for scoring extended-response questions.

Credit	Score	Criteria
Full	4	Full credit: A correct solution is given that is supported by well-developed, accurate explanations.
Partial	3, 2, 1	Partial credit: A generally correct solution is given that may contain minor flaws in reasoning or computation or an incomplete solution. The more correct the solution, the greater the score.
None	0	No credit: An incorrect solution is given indicating no mathematical understanding of the concept, or no solution is given.

> On some standardized tests, no credit is given for a correct answer if your work is not shown.

Make sure that when the problem says to *Show your work,* you show every part of your solution including figures, sketches of graphing calculator screens, or the reasoning behind your computations.

Example

1 Polygon *WXYZ* with vertices *W*(−3, 2), *X*(4, 4), *Y*(3, −1), and *Z*(−2, −3) is a figure represented on a coordinate plane to be used in the graphics for a video game. Various transformations will be performed on the polygon to use for the game.

a. Graph *WXYZ* and its image *W'X'Y'Z'* under a reflection in the *y*-axis. Be sure to label all of the vertices.

b. Describe how the coordinates of the vertices of *WXYZ* relate to the coordinates of the vertices of *W'X'Y'Z'*.

c. Another transformation is performed on *WXYZ*. This time, the vertices of the image *W'X'Y'Z'* are *W'*(2, −3), *X'*(4, 4), *Y'*(−1, 3), and *Z'*(−3, −2). Graph *WXYZ* and its image under this transformation. What transformation produced *W'X'Y'Z'*?

Strategy

Make a List
Write notes about what to include in your answer for each part of the question.

Full Credit Solution

Part a A complete graph includes labels for the axes and origin and labels for the vertices, including letter names and coordinates.

- The vertices of the polygon should be correctly graphed and labeled.
- The vertices of the image should be located such that the transformation shows a reflection in the *y*-axis.
- The vertices of the polygons should be connected correctly. Optionally, the polygon and its image could be graphed in two contrasting colors.

(continued on the next page)

PREPARING FOR STANDARDIZED TESTS

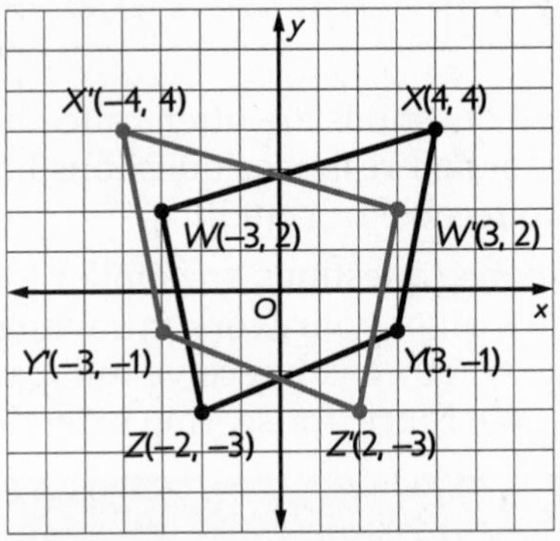

The first step of doubling the square footage for two coats of paint was left out.

Part b

The coordinates of W and W' are $(-3, 2)$ and $(3, 2)$. The x-coordinates are the opposite of each other and the y-coordinates are the same. For any point (a, b), the coordinates of the reflection in the y-axis are $(-a, b)$.

Part c

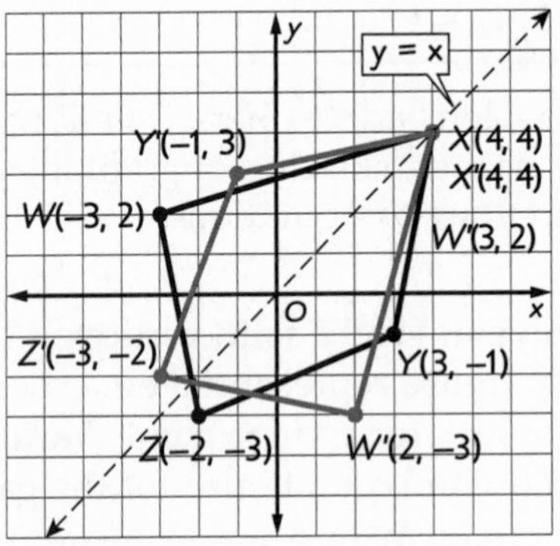

For full credit, the graph in Part C must also be accurate, which is true for this graph.

The coordinates of Z and Z' have been switched. In other words, for any point (a, b), the coordinates of the reflection in the y-axis are (b, a). Since X and X' are the same point, the polygon has been reflected in the line $y = x$.

Partial Credit Solution

Part a This sample graph includes no labels for the axes and for the vertices of the polygon and its image. Two of the image points have been incorrectly graphed.

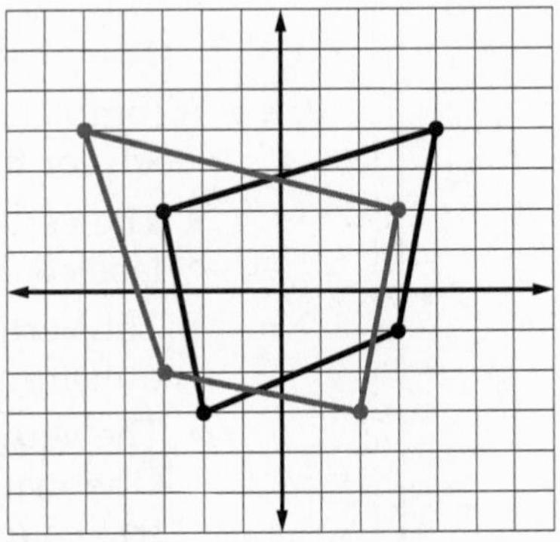

More credit would have been given if all of the points were reflected correctly. The images for *X* and *Y* are not correct.

Part b Partial credit is given because the reasoning is correct, but the reasoning was based on the incorrect graph in Part a.

> For two of the points, W and Z, the y-coordinates are the same and the x-coordinates are opposites. But, for points X and Y, there is no clear relationship.

Part c Full credit is given for Part c.

> I noticed that point X and point X' were the same. I also guessed that this was a reflection, but not in either axis. I played around with my ruler until I found a line that was the line of reflection. The transformation from WXYZ to W'X'Y'Z' was a reflection in the line y = x.

This sample answer might have received a score of 2 or 1, depending on the judgment of the scorer. Had the student graphed all points correctly and gotten Part b correct, the score would probably have been a 3.

No Credit Solution

Part a The sample answer below includes no labels on the axes or the coordinates of the vertices of the polygon. The polygon WXYZ has three vertices graphed incorrectly. The polygon that was graphed is not reflected correctly either.

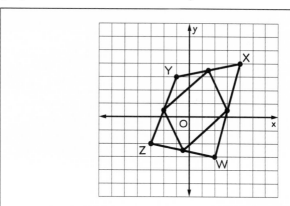

Part b
> I don't see any way that the coordinates relate.

Part c
> It is a reduction because it gets smaller.

In this sample answer, the student does not understand how to graph points on a coordinate plane and also does not understand the reflection of figures in an axis or other line.

Extended Response Practice

Solve each problem. Show all your work.

Number and Operations

1. Refer to the table.

Population		
City	1990	2000
Phoenix, AZ	983,403	1,321,045
Austin, TX	465,622	656,562
Charlotte, NC	395,934	540,828
Mesa, AZ	288,091	396,375
Las Vegas, NV	258,295	478,434

Source: census.gov

a. For which city was the increase in population the greatest? What was the increase?

b. For which city was the percent of increase in population the greatest? What was the percent increase?

c. Suppose that the population increase of a city was 30%. If the population in 2000 was 346,668, find the population in 1990.

2. Molecules are the smallest units of a particular substance that still have the same properties as that substance. The diameter of a molecule is measured in angstroms (Å). Express each value in scientific notation.

a. An angstrom is exactly 10^{-8} centimeter. A centimeter is approximately equal to 0.3937 inch. What is the approximate measure of an angstrom in inches?

b. How many angstroms are in one inch?

c. If a molecule has a diameter of 2 angstroms, how many of these molecules placed side by side would fit on an eraser measuring $\frac{1}{4}$ inch?

Algebra

3. The Marshalls are building a rectangular in-ground pool in their backyard. The pool will be 24 feet by 29 feet. They want to build a deck of equal width all around the pool. The final area of the pool and deck will be 1800 square feet.

a. Draw and label a diagram.

b. Write an equation that can be used to find the width of the deck.

c. Find the width of the deck.

4. The depth of a reservoir was measured on the first day of each month. (Jan. = 1, Feb. = 2, and so on.)

Depth of the Reservoir

a. What is the slope of the line joining the points with x-coordinates 6 and 7? What does the slope represent?

b. Write an equation for the segment of the graph from 5 to 6. What is the slope of the line and what does this represent in terms of the reservoir?

c. What was the lowest depth of the reservoir? When was this depth first measured and recorded?

Geometry

5. The Silver City Marching Band is planning to create this formation with the members.

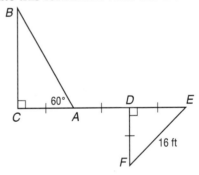

a. Find the missing side measures of $\triangle EDF$. Explain.

b. Find the missing side measures of $\triangle ABC$. Explain.

c. Find the total distance of the path: A to B to C to A to D to E to F to D.

d. The director wants to place one person at each point A, B, C, D, E, and F. He then wants to place other band members approximately one foot apart on all segments of the formation. How many people should he place on each segment of the formation? How many total people will he need?

Measurement

6. Two containers have been designed. One is a hexagonal prism, and the other is a cylinder.

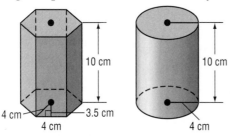

a. What is the volume of the hexagonal prism?

b. What is the volume of the cylinder?

c. What is the percent of increase in volume from the prism to the cylinder?

7. Kabrena is working on a project about the solar system. The table shows the maximum distances from Earth to the other planets in millions of miles.

Distance from Earth to Other Planets			
Planet	**Distance**	**Planet**	**Distance**
Mercury	138	Saturn	1031
Venus	162	Uranus	1962
Mars	249	Neptune	2913
Jupiter	602	Pluto	4681

Source: *The World Almanac*

a. The maximum speed of the Apollo moon missions spacecraft was about 25,000 miles per hour. Make a table showing the time it would take a spacecraft traveling at this speed to reach each of the four closest planets.

b. Describe how to use scientific notation to calculate the time it takes to reach any planet.

c. Which planet would it take approximately 13.3 years to reach? Explain.

Test-Taking Tip   Ⓐ Ⓑ Ⓒ Ⓓ

Question 6
While preparing to take a standardized test, familiarize yourself with the formulas for surface area and volume of common three-dimensional figures.

8. The table shows the average monthly temperatures in Barrow, Alaska. The months are given numerical values from 1-12. (Jan. = 1, Feb. = 2, and so on.)

Average Monthly Temperature			
Month	**°F**	**Month**	**°F**
1	−14	7	40
2	−16	8	39
3	−14	9	31
4	−1	10	15
5	20	11	−1
6	35	12	−11

a. Make a scatter plot of the data. Let x be the numerical value assigned to the month and y be the temperature.

b. Describe any trends shown in the graph.

c. Find the mean of the temperature data.

d. Describe any relationship between the mean of the data and the scatter plot.

9. A dart game is played using the board shown. The inner circle is pink, the next ring is blue, the next red, and the largest ring is green. A dart must land on the board during each round of play.

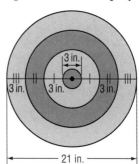

a. What is the probability that a dart landing on the board hits the pink circle?

b. What is the probability that the first dart thrown lands in the blue ring and the second dart lands in the green ring?

c. Suppose players throw a dart twice. For which outcome of two darts would you award the most expensive prize? Explain your reasoning.

Graphing Calculator Tutorial

General Calculator Information

Making the Display Lighter or Darker

To lighten or darken the display, turn the calculator on. Then hold down the 2nd key and press ▼ to lighten or ▲ to darken the display.

Turning the Calculator Off

Press 2nd [OFF].

Using Apps

TI Graphing Calculator Software Applications (apps) are pieces of software for your graphing calculator. Apps allow you to customize your TI calculator.

Accessing the Apps

Press APPS to display a list of apps that are available on your calculator. Use the arrow keys to scroll through the list. Press ENTER to open an app.

Downloading Apps

If you have Internet access and a TI GRAPH™ LINK cable, you can download apps to your calculator.

Using the Cabri Jr. Application

To use the Cabri Jr. Application on a TI-83 Plus or TI-84 Plus, you will select tools and options from a set of menus. The menus are labeled F1–F5.

Using Menus

- **Access the menus** using the graphing keys that are directly below the screen. To open a menu, press ALPHA plus the function key F1, F2, F3, F4, or F5. You can also open a menu by pressing Y=, WINDOW, ZOOM, TRACE, or GRAPH for the corresponding function key.

- **Move from one menu item to another** by pressing ▲ or ▼.

- **Select a menu item** by highlighting it and pressing ENTER.

- **Deactivate a tool or close a menu** by pressing CLEAR. This returns you to the drawing screen.

Opening a Geometry Session

To open a geometry session, press APPS choose CabriJr. Choose New from the menu to begin a new construction.

Saving a Session

Press ALPHA F1. Then choose Save. Type in a name up to eight characters long and press ENTER.

Dragging an Object

To select a point, use the arrow keys to move the cursor close to the point until the arrow turns clear. To drag an object, press ALPHA .

Key Skills

Each Graphing Calculator Exploration in the Student Edition requires the use of certain key skills. Use this section as a reference when you need further instruction on these skills.

Chapter	Page(s)	Key Skills
1	32	A, E, M, S
2	79	A, B, C, D, E
3	112	A, F, G, H, I
4	170	A, B, J, K, P
5	193	I, L
6	246–247	A, E, H, I, L, M
7	290	E, I, L, N, O
8	316–317	A, E, I, J, M
9	371	E, G, J, L, M, P
11	478	A, E
14	608	A, I, P
16	700	C, I, L, Q, R

Deleting an Object

Press ALPHA F5 . Choose Clear and then Object from the menu. Use the arrow keys to select the object and press ENTER .

A: *Creating a Line, Segment, or Circle*

Open the F2 menu, select the object type that you want to construct and press ENTER . Move the pointer to a location for the first point of the object and press ENTER to draw the point. Then move the pointer to a location for the second point of the object and press ENTER to draw the object.

Labeling an Object

First create the object. Then press ALPHA F5 . Choose Alph-Num from the menu. Move the pointer to the object and press ENTER . The object blinks when the pointer is close enough to the point to select it. Then type the letters.

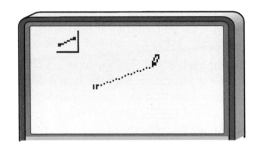

Hiding an Object

First create the object. Then press ALPHA F5 . Choose Hide/Show and then Objects from the menu. Point to each object you wish to hide, and press ENTER . The hidden object appears in dotted outline. *Use the same method to show the hidden objects again.*

B: *Show Coordinate Axes*

To show the axes, open the F5 menu. Select Hide/Show and then select Axes. *Use the same method to hide the axes again.*

C: *Finding Coordinates of Vertices*

To display the coordinates of a point in the underlying system of axes, open the F5 menu. Select Coord. & Eq. Then use the arrow keys to select the object.

Quitting the Application

Press 2nd QUIT or press ALPHA then select QUIT .

Graphing Calculator Tutorial 783

GRAPHING CALCULATOR TUTORIAL

D: *Finding a Midpoint*

Open the F3 menu and select Midpoint. Then move the pointer to the line segment and press ENTER . The midpoint is drawn.

You may also draw a midpoint by placing the cursor on one of the endpoints and pressing ENTER . Then move to the other endpoint and press ENTER .

E: *Finding a Distance or Length*

Use the Distance and Length tool to find the distance between two points, the length of a line segment, the perimeter of a triangle or a quadrilateral, or the circumference of a circle. Open the F5 menu and select Measure. Then select D. & Length. Select the two points the distance between which you want to find or the object for which you want the perimeter or circumference.

F: *Creating a Line through a Point*

To create a line through a point P, first create and label a point P. Open the F2 menu, select the object type that you want to construct and press ENTER . Move the pointer to the existing point, and then press ENTER to select that point. Move the cursor and the line is drawn in the same direction that you moved the cursor. Use the cursor keys to change the slope of the line, if desired. Then press ENTER to complete the construction.

G: *Marking Points on a Line*

Open the F2 menu and select Point. Highlight Point On, and then press ENTER . Move the pointer to the line or other object on which you want to draw a point, and then press ENTER to draw the point.

H: *Creating an Angle Bisector*

Open the F3 menu and select Angle Bis. Move the pointer to each of the three points that create the angle, pressing ENTER at each point. The second point must be the vertex of the angle. When you select the third point, the angle bisector is drawn.

I: *Measuring an Angle*

Open the F5 menu, select Measure, and then select Angle. Select three points to define the angle you wish to measure. The second point must be the vertex of the angle. The measure appears after you have selected the third vertex. Move the cursor where you want the label to appear and press ENTER .

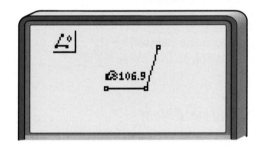

J: *Creating Perpendicular or Parallel Lines*

Create a line. Then open the F3 menu and select Perp or Parallel. Move the pointer to the line or segment for which you want to draw a perpendicular or parallel line and press ENTER . The line is drawn. Move the pointer to a point through which you want the perpendicular or parallel line to pass. Then press ENTER . The line is drawn through the point.

K: *Finding the Slope of a Line*

Open the F5 menu, select Measure and then select Slope. Select the line or segment. The measurement is displayed.

L: *Creating a Triangle or Quadrilateral*

Open the F2 menu and select Triangle or Quad. Move the cursor to each location at which you want a vertex and press `ENTER`.

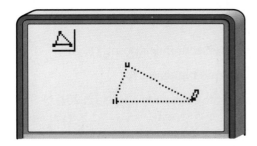

M: *Creating a Point of Intersection*

Create two intersecting objects. Then open the F2 menu, select Point and then select Intersection. If the objects intersect within the display, move the pointer to the intersection point. The objects that intersect at the pointer position blink. Press `ENTER` to create the intersection point. If the objects intersect outside of the display, move the pointer to the first object and press `ENTER`. Then move the pointer to the second object, and press `ENTER` to create the point.

N: *Creating a Perpendicular Bisector*

Create a segment or triangle. Open the F3 menu and select Perp. Bis. Move the pointer to the segment or triangle side for which you want to draw the perpendicular bisector and press `ENTER`.

O: *Creating a Comment*

A comment is similar to a label, but it is not attached to an object. Open the F5 menu and select Alpha-Num. Move the pointer where you want to type text Type the comment on the keyboard, and then press `ENTER`.

P: *Using the Calculate Feature*

Use the Calculate tool to perform calculations using values on the drawing screen. First find measurements of objects or place numeric labels on the drawing screen. Then open the F5 menu and select Calculate. Select the measurement and then select the operation you want to perform. Press `ENTER` to perform the calculation. The result is displayed.

Q: *Rotating an Object About a Point*

Create an object and a point around which the object will be rotated. Then draw three points whose angle will determine the angle of rotation. Open the F4 menu and then select Rotation. Select the three points that determine the angle of rotation. A new rotated object is created.

R: *Making Outlines Solid or Dotted*

Create an object. Open the F5 menu and select Display. Move the pointer to the object. The pointer changes from a solid arrow to a hollow arrow. Press `ENTER` to select the item and change the display.

S: *Using the Compass Tool*

When you draw a circle using the compass tool, a line segment or the distance between two points is the setting for the compass. Open the F3 menu and select Compass. If the segment that should be the compass setting is drawn, then move the pointer to the line segment and press `ENTER`. A dotted circle is drawn. If the segment is not already drawn, move the pointer to the first point and press `ENTER` then move to the second point and press `ENTER`. A dotted circle is drawn. Use the arrow keys to move the dotted circle (if necessary), and then press `ENTER` to finish.

Postulates and Theorems

Chapter 5   Triangles and Congruence

Postulate 5–1   **SSS Postulate**   If three sides of one triangle are congruent to three corresponding sides of another triangle, then the triangles are congruent. *(p. 211)*

Postulate 5–2   **SAS Postulate**   If two sides and the included angle of one triangle are congruent to the corresponding sides and included angle of another triangle, then the triangles are congruent. *(p. 212)*

Postulate 5–3   **ASA Postulate**   If two angles and the included side of one triangle are congruent to the corresponding angles and included side of another triangle, then the triangles are congruent. *(p. 215)*

Theorem 5–1   **Angle Sum Theorem**   The sum of the measures of the angles of a triangle is 180. *(p. 193)*

Theorem 5–2   The acute angles of a right triangle are complementary. *(p. 195)*

Theorem 5–3   The measure of each angle of an equiangular triangle is 60. *(p. 195)*

Theorem 5–4   **AAS Theorem**   If two angles and a nonincluded side of one triangle are congruent to the corresponding two angles and nonincluded side of another triangle, then the triangles are congruent. *(p. 216)*

Chapter 6   More About Triangles

Postulate 6–1   **HL Postulate**   If the hypotenuse and a leg of one right triangle are congruent to the hypotenuse and corresponding leg of another right triangle, then the triangles are congruent. *(p. 252)*

Theorem 6–1   The length of the segment from the vertex to the centroid is twice the length of the segment from the centroid to the midpoint. *(p. 230)*

Theorem 6–2   **Isosceles Triangle Theorem**   If two sides of a triangle are congruent, then the angles opposite those sides are congruent. *(p. 247)*

Theorem 6–3   The median from the vertex angle of an isosceles triangle lies on the perpendicular bisector of the base and the angle bisector of the vertex angle. *(p. 247)*

Theorem 6–4   **Converse of Isosceles Triangle Theorem**   If two angles of a triangle are congruent, then the sides opposite those angles are congruent. *(p. 248)*

Theorem 6–5   A triangle is equilateral if and only if it is equiangular. *(p. 249)*

Theorem 6–6   **LL Theorem**   If two legs of one right triangle are congruent to the corresponding legs of another right triangle, then the triangles are congruent. *(p. 251)*

Theorem 6–7   **HA Theorem**   If the hypotenuse and an acute angle of one right triangle are congruent to the hypotenuse and corresponding angle of another right triangle, then the triangles are congruent. *(p. 252)*

Theorem 6–8   **LA Theorem**   If one leg and an acute angle of one right triangle are congruent to the corresponding leg and angle of another right triangle, then the triangles are congruent. *(p. 252)*

Theorem 6–9   **Pythagorean Theorem**   In a right triangle, the square of the length of the hypotenuse c is equal to the sum of the squares of the lengths of the legs a and b. *(p. 256)*

Theorem 10–2	In any convex polygon, the sum of the measures of the exterior angles, one at each vertex, is 360. *(p. 410)*
Theorem 10–3	**Area of a Triangle** If a triangle has an area of A square units, a base of b units, and a corresponding altitude of h units, then $A = \frac{1}{2}bh$. *(p. 419)*
Theorem 10–4	**Area of a Trapezoid** If a trapezoid has an area of A square units, bases of b_1 and b_2 units, and an altitude of h units, then $A = \frac{1}{2}h(b_1 + b_2)$. *(p. 421)*
Theorem 10–5	**Area of a Regular Polygon** If a regular polygon has an area of A square units, an apothem of a units, and a perimeter of P units, then $A = \frac{1}{2}aP$. *(p. 426)*

Chapter 11   Circles

Postulate 11–1	**Arc Addition Postulate** The sum of the measures of two adjacent arcs is the measure of the arc formed by the adjacent arcs. *(p. 463)*
Theorem 11–1	All radii of a circle are congruent. *(p. 455)*
Theorem 11–2	The measure of the diameter d of a circle is twice the measure of the radius r of the circle. *(p. 455)*
Theorem 11–3	In a circle or in congruent circles, two minor arcs are congruent if and only if their corresponding central angles are congruent. *(p. 464)*
Theorem 11–4	In a circle or in congruent circles, two minor arcs are congruent if and only if their corresponding chords are congruent. *(p. 468)*
Theorem 11–5	In a circle, a diameter bisects a chord and its arc if and only if it is perpendicular to the chord. *(p. 469)*
Theorem 11–6	In a circle or in congruent circles, two chords are congruent if and only if they are equidistant from the center. *(p. 475)*
Theorem 11–7	**Circumference of a Circle** If a circle has a circumference of C units and a radius of r units, then $C = 2\pi r$ or $C = \pi d$. *(p. 479)*
Theorem 11–8	**Area of a Circle** If a circle has an area of A square units and a radius of r units, then $A = \pi r^2$. *(p. 483)*
Theorem 11–9	**Area of a Sector of a Circle** If a sector of a circle has an area of A square units, a central angle measurement of N degrees, and a radius of r units, then $A = \frac{N}{360}(\pi r^2)$. *(p. 485)*

Chapter 12   Surface Area and Volume

Theorem 12–1	**Lateral Area of a Prism** If a prism has a lateral area of L square units and a height of h units and each base has a perimeter of P units, then $L = Ph$. *(p. 504)*
Theorem 12–2	**Surface Area of a Prism** If a prism has a surface area of S square units and a height of h units and each base has a perimeter of P units and an area of B square units, then $S = Ph + 2B$. *(p. 504)*
Theorem 12–3	**Lateral Area of a Cylinder** If a cylinder has a lateral area of L square units and a height of h units and the bases have radii of r units, then $L = 2\pi rh$. *(p. 507)*
Theorem 12–4	**Surface Area of a Cylinder** If a cylinder has a surface area of S square units and a height of h units and the bases have radii of r units, then $S = 2\pi rh + 2\pi r^2$. *(p. 507)*

Theorem 12–5 **Volume of a Prism** If a prism has a volume of V cubic units, a base with an area of B square units, and a height of h units, then $V = Bh$. *(p. 511)*

Theorem 12–6 **Volume of a Cylinder** If a cylinder has a volume of V cubic units, a radius of r units, and a height of h units, then $V = \pi r^2 h$. *(p. 512)*

Theorem 12–7 **Lateral Area of a Regular Pyramid** If a regular pyramid has a lateral area of L square units a base with a perimeter of P units, and a slant height of ℓ units, then $L = \frac{1}{2}P\ell$. *(p. 517)*

Theorem 12–8 **Surface Area of a Regular Pyramid** If a regular pyramid has a total surface area of S square units, a slant height of ℓ units, and a base with perimeter of P units and an area of B square units, then $S = \frac{1}{2}P\ell + B$. *(p. 517)*

Theorem 12–9 **Lateral Area of a Cone** If a cone has a lateral area of L square units, a slant height of ℓ units, and a base with a radius of r units, then $L = \pi r\ell$. *(p. 519)*

Theorem 12–10 **Surface Area of a Cone** If a cone has a surface area of S square units, a slant height of ℓ units, and a base with a radius of r units, then $S = \pi r\ell + \pi r^2$. *(p. 519)*

Theorem 12–11 **Volume of a Pyramid** If a pyramid has a volume of V cubic units, and a height of h units, and the area of the base is B square units, then $V = \frac{1}{3}Bh$. *(p. 523)*

Theorem 12–12 **Volume of a Cone** If a cone has a volume of V cubic units, a radius of r units, and a height of h units, then $V = \frac{1}{3}\pi r^2 h$. *(p. 523)*

Theorem 12–13 **Surface Area of a Sphere** If a sphere has a surface area of S square units and a radius of r units, then $S = 4\pi r^2$. *(p. 529)*

Theorem 12–14 **Volume of a Sphere** If a sphere has a volume of V cubic units and a radius of r units, then $V = \frac{4}{3}\pi r^3$. *(p. 529)*

Theorem 12–15 If two solids are similar with a scale factor of $a{:}b$, then the surface areas have a ratio of $a^2{:}b^2$ and the volumes have a ratio of $a^3{:}b^3$. *(p. 536)*

Chapter 13 Right Triangles and Trigonometry

Theorem 13–1 **45°-45°-90° Triangle Theorem** In a 45°-45°-90° triangle, the hypotenuse is $\sqrt{2}$ times as long as a leg. *(p. 555)*

Theorem 13–2 **30°-60°-90° Triangle Theorem** In a 30°-60°-90° triangle, the hypotenuse is twice as long as the shorter leg, and the longer leg is $\sqrt{3}$ times as long as the shorter leg. *(p. 560)*

Theorem 13–3 If x is the measure of an acute angle of a right triangle, then $\frac{\sin x}{\cos x} = \tan x$. *(p. 574)*

Theorem 13–4 If x is the measure of an acute angle of a right triangle, then $\sin^2 x + \cos^2 x = 1$. *(p. 577)*

Chapter 14 Circle Relationships

Theorem 14–1 The degree measure of an inscribed angle equals one-half the degree measure of its intercepted arc. *(p. 587)*

Theorem 14–2 If inscribed angles intercept the same arc or congruent arcs, then the angles are congruent. *(p. 588)*

Theorem 14–3 If an inscribed angle of a circle intercepts a semicircle, then the angle is a right angle. *(p. 589)*

Theorem 14–4 In a plane, if a line is tangent to a circle, then it is perpendicular to the radius drawn to the point of tangency. *(p. 592)*

Theorem 14–5 In a plane, if a line is perpendicular to a radius of a circle at its endpoint on the circle, then the line is a tangent. *(p. 592)*

Theorem 14–6 If two segments from the same exterior point are tangent to a circle, then they are congruent. *(p. 594)*

Theorem 14–7 A line or line segment is a secant to a circle if and only if it intersects the circle in two points. *(p. 600)*

Theorem 14–8 If a secant angle has its vertex inside a circle, then its degree measure is one-half the sum of the degree measures of the arcs intercepted by the angle and its vertical angle. *(p. 601)*

Theorem 14–9 If a secant angle has its vertex outside a circle, then its degree measure is one-half the difference of the degree measures of the intercepted arcs. *(p. 601)*

Theorem 14–10 If a secant-tangent angle has its vertex outside the circle, then its degree measure is one-half the difference of the degree measures of the intercepted arcs. *(p. 606)*

Theorem 14–11 If a secant-tangent angle has its vertex on the circle, then its degree measure is one-half the degree measure of the intercepted arc. *(p. 606)*

Theorem 14–12 The degree measure of a tangent-tangent angle is one-half the difference of the degree measures of the intercepted arcs. *(p. 607)*

Theorem 14–13 If two chords of a circle intersect, then the product of the measures of the segments of one chord equals the product of the measures of the segments of the other chord. *(p. 612)*

Theorem 14–14 If two secant segments are drawn to a circle from an exterior point, then the product of the measures of one secant segment and its external secant segment equals the product of the measures of the other secant segment and its external secant segment. *(p. 613)*

Theorem 14–15 If a tangent segment and a secant segment are drawn to a circle from an exterior point, then the square of the measure of the tangent segment equals the product of the measures of the secant segment and its external secant segment. *(p. 614)*

Theorem 14–16 **General Equation of a Circle** The equation of a circle with center at (h, k) and a radius of r units is $(x - h)^2 + (y - k)^2 = r^2$. *(p. 618)*

Problem-Solving Strategy Workshops

Look for a Pattern

Project

When you first look at a section of a nautilus shell, you may not think of a number pattern. But if you examine the figure at the right, you'll discover a famous pattern called the *Fibonacci sequence*, which is named for Italian mathematician Leonardo Fibonacci (1170–1250). Explain the pattern in the Fibonacci sequence and tell how it is shown in the nautilus shell.

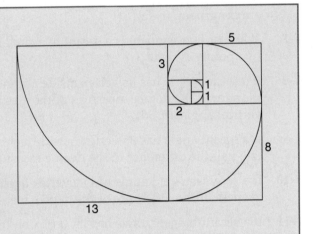

Working on the Project

Work with a partner and choose a strategy to help analyze the pattern. Develop a plan. Here are some suggestions to help you get started.

- Start at the innermost part of the spiral. As you go clockwise around the spiral, write the numbers.
- Look for a pattern and write five more numbers in the sequence.
- Do research about Fibonacci and his contributions to mathematics.

Technology Tools

- Use an **electronic encyclopedia** to do your research.
- Use **word processing software** to write your report.

 Research For more information about Fibonacci, visit: www.geomconcepts.com

Presenting the Project

Write a report about Fibonacci and the Fibonacci sequence. Make sure your report contains the following:

- a discussion of the pattern in the Fibonacci sequence and examples of the Fibonacci sequence in nature, and
- an explanation of how squares and rectangles are used when drawing the spiral in the nautilus shell.

Analyze Data and Make a Graph

Project

You are a reporter for your school newspaper. Your assignment is to conduct a survey about favorite television shows. The results of the survey must be shown in a circle graph. The angles in the circle graph must have the correct measure. How can you make an accurate circle graph that reflects your classmates' opinions?

Working on the Project

Work with a partner and choose a strategy to help analyze and solve the problem. Develop a plan. Here are some suggestions to help you get started.

- Do research to find the top six prime time television shows from last week according to the Nielsen ratings.

- Conduct a poll and ask each person to pick his or her favorite show from the list.

- What percent of people picked each show?

- Determine the angle measure to represent each show in the circle graph. (*Hint:* Multiply the percent by 360.)

Technology Tools

- Use **computer software** to design your circle graph.

- Use **word processing software** to write a paragraph explaining how angles are used to create circle graphs.

 Research   For more information on the Nielsen ratings, visit: www.geomconcepts.com

Presenting the Project

Draw your circle graph on unlined paper. Use color to enhance your graph and include labels. Make sure your paragraph contains the following information:

- the number of people polled,

- the number and percent of people who voted for each show, and

- an explanation of how you determined what portion of the circle graph to use for each show.

Act It Out

Project

A local furniture store is having a contest to see who can design the best dream bedroom for a teenager. Design a floor plan with furniture using a scale where $\frac{1}{2}$ inch represents one foot.

Working on the Project

Work with a partner and discuss what the room will need. Here are a few questions to get you started.

- What furniture do you want in the room?
- What size will you make your dream room?

Technology

- Use **software** to create your design.
- Use **word processing software** to write a paragraph explaining your floor design.
- Use **presentation software** to present your project.

*inter*NET
CONNECTION   **Research**   For more information about buying furniture, visit:
www.geomconcepts.com

Presenting the Project

Draw the floor plan of your dream room. Use colors to enhance your design. Along with your floor plan, include a paragraph that contains the following information:

- a description of the furniture you chose to include in your dream room, and
- an explanation of how you used proportions in this project.

Guess and Check

Project

You are applying for a job at a greeting card company to design wrapping paper. To get the job, you must create a sample design for the wrapping paper and submit a proposal to the company convincing them to choose your work. Your research shows that this company prefers wrapping paper designs that include tessellations. How can you include tessellations in your design and increase your chances of getting the job?

Working on the Project

Work with a partner and choose a strategy to help you get started. Here are some suggestions.

- Research the works of artist M.C. Escher.
- Make a list of which geometric shapes tessellate and which do not.
- Choose one or more shapes to tessellate.
- Decide what colors to use.

Technology Tools

- Use an **electronic encyclopedia** to do your research.
- Use **design software** to design your wrapping paper.
- Use **word processing software** to write your proposal.

 Research For more information about M.C. Escher and tessellations, visit: www.geomconcepts.com

Presenting the Project

Draw your design on large unlined paper, such as newsprint. Use the colors that you chose. Write a proposal for the greeting card company and explain why they should choose your design. Include the following information in your proposal:

- the shape or shapes you chose to tessellate,
- an explanation of why you chose that shape,
- the estimated area of one complete unit of your tessellation, and
- your list of which shapes tessellate and which do not.

Use a Formula

Project

Before calculators, mathematicians were able to calculate the value of pi (π) with surprising accuracy. Suppose your science or math teacher asks you to conduct an experiment to investigate ways to approximate π. How can you calculate the value of π to three decimal places by making direct measurements?

Working on the Project

Work with one or two other people to develop a strategy to solve this problem. Here are some suggestions to get you started.

- Research the history of π.
- Use string to find the circumference of, or distance around, several circular objects. Also, measure the diameters of the objects.
- Use your measurements and the formula $C = \pi d$ to calculate π to the nearest thousandth.

Technology

- Use a **scientific calculator** to do your calculations.
- Use **word processing software** to write a report on what you have discovered.
- Use **presentation software** to present your report.

 interNET
CONNECTION   **Research**   For more information about the history of π, visit: www.geomconcepts.com

Presenting the Project

Write a report on what you have discovered through this experiment. Be sure to include the following:

- the information you discovered in your research about π,
- a table of your measurements of the circular objects, and
- the answers to the following questions.
  - (1) How did your approximations compare to the actual value of π?
  - (2) Did this comparison change when you increased the circumferences of the object you were measuring? If so, how?

Use a Table

Project

Coastal Soup Company hires you to design a new soup container. They want the container to be a right cylinder that will hold 150 cubic centimeters using as little material as possible. How would you design the container that will meet their needs?

Working on the Project

Work with a partner and choose a strategy to solve the problem. Here are a few suggestions to get you started.

- Choose various values for the radius *r* and the height *h* of your container that will give you a volume of about 150 cubic centimeters. Also, find and list the surface area for each set of values.

- Make a table for the values of radius, height, volume, and surface area.

Technology

- Use the table feature on a **graphing calculator** to make the table.
- Use a **spreadsheet** to make your calculations.
- Use **word processing software** to write a report to the company.
- Use **presentation software** to present your project.

*inter*NET
CONNECTION   **Research**   For more information about packaging, visit: www.geomconcepts.com

Presenting the Project

Draw your design for the soup container on unlined paper. Include the radius, height, surface area, and the volume, which should be close to 150 cubic centimeters. Write a report for the company explaining why your design will suit their needs. Your report should include the following information:

- a description of the work you did to find the final dimensions,
- a copy of the table that you made,
- the formulas that you used, and
- an explanation of why your container will use the least amount of material.

Draw a Diagram

Project

The next time you scream in fear or excitement on a roller coaster, think of right triangles. Right triangles can show the vertical drop and angle of elevation that make a great roller coaster ride. Compare the vertical drops and angles of elevation for several different roller coasters.

Working on the Project

Work with a partner and choose a strategy to help analyze and solve the problem. Develop a plan. Here are some suggestions to help you get started.

- The first hill of the *Mean Streak* at an amusement park in Sandusky, Ohio, has a vertical drop of 155 feet and a 52° angle of elevation. Make a sketch of the first hill.

- Research other roller coasters.

Technology Tools

- Use an **electronic encyclopedia** to do your research.
- Use **drawing software** to make your drawings.
- Use **presentation software** to present your project.

 **interNET**
CONNECTION   **Research**   For more information about roller coasters, visit:
www.geomconcepts.com

Presenting the Project

Make a visual display that shows the vertical drop and angle of elevation for several different roller coasters. In your presentation, include the following:

- scale drawings of the right triangles,

- the vertical drop (rise), horizontal change (run), length of track (hypotenuse), and angle of elevation for each roller coaster, and

- an explanation of how you used trigonometry to find measures in your display.

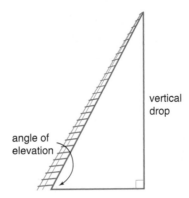

vertical drop

angle of elevation

Use Logical Reasoning

Project

Every time you look at a magazine, watch television, ride in a bus, or surf the Internet you are bombarded with advertisements. Sometimes advertisements contain faulty logic. Find five different advertisements and analyze the claims that are made in each of them. Identify the hypothesis, conclusion, and rules of logic that are used.

Working on the Project

Work with a partner and choose a strategy to help analyze each advertisement. Develop a plan. Here are some suggestions to help you get started.

- Write each statement in if-then form. You may want to review conditional statements in Lesson 1–4.

- Write the converse of each statement. Ask yourself whether the advertiser wants you to believe the conditional statement is true or its converse is true.

Technology Tools

- Surf the **Internet** to do some of your research.

- Use **word processing software** to write a report.

- Use **presentation software** to present your report.

 **Research** For more information about advertising, visit: www.geomconcepts.com

Presenting the Project

Write a report about your advertisements. Make sure your report contains the following:

- a discussion of the hypothesis, conclusion, and rules of logic that are used in each advertisement,

- an explanation of how inductive or deductive reasoning is used, and

- a discussion about whether the advertisement is misleading.

Glossary/Glosario

A mathematics multilingual glossary is available at www.math.glencoe.com.

The glossary is available in the following languages.

Arabic	Haitian Creole	Russian	Urdu	Bengali	Hmong
Spanish	Vietnamese	Cantonese	Korean	Tagalog	

absolute value   The number of units that a number is from zero on a number line. *(p. 52)*

valor absoluto   El número de unidades que un número dista de cero en una recta numérica. *(pág. 52)*

acute angle   An angle whose measure is less than 90. *(p. 98)*

ángulo agudo   Ángulo que mide menos de 90. *(pág. 98)*

$$0 < m\angle A < 90$$

acute triangle   A triangle with all acute angles. *(p. 188)*

triángulo acutángulo   Triángulo cuyos ángulos son todos agudos. *(pág. 188)*

three acute angles
tres ángulos agudos

adjacent angles   Two angles that share a common side and have the same vertex, but have no interior points in common. *(p. 110)*

ángulos adyacentes   Dos ángulos que comparten un lado, poseen el mismo mismo vértice, pero sin puntos interiores comunes. *(pág. 110)*

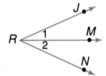

$\angle 1$ and $\angle 2$ are adjacent angles.
$\angle 1$ y $\angle 2$ son adyacentes.

adjacent arcs   Arcs of a circle with one point in common. *(p. 463)*

arcos adyacentes   Arcos de un círculo que sólo tienen un punto en común. *(pág. 463)*

$\overset{\frown}{PQ}$ and $\overset{\frown}{QR}$ are adjacent arcs.
$\overset{\frown}{PQ}$ y $\overset{\frown}{QR}$ son arcos adyacentes.

alternate exterior angles   See *transversal.*

ángulos alternos externos   Ver *transversal.*

alternate interior angles   See *transversal.*

ángulos alternos internos   Ver *transversal.*

altitude of a trapezoid   See *trapezoid.*

altura de un trapecio   Ver *trapecio.*

altitude of a triangle   A perpendicular segment in which one endpoint is a vertex of the triangle and the other is a point on the side opposite the vertex. *(p. 234)*

altura de un triángulo   Segmento perpendicular, uno de cuyos extremos es un vértice del triángulo y el otro es un punto en el lado opuesto al vértice. *(pág. 234)*

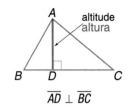

$$\overline{AD} \perp \overline{BC}$$

angle   A figure formed by two noncollinear rays that have a common endpoint and are not opposite rays. The rays are the *sides* of the angle. The endpoint is the *vertex* of the angle. An angle separates a plane into three parts, the *interior* of the angle, the *exterior* of the angle, and the angle itself. *(p. 90)*

ángulo   Figura formada por dos rayos no colineales y no opuestos que poseen un extremo común. Los rayos son los *lados* del ángulo y el extremo común es su *vértice*. Un ángulo separa el plano en tres partes: el *interior* del ángulo, el *exterior* del ángulo y el ángulo mismo. *(pág. 90)*

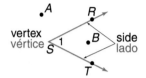

∠RST, ∠TSR, ∠S, ∠1
A is in the exterior of ∠1.
B is in the interior of ∠1.
A es el exterior del ∠1.
B es el interior del ∠1.

angle bisector   A ray whose endpoint is the vertex and is located in the interior of the angle that separates a given angle into two angles with equal measure. *(p. 106)*

bisectriz de un ángulo   Un rayo cuyo extremo es el vértice de un ángulo dado, está situado en su interior y lo divide en dos ángulos de igual medida. *(pág. 106)*

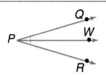

$\overrightarrow{PW}$ is the bisector of ∠P.
$\overrightarrow{PW}$ es la bisectriz del ∠P.

angle bisector of a triangle   A segment that separates an angle of a triangle into two congruent angles. *(p. 240)*

bisectriz de un ángulo de un triángulo   Segmento que divide un ángulo de un triángulo en dos ángulos congruentes. *(pág. 240)*

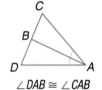

∠DAB ≅ ∠CAB

angle of depression   The angle formed by the line of sight and a horizontal line when looking down. *(p. 566)*

ángulo de depresión   Al mirar hacia abajo, ángulo formado por la línea visual y una recta horizontal. *(pág. 566)*

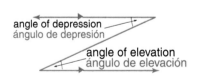

angle of depression
ángulo de depresión

angle of elevation   The angle formed by the line of sight and a horizontal line when looking up. *(p. 566)*

ángulo de elevación   Al mirar hacia arriba, ángulo formado por la línea visual y una recta horizontal. *(pág. 566)*

angle of elevation
ángulo de elevación

apothem   A segment from the center of a polygon perpendicular to a side of the polygon. *(p. 425)*

apotema   Segmento trazado del centro de un polígono y que es perpendicular a uno de sus lados. *(pág. 425)*

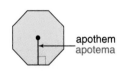

apothem
apotema

arc   A set of points along a circle defined by a central angle. *(p. 462)*

arco   Conjunto de puntos de un círculo determinado por un ángulo central. *(pág. 462)*

area   The number of square units in a polygonal region needed to cover its surface. *(p. 36)*

área   Número de unidades cuadradas que se requieren para cubrir la superficie de una región poligonal. *(pág. 36)*

axis of a cylinder   See *cylinder* and *oblique cylinder.*

eje de un cilindro   Ver *cilindro* y *cilindro oblicuo.*

B

base angles of an isosceles triangle   See *isosceles triangle.*

ángulos basales de un triángulo isósceles   Ver *triángulo isósceles.*

base angles of a trapezoid   See *trapezoid.*

bases de un trapecio   Ver *trapecio.*

base of an isosceles triangle   See *isosceles triangle.*

base de un triángulo isósceles   Ver *triángulo isósceles.*

bases of a trapezoid   See *trapezoid.*

bases de un trapecio   Ver *trapecio.*

betweenness   Point R is between points P and Q if and only if R, P, and Q are collinear and $PR + RQ = PQ$. *(p. 56)*

estar entre   El punto R está entre los puntos P y Q si y sólo si R, P y Q son colineales y $PR + RQ = PQ$. *(pág. 56)*

bisect   To separate a geometric figure into congruent parts using a point, line, ray, segment, or plane. *(p. 64)*

bisecar   Dividir una figura geométrica en partes congruentes usando un punto, una recta, un rayo, un segmento o un plano. *(pág. 64)*

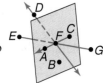

Point F, $\overrightarrow{FD}$, $\overrightarrow{FA}$, $\overline{AC}$, and plane ABC all bisect $\overline{EG}$.
El punto F, $\overrightarrow{FD}$, $\overrightarrow{FA}$, $\overline{AC}$ y el plano ABC todos bisecan $\overline{EG}$.

C

center of a circle   See *circle.*

centro de un círculo   Ver *círculo.*

center of a regular polygon   A unique point that is equidistant from all the vertices. *(p. 425)*

centro de un polígono regular   El punto único que equidista de todos sus vértices. *(pág. 425)*

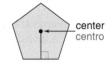

center
centro

center of rotation   The fixed point about which a figure is rotated. *(p. 697)*

centro de rotación   Punto fijo alrededor del cual gira una figura. *(pág. 697)*

central angle An angle whose vertex is the center of a circle and whose sides intersect the circle. *(p. 462)*

ángulo central Ángulo cuyo vértice es el centro de un círculo y cuyos lados lo intersecan. *(pág. 462)*

∠*MTD* is a central angle of ⊙*T*.
∠*MTD* es un ángulo central del ⊙*T*.

---

centroid of a triangle The point of intersection of the three medians of a triangle. *(p. 230)*

centroide de un triángulo Punto de intersección de las tres medianas de un triángulo. *(pág. 230)*

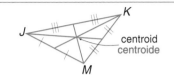

X is the centroid of △*JKM*.
X es el centroide del △*JKM*.

---

chord A segment of a circle whose endpoints are on the circle. *(p. 454)*

cuerda Segmento de un círculo cuyos extremos yacen en él. *(pág. 454)*

$\overline{JR}$ is a chord of ⊙*K*.
$\overline{JR}$ es una cuerda del ⊙*K*.

---

circle The set of all points in a plane that are a given distance from a given point in the plane, called the *center* of the circle. *(p. 454)*

círculo Conjunto de todos los puntos en un plano que equidistan de un punto dado del plano, llamado *centro* del círculo. *(pág. 454)*

P is the center of the circle.
P es el centro del círculo.

---

circumference The distance around a circle. *(p. 478)*

circunferencia Longitud del contorno de un círculo. *(pág. 478)*

---

circumscribed polygon A polygon with each side tangent to a circle. *(p. 474)*

polígono circunscrito Un polígono cuyos lados son tangentes a un círculo. *(pág. 474)*

---

collinear points Three or more points that lie on the same line. *(p. 13)*

puntos colineales Tres o más puntos que yacen en la misma recta. *(pág. 13)*

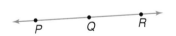

P, *Q*, and *R* are collinear.
P, *Q* y *R* son colineales.

---

compass An instrument used to draw circles and arcs of circles. *(p. 30)*

compás Instrumento que se utiliza para trazar círculos y arcos de círculos *(pág. 30)*

---

complementary angles Two angles whose degree measures have a sum of 90. Each angle is a *complement* of the other. *(p. 116)*

ángulos complementarios Dos ángulos cuyas medidas angulares suman 90. Cada ángulo es *complemento* del otro. *(pág. 116)*

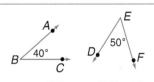

m∠*ABC* + *m*∠*DEF* = 90

compound statement   Two or more logic statements joined by *and* or *or*. (p. 633)

enunciado compuesto   Dos o más enunciados lógicos unidos por *y* o *o*. (*pág. 633*)

---

concave polygon   A polygon such that a point on at least one of its diagonals lies outside the polygon. (p. 404)

polígono cóncavo   Polígono para el que existe un punto en una de sus diagonales que yace fuera del polígono. (*pág. 404*)

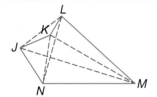

Diagonal $\overline{JL}$ lies outside polygon *JKLMN*.
La diagonal $\overline{JL}$ yace fuera del polígono *JKLMN*.

---

concentric circles   Circles that lie in the same plane, have the same center, and have radii of different lengths. (p. 456)

círculos concéntricos   Círculos que yacen en el mismo plano y tienen el mismo centro, pero radios distintos. (*pág. 456*)

$\odot R$ with radius $\overline{RS}$ and $\odot R$ with radius $\overline{RT}$ are concentric circles.
El $\odot R$ de radio $\overline{RS}$ y el $\odot R$ de radio $\overline{RT}$ son círculos concéntricos.

---

conclusion   See *conditional statement*.

conclusión   Ver *enunciado condicional*.

---

concurrent   Three or more lines or segments that meet at a common point. (p. 230)

concurrente   Tres o más rectas o segmentos que se intersecan en un punto. (*pág. 230*)

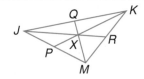

$\overline{JR}$, $\overline{KP}$, and $\overline{MQ}$ are concurrent.
$\overline{JR}$, $\overline{KP}$ y $\overline{MQ}$ son concurrentes.

---

conditional statement   A statement written in if-then form. The part following *if* is the *hypothesis*. The part following *then* is the *conclusion*. (p. 24)

enunciado condicional   Enunciado de la forma si-entonces. La parte entre el *si* y el *entonces* es la *hipótesis* y la que sigue al *entonces* es la *conclusión*. (*pág. 24*)

---

cone   A solid figure in which the base is a circle and the lateral surface is a curved surface. (p. 497)

cono   Sólido de base circular cuya superficie lateral es curva. (*pág. 497*)

vertex
vértice

base
base

---

congruent angles   Angles that have the same degree measure. (p. 122)

ángulos congruentes   Ángulos que tienen la misma medida. (*pág. 122*)

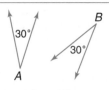

$\angle A \cong \angle B$

---

congruent segments Segments that have the same length. *(p. 62)*

segmentos congruentes Segmentos que tienen la misma longitud. *(pág. 62)*

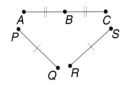

$$\overline{AB} \cong \overline{BC} \text{ and } \overline{PQ} \cong \overline{RS}$$

congruent triangles Triangles whose corresponding parts are congruent. *(p. 203)*

triángulos congruentes Triángulos cuyas partes correspondientes son congruentes. *(pág. 203)*

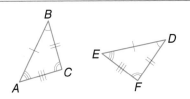

$$\triangle ABC \cong \triangle EDF$$

conjecture A conclusion reached based on inductive reasoning. *(p. 6)*

conjetura Una conclusión a la que se llega mediante razonamiento inductivo. *(pág. 6)*

conjunction A compound statement formed by joining two statements with the word *and*. *(p. 633)*

conjunción Enunciado compuesto formado al unir dos enunciados por la palabra *y*. *(pág. 633)*

consecutive interior angles See *transversal*.

ángulos internos consecutivos Ver *transversal*.

consecutive sides Sides of a polygon that share a vertex. *(p. 311)*

lados consecutivos Lados de un polígono con un vértice común. *(pág. 311)*

$\overline{PS}$ and $\overline{PQ}$ are consecutive sides.
$\overline{PS}$ y $\overline{PQ}$ son lados consecutivos.

construction The process of drawing a figure using only a compass and a straightedge. *(p. 30)*

construcción El proceso de trazar figuras sólo con regla y compás. *(pág. 30)*

converse The converse of a conditional statement is formed by exchanging the hypothesis and the conclusion in the conditional. *(p. 25)*

recíproca La recíproca de un enunciado condicional se forma intercambiando la hipótesis y la conclusión del condicional. *(pág. 25)*

convex polygon If all diagonals of a polygon are located in the interior of the figure, the polygon is convex. *(p. 404)*

polígono convexo Si todas las diagonales de un polígono están situadas dentro de éste, el polígono es convexo. *(pág. 404)*

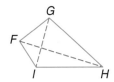

All diagonals lie inside polygon *FGHI*.
Todas las diagonales yacen dentro del polígono *FGHI*.

coordinate A number associated with a point on a number line. *(p. 52)*

coordenada Número asociado con un punto de una recta numérica. *(pág. 52)*

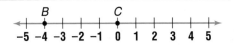

The coordinate of *B* is -4.
La coordenada de *B* es -4.

coordinate plane   The number plane formed by two perpendicular number lines that intersect at their zero points to form a grid. The vertical number line is called the *y-axis*. The horizontal number line is called the *x-axis*. The point of intersection of the two axes is called the *origin*, *O*. The two axes separate the plane into four regions called *quadrants*. (p. 68)

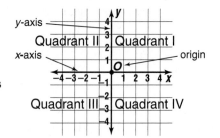

plano coordenado   Plano en el que se han trazado dos rectas numéricas perpendiculares, que se intersecan en sus puntos cero, formando un cuadriculado. La recta numérica vertical se llama *eje y* y la horizontal se llama *eje x*. El punto de intersección de los ejes se llama *origen*, *O*. Los ejes dividen el plano en cuatro regiones llamadas *cuadrantes*. (pág. 68)

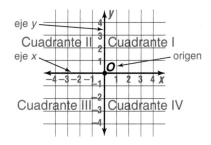

coordinate proof   A geometric proof that uses figures on a coordinate plane. (p. 660)

demostración por coordenadas   Demostración geométrica que usa figuras en un plano coordenado. (pág. 660)

coordinates   An *ordered pair* of real numbers used to locate a point on the coordinate plane. The point is called the *graph* of the ordered pair. In an ordered pair, the first component is called the *x-coordinate* and the second component is called the *y-coordinate*. (p. 68)

coordenadas   Un *par ordenado* de números reales que se usa para ubicar un punto en el plano coordenado. El punto se llama la *gráfica* del par ordenado. En un par ordenado, la primera coordenada se llama *coordenada x* y la segunda se llama *coordenada y*. (pág. 68)

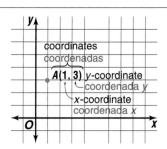

coplanar   See *plane*.

coplanario   Ver *plano*.

corresponding angles   See *transversal*.

ángulos correspondientes   Ver *transversal*.

corresponding parts   See *congruent triangles*.

partes correspondientes   Ver *triángulos congruentes*.

cosine   See *trigonometric ratio*.

coseno   Ver *razón trigonométrica*.

counterexample   An example that shows that a conjecture is not true. (p. 6)

contraejemplo   Un ejemplo que demuestra que una conjetura no es verdadera. (pág. 6)

cross products   See *proportion*.

productos cruzados   Ver *proporción*.

cube   A rectangular prism in which all of the faces are squares. (p. 497)

cubo   Prisma rectangular cuyas caras son cuadrados. (pág. 497)

cylinder   A solid figure whose bases are formed by congruent circles in parallel planes and whose lateral surface is curved. The segment whose endpoints are the centers of the circular bases is called the *axis* of the cylinder. The *altitude* is a segment perpendicular to the base planes with an endpoint in each plane. (p. 497)

cilindro   Sólido cuyas bases son círculos congruentes que yacen en planos paralelos y cuya superficie lateral es curva. El segmento cuyos extremos son los centros de las bases circulares se llama *eje* del cilindro. La *altura* es un segmento perpendicular al plano de las bases con un extremo en cada plano. (pág. 497)

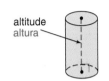

altitude
altura

D

deductive reasoning   The process of using facts, rules, definitions, or properties in logical order to reach a conclusion. (p. 638)

razonamiento deductivo   El proceso de usar lógicamente hechos, reglas, definiciones o propiedades para llegar a una conclusión. (pág. 638)

degree   A unit of measure used when measuring angles. (p. 96)

grado   Unidad de medida que se usa para medir ángulos. (pág. 96)

diagonal   A segment joining two nonconsecutive vertices of a polygon. (p. 311)

diagonal   Segmento que une dos vértices no consecutivos de un polígono. (pág. 311)

$\overline{SQ}$ is a diagonal.
$\overline{SQ}$ es una diagonal.

diameter   A chord of a circle that contains the center of the circle. (p. 454)

diámetro   Cuerda de un círculo que contiene su centro. (pág. 454)

$\overline{TG}$ is a diameter of $\odot K$.
$\overline{TG}$ es un diámetro del $\odot K$.

dilation   A transformation that alters the size of a figure, but not its shape. (p. 703)

dilatación   Transformación que altera el tamaño de una figura, pero no su forma. (pág. 703)

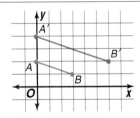

Segment $A'B'$ is a dilation of segment AB.
El segmento $A'B'$ es una dilatación del segmento AB.

disjunction   A compound statement formed by joining two statements with the word *or*. (p. 633)

disjunción   Enunciado compuesto formado al unir dos enunciados por la palabra *o*. (pág. 633)

edge   See *polyhedron.*

arista   Ver *poliedro.*

endpoint   See *line segment* or *ray.*

extremo   Ver *segmento de recta* o *rayo.*

equation   A statement that includes the symbol =. *(p. 57)*

ecuación   Enunciado que contiene el signo =. *(pág. 57)*

equiangular triangle   A triangle with three congruent angles. *(p. 195)*

triángulo equiangular   Triángulo con tres ángulos congruentes. *(pág. 195)*

equilateral triangle   A triangle with three congruent sides. *(p. 189)*

triángulo equilátero   Triángulo con tres lados congruentes. *(pág. 189)*

$$\overline{AB} \cong \overline{BC} \cong \overline{AC}$$
$$\angle A \cong \angle B \cong \angle C$$

exterior angles   See *transversal.*

ángulos externos   Ver *transversal.*

exterior of an angle   See *angle.*

exterior de un ángulo   Ver *ángulo.*

exterior angle of a triangle   An angle that forms a linear pair with an angle of a triangle. *(p. 282)*

ángulo exterior de un triángulo   Ángulo que forma un par lineal con un ángulo de un triángulo. *(pág. 282)*

exterior angles: $\angle 4, \angle 5, \angle 6, \angle 7, \angle 8, \angle 9$
ángulos exterioes: $\angle 4, \angle 5, \angle 6, \angle 7, \angle 8, \angle 9$

external secant segment   See *secant segment.*

segmento secante externo   Ver *segmento secante.*

extremes   See *proportion.*

extremos   Ver *proporción.*

F

face   See *polyhedron.*

cara   Ver *poliedro.*

formula   An equation that shows how certain quantities are related. *(p. 35)*

fórmula   Ecuación que muestra una relación entre ciertas cantidades. *(pág. 35)*

45°-45°-90° triangle   A special right triangle with two 45° angles. *(p. 554)*

triángulo 45°-45°-90°   Triángulo especial con dos ángulos de 45°. *(pág. 554)*

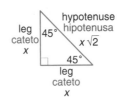

H

hypotenuse   See *right triangle.*

hipotenusa   Ver *triángulo rectángulo.*

hypothesis   See *conditional statement.*

hipótesis   Ver *enunciado condicional.*

I

if-then statement   See *conditional statement. (p. 24)*

enunciado si-entonces   Ver *enunciado condicional. (pág. 24)*

image   See *transformation.*

imagen   Ver *transformación.*

included angle   An angle formed by two given sides of a triangle. *(p. 211)*

ángulo incluido   Ángulo formado por dos lados de un triángulo. *(pág. 211)*

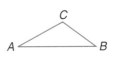

included side   A side common to two given angles of a triangle. *(p. 215)*

lado incluido   Lado de un triángulo común a dos de sus ángulos. *(pág. 215)*

$\angle A$ is the included angle of $\overline{AB}$ and $\overline{AC}$.
$\overline{AB}$ is the included side of $\angle A$ and $\angle B$.
El $\angle A$ es el ángulo incluido de $\overline{AB}$ y $\overline{AC}$.
$\overline{AB}$ es el lado incluido del $\angle A$ y del $\angle B$.

inductive reasoning   Making a conclusion based on a pattern of examples or past events. *(p. 4)*

razonamiento inductivo   Conclusión basada en un patrón de ejemplos o de sucesos pasados. *(pág. 4)*

inequality   A statement used to compare nonequal measures. *(p. 276)*

desigualdad   Enunciado que se usa para comparar dos magnitudes desiguales. *(pág. 276)*

inscribed angle   An angle whose vertex lies on a circle and whose sides contain chords of the circle. *(p. 586)s*

ángulo inscrito   Ángulo cuyo vértice yace en un círculo y cuyos lados contienen cuerdas de éste. *(pág. 586)*

$\angle RST$ is inscribed in $\odot P$.
El $\angle RST$ está inscrito en el $\odot P$.

inscribed polygon   A polygon in which every vertex of the polygon lies on the circle. *(p. 474)*

polígono inscrito   Un polígono cuyos vértices yacen en un círculo. *(pág. 474)*

---

integers   The set of real numbers $\{\ldots, -3, -2, -1, 0, 1, 2, 3, \ldots\}$. *(p. 50)*

enteros   El conjunto de números reales $\{\ldots, -3, -2, -1, 0, 1, 2, 3, \ldots\}$. *(pág. 50)*

---

intercepted arc   An angle intercepts an arc *if and only if* each of the following conditions holds. *(p. 586)*

1. The endpoints of the arc lie on the angle.
2. All points of the arc, except the endpoints, are in the interior of the angle.
3. Each side of the angle contains an endpoint of the arc.

arco interceptado   Un ángulo intercepta un arco *si y sólo si* se cumplen todas estas condiciones. *(pág. 586)*
1. Los extremos del arco yacen en el ángulo.
2. Todos los puntos del arco, salvo sus extremos, yacen en el interior del ángulo.
3. Cada lado del ángulo contiene un extremo del arco.

$\angle PQR$ intercepts $\overparen{PR}$.
El $\angle PQR$ intercepta $\overparen{PR}$.

---

interior angles   See *transversal*.

ángulos internos   Ver *transversal*.

---

interior of an angle   See *angle*.

interior de un ángulo   Ver *ángulo*.

---

irrational number   A real number that is a nonterminating and nonrepeating decimal. *(p. 51)*

número irracional   Número real que no es un decimal exacto o periódico. *(pág. 51)*

---

isosceles trapezoid   A trapezoid with two congruent legs. *(p. 334)*

trapecio isósceles   Trapecio con dos catetos congruentes. *(pág. 334)*

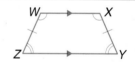

$\angle W \cong \angle X, \angle Z \cong \angle Y; \overline{WZ} \cong \overline{XY}$

---

isosceles triangle   A triangle with two congruent sides. The congruent sides are called *legs*. The angle formed by the congruent sides is called the *vertex angle*. The side opposite the vertex angle is called the *base*. The angles formed by the base and each of the legs are called *base angles*. They are congruent. *(p. 189)*

triángulo isósceles   Triángulo que tiene dos lados congruentes. Los lados congruentes se llaman *catetos*. El ángulo formado por los catetos se llama *ángulo del vértice*. El lado opuesto al ángulo del vértice se llama *base*. Los ángulos formados por la base y cada uno de los catetos se llaman *ángulos basales*, los cuales son congruentes. *(pág. 189)*

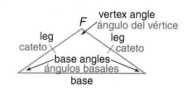

---

lateral area   The sum of the areas of the lateral faces of a solid. *(p. 504)*

área lateral   La suma de las áreas de las caras laterales de un sólido. *(pág. 504)*

lateral face   See *prism* and *pyramid*.

cara lateral   Ver *prisma* y *pirámide*.

legs of an isosceles triangle   See *isosceles triangle*.

catetos de un triángulo isósceles   Ver *triángulo isósceles*.

legs of a right triangle   See *right triangle*.

catetos de un triángulo rectángulo   Ver *triángulo rectángulo*.

legs of a trapezoid   See *trapezoid*.

catetos de un trapecio   Ver *trapecio*.

line   A basic undefined term of geometry. Lines extend indefinitely and have no thickness or width. *(p. 12)*

recta   Uno de los términos primitivos en geometría. Las rectas se extienden indefinidamente en ambos sentidos y no tienen espesor o ancho. *(pág. 12)*

linear equation   An equation whose graph is a straight line. *(p. 174)*

ecuación lineal   Ecuación cuya gráfica es una recta. *(pág. 174)*

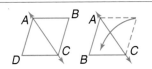

linear pair   Two angles that are adjacent and whose noncommon sides are opposite rays. *(p. 111)*

par lineal   Dos ángulos adyacentes cuyos lados no comunes son rayos opuestos. *(pág. 111)*

∠1 and ∠2 are a linear pair.
Los ∠1 y ∠2 forman un par lineal.

line of symmetry   A line that can be drawn through a plane figure so that part of the figure on one side of the line is the congruent reflected image of the part on the other side of the line. *(p. 434)*

eje de simetría   Recta que puede trazarse por una figura plana de modo que parte de la figura en un lado de la recta es la imagen congruente reflejada de la parte al otro lado de la recta. *(pág. 434)*

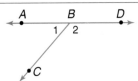

$\overrightarrow{AC}$ is a line of symmetry.
$\overrightarrow{AC}$ es un eje de simetría.

line segment   Part of a line containing two endpoints and all points between them. *(p. 13)*

segmento de recta   Parte de una recta que contiene dos extremos y todos los puntos entre éstos. *(pág. 13)*

line symmetry   Each half of a figure is a mirror image of the other half when a line of symmetry is drawn. *(p. 434)*

simetría lineal   Al trazar una recta de simetría, cada mitad de una figura es una imagen especular de la otra mitad. *(pág. 434)*

major arc   A part of the circle in the exterior of a central angle that measures greater than 180. *(p. 462)*

arco mayor   Parte de un círculo en el exterior de un ángulo central que mide más de 180. *(pág. 462)*

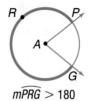

$m\overset{\frown}{PRG} > 180$

---

means   See *proportion.*

medios   Ver *proporción.*

---

measure   See *measurement.*

medida   Ver *medición.*

---

measurement   A measurement consists of a number called a *measure* and the *unit of measure.* *(p. 57)*

medición   Una medición consta de un número llamado *medida* y la *unidad de medida.* *(pág. 57)*

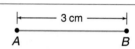

---

median of a trapezoid   A segment joining the midpoints of the legs of a trapezoid. *(p. 334)*

mediana de un trapecio   Segmento que une los puntos medios de los catetos de un trapecio. *(pág. 334)*

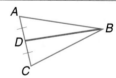

$\overline{DE} \parallel \overline{MN}, \overline{GF} \parallel \overline{MN}$

---

median of a triangle   A segment in which one endpoint is the vertex of a triangle and the other endpoint is the midpoint of the side opposite that vertex. *(p. 228)*

mediana de un triángulo   Segmento que une el vértice de un triángulo con el punto medio del lado opuesto a dicho vértice. *(pág. 228)*

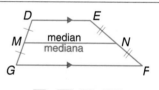

median $\overline{BD}$
mediana $\overline{BD}$

---

midpoint   A point M is the midpoint of segment ST if and only if M is between S and T, and $SM = MT$. *(pp. 31, 63)*

punto medio   Un punto M es el punto medio del segmento ST si y sólo si M está entre S y T y $SM = MT$. *(págs. 31, 63)*

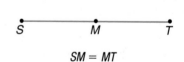

$SM = MT$

---

minor arc   A part of a circle in the interior of a central angle that measures less than 180. *(p. 462)*

arco menor   Parte de un círculo en el interior de un ángulo central que mide menos de 180. *(pág. 462)*

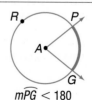

$m\overset{\frown}{PG} < 180$

natural numbers   The set of real numbers {1, 2, 3, . . .}. These are also called *counting numbers*. *(p. 50)*

números naturales   El conjunto de números reales {1, 2, 3, . . .}. Éstos también se llaman *números de contar*. *(pág. 50)*

negation   The negative of a statement. *(p. 632)*

negación   Negativa de un enunciado. *(pág. 632)*

net   A two-dimensional pattern that folds to form a solid. *(p. 507)*

red   Patrón bidimensional que una vez plegado forma un sólido. *(pág. 507)*

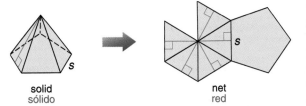

solid
sólido

net
red

noncollinear points   Three or more points that do not lie on the same line. *(p. 13)*

puntos no colineales   Tres o más puntos que no yacen en la misma recta. *(pág. 13)*

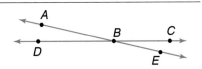

A, *B*, and *C* are noncollinear.
A, *B* y *C* no son colineales.

nonconsecutive sides   Sides of a polygon that do not share a vertex. *(p. 311)*

lados no consecutivos   Lados de un polígono que no tienen un vértice común. *(pág. 311)*

$\overline{PS}$ and $\overline{QR}$ are nonconsecutive sides.
$\overline{PS}$ y $\overline{QR}$ son lados no consecutivos.

noncoplanar   See *plane*.

no coplanario   Ver *plano*.

nonterminating decimal   An infinite number of digits either with a repeating pattern or not repeating. *(p. 51)*

decimal no terminal   Un número infinito de dígitos con un patrón repetitivo o sin él. *(pág. 51)*

oblique cone   A cone in which the altitude is perpendicular to the base at a point other than its center. *(p. 516)*

cono oblicuo   Cono en que la altura es perpendicular a la base en un punto distinto de su centro. *(pág. 516)*

oblique cylinder   A cylinder in which the axis is not an altitude. *(p. 506)*

cilindro oblicuo   Cilindro en el que su eje no es una altura. *(pág. 506)*

oblique prism   A prism in which a lateral edge is not an altitude. *(p. 504)*

prisma oblicuo   Prisma en el que una arista lateral no es una altura. *(pág. 504)*

oblique pyramid   A pyramid in which the altitude is perpendicular to the base at a point other than its center. *(p. 516)*

pirámide oblicua   Pirámide en que la altura es perpendicular a la base en un punto distinto de su centro. *(pág. 516)*

obtuse angle   An angle whose measure is greater than 90 but less than 180. *(p. 98)*

ángulo obtuso   Ángulo que mide más de 90 y menos de 180. *(pág. 98)*

$90 < m\angle A < 180$

obtuse triangle   A triangle with one obtuse angle. *(p. 188)*

triángulo obtusángulo   Triángulo que tiene un ángulo obtuso. *(pág. 188)*

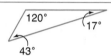

one obtuse angle
un ángulo obtuso

opposite rays   Two rays that are part of the same line and have only their endpoints in common. *(p. 90)*

rayos opuestos   Dos rayos que forman parte de la misma recta y que sólo poseen sus extremos en común. *(pág. 90)*

$\overrightarrow{YX}$ and $\overrightarrow{YZ}$ are opposite rays.
$\overrightarrow{YX}$ y $\overrightarrow{YZ}$ son rayos opuestos.

ordered pair   See *coordinates*.

par ordenado   Ver *coordenadas*.

origin   See *coordinate plane*.

origen   Ver *plano coordenado*.

paragraph proof   A logical argument used to validate a conjecture in paragraph form. *(p. 644)*

demostración de párrafo   Un argumento lógico en forma de párrafo que se usa para validar una conjetura. *(pág. 644)*

parallel lines   Two lines that lie in the same plane and do not intersect. *(p. 142)*

rectas paralelas   Rectas que yacen en un mismo plano y que no se intersecan. *(pág. 142)*

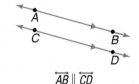

$\overleftrightarrow{AB} \parallel \overleftrightarrow{CD}$

parallel planes Planes that do not intersect. *(p. 142)*

planos paralelos Planos que no se intersecan. *(pág. 142)*

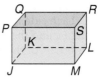

plane *PQR* ∥ plane *JKL*
plano *PQR* ∥ plano *JKL*

---

parallelogram A quadrilateral with two pairs of parallel sides. *(p. 316)*

paralelogramo Cuadrilátero con dos pares de lados paralelos. *(pág. 316)*

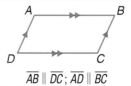

$\overline{AB} \parallel \overline{DC}$; $\overline{AD} \parallel \overline{BC}$

---

perfect square A number multiplied by itself. *(p. 548)*

cuadrado perfecto Número multiplicado por sí mismo. *(pág. 548)*

---

perimeter The sum of the lengths of the sides of a polygon. *(p. 35)*

perímetro La suma de las longitudes de los lados de un polígono. *(pág. 35)*

---

perpendicular bisector A segment that is perpendicular to another segment and passes through that segment's midpoint. *(p. 235)*

mediatriz Segmento perpendicular a otro y que pasa por su punto medio. *(pág. 235)*

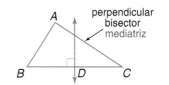

perpendicular bisector
mediatriz

D is the midpoint of $\overline{BC}$.
D es el punto medio de $\overline{BC}$.

---

perpendicular lines Lines that intersect to form right angles. *(p. 128)*

rectas perpendiculares Rectas que al intersecarse forman un ángulo recto. *(pág. 128)*

line *m* ⊥ line *n*
recta *m* ⊥ recta *n*

---

pi (π) A Greek letter that represents the ratio of the circumference of a circle to its diameter. *(p. 479)*

pi (π) Letra griega que representa la razón de la circunferencia de un círculo a su diámetro. *(pág. 479)*

---

plane A flat surface that extends in all directions containing at least three noncollinear points. Points or lines that lie in the same plane are *coplanar*. Points or lines that do not lie in the same plane are *noncoplanar*. *(p. 14)*

plano Superficie llana que se extiende en todas direcciones y que contiene por lo menos tres puntos no colineales. Son *coplanarios* los puntos o rectas que yacen en el mismo plano; son *no coplanarios* los que no yacen en el mismo plano. *(pág. 14)*

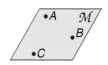

---

point A basic undefined term of geometry. Points have no size. *(p. 12)*

punto Uno de los términos primitivos en geometría. Los puntos no tienen tamaño. *(pág. 12)*

---

point of tangency See *tangent*.

punto de tangencia Ver *tangente*.

polygon   A geometric figure formed by three or more coplanar segments called *sides*. Each side intersects exactly two other sides, but only at their endpoints, and the intersecting sides must be noncollinear. The intersection points of the sides are the *vertices* of the polygon. *(p. 356)*

polígono   Figura geométrica formada por tres o más segmentos coplanarios llamados *lados*. Cada lado interseca exactamente dos lados y sólo en sus extremos y los lados que se intersecan no son colineales. Los puntos de intersección de los lados son los *vértices* del polígono. *(pág. 356)*

---

polygonal region   Any polygon and its interior form a polygonal region. *(p. 413)*

región poligonal   Cualquier polígono y su interior. *(pág. 413)*

---

polyhedron   A solid with flat surfaces that are polygonal regions. The flat surfaces formed by the polygons and their interiors are called *faces*. Pairs of faces intersect at *edges*. Three or more edges intersect at a *vertex*. *(p. 496)*

poliedro   Sólido de superficies llanas que son regiones poligonales. Las superficies llanas formadas por los polígonos y sus interiores se llaman *caras*. Pares de caras se intersecan en *aristas* y tres o más aristas se intersecan en un *vértice*. *(pág. 496)*

---

postulate   A rule of geometry that is accepted as being true without proof. *(p. 18)*

postulado   Una regla de geometría que se acepta como verdadera sin necesidad de demostración. *(pág. 18)*

---

preimage   See *transformation*.

preimagen   Ver *transformación*.

---

prism   A solid with the following characteristics:
1. Two faces, called *bases*, are formed by congruent polygons that lie in parallel planes.
2. The faces that are not bases, called *lateral faces*, are formed by parallelograms.
3. The intersection of two adjacent lateral faces are called *lateral edges* and are parallel segments. *(p. 496)*

prisma   Sólido con las siguientes características:
1. Dos caras, llamadas *bases*, están formadas por polígonos congruentes que yacen en planos paralelos.
2. Las caras que no son bases, llamadas *caras laterales*, están formadas por paralelogramos.
3. Las intersecciones de dos caras laterales adyacentes se llaman *aristas laterales* y son todas segmentos paralelos. *(pág. 496)*

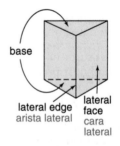

base
lateral edge
arista lateral
lateral face
cara lateral

triangular prism
prisma triangular

---

proof   A logical argument used to validate a conjecture. *(p. 644)*

demostración   Argumento lógico que se usa para validar una conjetura. *(pág. 644)*

proportion An equation of form $\frac{a}{b} = \frac{c}{d}$ that states that two ratios are equivalent. The *extremes* are a and d, and the *means* are b and c. The *cross products* are the product of the extremes and the product of the means. *(p. 351)*

proporción Ecuación de la forma $\frac{a}{b} = \frac{c}{d}$ que afirma la igualdad de dos razones. Los *extremos* son a y d y los *medios* son b y c. Los *productos cruzados* son el producto de los extremos y el de los medios. *(pág. 351)*

---

protractor An instrument used to measure angles in degrees. *(p. 96)*

transportador Instrumento que se usa para medir ángulos en grados. *(pág. 96)*

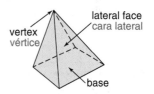

---

pyramid A solid with the following characteristics:
1. All the faces, except one, intersect at a common point called the *vertex*.
2. The face that does not intersect at the vertex is called the *base*. The base is formed by a polygon.
3. The faces meeting at the vertex are called *lateral faces*. They are formed by triangles. *(p. 496)*

pirámide Sólido con las siguientes características:
1. Todas las caras, excepto una, se intersecan en un punto llamado *vértice*.
2. La cara que no contiene el vértice se llama *base* y es un polígono.
3. Las caras que se encuentran en el vértice se llaman *caras laterales* y están formadas por triángulos. *(pág. 496)*

vertex / vértice · lateral face / cara lateral · base

rectangular pyramid
pirámide rectangular

Pythagorean Theorem In a right triangle, the sum of the squares of the measures of the legs equals the square of the measure of the hypotenuse. A *Pythagorean triple* is a group of three whole numbers that satisfies the Pythagorean Theorem. *(p. 256)*

teorema de Pitágoras En un triángulo rectángulo, la suma de los cuadrados de las longitudes de los catetos es igual al cuadrado de la longitud de la hipotenusa. Un *triple Pitágorico* es un grupo de tres enteros que satisfacen el teorema de Pitágoras. *(pág. 256)*

 Q

quadrant See *coordinate plane*.

cuadrante Ver *plano coordenado*.

---

quadrilateral A four-sided closed figure with four vertices. *(pp. 103, 310)*

cuadrilátero Figura cerrada de cuatro lados y cuatro vértices. *(págs. 103, 310)*

 R

radical expression An expression that contains a square root. *(p. 549)*

expresión radical Expresión que contiene una raíz cuadrada. *(pág. 549)*

radical sign A symbol used to indicate the positive square root. *(p. 548)*

signo radical Símbolo que indica la raíz cuadrada positiva. *(pág. 548)*

radicand The number under a radical sign. *(p. 549)*

radicando Número bajo el signo radical. *(pág. 549)*

radius A segment of a circle whose endpoints are the center of the circle and a point on the circle. *(p. 454)*

radio Segmento de un círculo cuyos extremos son el centro del círculo y un punto en él. *(pág. 454)*

$\overline{KA}$ is a radius of $\odot K$.
$\overline{KA}$ es un radio del $\odot K$.

ratio A comparison of two numbers by division. *(p. 350)*

razón Comparación de dos números por división. *(pág. 350)*

rational number Any real number that can be expressed in the form $\frac{a}{b}$, where a and b are integers and b cannot equal 0. *(p. 50)*

número racional Número real que puede escribirse como $\frac{a}{b}$, con a y b enteros y b distinto de 0. *(pág. 50)*

ray A part of a line that has an endpoint and contains all the points of the line without end in one direction. *(p. 13)*

rayo Parte de una recta que posee un extremo y que contiene todos los puntos de la recta en una dirección. *(pág. 13)*

real numbers The union of the sets of rational and irrational numbers. *(p. 51)*

números reales Unión del conjunto de números racionales con el de números irracionales. *(pág. 51)*

rectangle A parallelogram with four right angles. *(p. 327)*

rectángulo Paralelogramo que tiene cuatro ángulos rectos. *(pág. 327)*

reflection The flip of a figure over a line to produce a mirror image. *(pp. 198, 692)*

reflexión Tipo de transformación en la que una figura se voltea al otro lado de una recta para producir una imagen especular. *(págs. 198, 692)*

line
recta

regular polygon A convex polygon that is both equilateral and equiangular. *(p. 402)*

polígono regular Polígono convexo que es equilátero y equiangular. *(pág. 402)*

regular pyramid A pyramid whose base is a regular polygon and in which the segment from the vertex to the center of the base is the altitude. *(p. 516)*

pirámide regular Pirámide cuya base es un polígono regular y en la que el segmento del vértice al centro de la base es la altura. *(pág. 516)*

regular tessellation See *tessellation*.

teselado regular Ver *teselado*.

remote interior angles   The angles in a triangle that are not adjacent to a given exterior angle of the triangle. *(p. 282)*

ángulos interiores no adyacentes   Ángulos de un triángulo que no son adyacentes a un ángulo exterior dado del triángulo. *(pág. 282)*

∠2 and ∠3 are remote interior angles for exterior angle 4.
Los ∠2 y ∠3 son ángulos interiores no adyacentes al ángulo exterior 4.

---

rhombus   A parallelogram with four congruent sides. *(p. 327)*

rombo   Paralelogramo de cuatro lados congruentes. *(pág. 327)*

---

right angle   An angle with a degree measure of 90. *(p. 98)*

ángulo recto   Ángulo que mide 90°. *(pág. 98)*

$m\angle A = 90$

---

right triangle   A triangle with one right angle. *(p. 188)* The side opposite the right angle is called the *hypotenuse*. The two sides that form the right angle are called *legs*. *(p. 252)*

triángulo rectángulo   Triángulo que tiene un ángulo recto. *(p. 188)* El lado opuesto al ángulo recto se llama *hipotenusa*. Los otros dos lados, que forman el ángulo recto, se llaman *catetos*. *(pág. 252)*

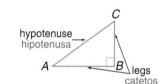

hypotenuse
hipotenusa

legs
catetos

one right angle
un ángulo recto

---

rotation   A geometric turn of a figure around a fixed point. *(pp. 198, 697)*

rotación   Un giro geométrico de una figura en torno a un punto fijo. *(págs. 198, 697)*

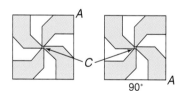

fixed point
punto fijo

---

rotational symmetry   A figure that can be turned or rotated less than 360° about a fixed point so that the figure looks exactly as it does in its original position has rotational or *turn* symmetry. *(p. 435)*

simetría rotacional   Una figura posee simetría rotacional si se la puede girar en menos de 360° en torno a un punto fijo sin que esto cambie su apariencia con respecto a la figura original. *(pág. 435)*

---

scale drawing   A drawing that represents something proportionally that is too large or too small to be drawn actual size. *(p. 358)*

dibujo a escala   Dibujo que representa proporcionalmente algo que es demasiado grande o pequeño como para ser dibujado de tamaño natural. *(pág. 358)*

---

scale factor   The ratio found by comparing the measures of corresponding sides of similar triangles. The scale factor is also called the *constant of proportionality*. *(p. 389)*

factor de escala   Razón que se halla al comparar las medidas de lados correspondientes de triángulos semejantes. El factor de escala también se llama *constante de proporcionalidad*. *(pág. 389)*

scalene triangle   A triangle with no congruent sides. *(p. 189)*

triángulo escaleno   Triángulo sin lados congruentes. *(pág. 189)*

---

secant   A line that intersects a circle in two points. *(p. 600)*

secante   Recta que interseca un círculo en dos puntos. *(pág. 600)*

$\overleftrightarrow{CD}$ is a secant of $\odot P$.
$\overleftrightarrow{CD}$ es una secante del $\odot P$.

---

secant angles   The angles formed when two secants intersect. *(p. 600)*

ángulos secantes   Ángulos que se forman al intersecarse dos secantes. *(pág. 600)*

secant angle *CAB*, secant angle *DHG*, secant angle *JQL*
ángulo secante *CAB*, ángulo secante *DHG*, ángulo secante *JQL*

---

secant segment   A segment that contains a chord of a circle. An *external secant segment* is the part of a secant segment that lies outside of the circle. *(p. 600)*

segmento secante   Segmento que contiene una cuerda de un círculo. Un *segmento secante externo* es la parte del segmento secante que yace fuera del círculo. *(pág. 600)*

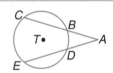

$\overline{CA}$ and $\overline{EA}$ are secant segments.
$\overline{AB}$ and $\overline{AD}$ are external secant segments.
$\overline{CA}$ y $\overline{EA}$ son segmentos secantes.
$\overline{AB}$ y $\overline{AD}$ son segmentos secantes externos.

---

secant-tangent angle   An angle formed by a secant segment and a tangent to a circle. *(p. 606)*

ángulo secante-tangente   Ángulo formado por un segmento secante y una tangente a un círculo. *(pág. 606)*

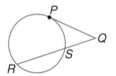 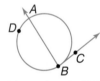

secant-tangent angle *PQR*, secant-tangent angle *ABC*
ángulo secante-tangente *PQR*, ángulo secante-tangente *ABC*

---

sector   A region of a circle bounded by a central angle and its corresponding arc. *(p. 485)*

sector   Región de un círculo acotada por un ángulo central y su arco correspondiente. *(pág. 485)*

---

segment   See *line segment*.

segmento   Ver *segmento de recta*.

---

semicircle   An arc whose endpoints lie on a diameter of a circle. *(p. 462)*

semicírculo   Un arco cuyos extremos yacen en un diámetro de un círculo. *(pág. 462)*

$m\overset{\frown}{PRW} = 180$

---

semi-regular tessellation   See *tessellation*.

teselado semirregular   Ver *teselado*.

side of an angle   See *angle*.

lado de un ángulo   Ver *ángulo*.

sides of a polygon   See *polygon*.

lados de un polígono   Ver *polígono*.

similar polygons   Two polygons whose corresponding angles are congruent and whose corresponding sides have measures that are proportional. *(p. 356)*

polígonos semejantes   Dos polígonos cuyos ángulos correspondientes son congruentes y cuyos lados correspondientes tienen medidas que son proporcionales. *(pág. 356)*

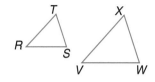

$\angle T \cong \angle X, \angle S \cong \angle W, \angle R \cong \angle V;$
$$\frac{RT}{VX} = \frac{ST}{WX} = \frac{RS}{VW}$$

similar solids   Solids that have the same shape but are not necessarily the same size. *(p. 534)*

sólidos semejantes   Sólidos que tienen la misma forma, pero no necesariamente el mismo tamaño. *(pág. 534)*

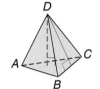

The pyramids are similar.
Estas pirámides son semejantes.

sine   See *trigonometric ratio*.

seno   Ver *razón trigonométrica*.

skew lines   Two nonparallel lines that do not intersect. *(p. 143)*

rectas alabeadas   Rectas que no se intersecan y que no son paralelas. *(pág. 143)*

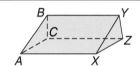

$\overline{AX}$ and $\overline{BC}$ are skew segments.
$\overline{AX}$ y $\overline{BC}$ son segmentos alabeados.

slant height   The height of each lateral face of a regular pyramid or the length of any segment joining the vertex to the base of a circular cone. *(p. 516)*

altura oblicua   La altura de cada cara lateral de una pirámide regular o la longitud de cualquier segmento que une el vértice a la base de un cono circular. *(pág. 516)*

slope   The ratio of the *rise*, or vertical change, to the *run*, or horizontal change. *(p. 168)*

pendiente   Razón de la *elevación* o cambio vertical al *tramo* o cambio horizontal. *(pág. 168)*

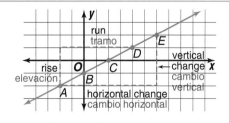

slope-intercept form   The form of a linear equation written as $y = mx + b$. *(p. 174)*

forma pendiente-intersección   Ecuación lineal de la forma $y = mx + b$. *(pág. 174)*

solid figure   A figure that encloses a part of space. *(p. 496)*

sólido   Figura que encierra una parte del espacio. *(pág. 496)*

sphere   A sphere is the set of all points in space that are a given distance from a given point, called the *center*. It has the following characteristics.

1. A *radius* is a segment whose endpoints are the center and a point on the sphere.

2. A *chord* is a segment whose endpoints are points on the sphere.

3. A *diameter* is a chord of the sphere that contains the center.

4. A *tangent* to a sphere is a line that intersects the sphere at exactly one point. *(p. 528)*

esfera   Una esfera es el conjunto de todos los puntos en el espacio que equidistan de un punto dado llamado *centro*. Tiene las siguientes características.

1. Un *radio* es un segmento cuyos extremos son el centro y un punto en la esfera.

2. Una *cuerda* es un segmento cuyos extremos son puntos de la esfera.

3. Un *diámetro* es una cuerda de la esfera que pasa por su centro.

4. Una *tangente* a una esfera es una recta que la interseca en un único punto. *(pág. 528)*

---

square   **1.** A parallelogram with four congruent sides and four right angles. *(p. 327)* **2.** See *perfect square*.

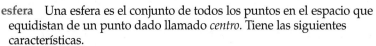

cuadrado   **1.** Paralelogramo de cuatro lados congruentes y cuatro ángulos rectos. *(pág. 327)* **2.** Ver *cuadrado perfecto*.

---

square root   One of two identical factors of a number. *(p. 548)*

raíz cuadrada   Uno de dos factores iguales de un número. *(pág. 548)*

---

statement   A sentence that is either true or false, but not both. *(p. 632)*

enunciado   Una oración que es verdadera o falsa, pero no ambas. *(pág. 632)*

---

straightedge   Any object that can be used as a guide to draw a straight line. *(p. 29)*

regla recta   Cualquier objeto que se usa para trazar rectas. *(pág. 29)*

---

supplementary angles   Two angles whose angle measures have a sum of 180. Each angle is called the *supplement* of the other. *(p. 116)*

ángulos suplementarios   Dos ángulos cuyas medidas angulares suman 180. Cada ángulo se llama *suplemento* del otro. *(pág. 116)*

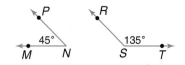

$$m\angle MNP + m\angle RST = 180$$

---

surface area   The sum of the areas of the surfaces of a solid figure. *(p. 504)*

área de superficie   Suma de las áreas de las superficies de un sólido. *(pág. 504)*

---

symmetry   See *line symmetry* and *rotational symmetry*.

simetría   Ver *simetría lineal* and *simetría rotacional*.

---

system of equations   A set of two or more unique equations. *(p. 676)*

sistema de ecuaciones   Conjunto de dos o más ecuaciones. *(pág. 676)*

tangent **1.** In a plane, a line that intersects a circle at exactly one point. The point of intersection is the *point of tangency*. (p. 592) **2.** See *trigonometric ratio*.

tangente **1.** En el plano, recta que interseca un círculo en un solo punto. Éste es el *punto de tangencia*. (pág. 592) **2.** Ver *razón trigonométrica*.

Line ℓ is tangent to $\odot P$.
T is the point of tangency.
La recta ℓ es tangente al $\odot P$.
T es el punto de tangencia.

tangent-tangent angle The angle formed by two tangents to a circle. (p. 607)

ángulo tangente-tangente El ángulo formado por dos tangentes a un círculo. (pág. 607)

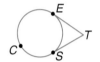

tangent-tangent angle *ETS*
ángulo tangente-tangente *ETS*

terminating decimal A decimal with a finite number of digits. (p. 51)

decimal terminal Decimal con un número finito de dígitos. (pág. 51)

tessellation A tiled pattern formed by repeating figures to fill a plane without gaps or overlaps. In a *regular tessellation*, a single regular polygon is used to form the pattern. In a *semi-regular tessellation*, two or three regular polygons are used. (p. 440)

teselado Un patrón de teselas formado por figuras que se repiten y que cubren el plano sin dejar huecos o traslapos. En un *teselado regular*, el patrón se hace con un solo polígono regular. En un *teselado semirregular* se usan dos o tres polígonos regulares. (pág. 440)

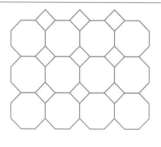

regular tessellation
teselado regular

semi-regular tessellation
teselado semirregular

tetrahedron A triangular pyramid. (p. 497)

tetraedro Una pirámide triangular. (pág. 497)

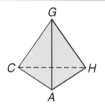

theorem A statement that can be justified by logical reasoning. It must be proven before it is accepted as true. (p. 62)

teorema Enunciado que se puede justificar por razonamiento lógico. Su validez debe demostrarse. (pág. 62)

30°-60°-90° triangle A special right triangle with a 30° angle and a 60° angle. (p. 559)

triángulo 30°-60°-90° Triángulo especial con un ángulo de 30° y uno de 60°. (pág. 559)

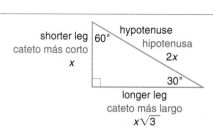

shorter leg
cateto más corto
x

hypotenuse
hipotenusa
$2x$

longer leg
cateto más largo
$x\sqrt{3}$

transformation   The moving of each point of an original figure called the *preimage* to a new figure called the *image*. *(p. 199)*

transformación   Movimiento de cada punto de una figura, llamada *preimagen*, a una nueva figura, llamada *imagen*. *(pág. 199)*

translation   The slide of a figure from one position to another. *(pp. 198, 687)*

traslación   El deslizamiento de una figura de una posición a otra. *(págs. 198, 687)*

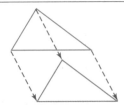

transversal   In a plane, a line that intersects two or more lines, each at a different point. *(p. 148)*

alternate exterior angles:	∠1 and ∠7; ∠2 and ∠8
alternate interior angles:	∠4 and ∠6; ∠3 and ∠5
consecutive interior angles:	∠3 and ∠6; ∠4 and ∠5
corresponding angles:	∠1 and ∠5; ∠2 and ∠6; ∠3 and ∠7; ∠4 and ∠8
exterior angles:	∠1, ∠2, ∠7, ∠8
interior angles:	∠3, ∠4, ∠5, ∠6

transversal   En un plano, recta que interseca dos o más rectas en puntos distintos. *(pág. 148)*

ángulos alternos externos:	∠1 y ∠7; ∠2 y ∠8
ángulos alternos internos:	∠4 y ∠6; ∠3 y ∠5
ángulos conjugados internos:	∠3 y ∠6; ∠4 y ∠5
ángulos correspondientes:	∠1 y ∠5; ∠2 y ∠6; ∠3 y ∠7; ∠4 y ∠8
ángulos externos:	∠1, ∠2, ∠7, ∠8
ángulos internos:	∠3, ∠4, ∠5, ∠6

trapezoid   A quadrilateral with exactly one pair of parallel sides called *bases*, and nonparallel sides called the *legs*. *(p. 333)* The *altitude* is a segment perpendicular to the lines containing the bases. *(p. 420)*

trapecio   Cuadrilátero con sólo un par de lados paralelos, llamados *bases*, y lados no paralelos, llamados *catetos*. *(p. 333)* La *altura* es un segmento perpendicular a las rectas que contienen las bases. *(pág. 420)*

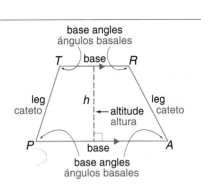

triangle   A figure formed by three noncollinear points connected by segments. *(p. 188)*

triángulo   Figura formada por tres puntos no colineales y los segmentos que los unen. *(pág. 188)*

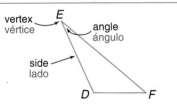

trigonometric identity   An equation involving trigonometric ratios that is true for all values of the angle. *(p. 574)*

identidad trigonométrica   Ecuación que contiene razones trigonométricas y que se cumple para todos los valores del ángulo. *(pág. 574)*

trigonometric ratio A ratio of the measure of two sides of a right triangle. *(p. 564)* The *cosine* is the ratio of the measure of the leg adjacent to the acute angle to the measure of the hypotenuse. *(p. 572)* The *sine* is the ratio of the measure of the leg opposite the acute angle to the measure of the hypotenuse. *(p. 572)* The *tangent* is the ratio of the measure of the leg opposite the acute angle to the measure of the leg adjacent to the acute angle. *(p. 564)*

razón trigonométrica Razón de las longitudes de dos lados de un triángulo rectángulo. *(pág. 564)* El *coseno* es la razón de la longitud del cateto adyacente al ángulo agudo a la longitud de la hipotenusa. *(pág. 572)* El *seno* es la razón de la longitud del cateto opuesto al ángulo agudo a la longitud de la hipotenusa. *(pág. 572)* La *tangente* es la razón de la longitud del lado opuesto al ángulo agudo al lado adyacente al ángulo agudo. *(pág. 564)*

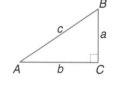

$$\tan A = \frac{a}{b},\ \sin A = \frac{a}{c},\ \cos A = \frac{b}{c}$$

trigonometry The study of the properties of triangles using ratios of angle measures and measures of sides. *(p. 564)*

trigonometría Estudio de las propiedades de los triángulos usando las razones de las medidas angulares y medidas de los lados. *(pág. 564)*

truth table A convenient way to organize truth values. *(p. 633)*

tabla de verdad Una manera conveniente de organizar los valores de verdad. *(pág. 633)*

truth value The true or false nature of a statement. *(p. 632)*

valor de verdad Verdad o falsedad de un enunciado. *(pág. 632)*

turn symmetry See *rotational symmetry*.

simetría de giro Ver *simetría rotacional*.

two-column proof A deductive argument that contains statements and reasons organized in two columns. *(p. 649)*

demostración a dos columnas Argumento deductivo que contiene enunciados y razones organizados en dos columnas. *(pág. 649)*

unit of measure See *measurement*.

unidad de medida Ver *medición*.

vertex See *angle, cone, isosceles triangle*, and *pyramid*.

vértice Ver *ángulo, cono, triángulo isósceles y pirámide*.

vertex angle See *isoceles triangle*.

ángulo del vértice Ver *triángulo isósceles*.

vertical angles Two nonadjacent angles formed by a pair of intersecting lines. *(p. 122)*

ángulos opuestos por el vértice Dos ángulos no adyacentes formados al intersecarse dos rectas. *(pág. 122)*

∠1 and ∠3 are vertical angles.
∠2 and ∠4 are vertical angles.
Los ∠1 y ∠3 son opuestos por el vértice.
Los ∠2 y ∠4 son opuestos por el vértice.

volume The measurement of the space occupied by a solid region. *(p. 510)*

volumen Medida del espacio que ocupa un sólido. *(pág. 510)*

whole numbers The set of real numbers {0, 1, 2, 3, . . .}. *(p. 50)*

números enteros El conjunto de números reales {0, 1, 2, 3, . . .}. *(pág. 50)*

x-axis See *coordinate plane.*

eje x Ver *plano coordenado.*

x-coordinate See *coordinates.*

coordenada x Ver *coordenadas.*

x-intercept The x value of the point where a line crosses the x-axis. *(p. 174)*

intersección x El valor de x del punto donde una recta cruza el eje x. *(pág. 174)*

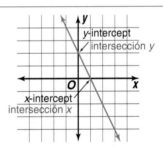

y-axis See *coordinate plane.*

eje y Ver *planos de coordenadas.*

y-coordinate See *coordinates.*

coordenada y Ver *coordenadas.*

y-intercept The y value of the point where a line crosses the y-axis. *(p. 174)* (See art for *x-intercept.*)

intersección y El valor de y del punto donde una recta cruza el eje y. *(pág. 174)* (Ver arte para *intersección x.*)

Chapter 1 Reasoning in Geometry

Page 3 Check Your Readiness
1. 24 **3.** 42 **5.** 22 **7.** 27 **9.** 84 **11.** 1.2 **13.** 5.5
15. 7.5 **17.** 11.5 **19.** 20.24 **21.** 86.1 **23.** $\frac{3}{5}$
25. $\frac{7}{12}$ **27.** $\frac{3}{10}$ **29.** $\frac{21}{40}$

Pages 7–9 Lesson 1–1
1. A conjecture is a conclusion you reach based on inductive reasoning. **5.** Subtract 3. **7.** Add 4.
9. $-3, -6, -9$ **11.** 23, 32, 43
13. **15.** 21, 25, 29 **17.** 48, 57, 66
19. 1875, 9375, 46,875 **21.** 14.6, 18.6, 22.6
23. 23, 28, 34 **25.** 40, 55, 73

27. **29.**

31.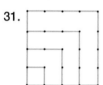

33. $\frac{7}{2}$ **35.** Sample answer: A golden retriever doesn't have spots. **37.** Someone's fingerprint is *not* an arch, loop, or whorl. **39.** C

Pages 15–17 Lesson 1–2
1. A line extends without end in two directions; a line segment has endpoints. **3.** D **5.** $\overleftrightarrow{DB}$
7. collinear: Denver, Colorado Springs, Pueblo; noncollinear: any three cities other than Denver, Colorado Springs, and Pueblo **9.** $\overrightarrow{BE}, \overrightarrow{AD}$
11. $\overrightarrow{AF}, \overrightarrow{BF}, \overrightarrow{FC}$ **13.** $\overleftrightarrow{AD}$ **15.** $\overline{EF}$ **17.** A, F, B
19. plane **21.** segment **23.** plane

25. Sample answer:

27. Sample answer:

29. Sample answer:

31a. close together **31b.** far apart
33. 80, 160, 320 **35.** $-7, -9, -11$
37. Sample answer:

Page 17 Quiz 1
1. $-1, -5, -9$ **3.** Sample answer: $\overleftrightarrow{AC}$
5. Sample answer: $\overleftrightarrow{BE}$

Pages 20–22 Lesson 1–3
1.

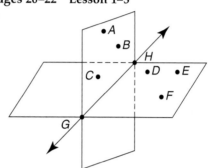

5. point C **7.** $\overleftrightarrow{AC}$ **9.** $\overleftrightarrow{AC}, \overleftrightarrow{AB}, \overleftrightarrow{BC}$ **11.** $\overleftrightarrow{KG}, \overleftrightarrow{KH}, \overleftrightarrow{KJ}$, $\overleftrightarrow{GH}, \overleftrightarrow{GJ}, \overleftrightarrow{HJ}$ **13.** point E **15.** planes QRS, QST, QTR, RST **17.** planes GHJ, DEF, GHD, HJE, JEF
19. $\overleftrightarrow{HG}$ **21.** planes EFG, AEF **23.** false **25.** false
27. true **29.** true **31.** According to the geometric definition of line, there can be only one line through any two points. In this case, there can be many lines through any two points. **33.** infinite number **35.** any segment **37.** A ray has a definite starting point and extends without end in one direction; a line is a series of points that extends without end in two directions.
39. Sample answer: 14, 17

Pages 26–28 Lesson 1–4
1. If there are clouds in the sky, then it may rain.
3. H: a figure is a quadrilateral; C: it has four sides
5. All people who are at least 18 years old can vote. If you are at least 18 years old, then you can vote.
7. If the ground is wet, then it is raining. **9.** If an animal is a cat, then it is a mammal. **11.** H: a set of points has two endpoints; C: it is a line segment
13. H: I finish my homework; C: I will call my friend **15.** H: you are a student; C: you should report to the gymnasium **17.** You will be healthy if you eat fruits and vegetables. All people who eat fruits and vegetables will be healthy. **19.** If you run the fastest, then you'll win the race. All people who run the fastest will win the race.
21. If you are over age 18, you can serve in the armed forces. You can serve in the armed forces if you are over age 18. **23.** If you do well in school, then you play a musical instrument. **25.** If you play softball, it stops raining. **27.** If points extend without end in two directions, then it is a line.

31a. I: If a figure does not have five sides, then it is not a pentagon. C: If a figure is not a pentagon, then it does not have five sides.   **33.** Sample answer: $\overrightarrow{QS}$ **35.** Sample answer: *P, Q, R*   **37.** B

Page 28   Quiz 2
1. $\overline{YZ}$ **3.** Today is Monday.   **5.** If I have band practice, then today is Monday.

Pages 32–34   Lesson 1–5
1. A construction is a precise drawing done with a compass and straightedge. Other drawings may be only rough sketches. Also, a construction does not use standard measurement units.   **3.** Curtis is correct. A ruler can be used as a straightedge, but a ruler has measurement units and a straightedge does not.   **5.** width   **7.** B   **9.** straight   **13.** Sample answer: The measuring tape is a fixed distance, like a compass. Making marks on the ground is like drawing arcs with the pencil.   **15.** If a figure has three sides, it is a triangle.   **17.** If you like the ocean, then you are a surfer.   **19.** B

Pages 38–40   Lesson 1–6
1. Sample answers:

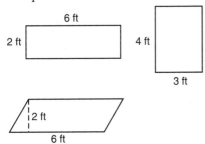

6 ft
2 ft
4 ft
3 ft
2 ft
6 ft

3. Explore, Plan, Solve, Examine   **5.** 12 ft^2
7. $P = 134$ cm, $A = 660$ cm^2 **9.** $P = 60$ ft, $A = 125$ ft^2 **11.** 5 rolls   **13.** $P = 40$ m, $A = 100$ m^2 **15.** $P = 22.4$ mm, $A = 15.36$ mm^2
17. $P = 24$ mi, $A = 36$ mi^2 **19.** $P = 50$ ft, $A = 150$ ft^2 **21.** $P = 72$ cm, $A = 324$ cm^2
23. $P = 33$ mi, $A = 65$ mi^2 **25.** 80 ft^2 **27.** 34.5 m^2
29. 5 yd   **31.** $1755   **33.** They are the same length.
35. $-3, -7, -11$ **37.** C

Pages 42–44   Chapter 1 Study Guide and Assessment
1. plane   **3.** hypothesis   **5.** perimeter
7. line segment   **9.** straightedge
11. 18, 27, 38
13.

15. Sample answer: $\overrightarrow{OM}, \overrightarrow{ON}$ **17.** Sample answer:

M, N, O   **19.** Sample answer: planes *ABG, CBG,* and *BDC*   **21.** H: a bus is a school bus; C: it is yellow   **23.** If you own a pet, then you will live a long life. You will live a long life if you own a pet.   **25.** If you like to play baseball, then you are a student.   **29.** $P = 82$ ft, $A = 414$ ft^2 **31.** 12 cm
33. $1215

Page 47   Preparing for Standardized Tests
1. C   **3.** C   **5.** A   **7.** E   **9.** 15

Chapter 2   Segment Measure and Coordinate Graphing

Page 49   Check Your Readiness
1. 4   **3.** 19   **5.** 13   **7.** 9   **9.** 15   **11.** 11   **13.** 7   **15.** 4
17. 8   **19.** 13   **21.** -5 **23.** 4   **25.** 5   **27.** 4   **29.** 5

Pages 53–55   Lesson 2–1
1. Sample answer: Negative numbers and positive numbers never stop. The arrows show that they continue without end.
3a. All three numbers are rational numbers. $0.34 = 0.34$; $0.3\overline{4} = 0.344444\ldots$; $0.\overline{34} = 0.34343434\ldots$
3b. 0.3400000000
0.3444444444 $\}$ $0.3\overline{4}$ is greatest.
0.3434343434
3c. Sample answers: Read 0.34 as *zero point three four.* Read $0.3\overline{4}$ as *zero point three four repeating.* Read $0.\overline{34}$ as *zero point three four, all repeating.*   **5.** Sample answer: 1.1211221112…   **7.** 3   **9.** 4   **11.** Sample answer: $-1.1212121212\ldots$ **13.** Sample answer: 3.1234123412…   **15.** Sample answer: 0.1211221112… and 0.3456789101…   **17.** 4
19. 1   **21.** $\frac{1}{4}$ **23.** $2\frac{1}{4}$ **25.** $3\frac{1}{2}$ **27.** $3\frac{3}{4}$ **29a.** 16 ft
29b. 13 ft   **31.** -12 and 2   **33.** $P = 32$ cm; $A = 60$ cm^2 **35.** compass   **37.** B

Pages 59–61   Lesson 2–2
1. The measure of a segment is the number of units and the measurement of a segment is the number of units and the units of measure.
3. Joseph; 2 lb is measured to the nearest pound, and 34 oz is measured to the nearest ounce. Therefore, 34 oz is more precise.   **5.** Y   **7.** 57
9. 5.1 cm; 2 in.   **11.** D   **13.** G   **15.** C   **17.** 52
19. 17   **21.** 7.4   **23.** 3.1 cm; $1\frac{1}{4}$ in.   **25.** 3.5 cm; $1\frac{3}{8}$ in.   **27.** 8.9 cm; $3\frac{1}{2}$ in.   **29a.** 10 mm
29b. 14 mm   **29c.** 8 mm   **31.** 11; 3; 16   **33.** 6

35a.

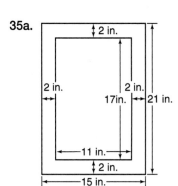

35b. 187 in² **37.** A

Page 61   Quiz 1
1. Sample answer: 4.1231231231 . . .   **3.** T
5. 8.2 cm; $3\frac{1}{4}$ in.

Pages 65–67   Lesson 2–3
1. Sample answers:

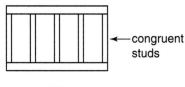

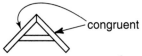

3. true; $AB = 3$ and $CD = 3$ **5.** False; an endpoint (Y) cannot also be the midpoint.   **7.** 12   **9.** false; $BF = 7$ and $EI = 8$ **11.** false; $BF = 7$ and $FI = 6$
13. true; $CD = 3$ and $DF = 3$ **15.** True; segment congruence is symmetric and transitive.   **17.** False; a plane can only bisect a segment in one point.
19. False; D may be between E and F.   **21a.** 9
21b. 27; 27   **21c.** 54   **23.** 2:1   **25.** T **27.** 7300

Pages 71–73   Lesson 2–4
1. Sample answer: The artist would use two sizes of grids and locate corresponding points on the two grids.   **3.** Sample answers: quadrant: one of *four* parts; quadriceps: muscle in the front of the thigh that is divided into *four* parts; quadrilateral: polygon with *four* sides; quadruple: to multiply by *four*; quadruplet: one of *four* children at one birth
5. $x = -3, y = -6$ **7.** $x = 11, y = 0$

8–10.

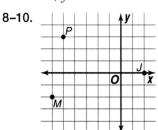

---

11. (2, 3)   **13.** (2, –2)

15–23.

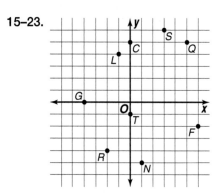

25. $(-1, -4)$ **27.** $(1, 5)$ **29.** $(0, -5)$ **31.** $(-5, 0)$
33. N **35a.** New Orleans   **35b.** 60°N, 30°E
35c. South Africa   **35d.** Answers will vary.

37.

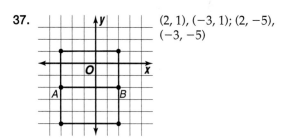

(2, 1), (−3, 1); (2, −5), (−3, −5)

39. true; $AC = 7$ and $CE = 7$ **41.** 39   **43.** If students do their homework, then they will pass the course.

Page 73   Quiz 2
1. true; $AC = 7$ and $EF = 7$
3. false; $AD = 9$ and $DF = 10$
4–5.

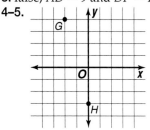

Pages 79–81   Lesson 2–5
1.

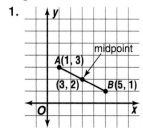

$$\left(\frac{x_1 + x_2}{2}, \frac{y_1 + y_2}{2}\right)$$
$$= \left(\frac{1 + 5}{2}, \frac{3 + 1}{2}\right)$$
$$= \left(\frac{6}{2}, \frac{4}{2}\right)$$
$$= (3, 2)$$

3. Both are correct; adding the number to the coordinate of the left endpoint will give the same answer as subtracting the number from the coordinate of the right endpoint. **5.** 2

7. -7 **9.** $-\frac{1}{4}$ **11.** $\left(-3\frac{1}{2}, 1\frac{1}{2}\right)$ **13.** $(-1, -8)$ **15.** $1\frac{1}{2}$

17. $-2\frac{1}{2}$ **19.** $-3\frac{3}{4}$ **21.** $(-2, -4)$ **23.** $\left(1, 1\frac{1}{2}\right)$

25. $\left(2\frac{1}{2}, -3\right)$ **27.** $\left(-10\frac{1}{2}, -5\frac{1}{2}\right)$ **29.** $\left(\frac{a+c}{2}, \frac{b+d}{2}\right)$

31. (a, b) **33.** 156 **35.** Sample answers: (3, 5), (9, 11); (4, 3), (8, 13); (−2, 0), (14, 16); (−5, 8), (17, 8); (18, 4), (−6, 20) **37.** (0, −4) **39.** (−2, −5) **41.** $\overrightarrow{AB}$

Pages 82–84 Chapter 2 Study Guide and Assessment
1. whole **3.** irrational **5.** absolute value
7. congruent **9.** bisect **11.** 3 **13.** 5 **15.** 10.5
17. true; $BD = 5$ and $EG = 5$ **19.** true; $AC = 5$ and $CE = 5$ **21.** False; since $\overline{LM}$ and $\overline{ML}$ have the same length, $\overline{LM} \cong \overline{ML}$ by the definition of congruent segments. **23.** (0, −5) **25.** (2, 2)

27–30.

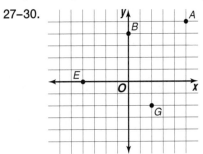

31. $\frac{1}{2}$ **33.** (1, −4) **35.** 153°C

37a.

Municipal Waste (kg per person) vs GNP ($ per person)

37b. Sample answer: The graph shows that as the x-values increase, the y-values also increase, indicating a tendency for countries with higher GNP per person to produce more waste per person. Japan, however, does not follow this tendency.

Page 87 Preparing for Standardized Tests
1. C **3.** B **5.** B **7.** C **9.** 12, 18, 21, 24, or 27

Chapter 3 Angles

Page 89 Check Your Readiness
1. Sample answer: M, S, N **3.** Sample answer: Q
5. 45 **7.** 21 **9.** 78 **11.** 29 **13.** 54 **15.** 33 **17.** 14
19. 31

Pages 92–94 Lesson 3–1
1.

3. There is more than one angle with T as its vertex.
5. $\angle 1, \angle 2, \angle DKF$ **7.** exterior **9.** $\angle DEF, \angle FED,$ $\angle E, \angle 2$; E; $\overrightarrow{ED}, \overrightarrow{EF}$ **11.** $\angle HIJ, \angle JIH, \angle I, \angle 4$; I; $\overrightarrow{IJ},$ $\overrightarrow{IH}$ **13.** $\angle 4, \angle 5, \angle MJP$ **15.** exterior **17.** on
19. on **21.** true **23.** false **25.** $\angle ADB; \angle BDC;$ $\angle ADC; \overrightarrow{DA}, \overrightarrow{DB}; \overrightarrow{DB}, \overrightarrow{DC}; \overrightarrow{DA}, \overrightarrow{DC}$ **27.** (3, 4)
29. 26 ft **31.** $\overrightarrow{PQ}, \overrightarrow{PR}, \overrightarrow{PS}, \overrightarrow{QR}, \overrightarrow{QS}, \overrightarrow{RS}$

Pages 100–101 Lesson 3–2
1.

70°

3. Sample answer: Rulers are used to measure segments. Protractors are used to measure angles. **5.** 105; obtuse **7.** 60; acute **9.** obtuse **11.** 110; obtuse **13.** 30; acute **15.** 90; right **17.** 40; acute
19. 140; obtuse **21.** 90; right **23.** acute **25.** acute
27. acute **29a.** Algebra-150; Calculus-20; Trigonometry-25; Advanced Algebra-35; Geometry-130 **29b.** Algebra-obtuse; Calculus-acute; Trigonometry-acute; Advanced Algebra-acute; Geometry-obtuse **29c.** To the nearest degree, the greatest measure of an acute angle is 89°. The total number of degrees in a circle is 360°. So, the greatest percentage is (89° ÷ 360°) × 100 or about 24.7%. **31.** $m\angle ABC = 62; m\angle EFG = 28$
33. Sample answer:

35. (5, 4) **37.** Sample answer: 1.2, 7.2, 13.2, 19.2, . . .

Pages 108–109 Lesson 3–3
1. For any angle ABC, if X is in the interior of $\angle ABC$, then $m\angle ABX + m\angle XBC = m\angle ABC$.

3. Brandon is correct. Since the measure of any angle is between 0 and 180, bisecting the angle with the greatest possible measure will produce two smaller angles with a measure less than 90. These two angles would be classified as acute. **5.** 79 **7.** 64 **9.** 58 **11.** 42 **13.** 103 **15.** 15 **17.** 63 **19.** 72 **21.** acute **23.** 4 **25.** Definition of betweenness **27.** $\angle 1, \angle 2, \angle OPQ$ **29.** $\overline{NK}$

Pages 112–114   Lesson 3–4
1. Sample answer:

3. adjacent angles; linear pair **5.** $\angle XUY, \angle XUZ$ **7.** adjacent angles **9.** neither **11.** adjacent angles **13.** adjacent angles **15.** No, they are not adjacent angles. **17.** $\angle AGB, \angle DGE$ **19.** Yes, their noncommon sides are opposite rays. **21.** 4 **23.** 68
25. Sample answer:

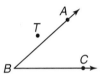

27. D

Page 114   Quiz 1
1. $\angle FGH, \angle HGF, \angle G, \angle 1$; G; $\overrightarrow{GF}, \overrightarrow{GH}$ **3.** acute **5.** adjacent angles; linear pair

Pages 119–121   Lesson 3–5
1.

3. neither **5.** 48; 138 **7.** 35; 125 **9.** Sample answers: $\angle AGB, \angle DGE$ **11.** 62; 118 **13.** $\angle MNK$, $\angle KNJ$; $\angle KNJ, \angle HNI$ **15.** Sample answer: $\angle HNI$, $\angle INJ$ **17.** Sample answer: $\angle QWV, \angle SWT$ **19.** 95 **21.** $\angle DGE, \angle AGF$ **23.** 53 **25.** No; for two angles to be supplementary, their sum must be equal to 180°. To the nearest degree, the greatest measure an acute angle can have is 89. $89 + 89 = 178$. So, two acute angles cannot be supplementary. **27.** right **29.** 120; 60 **31.** Their measures are the same. **33.** 130 **35.** If he will go skiing, then it is snowing. **37.** B

Pages 125–127   Lesson 3–6
3. Roberta is correct. To say that the angles have the same measure, it is correct to write $m\angle A = m\angle B$. Keisha is incorrect. To say that the angles are congruent, it is correct to write $\angle A \cong \angle B$ not $m\angle A \cong m\angle B$. **5.** 96 **7.** 64; 116 **9.** 70 **11.** 65 **13.** 14 **15.** 75 **17.** 125 **19.** 47 **21.** 11

23. You can show that Theorem 3–6 is true by using algebra. Let x = the measure of the first angle. Since the angles are congruent, the measure of the second angle also equals x. The angles are supplementary. So, their sum equals 180.

$$x + x = 180$$
$$2x = 180$$
$$\frac{2x}{2} = \frac{180}{2}$$
$$x = 90$$

Thus, if two angles are congruent and supplementary, then each is a right angle.
25. neither **27.** Sample answer: 2.1646646664 . . .

Page 127   Quiz 2
1. Sample answer:

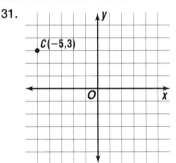

3. 37 **5.** 45; 135

Pages 131–133   Lesson 3–7
1. c **3.** false **5.** true **7.** 48; 42 **9.** false **11.** true **13.** false **15.** false **17.** false **19.** true **21.** 42 **23.** Sample answer: $\angle YPT$ and $\angle TPL$ **25.** 65; 25 **27.** Sample answer: two walls of the classroom that meet in a corner **29.** 47
31.

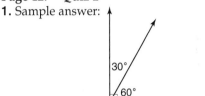

33. B

Pages 134–136   Chapter 3 Study Guide and Assessment
1. true **3.** false; protractor **5.** false; congruent **7.** false; right **9.** true **11.** $\angle FGH, \angle HGF, \angle G, \angle 5$; G; $\overrightarrow{GF}, \overrightarrow{GH}$ **13.** $\angle 2, \angle 3, \angle NPO$ **15.** 155; obtuse **17.** 90; right **19.** 83 **21.** 48 **23.** Sample answer: $\angle UTR, \angle STV$ **25.** $\angle LAS, \angle DAF$ **27.** 126 **29.** 115 **31.** 17 **33.** false **35.** true **37.** true **39.** 60, 60, 60

Page 139   Preparing for Standardized Tests
1. B **3.** B **5.** A **7.** D **9.** 2/3

Chapter 4 Parallels

Page 141 Check Your Readiness

1. Sample answer: $\overline{AB}$ **3.** Sample answer: D
5. 2 **7.** $\frac{1}{4}$ **9.** $-\frac{5}{2}$ **11.** 2 **13.** $\frac{7}{2}$

15, 17, 19.

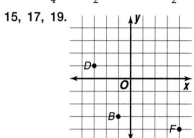

Pages 144–147 Lesson 4–1

1.

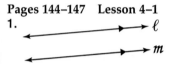

3. Lines ℓ and m are skew. **5.** intersecting
7. plane ZYR **9.** $\overline{XQ}$, $\overline{YR}$, $\overline{WX}$, $\overline{ZY}$ **11.** Sample answers: The slats on the chair back are parallel; the seat and the back are parts of intersecting planes; the front edge of the seat and a back leg are skew. **13.** parallel **15.** parallel **17.** skew
19. skew **21.** skew **23.** plane ABC ∥ plane EGF, plane ABF ∥ plane CDG, plane EDA ∥ plane BCG
25. $\overline{BF}$, $\overline{AD}$, $\overline{BC}$, $\overline{AE}$ **27.** $\overline{CD}$, $\overline{GH}$, $\overline{EH}$, $\overline{AD}$
29. $\overline{ST}$ ∥ $\overline{PO}$, $\overline{PS}$ ∥ $\overline{OT}$ **31.** $\overline{SP}$, $\overline{ST}$ and $\overline{SM}$; $\overline{ST}$, $\overline{TO}$, and $\overline{TM}$; $\overline{TO}$, $\overline{OP}$, and $\overline{OM}$; $\overline{OP}$, $\overline{PS}$, and $\overline{PM}$; $\overline{SM}$, $\overline{MT}$, and $\overline{MO}$; $\overline{SM}$, $\overline{MT}$, and $\overline{MP}$; $\overline{SM}$, $\overline{MO}$, and $\overline{MP}$; $\overline{MT}$, $\overline{MO}$, and $\overline{MP}$ **41.** never **43.** sometimes
45. always **47.** 12.25 ft **49.** The rails of the railroad track are parallel and thus never cross; the character is saying that her life's path and Mr. Right's life path are parallel and thus they will never meet. **51a.** $\angle AXB$, $\angle BXD$, $\angle DXE$, $\angle EXA$ **51b.** Sample answer: $\angle AXC$ and $\angle CXD$
51c. Sample answer: $\angle BXC$ and $\angle CXD$ **53.** 48
55. obtuse **57.** Carlos **59.** C

Pages 151–153 Lesson 4–2

1. Theorem 4–1 **3.** transversal c: s, t; transversal s: c, t, d; transversal d: s, t; transversal t: c, s, d
5. alternate exterior **7.** consecutive interior
9. 48; $\angle 1$ and $\angle 3$ are vertical angles and are congruent. **11.** 48; $\angle 1$ and $\angle 7$ are alternate exterior angles and are congruent. **13.** alternate interior **15.** vertical **17.** consecutive interior
19. consecutive interior **21.** consecutive interior
23. vertical **25.** Vertical angles are congruent.
27. Consecutive interior angles are supplementary.
29. 76; Linear pairs are supplementary. **31.** 76; Alternate interior angles are congruent ($\angle 13$ and $\angle 11$); linear pairs are supplementary ($\angle 11$ and $104°$ angle). **33.** 98; Linear pairs are supplementary.
35. 82; Alternate interior angles are congruent.

37. $x = 12$, $y = 10$ **39.** $x = 14$, $y = 14$
41. $\angle XAC \cong \angle XBD$, $\angle XCA \cong \angle XDB$
47–49.

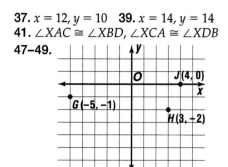

Pages 158–161 Lesson 4–3

1a. $\angle 1$ and $\angle 3$, $\angle 6$ and $\angle 4$ **1b.** Postulate 4–1
5. 13 **7.** $\angle 3$, $\angle 6$, $\angle 8$, $\angle 9$, $\angle 11$, $\angle 14$, $\angle 16$; $\angle 3$, corresponding; $\angle 6$, vertical; $\angle 8$, alternate exterior; $\angle 9$, corresponding; $\angle 11 \cong \angle 9$ (corresponding); $\angle 14$ alternate exterior; $\angle 16 \cong \angle 14$ (corresponding)
9. $m\angle 1 = 112$, $m\angle 2 = 68$, $m\angle 3 = 112$ **11.** 128
13. $\angle 4$, $\angle 6$, $\angle 8$; $\angle 4$, vertical; $\angle 6$, corresponding; $\angle 8$, alternate interior **15.** $\angle 2$, $\angle 4$, $\angle 6$; $\angle 2$, alternate interior; $\angle 4$, corresponding; $\angle 6$, vertical
17. $\angle 10$, $\angle 14$, $\angle 16$; $\angle 10$, vertical; $\angle 14$, alternate exterior; $\angle 16$, corresponding **19.** $m\angle 11 = 124$, $m\angle 12 = 98$, $m\angle 13 = 82$, $m\angle 14 = 124$, $m\angle 15 = 98$
21. $m\angle 23 = 120$, $m\angle 24 = 120$, $m\angle 25 = 120$
23. $m\angle 32 = 61$, $m\angle 33 = 78$, $m\angle 34 = 41$, $m\angle 35 = 61$
25. $x = 20$, $m\angle 4 = 47$, $m\angle 8 = 47$ **27.** $x = 10$, $m\angle 1 = 58$, $m\angle 4 = 122$ **29.** $\angle 1 \cong \angle 2$ and $\angle 3 \cong \angle 4$ by Postulate 4–1. **31.** $\angle 6 \cong \angle 4$ by Postulate 4–1. $\angle 6 \cong \angle 2$ by Theorem 4–1 only if $\overline{AM}$ ∥ $\overline{KI}$, which is not given. **33.** $\overleftrightarrow{AB}$ ∥ $\overleftrightarrow{DC}$; $\overleftrightarrow{AD}$ ∥ $\overleftrightarrow{BC}$ **39.** D

Page 161 Quiz 1

1. Sample answer: Stair railings have parallel posts; the railing is a transversal. **3.** 56 **5.** 41

Pages 165–167 Lesson 4–4

1. Sample answer: Neither $\angle A$ and $\angle R$ nor $\angle C$ and $\angle T$ are supplementary angles.
3. (1) $\angle 1 \cong \angle 2$; Given
(2) $\angle 2 \cong \angle 3$; Vertical angles are congruent.
(3) $\angle 1 \cong \angle 3$; Congruence of angles is transitive.
(4) ℓ ∥ n; Postulate 4–2 **5.** 12 **7.** p ∥ q **9.** 110

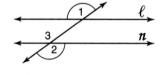

11. 64 **13.** 35 **15.** m ∥ n **17.** $\overline{ST}$ ∥ $\overline{YZ}$ **19.** p ∥ q
21a. 55 **21b.** yes **23.** Show $\angle AGE \cong \angle CHG$; show $\angle GAC$ and $\angle ACD$ are right angles; show $\angle AGH$ and $\angle CHG$ are supplementary. **25.** 68
27. true; definition of congruent segments **29.** D

Pages 171–173 Lesson 4–5

1. horizontal line, vertical line **3.** Sang Hee; see students' drawings. **5.** 0 **7.** perpendicular
9. Yes, the slope, $\frac{6}{11}$ or $0.\overline{54}$, is less than 0.88.
11. 5 **13.** -4 **15.** $-\frac{1}{7}$ **17.** 0 **19.** parallel

21. perpendicular **23.** neither **25.** $\frac{5}{9}$ **27.** 1800 feet
29. Yes; the product of the slopes of $\overline{AB}$ and $\overline{BC}$ is -1, so they are perpendicular. The same is true for $\overline{BC}$ and $\overline{CD}$, $\overline{CD}$ and $\overline{AD}$, and $\overline{AD}$ and $\overline{AB}$. Since all four angles are right angles, the figure is a rectangle. **31.** $m\angle 1 = 90$, $m\angle 2 = 125$, $m\angle 3 = 55$ **33.** $\angle AXB$ and $\angle BXC$, $\angle BXC$ and $\angle CXD$, $\angle AXC$ and $\angle CXD$, $\angle AXB$ and $\angle BXD$

Page 173 Quiz 2
1. 70 **3.** 36 **5.** perpendicular

Pages 177–179 Lesson 4–6
1. If you are given the slope and the y-intercept of a line, you can find an equation of the line using this form. **3.** $y = 6x - 3$ **5.** $y = \frac{5}{3}x - 3$ **7.** $-\frac{3}{2}$; 4

9.
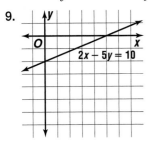
$2x - 5y = 10$

11. $y = -2x - 4$ **13.** 9; 1 **15.** $\frac{3}{2}$; -9 **17.** 0; 5

19.
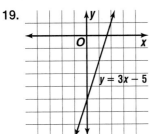
$y = 3x - 5$

21.
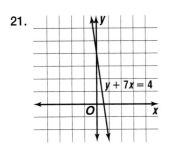
$y + 7x = 4$

23.
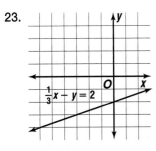
$\frac{1}{3}x - y = 2$

25. $y = 3x + 7$ **27.** $y = -4x - 5$
29. $x = -3$ **31.** d **33.** a

35a.

x	y
0	0.99
1	1.39
2	1.79
3	2.19
4	2.59
5	2.99

35b.
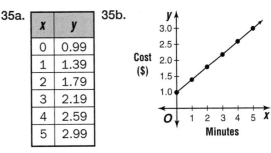

35c. 0.40; rate per minute **35d.** 0.99; base charge for making any call **37.** Sample answer: Use the points to find the slope, then choose one of the points and substitute in the slope-intercept form to find the y-intercept. **39.** The slope would be positive. **41.** $\angle UQT$, $\angle TQR$, $\angle VQR$, $\angle UQV$, $\angle VQT$, $\angle UQR$ **43.** $\overline{KR}$; $\overline{AJ}$ is longer.

Pages 180–182 Chapter 4 Study Guide and Assessment
1. c **3.** d **5.** b **7.** f **9.** h **11.** parallel **13.** intersecting **15.** $\overline{DF}$, $\overline{BH}$, $\overline{AD}$, $\overline{DC}$, $\overline{BC}$, $\overline{AB}$ **17.** alternate exterior **19.** alternate interior **21.** 124; Vertical angles are congruent. **23.** 124; Alternate exterior angles are congruent. **25.** $\angle 4$, alternate interior; $\angle 6$, corresponding **27.** $\angle 9$, vertical **29.** $m\angle 11 = 116$, $m\angle 12 = 51$, $m\angle 13 = 116$, $m\angle 14 = 129$, $m\angle 15 = 51$, $m\angle 16 = 116$, $m\angle 17 = 64$, $m\angle 18 = 129$, $m\angle 19 = 116$, $m\angle 20 = 129$ **31.** 17 **33.** $\overline{EF} \parallel \overline{ST}$ **35.** 0 **37.** neither

39. $-\frac{3}{5}$, 1
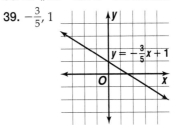
$y = -\frac{3}{5}x + 1$

41. $y = \frac{1}{2}x + 7$ **43.** 58

Page 185 Preparing for Standardized Tests
1. C **3.** D **5.** D **7.** B **9.** 2/15

Chapter 5 Triangles and Congruence

Page 187 Check Your Readiness
1. acute **3.** acute **5.** acute **7.** 56°, 146° **9.** 46°, 136° **11.** 24°, 114° **13.** 66 **15.** 18 **17.** 67 **19.** 27 **21.** 24.5 **23.** 92

Pages 190–192 Lesson 5–1
1. Sample answer:
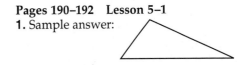

3. Yes; an equilateral triangle has at least two congruent sides, so it is also an isosceles triangle.   **5.** acute, equilateral   **7.** 5, 2   **9.** acute, equilateral   **11.** right, scalene   **13.** acute, isosceles   **15.** right, scalene   **17.** right   **19.** not possible   **21.** Sample answer:

23a. right, scalene   **23b.** acute, isosceles   **23c.** obtuse, isosceles   **25a.** right, isosceles   **25b.** acute isosceles, right scalene, right isosceles, obtuse scalene   **27.** 4, 8, 8   **29.** $y = -3x + 4$ **31.** $y = -2x - 3$ **33.** 0   **35.** 160°

Pages 196–197   Lesson 5–2
1. c   **3.** No; the sum of the measures of the angles of a triangle is 180. If a triangle has two obtuse angles, the sum of the measures of these two angles alone would be greater than 180.   **5.** 75   **7.** 40, 60, 80   **9.** 60   **11.** 27   **13.** $a = 25$, $b = 120$ **15.** $x = 70$, $y = 60$ **17.** 55   **19.** 51, 66   **21.** 98   **23.** The sum of the measures of the angles of each triangle is 180. By substitution, the measures of the third angles are equal. Therefore, the third angles are congruent.
25. $-\frac{1}{3}$ **27.** vertical   **29.** alternate exterior

Pages 200–202   Lesson 5–3
1. A translation involves moving a figure without changing its orientation; a rotation involves turning a figure in a circular motion.   **3.** translation
5. reflection   **7.** $\angle X$ **9.** reflection   **11.** translation
13. rotation   **15.** rotation   **17.** translation   **19.** $\overline{FG}$
21. point J **23.** $\overline{HJ}$ **25.** rotation   **27.** A translation, reflection, or rotation is the result of a single motion; a glide reflection is a combination of a translation and a reflection.   **29.** 12

31. 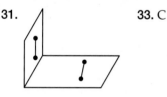   **33.** C

Page 202   Quiz 1
1. right, isosceles   **3.** acute, equilateral
5. reflection

Pages 205–207   Lesson 5–4
1. They have the same size and shape.
3. $\angle C$ **5.** $\overline{DF}$ **7.** $\angle X \cong \angle E$, $\angle Y \cong \angle D$, $\angle Z \cong \angle F$, $\overline{XY} \cong \overline{ED}$, $\overline{XZ} \cong \overline{EF}$, $\overline{YZ} \cong \overline{DF}$

7. $\angle X \cong \angle E$, $\angle Y \cong \angle D$, $\angle Z \cong \angle F$, $\overline{XY} \cong \overline{ED}$, $\overline{XZ} \cong \overline{EF}$, $\overline{YZ} \cong \overline{DF}$

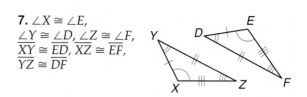

9. DEF **11.** $\angle A \cong \angle E$, $\angle B \cong \angle D$, $\angle C \cong \angle F$, $\overline{AB} \cong \overline{ED}$, $\overline{BC} \cong \overline{DF}$, $\overline{AC} \cong \overline{EF}$

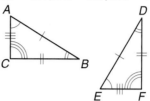

13. CDA **15.** CDB **17.** CAB **19.** $\angle C$ **21.** $\angle H$
23. $\angle B$ **25.** 2   **27.** They have the same size and shape.   **29.** rotation   **31.** translation   **33.** acute

Pages 212–214   Lesson 5–5
1.

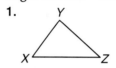

3. Sample answer: $\triangle RST \cong \triangle UVW$ **5.** Sample answer: $\triangle ABC \cong \triangle FDE$; SAS   **7.** There is only one triangle with three given measures. Therefore, the triangles in the truss will not shift into a different triangle.   **9.** Sample answer: $\triangle CBA \cong \triangle EFD$
11. Sample answer: $\triangle GHI \cong \triangle RTS$ **13.** Sample answer: $\triangle BCA \cong \triangle DFE$; SAS   **15.** Sample answer: $\triangle DBA \cong \triangle DBC$; SAS   **17.** yes   **19.** no
21. A triangle is formed by the tree trunk, the stake, and the ground. Since the triangle won't shift, it will provide support from the wind.
23. 97   **25.** $(-2, -5)$ **27.** $\left(\frac{1}{2}x, \frac{1}{2}y\right)$

Page 214   Quiz 2
1. The small isosceles triangles are congruent to each other as are the small equilateral triangles.
3. Sample answer: $\triangle XYZ \cong \triangle BAC$ **5.** Sample answer: $\triangle NML \cong \triangle QPR$; SAS

Pages 217–219   Lesson 5–6
1. $\angle X$ and $\angle Z$; Sample answer:

5. $\triangle ABC \cong \triangle XYZ$ **7.** $\overline{BA} \cong \overline{FE}$ or $\overline{CA} \cong \overline{DE}$
9. AAS   **11.** $\triangle QRS \cong \triangle TVU$ **13.** $\triangle RST \cong \triangle YXZ$
15. $\angle E \cong \angle C$ **17.** $\angle C \cong \angle E$ **19.** AAS   **21.** SAS
23. The triangle made by the ship and points P and Q is congruent to $\triangle PQT$ by ASA. Therefore, the distance from the ship to point Q is the same as the distance from point Q to point T by CPCTC.

25. $\triangle MNO \cong \triangle PQR$; SSS **27.** $\overline{MP}$ **29.** C

Pages 220–222   Chapter 5 Study Guide and Assessment

1. true **3.** false; 180 **5.** false; complementary
7. true **9.** false; nonincluded **11.** right, scalene
13. 35 **15.** $a = 25, b = 125$ **17.** $\angle CBD$
19. reflection **21.** RVW **23.** $\triangle FED \cong \triangle CBA$; SAS
25. AAS **27.** 21

Page 225   Preparing for Standardized Tests

1. C **3.** D **5.** B **7.** A **9.** 14

Chapter 6   More About Triangles

Page 227   Check Your Readiness

1. acute, isosceles **3.** equiangular, equilateral
5. 56 **7.** 7 **9.** 2 **11.** 10 **13.** 4

Pages 231–233   Lesson 6–1

1. Locate the midpoint of a side of the triangle. Then draw a segment from that point to the vertex opposite that side. **3.** Kim and Hector are both wrong. Medians of an equilateral triangle are the same length. But, medians of a scalene triangle are not the same length. **5.** 11 **7.** 6.7
9. 16 **11.** 11 **13.** 6.5 **15.** 5.3 **17.** 8.5
19. Sample answer:

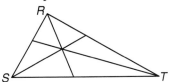

21. 14 **23.** Sample answer: Find the point that is two-thirds of the way from the vertex to the midpoint of the opposite side. The triangle should balance on that point because the centroid is the center of gravity. **25.** No; the pair of congruent angles is not included between the sides.
27. slope: 5, the cost per person; y-intercept: 3, the base cost **29.** A

Pages 237–239   Lesson 6–2

1. Sample answer:

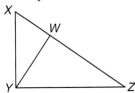

$\overline{XY}$ is the altitude from X. $\overline{ZY}$ is the altitude from Z. **3.** Sample answer: An altitude is a perpendicular segment in which one endpoint is at a vertex and the other is on the side opposite that vertex. A perpendicular bisector is a line that contains the midpoint of that side and is perpendicular to that side. **5.** neither **7.** both

9. both **11.** perpendicular bisector **13.** both
15. neither **17.** $\overline{DE}$ **19.** Yes; 3 perpendicular bisectors can be constructed, one to each of the 3 sides of the triangle. **21.** altitude
23.

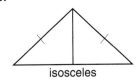

isosceles   equilateral

25. not possible **27.**

Pages 242–243   Lesson 6–3

1. An angle bisector of a triangle is a segment that separates an angle of the triangle into two congruent angles. **3.** 48 **5.** 84 **7.** 110 **9.** 22
11. 40 **13.** 17 **15.** 88 **17.** 124 **19a.** 60 **19b.** 96
21. altitude **23.** 30, 115, 35 **25.** B

Page 243   Quiz 1

1. 4.5 **3.** 36 **5.** 15

Pages 249–250   Lesson 6–4

1.

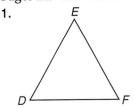

$\overline{ED} \cong \overline{EF}$; $\angle D \cong \angle F$; $\angle E$ is the vertex angle. $\angle D$ and $\angle F$ are base angles. **3.** $x = 75$;
$y = 15$ **5.** 37; 13; 13 **7.** $x = 60$; $y = 5$
9. $x = 68$; $y = 112$ **11.** $x = 86$; $y = 9$ **13.** 55
15. 38; 38 **17a.** isosceles **17b.** 35, 35 **19.** 66
21.

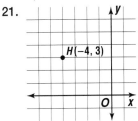

23. Sample answer: 48, 41, 34, 27, . . .

Pages 253–255   Lesson 6–5

1. SAS **3.** HA **5.** LA **7.** HA **9.** LL **11.** not possible **13.** $\overline{BC} \cong \overline{EF}$; $\overline{BA} \cong \overline{ED}$ **15.** $\overline{CA} \cong \overline{FD}$; $\overline{BC} \cong \overline{EF}$ or $\overline{BA} \cong \overline{ED}$ **17.** $\overline{ED} \cong \overline{BA}$; EDF; HL
19. $\overline{VW} \cong \overline{XY}$; $\overline{WZ} \cong \overline{YZ}$; VWZ; LL

21. LA; Since $\overline{AC}$ bisects $\angle BAD$, $m\angle BAC = m\angle DAC$. So, $\angle BAC \cong \angle DAC$. $\overline{AC} \cong \overline{AC}$. Since $\overline{AC} \perp \overline{BD}$, $\angle BCA$ and $\angle DCA$ are right angles. So, $\triangle ABC \cong \triangle ADC$ by the LA Theorem.

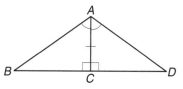

23. $x = 90$; $y = 38$ **25.** 75   **27.** B

Pages 259–261   Lesson 6–6
1. The square of the length of the hypotenuse is equal to the sum of the squares of the lengths of the legs.   **3.** Sample answer: If the sum of the squares of the lengths of the legs equals the square of the length of the hypotenuse, then the triangle is a right triangle.   **5.** 7.3   **7.** 11   **9.** 9.3   **11.** 19.3   **13.** 8.1   **15.** no   **17.** 20   **19.** 2.9   **21.** 10.9   **23.** 6.7   **25.** 21   **27.** 4   **29.** no   **31.** no   **33.** yes   **35.** yes; $30^2 + 40^2 = 50^2$ **37.** 176   **39.** 6.4 ft   **41.** HL

43. Sample answer:

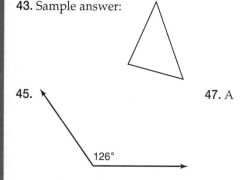

45.

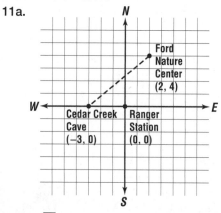

47. A

Pages 265–267   Lesson 6–7
1. $d = \sqrt{(x_2 - x_1)^2 + (y_2 - y_1)^2}$ **3.** Both are correct. Either point can be used as (x_1, y_1) or (x_2, y_2).   **5.** 50   **7.** 2.8   **9.** 9.4

11a.

11b. $\sqrt{41}$ or about 6.4 km   **13.** 5   **15.** 7.6   **17.** 4.5   **19.** 12   **21.** 4.2   **23.** 7.1   **25.** Yes; all three sides have different measures.

27a.

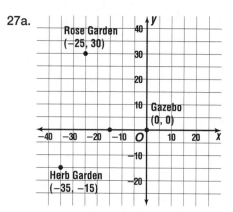

27b. 46.1 yd   **27c.** 39.1 yd
29. 6.3;

31. LL   **33.** obtuse

Page 267   Quiz 2
1. 38   **3.** 16.1   **5.** Yes; sides JK and JL have the same measure.

Pages 268–270   Chapter 6 Study Guide and Assessment
1. true   **3.** false; median   **5.** false; $(JK)^2 + (KL)^2 = (JL)^2$ **7.** false; obtuse   **9.** true   **11.** 6   **13.** 5.5   **15.** 38   **17.** both   **19.** neither   **21.** 31   **23.** $x = 51$; $y = 39$ **25.** LL   **27.** 20   **29.** yes   **31.** 10   **33.** 11   **35.** both

Page 273   Preparing for Standardized Tests
1. D   **3.** D   **5.** B   **7.** C   **9.** 0.6

Chapter 7   Triangle Inequalities

Page 275   Check Your Readiness
1. 2   **3.** 3   **5.** 3
7. $n < 5$

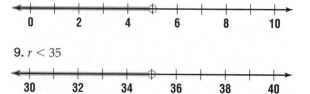

9. $r < 35$

11. $w > 30$

13. Sample answer: $\angle BFA$ and $\angle BFC$; $\angle EFD$ and $\angle DFC$ **15.** 130°

Pages 280–281   Lesson 7–1
1. The measure of angle J is not less than or equal to the measure of angle T. The measure of angle J is greater than the measure of angle T.   **3.** Mayuko; the Transitive Property of Inequality states that if $a > 7$ and $7 > b$, then $a > b$.   **5.** no   **7.** >   **9.** true   **11.** If $q > d$ and $d > w$, then $q > w$.   **13.** >   **15.** true   **17.** true   **19.** =   **21.** true   **23.** false   **25.** true   **27.** true   **29.** $3 \le b \le 22\frac{1}{3}$ **31.** -13.6 **33.** This is not always true: $-5 < 8$ and $-2 < -1$, but $10 > -8$.   **35.** 8   **39.** C

Pages 285–287   Lesson 7–2
3. Maurice; the exterior angles are vertical angles and vertical angles are congruent.   **5.** 77   **7.** <   **9.** $\angle JET$ or $\angle BES$ **11.** $\angle 1$ or $\angle 5$ **13.** 61   **15.** 32   **17.** 32   **19.** <   **21.** $m\angle BAC < m\angle ACD$ **23.** no; $x = 109$, $y = 110$ **25.** 76   **27.** yes; $XY = XZ$ **29.** 9.4 m; 4.2 m²

Pages 292–295   Lesson 7–3
1. $\angle G$ **3.** $\overline{PD}$; the perpendicular segment is the shortest segment from a point to a line.   **5.** $\overline{PR}$, $\overline{RQ}$, $\overline{QP}$ **7.** $\overline{MN}$ **9.** $\angle F$, $\angle E$, $\angle D$ **11.** $\angle Z$, $\angle X$, $\angle Y$ **13.** $\overline{PQ}$, $\overline{QN}$, $\overline{NP}$ **15.** $\angle L$ **17.** $\angle T$ **19.** $\overline{DS}$ **21.** $\overline{PR}$ **23.** Less than; the measure of the side opposite $\angle E$ is less than the measure of the side opposite $\angle G$.   **25.** The obtuse angle is the largest angle of a triangle since the other two angles must be acute.   **27.** $5 < x \le 29$ **29.** DFE

Page 295   Quiz
1. <   **3.** 101   **5.** Perth and Sydney

Pages 298–300   Lesson 7–4
1. Any number between 8 and 26 is correct.   **5.** yes; $100 + 100 > 8$, $100 + 8 > 100$ **7.** $22 < x < 102$ **9.** yes; $7 + 12 > 8$, $7 + 8 > 12$, $12 + 8 > 7$ **11.** no; $1 + 2 \not> 3$ **13.** no; $5 + 10 \not> 20$ **15.** $4 < x < 20$ **17.** $1 < x < 43$ **19.** $6 < x < 82$ **21.** LM **23.** $22 < x < 144$ **25.** 3 triangles having the following side measures in units: $2, 5, 5$; $3, 4, 5$; $4, 4, 4$ **27a.** $\overline{LM}$, $\overline{MN}$, $\overline{NL}$ **27b.** $\overline{UV}$, $\overline{WU}$, $\overline{VW}$ **27c.** $\overline{CD}$, $\overline{BC}$, $\overline{DB}$ **29.** 30   **31.** B

Pages 302–304   Chapter 7 Study Guide and Assessment
1. false; inequality   **3.** false; greater than or equal to   **5.** true   **7.** false; less than   **9.** false; greater than   **11.** <   **13.** <   **15.** false   **17.** $\angle 8$, $\angle 5$, or $\angle 1$

19. $\angle 7$ or $\angle ZQJ$ **21.** 65   **23.** <   **25.** 28   **27.** $\angle Y$, $\angle X$, $\angle W$ **29.** $\overline{TP}$ **31.** yes; $12 + 5 > 13$, $5 + 13 > 12$, and $12 + 13 > 5$ **33.** no; $15 + 45 \not> 60$ **35.** $20 < x < 40$ **37.** 10.5   **39.** False; in $\triangle KAT$, $TA < KT$ by Theorem 7–7.

Page 307   Preparing for Standardized Tests
1. A   **3.** C   **5.** A   **7.** C   **9.** 16.50

Chapter 8   Quadrilaterals

Page 309   Check Your Readiness
1. 36   **3.** 38, 124   **5.** 54°   **7.** 126°   **9.** 44, 122

Pages 313–315   Lesson 8–1
1. Sample answer:

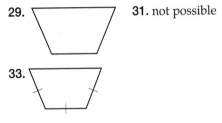

3. 100   **5.** 126   **7.** Sample answer: $\angle M$, $\angle Q$ **9.** $\overline{MP}$ or $\overline{NQ}$ **11.** 150   **13.** $\overline{ST}$ or $\overline{RQ}$ **15.** Sample answer: Q and R **17.** $\overline{QS}$, $\overline{RT}$ **19.** $\overline{FH}$ or $\overline{GJ}$ **21.** $\angle H$ **23.** 40   **25.** 58, 116   **27.** 60, 120

29.   **31.** not possible

33.

35. $m\angle R = 90$, $m\angle S = 100$, $m\angle T = 120$ **37.** no, $4 + 2 + 1 < 8$ **39.** yes   **41.** $\overline{LN}$ **43.** $\angle P \cong \angle T$

Pages 319–321   Lesson 8–2
1. Opposite sides are congruent, opposite angles are congruent, consecutive angles are supplementary, diagonals bisect each other, and a diagonal separates the parallelogram into two congruent triangles.   **3.** Karen; opposite angles are congruent and consecutive angles are supplementary.   **5.** 110   **7.** 60   **9a.** $\overline{DE}$, $\overline{GF}$ **9b.** $\overline{AD}$, $\overline{CF}$ **9c.** $\overline{EF}$, $\overline{BC}$ **11.** 40   **13.** 9   **15.** 25   **17.** 110   **19.** 12   **21.** 24   **23.** 35, 145, 145   **25.** true   **29.** They decrease by the same amount.   **31.** 74   **33.** reflection

Page 321   Quiz 1
1. 163   **3.** 166   **5.** 16

Pages 324–326   Lesson 8–3
1a. Sample answer:

1b. Sample answer:

1c. Sample answer:

3. yes, Theorem 8–7 **5a.** congruent alternate interior angles **5b.** definition of parallelogram **7.** yes, Theorem 8–8 **9.** yes, definition of parallelogram **11.** no **13a.** vertical angles **13b.** ASA **13c.** CPCTC **13d.** congruent alternate interior angles **13e.** Theorem 8–8 **15.** No; in order to use Theorem 8–8, the same pair of sides must be parallel and congruent. In this case, one pair of sides is congruent and the other pair is parallel. **17.** Sample answer: The quilt pieces fit together because opposite sides are congruent, opposite angles are congruent, and consecutive angles are supplementary. **19.** 62 **21.** 45 **23.** 25 in. **25.** Sample answer: $\angle 2$ and $\angle 8$ are alternate interior angles; $\angle 2$ and $\angle 6$ are corresponding angles.

Pages 330–332 Lesson 8–4
1. Sample answer:

3. Teisha; every square has four congruent sides, which is the definition of a rhombus, but every rhombus does not have right angles. **5.** parallelogram, rectangle, square, rhombus **7.** rectangle, rhombus, square **9.** 24 **11.** 90 **13.** 6 **15.** Sample answer: soccer, tennis; ice hockey, golf **17.** rectangle **19.** rectangle, rhombus, square **21.** none of these **23.** 16 **25.** 32 **27.** 90 **29.** 45 **31.** 30 **33.** 124 **35.** 17 **37.** 62 **39.** true **41.** false **43.** true **45.** 12 ft **47a.** isosceles **47b.** right **47c.** Yes; sample answer: $\overline{PE} \cong \overline{EA}$ and $\overline{NE} \cong \overline{EL}$ because the diagonals of a rhombus bisect each other. Also, $\angle PEN \cong \angle LEA$ because they are vertical angles. Therefore, $\triangle PEN \cong \triangle AEL$ by SAS. **49.** no **51.** true **53.** false **55.** A

Page 332 Quiz 2
1. Yes; two sides are parallel and congruent. **3.** $\angle BIT$, $\angle TIL$, $\angle LIE$ **5.** EL, TL, BT

Pages 336–338 Lesson 8–5
1. Sample answer:

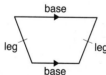

3. parallelogram: yes, yes, yes, yes, yes, no, no, no; rectangle: yes, yes, yes, yes, yes, yes, no, no; rhombus: yes, yes, yes, yes, yes, no, yes, yes; square: yes, yes, yes, yes, yes, yes, yes, yes; trapezoid: no, no, no, no, no, no, no, no **5.** 37 ft

7. 65, 115, 115 **9.** $\overline{VT}$, $\overline{SR}$; $\overline{VS}$, $\overline{TR}$; $\angle V$ and $\angle T$, $\angle S$ and $\angle R$ **11.** $\overline{GH}$, $\overline{JK}$; $\overline{GK}$, $\overline{HJ}$; $\angle J$ and $\angle K$, $\angle G$ and $\angle H$ **13.** 20 yd **15.** 40 mm **17.** 26.5 ft **19.** 85, 95, 95 **21.** 19 m
23. yes; **25.** no

27. yes;

29. The support cables form one pair of parallel sides. The other two sides are not parallel. **31a.** $2 \cdot 10 + 1$ or 21 **31b.** $2n + 1$ **33.** rectangle, square
35. Sample answer:

Pages 342–344 Chapter 8 Study Guide and Assessment
1. parallelogram **3.** rhombus **5.** quadrilateral **7.** base angles **9.** square **11.** $\overline{MA}$, $\overline{NY}$ or $\overline{AY}$, $\overline{MN}$ **13.** $\angle Y$ **15.** 74 **17.** 5 **19.** 80 **21.** 6 **23.** 28, 152, 152 **25.** yes, Theorem 8–8 **27.** none of these **29.** rectangle **31.** $\overline{CD}$, $\overline{HJ}$; $\overline{CH}$, $\overline{DJ}$; $\angle C$ and $\angle D$, $\angle H$ and $\angle J$ **33.** 74, 106, 106 **35.** 80, 80 **37.** The sides of the quadrilateral formed by the four metal pieces have equal lengths. By Theorem 8–7, quadrilateral $ABCD$ is a parallelogram. By definition, opposite sides of a parallelogram are parallel.

Page 347 Preparing for Standardized Tests
1. A **3.** A **5.** B **7.** B
9.

Chapter 9 Proportions and Similarity
Page 349 Check Your Readiness
1. 54 **3.** 3.8 **5.** $\frac{1}{2}$ **7.** $1\frac{1}{2}$ **9.** $\angle 6$, vertical; $\angle 9$, corresponding; $\angle 14$, alternate exterior **11.** $\angle 2$, vertical; $\angle 10$, alternate interior; $\angle 13$, corresponding **13.** 16.5 cm **15.** 288 mm

Pages 352–355 Lesson 9–1
1. Sample answers: $\frac{1}{2} = \frac{2}{4}$; $\frac{1}{2} \neq \frac{1}{3}$

3. Lawanda; if $\frac{7}{8} = \frac{x}{y}$, then $7(y) = 8(x)$. Using the Symmetric Property, $8(x) = 7(y)$. If $8(x) = 7(y)$, then $\frac{8}{7} = \frac{y}{x}$. **5.** $\frac{3}{11}$ **7.** $\frac{21}{16}$ **9.** $\frac{3}{5}$ **11.** $\frac{18}{25}$ **13.** 2 **15.** 5 **17.** $\frac{1}{5}$ **19.** $\frac{5}{11}$ **21.** $\frac{15}{7}$ **23.** $\frac{11}{50}$ **25.** $\frac{12}{1}$ or 12 **27.** 10 **29.** 1 **31.** 9 **33.** 12 **35.** 12 **37.** 2 **39.** yes **41.** yes **43.** 3125 ft^2 **45.** 2.5 mL **47.** 17.5 in. **49.** yes **51.** neither **53.** both

Pages 359–361 Lesson 9–2

1. Congruent polygons are the same shape and the same size. The corresponding angles and sides of congruent polygons are congruent. Similar polygons are the same shape, but they may be a different size. The corresponding angles of similar polygons are congruent, but the measures of corresponding sides have equivalent ratios.

3. Sample answer:

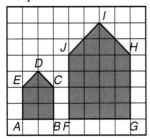

pentagon $ABCDE \sim$ pentagon $FGHIJ$; $\angle A \leftrightarrow \angle F$, $\angle B \leftrightarrow \angle G$, $\angle C \leftrightarrow \angle H$, $\angle D \leftrightarrow \angle I$, $\angle E \leftrightarrow \angle J$; $\overline{AB} \leftrightarrow \overline{FG}$, $\overline{BC} \leftrightarrow \overline{GH}$, $\overline{CD} \leftrightarrow \overline{HI}$, $\overline{DE} \leftrightarrow \overline{IJ}$, $\overline{EA} \leftrightarrow \overline{JF}$; $\frac{AB}{FG} = \frac{BC}{GH} = \frac{CD}{HI} = \frac{DE}{IJ} = \frac{EA}{JF}$ **5.** Yes; corresponding angles are congruent and $\frac{5}{4} = \frac{5}{4}$. **7.** $x = 10, y = 20.5$ **9.** Yes; corresponding angles are congruent and $\frac{9}{7} = \frac{9}{7}$. **11.** No; $\frac{4}{2} \neq \frac{14}{8}$. **13.** Yes; corresponding angles are congruent and $\frac{6.4}{4.8} = \frac{7.6}{5.7} = \frac{6}{4.5}$. **15.** $x = 10, y = 10$ **17.** $x = 3, y = 5$ **19.** $x = 27, y = 14$ **21.** sometimes

23.

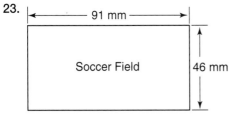

25a. 308 in. **25b.** 98 in. **25c.** 168 in. **27.** about 33 mi **29.** 68; 112; 68 **31.** $\overline{AP}$

Page 361 Quiz 1
1. 3 **3.** 11 **5.** $x = 13, y = 39$

Pages 365–367 Lesson 9–3
1.

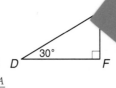

$\frac{AB}{DE} = \frac{BC}{EF} = \frac{CA}{FD}$
3. yes; FRT; SAS **5.** 36 m **7.** yes; VPK; SSS **9.** $x = 12$ **11.** $x = 17.5, y = 15$ **13a.** Given **13b.** Corresponding angles are congruent. **13c.** Reflexive Property of Congruent Angles **13d.** AA Similarity **13e.** Definition of Similar Polygons **15.** 4 ft **17.** $\triangle JKP \sim \triangle MNP$; since $\overline{JK} \parallel \overline{MN}$ and $\angle J$ and $\angle M$ are alternate interior angles, $\angle J \cong \angle M$. Likewise, $\angle K \cong \angle N$. The triangles are similar by AA Similarity. **19.** $\frac{1}{6}$ **21.** 72; 72

Pages 372–373 Lesson 9–4
1. $\angle 1$ and $\angle 2$ are congruent corresponding angles. $\angle N \cong \angle N$ by the Reflexive Property of Congruent Angles. $\triangle NRT \sim \triangle NPM$ by AA Similarity. **3.** Jacob; since $\triangle ADE \sim \triangle ABC$, $\frac{AD}{AB} = \frac{DE}{BC}$ and $AB = 5 + 6$ or 11. **5.** NR **7.** 10 **9.** GJ **11.** GM **13.** GM **15.** $x = 11$ **17.** $x = 20$ **19.** $x = 4\frac{4}{5}$, $y = 19\frac{1}{5}$ **21.** 8; 16; 12; 24 **23.** 12 in. **25.** Yes; if $\overline{KM} \parallel \overline{JN}$, $\triangle JLN \sim \triangle KLM$. Similar triangles are the same shape, so $\triangle KLM$ must also be equilateral.

27.

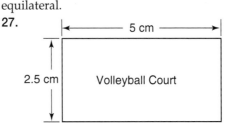

29.

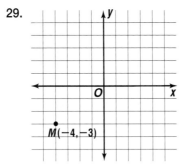

Pages 376–378 Lesson 9–5
1. If the collar tie divides the rafters proportionally, the collar tie is parallel to the joist. **3.** no **5.** $\overline{AC}$ **7.** s **9.** 24 in. **11.** no **13.** yes **15.** yes **17.** $\overline{ST}$ **19.** 22 **21.** 31 **23.** $40y$ **25.** $8b$ **27.** $x + 5$ **29.** 12; 6 **31.** 7 **33.** 8 **35.** yes; SAS Similarity; BCA **37.** 85

Sample answers: $\frac{DE}{EF} = \frac{GH}{HJ}$, $\frac{DE}{DF} = \frac{GH}{GJ}$, $\frac{EF}{DF} = \frac{HJ}{GJ}$

3e. Theorem 9–9   **5.** RS **7.** 2   **9.** AC **11.** AE
13. DB **15.** 12   **17.** 16   **19.** 9   **21.** $\frac{60}{13}$ or $4\frac{8}{13}$
23. about 501 m   **25.** Draw segment AB. Draw $\overrightarrow{AC}$
so that $\angle BAC$ is an acute angle. With a compass,
start at A and mark off six congruent segments on
$\overrightarrow{AC}$. Label points D, E, and F so that AD is 1 unit,
DE is 2 units, and EF is 3 units. Draw $\overline{BF}$. Construct
lines through D and E that are parallel to $\overline{BF}$. These
parallel lines will divide $\overline{AB}$ into three segments
with the ratio 1:2:3.

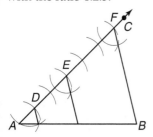

27. YB **29.** 48

Page 387   Quiz 2
1. $x = 12$, $y = 20$ **3.** $x = 17$ **5.** 36 ft

Pages 391–393   Lesson 9–7
1. scale factor $= \frac{4}{1}$, perimeter of $\triangle HJN = 48$,
perimeter of $\triangle MJK = 12$, $\frac{48}{12} = \frac{4}{1}$ **3.** $\frac{1}{7}$ **5.** $\frac{7}{2}$
7. $x = 4$, $y = 3$, $z = 2$ **9.** $\frac{4}{3}$ **11.** 38 in.   **13.** $x = 8$,
$y = 12$, $z = 9$ **15.** $x = 12$, $y = 15$, $z = 9$ **17.** $x = 24$,
$y = 27$, $z = 30$ **19.** $\frac{2}{3}$ **21.** $\frac{4}{1}$ **23.** $\frac{9}{5}$ **25.** 38 ft
27. 15 m   **29a.** $\frac{23}{9}$ **29b.** 4600 ft   **29c.** 1800 ft
29d. $\frac{23}{9}$ **29e.** They are equivalent.   **31.** 17.5
33. true   **35.** No; the angle is not included between
the two sides.

**Pages 394–396   Chapter 9 Study Guide and
Assessment**
1. cross products   **3.** proportion   **5.** extremes
7. means   **9.** similar figures   **11.** $\frac{1}{3}$ **13.** $\frac{5}{2}$ **15.** 2
17. 5   **19.** $x = 9$, $y = 6$ **21.** yes; SSS   **23.** MN
25. $\overline{ST}$ **27.** 12   **29.** 15   **31.** $\frac{4}{5}$ **33.** 3.5 ft

Page 399   Preparing for Standardized Tests
1. D   **3.** C   **5.** D   **7.** C   **9.** 420

Chapter 10   Polygons and Area

Page 401   Check Your Readiness
1. 116   **3.** 37   **5.** 30.8 cm; 48.4 cm^2 **7.** altitude
9. altitude

Pages 404–407   Lesson 10–1
1. Sample answer:

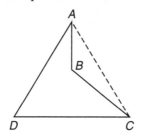

Diagonal AC lies outside of quadrilateral $ABCD$.
5. octagon, regular   **7.** Sample answer: $\overline{NT}$, $\overline{TA}$,
$\overline{AP}$ **9.** convex   **11a.** heptagon   **11b.** concave
13. quadrilateral, regular   **15.** heptagon, not
regular; angles and sides are not congruent
17. pentagon, regular   **19.** Sample answer: $\overline{MO}$,
$\overline{MP}$ **21.** Sample answer: $\overline{TS}$, $\overline{SR}$, $\overline{RQ}$ **23.** convex
25. concave   **27.** convex   **29.** 100 in.   **31.** 34.2 mm
35. cycloheptene; cyclooctene   **37.** 4 to 3   **39.** yes
41. no   **43.** D

Pages 411–412   Lesson 10–2
1. Use Theorem 10–1 to find the sum of measures
of the interior angles. Then divide the sum by n.
3. No; the segments forming the triangles are not
diagonals.   **5.** 90   **7.** 720   **9.** 540   **11.** 108, 72
13. 144, 36   **15.** 90   **17.** 35; 45   **19a.** 5 turns
19b. 360   **19c.** 540   **21.** pentagon, not regular
23. 96   **25.** A

Pages 416–418   Lesson 10–3
3. Kevin; Figure 1 has an area of 6 square units and
a perimeter of 10 units. Figure 2 has twice the area
of Figure 1, or 12 square units, but not twice the
perimeter (14 units).

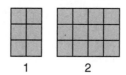

5. 5 units^2 **7.** 28 units^2 **9.** 5 units^2 **11.** 7 units^2
13. 6.5 units^2 **15.** 5 units^2 **17.** 6.5 units^2
19. Sample answer: 30 units^2
21. Sample answer:

23. Sample answer: 82 in^2

25. 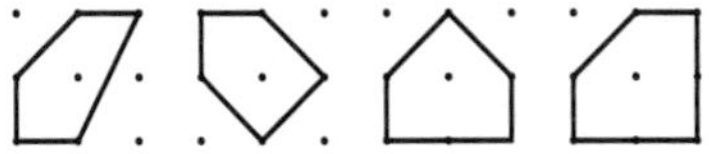

2.5 2.5 3 3.5
(units^2)

27. 21 ft 29. 69; 21

Page 418 Quiz 1
1. pentagon, convex 3. 150 5. 384 ft^2

Pages 422–424 Lesson 10–4
1. The new area is 4 times the original area.

3a. 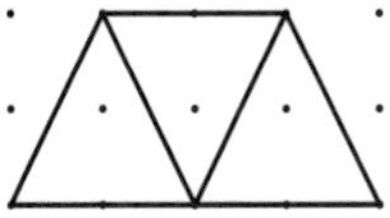

3b. 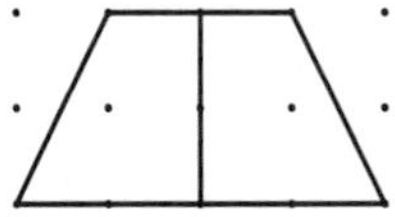

3c. 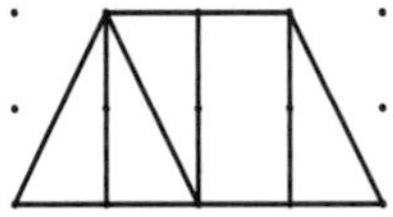 4 triangles, 1 rectangle

5. 23 7. 30 m^2 9. 20 yd^2 11. 20 in^2 13. 12 km^2
15. 522 in^2 17. $136\frac{1}{8}$ ft^2 19. 9 yd^2 21. 29 m
23a. 15.6 cm^2 23b. 35 cm^2 23c. 105 cm^2
25. 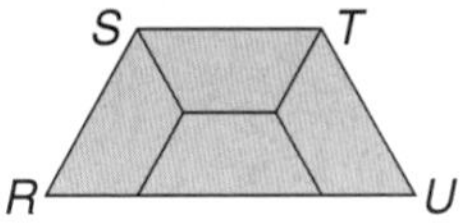

27. 720 29. 24.5 ft

Pages 428–430 Lesson 10–5
1. More significant digits in the measures will increase the precision of the calculation and the number of significant digits in the measure of its area. 3. Sample answer: Construct a perpendicular bisector to each side of the figure. The point where the perpendicular bisectors intersect is the center.
5. 110.4 m^2 7. 504 yd^2 9. 186 ft^2 11. 106 in^2
13a. $155\frac{1}{4}$ in^2 13b. $157\frac{1}{2}$ in^2; It is $2\frac{1}{4}$ in^2 greater.
15. 549,240 ft^2 17a. 3020 ft^2 17b. 696 ft^2
17c. 2324 ft^2 19. 22 in^2 21. (1, −2) 23. B

Pages 436–439 Lesson 10–6
1. 4 ways; 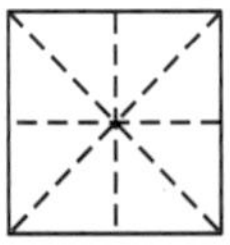

5. no 7. no 9. no 11. yes 13. yes 15. no
17. yes 19. no 21. yes 23. no 25. isosceles, equilateral 27a. 1 time 27b. 2 times
27c. 3 times 29. $\frac{3}{7}$ 31. 48 cm^2 33. no; $\frac{8}{7} \neq \frac{9}{7}$
35. D

Page 439 Quiz 2
1. 34.2 cm^2 3. 1599.67 km^2 5. 429 yd^2
7. 952.56 cm^2 9a. yes; 2 lines 9b. yes

Pages 441–444 Lesson 10–7
3. Hexagons form tessellations. Since there is no space in between, more hexagonal pencils can be made from the same amount of wood than round pencils. Also, packaging is less expensive.

5. Sample answer:

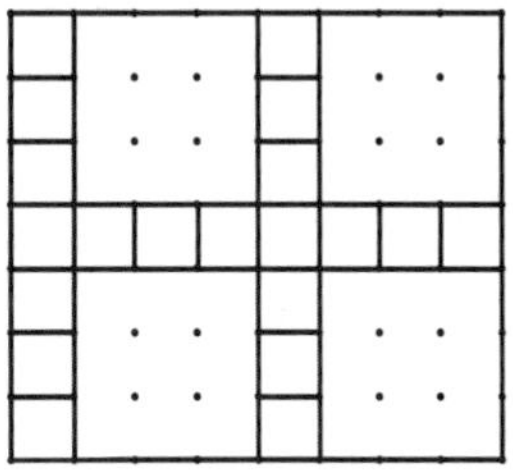

7. regular hexagons; regular 9. regular hexagons, equilateral triangles; semi-regular

11. Sample answer:

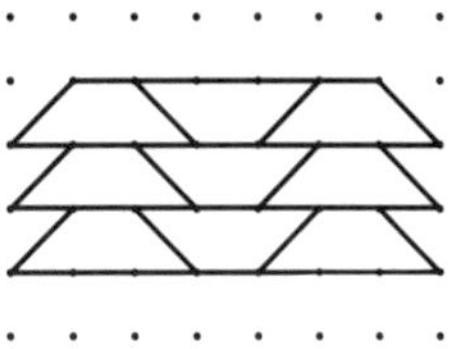

13. Sample answer:

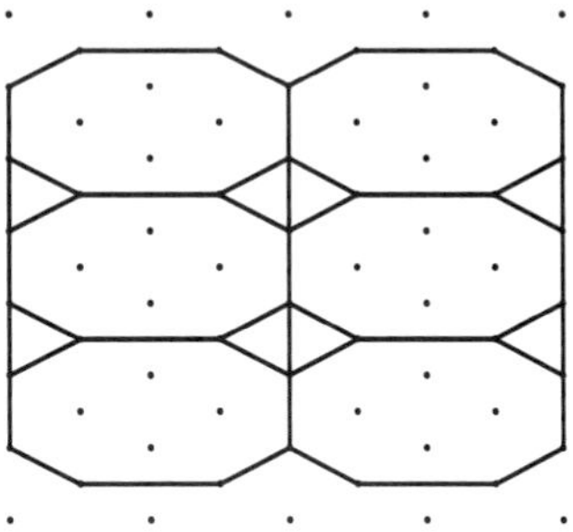

15. neither

21a. 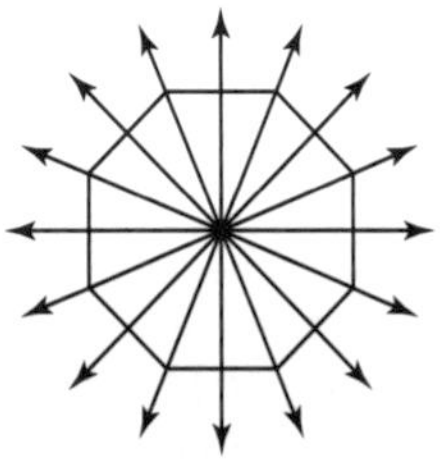

21b. yes; point where lines intersect 23. Yes; the diagonals bisect each other (Theorem 8–9).
25. $7 < y < 22.5$

Pages 394–396   Chapter 10 Study Guide and Assessment
1. convex   **3.** altitude   **5.** tessellation
7. line symmetry   **9.** regular tessellation
11. quadrilateral, not regular   **13.** convex
15. 720   **17.** 135, 45   **19.** 5.5 units^2 **21.** 45 cm^2
23. 615 in^2 **25.** 130.5 cm^2 **27.** both   **29.** neither
31. squares, rectangles; neither
33. Sample answer:

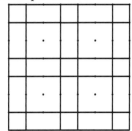

35. 120 ft^2

Page 451   Preparing for Standardized Tests
1. B   **3.** E   **5.** B   **7.** D   **9.** 7

Chapter 11   Circles

Page 453   Check Your Readiness
1. 14   **3.** 8   **5.** 87   **7.** 43   **9.** 8   **11.** 16.1

Pages 456–458   Lesson 11–1
1. A radius can be formed between any point on the circle and the center of the circle. Thus, a circle has an infinite number of radii.   **3.** Jason; a diameter is a chord through the center, and some chords are not diameters.   **5.** $1\frac{3}{4}$ **7.** true
9. true   **11.** 38   **13.** false   **15.** true   **17.** false
19. true   **21.** false   **23.** 2   **25.** 15   **27.** 4x
29. 31.5 units   **31a.** Triangle Inequality Theorem
31b. All radii are congruent.   **31c.** Substitution
31d. Segment Addition Postulate   **33.** yes, yes
35. $\frac{3}{4}$ **37a.** Water boils down to nothing; snow boils down to nothing; ice boils down to nothing.
37b. Everything boils down to nothing.

Pages 465–467   Lesson 11–2
1a. $\overarc{PNH}$ is a semicircle. By the Definition of Arc Measure, the degree measure of a semicircle is 180.
1b. No, $\overarc{PNH}$ is a semicircle and $\overarc{PHN}$ is not.
1c. If $m\angle NRH = 35$, then by the Definition of Arc Measure, the degree measure of $\overarc{HPN}$ is $360 - 35$ or 325.   **1d.** Diameter $\overline{PH}$ separates the circle into two congruent arcs called semicircles. By the Definition of Arc Measure, the degree measure of a semicircle is 180.   **3.** Marisela; arcs having the same measure are congruent only if they are part of the same circle or congruent circles.
5. major   **7.** 30   **9.** 210   **11.** 150   **13.** 38   **15.** 28

17. 180   **19.** 180   **21.** 114   **23.** 246   **25.** false
27. true   **29.** false   **31.** 112   **33.** 32   **35.** 216
37a. 15   **37b.** 150   **39.** 9 in.   **41.** 5   **43.** 130

Page 467   Quiz 1
1. true   **3.** false   **5.** false   **7.** true   **9.** true

Pages 471–473   Lesson 11–3
1a. congruent   **1b.** perpendicular, arc   **1c.** arcs
3. $\overline{ST}$ **5.** 5   **7.** 8.66 cm   **9.** $\overarc{DE}$ **11.** $\overarc{BD}$ **13.** $\overline{GC}$
15. 8   **17.** 41   **19.** 5   **21.** 13   **23.** 31   **25a.** $AK > AT$; the perpendicular segment from a point to a line is the shortest segment from the point to the line.   **25b.** T
25c. $\overarc{ED}$ **25d.** no   **27.** $\overline{AB}, \overline{FC}$ **29.** 13   **31.** A

Pages 476–477   Lesson 11–4
1. circle   **3.** Construct the six arcs for an inscribed hexagon. Connect every other arc to form the equilateral triangle.   **5.** 5   **7.** Use the construction of an inscribed square from Example 1. Then construct the perpendicular bisectors of each side. The intersections of the bisectors and the circle determine the additional four vertices   **9.** 7   **11.** 10
13. 4   **15a.** 90   **15b.** $12\sqrt{2}$ cm   **15c.** isosceles right   **15d.** $6\sqrt{2}$ cm   **15e.** yes   **17.** chord $\overline{AT}$
19. the distance from one corner to the opposite corner   **21.** no   **23.** 1800   **25.** B

Pages 480–482   Lesson 11–5
1. because $d = 2r$ **3.** 4 m; 25.1 m   **5.** 13.5 ft; 27 ft
7. $11\frac{1}{2}$ yd; 36.1 yd   **9.** 75.4 cm   **11.** 20.4 in.
13. 12.6 mm   **15.** 106.8 yd   **17.** 1.0 in.   **19.** 1.3 ft
21. 56.55 in.   **23.** 47.12 cm   **25.** 1257 ft   **27.** 6.28 cm   **29.** about 52.65 cm   **31.** 110

Page 482   Quiz 2
1. $\overarc{QR}$ or $\overarc{PR}$ **3.** 24   **5.** 23 units

Pages 485–487   Lesson 11–6
1. $A = \frac{N}{360}(\pi r^2)$ *Theorem 11–9*
$= \frac{90}{360}[\pi(14)^2]$ *Replace N with 90 and r with 14.*
$= \frac{1}{4}(196\pi)$ $\frac{90}{360} = \frac{1}{4}$ *and* $14^2 = 196$
$= 49\pi$ $\frac{1}{4}(196) = 49$
≈ 153.93804
3. 5.50 m; 17.28 m; 23.76 m^2 **5.** 3.46 in.; 6.93 in.; 37.61 in^2 **7.** $\frac{21}{25}$ **9.** 19.63 ft^2 **11.** 176.71 mi^2
13. 14.45 m^2 **15.** 109.35 m^2 **17.** 254.47 cm^2
19. 6.28 cm^2 **21.** 37.70 cm^2 **23.** 7.75 m   **25.** 0.21
27. 93.73 in^2 **29.** about 38 m   **31.** \$325.50   **33.** D

Pages 488–490   Chapter 11 Study Guide and Assessment
1. true   **3.** false; circumference   **5.** false; radius
7. true   **9.** true   **11.** chord   **13.** radii   **15.** 282
17. 78   **19.** 247   **21.** 6   **23.** 6   **25.** true   **27.** false

29. 100.5 ft   **31.** 138.2 in.   **33.** 2.8 cm   **35.** 1963.50 in^2
37. 764.54 ft^2 **39.** 156   **41.** about 46 in^2

Page 493   Preparing for Standardized Tests
1. C   **3.** B   **5.** C   **7.** A   **9.** 22/8, 11/4, or 2.75

Chapter 12   Surface Area and Volume

Page 495   Check Your Readiness
1. 32 m, 48 m^2 **3.** 106 ft, 520 ft^2 **5.** 20 ft^2
7. 76 in^2 **9.** 585 m^2 **11.** 25.13 m, 50.27 m^2
13. 20.11 cm, 32.17 cm^2

Pages 498–501   Lesson 12–1
1a.
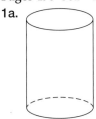
1b.
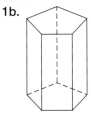

1c.

1d.
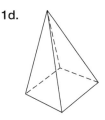

3. Both; a tetrahedron is a triangular pyramid.
5. squares   **7.** triangles   **9.** triangles
11. rectangular prism   **13.** rectangular prisms
15. faces: quadrilaterals *FGJI*, *GHKJ*, *HFIK* and
triangles *FGH*, *IJK*; edges: $\overline{FI}$, $\overline{GJ}$, $\overline{HK}$, $\overline{FG}$, $\overline{GH}$, $\overline{HF}$,
$\overline{IJ}$, $\overline{JK}$, $\overline{KI}$; vertices: *F*, *G*, *H*, *I*, *J*, *K*   **17.** cylinder
19. cone   **21.** square pyramid   **23.** false   **25.** true
27. true   **29.** true   **31.** triangular prism

33a.

V	*F*	*E*
4	4	6
8	6	12
5	5	8
10	7	15

33b. $V + F - 2 = E$ **33c.** 16 vertices   **35.** 695
37. yes;

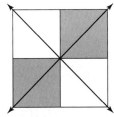

39. no

Pages 508–509   Lesson 12–2
1. Lateral area is the sum of the areas of the
lateral surfaces. Surface area is the sum of
the areas of the lateral surfaces and the bases.
3. 60 cm^2; 72 cm^2 **5.** 282.74 ft^2; 339.29 ft^2
7. 558 m^2; 858 m^2 **9.** 36 m^2; 54 m^2 **11.** 144 in^2;
192 in^2 **13.** 439.82 cm^2; 747.70 cm^2 **15.** 15.71 in^2;
21.99 in^2 **17a.** about 2010.62 ft^2 **17b.** about
3619.11 ft^2 **19.** 2 gal
21. Sample answer:

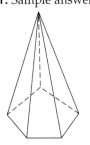

23. about 22.0 in.   **25.** C

Pages 513–515   Lesson 12–3
1. Sample answer:

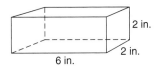

3. Caitlin; the volume of the new tank would be
8 times more than the original tank.   **5.** 38.48 m^2
7. 840 m^3 **9.** 4071.50 mm^3 **11.** 48 m^3 **13.** 7920 ft^3
15. 1508.75 m^3 **17.** 72 ft^3 **19.** 336 cm^3

21. 96 cm^3;

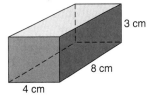

23. about 62.83 in^3 **25.** 96 in^3; The dimensions
of the prism are found by examining the factors
of 12, 32, and 24. Adjacent sides must have a
dimension in common. The dimensions are 8 in.
by 3 in. by 4 in.   **27.** rectangular prism   **29.** 186 ft^2

Page 515   Quiz 1
1. rectangular pyramid   **3.** hexagonal prism
5. 200 m^2; 392 m^2 **7.** 351.86 in^2; 753.98 in^2
9. about 12,666.90 m^3

Pages 520–521   Lesson 12–4
1. The slant height of a regular pyramid is the height of a lateral face. The altitude of a regular pyramid is the segment from the vertex perpendicular to the plane containing the base. **3.** To find the surface area of a prism, cylinder, pyramid, or cone, you find the sum of the areas of the surfaces. In a prism or pyramid, all of the surfaces are polygons. In a cylinder or cone, the lateral surface is curved. The lateral area of a right prism or a right cylinder equals the perimeter or circumference of the base times the height. To find the surface area, you add the areas of the two bases to the lateral area. The lateral area of a regular pyramid or a right circular cone equals $\frac{1}{2}$ times the perimeter or circumference of the base times the slant height. To find the surface area, you add the area of the one base to the lateral area.
5. 753.98 in^2; 1206.37 in^2 **7.** 36 ft^2; 45 ft^2
9. 89.44 m^2; 116.48 m^2 **11.** 52.28 mm^2; 84.45 mm^2
13. 96 ft^2 **15.** 160 ft^2 **17a.** 477.52 cm^2

19. 52 cm^2;

4 cm
2 cm
3 cm

21. perpendicular

Pages 525–527   Lesson 12–5
1. In both formulas, the area of the base is multiplied by the height. The volume of a prism equals the area of the base times the height. The volume of a pyramid equals $\frac{1}{3}$ times the area of the base times the height.   **3.** Darnell; since the radius is squared and the height is not squared, doubling the radius increases the volume more than doubling the height.   **5.** 20 ft^3 **7.** about 10,000 m^3
9. 20.94 cm^3 **11.** 149.33 in^3 **13.** 528 m^3
15. 2412.74 in^3 **17.** 30 cm^3 **19.** about 1017.88 ft^3
21. about 2.36 ft^3 **23.** The height of the cone is three times the height of the cylinder.   **25.** 540 in^3
27. 23.65

Page 527   Quiz 2
1. 60 in^2; 85 in^2 **3.** 197.92 m^2; 351.86 m^2
5. about 21,160 ft^2 **7.** 63 in^3 **9.** 402.12 m^3

Pages 531–533   Lesson 12–6
1. Both a circle and a sphere are a set of points that are a given distance from a given point. A circle is the set of all points in a plane that are a given distance from a given point in the plane. It is a two-dimensional figure. A sphere is the set of all points in space that are a given distance from a given point. It is a three-dimensional figure.

3. Sample answers: (1) prism; a shoe box; $V = Bh$; (2) cylinder; can of soup; $V = \pi r^2 h$; (3) pyramid; a Egyptian pyramid; $V = \frac{1}{3}Bh$; (4) cone; an ice cream cone; $V = \frac{1}{3}\pi r^2 h$; (5) sphere; a basketball; $V = \frac{4}{3}\pi r^3$ **5.** 6082.12 cm^2; 44,602.24 cm^3
7. 2827.43 in^2; 14,137.17 in^3 **9.** 201.06 m^2; 268.08 m^3
11. 1520.53 cm^2; 5575.28 cm^3 **13.** 1809.56 m^2
15. 53.21 yd^3 **17.** No; the volume of the cone is about 41.89 cm^3; and the volume of the ice cream is about 33.51 cm^3.   **19.** about 526.47 m^3
21a. 196,066,800 mi^2 **21b.** 258,154,616,700 mi^3
21c. 9,927,956,400 mi^3 **23.** 153.94 in^3 **25.** regular hexagons and rhombi; neither   **27.** $\frac{3}{2}$

Pages 537–539   Lesson 12–7
1. Similar solids have the same shape but not necessarily the same size. Yes, two solids with the same size and shape are similar. Their scale factor is 1:1.   **3.** no   **5.** $\frac{3}{2}$; $\frac{9}{4}$; $\frac{27}{8}$ **7.** yes   **9.** no
11. yes   **13.** $\frac{3}{1}$; $\frac{9}{1}$; $\frac{27}{1}$ **15.** $\frac{5}{1}$; $\frac{25}{1}$; $\frac{125}{1}$ **17a.** It is 4 times greater.   **17b.** It is 8 times greater.   **19a.** 3:4
19b. 27:64   **21.** 216,000:1   **23.** Since all linear measures for each cube are the same, the ratio of corresponding parts of any two cubes will be equivalent. Sample answer: sphere   **25.** 84 ft^3 **27.** 60

Pages 540–542   Chapter 12 Study Guide and Assessment
1. d   **3.** b   **5.** i   **7.** j   **9.** f   **11.** The faces are $ABCD$, $ABFE$, $BCGF$, $CDHG$, $ADHE$, and $EFGH$. The edges are $\overline{AB}$, $\overline{BC}$, $\overline{CD}$, $\overline{AD}$, $\overline{AE}$, $\overline{BF}$, $\overline{CG}$, $\overline{DH}$, $\overline{EF}$, $\overline{FG}$, $\overline{GH}$, and $\overline{EH}$. The vertices are A, B, C, D, E, F, G, and H.   **13.** true   **15.** 228 in^2; 396 in^2
17. 405 in^3 **19.** 32 m^3 **21.** 1055.58 cm^2; 1859.82 cm^2
23. 12.83 m^3 **25.** 615.75 cm^2; 1436.76 cm^3 **27.** yes
29. 1620 in^3 **31.** 14,657,415 mi^2; 5,276,669,286 mi^3

Page 545   Preparing for Standardized Tests
1. A   **3.** C   **5.** A   **7.** B   **9.** 145

Chapter 13   Right Triangles and Trigonometry
Page 547   Check Your Readiness
1. 16.1   **3.** 25.5   **5.** 17.3   **7.** 45   **9.** 34   **11.** 1.5
13. 2.5   **15.** 7.44   **17.** 12.32

Pages 552–553   Lesson 13–1
1. $\sqrt{100}$ **3.** Talisa is correct. $\sqrt{30}$ is not a fraction; nor does it contain any perfect square factors in the radicand.   **5.** 2   **7.** 6   **9.** $3\sqrt{3}$
11. $\sqrt{3}$ **13.** $\frac{\sqrt{21}}{3}$ **15.** about 58 mi   **17.** 11
19. $4\sqrt{2}$ **21.** $4\sqrt{3}$ **23.** $10\sqrt{2}$ **25.** $5\sqrt{3}$

27. $6\sqrt{2}$ **29.** $\sqrt{6}$ **31.** $\frac{4}{9}$ **33.** $\frac{\sqrt{6}}{2}$ **35.** $\frac{4\sqrt{7}}{7}$

37. $\frac{\sqrt{2}}{4}$ **39.** $\sqrt{38}$ **41.** 15 **43.** 97.12 ft/s

45. 27,000:1 **47.** $\overline{CE}$ **49.** 50; $AB > BC$ and
$BC > DC$. So, $AB > DC$. **51.** 11

Pages 556–558 Lesson 13–2
1. Sample answer:

3. Kyung; a leg of a right triangle is always shorter
than the hypotenuse. **5.** $x = 4\sqrt{2}, y = 4\sqrt{2}$
7. $x = 8, y = 8\sqrt{2}$ **9.** $x = 9\sqrt{2}, y = 9\sqrt{2}$
11. $x = 1, y = \sqrt{2}$ **13.** 6 ft **15.** 2.8 cm **17a.** $\sqrt{2}$
17b. $\sqrt{3}$ **17c.** 2 **17d.** $\sqrt{5}$ **17e.** $\sqrt{6}$ **17f.** $\sqrt{7}$
19. $5\sqrt{3}$ **21.** 113.1 cm^2

Page 558 Quiz 1
1. $2\sqrt{3}$ **3.** $3\sqrt{2}$ **5a.** 4 in. **5b.** 16 in.

Pages 562–563 Lesson 13–3
1. Sample answer:

3. $x = 4\sqrt{3}, y = 8$ **5.** $x = 4\sqrt{3}, y = 2\sqrt{3}$
7. $x = 30, y = 15\sqrt{3}$ **9.** $x = 14\sqrt{3}, y = 14$
11. $x = 0.6\sqrt{3}, y = 1.2$ **13.** $x = 6\sqrt{3}, y = 3\sqrt{3}$
15. $x = \frac{2\sqrt{3}}{3}, y = \frac{4\sqrt{3}}{3}$ **17.** 42 ft **19.** $24\sqrt{3}$ ft^2
21. $5\sqrt{2}$; 10 **23.** 5

Pages 567–569 Lesson 13–4
1. the ratio of the measure of the leg opposite an
acute angle to the measure of the leg adjacent to
the acute angle **3.** 0.6 **5.** 105.6 **7.** 42.5 m
9. 1.3333 **11.** 2.4 **13.** 67.2 **15.** 43.6 **17.** 7.5
19. 140.0 m **21.** They are equal. **23.** $40\sqrt{2}$ or
about 56.6 ft **25.** no

Page 569 Quiz 2
1. $x = 3, y = 3\sqrt{3}$ **3.** $x = 3.9$ **5.** 5.2

Pages 575–577 Lesson 13–5
1. They are the same because both ratios use the
hypotenuse. They are different because the sine
uses the opposite leg and the cosine uses the
adjacent leg. **5.** $\overline{ST}$ **7.** 0.4706 **9.** 8.2 **11.** 37.5 ft

13. 0.6 **15.** 0.9459 **17.** 195.4 **19.** 84.8 **21.** 24.5
23. 0.8660 **25.** 0.8660 **27.** 0.7071 **29.** 1.7321
31. 1.9 ft **33.** 19.3 ft **35a.** Definition of sine
and cosine **35b.** $\sin^2 x = (\sin x)^2$ **35c.** Adding
like terms **35d.** Pythagorean Theorem
35e. Substitution **35f.** $\frac{4}{5}$ **35g.** $\frac{5}{13}$ **37.** 79.8° **41.** B

**Pages 578–580 Chapter 13 Study Guide and
Assessment**
1. trigonometric ratio **3.** Trigonometry
5. square root **7.** angle of depression
9. angle of depression **11.** 6 **13.** $\frac{\sqrt{10}}{2}$
15. $4\sqrt{3}$ **17.** $x = 3\sqrt{2}, y = 3\sqrt{2}$ **19.** $x = 2$,
$y = 2\sqrt{2}$ **21.** $x = 7\sqrt{3}, y = 14\sqrt{3}$
23. $x = 9\sqrt{3}, y = 9$ **25.** 1.3333 **27.** 60.9
29. 0.3846 **31.** 24.8 **33.** 116 in. **35.** 4.8°

Page 583 Preparing for Standardized Tests
1. B **3.** C **5.** B **7.** A **9.** 471

Chapter 14 Circle Relationships

Page 585 Check Your Readiness
1. false **3.** true **5.** 36 **7.** 120 **9.** 204 **11.** 17
13. 14.1

Pages 589–591 Lesson 14–1
1. Sample answer: It is the part of the circle that lies
inside the angle. Its measure is twice the measure
of the inscribed angle. **3.** yes; $\overarc{WS}$ **5.** 30 **7.** 55
9. yes; $\overarc{DGF}$ **11.** no; $\overarc{JS}$ **13.** 38 **15.** 118 **17.** 30
19. 17 **21.** 12 **23.** 31 **25.** No; Dante's suggestion
is impossible if a triangle is inscribed in a semicircle
and one of its angles intercepts a semicircle.
27. Sample answer: $\angle M$ intercepts $\overarc{ATH}$, so $m\angle M$
$= \frac{1}{2}m\overarc{ATH}$. $\angle T$ intercepts $\overarc{HMA}$, so $m\angle T = \frac{1}{2}m\overarc{HMA}$.
$m\angle M + m\angle T = \frac{1}{2}m\overarc{ATH} + \frac{1}{2}m\overarc{HMA}$
$m\angle M + m\angle T = \frac{1}{2}(m\overarc{ATH} + m\overarc{HMA})$
$m\angle M + m\angle T = \frac{1}{2}(360)$ or 180

The same can be shown for angles H and A.
Thus, opposite angles of the quadrilateral are
supplementary. **29.** 552.92 cm^2 **31.** 0.33

Pages 595–597 Lesson 14–2
1. 2: Theorem 14–6 **3.** 6.4 **5.** 14.7 **7.** 24 in.
9. 15 cm **11.** 72 **13.** 13 ft **15.** 90 **17.** 8
19. $42\frac{1}{2}$ ft; Theorem 14–6 **21.** 29 **23.** All radii
of a circle are congruent **25.** SSS **27a.** 21
27b. By Theorem 14–6, segments from a vertex
to the tangent points are congruent. Since this is
true for all vertices, it can be shown that all these
segments are congruent. Therefore, the points of
tangency are the midpoints of each side. **29.** 86

31. 98.7 m   **33.** C

Pages 603–605   Lesson 14–3
1. $m\widehat{CD}$ **3.** *Secare* means to cut. A secant cuts a circle into two parts.   **5.** 130   **7.** 20.5; 36   **9.** 14
11. 21   **13.** 16   **15.** 60; 44   **17.** 13; 31   **19.** 12; 42
21. 114   **23.** $x = 14$, $m\widehat{AB} = 100$, $m\widehat{CD} = 30$
25. 116   **27.** 22   **29.** 81.7 m   **31.** B

Page 605   Quiz 1
1. yes; $\widehat{NP}$ **3.** 13   **5.** 24; 76

Pages 609–611   Lesson 14–4
1. Find one-half the difference of the measures of the intercepted arcs.   **3.** Yes; see students' drawings.   **5.** 120   **7.** 52   **9.** 38   **11.** 90
13. 30   **15.** 145   **17.** 84   **19.** $270 - 4x$ **21a.** 120
21b. 24.9 cm; The shard is a 120° arc, which is one third of a circle. Therefore, the circumference of the original plate was $3 \cdot 8.3$ or 24.9 centimeters.
23. 77.5   **25.** 1:12   **27.** A

Pages 615–617   Lesson 14–5
1. Sample answer:

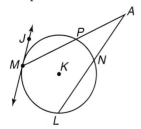

3. Yoshica, by Theorem 14–14   **5.** 7.1   **7.** 8.2 in.   **9.** 8
11. 2   **13.** 1.2   **15.** 15   **17.** 6.9   **19.** about 1087 mi
21a. 8   **21b.** 8   **23.** $\frac{\sqrt{2}}{3}$ **25.** line symmetry

Page 617   Quiz 2
1. 40   **3.** 4.6   **5.** 14

Pages 620–622   Lesson 14–6
3. Sample answer: Graph the circle on grid paper. Draw a radius and label its endpoint on the circle P. Find the slope of the line containing the radius. Use the opposite inverse of that slope and the coordinates of P to write an equation of a line perpendicular to the radius. By Theorem 14–5, this line will be tangent to the circle at P.   **5.** 26   **7.** $\frac{1}{25}$
9. $(x - 1)^2 + (y + 5)^2 = 16$ **11.** $(7, -5)$, 2
13. $x^2 + y^2 = 210.25$ **15.** $(x + 4)^2 + (y - 2)^2 = 1$
17. $(x - 6)^2 + y^2 = \frac{4}{9}$ **19.** $(x + 5)^2 + (y - 9)^2 = 20$
21. $(0, -5)$, 10   **23.** $\left(-\frac{1}{2}, -\frac{1}{3}\right), \frac{4}{5}$ **25.** $(24, -8.1), 2\sqrt{3}$

27.

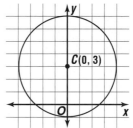

29. $(x - 5)^2 + (y + 13)^2 = 25$ **31.** $x^2 + (y + 3)^2 = 13.69$ **33.** 21.4   **35.** 110 in^2 **37.** D

Pages 624–626   Chapter 14 Study Guide and Assessment
1. secant angles   **3.** tangents   **5.** intercepted arc
7. secant-tangent angle   **9.** tangent-tangent angle
11. 96   **13.** 22   **15.** 12 in.   **17.** 35; 10   **19.** 72
21. 28   **23.** 9.8   **25.** 10.1   **27.** $(x + 3)^2 + (y - 2)^2 = 25$
29. $(x - 5)^2 + (y + 5)^2 = 4$ **31.** $(9, -6)$, 4   **33.** 6.5 ft

Page 629   Preparing for Standardized Tests
1. B   **3.** C   **5.** C   **7.** C   **9.** 3

Chapter 15   Formalizing Proof

Page 631   Check Your Readiness
1. H: it rains; C: we will not have soccer practice
3. H: I win this game; C: I will advance to the semi-finals   **5.** true; definition of congruent segments
7. False; *J*, *K*, and *L* may not be collinear.   **9.** 5
11. 4.2   **13.** 4.1

Pages 636–637   Lesson 15–1
1. A conjunction is a compound statement joined with *and*. A disjunction is a compound statement joined with *or*.   **3.** false   **5.** true   **7.** Mark Twain is not a famous author.   **9.** True; a square has congruent sides, or a parallelogram has parallel sides.

11.

p	q	~q	p ∨ ~q
T	T	F	T
T	F	T	T
F	T	F	F
F	F	T	T

13. false   **15.** Memorial Day is not in July.
17. A pentagon does not have five sides.
19. Water freezes at 32°F, and Memorial Day is in July; false.   **21.** Water freezes at 32°F, and a pentagon has five sides; true.   **23.** Water does not freeze at 32°F, and a pentagon does not have five sides; false.   **25.** Memorial Day is not in July, and $20 \times 5 \neq 90$; true.

27.

p	q	(p ∨ q)	~(p ∨ q)
T	T	T	F
T	F	T	F
F	T	T	F
F	F	F	T

29.

p	q	~p	~q	~(p ∧ q)
T	T	F	F	F
T	F	F	T	F
F	T	T	F	F
F	F	T	T	T

31.

p	q	~p	~p → q
T	T	F	T
T	F	F	T
F	T	T	T
F	F	T	F

33.

p	q	~p	~p ∧ q	~(~p ∧ q)
T	T	F	F	T
T	F	F	F	T
F	T	T	T	F
F	F	T	F	T

35a. If you use Skin-So-Clear, you want clear skin. **35b.** Using Skin-So-Clear will result in clear skin. **35c.** No; a true statement does not always have a true converse. **37.** p and q are both true or both false. **39.** 2 **41.** 4 **43.** 0.6 **45.** 0.8

Pages 640–643   Lesson 15–2
1. Inductive reasoning is the process of using a pattern of examples or experiments to reach a conclusion; deductive reasoning is the process of using facts, rules, definitions, and properties to reach a conclusion.   **3.** Candace; the Law of Detachment does not apply if the hypothesis is negated.   **5.** no valid conclusion   **7.** If two angles are vertical angles, their supplements are congruent.   **9.** no valid conclusion   **11.** The sum of 5 and 3 is an even number.   **13.** Angle B is an acute angle.
15. If a parallelogram has four congruent sides, the diagonals are perpendicular.   **17.** no valid conclusion   **19.** no valid conclusion   **21.** inductive
23. deductive   **25.** The owner is a physician.
27. Sample answer: Babies cannot manage crocodiles.   **29.** True; dogs are mammals, or snakes are reptiles.   **31.** (3, 5); 1   **33.** 11.3

Pages 646–648   Lesson 15–3
1. definitions, postulates, previously proven theorems

3. Given: T bisects $\overline{PN}$ and $\overline{RM}$.
   Prove: $\angle M \cong \angle R$
You know that T bisects $\overline{PN}$ and $\overline{RM}$. So, $\overline{PT} \cong \overline{NT}$ and $\overline{MT} \cong \overline{RT}$. Also, $\angle PTM \cong \angle NTR$ because vertical angles are congruent. $\triangle PTM \cong \triangle NTR$ by SAS. Therefore, $\angle M \cong \angle R$ because CPCTC.

5. Given: $\overline{EF} \parallel \overline{DB}$, $\overline{ED} \parallel \overline{CA}$, $m\angle CAB = 25$
   Prove: $m\angle FED = 25$
You know that $\overline{ED} \parallel \overline{CA}$. These lines are cut by transversal $\overline{DB}$. You also know that $m\angle CAB = 25$. Since $\angle CAB$ and $\angle EDB$ are corresponding angles, $\angle CAB \cong \angle EDB$. So, $m\angle CAB = m\angle EDB$, and $m\angle EDB = 25$. You also know that $\overline{EF} \parallel \overline{DB}$. These lines are cut by transversal $\overline{ED}$. Since $\angle FED$ and $\angle EDB$ are alternate interior angles, $\angle FED \cong \angle EDB$. So, $m\angle FED = m\angle EDB$. Since $m\angle EDB = 25$, $m\angle FED = 25$.

7. Given: $\overline{MQ} \parallel \overline{NP}$, $m\angle 4 = m\angle 3$
   Prove: $m\angle 1 = m\angle 5$
You know that $\overline{MQ} \parallel \overline{NP}$. These lines are cut by transversal $\overline{NQ}$. So, $\angle 3 \cong \angle 5$ because alternate interior angles are congruent, and $m\angle 3 = m\angle 5$. Similarly, $\overline{MQ}$ and $\overline{NP}$ are cut by transversal $\overline{MO}$. So, $\angle 1 \cong \angle 4$ because corresponding angles are congruent, and $m\angle 1 = m\angle 4$. You also know that $m\angle 4 = m\angle 3$. Therefore $m\angle 4 = m\angle 5$ by substitution and $m\angle 1 = m\angle 5$ by substitution.

9. Given: $\triangle GMK$ is an isosceles triangle with vertex $\angle GMK$.
   $\angle 1 \cong \angle 6$
   Prove: $\triangle GMH \cong \triangle KMJ$
You know that $\triangle GMK$ is an isosceles triangle. Therefore, $\overline{MG} \cong \overline{MK}$ by the definition of isosceles triangle. Also, $\angle G \cong \angle K$ because base angles of an isosceles triangle are congruent. You also know that $\angle 1 \cong \angle 6$. Therefore, $\triangle GMH \cong \triangle KMJ$ by ASA.

11. Given: $\angle 5 \cong \angle 6$, $\overline{FR} \cong \overline{GS}$
   Prove: $\angle 4 \cong \angle 3$
You know that $\angle 5 \cong \angle 6$ and $\overline{FR} \cong \overline{GS}$. $\angle 1 \cong \angle 2$ because vertical angles are congruent. Therefore, $\triangle FXR \cong \triangle GXS$ by AAS. $\overline{FX} \cong \overline{GX}$ by CPCTC, and $\triangle FXG$ is isosceles. Therefore, $\angle 3 \cong \angle 4$ because base angles of an isosceles triangle are congruent.

13. Given: $\overline{AD}$ is an angle bisector and an altitude of $\triangle ABC$.
   Prove: $\triangle ABC$ is isosceles.

You know that $\overline{AD}$ is an angle bisector. So, $\angle 1 \cong \angle 2$. You also know that $\overline{AD}$ is an altitude. So, $\angle ADC$ and $\angle ADB$ are right triangles, and $\triangle ADC$ and $\triangle ADB$ are right triangles. $\overline{AD} \cong \overline{AD}$ because congruence of segments is reflexive. Therefore, $\triangle ADC \cong \triangle ADB$ by LA. $\overline{AB} \cong \overline{AC}$ by CPCTC. Therefore, $\triangle ABC$ is isosceles by definition of isosceles triangle.

15. Sample answer: In a closing argument, the attorney presents the evidence in a logical order, tells how the evidence is related, and gives reasons to find the defendant guilty.   **17.** $m\angle 1 + m\angle 4 = 90$

19.

p	q	~p	~q	~p ∧ ~q
T	T	F	F	F
T	F	F	T	F
F	T	T	F	F
F	F	T	T	T

21.

p	q	~q	p ∨ ~q
T	T	F	T
T	F	T	T
F	T	F	F
F	F	T	T

23. C

Page 648   Quiz 1
1. false   **3.** no valid conclusion

5. Given: $\triangle CAN$ is an isosceles triangle with vertex $\angle N$.
$\overline{CA} \parallel \overline{BE}$

Prove: $\triangle NEB$ is an isosceles triangle.

You know that $\triangle CAN$ is an isosceles triangle with vertex $\angle N$. Therefore, $\angle A \cong \angle C$ because the base angles of an isosceles triangle are congruent. You also know that $\overline{CA} \parallel \overline{BE}$. So, $\angle A \cong \angle NEB$ and $\angle C \cong \angle NBE$ because they are corresponding angles. $\angle NEB \cong \angle NBE$ by substitution. Therefore, $\overline{NE} \cong \overline{NB}$ because if two angles of a triangle are congruent, then the sides opposite those angles are congruent. $\triangle NEB$ is an isosceles triangle by definition.

Pages 651–653   Lesson 15–4
1. given statement, prove statement, figure, statements, reasons   **3.** Paragraph proofs and two-column proofs both have a given statement, a prove statement, and usually have a figure. In a paragraph proof, the statements and reasons are written in paragraph form. In a two-column proof, the statements and reasons are listed in two columns.

5. Given: $-2x + 5 = -13$
   Prove: $x = 9$

Statements	Reasons
1. $-2x + 5 = -13$	**1.** Given
2. $-2x + 5 - 5 = -13 - 5$	**2.** Subtraction Property, =
3. $-2x = -18$	**3.** Substitution Property, =
4. $\frac{-2x}{-2} = \frac{-18}{-2}$	**4.** Division Property, =
5. $x = 9$	**5.** Substitution Property, =

7a. Given   **7b.** Multiplication, =
7c. Division, =

9. Given: $F = ma$
   Prove: $m = \frac{F}{a}$

Statements	Reasons
1. $F = ma$	**1.** Given
2. $\frac{F}{a} = \frac{ma}{a}$	**2.** Division Property, =
3. $\frac{F}{a} = m$	**3.** Substitution Property, =
4. $m = \frac{F}{a}$	**4.** Symmetric Property, =

11. Given: $\overline{VR} \perp \overline{RS}, \overline{UT} \perp \overline{SU}, \overline{RS} \cong \overline{US}$
   Prove: $\overline{VR} \cong \overline{TU}$

You know that $\overline{VR} \perp \overline{RS}$ and $\overline{UT} \perp \overline{SU}$. So, $\triangle VRS$ and $\triangle TUS$ are right triangles. You also know that $\overline{RS} \cong \overline{US}$. $\angle VSR \cong \angle TSU$ because vertical angles are congruent. So, $\triangle VSR \cong \triangle TSU$ by LA. Therefore, $\overline{VR} \cong \overline{TU}$ by CPCTC.

13. 300   **15.** 7   **17.** B

Pages 656–659   Lesson 15–5
1. Yes; $\angle 1 \cong \angle 4$ because $\overline{RS} \parallel \overline{WT}$ and $\angle 1$ and $\angle 4$ are alternate interior angles.

3. Given: $\overline{EF} \cong \overline{GH}, \overline{EH} \cong \overline{GF}$
   Prove: $\triangle EFH \cong \triangle GHF$

Statements	Reasons
1. $\overline{EF} \cong \overline{GH}, \overline{EH} \cong \overline{GF}$	**1.** Given
2. $\overline{HF} \cong \overline{HF}$	**2.** Congruence of segments is reflexive.
3. $\triangle EFH \cong \triangle GHF$	**3.** SSS

5. Given: $\overline{AC} \perp \overline{BD}, \overline{AF} \cong \overline{CF}$
Prove: $\overline{AB} \cong \overline{CB}, \overline{CD} \cong \overline{AD}$

Statements	Reasons
1. $\overline{AC} \perp \overline{BD}$	1. Given
2. $\angle AFB, \angle CFB, \angle AFD,$ and $\angle CFD$ are right angles.	2. Perpendicular lines form four right angles.
3. $\triangle AFB, \triangle CFB, \triangle AFD,$ and $\triangle CFD$ are right triangles.	3. Definition of right triangle
4. $\overline{AF} \cong \overline{CF}$	4. Given
5. $\overline{BF} \cong \overline{BF}, \overline{DF} \cong \overline{DF}$	5. Congruence of segments is reflexive.
6. $\triangle AFB \cong \triangle CFB;$ $\triangle AFD \cong \triangle CFD$	6. LL
7. $\overline{AB} \cong \overline{CB}; \overline{CD} \cong \overline{AD}$	7. CPCTC

7. Given: $\overline{AB} \cong \overline{CD}, \angle 1 \cong \angle 2$
Prove: $\overline{AD} \cong \overline{CB}$

Statements	Reasons
1. $\overline{AB} \cong \overline{CD}, \angle 1 \cong \angle 2$	1. Given
2. $\overline{AC} \cong \overline{AC}$	2. Congruence of segments is reflexive.
3. $\triangle ABC \cong \triangle CDA$	3. SAS
4. $\overline{AD} \cong \overline{CB}$	4. CPCTC

9. Given: $HJLM$ is a rectangle, $\overline{KJ} \cong \overline{NM}$.
Prove: $\overline{HK} \cong \overline{LN}$

Statements	Reasons
1. $HJLM$ is a rectangle.	1. Given
2. $\angle J$ and $\angle M$ are right angles.	2. Definition of rectangle.
3. $\triangle HJK$ and $\triangle LMN$ are right triangles.	3. Definition of right triangle
4. $\overline{HJ} \cong \overline{LM}$	4. Opposite sides of a rectangle are $\cong$.
5. $\overline{KJ} \cong \overline{NM}$	5. Given
6. $\triangle HJK \cong \triangle LMN$	6. LL
7. $\overline{HK} \cong \overline{LN}$	7. CPCTC

11. Given: $\overline{CD}$ is a diameter of $\odot E, \overline{CD} \perp \overline{AB}$.
Prove: $\overline{AF} \cong \overline{BF}$

Statements	Reasons
1. $\overline{CD} \perp \overline{AB}$	1. Given
2. $\angle EFB$ and $\angle EFA$ are right angles.	2. Definition of perpendicular lines
3. $\triangle EFB$ and $\triangle EFA$ are right triangles.	3. Definition of right triangle

4. $\overline{EB} \cong \overline{EA}$	4. Radii of a circle are congruent.
5. $\overline{EF} \cong \overline{EF}$	5. Congruence of segments is reflexive.
6. $\triangle EBF \cong \triangle EAF$	6. HL
7. $\overline{AF} \cong \overline{BF}$	7. CPCTC

13. Given: isosceles trapezoid $QRST$ with bases $\overline{QR}$ and $\overline{TS}$ and diagonals $\overline{QS}$ and $\overline{RT}$
Prove: $\overline{QS} \cong \overline{RT}$

Statements	Reasons
1. isosceles trapezoid $QRST$ with bases $\overline{QR}$ and $\overline{TS}$	1. Given
2. $\overline{QT} \cong \overline{RS}$	2. Definition of isosceles trapezoid
3. $\angle QTS \cong \angle RST$	3. Base angles of an isosceles trapezoid are $\cong$.
4. $\overline{TS} \cong \overline{TS}$	4. Congruence of segments is reflexive.
5. $\triangle QST \cong \triangle RTS$	5. SAS
6. $\overline{QS} \cong \overline{RT}$	6. CPCTC

15a. If the diagonals of a parallelogram are congruent, the parallelogram is a rectangle.

15b. Given: parallelogram $ABCD$ with diagonals $\overline{AC}$ and $\overline{BD}$
$\overline{AB} \cong \overline{DC}, \overline{AD} \cong \overline{BC}, \overline{AC} \cong \overline{BD}$
Prove: $ABCD$ is a rectangle.

Statements	Reasons
1. $ABCD$ is a parallelogram. $\overline{AB} \cong \overline{DC}, \overline{AD} \cong \overline{BC}, \overline{AC} \cong \overline{BD}$	1. Given
2. $\triangle ACD \cong \triangle BDC$	2. SSS
3. $\angle ADC \cong \angle BCD$	3. CPCTC
4. $m\angle ADC = m\angle BCD$	4. Definition of congruent angles
5. $m\angle ADC + m\angle BCD = 180$	5. Adjacent angles of a parallelogram are supplementary.
6. $m\angle ADC + m\angle ADC = 180$	6. Substitution Property, $=$
7. $2(m\angle ADC) = 189$	7. Substitution Property, $=$

8. $m\angle ADC = 90$

9. $m\angle BCD = m\angle ADC$
$= 90$

10. $ABCD$ is a rectangle.

8. Division Property, $=$

9. Substitution Property, $=$

10. Definition of rectangle

17. Given: $-4x + 5 = -15$

Prove: $x = 5$

Statements	Reasons
1. $-4x + 5 = -15$	**1.** Given
2. $-4x + 5 - 5 = -15 - 5$	**2.** Subtraction Property, $=$
3. $-4x = -20$	**3.** Substitution Property, $=$
4. $\dfrac{-4x}{-4} = \dfrac{-20}{-4}$	**4.** Division Property, $=$
5. $x = 5$	**5.** Substitution Property, $=$

19. 20   **21.** 110

Page 659   Quiz 2
1. Given   **3.** Substitution, $=$

5. Given: T is the midpoint of $\overline{BQ}$. $\triangle ABT$ and $\triangle PQT$ are right triangles.
$\angle 1 \cong \angle 2$

Prove: $\overline{AT} \cong \overline{PT}$

Statements	Reasons
1. T is the midpoint of $\overline{BQ}$.	**1.** Given
2. $BT = QT$	**2.** Definition of midpoint
3. $\overline{BT} \cong \overline{QT}$	**3.** Definition of congruent segments
4. $\triangle ABT$ and $\triangle PQT$ are right triangles.	**4.** Given
5. $\angle 1 \cong \angle 2$	**5.** Given
6. $\triangle ATB \cong \triangle PTQ$	**6.** LA
7. $\overline{AT} \cong \overline{PT}$	**7.** CPCTC

Pages 663–665   Lesson 15–6
1. Use the origin as a vertex or center, place at least one side of a polygon on an axis, and try to keep the figure within the first quadrant.

3. $(4, -1)$ **5.** (e, f)

7. Sample answer:

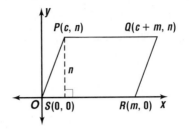

9. Given: rectangle $ABCD$ with diagonals $\overline{AC}$ and $\overline{BD}$

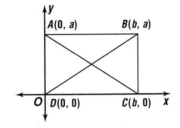

Prove: $\overline{AC}$ and $\overline{BD}$ bisect each other.
Find the midpoint of $\overline{AC}$.
$$\left(\frac{0 + b}{2}, \frac{a + 0}{2}\right) = \left(\frac{b}{2}, \frac{a}{2}\right)$$
Find the midpoint of $\overline{BD}$.
$$\left(\frac{b + 0}{2}, \frac{a + 0}{2}\right) = \left(\frac{b}{2}, \frac{a}{2}\right)$$
The midpoints of the diagonals have the same coordinates. Therefore, they name the same point, and $\overline{AC}$ and $\overline{BD}$ bisect each other.

11. Sample answer:

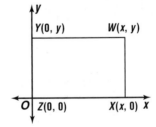

13. Sample answer:

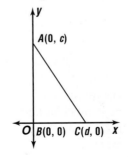

15. Sample answer:

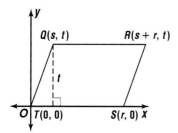

17. Given: right triangle ABC

M is the midpoint of the hypotenuse $\overline{BC}$.

Prove: $BM = CM = AM$

First, use the Midpoint Formula to find the coordinates of M.

$\left(\dfrac{0 + 2c}{2}, \dfrac{2b + 0}{2}\right) = \left(\dfrac{2c}{2}, \dfrac{2b}{2}\right)$ or (c, b)

Next, use the Distance Formula to find BM, CM, and AM.

$BM = \sqrt{(c - 0)^2 + (b - 2b)^2} = \sqrt{c^2 + b^2}$

$CM = \sqrt{(c - 2c)^2 + (b - 0)^2} = \sqrt{c^2 + b^2}$

$AM = \sqrt{(c - 0)^2 + (b - 0)^2} = \sqrt{c^2 + b^2}$

Therefore, $BM = CM = AM$.

19. Given: isosceles triangle XYZ with medians $\overline{MZ}$ and $\overline{NX}$

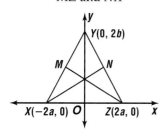

Prove: $\overline{MZ} \cong \overline{NX}$

First, use the Midpoint Formula to find the coordinates of M and N.

$M: \left(\dfrac{-2a + 0}{2}, \dfrac{0 + 2b}{2}\right) = \left(\dfrac{-2a}{2}, \dfrac{2b}{2}\right)$ or $(-a, b)$

$N: \left(\dfrac{2a + 0}{2}, \dfrac{0 + 2b}{2}\right) = \left(\dfrac{2a}{2}, \dfrac{2b}{2}\right)$ or (a, b)

Next, use the Distance Formula to find MZ and NX.

$MZ = \sqrt{[2a - (-a)]^2 + (0 - b)^2}$

$\quad = \sqrt{(3a)^2 + b^2}$ or $\sqrt{9a^2 + b^2}$

$NX = \sqrt{(-2a - a)^2 + (0 - b)^2}$

$\quad = \sqrt{(-3a)^2 + b^2}$ or $\sqrt{9a^2 + b^2}$

Since the medians have the same measure, $\overline{MZ} \cong \overline{NX}$.

21. Given: isosceles trapezoid $PQRS$ with diagonals $\overline{PR}$ and $\overline{QS}$

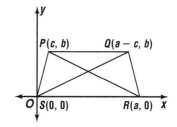

Prove: $\overline{PR} \cong \overline{QS}$

Use the Distance Formula to find PR and QS.

$PR = \sqrt{(c - a)^2 + (b - 0)^2}$

$\quad = \sqrt{c^2 - 2ac + a^2 + b^2}$

$QS = \sqrt{[(a - c) - 0]^2 + (b - 0)^2}$

$\quad = \sqrt{a^2 - 2ac + c^2 + b^2}$

Since the diagonals have the same measure, $\overline{PR} \cong \overline{QS}$.

23. Sample answer:

Given: parallelogram $PQRS$ with $P(20, 50)$, $Q(60, 50)$, $R(60, 20)$, and $S(20, 20)$

Prove: $PQRS$ is a rectangle.

Use the Distance Formula to find PR and QS.

$PR = \sqrt{(60 - 20)^2 + (20 - 50)^2} = \sqrt{40^2 + (-30)^2}$ or 50

$QS = \sqrt{(20 - 60)^2 + (20 - 50)^2} =$

$\quad \sqrt{(-40)^2 + (-30)^2}$ or 50

Since the diagonals have the same measure, they are congruent. Therefore, $PQRS$ is a rectangle.

25. Given: $ACDE$ is a rectangle. $ABCE$ is a parallelogram.

Prove: $\triangle ABD$ is isosceles.

Statements	Reasons
1. $ACDE$ is a rectangle. $ABCE$ is a parallelogram.	1. Given
2. $\overline{AB} \cong \overline{EC}$	2. Opposite sides of a parallelogram are $\cong$.
3. $\overline{EC} \cong \overline{AD}$	3. Diagonals of a rectangle are $\cong$.
4. $\overline{AB} \cong \overline{AD}$	4. Congruence of segments is transitive.
5. $\triangle ABD$ is isosceles.	5. Definition of isosceles triangle

27. 205 in² **29.** $a = 65$, $b = 72$ **31.** 7

1. e **3.** i **5.** a **7.** b **9.** h **11.** true **13.** false
15. false **17.** false **19.** no valid conclusion

21. Given: $m\angle BCD = m\angle EDC$, $\overline{AC}$ bisects $\angle BCD$, $\overline{AD}$ bisects $\angle EDC$.

Prove: $\triangle ACD$ is isosceles.

You know that $m\angle BCD = m\angle EDC$, $\overline{AC}$ bisects $\angle BCD$, and $\overline{AD}$ bisects $\angle EDC$. By the definition of an angle bisector, $m\angle 1 = \frac{1}{2}m\angle BCD$ and $m\angle 2 = \frac{1}{2}m\angle EDC$. Halves of equal quantities are equal by the Division Property of Equality. So, $m\angle 1 = m\angle 2$. Therefore, $\overline{AC} \cong \overline{AD}$ because if two angles of a triangle are congruent, then the sides opposite those angles are congruent. $\triangle ACD$ is isosceles by the definition of isosceles triangle.

23. Given: $m\angle AEC = m\angle DEB$

Prove: $m\angle AEB = m\angle DEC$

Statements	Reasons
1. $m\angle AEC = m\angle DEB$	**1.** Given
2. $m\angle AEC = m\angle AEB$ $+ m\angle BEC$ $m\angle DEB = m\angle DEC$ $+ m\angle BEC$	**2.** Angle Addition Postulate
3. $m\angle AEB + m\angle BEC =$ $m\angle DEC + m\angle BEC$	**3.** Substitution Property of Equality
4. $m\angle AEB = m\angle DEC$	**4.** Subtraction Property of Equality

25. Sample answer:

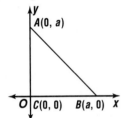

27. Given: square $WXYZ$ with diagonals $\overline{WY}$ and $\overline{XZ}$ intersecting at T

Prove: $\overline{WY}$ and $\overline{XZ}$ bisect each other.

Find the midpoint of $\overline{WY}$.
$\left(\frac{0+a}{2}, \frac{a+0}{2}\right) = \left(\frac{a}{2}, \frac{a}{2}\right)$

Find the midpoint of $\overline{XZ}$.
$\left(\frac{a+0}{2}, \frac{a+0}{2}\right) = \left(\frac{a}{2}, \frac{a}{2}\right)$

The midpoints of the diagonals have the same coordinates. Therefore, they name the same point, and $\overline{WY}$ and $\overline{XZ}$ bisect each other.

29. Julia earned an A.

1. D **3.** B **5.** D **7.** A **9.** 20

Chapter 16 More Coordinate Graphing and Transformations

Page 675 Check Your Readiness

1.

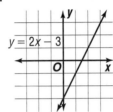

3.

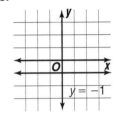

5.

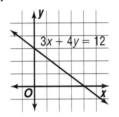

7.

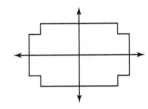

9.

11. yes
13. No; $\frac{4}{5} \neq \frac{9}{10}$

1. Sample answer:

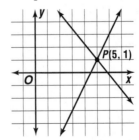

Point P is the solution of the system of equations because it lies on both graphs.

3a.

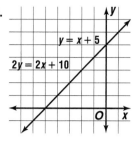

$y = x + 5$

$2y = 2x + 10$

3b. Sample answer: $(-2, 3)$, $(-1, 4)$, $(0, 5)$, $(1, 6)$, $(2, 7)$

3c. When both graphs are the same line, the system of equations has infinitely many solutions.

5. $y = -2x + \dfrac{7}{2}$

7. $(4, 2)$

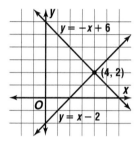

$y = -x + 6$

$(4, 2)$

$y = x - 2$

9. no solution

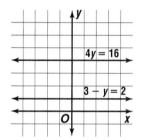

$4y = 16$

$3 - y = 2$

11. $(3, -3)$

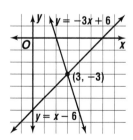

$y = -3x + 6$

$(3, -3)$

$y = x - 6$

13. $(4, 2)$

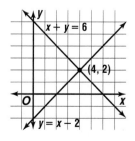

$x + y = 6$

$(4, 2)$

$y = x - 2$

15. $(8, 2)$

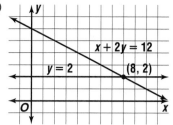

$x + 2y = 12$

$y = 2$ $(8, 2)$

17. $(-6, 5)$

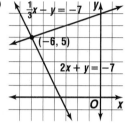

$\frac{1}{3}x - y = -7$

$(-6, 5)$

$2x + y = -7$

19. $(1, -5)$

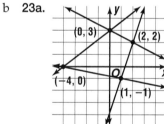

$x + 2y = -9$ $x - y = 6$

$(1, -5)$

21. b   **23a.**

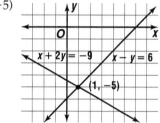

$(0, 3)$ $(2, 2)$

$(-4, 0)$

$(1, -1)$

23b. $(2, 2)$, $(1, -1)$, $(-4, 0)$, $(0, 3)$

25a.

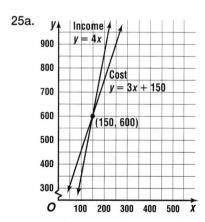

Income

$y = 4x$

Cost

$y = 3x + 150$

$(150, 600)$

25b. $(150, 600)$; If 150 gadgets are produced and sold, the cost and the income both equal \$600.

27. Sample answer:

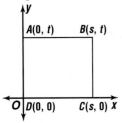

29. 155   **31.** A

Pages 684–686   Lesson 16–2
1. (3, −3) satisfies each equation.
3. Sample answer: In the substitution method, one equation is substituted into the other to solve for a variable. It is easy to use this method when one of the equations is already solved for a variable. Elimination uses addition or subtraction to eliminate one of the variables to solve for the other variable. This method is easier to use when the same variable in both equations has the same coefficient.
5. $\frac{4}{5}$ **7.** (3, 1)   **9.** (2, 0)   **11a.** (114, 136)   **11b.** 114 creamy Italian and 136 garlic herb   **13.** (9, −2)
15. (−1, −1)   **17.** (2, 1)   **19.** $\left(\frac{13}{2}, \frac{1}{2}\right)$ **21.** $\left(-\frac{3}{5}, \frac{7}{5}\right)$
23. (7, 3)   **25.** Sample answer: substitution, (−11, −4)   **27.** (2.74, −0.16)   **29a.** Set up a system of equations and use substitution to find x.   **29b.** 10 min   **29c.** The y value is the number of miles Josh and his mother travel before Josh's mother catches up to him.   **31a.** 0 = 0; There is an infinite number of solutions.   **31b.** 0 = 33; There is no solution.
33. Given: $\overline{SL} \cong \overline{VR}$, $\overline{LT} \cong \overline{RN}$, $\angle L \cong \angle R$; Prove: $\angle S \cong \angle V$ **35.** 41.41 in³
37. Sample answer:

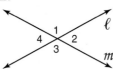

$\angle 1$ and $\angle 2$; $\angle 2$ and $\angle 3$; $\angle 3$ and $\angle 4$; $\angle 4$ and $\angle 1$

Page 686   Quiz 1
1. (4, 5)

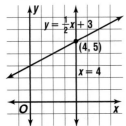

3. $\left(\frac{7}{3}, \frac{8}{3}\right)$ **5a.** (29, 9)   **5b.** 20

Pages 688–690   Lesson 16–3
1. A figure is moved 2 units left and 4 units down.
3. Nicole; $\triangle ABC$ is translated 3 units right and 1 unit up.

5. $R'(6, -2)$, $S'(-1, 0)$, $T'(0, -3)$

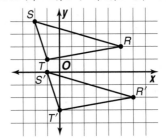

7. $E'(-1, 4)$, $F'(1, 3)$, $G'(1, 6)$

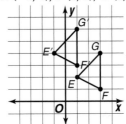

9. $P'(-3, -4)$, $Q'(-5, -5)$, $R'(-4, 0)$

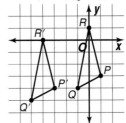

11. $X'(1, 0)$, $Y'(4, 9)$, $Z'(7, -4)$

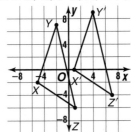

13. $T'(11, -1)$, $U'(3, -9)$, $V'(-7, -5)$, $W'(-2, 0)$

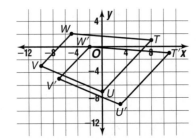

17. The figure moves 3 units right, then 3 units left. It also moves 2 units down, then 2 units up. So, the final position is the same as its original position.
19. $(x - 6)^2 + (y + 21)^2 = 20.25$ **21.** $x = 16$, $y = 16\sqrt{2}$

Pages 694–696   Lesson 16–4

1. Yes, if the original figure is symmetrical. Sample answer: In the figure, △ABC → image by reflection over the y-axis, or by translation of (4, 0).

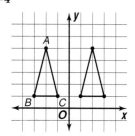

3. Manuel; reflections do not change the size or the shape of a figure.

5. H'(−7, 2), I'(−6, 4), J'(3, −4), K'(−5, −3)

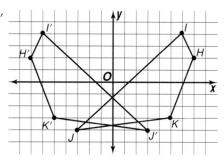

7. H'(3, 2), I'(1, 3), J'(1, −2)

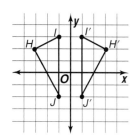

9. S'(−1, −3), T'(3, −2), U'(3, 2), V'(−3, −2)

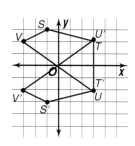

11. P'(1, −2), Q'(4, −4), R'(2, 3)

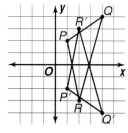

13a. C'(4, 3), D'(6, −2), E'(3, −1)

13b.

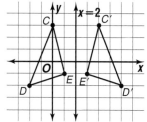

15. H'(−5, 7), J'(0, 4), K'(7, 12)

17.

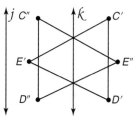

19. ∠A ≅ ∠D   **21.** 3√6

Pages 700–702   Lesson 16–5

1. The image of a 90° rotation reverses the values of the coordinates and uses the appropriate sign for the quadrants.   **3.** Yes; 55° in one direction is the same as 305° in the opposite direction because 55 + 305 = 360, a complete circle.

5.

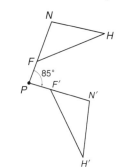

7.

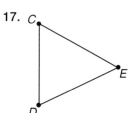

9.

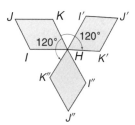

11.

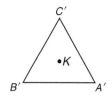

13. $X'(0, 0)$, $Y'(3, 1)$, $Z'(1, 4)$

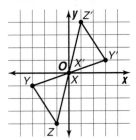

17. $50°$ **19.** $\left(-\dfrac{1}{4}, -\dfrac{17}{2}\right)$

21. $\overline{MN}$

Page 702   Quiz 2

1. $H'(0, 5)$, $I'(4, 7)$, $J'(2, 4)$

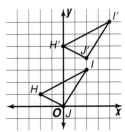

3. $A'(-4, 2)$, $B'(-1, 4)$, $C'(2, 2)$

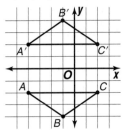

5. Sample answer: The outer portion was created using 45° rotation; the inner portion was created using 90° rotation.

Pages 704–707   Lesson 16–6

1. If the scale factor is between 0 and 1, it is a reduction. If the scale factor is greater than 1, it is an enlargement.

5. $\left(-2, \dfrac{1}{2}\right)$

7. $A'(-2, 0)$, $B'(-4, 4)$, $C'(4, 4)$, $D'(2, 0)$

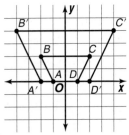

9. $F'(-6, 3)$, $G'(3, -6)$

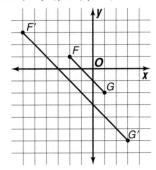

11a. $A'\left(\dfrac{9}{4}, 6\right)$, $B'\left(6, \dfrac{3}{4}\right)$, $C'\left(\dfrac{3}{4}, 3\right)$

11b.

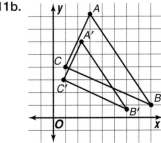

13. $R'(0, 0)$, $S'(0, 2)$, $T'(-1, 2)$, $U'(-2, 0)$

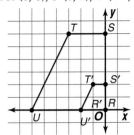

15. $K'(-3, 0)$, $P'(-3, 6)$, $Q'(3, 6)$

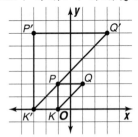

17. $W'(4, 0)$, $X'(8, 6)$, $Y'(0, 6)$

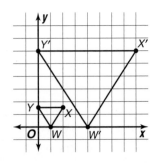

19. $A'\left(\frac{1}{2}, 0\right)$, $B'\left(0, -\frac{3}{2}\right)$, $C'(-1, -1)$

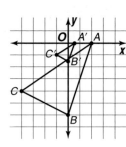

21. $J'(0, 0)$, $K'\left(\frac{15}{4}, \frac{9}{4}\right)$, $L'\left(\frac{21}{4}, -\frac{3}{2}\right)$, $M'(3, -3)$

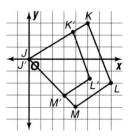

23. 5 yd   **25.** Reflection, dilation, translation; the artist is reflected in the mirror and a dilated image of him is translated to the canvas.
26b. The perimeter of the image is twice that of the preimage.

27.

29. 23.5   **31.** D

Pages 710–712   Chapter 16 Study Guide and Assessment
1. true   **3.** false; rotation   **5.** false; translation
7. true   **9.** true
11. (1, 3)

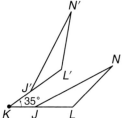

13. no solution

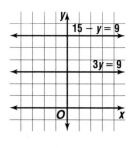

15. Sample answer: elimination; (3, −4)

17. Sample answer: substitution; (2, 24)

19. Sample answer: substitution; (2, −1)

21. $L'(-2, 2)$, $M'(1, 4)$, $N'(2, 0)$

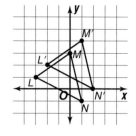

23. $S'(0, 2)$, $T'(-4, 1)$, $U'(-2, -1)$, $V'(1, -2)$

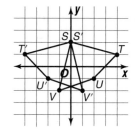

25. $R'(1, -1)$, $S'(2, -1)$, $T'(1, -2)$

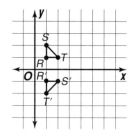

27. $Q'(0, 3)$, $R'(0, 0)$, $S'(-3, -3)$, $T'(-6, 3)$

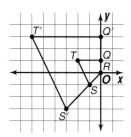

29. 148; 72

Page 715   Preparing for Standardized Tests
1. D   **3.** D   **5.** B   **7.** D   **9.** 40

Photo Credits

Cover: (l)Douglas Hill/Getty Images, (r)Stephan Simpson/Getty Images; **v** file photo; **viii** Georges Seurat, French, 1859–1891, *A Sunday on La Grande Jatte, 1884,* oil on canvas, 1884–1886, 207.6 × 308 cm, Helen Birch Bartlett Memorial Collection, 1926.224 © 1999, the Art Institute of Chicago. All rights reserved; **ix** Steve Prezant/CORBIS; **x** Frank Cezus; **xi** William Whitehurst/CORBIS; **xii** Tom Tallant; **xiii** Jeff Greenberg/PhotoEdit; **xiv** Oberon, courtesy Paul Newman, Australian Centre for Field Robotics; **xv** Timothy Hursley; **xvi** Michael Newman/PhotoEdit; **xvii** Dr. Jeremy Burgess/Science Photo Library/Photo Researchers; **xviii** David Aubrey/CORBIS; **xix** Paul A. Souders/CORBIS; **xx** Michelle Burgess/CORBIS; **xxi** Ellen Knight, Sneed Middle School, Florence SC; **xxii** Michael P. Gadomski/Photo Researchers; **xxiii** P. Ridenour/Getty Images; **2** Eric Kamp/Index Stock Imagery/PictureQuest; **4** A & J Verkaik/Skyart; **9** Michael Newman/PhotoEdit; **12** Georges Seurat, French, 1859–1891, *A Sunday on La Grande Jatte, 1884,* oil on canvas, 1884–1886, 207.6 × 308 cm, Helen Birch Bartlett Memorial Collection, 1926.224 ©1999, the Art Institute of Chicago. All rights reserved; **16** Randy Trina; **17** Christie's Images/CORBIS; **21** John D. Pearce; **22** Irene Rice Pereira. *Untitled,* 1951, oil on board, 101.6 × 61 (40 × 24"). Solomon R. Guggenheim Museum, New York, NY. Gift of Jerome B. Lurie, 1981; **25** Donald C. Johnson/CORBIS; **27** Tribune Media Services, Inc. All rights reserved. Reprinted with permission; **29** (t)Mary Lou Uttermohlen, (b)Amanita Pictures; **30** Aaron Haupt Photography; **34** A. Schoenfeld/Photo Researchers; **38** Aaron Haupt Photography; **41** Paul Barton/CORBIS; **48** Getty Images; **53** Aaron Haupt Photography; **54** Jim Brown/CORBIS; **55** Mark E. Gibson; **60** Donna Terek/PEOPLE Weekly; **61** Matt Meadows; **67** Jon Feingersh/CORBIS; **72** Steve Prezant/CORBIS; **73 81** Matt Meadows; **88** Kayte M. Deioma/PhotoEdit; **95** Jerry Ledriguss/Photo Researchers; **102** Matt Meadows; **111** Frank Cezus; **113** Jim Corwin/Photo Researchers; **115** Ariel Skelly/CORBIS; **116** Kunio Owaki/CORBIS; **127** KS Studio; **136** Mehau Kulyk/Science Photo Library/Photo Researchers; **140** Steve Craft/Masterfile; **142** Frank Whitney/Getty Images; **144** Ira Montgomery/Getty Images; **146** (t)Mike Chew/CORBIS, (b)By permission of Johnny Hart and Creators Syndicate, Inc.; **147** William Manning/CORBIS; **153** Paul Barton/CORBIS; **154** William Whitehurst/CORBIS; **160** Tim Courlas **163** A. Ramey/PhotoEdit; **166** Roger K. Burnard; **171** Zefa Germany/CORBIS; **173 177** Aaron Haupt Photography; **186** Gunter Marx Photography/CORBIS; **191** ML Sinibaldi/CORBIS; **192** Vandysadt/Getty Images; **197** file photo; **200** (t) M.C. Escher ©Cordon Art, Baarn, Holland. All rights reserved, (b)Tom Tallant; **202** M.C. Escher ©Cordon Art, Baarn, Holland. All rights reserved; **207** Henley & Savage/CORBIS; **208** Aaron Haupt Photography; **213** Novastock/PhotoEdit; **226** Jacob Halaska/Index

Stock Imagery; **237** (t)Max B. McCullough, (b)Matt Meadows; **238** Vince Streano/CORBIS; **243** Jeff Greenberg/PhotoEdit; **244** Matt Meadows; **254** KS Studio; **256** Bud Fowle; **261** Larry Lefever for Grant Heilman Photography; **265** Taryn Howard/CORBIS; **266** Larry Lefever for Grant Heilman Photography; **270** Amy C. Etra/PhotoEdit; **274** Firefly Productions/CORBIS; **279** Rick Gayle/CORBIS; **281 286** Aaron Haupt Photography; **288–289 290** KS Studio; **292** Oberon, courtesy of Paul Newman, Australian Centre for Field Robotics; **293** Spencer Grant/PhotoEdit; **294** Gianni Dagli Orti/CORBIS; **298** North Wind Picture Archives; **299** Shearn Benjamin/Getty Images; **300** Musee d'Orsay, Paris/Lauros-Giraudon, Paris/SuperStock; **301** G. Contorakes/CORBIS; **308** Stefano Amantini/Bruce Coleman, Inc.; **310** Timothy Hursley; **315** Daniel Chester Grench/CORBIS; **320** M.C. Escher ©Cordon Art, Baarn, Holland. All rights reserved; **321** Wolfgang Kaehler/CORBIS; **326** Faith Ringgold; **328** Diana Ong/SuperStock; **337** Icon Images; **338** Richard Berenholtz/CORBIS; **339** Aaron Haupt Photography; **340 341** Matt Meadows; **348** Victor Fraile/Reuters/CORBIS; **350** Aaron Haupt Photography; **352** Ron Kimball; **355** Dominic Oldershaw; **358** KS Studio; **361** Michael S. Yamashita/CORBIS; **362** Telegraph Colour Library/Getty Images; **376** Brownie Harris/CORBIS; **379** Michael Newman/PhotoEdit; **380** Harald Sund/Getty Images; **381** Scala/Art Resource, NY; **385** Wes Thompson/CORBIS; **393** Photo Researchers; **400** Darlyne A. Murawski/Getty Images; **403** Raphael Gaillarde/Getty Images News Services; **407** Jeffery Coolidge/Getty Images; **408** John Foster/Photo Researchers; **409** Larry Lefever for Grant Heilman Photography; **411** Dr. Jeremy Burgess/Science Photo Library/Photo Researchers; **416** Michael Frye/Getty Images; **421** Robert Nelson; **423 426** Aaron Haupt Photography; **430** Icon Images; **431** Alan Levenson/Getty Images; **432** Aaron Haupt Photography; **434** Gerben Oppermans/Getty Images; **437** Darryl Torckler/Getty Images; **440** Michael St. Maur Sheil/CORBIS; **442** Kevin Schafer/Getty Images; **443** Aaron Haupt Photography; **452** Jim Erickson/CORBIS; **457** Ken Frick; **458** ©Saturday Review, reproduced by permission of Ed Fisher; **460** David Aubrey/CORBIS; **461** Kjell B. Sandved/Photo Researchers; **473** Joel Warren/Photofest; **476** Gerolimetto/CORBIS; **478** Aaron Haupt Photography; **494** John Lamb/Getty Images; **498** Janice Fullman/Index Stock Imagery; **499** (c)Collection Walker Art Center, Minneapolis, MN; Gift of the T.B. Walker Foundation, 1966/2001 © Estate of David Smith/Licensed by VAGA, New York, NY, (br) Morton & White, (others) Aaron Haupt Photography; **500** (tl tr)Aaron Haupt Photography, (tc)Deni McIntyre/Photo Researchers, (b)Gail Mooney/CORBIS; **502–503 504** Aaron Haupt Photography; **510** Dominic Oldershaw; **511** Aaron Haupt Photography; **512** T. Anderson/Getty

PHOTO CREDITS

Index

INDEX

rational, 50, 51, 54, 61, 85, 642
real, 51–52, 57, 83, 85, 278,
    641–642
squares, 256, 577, 661
whole, 50, 297

number of sides of a convex polygon, 408

number theory, 46, 47, 50–55, 61,
    87, 214, 225, 272, 273, 493, 583,
    629

Number Theory Link, 6

number of triangles formed, 408

O

obelisk, 344

Oberon, Undersea Robot Vehicle, 292

oblique cones, 516

oblique cylinders, 506

oblique prisms, 504

oblique pyramid, 516

obtuse, 188, 207, 267

obtuse angles, 98, 100–101, 109,
    119, 135, 137, 196, 243, 315, 326,
    569

obtuse triangles, 188, 191, 235

octagons, 360, 405–409, 412, 430,
    439, 447, 487

one-to-one correspondence, 199

Open-Ended Test Practice, 22,
    81, 101, 153, 197, 250, 295, 326,
    355, 387, 424, 467, 527, 558, 617,
    637, 659

opposite angles, 317

opposite rays, 90, 132, 135, 137,
    173, 387

opposite sides, 317, 320, 322, 327

optical illusions, 29

ordered pairs, 68–72, 81, 83, 84,
    85, 101, 147, 174, 202, 346, 399,
    515

order of operations, 86

origin, 52, 68, 661, 670

orthocenter, 245

orthographic drawing, 74–75, 500

P

pantograph, 322, 325

paragraph (informal) proofs,
    644–647, 649, 659, 669–671

parallel, 145, 161, 162, 163, 164,
    165, 172, 173, 176, 178, 182, 183,
    185, 324, 374, 382, 385, 424, 521,
    545, 640, 667

parallel lines, 141, 142, 144, 148,
    150, 153, 157, 158, 161, 166, 167,
    170, 173, 181, 183, 316, 383, 384,
    396, 461, 487, 539, 544, 545, 566
    alternate exterior angles, 148,
        149, 150, 152, 156, 163, 164,
        181, 183, 197
    alternate interior angles, 148,
        150, 152, 163, 164, 181, 183,
        197, 319, 323, 544, 566, 645,
        646, 667
    consecutive interior angles,
        148, 149, 150, 152, 156, 164,
        183, 197, 335
    corresponding angles, 156, 157,
        158, 164, 183, 356, 645
    proving, 163
    transversals, 141, 148, 149, 150,
        151, 152, 153, 156, 157, 158,
        161, 162, 163, 164, 180, 181,
        255, 261, 323, 376, 382, 383,
        384, 640, 646, 667

parallelograms, 36–40, 74–75,
    223, 316–329, 331–332, 335–336,
    338, 341–343, 345, 355–357,
    378, 405, 420, 439, 444, 446, 450,
    451, 582–583, 597, 605, 636, 639,
    648, 657, 661–665, 670, 686
    area, 36, 37, 39, 40, 44, 45, 420,
        421
    diagonals, 74, 318, 658
    rectangles, 25, 35–40, 67, 121,
        137, 327–331, 335–336,
        343–345, 355, 359, 360, 395,
        405, 419, 435, 439, 450, 459,
        527, 545, 583, 622, 629, 635,
        642, 647, 656–657, 662, 664,
        665, 671, 673
    rhombi, 327, 328, 329, 330, 331,
        332, 335, 336, 343, 345, 355,
        434, 437, 642
    tests for, 323, 324, 343

parallel planes, 142, 144, 146,
    147, 183, 202

Parallel Postulate, 167

parallel segments, 153, 173, 319,
    333, 376

parallel sides, 333, 344

Pascal's Triangle, 10–11

patterns, 7–11, 17, 51, 54, 73, 133,
    153, 161, 283, 493, 638

pentagonal prism, 498–501

pentagonal pyramid, 497, 520

pentagons, 380–381, 402,
    404–408, 411–412, 418, 425,
    429, 430, 437, 439, 446–449,
    482, 597

percent of error, 58

Percent Review, 55, 109, 147, 239,
    367, 509, 553, 702

percents, 55, 87, 89, 147, 184, 185,
    225, 239, 273, 307, 315, 367, 399,
    493, 509, 553, 629, 702

Pereira, Irene Rice, 22

perfect square factors, 549

perfect squares, 548

perimeters, 35, 38–39, 44, 45, 47,
    55, 94, 121, 250, 275, 287, 307,
    377, 388–393, 396–397, 403,
    405–406, 416–418, 426, 429,
    430, 432–433, 447–478, 491,
    505–506, 509, 511, 515, 517–518,
    558, 563, 580, 582–583, 596, 611,
    622, 629, 677–678

perpendicular, 128, 155, 157, 164,
    171–172, 176, 178, 197, 336, 407,
    420, 425, 467, 469, 470, 475, 502,
    520–521, 582, 592, 629, 642, 648

perpendicular bisector, 235–239,
    243, 245, 247, 250, 269–271, 290,
    412, 658

perpendicular lines, 128, 129,
    130, 133, 170, 183
    constructing, 131, 246, 474
    from a point to a line, 131
    slopes of, 170, 545

perpendicular planes, 133

perpendicular segments, 235, 271

perspective drawings, 23, 333

pi (π), 453, 479

planes, 14, 22, 29, 34, 64, 121, 179,
    180, 533, 673
    coordinate, 49, 68, 72, 73, 79,
        81, 83, 133, 250, 262, 266, 623,
        662, 664, 671
    intersecting, 202, 233
    naming, 19, 21
    parallel, 142, 144, 146, 147, 183,
        202

Platonic solid, 497

Plumbing Link, 508

points, 12, 16–17, 22, 28, 64, 147,
    207
    betweenness, 56
    centers, 425, 454–455, 459, 528,
        622, 626, 637
    collinear, 15, 16, 22, 25, 28, 43, 45
    coordinate planes, 68, 72, 83
    coplanar, 14–16

T

Formulas

Midpoint	on a number line	$M = \frac{a+b}{2}$
	on a coordinate plane	$M = \left(\frac{x_1 + x_2}{2}, \frac{y_1 + y_2}{2}\right)$
Distance	on a coordinate plane	$d = \sqrt{(x_2 - x_1)^2 + (y_2 - y_1)^2}$
Perimeter	square	$P = 4s$
	rectangle	$P = 2\ell + 2w$ or $P = 2(\ell + w)$
Circumference	circle	$C = 2\pi r$ or $C = \pi d$
Area	square	$A = s^2$
	rectangle	$A = \ell w$
	parallelogram	$A = bh$
	triangle	$A = \frac{1}{2}bh$
	trapezoid	$A = \frac{1}{2}h(b_1 + b_2)$
	circle	$A = \pi r^2$
Surface Area	cube	$S = 6s^2$
	prism	$S = Ph + 2B$
	cylinder	$S = 2\pi rh + 2\pi r^2$
	regular pyramid	$S = \frac{1}{2}P\ell + B$
	cone	$S = \pi r\ell + \pi r^2$
	sphere	$S = 4\pi r^2 h$
Lateral Area	cube	$L = 4s^2$
	prism	$L = Ph$
	cylinder	$L = 2\pi rh$
	regular pyramid	$L = \frac{1}{2}P\ell$
	cone	$L = \pi r\ell$
Volume	cube	$V = s^3$
	prism	$V = Bh$
	cylinder	$V = \pi r^2 h$
	regular pyramid	$V = \frac{1}{3}Bh$
	cone	$V = \frac{1}{3}\pi r^2 h$
	sphere	$V = \frac{4}{3}\pi r^3$